Handbook of Neurochemistry and Molecular Neurobiology

Degenerative Diseases of the Nervous System

Abel Lajtha (Ed.)

Handbook of Neurochemistry and Molecular Neurobiology

Degenerative Diseases of the Nervous System

Volume Editors: Moussa B. H. Youdim, Peter Riederer,
Silvia A. Mandel and Leontino Battistin

With 39 Figures and 32 Tables

 Springer

Editor
Abel Lajtha
Director
Center for Neurochemistry
Nathan S. Kline Institute for Psychiatric Research
140 Old Orangeburg Road
Orangeburg
New York, 10962
USA

Volume Editors
Moussa B. H. Youdim
Professor of Pharmacology
Professor of Life Sciences
Director of Eve Topf and NPF Centers of Excellence
for Neurodegenerative Diseases Research
Technion-Rappaport Family Faculty of Medicine
Efron St. POBox 9697
Haifa 31096
Israel

Peter Riederer
Clinic and Policlinic of Psychiatry and Psychotherapy
Dept. Clinical Neurochemistry
University of Würzburg
Füchsleinstr. 15
97080 Würzburg
Germany

Silvia A. Mandel
Eve Topf Center for Neurodegenerative Diseases
Research and Dept. of Pharmacology
Faculty of Medicine
Technion
Efron St.P.O.B. 9697
Haifa 31096
Israel

Leontino Battistin
Department of Neurosciences
University of Padova Medical School
Via Giustiniani 5
35128, Padova
Italy

Library of Congress Control Number: 2006922553

ISBN: 978–0–387–30344–4

Additionally, the whole set will be available upon completion under ISBN: 978–0–387–35443–9
The electronic version of the whole set will be available under ISBN: 978–0–387–30426–7
The print and electronic bundle of the whole set will be available under ISBN: 978–0–387–35478–1

springer.com

Printed on acid-free paper   SPIN: 11417842 2109 - 5 4 3 2 1 0

Preface

This volume of the *Handbook of Neurochemistry and Molecular Neurobiology* is a collection of papers that describe recent developments in various pathological aspects of neurological disorders with special emphasis on neurodegenerative diseases, molecular mechanism of neuronal death in the various diseases and their models, prospects for neuroprotection, and development of disease-modifying drugs. Although the goal of this volume was not to include all neurological disorders, we feel that the selective topics have significantly advanced the field. The first seven chapters including those of Berg et al., Müller, Kruger et al., Toft et al., Weinreb et al., and Huber et al. concentrate on the various aspects of Parkinson's disease (PD). These include pathology, diagnosis, and prospects for neuroprotection. The familial LRRK2 (leucine-rich repeats kinase 2)-associated parkinsonism is described at length by Mathias Toft et al. and its role in PD. Although this gene accounts for a minority of PD subjects, it may offer the chance of identifying patients in a presymptomatic state. Mutations in LRRK2 constitute the most common cause of autosomal-dominant inherited PD so far discovered. Another aspect of determining the molecular pathology of neurodegenerative diseases is the introduction of cutting-edge technology of transcriptomic and proteomic profiling. These have brought a new dimension to determine the global view of molecular cascade of events in neurodegeneration in the various neurological diseases and their animal models. Such approaches may contribute to the development of novel drugs with disease-modifying activity. The chapters on Parkinson's disease are supplemented with those on progressive supranuclear palsy and corticobasal degeneration (Geser and Wenning) as well as with a chapter on central tremors (Glöckler and Reichmann). The chapters by Weinreb et al., Jayanthi et al., Butterfield and Sultana, and Golubnitschaja and Fountoulakis go into extensive description of various genes and proteins as potential candidates as biomarkers for disease progression, early detection, and drug development in experimental models of Huntington's disease and Alzheimer's dementia. Unraveling the pathophysiology of the familial disorder Huntington disease is illuminated in the chapter by Kosinski et al., wherein the various genes and proteins that may be involved in this genetic disorder and the contribution of transgenic gene-based models developed are described.

Various other aspects of dementia are described in three chapters: Hoyer and Frölich describe the significance of cerebral metabolic disturbance in Alzheimer's disease, while Cagnin and Battistin discuss at length vascular dementia and its relation to this disease. HIV dementia as a viral neurodegenerative disorder is discussed by Koutsilieri et al., while Jellinger describes in all facets the Lewy body disorders and gives a neuropathological differentiation to Alzheimer's disease and Parkinson's disease dementia.

Finally, multiple sclerosis is described in its detailed pathology by Lassmann, while the latest news on neuropathic pain is reviewed by Pazzaglia and Battistin.

This volume is intended to serve as a reference source for basic researchers and clinicians involved in the neurodegenerative field in particular and in the neuroscience field in general.

We thank all those who contributed to this volume: the series editor, Abel Lajtha, the managing secretary, Kristine Immediato, and the publisher.

Table of Contents

Contributors

J. O. Aasly
Department of Neurology, St. Olav's Hospital,
Trondheim, Norway

T. Amit
Eve Topf Center for Neurodegenerative Diseases
Research and Department of Pharmacology,
Faculty of Medicine, Technion, Haifa, Israel

G. Arendt
Department of Neurology, University of Düsseldorf,
Germany

L. Battistin
Department of Neurosciences, University of Padova
Medical School, Via Giustiniani 5, 35128, Padova, Italy,
and I.R.C.C.S. San Camillo Hospital, Venice, Italy

D. Berg
Center of Neurology, Hertie-Institute for Clinical Brain
Research, University of Tübingen, 72076 Tübingen,
Germany

C. Briani
Department of Neurosciences,
University of Padova, Italy

A. Bürkle
Molecular Toxicology Group, Department of Biology,
Box X911, University of Konstanz, D–78457 Konstanz,
Germany

D. A. Butterfield
Department of Chemistry, Sanders-Brown Center on
Aging, Center of Membrane Sciences, University of
Kentucky, Lexington, KY 40506, USA

J. L. Cadet
Molecular Neuropsychiatry Branch, NIH/NIDA,
Intramural Research Program,5500 Nathan Shock Drive,
Baltimore, Maryland 21224, USA

A. Cagnin
Department of Neurosciences, University of
Padova Medical School, Via Giustiniani 5, 35128,
Padova, Italy

D. W. Dickson
Department of Neuroscience, Mayo Clinic, Jacksonville,
Florida, USA, and Department of Pathology, Mayo
Clinic, Jacksonville, Florida, USA

M. J. Farrer
Department of Neuroscience, Mayo Clinic, Jacksonville,
Florida, USA

M. Fountoulakis
Foundation for Biomedical Research of the Academy of
Athens, Greece

L. Frölich
Division of Geriatric Psychiatry, Central Institute of
Mental Health Mannheim, University of Heidelberg, J 5,
D–68159 Mannheim, Germany

F. Geser
Clinical Department of Neurology, Innsbruck Medical
University, A–6020 Innsbruck, Austria

T. Glöckler
Klinik und Poliklinik für Neurologie,
Universitätsklinikum Carl Gustav Carus Dresden,
Fetscherstrasse 74, 01307 Dresden, Germany

O. Golubnitschaja
Department of Radiology,
Division of Molecular/Experimental Radiology,
University of Bonn, Germany

E. Grünblatt
Clinical Neurochemistry (National Parkinson
Foundation Center of Excellence Research Laboratory),
Psychiatry and Psychotherapy, University of Würzburg,
Germany

S. Hoyer
Institute of Pathology, University of Heidelberg,
Im Neuenheimer Feld 220/221, D–69120 Heidelberg,
Germany

A. Huber
Comparative Genomics Centre,
James Cook University, Townsville, 4811,
Queensland, Australia

S. Jayanthi
Molecular Neuropsychiatry Branch, NIH/NIDA,
Intramural Research Program,
5500 Nathan Shock Drive, Baltimore,
Maryland 21224, USA

K. A. Jellinger
Institute of Clinical Neurobiology, Kenyongasse 18,
A–1070 Vienna, Austria

C. M. Kosinski
Department of Neurology, University Hospital RWTH
Aachen, Pauwelsstr. 30, D–52074 Aachen, Germany

E. Koutsilieri
Clinical Neurochemistry (National Parkinson
Foundation Center of Excellence Research
Laboratory), Psychiatry and Psychotherapy,
University of Würzburg, Germany, and Institute
of Virology and Immunobiology, University of
Würzburg, Germany

R. Krüger
Center of Neurology, Hertie-Institute for Clinical Brain
Research, University of Tübingen, 72076 Tübingen,
Germany

B. Landwehrmeyer
Department of Neurology, University of Ulm,
Oberer Eselsberg 45/1, D–89081 Ulm, Germany

H. Lassmann
Center for Brain Research, Medical University of Vienna,
Austria

A. Ludolph
Department of Neurology, University of Ulm,
Oberer Eselsberg 45/1, D–89081 Ulm, Germany

M. T. McCoy
Molecular Neuropsychiatry Branch,
NIH/NIDA, Intramural Research Program,
5500 Nathan Shock Drive, Baltimore,
Maryland 21224, USA

S. A. Mandel
Eve Topf Center for Neurodegenerative
Diseases Research and Department of
Pharmacology, Faculty of Medicine,
Technion, Haifa, Israel

T. Müller
Department of Neurology, St. Josef Hospital, Ruhr
University Bochum, Gdrunstrasse 56, 44791 Bochum,
Germany

G. Münch
Comparative Genomics Centre, James Cook University,
Townsville, 4811, Queensland, Australia

E. Neuen-Jacob
Institute of Neuropathology, University of Düsseldorf,
Germany

L. Padua
Institute of Neurology, Cattolica University, Rome, and
Fondazione Don C. Gnocchi, Rome, Italy

C. Pazzaglia
Institute of Neurology, Cattolica University, Rome, Italy

H. Reichmann
Klinik und Poliklinik für Neurologie,
Universitätsklinikum Carl Gustav Carus Dresden,
Fetscherstrasse 74, 01307 Dresden, Germany

P. Riederer
Department of Clinical Neurochemistry,
University of Würzburg, 97080 Würzburg,
Germany

O. Rieß
Medical Genetics, University of Tübingen, 72076
Tübingen, Germany

C. Scheller
Institute of Virology and Immunobiology,
University of Würzburg, Germany

S. W. Scholz
Clinical Department of Neurology,
Innsbruck Medical University, A–6020 Innsbruck, Austria

R. Sultana
Department of Chemistry, Sanders-Brown Center
on Aging, University of Kentucky, Lexington,
KY 40506, USA

M. Toft
Department of Neuroscience, Norwegian University
of Science and Technology, Trondheim, Norway,
and Department of Neuroscience, Mayo Clinic,
Jacksonville, Florida, USA

O. Weinreb
Eve Topf Center for Neurodegenerative Diseases
Research and Department of Pharmacology,

Faculty of Medicine, Technion, Haifa,
Israel

G. K. Wenning
Clinical Department of Neurology, Innsbruck Medical
University, A–6020 Innsbruck, Austria

Z. K. Wszolek
Department of Neurology, Mayo Clinic, Jacksonville,
Florida, USA

M. B. H. Youdim
Eve Topf Center for Neurodegenerative
Diseases Research and Department of
Pharmacology, Faculty of Medicine,
Technion, Haifa, Israel

1 Parkinson's Disease

D. Berg · R. Krüger · O. Rieß · P. Riederer

Abstract: The causes for the initiation of neurodegeneration in idiopathic Parkinson's disease (iPD) are still obscure. However, research of the past decades has shed much light into pathomechanisms involved in the disease. Defined clinical symptoms, new imaging techniques, and additional functional investigations allow diagnosis of most PD cases in the course of the disease. Understanding of the pathoanatomy is essential for realization of the involvement of different brain areas and neurotransmitter systems that explain the individual development of symptoms. Findings of different factors involved in the pathogenesis have led to the hypothesis that iPD is a multifactorial disorder. It is hoped that further elucidation of these and other pathogenetic factors will lead to an earlier diagnosis and allow a more causative, neuroprotective treatment.

List of Abbreviations: CBGD, corticobasal ganglionic degeneration; CCT, cranial computer tomography; ILBD, Lewy body disease; iPD, idiopathic Parkinson's disease; MRI, magnetic resonance imaging; MSA, multiple system atrophy; PD, Parkinson's disease; PSP, progressive supranuclear palsy; SN, substantia nigra; SNc, substantia nigra pars compacta; SNr, substantia nigra pars reticulata; TCS, transcranial sonography; UPDRS, Unified Parkinson's Disease Rating Scale

1   Introduction

With a prevalence of 1:1000, Parkinson's disease (PD) is one of the most common neurological diseases and the second most common movement disorder. An age-dependent increase in incidence is known, affecting 1.4% of the 55-year old group and 3.4% of the population of 75 years of age (De Rijk et al., 1997). According to growing life expectancy, a four times increase in prevalence is expected for the next 20 years in industrialized countries (World Health Organisation, 1998) posing great medical, social, and economical challenges.

Although the primary pathology and key defects of neurotransmission leading to the clinical picture of PD are known, the cause of initiation and progression of neurodegeneration in the disease process is still obscure. However, extensive research has elucidated biochemical alterations of the affected brain areas and knowledge about a genetic contribution to and molecular biological mechanism of the disease (see chapter by Krüger et al.) is still growing, giving promising perspectives for future research and therapeutic strategies.

2   History

The first clinical description of PD was presented by Parkinson in 1817, who believed that the disorder originated from disturbances in the upper cervical spinal cord. More than 70 years later, Blocq and Marinesco (1893) discovered the destruction of the substantia nigra (SN) in Parkinsonian patients. The identification of the typical intracellular and extraneuronal eosinophil inclusions in the dorsal nucleus of the vagal nerve and the nucleus basalis Meynert in PD patients by Lewy in 1912 was followed by the discovery of the same inclusion in the SN by Trétiakoff a few years later. These findings initiated further research on the impact of SN alterations for the clinical manifestation of PD (Trétiakoff, 1919; Hassler, 1938; Greenfield and Bosanquet, 1953) cumulating in two pioneering findings: (i) Carlsson et al. (1957) realized that dopamine is a neurotransmitter and (ii) Ehringer and Hornykiewicz (1962) described dopamine deficiency in the striatum as the cause for the clinical manifestation of the disease. First administrations of intravenous (Birkmayer and Hornykiewicz, 1961) and oral (Barbeau et al., 1962) dopamine constituted the basis of the still most effective symptomatic therapy: Substitution of dopamine and implementation of dopamine agonists. The discoveries that toxins like MPP^+, derived from the drug MPTP might lead to the clinical picture of PD with similar histological findings (Davis et al., 1979; Langston et al., 1983) and that disturbances in the trace metal content like increased iron levels of the SN may induce and accelerate oxidative stress (Sofic et al., 1988) initiated extensive research on the influence of exo- and endotoxins in disease development. Discovery of monogenetically caused forms of PD (chapter 3 of this book) and elucidation of the underlying pathomechanisms as well as of genetic susceptibility factors for the disease have now lead to the consensus of multifactorial causes of sporadic PD.

3   Clinical Manifestation

For the diagnosis of PD it is essential to differentiate between idiopathic or primary PD and secondary Parkinsonian syndromes. In contrast to idiopathic PD and Parkinsonian syndromes associated with other neurodegenerative diseases, the etiology of secondary Parkinsonian syndromes is known (❱ *Table 1-1*). These secondary Parkinsonisms need to be excluded as they require different therapeutical strategies.

❑ Table 1-1
Symptomatic causes for parkinsonian syndromes, also called secondary PD

Diagnosis	Indices
Drug induced parkinsonism	Treatment with neuroleptics, other dopamine receptor blockers, lithium, calcium-antagonists, methyldopa
Tumors or hydrocephalus	Other focal symptoms, CCT, MRI
Recurring ischemic insults	Vascular risk factors, stepwise worsening and additional symptoms
Infections, postencephalic	History, systemic symptoms
Metabolic causes	Changes in copper or calcium metabolism
Hereditary neurodegenerative disorders	Family history, additional symptoms
Severe brain traumas	History, CCT, MRI
Seldom intoxications	History, systemic symptoms
Psychogenic	Remission for longer periods, additional symptoms

Idiopathic PD is characterized by disturbances of voluntary and involuntary movements. Four cardinal symptoms, usually beginning at one side, can be differentiated: *Akinesia,* which refers to a slowing of movement (bradykinesia) and a reduction in amplitude (hypokinesia) that may lead to a complete loss of voluntary and involuntary movements (akinesia). In the beginning, especially the fine and rapid alternating movements are affected. Mikrographia occurs and blinking is infrequent. The face lacks expressive mobility (hypomimea), and speech gets increasingly monotonic. In later stages saliva is not swallowed as fast as it is produced resulting in siallorrhea. Posture gets increasingly flexed and gait turns slow, shuffling with small steps and there is reduced movement of arms. A limited capacity to initiate and stop movements leads to falling. *Rigidity* due to tonic innervation of agonists and antagonists is usually more pronounced in muscles that maintain flexed postures and somewhat greater in large muscle groups. The patients notice firmness and tenseness of the muscles, which is encountered as a ratchet-like resistance to passive movements by the examiner interrupted by a rhythmical interruption, the cogwheel phenomenon. The permanent stiffness of muscles often leads to pain, clinically manifesting as lumbago or shoulder-arm-syndrome. *Resting tremor* is a coarse, rhythmic tremor with a frequency of 3–5 Hz, electromyographically characterized by bursts of activity that alternate between opposing muscle groups. It is most often localized in one or both hands with the typical sign of pill-rolling, less frequently in the feet, jaw, or tongue and usually occurs when the limb is in an attitude of response. It is diminished by voluntary movements and usually disappears when the limb is completely relaxed, however, the patient rarely achieves this state. Frequently, a fine 7–8 per/sec slightly irregular action tremor of the outstretched fingers and hands may also be noted. Usually, later in the disease process *postural instability* occurs, which includes the inability of a patient to make appropriate postural adjustments to tilting or falling as well as the inability to change from the reclining to the standing position. Impairment of anticipatory and compensatory rigthening reflexes is the cause of these symptoms leading to retropulsion, propulsion, and festination.

According to the prevalence of these symptoms PD is classified in subgroups like a tremor dominant or rigid-akinetic type. Besides these cardinal symptoms a number of vegetative symptoms like seborrhea, obstipation, bladder dysfunction, orthostatic dysregulation, as well as somatosensoric disturbances like impairment of olfaction, color discrimination, and paraestesia may occur. About 40% of PD patients suffer from a concomitant depression (Cummings, 1992); in 15–20% dementia can be recognized (Biggins et al., 1992). Gastrointestinal and brainstem associated deficits may precede the cardinal symptoms (Przuntek et al., 2004) in line with the histopathological staging of Braak (Braak et al., 2003, 2004).

Because of the available symptomatic medication, life expectancy of patients with PD is almost normal. However, with this medication and partly because of this medication later complications may occur. These include fluctuation of medication effects like "wearing-off," "end-of-dose-akinesia" and paroxymal "on-off-phenomenons" as well as L-Dopa-induced dyskinesia manifesting as choreatic "peak-of-dose-," "biphasic-" or "off dyskinesias."

4   Diagnosis

Diagnosis of PD is primarily clinical. According to the cardinal symptoms, a parkinsonian syndrome (PS) is diagnosed when hypo- and bradykinesia are accompanied by at least one of the other cardinal symptoms rigor, resting tremor, or postural instability. Supporting criteria for idiopathic PD are given in ❷ *Table 1-2*. Symptomatic PS (❷ *Table 1-1*) must be ruled out. The main diagnostic difficulties occur in the early disease stages when additional symptoms arguing for an atypical PS (❷ *Table 1-3*) cannot be delineated clearly.

◘ Table 1-2
If three of the supporting criteria are fulfilled diagnosis is clinically reliable

Supporting criteria for the diagnosis of idiopathic PD
Unilateral manifestation and/or persisting lateralization in the course of the disease
Resting tremor
Positive response (at least 30% improvement) to L-dopa treatment
No additional symptoms indicating involvement of other systems for ten or more years

◘ Table 1-3
Atypical PS

Atypical PS	Most frequent additional symptoms
Parkinsonian multiple system atrophy (MSAp)	Autonomic dysfunction, especially orthostatic hypotension, male impotence, urinary difficulties, dystonia
Cerebellar multiple system atrophy (MSAp)	Cerebellar signs, pyramidal signs, autonomic dysfunction
Progressive supranuclear palsy (PSP)	Postural instability, vertical gaze palsy, dementia, dysphagia
Cortico basal ganglionic degeneration (CBGD)	Cortical signs, alien limb syndrome, myoclonus, pyramidal signs
Diffuse Lewy body dementia (DLBD)	Dementia, fluctuation of vigilance, early and frequent hallucinations on dopaminergic treatment

With progression of the disease, however, it needs to be considered that symptoms which in the beginning of the disease would have been ascribed to an atypical PD like dementia and dystonic postures become frequent in idiopathic PD. Also, postural instability and autonomic symptoms occur in the majority of advanced PD cases. Therefore, it is important to document a careful history including age of and symptoms at onset, duration of the disease, and manifestation of additional symptoms. Response to the symptomatic medication may also vary as the disease progresses. However, in the beginning there should be a good and sustained response to dopaminergic agents. If the response to dopaminergic medication is not clear an L-dopa or apomorphine test should be applied to evaluate therapeutic effects. Severity of PD should be documented by standardized scales like Hoehn and Yahr (1969) or the Unified Parkinson's Disease Rating Scale (UPDRS) (Fahn et al., 1987), which is helpful for estimating the individual progression and complications.

Different methods are useful to rule out symptomatic or atypical PS. Structural neuroimaging techniques like cranial computer tomography (CCT) or magnetic resonance imaging (MRI) are helpful in ruling out symptomatic PS like normal pressure hydrocephalus, a frontal tumor, or ischemic lesions. MRI may also be helpful in the differentiation of the atypical PSs multiple system atrophy (MSA), progressive supranuclear palsy (PSP), and corticobasal ganglionic degeneration (CBGD) (Warmuth-Metz et al., 2001). While normal MRI weightings of PD patients do not show any disease-related abnormalities, there are typical MRI signs indicating MSA-like diminished T2-time signal in the dorsolateral putamen, hyperintense signals at the border of the lateral putamen, and the external capsule as well as cerebellar atrophy. PSP is characterized on MRI by lower anteroposterior midbrain diameters, whereas patients with CBGD show a mainly unilateral pronounced atrophy of the parietal cortex (Gimenez-Roldan et al., 1994; Warmuth-Metz et al., 2001).

As a new, noninvasive and quickly performable method, transcranial sonography (TCS) has become increasingly important in the diagnosis and differential diagnosis of PD (Becker et al., 1995; Berg et al., 2001c; Walter et al., 2002). It enables the visualization of the butterfly-shaped mesencephalic brainstem as a normally hypoechogenic structure. In PD, however, hyperechogenic signals can be delineated at the anatomical side of the SN, which exceed the area of increased echogenicity in the healthy population (❯ Figure 1-1). Studies of different groups demonstrated that this echofeature occurring in more than 90%

◘ Figure 1-1

Transcranial sonography of the mesencephalic brainstem. The butterfly-shaped mesencephalic brainstem is normally hypoechogenic (*left*). In Parkinson's disease areas of hyperechogenicity (*encircled and arrows*) may be displayed on both sides at the anatomical area of the substantia nigra

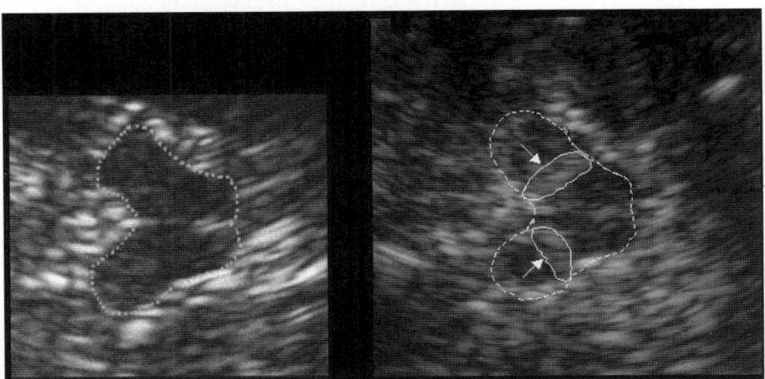

of patients with idiopathic PD, may also be used to differentiate between idiopathic PD and atypical PS (Walter et al., 2004; Behnke et al., 2005). Interestingly, this maker indicates a subgroup of the population at risk for the development of signs of nigrostriatal dysfunction for example, in the process of aging (Berg et al., 2001d) or under neuroleptic therapy (Berg et al., 2001b). Moreover, it has been shown that increased echogenicity occurring very early in life in a small proportion of healthy individuals may be associated with impaired nigrostriatal function detected by 18F-dopa PET (Berg et al., 1999a, 2002). An association between increased echogenicity and increased iron levels has been established (Berg et al., 1999b, 2002).

Functional neuroimaging methods include Single Photon Emission Tomography (SPECT) and Positron Emission Tomography (PET) (Thobois et al., 2004). To visualize the presynaptic function, radioactive markers like ^{123}I-FP-CIT for SPECT and ^{18}F-Dopa are used for examinations. Postsynaptically, IBZM is applied in SPECT examinations to visualize primarily D2 receptors. A number of other radionucleotides like ^{11}C-Nomifensin, ^{11}C-WIN, and ^{11}C-DTBZ for presynaptic PET; and ^{123}I-β-CIT, ^{123}I-IPT, and ^{99m}TC-TRODAT-1 for presynaptic SPECT studies as well as ^{11}C-Racloprid (D2); and ^{18}F-Desmethoxy-Fallpyrid (D2) and ^{11}C-Sch23390 (D1) for postsynaptic PET examinations are only used in specific studies (Brooks et al., 2003).

The Schellong test, urodynamic examinations, and sympathetic skin response may be helpful to test for autonomic dysfunction. Electrophysiological investigations like electromyography of the sphincter muscle and long-latency reflexes may be helpful in the differentiation of atypical PS. Smelling tests are also used in the differential diagnosis of PS, as more than 80% of patients with idiopathic PD show alterations in quantitative smelling tests (Katzenschlager and Lees, 2004).

5   Pathoanatomy

5.1   Histopathology

Idiopathic PD is a multisystem disorder, affecting predisposed nerve cell types in specific regions not only of the central but also of the peripheral and enteric nervous system. Essential for the neuropathological diagnosis of PD are alpha-synuclein positive Lewy bodies (LB) (❷ *Figure 1-2*) and Lewy neurites.

❑ **Figure 1-2**

Two Lewy bodies (*arrows*) in a partly depigmented neuron in the substantia nigra of a patient with Parkinson's disease (HE-staining) (kindly provided by A. Bornemann, Institute of Neuropathology, University of Tübingen)

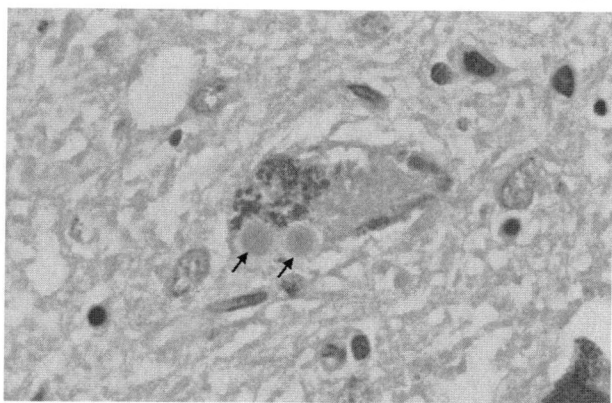

LBs are typical but not specific for PD, as they or Lewy body-like inclusions are also found in other neurodegenerative diseases like multiple system atrophy, progressive supranuclear palsy, Alzheimer's disease, and other diseases affecting the brain like prion diseases, Down syndrome or subacute progressive panencephalitis (Lowe et al., 1997; Goedert et al., 1998; Trojanowski and Lee, 2001). More than 25 different protein components of Lewy bodies are known. One main group of components consists of neurofilaments, α-synuclein, β-amyloid, actin-like protein, and many others; another group of proteins is expressed as a cellular response to the abnormal protein aggregation like ubiquitin, proteasome subunits, and chaperones (Pollanen et al., 1993; Galvin et al., 1999; Fornai et al., 2003).

In PD, brain areas with the highest levels of α-synuclein are those associated with Lewy bodies. It has therefore been speculated that Lewy bodies result from altered handling of oxidized proteins and may at least initially represent a protective mechanism of the cell from the toxicity of protein accumulation (Jenner and

Olanow, 1998; Tanaka et al., 2004). Also, the sequestration of toxic iron by Lewy bodies indicates a protective role of these inclusions (Castellani et al., 2000). New lines of evidence indicate, that by increased expression even α-synuclein itself might protect cells from oxidative stress by inactivation of stress-signaling pathways (Hashimoto et al., 2002). Some investigations, however, indicate that Lewy bodies might contribute to the pathological cascade. In analogy to a model with transgenic mice (Tu et al., 1997), it has been hypothesized that proteins usually transported by axons are accumulated in Lewy bodies leading to a lack of function of neurons and ultimately to degeneration (Trojanowski, 2001). Moreover, it has been shown that α-synuclein overexpression may contribute to the formation of free radicals, thereby exaggerating the vulnerability of neurons to dopamine-induced cell death (Turnbull et al., 2001; Junn and Mouradian, 2002). However, it needs to be noted that the majority of degenerating neurons do not contain Lewy bodies. Therefore, these inclusions do not seem to be necessary for neurodegeneration (Tompkins and Hill, 1997).

According to the spreading of these histopathological hallmarks of the disease, Braak et al. suggest that the disease starts at defined sites and advances in a topographically predictable sequence. During the presymptomatic stages 1–2 Lewy body pathology is confined to the medulla oblongata with lesions occurring in the dorsal motor nuclei of the glossopharyngeal and vagal nerves as well as in the olfactory bulb and nucleus. In the stages 3–4, the SN and other nuclei of the midbrain and forebrain show increasing signs of neurodegeneration reaching the anteromedial temporal mesocortex in stage 4. Involvement of the mature neocortex in the end-stages 5–6 starts with high-order sensory association and prefrontal areas, followed by first-order sensory and association/premotor area and primary sensory/motor fields (Braak et al., 2003, 2004). Most of the leading clinical symptoms can be attributed to degeneration of the SN. Clinical symptoms usually occur when more than 60% of the dopaminergic cells of the SN have degenerated (Bernheimer et al., 1973). Neuronal loss occurs in a characteristic distribution affecting predominantly the ventrolateral and to a lesser extent the medioventral and dorsal parts with a slight concomitant reactive gliosis and appearance of the characteristic Lewy bodies (Fearnley and Lees, 1991). The reason why it is the SN, which is the target of the high degree of oxidative stress-related neurodegeneration in PD, may lie in its high energy metabolism and the high content of dopamine. In the case of the disease, the dopaminergic cells cannot be protected by the normally huge range of protective mechanisms. The fact that primarily long fibers with scarce myelinization needing more energy degenerate (Braak et al., 2003) indicates that a dysbalance of oxidative stress and the physiologically normal lack of antioxidant capacities of these fibers make them especially vulnerable to neurodegeneration.

5.2 Pathoanatomy of the Basal Ganglia in PD

PD has been recognized as a basal ganglia disorder affecting a multitude of projection areas for a long time. Specific clinical symptoms could be derived from the physiological function of the basal ganglia like control of movement amplitude (Desmurget et al., 2004) or reinforcement-based learning (Saint-Cyr, 2003). However, the complexity and parallel processing of the information processes have prevented the establishment of a model embracing the underlying pathoanatomy completely. Moreover, altered representation of different cortical areas as adaptions to the impairments of striatocortical circuits in PD patients renders the understanding of the circuits involved even more difficult (Kagerer et al., 2003). Still most frequently used is the basal ganglia model developed by Alexander et al. (1986). Although there is increasing evidence that this model is not sufficient for all aspects (Foley and Riederer, 2000), it enables a good understanding of the main pathomechanisms of the disease. According to this model the basal ganglia plays a central role in integrating and reorganizing information from cortical and subcortical areas. Signals from the cortex, thalamus, and brainstem nuclei are taken up by the caudate nucleus and dorsal putamen. Within the basal ganglia the information is processed by feedback mechanisms induced by the SN and pallidum. Either directly via the monosynaptic striatal pallidal or indirectly via the polysynaptic striatopallidal way information is then lead to the pallidum and substantia nigra pars reticulata (SNr) and projected via the thalamus to projection areas outside the basal ganglia. The dopaminergic system of the substantia nigra pars compacta (SNc) as well as dopaminergic cell groups of the of the SNr project back to the striatum and to the prefrontal cortex. According to Alexander et al. (1990), there are at least 5 parallel processing basal-ganglia-thalamocortical feedback circuits to be discerned, projecting to the respective cortical areas including a motor, an oculomotor, two prefrontal, and a limbic circuit. These circuits maintain the somatotopic organization of movement related neurons (Herrero et al., 2002). The inhibitory influence in each of these circuits is modulated by two counteracting systems. In the direct system, activation of the striatum is the source of direct inhibition of the internal pallidum and SNr, reducing the normally inhibitory influence of these nuclei on the thalamus. Neurons involved are biochemically characterized by containing mainly D_1 dopamine receptors and coexpress the peptide substance P and dynorphin (Steiner and Gerfen, 1998). Contrary, in the indirect system inhibitory projection to the external pallidum leads via inhibition of the subthalamic nucleus to enhanced activation of the inhibitory influence of the internal pallidum and SNr resulting in enhanced inhibitory function of the thalamus. Striatal neurons involved in this pathway express mainly D_2 dopamine receptors and the peptide enkephalin. In PD especially the indirect system is thought to be involved. Diminished dopaminergic projection of the SNc to the striatum is supposed to result in reduced activation of the inhibitory influence of this basal ganglia area and to be the causative for many of the clinical symptoms (Mitchell et al., 1989).

Additionally, the basal ganglia are intimately interconnected with brain stem nuclei like the locus ceruleus or raphe nuclei (Obeso et al., 2002). Besides, neurodegeneration and Lewy body pathology are found in many other cortical and subcortical areas involved in these circuits (Braak et al., 1995; Jellinger, 1999). Involvement of the limbic system results in a loss of emotional modulation of movements (Braak and Braak, 2000) and additional endocrine, autonomic, and mnestic disturbances, when the amygdala are affected (Braak et al., 1998). When the met- and leu-encephalin systems, which normally modulate the dopaminergic system, are affected it leads to further dysregulation of this neurotransmitter. Degeneration of nuclei involved in the synthesis of other neurotransmitters like noradrenalin (locus ceruleus), acetylcholine (nucleus basalis Meynert), and serotonin (raphe nuclei) leads to symptoms like cognitive decline and depression (Jellinger, 1999, 2001). Interestingly, vulnerability for neurodegeneration is very selective, leaving even specific dopaminergic subgroups of the SN unaffected (Braak and Braak, 2000). The reason for this selective vulnerability is not clear, yet, although there are a number of models referring to known mechanisms to explain this selective neurodegeneration.

6 Factors Contributing to the Pathogenesis of PD

Only a small percentage of PD patients suffer from a monogenetically caused disease (chapter 3 of this book). However, understanding derived from pathomechansims of these genetically determined metabolic pathways, like disturbance of the ubiquitin-proteasome system, have greatly extended knowledge of possible causes of the disorder, that may also contribute to sporadic PD. The main group of disease cases seems to be caused by a variable combination of factors including endogenous production of oxidative stress on the one hand and a variable chronic exposure to environmental agents, in addition to a genetic vulnerability. These factors may influence gene expression in the substantia nigra (Grunblatt et al., 2004). Extensive research of the past years has discovered some main factors contributing to the pathogenetic cascade of the disease.

6.1 Aging

One of the main risk factors for sporadic PD is aging. It has been estimated that 10,000 oxidative interactions occur between DNA and endogenously generated free radicals per human cell per day, and that in older animals at least one of every three proteins in the cell is dysfunctional as an enzyme or structural protein, due to oxidative modification, indicating a key role of free radicals in the process of aging (Poon et al., 2004). Consequences of this aging-related oxidative modifications are increased levels of lipid peroxidation metabolites (Weisser et al., 1997; Volchegorskii et al., 2004) and a reduction in cytoprotective mechanisms, like decreased levels of hyaluronic acid, which is known to inhibit expression of proinflammatoric cytokines. In addition aging-related changes in the synthesis and metabolism of neurotransmitters and their receptors, like a reduction in the density of dopamine receptors (Rinne et al., 1990) in contrast to an increase in serotonergic receptors (Marcusson et al., 1984) or changes in the metabolism of essential heavy metals like iron (Zecca et al., 2001) may induce or accelerate pathological cascades of neurodegeneration. The time of manifestation of clinical symptoms seems to depend on the genetic constellation of each subject, which is responsible for the increased vulnerability to each of these factors.

However, PD can not be explained as an accelerated process of aging. While there is a selective degeneration of dopaminergic neurons of the SNc in PD, aging is accompanied by loss of dopaminergic neurons of the dorsal tier of the SN (Fearnley and Lees, 1991; Halliday et al., 1990). Moreover, aging is not accompanied by the formation of Lewy bodies, the pathologic hallmark of PD.

6.2 Oxidative Stress

6.2.1 Definition and Findings

Reactive oxygen species (ROS) and reactive nitrogen species (RNS) are generated as a result of normal metabolism. However, the deleterious condition termed oxidative stress (OS) occurs when ROS or RNS such as the superoxide radical ($O_2^{\bullet-}$), nitric oxide (NO) and especially the hydroxyl radical ($^{\bullet}OH$) due to an excessive production or diminished decontamination overwhelm the protective defense mechanisms of a cell resulting in functional disruption and ultimately in cell death.

Although it is no question that OS it as key event in the neurodegenerative process of all forms of PD (Berg et al., 2004), it is not clear, yet, whether it constitutes a primary event or a consequence of other pathogenetic factors. Studies in patients with incidental Lewy body disease (ILBD), the supposed pre-symptomatic form of PD, implicated that there is no conclusive evidence of markers of OS at an early stage of neurodegeneration besides a reduction in GSH levels (Jenner, 2003). However, only few samples have been examined with tissue homogenates rather than detailed studies on dopaminergic neurons. Therefore, no final conclusion can be derived from this observation. On the other hand, it has been shown that oxidative dimer formation constitutes the critical rate-limiting step for fibrillogenesis of α-synuclein, providing an explanation that overproduction of ROS and/or impairment of cellular antioxidative mechanisms could be a primary event both in the initiation and in the progression of PD (Krishnan et al., 2003).

Because of their high reactivity free radicals cannot be measured directly. However, there are a number of indices for OS and its deleterious consequences in the SN of PD patients:

- *Lipid peroxidation of membranes* – Membranes are crucial for cell viability. As a consequence of OS reaction of fatty acids of polar lipids with oxygen or a reaction catalyzed by either metals like iron or by NADPH cytochrome P-450 reductase leads to the formation of free radical intermediates and semi-stable peroxide. Chemically and metabolically stable oxidation products like isoprostanes and increased levels of secondary products like conjugated dienes, hydrocarbon gases (e.g., ethane) and carbonyl compounds (e.g., malondialdehyde), and decreased levels of polyunsaturated fatty acid can be measured (Dexter et al., 1989; Practico, 2001; Montine et al., 2004).
- *Oxidation of proteins* – ROS may directly oxidize amino acids leading to a loss of function of proteins and a deactivation of enzymes (Stadtman, 2001). Increase in malondialdehyde has been suggested to lead to intra- and intermolecular cross-links of proteins (Goetz and Gerlach, 2004). Moreover, conformational changes and adducts of lipid peroxidation (Marnett, 2000) may result in aggregation or precipitation of undegradable proteins, a hypothesis that is underlined by the detection of OS-dependent protein aggregation in the form of advanced glycation end products (Münch et al., 1999) which are already detectable within Lewy bodies, when no clinical sign of PD is obvious (Münch et al., 2000).
- *DNA damage caused by ROS* – Increased levels of 8-hydroxy-2′hydroxyguanine, and thymidine glycol in the SN and striatum of Parkinson's disease brain indicate DNA base damage (Sanchez-Ramos et al., 1994; Alam et al., 1997; Poon et al., 2004) as a result of strand breakage, nucleic acid-protein cross-linking, and nucleic base modification by ROS. These processes also affect the mitochondrial DNA (mtDNA), which is transiently attached to the inner mitochondrial membrane where a large amount of ROS is produced (Goetz and Gerlach, 2004).

- *Further markers of OS* – 3-Nitrotyrosine, the marker of OS induced by peroxynitrite, has been detected in Lewy bodies (Giasson et al., 2000). Decreased levels of reduced glutathione in the SN of PD (Sofic et al., 1992) and low activity of phospholipid-catabolising enzymes in normal SN compared with other regions of the human brain, indicate a reduced repair capacity of oxidative damage (Ross et al., 1998). On the other hand up-regulation of antioxidant enzymes like Mn-dependent superoxide dismutase and nonselenium glutathione peroxidase has been described (Martilla et al., 1988; Yoshida et al., 1994; Power et al., 2002).

Very importantly, the 30–40% changes in parameters of oxidative stress found in homogenates of nigral tissue cannot be restricted to the dopaminergic cells accounting for only 1–2% of the whole nigral cell population (Jenner, 2003). It needs to be considered that changes occur also in other cell types, predominantly in glial cells, implying a concept of general metabolic failure in the SN of PD patients.

Moreover, peripheral markers for oxidative stress indicate a possible systemic affection related to the oxidative stress in the brain. These include altered SOD activity, reductions in glutamate uptake, increased levels of malondialdehyde as well as thiobarbituric acid reactive substances detected in erythrocytes, serum, and plasma of PD patients and primary and oxidative DNA damage in lymphocytes of untreated Parkinson patients (Bostantjopoulou et al., 1997; Nagatsu et al., 1999; Ferrarese et al., 2001; Serra et al., 2001; Migliore et al., 2002).

6.2.2 Factors Contributing to the Generation of Oxidative Stress

A number of factors are known to exert toxic influences in neurodegenerative diseases (for details see the chapter by Gerlach and Double). For the generation of oxidative stress in PD some factors have been shown to be responsible.

1. *Dopamine* – Metabolism of this main neurotransmitter of the neurons degenerating in PD produces ROS and might therefore account at least in part for the selective vulnerability of the SNc. Already during the process of dopamine synthesis cytotoxic products like reactive dopamine quinone products may be formed (Choi et al., 2003). Then, autooxidation leading to the production of dopaquinone and $O_2^{\bullet-}$ (Lotharius and Brundin, 2002) or enzymatic metabolism forming H_2O_2, which is converted into $^\bullet OH$ via the iron-mediated Fenton reaction (Götz et al., 1994; Maruyama and Naoi, 2002) are the main sources of ROS derived from dopamine metabolism. Moreover, dopamine and various of its metabolites can inhibit complex I of the electron-transport chain (see later). An additional toxic effect of the antiparkinsonian drug L-Dopa has been the matter of discussion for many years. In vitro experiments seem to indicate an increase in OS and neurodegeneration after L-Dopa administration (Spencer et al., 1994; Walinshaw and Waters, 1995). In animal experiments, however, no increased toxicity was seen after L-Dopa application (Perry et al., 1984). Also, the ELL-Dopa clinical trial has not substantiated an L-Dopa-induced toxicity (Fahn, 1999).

2. *Transition metals iron, copper, and manganese* – By their ability to undergo one-electron transfer, these essential metals are also potentially dangerous, enabling autooxidation (e.g., of dopamine and ascorbate), conversion of H_2O_2 to $(HO)^\bullet$, or decomposition of lipid peroxides to reactive peroxyl and alkoxyl radicals. Most evidence exists for the contribution of iron to OS in PD. Iron content of the basal ganglia and SN is already under physiologic conditions higher than in most other regions of the brain (Hallgren and Sourander, 1958). This is important, as iron is required as a cofactor for the synthesis of dopamine by the enzyme tyrosine hydroxylase. The physiologically high iron content of the SN is even 35% higher in PD, with an elevation of the Fe(III)/Fe(II)-ratio from 2:1 to almost 1:2 (Riederer et al., 1985, 1988, 1989; Sofic et al., 1988; Dexter et al., 1993; Gerlach et al., 1994). Increased levels of iron and Fe(II) enhance the conversion of H_2O_2 to $^\bullet OH$ via the Fenton reaction and favor a greater turnover in the Haber–Weiss cycle, which leads to an amplification of ROS (Riederer and Youdim, 1993; Berg et al., 2001a) with all detrimental consequences. Moreover, by catalyzing oxidative reactions iron enhances the conversion of unfolded or α-helical conformation of α–synuclein to β-pleated sheet conformation, the primary form in Lewy bodies (Hashimoto et al., 1998; Münch et al., 2000; Golts et al., 2002). It is not entirely clear yet, at what time in the pathophysiological cascade of PD iron accumulation occurs.

Data from recent transcranial ultrasound studies also imply iron accumulation to occur very early in the disease process, constituting a rather primary cause of the disease(Becker et al., 1995; Berg et al., 1999a, 2002). Continuous and unlimited iron uptake through a disturbed blood–brain barrier may be one reason for the accumulation (Riederer, 2004; Kortekaas et al., 2005). Moreover, an association of sequence variations in some genes encoding for iron metabolizing proteins within the brain and PD has been established (Borie et al., 2002; Hochstrasser et al., 2004).

3. *Neuromelanin* – This excellent chelator of metal ions, especially iron (Ben-Shachar et al., 1991), has been discussed to be neuroprotective (Gerlach et al., 1994). However, it seems that the amount of iron determines the role of NM: In the situation of normal iron levels, this redox active metal is sequestered. In the presence of excess iron, however, NM promotes the formation of reactive oxygen species and fosters the release of iron into the cytoplasm (Zareba et al., 1995; Zecca et al., 1996; Double et al., 1999). Additionally, NM can bind a variety of potentially toxic substances like MPP+, the neurotoxic metabolite of MPTP or pesticides suggesting a contribution to neurotoxin-mediated neurodegeneration (D'Amato et al., 1986; Gerlach et al., 1994).

4. *Nitric Oxide* – Nitric oxide (NO) is produced in the reaction of arginine with molecular oxygen to generate citrulline and NO (Bredt, 1999). In PD enhanced production of NO by activated glial cells has been observed (Hunot et al., 1999). As a free radical, NO contributes to OS by reacting with proteins and lipids (Beckmann, 1996; Iravani et al., 2002). Additionally, highly reactive intermediates like peroxinitrite are formed in the presence of superoxide, inducing DNA damage and lipid peroxidation (Liu et al., 2002). NO also mediates iron release from ferritin (Reif and Simmons, 1990) and interacts directly with the respiratory chain (Antunes et al., 2002).

6.2.3 Factors Diminishing Oxidative Stress

The cell protects itself against oxidative damage using either enzymatic reactions by superoxide dismutase (SOD), glutathione peroxidase, or catalase, or interaction with substances binding the toxic metabolites like the antioxidants vitamin C and E, glutathione (GSH), cystein, or coenzyme Q (Gerlach et al., 2001). However, although expected to counteract the increased oxidative stress no difference in the levels of antioxidative vitamins was found between PD patients and controls (Riederer et al., 1989). For catalase and glutathione peroxidase almost equal or only slightly reduced levels and activities were found (Ambani et al., 1975; Kish et al., 1985; Sian et al., 1994). However, increase in the antioxidant nonselenium glutathione peroxidase in astrocytes in PD brains (Power et al., 2002) and a specific and unique increase in SOD in the SNc were detected (Martilla et al., 1988; Yoshida et al., 1994). However, this increase is not accompanied by an increase in the subsequent enzymes of the detoxification process. As H_2O_2 is generated in the process of detoxifying free radicals by SOD, lack of the enzymes catalase and glutathione needed for further conversion leads to a deleterious increase in H_2O_2 (Foley and Riederer, 1999). Moreover, a reduction of total and reduced GSH, a very important antioxidant synthesized in every brain cell (Wu et al., 2004), is detectable in the SNc in PD (Riederer et al., 1989; Sofic et al., 1992) in the very early pathogenetic process, rendering neurons more vulnerable to endo- or exotoxins (Jenner and Olanow, 1996). Especially for PD GSH depletion has been shown to cause a cascade of deleterious events like promoting the formation of defective proteins and thereby impairing the normal ubiquitin-proteasome pathway of protein degradation (Bharat et al., 2002). Additionally, release of arachidonic acid by GSH depletion may cause cellular damage through its metabolism by lipooxygenase (Kramer et al., 2002; Mytilineou et al., 2002). GSH depletion may also have an inhibitory affect of mitochondrial complex I activity (Bharat et al., 2002).

6.3 Mitochondrial Dysfunction

Activity of complex I, the first of five complexes of the mitochondrial respiratory chain, has been found to be reduced by about 30% in SNc of PD patients (Mizuno et al., 1989; Reichmann and Riederer, 1989;

Janetzky et al., 1994). Decline of complex I activity does not only lead to reduced ATP production resulting in decreased energy for the cell. Also, less NAD is formed from NADH with the consequence of an impairment of proton pumping and electron transport, which in turn leads to an increase in reactive oxygen species, proteasome inhibition, and cell death (Reichmann et al., 1993, Betarbet et al., 2000; Liu et al., 2002). In addition to the cellular energy supply, the specific role of determining survival and death in apoptosis induced by endogenous neurotoxins, which has been attributed to mitochondria (Naoi et al., 2000, 2002), is also impaired by the reduced activity. It is not clear, yet, whether complex I impairment is a secondary phenomenon for example, resulting from radical formation. On the other hand, sequence variations in the mitochondrial genome in the brain of PD patients with an increase in number and variety of mtDNA deletions/rearrangements compared to patients with other movement disorders or age-matched controls (Gu et al., 2002) and a variety of amino acid variations detected in cytochrome *b* of the mitochondrial genome in PD patients compared to only few variations in centenarians (Tanaka, 2002) argue for a primary contribution. Besides an additional impairment of complex IV activity (Itoh et al., 1997) no alterations in the functions of the other complexes of the respiratory chain have been detected, yet.

6.4　Impairment of Calcium Homeostasis and the Role of Glial Cells

Increase in ROS, overstimulation of a cell by "excitotoxic" inputs, and impairment of the respiratory chain may lead to an increase in intracellular calcium. Excitotoxicity means excitatory and toxic properties of a substance at the same time. Impaired calcium homeostasis with high levels of intracellular calcium may lead to uncontrolled stimulation of various calcium-dependent enzymes and stimulation of mitochondrial generation of ROS (Gerlach et al., 1996; Starkov et al., 2002). Moreover, calcium overload may uncouple mitochondria leading to a further enhancement of OS. Because of their need of high rates of energy, dopaminergic neurons are especially vulnerable to these processes. Moreover, studies on calcium binding to α-synuclein suggest that calcium ions may on the one hand participate in normal α-synuclein functioning, while on the other hand exercising pathological effects known to be involved in Lewy body formation (Nielsen et al., 2001).

The release of glutamate, the main excitatory transmitter in human central nervous system, is among others mediated by cytokines (Mogi et al., 1994; Hunot et al., 1999) produced by activated glia cells (McNaught and Jenner, 2000; Koutsilieri et al., 2002) that lead to an elevation of OS (Gao et al., 2002) and excitotoxicity. However, activation of glial cells does not only exert detrimental influences on dopaminergic neurons (Teismann and Schulz, 2004). They also produce glutathione peroxidase and pyruvate, which protect cells from highly toxic free radicals (Damier et al., 1993; Desagher et al., 1997), and neurotrophic factors exerting a neuroprotective influence-like GDNF (glial cell line-derived neurotrophic factor) and BDNF (brain-derived neurotrophic factor), which have proven neuroprotective against different neurotoxic substances in vitro and in vivo (Beck et al., 1993; Gash et al., 1996).

6.5　Exotoxins

Since the discovery that MPTP, an impurity formed during the preparation of meperidine, provokes severe, progressive parkinsonism (Davis et al., 1979; Langston et al., 1983), many epidemiological studies have focussed on the possible contribution of environmental factors, nutritional components, or exposition to certain toxins to the pathogenesis of PD. So far, substances known to cause or promote PD are CO, manganese, cyanid, and methanol (Glass 1983; Zayed et al., 1990). Exposition to these substances might be at home, as seen in the case of drinking water from wells, polluted with heavy metals (Rajput et al., 1986; Barbeau et al., 1987). However, not all investigations corroborate these findings (Rajput et al., 1987). At work pesticides and herbicides are discussed to be the main substances of exposition (Sanchez-Ramos et al., 1987; Sechi et al., 1992; Sherer et al., 2002a; McCarthy et al., 2004). Chronic treatment with the insecticide rotenone, a specific inhibitor of complex I, has been suggested to cause delayed oxidative stress (Sherer et al., 2002b;

Hoglinger et al., 2003). Other toxins like chloral hydrate or 1-trichloromethyl-1,2,3,4-tetrahydro-beta-carboline reacting with endogen amines (Bringmann et al., 2000; Riederer et al., 2002) and dieldrin (Kitazawa et al., 2001) are still investigated. As not all individuals exposed to the same environmental toxins develop PD, a genetically determined susceptibility is probable (see earlier; Jenner, 2001; Tsang and Soong, 2003).

The time point of exposure may be variable. Interestingly, there are indices that environmental factors, including viral and risk factors already associated with pregnancy and birth, together with genetically determined lability may increase the incidence of early hypokinesia and parkinsonism in particular and of Parkinson's disease in later life by disturbing the immature basal ganglia at critical developmental stages (Riederer and Foley, 2002; Iova et al., 2004).

Taken together, research of the past decades has elucidated many aspects concerning the pathoanatomy, histology, and some risk factors for PD. However, although some of the basic pathomechanisms for neuronal death in PD have been elucidated, the initial cause and reason for progression are still not entirely understood. Many helpful aspects concerning possible pathomechanisms have been revealed by insights derived from studies in monogenetically caused forms of the disease (see chapter 3 of this book).

Together with the understanding of the pathophysiology of the disease, methods for an earlier, in the best case preclinical diagnosis, need to be developed to enable the ingenious application of neuroprotective therapeutic strategies (chapter 6 of this book).

References

Alam ZI, Jenner A, Daniel SE, Lees AJ, Cairns N, et al. 1997. Oxidative DNA damage in the parkinsonian brain: A selective increase in 8-hydroxyguanine in substantia nigra? J Neurochem 69: 1196-1203.

Alexander GE, Crutcher MD, DeLong MR. 1990. Basal-ganglia-thalamocortical circuits: Parallel substrates for motor, oculomotor, "prefrontal" and "limbic"functions. Prog Brain Res 85: 119-146.

Alexander GE, DeLong MR, Strick PL. 1986. Parallel organisation of functionally segregated circuits linking basal ganglia and cortex. Annu Rev Neurobiol 9: 357-381.

Ambani LM, Van Woert MH, Murphy S. 1975. Brain peroxidase and catalase in Parkinson's disease. Arch Neurol 32: 114-118.

Antunes F, Han D, Rettori D, Cadenas E. 2002. Mitochondrial damage by nitric oxide potentiated by dopamine in PC12 cells. Biochim Biophys Acta 1556: 233-238.

Barbeau A, Roy M, Cloutier T, Plasse L, Paris S. 1987. Environmental and genetic factors in the etiology of Parkinson's disease. Adv Neurol 45: 299-306.

Barbeau AT, Sourkes L, Murphy G. 1962. Les catécholamines dans la maladie de Parkinson. Monoamines et Système Nerveux Central. De Ajuriaguerra J, editor. Paris: Georg, Geneva and Masson; pp. 247-262.

Beck KD, Knusel B, Hefti F. 1993. The nature of the trophic action of brain-derived neurotrophic factor, des(T-3)-insulin-like growth factor, and basic fibroblast growth factor on mesencephalic dopaminergic neurons developing in culture. Neuroscience 52: 855-866.

Becker G, Seufert J, Bogdahn U, Reichmann H, Reiners K. 1995. Degeneration of substantia nigra in chronic Parkinson's disease visualized by transcranial color-coded real-time sonography. Neurology 45: 443-454.

Beckmann JS. 1996. Oxidative damage and tyrosine nitration from peroxynitrite. Chem Res Toxicol 9: 836-844.

Behnke S, Berg D, Naumann M, Becker G. 2005. Differentiation of Parkinson's disease and atypical parkinsonian syndromes by transcranial ultrasound. J Neurol Neurosurg Psychiatry 76: 423-425.

Ben-Shachar D, Riederer P, Youdim MB. 1991. Iron-melanin interaction and lipid peroxidation: Implications for Parkinson's disease. J Neurochem 57: 1609-1614.

Berg D, Becker G, Zeiler B, Tucha O, Hofmann E, et al. 1999a. Vulnerability of the nigrostriatal system as detected by transcranial ultrasound. Neurology 53: 1026-1031.

Berg D, Grote C, Rausch WD, Mäurer M, Wesemann W, et al. 1999b. Iron accumulation of the substantia nigra in rats visualized by ultrasound. Ultrasound Med Biol 25: 901-904.

Berg D, Gerlach M, Youdim MBH, Doulbe KL, Zecca L, et al. 2001a. Brain iron pathways and their relevance to Parkinson's disease. J Neurochem 79: 225-236.

Berg D, Jabs B, Merschdorf U, Beckmann H, Becker G. 2001b. Echogenicity of substantia nigra determined by transcranial ultrasound correlates with severity of Parkinsonian symptoms induced by neuroleptic therapy. Biol Psychiatry 50: 463-467.

Berg D, Siefker C, Becker G. 2001c. Echogenicity of the substantia nigra in Parkinson's disease and its relation to clinical findings. J Neurol 248: 684-689.

Berg D, Siefker C, Ruprecht-Dörfler P, Becker G. 2001d. Relationship of substantia nigra echogenicity and motor function in elderly subjects. Neurology 56: 13-17.

Berg D, Roggendorf W, Schröder U, Klein R, Tatschner T, et al. 2002. Echogenicity of the substantia nigra—association with increased iron content and marker for susceptibility to nigrostriatal injury. Arch Neurol 59: 999-1005.

Berg D, Youdim MB, Riederer P. 2004. Redox imbalance. Cell Tissue Res 318: 201-213.

Bernheimer H, Birkmayer W, Hornykiewicz O, Jellinger K, Seitelberger F. 1973. Brain dopamine and the syndromes of Parkinson and Huntington. Clinical, morphological and neurochemical correlations. J Neurol Sci 20: 415-455.

Betarbet R, Sherer TB, Mac Kenzie G, Garcia-Osuna M, Panov AV, et al. 2000. Chronic systemic pesticide exposure reproduces features of Parkinson's disease. Nat Neurosci 3: 1301-1306.

Bharat S, Hsu M, Kaur D, Rajagopalan S, Andersen JK. 2002. Glutathione, iron and Parkinson's disease. Biochem Pharmacol 64: 1037-1048.

Biggins CA, Boyd JL, Harrop FM, Madeley P, Mindham RH, et al. 1992. A controlled, longitudinal study of dementia in Parkinson's disease. J Neurol Neurosurg Psychiatry 55: 566-571.

Birkmayer WO, Hornykiewicz O. 1961. Der L-Dioxyphenyla-lanin (=L-Dopa)-Effekt bei der Parkinson-Akinese. Wien Klein WschR 73: 787-788.

Blocq PO, Marinesco G. 1893. Sur en cas de tremblement parkinsonien hemiplégique symptomatique d'une tumeur du pédoncule cérébral. CR Soc de Biol 5: 105-111.

Borie C, Gasparini F, Verpillat P, Bonnet AM, Agid Y, et al. 2002. Association study between iron-related genes polymorphisms and Parkinson's disease. J Neurol 249: 801-804.

Bostantjopoulou S, Kyriazis G, Katsarou Z, Kiosseoglou G, Kazis A, et al. 1997. Superoxide dismutase activity in early and advanced Parkinson's disease. Funct Neurol 12: 63-68.

Braak H, Braak E. 2000. Pathoanatomy of Parkinson's disease. J Neurol 247: II/3-II/10.

Braak H, Braak E, Yilzmar D, Schlutz C, DeVos RAI, et al. 1995. Nigral and extranigral pathology in Parkinson's disease. J Neural Transm 46: 15-31.

Braak H, Del Tredici K, Rub U, de Vos RA, Jansen EN, et al. 2003. Staging of brain pathology related to sporadic Parkinson's disease. Neurobiol Aging 24: 197-211.

Braak H, DeVos RAI, Jansen ENH, Bratzke H, Braak E. 1998. Neuropathological hallmarks of Alzheimer's and Parkinson' disease. Prog Brain Res 117: 267-285.

Braak H, Ghebremedhin E, Rub U, Bratzke H, Del Tredici K. 2004. Stages in the development of Parkinson's disease-related pathology. Cell Tissue Res 318: 121-134

Bredt DS. 1999. Endogenous nitric oxide synthesis: Biological functions and pathophysiology. Free Radic Res 31: 577-596.

Bringmann G, Feineis D, Bruckner R, Blank M, Peters K, et al. 2000. Bromal-derived tetrahydro-beta-carbolines as neurotoxic agents: Chemistry, impairment of the dopamine metabolism, and inhibitory effects on mitochondrial respiration. Bioorg Med Chem 8: 1467-1478.

Brooks DJ, Frey KA, Marek KL, Oakes D, Paty D, et al. 2003. Assessment of neuroimaging techniques as biomarkers of the progression of Parkinson's disease. Exp Neurol 184: S68-S79.

Carlsson A, Lundqvist M, Magnusson T. 1957. 3,4-Dihydrox-yphenylalanine and 5-hydroxy-tryptophan as reserpine antagonists. Nature 180: 1200.

Castellani RJ, Siedlak SL, Perry S, Smith MA. 2000. Sequestration of iron by Lewy bodies in Parkinson's disease. Acta Neuropathol 100: 111-114.

Choi HJ, Kim SW, Lee SY, Hwang O. 2003. Dopamine-dependent cytotoxicity of tetrahydrobiopterin: A possible mechanism for selective neurodegeneration in Parkinson's disease. J Neurochem 86: 143-152.

Cummings JL. 1992. Depression and Parkinson's disease: A review. Am J Psychiatry 149: 443-454.

D'Amato RJ, Lipman ZP, Snyder SH. 1986. Selectivity of the Parkinson neurotoxin MPTP: Toxic metabolite MPP^+ binds to neuromelanin. Science 231: 987-989.

Damier P, Hirsch EC, Zhang P, Agid Y, Javoy-Agid F. 1993. Glutathione peroxidase, glial cells and Parkinson's disease. Neuroscience 52: 1-6.

Davis GC, Williams AC, Markey SP, Ebert MH, Caine ED, et al. 1979. Chronic Parkinsonism secondary to intravenous injection of meperidine analogues. Psychiat Res 1: 249-253.

De Rijk MC, Tzourio C, Breteler MMB, Dartigues JF, Amaducci L, et al. 1997. Prevalence of parkinsonism and Parkinson's disease in Europe: The EUROPARKINSON collaborative study. J Neurol Neurosurg Psychiatry 62: 10-15.

Desagher S, Glowinski J, Premont J. 1997. Pyruvate protects neurons against hydrogen peroxide-induced toxicity. J Neurosci 17: 9060-9067.

Desmurget M, Grafton ST, Vindras P, Grea H, Turner RS. 2004. The basal ganglia network mediates the planning of movement amplitude. Eur J Neurosci 19: 2871-2880.

Dexter DT, Carter CJ, Wells FR, Javoy-Agid F, Agid Y, et al. 1989. Basal lipid peroxidation in substantia nigra is increased in Parkinson's disease. J Neurochem 52: 381-389.

Dexter DT, Sian J, Jenner P, Marsden CD. 1993. Implications of alterations in trace element levels in brain in Parkinson's disease and other neurological disorders affecting the basal ganglia. Adv Neurol 60: 273-281.

Double KL, Riederer P, Gerlach M. 1999. Significance of neuromelanin for neurodegeneration in Parkinson's disease. Drug News Perspect 12: 333-340.

Ehringer H, Hornykiewicz O. 1962. Verteilung von Nor-adrenalin und Dopamin (3-Hydroxytyramin) im Gehirn des Menschen und ihr Verhalten bei Erkrankungen des extrapyramidalen Systems. Klin Wschr 38: 1236-1239.

Fahn S. 1999. Parkinson disease, the effect of levodopa, and the ELLDOPA trial. Earlier vs Later L-DOPA. Arch Neurol 56: 529-535.

Fahn S, Elton RL, Members of the UPDRS Development Committee. 1987. Unified Parkinson's disease rating scale. Fahn S, Marsden CD, Goldstein M, editors. et al. Recent Developments in Parkinson' s Disease. Macmillan; New York: pp. 153-163.

Fearnley JM, Lees AJ. 1991. Ageing and Parkinson's disease: Substantia nigra regional selectivity. Brain 114: 2283-2301.

Ferrarese C, Tremolizzo L, Rigoldi M, Sala G, Begni B. 2001. Decreased platelet glutamate uptake and genetic risk factors in patients with Parkinson's disease. Neurol Sci 22: 65-66.

Foley P, Riederer P. 2000. The motor circuit of the human basal ganglia reconsidered. J Neural Transm Suppl 58: 97-110.

Foley P, Riederer P. 1999. Pathogenesis and preclinical course of Parkinson's disease. J Neural Transm 56: S31-S74.

Fornai F, Lenzi P, Gesi M, Ferrucci M, Lazzeri G, et al. 2003. Fine structure and biochemical mechanisms underlying nigrostriatal inclusions and cell death after proteasome inhibition. J Neurosci 23: 8955-8966.

Galvin JF, Lee VMY, Schmidt L, Tu PH, Iwatsubo T, et al. 1999. Pathology of the Lewy body. Adv Neurol 80: 313-324.

Gao HM, Jiang J, Wilson B, Zhang W, Hong JS, et al. 2002. Microglia activation-mediated delayed and progressive degeneration of rat nigral dopaminergic neurons: Relevance to Parkinson's disease. J Neurochem 81: 1285-1297.

Gash DM, Zhang Z, Ovadia A, Cass WA, Yi A, et al. 1996. Functional recovery in parkinsonian monkeys treated with GDNF. Nature 380: 252-255.

Gerlach M, Ben-Shachar D, Riederer P, Youdim MB. 1994. Altered brain metabolism of iron as a cause of neurodegenerative diseases. J Neurochem 63: 793-807.

Gerlach M, Reichmann H, Riederer P. 2001. Die Parkinsonkrankheit, Grundlagen, Klinik, Therapie. New York: Springer, Wien.

Gerlach M, Riederer P, Youdim MBH. 1996. Molecular mechanisms for neurodegeneration: Synergism between reactive oxygen species, calcium and exitotoxic amino acids. Parkinson's Disease, Advances in Neurology, Vol. 69. Battistin L, Scarlato G, Caraceni T, Ruggieri S, editors. Philadelphia: Lippincott-Raven; pp. 177-194.

Giasson BI, Duda JE, Murray IV, Chen Q, Souza JM, et al. 2000. Oxidative damage linked to neurodegeneration by selective alpha-synuclein nitration in synucleinopathy lesions. Science 3: 985-989.

Gimenez-Roldan S, Mateo D, Benito C, Grandas F, Perez-Gilabert Y. 1994. Progressive supranuclear palsy and corticobasal ganglionic degeneration: Differentiation by clinical features and neuroimaging techniques. J Neural Transm Suppl 42: 79-90.

Glass J. 1983. Untersuchung zur Bedeutung chemischer Noxen in der Ätiologie des Parkinson Syndroms. Pathophysiologie, Klinik und Therapie des Parkinsonismus. Roches; Basel: pp. 103-107.

Goedert M, Spillantini MG, Davies SW. 1998. Filamentous nerve cell inclusions in neurodegenerative diseases. Curr Opin Neurobiol 8: 619-632.

Goetz ME, Gerlach M. 2004. Formation of radicals. Brain Damage and Repair. Herdegen T, Delgado-Garcia J, editors. London: Kluwer academic publishers; pp. 135-164.

Golts N, Snyder H, Frasier M, Theisler C, Choi P, et al. 2002. Magnesium inhibits spontaneous and iron-induced aggregation of alpha-synuclein. J Biol Chem 277: 16116-16123.

Götz ME, Künig G, Riederer P, Youdim MBH. 1994. Oxidative stress. Free radical production in neural degeneration. Pharmacol Ther 63: 37-122.

Greenfield JG, Bosanquet FD. 1953. The brain-stem lesions in Parkinsonism. J Neurol Neurosurg Psychiatry 16: 213-226.

Grunblatt E, Mandel S, Jacob-Hirsch J, Zeligson S, Amariglo N, et al. 2004. Gene expression profiling of parkinsonian substantia nigra pars compacta; alterations in ubiquitin-proteasome, heat shock protein, iron and oxidative stress regulated proteins, cell adhesion/cellular matrix and vesicle trafficking genes. J Neural Transm 111: 1543-1573.

Gu G, Reyes PE, Golden GT, Woltjer RL, Hulette C, et al. 2002. Mitochondrial DNA deletions/rearrangements in parkinson disease and related neurodegenerative disorders. J Neuropathol Exp Neurol 61: 634-639.

Hallgren B, Sourander P. 1958. The effect of age on non-haem iron in the human brain. J Neurochem 3: 41-51.

Halliday GM, Li YW, Blumbergs PC, Joh TH, Cotton RG, et al. 1990. Neuropathology of immunohistochemically identified brainstem neurons in Parkinson's disease. Ann Neurol 27: 373-385.

Hashimoto M, Hsu LJ, Rockenstein E, Takenouchi T, Mallory M, et al. 2002. alpha-Synuclein protects against oxidative stress via inactivation of the c-Jun N-terminal kinase stress-signaling pathway in neuronal cells. J Biol Chem 277: 11465-11472.

Hashimoto M, Hsu LJ, Sisk A, Xia Y, Takeda A, et al. 1998. Human recombinant NACP/alpha-synuclein is aggregated and fibrillated in vitro: Relevance for Lewy body disease. Brain Res 799: 301-306.

Hassler R. 1938. Zur Pathologie der Paralysis agitans und des postenzephalitischen Parkinsonismus. J Psychol Neurol 48: 387-476.

Herrero MT, Barcia C, Navarro JM. 2002. Functional anatomy of thalamus and basal ganglia. Childs Nerv Syst 18: 386-404.

Hochstrasser H, Bauer P, Walter U, Behnke S, Spiegel J, et al. 2004. Ceruloplasmin gene variations and substantia nigra hyperechogenicity in Parkinson disease. Neurology 63: 1912-1917.

Hoehn MM, Yahr MD. 1969. Parkinsonism: Onset, progression and mortality. Neurology 17: 427-442.

Hoglinger GU, Carrard G, Michel PP, Medja F, Lombes A, et al. 2003. Dysfunction of mitochondrial complex I and the proteasome: Interactions between two biochemical deficits in a cellular model of Parkinson's disease. J Neurochem 86: 1297-1307.

Hunot S, Dugas N, Faucheux B, Hartmann A, Tardieu M, et al. 1999. Fcepsilon-RII/CD23 is expressed in Parkinson's disease and induces, in vitro, production of nitric oxide and tumor necrosis factor-alpha in glial cells. J Neurosci 19: 3440-3447.

Iova A, Garmashov A, Andruchtchenko N, Kehrer M, Berg D, et al. 2004. Postnatal decrease in substantia nigra echogenicity–implications for the pathogenesis of Parkinson's disease. J Neurol 251: 1451-1454.

Iravani MM, Kashefi K, Mander P, Rose S, Jenner P. 2002. Involvement of inducible nitric oxide synthase in inflammation-induced dopaminergic neurodegeneration. Neuroscience 110: 49-58.

Itoh K, Weis S, Mehraein P, Muller-Hocker J. 1997. Defects of cytochrome c oxidase in the substantia nigra of Parkinson's disease: An immunohistochemical and morphometric study. Mov Disord 12: 9-16.

Jellinger KA. 1999. Post mortem studies in Parkinson's disease—is it possible to detect brain areas for specific symptoms? J Neural Transm Suppl. 56: 1-29.

Jellinger KA. 2001. The pathology of Parkinson's disease. Parkinson's Disease, Advances in Neurology, Vol. 86. Calne D, Calne S, editors. Philadelphia: Lippincott Williams & Wilkins; pp. 55-72.

Janetzky B, Hauck S, Youdim MB, Riederer P, Jellinger K, Pantucek F, Zochling R, Boissl KW, Reichmann H. 1994. Unaltered aconitase activity, but decreased complex I activity in substantia nigra pars compacta of patients with Parkinson's disease. Neurosci Lett 169(1-2): 126-128.

Jenner P. 2001. Parkinson's disease, pesticides and mitochondrial dysfunction. Trends Neurosci 24: 245-246.

Jenner P. 2003. Oxidative stress in Parkinson's disease. Ann Neurol 53(Suppl. 3): S26-S36.

Jenner P, Olanow CW. 1998. Understanding cell death in Parkinson's disease. Ann Neurol 44: S72-S84.

Junn E, Mouradian MM. 2002. Human alpha-synuclein over-expression increases intracellular reactive oxygen species levels and susceptibility to dopamine. Neurosci Lett 320: 146-150.

Kagerer FA, Summers JJ, Byblow WD, Taylor B. 2003. Altered corticomotor representation in patients with Parkinson's disease. Mov Disord 18: 919-927.

Katzenschlager R, Lees AJ. 2004. Olfaction and Parkinson's syndromes: Its role in differential diagnosis. Curr Opin Neurol 17: 417-423.

Kish SJ, Morito CH, Hornykiewics O. 1985. Glutathione peroxidase activity in Parkinson's disease brain. Neurosci Lett 58: 343-346.

Kitazawa M, Anantharam V, Kanthasamy AG. 2001. Dieldrin-induced oxidative stress and neurochemical changes contribute to apoptotic cell death in dopaminergic cells. Free Radic Biol Med 31: 1473-14885.

Kortekaas R, Leenders KL, van Oostrom JC, Vaalburg W, Bart J et al. 2005. Blood–brain barrier dysfunction in parkinsonian midbrain in vivo. Ann Neurol 57: 176-179.

Koutsilieri E, Scheller C, Tribl F, Riederer P. 2002. Degeneration of neuronal cells due to oxidative stress—microglial contribution. Parkinsonism Relat Disord 8: 401-406.

Kramer BC, Yabut JA, Cheong J, JnoBaptiste R, Robakis T, et al. 2002. Lipopolysaccharide prevents cell death caused by glutathione depletion: Possible mechanism of protection. Neuroscience 114: 361-372.

Krishnan S, Chi EY, Wood SJ, Kendrick BS, Li C, et al. 2003. Oxidative dimer formation is the critical rate-limiting step for Parkinson's disease α-synuclein fibrillogenesis. Biochemistry 42: 829-837.

Langston JW, Ballard P, Tetrud JW, Irwin I. 1983. Chronic parkinsonism in humans due to a product of meperidine-analog synthesis. Science 219: 989-980.

Lewy FH. 1912. Paralysis agitans: I. Pathologische Anatomie. Handbuch der Neruologie, Vol. 3. Lewandowsky M (Hrsg), editor. Berlin: Springer; pp. 920-933.

Liu Y, Fiskum G, Schubert D. 2002. Generation of reactive oxygen species by the mitochondrial electron transport chain. J Neurochem 80: 780-787.

Lotharius J, Brundin P. 2002. Impaired dopamine storage resulting from alpha-synuclein mutations may contribute to the pathogenesis of Parkinson's disease. Hum Mol Genet 11: 2395-2405.

Lowe J, Lennox G, Leigh PN. 1997. Disorders of movement and system degenerations. Greenfield's Neuropathology, 6th ed. Graham D, Lantos PL, editors. London: Edward Arnold; pp. 280-366.

Marcusson J, Oreland L, Winblad B. 1984. Effect of age on human brain serotonin (S-1) binding sites. J Neurochem 43: 1699-1705.

Marnett LJ. 2000. Oxyradicals and DNA damage. Carcinogenesis 21: 361-370.

Martilla RJ, Lorentz H, Rinne UK. 1988. Oxygen toxicity protecting enzymes in Parkinson's disease: Increase of superoxide-dismutase-like activity in the substantia nigra and basal nucleus. J Neurol Sci 86: 321-331.

Maruyama W, Naoi M. 2002. Cell death in Parkinson's disease. J Neurol 249: II/6-;II/10.

McCarthy S, Somayajulu M, Sikorska M, Borowy-Borowski H, Pandey S. 2004. Paraquat induces oxidative stress and neuronal cell death; neuroprotection by water-soluble Coenzyme Q(10). Toxicol Appl Pharmacol 201: 21-31.

McNaught KSP, Jenner P. 2000. Extracellular accumulation of nitric oxide, hydrogen peroxide and glutamate in astrocytic cultures following glutathione depletion, complex I inhibition and/or lipopolysaccharide-induced activation. Biochem Pharmacol 60: 979-988.

Migliore L, Petrozzi L, Lucetti C, Gambaccini G, Bernardini S, et al. 2002. Oxidative damage and cytogenetic analysis in leukocytes of Parkinson's disease patients. Neurology 58: 1809-1815.

Mitchell IJ, Clarke CE, Boyce S, Robertson RG, Peggs D, et al. 1989. Neural mechanisms underlying Parkinsonian symptoms based upon regional uptake of 2-deoxyglucose in monkeys exposed to 1-methyl-4-phenyl-1,2,3,6-tetrahydropyridine. Neuroscience 32: 213-226.

Mizuno Y, Ohta S, Tanaka M, Takamiya S, Suzuki K, et al. 1989. Deficiencies in complex I subunits of the respiratory chain in Parkinson's disease. Biochem Biophys Res Commun 163: 1450-1455.

Mogi M, Harada M, Kondo T, Riederer P, Inagaki H, et al. 1994. Interleukin-1 beta, interleukin-6, epidermal growth factor and transforming growth factor-alpha are elevated in the brain from parkinsonian patients. Neurosci Lett 180: 147-150.

Montine KS, Quinn JF, Zhang J, Fessel JP, Roberts LJ 2nd, et al. 2004. Isoprostanes and related products of lipid peroxidation in neurodegenerative diseases. Chem Phys Lipids 128: 117-124.

Münch G, Lüth HJ, Wong A, Arendt T, Hirsch E, et al. 2000. Crosslinking of α-synuclein by advanced glycation end products—an early pathophysiological step in Lewy body formation. J Clin Neuroanat 20: 253-257.

Münch G, Schicktanz D, Behme A, Gerlach M, Riederer P, et al. 1999. Amino acid specificity of glycation and protein—AGE crosslinking reactivities determined with a dipeptide SPOT library. Res Nature Biotech 17: 1006-1010.

Mytilineou C, Kramer BC, Yabut JA. 2002. Glutathione depletion and oxidative stress. Parkinsonism Relat Disord 8: 385-387.

Nagatsu T, Mogi M, Ichinose H, Togari A, Riederer P. 1999. Cytokines in Parkinson's disease. Neurosci News 2: 88-90.

Naoi M, Maruyama W, Akao Y, Yi H. 2002. Mitochondria determine the survival and death in apoptosis by an endogenous neurotoxin, N-methyl(R)salsolinol, and neuroprotection by propargylamines. J Neural Transm 109: 607-621.

Naoi M, Maruyama W, Akao Y, Zhang J, Parvez H. 2000. Apoptosis induced by an endogenous neurotoxin, N-methyl(R)salsolinol, in dopamine neurons. Toxicology 153: 123-141.

Nielsen MS, Vorum H, Lindersson E, Jensen PH. 2001. Ca^{2+} binding to alpha-synuclein regulates ligand binding and oligomerization. J Biol Chem 276: 22680-22684.

Obeso JA, Rodriguez-Oroz MC, Rodriguez M, Arbizu J, Gimenez-Amaya JM. 2002. The basal ganglia and disorders of movement: Pathophysiological mechanisms. News Physiol Sci 17: 51-55.

Parkinson J. 1817. Essay on the shaking palsy. London: Whittingham and Rowland.

Perry TL, Yong VW, Ito M, Foulks JG, Wall RA, et al. 1984. Nigrostriatal dopaminergic neurons remain undamaged in rats given high doses of L-Dopa and carbidopa chronically. J Neurochem 43: 990-993.

Pollanen MS, Dickson DW, Bergeron C. 1993. Pathology and biology of the Lewy Body. J Neuropathol Exp Neurol 52: 183-191.

Poon HF, Calabrese V, Scapagnini G, Butterfield DA. 2004. Free radicals and brain aging. Clin Geriatr Med 20: 329-359.

Power JH, Shannon JM, Blumbergs PC, Gai WP. 2002. Non-selenium glutathione peroxidase in human brain: Elevated levels in Parkinson's disease and dementia with Lewy bodies. Am J Pathol 161: 885-894.

Practico D. 2001. In vivo measurement of the redox state. Lipids 36: S45-S49.

Przuntek H, Müller T, Riederer P. 2004. Diagnostic staging of Parkinson's disease: Conceptual aspects. J Neural Transm 111: 201-216.

Rajput AH, Uitti RJ, Stern W, Laverty W. 1986. Early onset Parkinson's disease and childhood environment. Adv Neurol 45: 295-297.

Rajput AH, Uitti RJ, Stern W, Laverty W, O'Donnell K, et al. 1987. Geography, drinking water chemistry, pesticides and herbicides and the etiology of Parkinson's disease. Can J Neurol Sci 14: 414-418.

Reichmann H, Lestienne P, Jellinger K, Riederer P. 1993. Parkinson's disease and the electron transport chain in post mortem brain. Parkinson's Disease: From Basic Research to Treatment. Advances in Neurology, Vol. 60. Narabayashi H, Nagatsu T, Yanagisawa, Mizuno Y, editors. New York: Raven; pp. 297-299.

Reichmann H, Riederer P. 1989. Biochemische Analyse der Atmungskettenkomplexe verschiedener Hirnregionen von Patienten mit M. Parkinson. Morbus Parkinson und andere Basalganglienerkrankungen. Symposium zu einem Förderschwerpunkt des BMFT. Bad Kissingen, 23–25, pp. 44.

Reif DW, Simmons RD. 1990. Nitric oxide mediates iron release from ferritin. Arch Biochem Biophys 283: 537-541.

Riederer P, Youdim MBH. 1993. Iron in central nervous system disorders. Vienna: Springer.

Riederer P, Foley P. 2002. Mini-Review: Multiple developmental forms of parkinsonism. The basis for further research as to the pathogenesis of parkinsonism. J Neural Transm 109: 1469-1475.

Riederer P, Foley P, Bringmann G, Feineis D, Brückner R, et al. 2002. Biochemical and pharmacological characterization of 1-trichloromethyl-1,2,3,4-tetrahydro-beta-carboline: A biologically relevant neurotoxin? Eur J Pharmacol 442: 1-16.

Riederer P, Rausch WD, Schmidt B, Kruzik P, Konradi C, et al. 1988. Biochemical fundamentals of Parkinson's disease. Mt Sinai J Med 55: 21-28.

Riederer P, Sofic E, Rausch WD, Kruzik P, Youdim MBH. 1985. Dopaminforschung heute und morgen—L-Dopa in der Zukunft. Riederer P, Umek H, editors. L-Dopa-Substitution der Parkinson-Krankheit. Geschichte-Gegenwart-Zukunft. New York: Springer Verlag, Wien; pp. 127-144.

Riederer P, Sofic E, Rausch WD, Schmidt B, Reynolds GP, et al. 1989. Transition metals, ferritin, glutathione and ascorbic acid in Parkinsonian brains. J Neurochem 52: 515-520.

Riederer PF. 2004. Views on neurodegeneration as a basis for neuroprotective strategies. Med Sci Monit 10: RA287-RA290.

Rinne JO, Lonnberg P, Marjamaki P. 1990. Age-dependent decline in human brain dopamine D1 and D2 receptors. Brain Res 508: 349-352.

Ross BM, Moszczynska A, Ehrlich J, Kish SJ. 1998. Low activity of key phospholipid catabolic and anabolic enzymes in human substantia nigra: Possible implications for Parkinson's disease. Neuroscience 83: 791-798.

Saint-Cyr JA. 2003. Frontal-striatal circuit functions: Context, sequence, and consequence. J Int Neuropsychol Soc 9: 103-127.

Sanchez-Ramos JR, Hefti F, Weiner WJ. 1987. Paraquat and Parkinson's disease. Neurology 37: 728.

Sanchez-Ramos JR, Övervik E, Ames BN. 1994. A marker of oxyradical-mediated DNA damage (8-hydroxy-2′-deoxyguanosine) is increased in nigro-striatum of Parkinson's disease brain. Neurodegeneration 3: 197-204.

Sechi GP, Angetti V, Piredda M, Canu M, Deserta F, et al. 1992. Acute and persistent parkinsonism after use of diquat. Neurology 42: 261-263.

Serra JA, Domiguez RO, de Lustig ES, Guareschi EM, Famulari AL, et al. 2001. Parkinson's disease is associated with oxidative stress: Comparison of peripheral antioxidant profiles in living Parkinson's, Alzheimer's and vascular dementia patients. J Neural Transm 108: 1135-1148.

Sherer TB, Betarbet R, Greenamyre JT. 2002a. Environment, mitochondria, and Parkinson's disease. Neuroscientist 8: 192-197.

Sherer TB, Betarbet R, Stout AK, Lund S, Baptista M. 2002b. An in vitro model of Parkinson's disease: Linking mitochondrial impairment to altered α-synuclein metabolism and oxidative damage. J Neurosci 22: 7006-7015.

Sian J, Dexter DT, Lees AJ, Daniel S, Jenner P, et al. 1994. Glutathione-related enzymes in brain in Parkinson's disease. Ann Neurol 36: 356-361.

Sofic E, Lange KW, Jellinger K, Riederer P. 1992. Reduced and oxidized glutathione in the substantia nigra of patients with Parkinson's disease. Neurosci Lett 142: 128-130.

Sofic E, Riederer P, Heinsen H, Beckmann H, Reynolds GP, et al. 1988. Increased iron (III) and total iron content in post mortem substantia nigra of parkinsonian brain. J Neural Transm 74: 199-205.

Spencer JPE, Jenner A, Aruoma OI, Evans PJ, Kaur H, et al. 1994. Intense oxidative DNA damage promoted by L-Dopa and its metabolites: Implications for neurodegenerative disease. FEBS Lett 353: 246-250.

Stadtman ER. 2001. Protein oxidation in aging and age-related diseases. Ann NY Acad Sci 928: 22-38.

Starkov AA, Polster BM, Fiskum G. 2002. Regulation of hydrogen peroxide production by brain mitochondria by calcium and Bax. J Neurochem 83: 220-228.

Steiner H, Gerfen CR. 1998. Role of dynorphin and enkephalin in the regulation of striatal output pathways and behavior. Exp Brain Res 123: 60-76.

Tanaka M. 2002. Mitochondrial genotypes and cytochrome b variants associated with longevity or Parkinson's disease. J Neurol 249: III 1-III 8.

Tanaka M, Kim YM, Lee G, Junn E, Iwatsubo T, et al. 2004. Aggresomes formed by alpha-synuclein and synphilin-1 are cytoprotective. J Biol Chem 279: 4625-4631.

Teismann P, Schulz JB. 2004. Cellular pathology of Parkinson's disease: Astrocytes, microglia and inflammation. Cell Tissue Res 318: 149-161.

Thobois S, Jahanshahi M, Pinto S, Frackowiak R, Limousin-Dowsey P. 2004. PET and SPECT functional imaging studies in Parkinsonian syndromes: From the lesion to its consequences. Neuroimage 23: 1-16.

Tompkins MM, Hill WD. 1997. Contribution of somal Lewy bodies to neuronal death. Brain Res 775: 24-29.

Trétiakoff C. 1919. Contribution à l'étude de l'anatomie patholique du locus niger de Soemmering avec quelques déductions relatives à la pathogénie des troubles du tonus musculaire et de la maladie de Parkinson. Thèse, Faculté de Médecine, Paris: Univerisité de Paris.

Trojanowski JQ, Lee VM. 2001. Parkinson's disease and related neurodegenerative synucleinopathies linked to progressive accumulations of synuclein aggregates in brain. Parkinsonism Relat Disord 7(3): 247-251.

Tsang F, Soong TW. 2003. Interactions between environmental and genetic factors in the pathophysiology of Parkinson's disease. IUBMB Life 55: 323-327.

Tu PH, Robinson KA, de Snoo F, Eyer J, Peterson A, et al. 1997. Selective degeneration of Purkinje cells with Lewy body-like inclusions in aged NFHLACZ transgenic mice. J Neurosci 17: 1064-1074.

Turnbull S, Tabner BJ, El-Agnaf OMA, Moore S, Davies Y, et al. 2001. α-synuclein implicated in Parkinson's disease catalyses the formation of hydrogen peroxide in vitro. Free Radical Biol & Med 30: 1163-1170.

Volchegorskii IA, Shemyakov SE, Turygin VV, Malinovskaya NV. 2004. The age dynamics of monoamine oxidase activity and levels of lipid peroxidation products in the human brain. Neurosci Behav Physiol 34: 303-305.

Walinshaw G, Waters CM. 1995. Induction of apoptosis in catecholaminergic PC12 cells by L-Dopa: Implications for the treatment of Parkinson's disease. J Clin Invest 95: 2458-2464.

Walter U, Dressler D, Wolters A, Probst T, Grossmann A, et al. 2004. Sonographic discrimination of corticobasal degeneration vs progressive supranuclear palsy. Neurology 63: 504-509.

Walter U, Wittstock M, Benecke R, Dressler D. 2002. Substantia nigra echogenicity is normal in non-extrapyramidal disorders but increased in Parkinson's disease. J Neural Transm 109: 191-196.

Warmuth-Metz M, Naumann M, Csoti I, Solymosi L. 2001. Measurement of the midbrain diameter on routine magnetic resonance imaging: A simple and accurate method of differentiating between Parkinson disease and progressive supranuclear palsy. Arch Neurol 58: 1076-1079.

Weisser M, Vieth M, Stolte M, Riederer P, Pfeuffer R, et al. 1997. Dramatic increase of alpha-hydroxyaldehydes derived from plasmalogens in the aged human brain. Chem Phys Lipids 90: 135-142.

World Health Organization. 1998. World Health Report. Geneva: World Health Organization.

Wu G, Fang YZ, Yang S, Lupton JR, Turner ND. 2004. Glutathione metabolism and its implications for health. J Nutr 134: 489-492.

Yoshida E, Mokuno K, Aoki SI, Takahashi A, Riku S, et al. 1994. Cerebrospinal fluid levels of superoxide dismutase. J Neurol Sci 124: 25-31.

Zareba M, Bober A, Korytowski W, Zecca L, Sarna T. 1995. The effect of a synthetic neuromelanin on yield of free hydroxyl radicals generated in model systems. Biochim Biophys Acta 1271: 343-348.

Zayed J, Ducic S, Campanella G, Andre P, Masson H, et al. 1990. Facteurs environnementeaux dans la maladie de Parkinson. Can J Neurol Sci 17: 286-291.

Zecca L, Gallorini M, Schünemann V, Trautwein AX, Gerlach M, et al. 2001. Iron, neuromelanin and ferritin in substantia nigra of normal subjects at different ages, consequences for iron storage and neurodegenerative processes. J Neurochem 76: 1766-1773.

Zecca L, Shima T, Stroppolo A, Goj C, Battistron GA, et al. 1996. Interaction of neuromelanin and iron in the substantia nigra and other areas of human brain. Neuroscience 73: 407-415.

2 Diagnosis and Neuroprotection in Parkinson's Disease

T. Müller

Abstract: Insidious onset of mild, unspecific vegetative, psychopathological, cognitive and perceptive disturbances with a resulting change of personal behavior often precede the initially intermittend occurring motor symptoms in patients with Parkinson's disease (PD). Thus the current available diagnostic tools and concepts do not enable such an early diagnosis of Parkinson's disease. However they are necessary to perform an effective neuroprotective treatment regimes, since neuronal dopaminergic death is advanced at the clinical treshold of onset of motor symptoms. Treatment of these symptoms should be formed in a more continuous fashion, in order to imitate and provide more physiological conditions of dopaminergic neurotransmission. This delays onset of motor complications, which may be interpreted as a certain neuroprotective effect from the clinical point of view.

List of Abbreviations: 5-HT, 5-hydroxytryptamine; PD, Parkinson's disease

1 Introduction

James Parkinson's original 1817 description of the "shaking palsy" is a remarkable, accurate account of the disease, which now bears his name. He focused his essay on akinesia, tremor, and rigidity as cardinal motor symptoms of Parkinson's disease (PD), but he also reported on an array of further more secondary clinical signs, which are not directly related to the dysfunction of the motor system (Parkinson, 1817). These nonmotor symptoms also affect quality of life in PD patients. The self-description of onset of PD by Wilhelm von Humboldt or Michael J. Fox also initially mentions motor symptoms as a guiding clinical feature (Horowski et al., 1995; Fox, 2003). But both of them also report apathy and anhedonia-like depressive symptoms, which occur initially and later in the course of PD. These unspecific fatigue-like symptoms, disturbances of sleep, disturbances in emotional and personal behavior, reduced tolerance to stress, deficits of motivation and impulse, lack of concentration, restlessness, and anxiety clinically often appear before the often intermittent manifestation of motor symptoms. These signs of psychomotor retardation with reduced lust for life, anhedonia, and psychomotor retardation are often clinically diagnosed as depression of the elderly or as early signs of mild cognitive impairment or dementia in clinical practice (Müller, 2004). This view is confirmed by a trial, which uses a case-control design and investigates an association between preceding anxiety, depression, and later diagnosis of PD. In the case of depression, the significance of the association was lost when depressive disorders first recognized within the 5 years preceding onset of PD were excluded. However, this trial also reports significant relations between anxiety disorders and PD up to 20 years before onset of PD and that anxiety preceded depression in 72% of the PD patients who had both conditions (Shiba et al., 2000). These results are consistent with earlier studies and support the view that anxiety and depressive episodes are early nonmotor manifestations of the underlying disease process. This view is supported by the still hypothetical, revolutionary, and only partially confirmed approach of a pathological process with a somewhat stereotypic topographic expansion pattern of lesions in PD, which may even start outside the brain. These current neuropathological results also indicate that neurodegeneration not only takes place in nigral dopaminergic neurons but also appears in nondopaminergic neurotransmitter systems and in extranigral structures. One may even postulate that extranigral neurodegeneration may also play an essential role in the pathophysiology of PD, since it causes cognitive and behavioral disturbances, which precede the alteration of motor function (Wolters et al., 2000; Braak et al., 2003; Przuntek et al., 2004).

2 Do Some Mild Depressive Symptoms Indicate the Onset of Neurodegeneration?

The above mentioned predominant to date psychopathological symptoms clinically reflect the still hypothetical neuropathologic route of neurodegeneration in PD patients to a certain extent. In PD, this may be associated with a compensatory downregulation of 5-hydroxytryptamine (5-HT) and an associated development of mild depression, because 5-HT regulates dopamine turnover in the brain, i.e., in the frontal lobe or the brainstem (Birkmayer et al., 1972; Melamed et al., 1980; Agren et al., 1986). But, till date, research on relationships between neurotransmission, motor features of PD, and frequency and severity of depression or anxiety in predominant previously untreated PD patients in the early stage of PD is rare. The baseline data of the DATATOP trial revealed a higher frequency of depression in patients with postural instability, gait disability; and as another trial suggests, preponderant manifestation of akinesia and rigidity predisposes to the manifestation of depression in PD (Jankovic et al., 1990; Starkstein et al., 1998). In contrast to the various psychopathological features of an endogenous depression, PD patients more uniformly show a reduced frequency of guilt, feelings of failure, and self-blame behind their "iron" mask-like face, which may confound scoring of mood with subjective rating scales (Burn, 2002; Müller, 2004). However, depression in PD not only is a reaction to the chronic disabling illness and the resulting physical impairment but also is biologically or endogenously determined. The start of dopaminergic substitution therapy and the concomitant improvement of motor symptoms may reduce the intensity of the reactive components of mood disturbances. PD patients with predominant tremor appear to be more associated with the reduced neurotransmission of 5-HT in the raphe than with the nigrostriatal dopaminergic deficit (Doder et al., 2003). In later stages, these patients have a reduced risk for additional complications of PD, i.e., freezing of gait, and show less progress than the ones with akinesia and rigidity (Jellinger and Paulus, 1992).

3 The Need for Tools for Early Diagnosis Before Onset of Motor Symptoms in PD

There is a need for specific diagnostic tools for the early diagnosis of the neurodegenerative process, focusing on clinical symptoms associated with chronic decrease of 5-HT in certain brain areas, i.e., the brainstem or the frontal lobe (Mayeux, 1990; Starkstein et al., 1998; Foley and Riederer, 1999; Uekermann et al., 2003; Müller, 2004; Przuntek et al., 2004). Therefore, instruments for early diagnosis, i.e., a standardized interview technique, should include a special focus on mood-associated psychopathological features such as anhedonia, apathy, reduced stress tolerance, and transient appearance of early signs of disturbances of motor performance, i.e., onset of intermittent tremor in states of emotional distress or transient but longer-lasting dull pain syndromes of the neck, head, etc. (Shiba et al., 2000; Doder et al., 2003; Müller, 2004; Przuntek et al., 2004). Moreover, they should consist of a neuropsychologic tool for measuring a still unknown, biological, surrogate marker of the developing neurodegenerative process (Shiba et al., 2000; Uekermann et al., 2003; Müller, 2004). All of them should allow screening investigations to be performed, since the above-mentioned complaints are unspecific and occur in large cohorts, particularly in the elderly population. A further promising approach for research on more early premotor detection of PD would be the development of easy to perform, simple, and cheap objective instrumental tools for evaluation of disturbed fine motor behavior, which initially often intermittently occurs under stress conditions. Moreover, all these diagnostic procedures or combined test batteries should be simple and cheap (Müller et al., 2000). Such a concept of early diagnosis of neurodegeneration also suggests an associated modification of the diagnostic criteria and the staging of PD, in order to enable a "premotor" diagnosis of PD or neurodegeneration (Foley et al., 1999; Korczyn, 1999; Wolters et al., 2000; Przuntek et al., 2004).

4 Neuroprotection Should Start in the Premotor Phase of PD

When neuroprotective therapeutic interventions start more earlier in PD, they will be more effective, since nigral cell loss is about 50% and striatal dopamine content is reduced to approximately 80% at the threshold of occurrence of motor symptoms, which mostly lead to diagnosis (Foley et al., 1999; Przuntek et al., 2004). Thus, from an economical point of view, effective neuroprotection, which causes an effective delay of PD progression of about 10%, may reduce direct and indirect health care costs for PD patients of about 330 million USD per year according Bryson and coworkers (1992). Such an approach for a more earlier diagnosis of neurodegeneration must also consider nutrition habits, e.g., coffee consumption. Since caffeine also enhances dopamine synthesis or its synaptic release, which is also mediated by adenosine or nicotine, chronic caffeine intake may reduce psychomotor retardation, anhedonia-like symptoms, and apathy. This may hypothetically support the described decreased incidence of PD due to long-term coffee consumption in epidemiological studies. However, long-term caffeine not only may delay appearance of initial premotor mood disturbances and later motor symptoms related to neuronal dopaminergic deficits but also may exacerbate certain positive symptomatic effects from that stage onward to further more severe phases of PD with a possible positive neuroprotective impact on the progression of symptoms, i.e., akinesia and rigidity, or PD as a whole (Ross and Petrovitch, 2001; Gale and Martyn, 2003; Pedata et al., 2003; Tan et al., 2003; Ascherio et al., 2004). Considering all these aspects of neurotransmitter loss or compensatory downregulation of synthesis, several trials, which aimed to demonstrate neuroprotection, were rather questionable in the past years (Clarke, 2004a, b). These studies only focused on the dopaminergic system and measured progress of neurodegeneration with surrogate markers, which predominantly reflect striatal dopaminergic neurotransmission. Thus, the various, but only partially, convincing results of short- and long-term trials with various dopamine agonists and levodopa raises the question: whether the approach and design of these studies were sufficient to reflect the progress and the neuroprotective or even neuroregenerative therapeutic interventions in PD (Marek et al., 2002; Parkinson Study Group, 2002; Morrish, 2003a; Whone et al., 2003; Przuntek et al., 2004). Particularly, the definition of subjects, who were found negative for dopaminergic deficit using functional imaging techniques, as idiopathic PD patients and the results of the recent transplantation trials with embryonic dopaminergic cells support the view that increased dopaminergic neurotransmission, proved using [^{18}F]-Dopa PET, does not cause substantial clinical benefits for PD patients in the long term (Freed et al., 2001; Przuntek et al., 2004). If striatal death of dopaminergic neurons is a secondary phenomenon of the neurodegenerative process in PD, then the initial primary step of this process may be reduced neuronal and glial energy metabolism owing to different kinds of neurodegenerative processes, e.g., apoptosis, altered growth factor synthesis, or inflammatory mediators (Przuntek and Müller, 1999). Therefore, future neuroprotective agents should influence these initial steps of the subsequent cascade of neurodegeneration in various neurotransmitter systems in PD.

5 Neuroprotection

One goal of treatment in neurodegenerative disease is to slow progression, known as neuroprotection. This approach is still very important, since results of the neuroregenerative transplantation and curative growth factor trials were very disappointing and future stem cell therapy is far from clinical testing (Harrower and Barker, 2004). The essential problem of all these therapeutic approaches is their specific focus on the dopaminergic system and the still unknown regulation of physiologic dopamine release in the nigrostriatal system. Thus, the transplantation trials in idiopathic PD patients with onset of dyskinesia and cognitive disturbances due to dopamine overflow failed after a certain interval (Freed et al., 2001; Gill et al., 2003; Olanow et al., 2003; Clarke, 2004b). Since no curative treatment options exist in chronic neurodegenerative disorders, long-term neuroprotection still remains as the one essential option. In clinical and basic research, many trials, carried out with outstanding scientific interest and good financial support, investigated the issue of neuro-protection, particularly in the case of PD. Their more or less disappointing results were due to the inclusion of PD patients who exhibited early motor symptoms, which is rather late in the whole course of the neurodegen-erative disease. Moreover the durations of the trials were too short. The interpretations were confusing because the compounds investigated or the placebos used mimicked the symptoms of PD and also because of too short wash-out periods, which did not allow a consistent interpretation of the effect found in terms of modification of the disease progress (Parkinson Study Group, 1993; Sano et al., 1997; Foley et al., 1999; Marek et al., 2002; Whone et al., 2003; Fahn et al., 2004; Przuntek et al., 2004). Therefore, the promising and clinically relevant positive results of these compounds on neuroprotective and/or life expectancy increasing properties in clinical trials are still controversial and do not reflect the neuroprotective potential, which was shown in cell culture and animal models of PD. The applied biological surrogate marker, i.e., functional brain imaging techniques with various radiotracers that only focus on the dopaminergic system, did not consider that neurodegeneration in PD also affects other neurotransmitters, i.e., serotonin, norepinephrine, and acetylcholine with the corresponding onset of nonmotor features of PD (Foley et al., 1999; Wolters et al., 2000; Braak et al., 2003; Przuntek et al., 2004). This drawback was also observed in preclinical research in cell culture models especially in certain types of neuronal cells and in animal models exposed to selective toxins of the dopaminergic system.

6 Neuroprotection Within the Dopaminergic System may Depend on Its Modulation

The methodological design problems may play a major role in the negative outcomes of these trials in the select study populations, which try to overcome the gap between clinical practice and reality and the need for necessary GCP quality of data for consistent scientific outcomes. Since there is no simple, clinical and biological surrogate marker for measuring the progression of PD, the recently used expensive functional imaging techniques with various radiotracers, which visualize the function of the striatal dopaminergic system, are looked upon as a promising tool. However, their value is limited. There were earlier trials which postulated that levodopa/carbidopa or selegiline did not significantly affect presynaptic dopamine transporter imaging, although a small influence could not be excluded in the case of pergolide. These studies performed upon withdrawal from medication: in the case of levodopa/carbidopa for approximately 1 week, selegiline for 9 weeks, and pergolide for 4 weeks or levodopa for 2 weeks. But there are results which demonstrate that long-term stimulation of dopamine receptors decreases D_2 and D_3 receptor expression in regional and subtype-specific regions of the nigrostriatal system in animals (Ahlskog et al., 1999; Ekesbo et al., 1999; Innis et al., 1999; Stanwood et al., 2000; Marek et al., 2002; Ahlskog, 2003; Fahn et al., 2004). Moreover, an additional modulation of presynaptic endogenous dopamine synthesis and release of dopamine to the synaptic cleft takes place. Patients in earlier stages of PD may show an upregulated presynaptic inhibitory feedback regulation, particularly in the dorsal putamen, to maintain congruity within the dopaminergic system in response to antiparkinsonian medication. This inhibitory feedback regulation is diminished with the progression of nigrostriatal degeneration and chronic dopamine agonist treatment (Fearnley and Lees, 1990; Morrish et al., 1996; Ryoo et al., 1998; Ekesbo et al., 1999). Earlier studies on the impact of dopaminergic long-term treatment on functional imaging mainly included PD patients in more advanced stages of PD. Therefore, one cannot exclude that this loss of dopamine autoreceptor function in advanced stages of PD at least influenced the results and contributed to the conclusion that dopaminergic drugs did not have any impact on functional imaging techniques. This increasing loss of autoreceptor function could also have biased dopamine transporter imaging results in the CALM-PD study and their previously published discrepancy during the course of the whole trial (Ekesbo et al., 1999; Morrish, 2002, 2003b, 2005; Clarke, 2004b). However, since $[^{123}I]$-β-CIT SPECT only assesses presynaptic dopamine receptor activity, which is influenced by dopamine autoreceptor function particularly in early PD patients, the CALM-PD trial only provided evidence for a differing progression of PD under pramipexole compared with levodopa, but did not prove that pramipexole is neuroprotective, which is still only shown in laboratory and animal studies. One may also hypothesize that the alternating postsynaptic stimulation by levodopa administration did not downregulate the presynaptic autoreceptor mediated-dopamine synthesis and release via the dopamine transporter. In contrast, a more continuous modulation by the dopamine agonist pramipexole with its longer half-life caused a more persistent downregulation, followed by an upregulation during the washout period. This could have contributed to a better enrichment of the radiotracer. This was interpreted as a neuroprotective effect of the drug, which did not show the same clinical efficacy than levodopa. But this resulted in a discrepancy between clinical antiparkinsonian effects on motor symptoms and functional imaging results. This hypothesis holds also true for the functional imaging outcomes of the ELLDOPA trial with more distinct, alternating striatal dopamine levels in a dose-dependent manner with levodopa application, whereas placebo administration similar to a dopamine agonist also releases dopamine in a more continuous but less pronounced fashion via the dopamine-mediated reward pathway (Morrish et al., 1996; Ryoo et al., 1998; Stanwood et al., 2000; Fuente-Fernandez et al., 2001, 2004; Whone et al., 2003; Przuntek et al., 2004). The above-mentioned hypothetical considerations may also explain the observed discrepancy between less pronounced clinical benefit of the functional imaging outcome under ropinirole compared to levodopa in the REALPET study, which used $[^{18}F]$-Dopa PET imaging over a period of 5 years. This study also showed a better enrichment of $[^{18}F]$-Dopa under ropinirole compared with levodopa, which was interpreted as a neuroprotective benefit for ropinirole, whereas the clinical benefit was better in the levodopa arm. One may also argue that the more continuous stimulation during ropinirole treatment induced a long lasting downregulation of presynaptic endogenous dopamine synthesis followed by an upregulation, and accordingly increased long lasting presynaptic endogenous dopamine release during the washout period. This causes a delayed metabolism of $[^{18}F]$-Dopa in the synaptic cleft, which results in a more pronounced enrichment of the radiotracer in the PET scans (Müller et al., 1997; Torstenson et al., 1998). Moreover, there may be a further confounding factor. Administration of dopaminergic compounds influences brain function respectively, local perfusion patterns, as shown by fMRI trials (Buhmann et al., 2003; Peters et al., 2003; Schmid et al., 2004). In conclusion, this means that the role of presynaptic autoreceptor modulation of endogenous dopamine release is more important than originally assumed in more earlier courses of PD and its modulation by either more continuous or more intermittent kinds of dopamine drug administration influences outcomes of functional imaging techniques or even clinical effects of dopaminergic substitution.

7 Did the Recent Trials on Neuroprotection Compare Physiologic Conditions?

There is a further drawback in the design of all these trials, since they only compare initial monotherapy with dopamine agonists versus levodopa. Their designs did not consider that low dose of levodopa in combination with the dopamine agonist may represent the best treatment option in terms of tolerability and initial short-term efficacy due to levodopa with consecutive improved quality of life. Moreover, this combination may hypothetically better imitate physiological dopaminergic neurotransmission with already existing presynaptic neuronal synthesis of dopamine by the enzyme tyrosine hydroxylase because of increased alternating striatal dopamine levels. This demands a physiologic functioning of presynaptic dopaminergic autoreceptors (Müller et al., 1997; Torstenson et al., 1998). Thus, these trials compared two conditions that were not physiologic. One is caused by continuous delivery of dopaminergic compounds, i.e., dopamine agonists with long half-life or high receptor affinity, which results in an unphysiologic concomitant downregulation of endogenous presynaptic dopamine synthesis in the remaining healthy or less affected dopaminergic neurons. The other one is caused by nonphysiologic, intermittent stimulation of postsynaptic dopaminergic nigrostriatal uptake site receptors, i.e., with levodopa/DDC (Stanwood et al., 2000; Smith et al., 2005). This would explain the findings of the ELLDOPA-trial and its discrepancy between clinical benefit on motor symptoms even after a washout of 14 days and the functional imaging findings, which at least show a levodopa dose-dependent deteriorated $[^{123}I]$-β-CIT-uptake ($p = 0.036$), and which become most prominent in particular in the group of PD patients with 200 mg levodopa intake t.i.d. ($p = 0.015$) (Fahn et al., 2004). From this point of view, one could argue that all tested dopaminergic compounds are not toxic or neuroprotective in clinical practice. Only the mode of administration of the tested compound caused an impact on the progression of PD, which was clinically defined by the onset of motor complications. Even the trials with selegiline or rasagiline would also support the assumption of the impact of a more physiologic continuous long-term dopaminergic postsynaptic stimulation, since the MAO-B-inhibiting mode of action of both agents contributes to more continuous but less alternating dopamine levels in the synaptic cleft between pre- and postsynaptic nigrostriatal neurons. This results in a slower progression of the disease with a concomitant less pronounced clinical need for levodopa for adequate control of motor symptoms in terms of quality of life. This would even fit into trials with delayed treatment start designs. The observed but delayed beneficial effect of rasagiline in the group with initial placebo run consistently parallels the observed effect in patients who received the drug right from the beginning. Both consistently show a more smooth increase of disease intensity in the clinical rating when treated with rasagiline, which is interpreted as a neuroprotective effect of the compound (Parkinson Study Group, 1993, 2004, 2005; Clarke, 2005; Rascol et al., 2005). However, rasagiline and other compounds, which only modulate dopaminergic function at either pre- or postsynaptic striatal uptake sites, like NMDA antagonists, adenosine A_2 antagonists, or 5-HT$_{1A}$ agonists, reduce the effects of nonphysiologic alternating dopamine concentrations in the synaptic cleft, which clinically appear as a delayed onset of motor complications or as a reduction of the intensity of motor complications in the long run (Przuntek et al., 1999, 2002; Kulisevsky et al., 2002; Crosby et al., 2003; Hauser et al., 2003; Olanow et al., 2004; Thomas et al., 2004). This is reflected by the clinical rating scales and a better quality of life, but does this really mean neuroprotection in terms of less progression of neuronal death? Considering all these issues and the controversial results of the various trials performed, one might conclude that there is a need to reconsider the issue of neuroprotection in the treatment of PD.

8 Conclusion and Future Aspects

The neuroprotective treatment strategies of neurodegeneration suffer from a lack of timely diagnosis, which is too late in the course of disease onset. An effective neuroprotective therapeutic intervention has to start far more earlier. Therefore, there is a need for easy-to-handle, cheap diagnostic tools, which may be used for screening of the relatively nonspecific clinical signs of the premotor syndrome in PD or the clinically similar symptoms of mild cognitive impairment in Alzheimer's disease. Moreover, one should consider that diagnosis of idiopathic PD according to the current criteria only reflects a Parkinson syndrome and not the various types of the disease with its individually different course, symptoms, and response to treatment (Korczyn, 1999; Przuntek et al., 1999, 2004). Thus to date, the available diagnostic tools do not enable an early diagnosis of Parkinson's disease, which allows effective neuroprotective treatment regimes in order to delay the progression of the disease. This has been shown in preclinical research in various models. However, these models only focused on the dopaminergic system and on associated onset of motor symptoms. Treatment of these motor symptoms should be performed in a continuous fashion, in order to imitate and provide physiologic conditions of dopaminergic neurotransmission. This delays onset of motor complications, which may be interpreted as a certain neuroprotective effect from the clinical point of view.

References

Agren H, Mefford IN, Rudorfer MV, Linnoila M, Potter WZ. 1986. Interacting neurotransmitter systems. A non-experimental approach to the 5HIAA-HVA correlation in human CSF. J Psychiatr Res 20: 175-193.

Ahlskog JE. 2003. Slowing Parkinson's disease progression: Recent dopamine agonist trials. Neurology 60: 381-389.

Ahlskog JE, Uitti RJ, O'Connor MK, Maraganore DM, Matsumoto JY, et al. 1999. The effect of dopamine agonist therapy on dopamine transporter imaging in Parkinson's disease. Mov Disord 14: 940-946.

Ascherio A, Weisskopf MG, O'Reilly EJ, McCullough ML, Calle EE, et al. 2004. Coffee consumption, gender, and Parkinson's disease mortality in the cancer prevention study II cohort: The modifying effects of estrogen. Am J Epidemiol 160: 977-984.

Birkmayer W, Danielczyk W, Neumayer E, Riederer P. 1972. The balance of biogenic amines as condition for normal behaviour. J Neural Transm 33: 163-178.

Braak H, Rub U, Gai WP, Del Tredici K. 2003. Idiopathic Parkinson's disease: Possible routes by which vulnerable neuronal types may be subject to neuroinvasion by an unknown pathogen. J Neural Transm 110: 517-536.

Bryson HM, Milne RJ, Chrisp P. 1992. Selegiline: An appraisal of the basis of its pharmacoeconomic and quality-of-life benefits in Parkinson's disease. Pharmacoeconomics 2: 118-136.

Buhmann C, Glauche V, Sturenburg HJ, Oechsner M, Weiller C, et al. 2003. Pharmacologically modulated fMRI—cortical responsiveness to levodopa in drug-naive hemiparkinsonian patients. Brain 126: 451-461.

Burn DJ. 2002. Beyond the iron mask: Towards better recognition and treatment of depression associated with Parkinson's disease. Mov Disord 17: 445-454.

Clarke CE. 2004a. A "cure" for Parkinson's disease: Can neuroprotection be proven with current trial designs? Mov Disord 19: 491-498.

Clarke CE. 2004b. Neuroprotection and pharmacotherapy for motor symptoms in Parkinson's disease. Lancet Neurol 3: 466-474.

Clarke CE. 2005. Rasagiline for motor complications in Parkinson's disease. Lancet 365: 914-916.

Crosby NJ, Deane KH, Clarke CE. 2003. Amantadine for dyskinesia in Parkinson's disease. Cochrane Database Syst Rev CD003467.

Doder M, Rabiner EA, Turjanski N, Lees AJ, Brooks DJ. 2003. Tremor in Parkinson's disease and serotonergic dysfunction: An ^{11}C-WAY 100635 PET study. Neurology 60: 601-605.

Ekesbo A, Rydin E, Torstenson R, Sydow O, Laengstrom B, et al. 1999. Dopamine autoreceptor function is lost in advanced Parkinson's disease. Neurology 52: 120-125.

Fahn S, Oakes D, Shoulson I, Kieburtz K, Rudolph A, et al. 2004. Levodopa and the progression of Parkinson's disease. N Engl J Med 351: 2498-2508.

Fearnley JM, Lees AJ. 1990. Striatonigral degeneration. A clinicopathological study. Brain 113 (Pt 6): 1823-1842.

Foley P, Riederer P. 1999. Pathogenesis and preclinical course of Parkinson's disease. J Neural Transm Suppl 56: 31-74.

Fox MJ. 2003. Lucky Man: A Memoir. New York: Hyperion. 2002.

Freed CR, Greene PE, Breeze RE, Tsai WY, Du Mouchel W, et al. 2001. Transplantation of embryonic dopamine neurons for severe Parkinson's disease. N Engl J Med 344: 710-719.

Fuente-Fernandez R, Ruth TJ, Sossi V, Schulzer M, Calne DB, et al. 2001. Expectation and dopamine release: Mechanism of the placebo effect in Parkinson's disease. Science 293: 1164-1166.

Fuente-Fernandez R, Schulzer M, Stoessl AJ. 2004. Placebo mechanisms and reward circuitry: Clues from Parkinson's disease. Biol Psychiatry 56: 67-71.

Gale C, Martyn C. 2003. Tobacco, coffee, and Parkinson's disease. BMJ 326: 561-562.

Gill SS, Patel NK, Hotton GR, O'Sullivan K, McCarter R, et al. 2003. Direct brain infusion of glial cell line-derived neurotrophic factor in Parkinson disease. Nat Med 9: 589-595.

Harrower TP, Barker RA. 2004. Is there a future for neural transplantation? BioDrugs 18: 141-153.

Hauser RA, Hubble JP, Truong DD. 2003. Randomized trial of the adenosine A_{2A} receptor antagonist istradefylline in advanced PD. Neurology 61: 297-303.

Horowski R, Horowski L, Vogel S, Poewe W, Kielhorn FW. 1995. An essay on Wilhelm von Humboldt and the shaking palsy: First comprehensive description of Parkinson's disease by a patient. Neurology 45: 565-568.

Innis RB, Marek KL, Sheff K, Zoghbi S, Castronuovo J, et al. 1999. Effect of treatment with L-dopa/carbidopa or L-selegiline on striatal dopamine transporter SPECT imaging with [123I]β-CIT. Mov Disord 14: 436-442.

Jankovic J, McDermott M, Carter J, Gauthier S, Goe tz et al. 1990. Variable expression of Parkinson's disease: A base-line analysis of the DATATOP cohort. The Parkinson study group. Neurology 40: 1529-1534.

Jellinger KA, Paulus W. 1992. Clinico-pathological correlations in Parkinson's disease. Clin Neurol Neurosurg 94 Suppl: S86-S88.

Korczyn AD. 1999. Parkinson's disease: One disease entity or many? J Neural Transm Suppl 56: 107-111.

Kulisevsky J, Barbanoj M, Gironell A, Antonijoan R, Casas M, et al. 2002. A double-blind crossover, placebo-controlled study of the adenosine A2A antagonist theophylline in Parkinson's disease. Clin Neuropharmacol 25: 25-31.

Marek K, Seibyl J, Shoulson I, Holloway R, Kieburtz K, et al. 2002. Dopamine transporter brain imaging to assess the effects of pramipexole vs levodopa on Parkinson disease progression. JAMA 287: 1653-1661.

Mayeux R. 1990. The "serotonin hypothesis" for depression in Parkinson's disease. Adv Neurol 53: 163-166.

Melamed E, Hefti F, Wurtman RJ. 1980. Nonaminergic striatal neurons convert exogenous L-dopa to dopamine in parkinsonism. Ann Neurol 8: 558-563.

Morrish P. 2002. Is it time to abandon functional imaging in the study of neuroprotection? Mov Disord 17: 229-232.

Morrish P. 2005. The meaning of negative DAT SPECT and F-Dopa PET scans in patients with clinical Parkinson's disease. Mov Disord 20: 117-118.

Morrish PK. 2003a. REAL and CALM: What have we learned? Mov Disord 18: 839-840.

Morrish PK. 2003b. The harsh realities facing the use of SPECT imaging in monitoring disease progression in Parkinson's disease. J Neurol Neurosurg Psychiatry 74: 1447

Morrish PK, Sawle GV, Brooks DJ. 1996. Regional changes in [18F] dopa metabolism in the striatum in Parkinson's disease. Brain (Pt 6): 119 2097-2103.

Müller T. 2004. Mood disorders in early Parkinson's disease. Curr Opin Psychiatry 17: 191-196.

Müller T, Eising EG, Reiners C, Przuntek H, Jacob M, et al. 1997. 2-[123I]-iodolisuride SPET visualizes dopaminergic loss in de-novo parkinsonian patients: Is it a marker of striatal pre-synaptic degeneration? Nucl Med Commun 18: 1115-1121.

Müller T, Schafer S, Kuhn W, Przuntek H. 2000. Correlation between tapping and inserting of pegs in Parkinson's disease. Can J Neurol Sci 27: 311-315.

Olanow CW, Damier P, Goetz CG, Müller T, Nutt J, et al. 2004. Multicenter, open-label, trial of sarizotan in Parkinson disease patients with levodopa-induced dyskinesias (the SPLENDID Study). Clin Neuropharmacol 27: 58-62.

Olanow CW, Goetz CG, Kordower JH, Stoessl AJ, Sossi V, et al. 2003. A double-blind controlled trial of bilateral fetal nigral transplantation in Parkinson's disease. Ann Neurol 54: 403-414.

Parkinson J. 1817. An Essay on the shaking Palsy. London: Sherwood Neely and Jones.

Parkinson Study Group 1993. Effects of tocopherol and deprenyl on the progression of disability in early Parkinson's disease. The Parkinson Study Group. N Engl J Med 328: 176–183.

Parkinson Study Group. 2002. Dopamine transporter brain imaging to assess the effects of pramipexole vs levodopa on Parkinson disease progression. JAMA 287: 1653–1661.

Parkinson Study Group. 2004. A controlled, randomized, delayed-start study of rasagiline in early Parkinson disease. Arch Neurol 61: 561–566.

Parkinson Study Group. 2005. A randomized placebo-controlled trial of rasagiline in levodopa-treated patients with Parkinson disease and motor fluctuations: The PRESTO study. Arch Neurol 62: 241–248.

Pedata F, Pugliese AM, Melani A, Gianfriddo M. 2003. A2A receptors in neuroprotection of dopaminergic neurons. Neurology 61: S49-S50.

Peters S, Suchan B, Rusin J, Daum I, Koster O, et al. 2003. Apomorphine reduces BOLD signal in fMRI during voluntary movement in Parkinsonian patients. Neuroreport 14: 809-812.

Przuntek H, Müller T. 1999. Clinical efficacy of budipine in Parkinson's disease. J Neural Transm Suppl 56: 75-82.

Przuntek H, Bittkau S, Bliesath H, Büttner U, Fuchs G, et al. 2002. Budipine provides additional benefit in patients with

Parkinson disease receiving a stable optimum dopaminergic drug regimen. Arch Neurol 59: 803-806.

Przuntek H, Müller T, Riederer P. 2004. Diagnostic staging of Parkinson's disease: Conceptual aspects. J Neural Transm 111: 201-216.

Rascol O, Brooks DJ, Melamed E, Oertel W, Poewe W, et al. 2005. Rasagiline as an adjunct to levodopa in patients with Parkinson's disease and motor fluctuations (LARGO, lasting effect in adjunct therapy with rasagiline given once daily, study): A randomized, double-blind, parallel-group trial. Lancet 365: 947-954.

Ross GW, Petrovitch H. 2001. Current evidence for neuroprotective effects of nicotine and caffeine against Parkinson's disease. Drugs Aging 18: 797-806.

Ryoo HL, Pierrotti D, Joyce JN. 1998. Dopamine D3 receptor is decreased and D2 receptor is elevated in the striatum of Parkinson's disease. Mov Disord 13: 788-797.

Sano M, Ernesto C, Thomas RG, Klauber MR, Schafer K, et al. 1997. A controlled trial of selegiline, α-tocopherol, or both as treatment for Alzheimer's disease. The Alzheimer's Disease Cooperative Study. N Engl J Med 336: 1216-1222.

Schmid G, Suchan B, Rusin J, Daum I, Koster O, et al. 2004. Impact of apomorphine on BOLD signal during movement in normals. J Neural Transm Suppl 68: 69-78.

Shiba M, Bower JH, Maraganore DM, McDonnell SK, Peterson BJ, et al. 2000. Anxiety disorders and depressive disorders preceding Parkinson's disease: A case-control study. Mov Disord 15: 669-677.

Smith LA, Jackson MJ, Al Barghouthy G, Rose S, Kuoppamaki M, et al. 2005. Multiple small doses of levodopa plus entacapone produce continuous dopaminergic stimulation and reduce dyskinesia induction in MPTP-treated drug-naive primates. Mov Disord 20: 306-314.

Stanwood GD, Lucki I, McGonigle P. 2000. Differential regulation of dopamine D2 and D3 receptors by chronic drug treatments. J Pharmacol Exp Ther 295: 1232-1240.

Starkstein SE, Petracca G, Chemerinski E, Teson A, Sabe L, et al. 1998. Depression in classic versus akinetic-rigid Parkinson's disease. Mov Disord 13: 29-33.

Tan EK, Tan C, Fook-Chong SM, Lum SY, Chai A, et al. 2003. Dose-dependent protective effect of coffee, tea, and smoking in Parkinson's disease: A study in ethnic Chinese. J Neurol Sci 216: 163-167.

Thomas A, Iacono D, Luciano AL, Armellino K, Di Iorio A, et al. 2004. Duration of amantadine benefit on dyskinesia of severe Parkinson's disease. J Neurol Neurosurg Psychiatry 75: 141-143.

Torstenson R, Hartvig P, Langstrom B, Bastami S, Antoni G, et al. 1998. Effect of apomorphine infusion on dopamine synthesis rate relates to dopaminergic tone. Neuropharmacology 37: 989-995.

Uekermann J, Daum I, Peters S, Wiebel B, Przuntek H, et al. 2003. Depressed mood and executive dysfunction in early Parkinson's disease. Acta Neurol Scand 107: 341-348.

Whone AL, Watts RL, Stoessl AJ, Davis M, Reske S, et al. 2003. Slower progression of Parkinson's disease with ropinirole versus levodopa: The REAL-PET study. Ann Neurol 54: 93-101.

Wolters EC, Francot C, Bergmans P, Winogrodzka A, Booij J, et al. 2000. Preclinical (premotor) Parkinson's disease. J Neurol 247 Suppl 2: II103-II109.

3 Update on Parkinson's Disease Genetics

R. Krüger · D. Berg · O. Riess · P. Riederer

Abstract: Parkinson's disease (PD) is the most common neurodegenerative movement disorder with a multifactorial etiology, which includes environmental factors acting on genetically predisposed individuals. Although initially discounted, genetic factors causing PD become more and more important, allowing insight into molecular mechanisms leading to neuronal cell death. On the basis of families with Mendelian inheritance of PD, to date six genes could be identified that cause autosomal dominantly or autosomal recessively inherited forms of the disease. Interestingly, variants in some of these genes represent genetic risk factors also for the common sporadic form of PD. Therefore, the definition of novel disease-causing genes and the elucidation of molecular signaling pathways involved in neurodegeneration in PD provide the rationale for new strategies for disease prevention and causative therapy in the future.

List of Abbreviations: LRRK2, leucine-rich repeat kinase 2; NFM, neurofilament M; PD, Parkinson's disease; PINK1, PTEN-induced putative kinase 1; UCH-L1, ubiquitin C-terminal hydrolase L1

1   Introduction

The controversy on the role of genetic factors in Parkinson's desease (PD) has lasted for nearly 200 years after the initial description by James Parkinson (1817). The difficulties in defining a genetic contribution to the pathogenesis of PD may be at least in part explained by confounding variables including (1) late onset of the disease, (2) reduced penetrance of genetic traits, (3) genetic heterogeneity, and (4) environmental factors contributing to the disease. Initially, twin studies have been implemented as an established epidemiological approach to define a possible genetic contribution to PD. In this concept, clinical concordance of PD in monozygotic versus dizygotic twins is assessed. If the investigated trait is inherited, a significant overrepresentation of the trait in the monozygotic sample compared to dizygotic can be assumed indicating higher concordance rates. However, these studies remained inconclusive for several decades (Duvoisin et al., 1981; Marttila et al., 1988; Tanner et al., 1999). Most of the twin studies suffered from their cross-sectional concept that may be confounded by the late and variable age at disease onset. Because cases with subclinical disease were not detected, concordance was generally low and similar between monozygotic and dizygotic twins, arguing against a substantial genetic contribution to the disease (Marttila et al., 1988; Tanner et al., 1999). However, twin studies taking functional imaging into account allowed to detect preclinical cases of PD in asymptomatic co-twins (Burn et al., 1992; Holthoff et al., 1994; Laihinen et al., 2000). Thus, it was demonstrated that clinically discordant twin pairs could be already concordant for nigrostriatal dysfunction with concordance rates of 45% for monozygotic and 29% for dizygotic twins (Burn et al., 1992). A study using a longitudinal approach based on clinical assessment and positron emission tomography (PET) revealed highest concordance with combined levels for subclinical dopaminergic dysfunction and clinical PD of 75% in monozygotic twins and 22% in dizygotic twins (Piccini et al., 1999). These studies supported a substantial role of genetic factors in the pathogenesis of typical idiopathic PD and encouraged the search for disease-causing genes in families with inherited forms of PD. First evidence for genetic factors causing typical levodopa-responsive parkinsonism came from the identification of mutant α-synuclein in a large Italian family with autosomal dominant inheritance of the disease. This was the starting point for a large effort defining genes responsible for autosomal dominantly or autosomal recessively inherited parkinsonism that resulted in the identification of to date 11 known disease loci and five known disease genes (❷ *Table 3-1*). Subsequent functional characterization of these disease genes allowed insight into molecular pathways leading to neurodegeneration and dysfunction of the nigrostriatal system. There is increasing evidence that genes involved in monogenic forms of the disease may act as susceptibility factors also in the common sporadic form of PD. Thus, disturbance of the ubiquitin proteasome pathway came into focus of interest and underscored the relevance of protein misfolding and accumulation in neurodegeneration in PD. Other genes are involved in mitochondrial homeostasis and linked newly identified signaling pathways to the established paradigm of oxidative stress in PD. Today it is generally accepted that, apart from familial forms of PD that account for approximately 20% of the disease, environmental factors acting on genetically predisposed individuals play a role in the majority of sporadic PD patients (Bonifati et al., 1995). In the present chapter, we describe known genes and genetic susceptibility factors

◻ Table 3-1
Identified disease loci in inherited forms of typical levodopa-responsive parkinsonism

Locus	MIM No.	Inheritance	Chromosomal localization	Gene	Type of mutation	Clinical characteristics
PARK1	601508	AD	4q21-23	α-Synuclein	PM	Rapid disease progression, dementia, Lewy bodies
PARK2	600116	AR	6q25.2-27	Parkin	Del, Ins, Dupl, Tripl, PM	Early onset, dystonia, sleep benefit, variable Lewy body pathology
PARK3	602404	AD	2p13	–	–	Typical parkinsonism, dementia, Lewy bodies
PARK4	605543	AD	4q21-23	α-Synuclein	Dupl, Tripl	Typical parkinsonism, dementia, Lewy body pathology
PARK5	191342	AD	4p14	UCH-L1	PM	Typical parkinsonism, Lewy bodies
PARK6	605909	AR	1p35-36	PINK-1	PM	Early onset, slow disease progression
PARK7	606324	AR	1p36	DJ-1	Del, PM	Early onset, dystonia, psychiatric symptoms
PARK8	607060	AD	12cen	LRRK2	PM	Typical parkinsonism, variable pathology including Lewy bodies, neurofibrillary tangles, and senile plaques
PARK9	606693	AR	1p36	–	–	Supranuclear upgaze palsy, dementia
PARK10	606852	AD	1p32	–	–	Late age at disease onset
PARK11	607688	AD	2q36-37	–	–	Not specified

The mapping of the PARK4 locus has been revised and mutations in the α-synuclein gene (PARK1) were identified as disease causing in the respective family

and define their role in neurodegeneration. Transgenic animal models based on disease genes identified in monogenic forms of typical parkinsonism replicate important features of PD, including protein aggregation and progressive motor symptoms, giving options for novel neuroprotective approaches that might be beneficial to PD.

2 Identified Disease Loci in Familial Parkinsonism

2.1 α-Synuclein (PARK1/PARK4)

The first gene in the pathogenesis of PD was identified based on linkage analyses in a large Italian kindred with autosomal dominantly inherited parkinsonism (Polymeropoulos et al., 1997). The locus on chromosome 4q21-23 contained the α-synuclein gene that has been already implicated in the pathogenesis of another neurodegenerative disorder, Alzheimer disease (AD), because the encoded protein is part of the amyloid plaques in brains of AD patients. To date three point mutations in the α-synuclein gene have been identified: (1) an Ala53Thr mutation in the Contursi kindred and several families of Greek origin (Polymeropoulos et al., 1997), (2) an Ala30Pro mutation in a single family of German origin (Krüger et al., 1998), and (3) a Gly46Lys mutation in one Spanish family (Zarranz et al., 2004). For the more frequent Ala53Thr mutation, a founder effect has been shown by haplotype analyses in the Italian and seven Greek families indicating that the mutation spread in the Mediterranean area along ancient trading routes between Greece and Italy (Athanassiadou et al., 1999). Affected individuals carrying mutations in the α-synuclein gene display typical clinical signs of idiopathic PD, including an excellent response to levodopa therapy. The pathological hallmark of PD, Lewy bodies, was defined in affected brain regions of carriers of the Ala53Thr and Gly46Lys mutation. Functional imaging performed in carriers of the Ala30Pro mutation revealed dysfunction of the nigrostriatal system comparable to sporadic PD patients (Krüger et al., 2001; Zarranz et al., 2004). However, there is a large phenotypic variability between carriers of different mutations in the α-synuclein gene. Whereas carriers of the Ala30Pro mutation exhibit a typical late-onset parkinsonism with cognitive impairment occurring years after disease onset, carriers of the Ala53Thr mutation have an early age at disease onset and rapid progessive parkinsonism (Polymeropoulos et al., 1997; Krüger et al., 2001). Carriers of the Gly46Lys mutation display clinical and pathological features, i.e., cognitive decline in early disease stages and abundant cortical Lewy body pathology, which is in line with the diagnosis of dementia with Lewy bodies (Zarranz et al., 2004). All known point mutations in the α-synuclein gene are localized in the N-terminal portion of the α-synuclein protein that displays characteristic imperfect KTKEGV repeats. This region is highly conserved among different species and allows the formation of helical structures upon binding to synaptic vesicles (Davidson et al., 1998). However, the most intriguing property of α-synuclein is its ability to aggregate and form oligomeric and fibrillar structures in vitro (Conway et al., 1998; Giasson et al., 1999). Interestingly, the aggregation-forming capacity is more pronounced for the Ala53Thr mutant compared to Ala30Pro mutant and wild-type α-synuclein (Giasson et al., 1999). This might explain the more severe phenotype in Ala53Thr mutation carriers with an earlier age at disease onset and a more rapid disease progression compared to Ala30Pro carriers. The fact that also wild-type α-synuclein is able to form fibrils in vitro is in line with the observation that α-synuclein is the major component of Lewy bodies not only in brains of carriers of mutations in the α-synuclein gene but also in sporadic PD patients. This indicates a potential role of the wild-type protein in molecular mechanisms, leading to neurodegeneration in sporadic PD. Extensive mutation analyses in large samples of sporadic PD patients revealed no mutation in the coding sequence and in adjacent intronic sequences of the α-synuclein gene (Vaughan et al., 1998; Krüger et al., 1998; Berg et al., 2004). However, several studies observed an association of a dinucleotide repeat polymorphism in the promoter region of the α-synuclein gene with sporadic PD (Krüger et al., 1999; Tan et al., 2000; Tan et al., 2004). Although these results were not confirmed in other cohorts (Parsian et al., 1998; Spadafora et al., 2003), recent studies extending these analyses to haplotype studies in the promoter region support the genetic relevance of regulatory elements in the α-synuclein promoter (Farrer et al., 2001, 2004). Therefore, increased levels of wild-type α-synuclein might contribute to neurodegeneration via disturbed protein degradation and increased accumulation and aggregation of α-synuclein. Indeed, functional analyses of different polymorphisms in the promoter region of the α-synuclein gene revealed that risk alleles in PD patients lead to increased protein expression in vitro (Chiba-Falek and Nussbaum, 2001; Farrer et al., 2001; Holzmann et al., 2003). The pathogenic relevance of dose effects of wild-type α-synuclein was subsequently underscored by the identification of a triplication of the α-synuclein gene as the disease-causing mutation in the so-called Iowa kindred (formerly known as PARK4). The phenotype of this family with autosomal dominant inheritance of the disease shows a wide range of clinical symptom that includes typical PD as well as dementia with Lewy bodies (Farrer et al., 1999). The latter corresponds to the observation of Lewy bodies in affected brain regions including the cortex (Gwinn-Hardy et al., 2000). In blood and brain tissue of affected carriers, twofold increased levels of α-synuclein were demonstrated resulting from the expression of all four α-synuclein alleles (Farrer et al., 2004; Miller et al., 2004). Interestingly, a potential dose effect of α-synuclein expression is reflected by the milder phenotype of recently identified carriers of duplications of the α-synuclein gene (Chartier-Harlin et al., 2004). They display typical signs of PD and a later age at disease onset without prominent signs of dementia observed in carriers of triplications. When transferred to the animal model, it was shown that overexpression of human wild-type α-synuclein in mice is sufficient to cause motor deficits, loss of dopaminergic nerve terminals, and α-synuclein-positive ubiquitinated intracytoplasmic and intra-nuclear inclusions in vivo (Masliah et al., 2000).

2.2 Parkin (PARK2)

Mutations in the Parkin gene on chromosome 6q25.2-27 are the most common genetic cause for inherited parkinsonism known to date, accounting for 49% of familial and about 19% of sporadic early-onset PD cases (mean age at disease onset <45 years) (Lücking et al., 2000). Homozygous mutations were initially identified as responsible for an autosomal recessively inherited juvenile parkinsonism in Japanese families (ARJP). Subsequent studies in other ethnic groups revealed a variety of disease-causing mutations, including single base pair substitutions, deletions of variable size, and exon multiplications (Lücking et al., 2000). The Parkin gene encodes a ubiquitin E3 ligase implicating the ubiquitin-mediated protein degradation pathway in neurodegeneration in PD (Shimura et al., 2000; ❷ *Figure 3-1*). Mutations in the

❑ Figure 3-1

Genes identified in familial parkinsonism and their role in molecular mechanisms leading to neurodegeneration

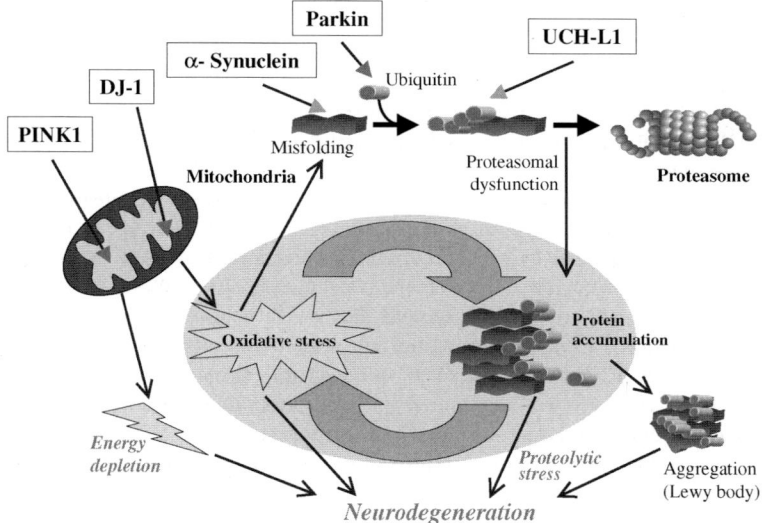

Parkin gene are thought to cause a loss of ligase function interfering with the ubiquitination and degradation of its substrates. A potential role of heterozygous point mutations in the Parkin gene came from studies in sporadic PD patients with typical late age at disease onset, because the authors identified the respective mutations as risk factors for typical PD (Foroud et al., 2003; Oliveira et al., 2003). Among these putative susceptibility alleles for late-onset PD, the R275W mutation has been functionally characterized in vitro. An altered subcellular localization compared to wild-type parkin and formation of cytoplasmic inclusions in cells overexpressing R275W mutant parkin was observed supporting a potential toxic gain-of-function (Cookson et al., 2003). Indeed, independent studies revealed an increased inclusion-forming capacity and preserved ligase function of R275W mutant parkin, indicating a possible different mechanism involved in parkin mediated late-onset parkinsonism (Wang et al., 2004).

2.3 UCH-L1 (PARK5)

Using a candidate approach focusing on genes involved in the ubiquitin-mediated protein degradation, the gene encoding ubiquitin C-terminal hydrolase L1 (UCH-L1) was defined as PARK5. In one German family with autosomal dominant parkinsonism, an Ile93Met mutation was identified cosegregating with the disease (Leroy et al., 1998). Affected family members displayed signs of idiopathic PD including the presence of Lewy bodies in affected brain regions. Subsequent mutation screenings failed to detect other disease-causing mutations in the UCH-L1 gene (Farrer et al., 2000; Wintermeyer et al., 2000). However, a Ser18Tyr polymorphism in the UCH-L1 gene was defined and the Tyr18 allele was found significantly underrepresented in sporadic PD patients compared to controls (Wintermeyer et al., 2000). This potential protective effect of the respective allele was recently confirmed by a pooled meta-analysis of all studies performed on the Ser18Tyr polymorphism in PD (Maraganore et al., 2004). Subsequent functional studies on UCH-L1 unserscored the relevance of the reported genetic findings. For wild-type UCH-L1, a balanced dual role acting as an ubiquitin-hydrolase and ubiquitin-ligase was found (Liu et al., 2002). The disease-causing Ile93Met mutation shifted the role of UCH-L1 in protein degradation toward the ubiquitin-ligase effect, suggesting the accumulation of ubiquitinated substrate. In contrast, the protective Ser18Tyr variant propagates the opposing hydrolase function possibly facilitating ubiquitin reutilization and protein clearance (❯ Figure 3-1). Thus, functional genetic studies strengthened the pathogenenic relevance of both variants of the UCH-L1 gene in familial and sporadic parkinsonism.

2.4 PINK1 (PARK6)

Mutations in the PINK1 gene turn out to represent the second most common cause for inherited PD known to date and are responsible for an autosomal recessively inherited form of the disease. The PARK6 locus has been initially mapped to chromosome 1p35-p36 in a large Italian family (Valente et al., 2001). Clinically affected individuals display signs of typical parkinsonism except for an early age at disease onset (range 32–48 years) and are characterized by early occurrence of levodopa-associated dyskinesia, slow disease progression and the absence of cognitive impairment (Valente et al., 2001, 2004a, b). Pathological data on the presence of Lewy bodies are not available to date. Disease-causing mutations identified in the PINK1 gene include point mutations (truncating mutations, missense mutations) and insertions (frameshift leading to truncation) that were identified in different ethnic groups (Hatano et al., 2004; Rohe et al., 2004; Valente et al., 2004a, b). As for Parkin, a potential role of heterozygous mutations in the PINK1 gene in the sporadic form of the disease has been suggested by a recent genetic study (Valente et al., 2004a, b) and is supported by functional brain imaging in asymptomatic carriers of PINK1 mutations exhibiting significant dopaminergic dysfunction (Khan et al., 2002). PINK1 encodes a PTEN-induced protein kinase in response to oxidative stress. The protein is encoded in the nucleus and seems to translocate from the cytoplasm to the mitochondria to exert its physiological function (Valente et al., 2004a, b). Overexpression of mutant PINK1 was shown to reduce cell viability in paradigms of cellular stress in vitro, suggesting an important protective function of the wild-type protein (Valente et al., 2004a, b). Future studies will define the substrates of PINK1 in mitochondria and whether PINK1 relates to dysfunction of the ubiquitin proteasome system.

2.5 DJ-1 (PARK7)

This locus was mapped to chromosome 1p36 in one Dutch family with autosomal recessively inherited early onset parkinsonism and independently confirmed in families from Italy and Uruguay (van Duijn et al., 2001). Mutation screenings in different ethnic groups indicate that mutations in the DJ-1 gene are only rare causes of early onset parkinsonism (Clark et al., 2004). Besides typical clinical symptoms of PD, affected individuals exhibit dystonic features, i.e., blepharospasm and laterocollis and psychiatric signs including psychotic episodes (van Duijn et al., 2001). A large deletion mutation encompassing exons 1–5 of the DJ-1 gene and a point mutation leading to a L166P exchange were identified as causative for the phenotype in the Dutch and Italian families (Bonifati et al., 2003). Functional characterization of the L166P mutation revealed decreased protein stability that is mediated, at least in part, by increased proteasomal degradation contributing to the loss of DJ-1 function in this recessively inherited form of parkinsonism (Miller et al., 2003).

Mutations in the DJ-1 gene are a rare cause of early-onset parkinsonism. To date two other homozygous point mutations, M26I and E64D, were observed in affected individuals with early onset parkinsonism (Abou-Sleiman et al., 2003; Hering et al., 2004a). Functional studies on these mutations show only slight differences in protein steady state levels of mutant compared to wild-type protein, indicating possible alternative pathways of DJ-1 pathogeneity (Moore et al., 2003; Hering et al., 2004a). An important feature of DJ-1 is the accumulation of an acidic form of the protein in response to oxidative stress. In this context, a specific acidification affecting the cysteine residue in position 106 was identified as responsible for mitochondrial relocalization of DJ-1 (Canet-Aviles et al., 2004). This modification and relocalization to mitochondria was found to be critical for the protective function of DJ-1 in cells in vitro (Canet-Aviles et al., 2004; ❷ *Figure 3-1*). Further functional studies are necessary to clarify the role of DJ-1 in neurodegeneration and define underlying molecular mechanisms.

2.6 LRRK2 (PARK8)

This locus for autosomal dominantly inherited PD was initially mapped on chromosome 12p11.2-q13.1 in a Japanese family (PARK8) (Funayama et al., 2002). The phenotype largely reflects common idiopathic PD including laterality of parkinsonism at onset, a mean age at disease onset of 51 ± 6 years, and a favorable response to dopaminergic medication. Reduced penetrance of the expected disease-causing mutation was reflected by the presence of unaffected family members sharing the risk haplotype. Subsequent studies revealed that the PARK8 locus is also responsible for PD in Caucasian families with autosomal dominant parkinsonism (Zimprich et al., 2004a). Recently, disease-causing mutations in a large gene encoding a multifunctional protein of the ROCO protein family were identified in PD families mapping to the PARK8 locus (Zimprich et al., 2004b). Brains of affected individuals reveal brainstem dopaminergic degeneration accompanied by strikingly diverse pathologies including Lewy bodies, tau-positive lesions, and amyloid plaques (Zimprich et al., 2004b). Based on genotyping in affected families, the clinical phenotype extended to dementia and amyotrophy as additional symptoms. Current data indicate that mutations in the LRKK2 gene are the most frequent cause for an autosomal dominantly inherited parkinsonism (Paisan-Ruiz et al., 2004; Zimprich et al., 2004b). Future studies will determine the relevance of the gene for the common sporadic form of the disease and elucidate the overlap in brain pathology with other major neurodegenerative disorders like AD and amyotrophic lateral sclerosis.

2.7 Other Loci (PARK3, PARK9, PARK10, PARK11)

Using a whole-genome mapping approach in six families with an autosomal dominantly inherited PD, the PARK3 locus has been initially mapped to a 10.6 cM region on chromosome 2p13 (Gasser et al., 1998). Four of these families showed positive lod scores and in two families (family B and C), which were traced back to the border region of Southern Denmark and Northern Germany, a common haplotype has been found. Clinically, patients in these families display typical signs of PD indistinguishable from sporadic forms of the disease, including an average age at disease onset of 61 years. Since the majority of individuals carrying the risk haplotype were not affected, a reduced penetrance of about 40% has been postulated for the disease-causing mutation. To date 15 genes in this candidate region have been screened for mutations in the coding region, however, no disease-causing mutation has been found (West et al., 2001). Further support for the relevance of the PARK3 locus came from independent linkage studies defining a disease-modulating factor for PD on chromosome 2p13 (DeStefano et al., 2002; Karamohamed et al., 2003; Pankratz et al., 2004).

The PARK9 locus has been mapped on chromosome 1p36 in one Arabian family with an autosomal recessively inherited atypical parkinsonian syndrome (Hampshire et al., 2001). The clinical syndrome has been named Kufor-Rakeb and includes an akinetic-rigid parkinsonian syndrome accompanied by atypical clinical features like spasticity, supranuclear upgaze palsy, and dementia. Although an excellent response of extrapyramidal symptoms to levodopa therapy has been reported, the relevance of the respective locus for typical PD remains unclear.

Another approach based on a genome-wide scan on 117 Icelandic PD patients from 51 families with more than one affected individual identified a susceptibility gene for PD on chromosome 1p32 (PARK10; Hicks et al., 2002). Patients present with typical signs of idiopathic PD including a late disease onset and a good and sustained response to levodopa therapy.

The PARK11 locus was identified in sib pairs with typical PD on chromosome 2q36-q37 (Pankratz et al., 2003). Expanding the initial sample to 150 families with strictest diagnostic criteria for PD, the authors confirmed significant linkage to the PARK11 locus assuming an autosomal dominant mode of inheritance (Pankratz et al., 2003).

3 Susceptibility Genes

Apart from genes identified on genetically defined chromosomal loci responsible for monogenic forms of parkinsonism, several susceptibility genes were identified based on a candidate approach, contributing to inherited and/or sporadic forms of the disease.

3.1 Tau

Mutations in the microtubule-associated protein tau (MAPT) gene are responsible for frontotemporal dementia with parkinsonism (FTDP-17), an atypical form of parkinsonism. However, a potential role of the MAPT gene in PD has been deducted from several genetic and functional studies. Using genetic association studies, a common H1 haplotype variability within MAPT gene was defined as a genetic risk factor for common sporadic PD (Healy et al., 2004). Using functional studies to define the transcriptional activity of different promoter haplotypes, the H1 haplotype lead to increased expression of the reporter gene in vitro compared to the H2 haplotype (Kwok et al., 2004). These data suggest a similar effect of increased tau expression as observed for α-synuclein promoter variants (see ➋ *Sect. 3.1*). Indeed, interference of both proteins in the pathogenesis of PD has been deducted from in vitro observations: at low protein concentrations the fibrillization of α-synuclein is promoted by the interacting tau protein (Giasson et al., 2003). Indeed in pathological conditions in brains of PD patients co-occurrence of α-synuclein and tau pathology has been observed (Duda et al., 2002). Moreover, pathological findings in affected individuals of PARK1 and PARK8 families indicate tau-positive neurofibrillary tangles (Kotzbauer et al., 2004; Wszolek et al., 2004).

3.2 Neurofilament

A linkage study defining susceptibility loci for PD found strongest evidence for linkage on chromosome 8p, a chromosomal region harboring the genes for the neurofilament (NF) subunits L and M (Scott et al., 2001). NF is a major component of Lewy bodies in PD brains (Hill et al., 1991). In one 22-year-old index patient of a family with autosomal recessively inherited juvenile parkinsonism, a novel Gly336Ser mutation in the NF-M gene was identified (Lavedan et al., 2002). In an independent mutation screening a Pro725Gln substitution and a novel deletion of valine in position 829 were identified in two PD patients (Krüger et al., 2003). All identified amino acid substitutions affect residues of the NF-M protein that are highly conserved among different species and therefore implicate functional relevance. These results might indicate that rare variants of the NF-M protein may act as susceptibility factors for PD. First functional analyses of the Gly336Ser mutation in the NF-M gene failed to demonstrate an effect of mutant NF-M on the intermediate filament network or intracellular distribution of cellular organelles or proteins (Perez-Olle et al., 2004). Further studies including Pro725Gln and del829Val NF-M are necessary to define a potential pathogenic relevance of mutations in the NF-M gene in neurodegeneration.

3.3 Nurr1

The transcription factor Nurr1 (NR4A2) is highly expressed in dopaminergic cells of the substantia nigra. Nurr1 knockout mice show a loss of dopaminergic neurons reflecting its relevance for the development and survival of dopaminergic neuron (Zetterstrom et al., 1997). Based on association studies using a polymorphism in intron 6 of the Nurr1 gene, first evidence for a possible contribution of variants in the Nurr1 gene to the pathogenesis of typical PD has been established (Xu et al., 2002; Zheng et al., 2003). Subsequently, two mutations, del-291T and T-245G, in the 5′-region of the Nurr1 gene have been found in patients with familial parkinsonism (Le et al., 2003). Functional characterization of the mutations revealed a marked decrease in Nurr1 mRNA levels in vitro and in vivo. Moreover, a downregulation of the gene encoding the dopamine biosynthesis enzyme tyrosine hydroxylase was found (Le et al., 2003). However the relatively high frequency of 10% mutations in the Nurr1 gene in the initial study could not be confirmed by other groups screening large samples of familial and sporadic PD patients (Hering et al., 2004b).

3.4 Synphilin-1

Synphilin-1 has been functionally linked to the pathogenesis of PD based on its identification as an α-synuclein (PARK1) and parkin (PARK2) interacting protein (Engelender et al., 1999; Chung et al., 2001). First genetic evidence for a contribution of synphilin-1 to the pathogenesis of PD came from linkage studies that identified the long arm of chromosome 5 containing the synphilin-1 gene as a candidate locus for PD (Scott et al., 2001). First studies sequencing the synphilin-1 gene in index patients from a total of 69 PD families found no coding changes and failed to provide genetic evidence for a role of synphilin-1 in PD (Bandopadhyay et al., 2001; Farrer et al., 2001). The first coding substitution leading to a novel R621C mutation in the synphilin-1 protein was identified in two apparently independent patients with sporadic PD (Marx et al., 2003). Genotyping polymorphisms in the chromosomal region containing the synphilin-1 gene demonstrated that both mutation carriers share rare alleles of the respective genetic markers, suggesting a common ancestor carrying the mutation. Subsequent functional in vitro studies supported the pathogenic relevance of the R621C mutation in the synphilin-1 gene, because overexpression of mutant synphilin-1 sensitized cells to toxic stress in vitro (Marx et al., 2003). In cultured cells, overexpression of synphilin-1 leads to the formation of intracytoplasmic inclusions, which is increased after proteasomal inhibition (O'Farrell et al., 2001; Marx et al., 2003). Interestingly, mutant synphilin-1 displayed a reduced inclusion-forming capacity, suggesting a protective function of inclusions in vitro (Marx et al., 2003). Indeed an independent study on synphilin-1 overexpression in cultured cells argues in favor of a cytoprotective function of inclusions formed by synphilin-1 (Tanaka et al., 2003).

4   Conclusions

Based on rare, monogenic forms of PD, substantial progress has been made in the understanding of the molecular signaling pathways leading to neurodegeneration in PD. Functional characterization of proteins encoded by the identified disease genes revealed the importance of proteolytic stress and centered on the ubiquitin-mediated protein degradation machinery. Genes mutated in inherited forms of parkinsonism encode proteins that are able to interfere with proteasomal function due to (1) misfolding, (2) degradation via the ubiquitin-proteasome system, or (3) as important key players in the degradation pathway. Recently identified genes encoding proteins involved in mitochondrial homeostasis and oxidative stress response link proteolytic stress to known biochemical markers of neurodegeneration in PD, namely energy depletion and oxidative stress. Thus genes defined in rare inherited forms of the disease open views for the understanding of the sporadic form of the disease. Indeed genetic variants in the α-synuclein and UCH-L1 gene have been identified as disease modifiers in sporadic PD patients (Krüger et al., 1999;

Maraganore et al., 2004). Moreover, previously identified biochemical features in sporadic PD as complex I deficiency and mitochondrial dysfunction may be related to genetic risk factors. Further support for an involvement of the ubiquitin-proteasome system, cytoskeletal proteins, and oxidative stress conditions in neurodegeneration in PD came from gene expression studies in brains of sporadic PD patients (Grünblatt et al., 2004). In general, genetic and biochemical data support the idea of a multifactorial etiology of sporadic PD with environmental factors acting on genetically predisposed individuals and might indicate first steps toward a unifying concept in the pathogenesis of PD.

References

Abou-Sleiman PM, Healy DG, Quinn N, Lees AJ, Wood NW. 2003. The role of pathogenic DJ-1 mutations in Parkinson's disease. Ann Neurol 54: 283-286.

Athanassiadou A, Voutsinas G, Psiouri L, et al. 1999. Genetic analysis of families with Parkinson disease that carry the Ala53Thr mutation in the gene encoding α-synuclein. Am J Hum Genet 65: 555-558.

Bandopadhyay R, de Silva R, Khan N, Graham E, Vaughan J, et al. 2001. No pathogenic mutations in the synphilin-1 gene in Parkinson's disease. Neurosci Lett 307: 125-127.

Berg D, Niwar N, Maass S, Zimprich A, Moeller C, et al. 2005. Mutations in the α-synuclein gene and Parkinson's disease– implications from the screening of more than 1900 patients. Mov Disord 20: 1191-1194.

Bonifati V, Oostra BA, Heutink P. 2004. Unraveling the pathogenesis of Parkinson's disease–the contribution of monogenic forms. Cell Mol Life Sci 61: 1729-1750.

Bonifati V, Rizzu P, van Baren MJ, Schaap O, Breedveld GJ, et al. 2003. Mutations in the DJ-1 gene associated with autosomal recessive early-onset parkinsonism. Science 299: 256-259.

Burn DJ, Mark MH, Playford ED, Maraganore DM, Zimmerman TR Jr, et al. 1992. Parkinson's disease in twins studied with 18F-dopa and positron emission tomography. Neurology 42: 1894-1900.

Canet-Aviles RM, Wilson MA, Miller DW, Ahmad R, McLendon C, et al. 2004. The Parkinson's disease protein DJ-1 is neuroprotective due to cysteine-sulfinic acid-driven mitochondrial localization. PNAS 101: 9103-9108.

Chartier-Harlin MC, Kachergus J, Roumier C, Mouroux V, Douay X, et al. 2004. Alpha-synuclein locus duplication as a cause of familial Parkinson's disease. Lancet 364: 1167-1169.

Chiba-Falek O, Nussbaum RL. 2001. Effect of allelic variation at the NACP-Rep1 repeat upstream of the α-synuclein gene (SNCA) on transcription in a cell culture luciferase reporter system. Hum Mol Genet 10: 3101-3109.

Chung KK, Zhang Y, Lim KL, Tanaka Y, Huang H, et al. 2001. Parkin ubiquitinates the alpha-synuclein-interacting protein, synphilin-1: Implications for Lewy-body formation in Parkinson disease. Nat Med 7: 1144-1150.

Clark LN, Afridi S, Mejia-Santana H, Harris J, Louis ED, et al. 2004. Analysis of an early-onset Parkinson's disease cohort for DJ-1 mutations. Mov Disord 19: 796-800.

Conway KA, Harper JD, Lansbury PT. 1998. Accelerated in vitro fibril formation by a mutant alpha-synuclein linked to early-onset Parkinson disease. Nat Med 4: 1318-1320.

Cookson MR, Lockhart PJ, McLendon C, O'Farrell C, Schlossmacher M, et al. 2003. RING finger 1 mutations in parkin produce altered localization of the protein. Hum Mol Genet 12: 2957-2965.

Davidson WS, Jonas A, Clayton DF, George JM. 1998. Stabilization of alpha-synuclein secondary structure upon binding to synthetic membranes. J Biol Chem 273: 9443-9449.

De Stefano AL, Lew MF, Golbe LI, Mark MH, Lazzarini AM, et al. 2002. PARK3 influences age at onset in Parkinson disease: A genome scan in the GenePD study. Am J Hum Genet 70: 1089-1095.

Duda JE, Giasson BI, Mabon ME, Miller DC, et al. 2002. Concurrence of α-synuclein and tau brain pathology in the Contursi kindred. Acta Neuropathol 104: 7-11.

Duvoisin RC, Eldridge R, Williams A, Nutt J, Calne D. 1981. Twin study of Parkinson disease. Neurology 31: 77-80.

Engelender S, Kaminsky Z, Guo X, Sharp AH, Amaravi RK, et al. 1999. Synphilin-1 associates with alpha-synuclein and promotes the formation of cytosolic inclusions. Nat Genet 22: 110-114.

Farrer M, Gwinn-Hardy K, Muenter M, De Vrieze FW, Crook R, et al. 1999. A chromosome 4p haplotype segregating with Parkinson's disease and postural tremor. Hum Mol Genet 8: 81-85.

Farrer M, Destee T, Becquet E, Wavrant-De Vrieze F, et al. 2000. Linkage exclusion in French families with probable Parkinson's disease. Mov Disord 15: 1075-1083.

Farrer M, Kachergus J, Forno L, Lincoln S, Wang DS, et al. 2004. Comparison of kindreds with parkinsonism and alpha-synuclein genomic multiplications. Ann Neurol 55: 174-179.

Farrer M, Maraganore DM, Lockhart P, et al. 2001. α-Synuclein gene haplotypes are associated with Parkinson's disease. Hum Mol Genet 10: 1847-1851.

Foroud T, Uniacke SK, Liu L, Pankratz N, Rudolph A, et al. 2003. Heterozygosity for a mutation in the parkin gene leads to later onset Parkinson's disease. Neurology 60: 796-801.

Funayama M, Hasegawa K, et al. 2002. A new locus for Parkinson's disease (PARK8) maps to chromosome 12p11.2-q13.1. Ann Neurol 51: 296-301.

Gasser T, Muller-Myhsok B, Wszolek ZK, Oehlmann R, Calne DB, et al. 1998. A susceptibility locus for Parkinson's disease maps to chromosome 2p13. Nat Genet 18: 262-265.

Giasson BI, Uryu K, Trojanowski JQ, Lee VM. 1999. Mutant and wild type human alpha-synucleins assemble into elongated filaments with distinct morphologies in vitro. J Biol Chem 274: 7619-7622.

Giasson BI, Forman MS, Higuchi M, Golbe LI, Graves CL, et al. 2003. Initiation and synergistic fibrillization of tau and alpha-synuclein. Science 300: 636-640.

Grünblatt E, Mandel S, Jacob-Hirsch J, Zeligson S, Amariglo N, et al. 2004. J Neural Transm 111: 1543-1573.

Gwinn-Hardy K, Mehta ND, Farrer M, Maraganore D, Muenter M, et al. 2000. Distinctive neuropathology revealed by alpha-synuclein antibodies in hereditary parkinsonism and dementia linked to chromosome 4p. Acta Neuropathol 99: 663-672.

Hampshire DJ, Roberts E, Crow Y, Bond J, Mubaidin A, et al. 2001. Kufor-Rakeb syndrome, pallido-pyramidal degeneration with supranuclear upgaze paresis and dementia, maps to 1p36. J Med Genet 38: 680-682.

Hatano Y, Li Y, Sato K, et al. 2004. Novel PINK1 mutations in early-onset parkinsonism. Ann Neurol 56: 424-427.

Healy DG, Abou-Sleiman PM, Lees AJ, Casas JP, Quinn N, et al. 2004. Tau gene and Parkinson's disease: A case-control study and meta-analysis. J Neurol Neurosurg Psychiatry 75: 962-965.

Hering R, Strauss K, Tao X, Bauer A, Mietz EM, et al. 2004a. Novel E64D mutation in DJ-1 gene is causative of early onset Parkinson's Disease. Hum Mutat 24: 321-329.

Hering R, Petrovic S, Mietz EM, Holzmann C, Berg D, et al. 2004b. Extended mutation analysis and association studies of Nurr1 (NR4A2) in Parkinson's Disease. Neurology 62: 1231-1232.

Hicks A, Petursson H, Jonsson T, Stefansson H, Johannsdottir H, et al. 2002. A susceptibility gene for late-onset idiopathic Parkinson's disease successfully mapped. Ann Neurol 52: 549-555.

Hill WD, Lee VM, Hurtig HI, Murray JM, Trojanowski JQ. 1991. Epitopes located in spatially separate domains of each neurofilament subunit are present in Parkinson's disease Lewy bodies. J Comp Neurol 309: 150-160.

Holthoff VA, Vieregge P, Kessler J, Pietrzyk U, Herholz K, et al. 1994. Discordant twins with Parkinson's disease: Positron emission tomography and early signs of impaired cognitive circuits. Ann Neurol 36: 176-182.

Holzmann C, Krüger R, Saecker AMM, Schmitt I, Schöls L, et al. 2003. Characterization of α-synuclein promoter polymorphisms in Parkinson's disease. J Neural Transm 110: 67-76.

Karamohamed S, De Stefano AL, Wilk JB, Shoemaker CM, Golbe LI, et al. 2003. A haplotype at the PARK3 locus influences onset age for Parkinson's disease: The GenePD study. Neurology 61: 1557-1561.

Khan NL, Valente EM, Bentivoglio AR, Wood NW, Albanese A, et al. 2002. Clinical and subclinical dopaminergic dysfunction in PARK6-linked parkinsonism: An 18F-dopa PET study. Ann Neurol 52: 849-853.

Kotzbauer PT, Giasson BI, Kravitz AV, Golbe LI, Mark MH, et al. 2004. Fibrillization of alpha-synuclein and tau in familial Parkinson's disease caused by the A53T alpha-synuclein mutation. Exp Neurol 187: 279-288.

Krüger R, Kuhn W, Muller T, Woitalla D, Graeber M, et al. 1998. Ala30Pro mutation in the gene encoding alpha-synuclein in Parkinson's disease. Nat Genet 18: 106-108.

Krüger R, Kuhn W, Müller T, Kühnl N, Fuchs GA, et al. 1999. Increased susceptibility to sporadic Parkinson's disease by a certain combined alpha-synuclein/apolipoprotein E genotype. Ann Neurol 45: 611-617.

Krüger R, Kuhn W, Leenders KL, Sprengelmeyer R, Müller T, et al. 2001. Familial parkinsonism with synuclein pathology: Clinical and PET studies of A30P mutation carriers. Neurology 56: 1355-1362.

Krüger R, Fischer C, Strauss K, Schulte T, Riess O, et al. 2003. Mutations in the neurofilament M gene in Parkinson's disease. Neurosci Lett 351: 125-129.

Kwok JB, Teber ET, Loy C, Hallupp M, Nicholson G, et al. 2004. Tau haplotypes regulate transcription and are associated with Parkinson's disease. Ann Neurol 55: 329-334.

Laihinen A, Ruottinen H, Rinne JO, Haaparanta M, Bergman J, et al. 2000. Risk for Parkinson's disease: Twin studies for the detection of asymptomatic subjects using [18F]6-fluorodopa PET. Neurology 247 (Suppl. 2): II110-II113.

Lavedan C, Buchholtz S, Nussbaum RL, Albin RL, Polymeropoulos MH. 2002. A mutation in the human neurofilament M gene in Parkinson's disease that suggests a role for the cytoskeleton in neuronal degeneration. Neurosci Lett 322: 57-61.

Le WD, Xu P, Jankovic J, Jiang H, Appel SH, et al. 2003. Mutations in NR4A2 associated with familial Parkinson disease. Nat Genet 33: 85-89.

Leroy E, Boyer R, Auburger G, Leube B, Ulm G, et al. 1998. The ubiquitin pathway in parkinson's disease. Nature 395: 451-452.

Liu Y, Fallon L, Lashuel HA, Liu Z, Lansbury PT. 2002. The UCH-L1 gene encodes two opposing enzymatic activities that affect alpha-synuclein degradation and Parkinson's disease susceptibility. Cell 111: 209-218.

Lücking CB, Dürr A, Bonifati V, Vaughan J, De Michele G, et al. 2000. Association between early-onset Parkinson's disease and mutations in the Parkin gene. N Engl J Med 342: 1560-1567.

Maraganore DM, Lesnick TG, Elbaz A, Chartier-Harlin MC, Gasser T, et al. 2004. Collaborative reanalysis of the ubiquitin carboxy-terminal hydrolase L1 (UCHL-1) gene S18 variant and its association with Parkinson's disease (PD). Ann Neurol 55: 512-521.

Marttila RJ, Kaprio J, Koskenvuo MD, Rinne UK. 1988. Parkinson's disease in a nationwide twin cohort. Neurology 38: 1217-1219.

Marx FP, Holzmann C, Strauss KM, Li L, Eberhardt O, et al. 2003. Identification and functional characterization of a novel R621C mutation in the synphilin-1 gene in Parkinson's disease. Hum Mol Gen 12: 1223-1231.

Masliah E, Rockenstein E, Veinbergs I, Mallory M, Hashimoto M, et al. 2000. Dopaminergic loss and inclusion body formation in alpha-synuclein mice: Implications for neurodegenerative disorders. Science 287: 1265-1269.

Miller DW, Ahmad R, Hague S, Baptista MJ, Canet-Aviles R, et al. 2003. L166P mutant DJ-1, causative for recessive Parkinson's disease, is degraded through the ubiquitin-proteasome system. J Biol Chem 278: 36588-36595.

Miller DW, Hague SM, Clarimon J, Baptista M, Gwinn-Hardy K, et al. 2004. Alpha-synuclein in blood and brain from familial Parkinson disease with SNCA locus triplication. Neurology 62: 1835-1838.

Moore DJ, Zhang L, Dawson TM, Dawson VL. 2003. A missense mutation (L166P) in DJ-1, linked to familial Parkinson's disease, confers reduced protein stability and impairs homo-oligomerization. J Neurochem 87: 1558-1567.

O'Farrell C, Murphy DD, Petrucelli L, Singleton AB, Hussey J, et al. 2001. Transfected synphilin-1 forms cytoplasmic inclusions in HEK293 cells. Brain Res Mol Brain Res 97: P94-P102.

Oliveira SA, Scott WK, Martin ER, Nance MA, Watts RL, et al. 2003. Parkin mutations and susceptibility alleles in late-onset Parkinson's disease. Ann Neurol 53: 624-629.

Paisan-Ruiz C, Hain S, Evans EW, Gilks WP, Simon J, et al. 2004. Cloning the gene containing mutations that cause PARK8–linked Parkinson's disease. Neuron 44: 595-600.

Pankratz N, Nichols WC, Uniacke SK, Halter C, Rudolph A, et al. 2003. Significant linkage of Parkinson disease to chromosome 2q36–37. Am J Hum Genet 72: 1053-1057.

Pankratz N, Uniacke SK, Halter CA, Rudolph A, Shults CW, et al. 2004. Genes influencing Parkinson disease onset: Replication of PARK3 and identification of novel loci. Neurology 62: 1616-1618.

Parsian A, Racette B, Zhang ZH, et al. 1998. Mutation, sequence analysis, and association studies of α-synuclein in Parkinson's disease. Neurology 51: 1757-1759.

Perez-Olle R, Lopez-Toledano MA, Liem RKH. 2004. The G336S variant in the human neurofilament-M gene does not affect its assembly or distribution: Importance of the functional analysis of neurofilament variants. J Neuropathol Exp Neurol 63: 759-774.

Piccini P, Burn DJ, Ceravolo R, Maraganore D, Brooks. 1999. The role of inheritance in sporadic Parkinson's disease: Evidence from a longitudinal study of dopaminergic function in twins. Ann Neurol 45: 577-582.

Polymeropoulos MH, Lavedan C, Leroy E, Die SE, Dehejia A, et al. 1997. Mutation in the α-synuclein gene identified in families with Parkinson's disease. Science 276: 2045-2047.

Rohe CF, Montagna P, Breedveld G, Cortelli P, Oostra BA, et al. 2004. Homozygous PINK1 C-terminus mutation causing early-onset parkinsonism. Ann Neurol 56: 427-431.

Scott WK, Nance MA, Watts RL, Hubble JP, Koller WC, et al. 2001. Complete genomic screen in Parkinson disease: Evidence for multiple genes. JAMA 286: 2239-2244.

Shimura H, Hattori N, Kubo S-I, Mizuno Y, Asakawa S, et al. 2000. Familial Parkinson disease gene product, parkin, is a ubiquitin-protein ligase. Nat Genet 25: 302-305.

Spadafora P, Annesi G, Pasqua AA, et al. 2003. NACP-REP1 polymorphism is not involved in Parkinson's disease: A case-control study in a population sample from southern italy. Neurosci Lett 351: 375-378.

Tan EK, Matsuura T, Nagamitsu S, Khajavi M, Jankovic J, et al. 2000. Polymorphism of NACP-Rep1 in Parkinson's disease: An etiologic link with essential tremor? Neurology 54: 1195-1198.

Tan EK, Chai A, Teo YY, Zhao Y, Tan C, et al. 2004. Alpha-synuclein haplotypes implicated in risk of Parkinson's disease. Neurology 62: 128-131.

Tanaka M, Kim YM, Lee G, Junn E, Iwatsubo T, et al. 2003. Aggresomes formed by alpha-synuclein and synphilin-1 are cytoprotective. J Biol Chem 279: 4625-4631.

Tanner CM, Ottman R, Goldman SM, Ellenberg J, Chan P, et al. 1999. Parkinson's disease in twins: An etiologic study. JAMA 281: 341-246.

Valente EM, Bentivoglio AR, Dixon PH, Ferraris A, Ialongo T, et al. 2001. Localization of a novel locus for autosomal recessive early-onset parkinsonism, PARK6, on human chromosome 1p35–p36. Am J Hum Genet 68: 895-900.

Valente EM, Abou-Sleiman PM, Caputo V, Muqit MMK, Harvey K, et al. 2004a. Hereditary early-onset Parkinson's disease is caused by mutations in PINK1. Science 304: 1158-1160.

Valente EM, Salvi S, Ialongo T, Marongiu R, Elia AE, et al. 2004b. PINK1 mutations are associated with sporadic early-onset parkinsonism. Ann Neurol 56: 336-341.

Van Duijn CM, Deker MCJ, Bonifati V, Galjaard RJ, Houwing-Duistermaat JJ, et al. 2001. PARK7, a novel locus for autosomal recessive early-onset parkinsonism, on chromosome 1p36. Am J Hum Genet 69: 629-634.

Vaughan JR, Farrer MJ, Wszolek ZK, Gasser T, Durr A, et al. 1998. Sequencing of the a-synuclein gene in a large series of familial Parkinson's disease fails to reveal any further mutations. Hum Mol Genet 7: 751-753.

Wang C, Tan JMM, Ho MWL, Zaiden N, Wong SH, et al. 2004. The solubility and intracellular localization of parkin are altered by several familial Parkinson's disease point mutations. Progr No. 558.4 Abstract Viewer/Itinerary Planner. Washington, DC: Society for Neuroscience.

West AB, Zimprich A, Lockhart PJ, Farrer M, Singleton A, et al. 2001. Refinement of the PARK3 locus on chromosome 2p13 and the analysis of 14 candidate genes. Eur J Hum Genet 9: 659-666.

Wintermeyer P, Krüger R, Kuhn W, Müller T, Woitalla D, et al. 2000. Mutation analysis and association studies of the UCHL1 gene in German Parkinson's disease patients. Neuroreport 11: 2079-2082.

Wszolek ZK, Pfeiffer RF, Tsuboi Y, Uitti RJ, McComb RD, et al. 2004. Autosomal dominant parkinsonism associated with variable synuclein and tau pathology. Neurology 62: 1619-1622.

Xu P-Y, Liang R, Jankovic J, Hunter C, Zeng Y-X, et al. 2002. Association of homozygous 7048G7049 variant in the intron six of Nurr1 gene with Parkinson's disease. Neurology 58: 881-884.

Zarranz JJ, Alegre J, Gomez-Estaban JC, Lezcano E, Ros R, et al. 2004. The new mutation, E64K, of alpha-synuclein causes Parkinson and Lewy body dementia. Ann Neurol 55: 164-173.

Zetterstrom RH, Solomin L, Jansson L, Hoffer BJ, Olson L, et al. 1997. Dopamine neuron agenesis in Nurr1–deficient mice. Science 276: 248-250.

Zheng K, Heydari B, Simon DK. 2003. A common NURR1 polymorphism associated with Parkinson disease and diffuse Lewy body disease. Arch Neurol 60: 722-725.

Zimprich A, Müller-Myhsok B, Farrer M, Leitner P, Sharma M, et al. 2004a. The PARK8 locus in autosomal dominant parkinsonism: Confirmation of linkage and further delineation of the disease-containing interval. Am J Hum Genet 74: 11-19.

Zimprich A, Biskup S, Leitner P, Lichtner P, Farrer M, et al. 2004b. Mutations in a large multifunctional protein cause autosomal dominant parkinsonism with pleiomorphic α-synuclein and tau pathology (PARK8). Neuron 44: 601-607.

4 *LRRK2*-Associated Parkinsonism

M. Toft · M. J. Farrer · J. O. Aasly · D. W. Dickson · Z. K. Wszolek

Abstract: A major breakthrough in Parkinson's disease (PD) research has been the discovery of a number of genes causing familial parkinsonism. Recently, several pathogenic mutations in the *Leucine-rich repeat kinase 2* (*LRRK2*) gene have been identified in both familial parkinsonism and sporadic PD. The novel *LRRK2* gene encodes a multifunctional protein of the ROCO class, which might play a crucial role in the pathogenesis of PD. This seminal finding opens new avenues for investigating the mechanisms of brain dysfunction and degeneration. The discovery also raises new questions, as both clinical presentation and neuropathologic findings in *LRRK2* vary. We review the latest insights into *LRRK2*-associated parkinsonism.

List of Abbreviations: AD, Alzheimer's disease; GTP, guanosine triphosphate; *LRRK2*, leucine-rich repeat kinase 2 (gene); *MAPT*, microtubule-associated protein tau (gene); PD, Parkinson's disease; PSP, progressive supranuclear palsy

1   Introduction

Parkinsonism is a clinical syndrome characterized by bradykinesia, resting tremor, muscle rigidity, and postural instability. The most common cause of parkinsonism is Parkinson's disease (PD), a neurodegenerative disorder affecting more than 1% of the population older than 55 years (de Rijk et al., 1995). Parkinsonism is also frequent in a large number of other neurodegenerative disorders (❷ *Table 4-1*), and symptomatic parkinsonism occurs in vascular, toxic, metabolic, infectious, and postinfectious disorders. The clinical diagnostic accuracy of PD can be improved with the use of published and validated criteria

◼ Table 4-1
Neurodegenerative disorders manifesting parkinsonism

α-Synucleinopathies
  Lewy body disease
    Parkinson's disease (brainstem form of Lewy body disease)
      Sporadic
      Familial
    Dementia with Lewy bodies
      Limbic and neocortical form
      Lewy body variant of Alzheimer's disease
    Pure autonomic failure
  Glial inclusion body disorders
    Multiple system atrophy
  Other α-synucleinopathies
    Pantothenate kinase associated neurodegeneration
    Pallidonigroluysian atrophy
Tauopathies
  Progressive supranuclear palsy
  Corticobasal degeneration
  Frontotemporal dementia with parkinsonism linked to chromosome 17
  Alzheimer's disease
  Postencephalitic parkinsonism
  Guam amyotrophic lateral sclerosis/parkinsonism dementia complex
Other neurodegenerations
  Spinocerebellar ataxia 2
  Spinocerebellar ataxia 3
  Dentatopallidoluysian dystrophy
  X-linked dystonia-parkinsonism (Lubag)

(Gelb et al., 1999). However, because of the overlapping clinical features of parkinsonian disorders, histopathologic confirmation is still required for the definite diagnosis of PD and other parkinsonian disorders (Hughes et al., 1992; Braak et al., 2003).

The neuropathologic hallmarks of PD are loss of dopaminergic neurons in the substantia nigra and the presence of intraneuronal inclusions called Lewy bodies within surviving neurons. Lewy bodies are eosinophilic cytoplasmic fibrillar aggregates, containing α-synuclein and various other proteins, and are found in affected brain regions. α-Synuclein aggregation is a pathologic feature common to sporadic and inherited forms of PD, as well as to other neurodegenerative disorders (❷ Table 4-1). These disorders have collectively been called α-synucleinopathies (Farrer et al., 1999). Other forms of parkinsonism are characterized neuropathologically by prominent intracellular accumulations of abnormal filaments of the microtubule-associated protein tau (MAPT). Mutations in this gene (MAPT) are found in families with frontotemporal dementia with parkinsonism linked to chromosome 17 (Hutton et al., 1998). Common variants in the MAPT gene are associated with progressive supranuclear palsy (PSP) (Baker et al., 1999) and possibly also with corticobasal degeneration (Houlden et al., 2001). Because the tau protein has a central role in these diseases, they are collectively known as neurodegenerative tauopathies (Lee et al., 2001). Specific MAPT polymorphisms have most recently been implicated in PD (Skipper et al., 2004).

A major breakthrough in recent years has been the mapping of a number of loci linked to familial parkinsonism and the cloning of several genes causing monogenic forms of the syndrome (❷ Table 4-2).

◻ Table 4-2

Familial parkinsonism (autosomal dominant) with reported mutations and loci[a]

Locus	Chromosome	Gene	Clinical phenotype	Reference
Autosomal dominant				
PARK1/PARK4	4q21	*SNCA*	Early-onset PD and DLB	Polymeropoulos et al. (1997); Singleton et al. (2003)
PARK3	2p13	*Unknown*	PD	Gasser et al. (1998)
PARK5	4p14	*UCHL1*	PD	Leroy et al. (1998)
PARK8	12q12	*LRRK2*	Predominantly PD, dementia, PSP-like syndrome, and ALS described	Paisan-Ruiz et al. (2004); Zimprich et al. (2004a)
Autosomal recessive				
PARK2	6q25-27	*Parkin*	Early-onset PD	Kitada et al. (1998)
PARK6	1p35-36	*PINK1*	Early-onset PD	Valente et al. (2004)
PARK7	1p36	*DJ-1*	Early-onset PD	Bonifati et al. (2003)
Unknown				
PARK10	1p32	*Unknown*	PD	Hicks et al. (2002)
PARK11	2q36-37	*Unknown*	PD	Pankratz et al. (2003)

Note: ALS, amyotrophic lateral sclerosis; DLB, dementia with Lewy bodies; PD, Parkinson's disease; PSP, peripheral supranuclear palsy

[a]PARK9 has been assigned to a locus on 1p36 linked to Kufor-Rakeb syndrome, an autosomal recessive parkinsonian syndrome described in a consanguineous Jordanian family

Missense mutations in and genomic multiplication of the α-synuclein (SNCA) gene have been identified in a small number of families with autosomal dominant parkinsonism (PARK1/PARK4) (Polymeropoulos et al., 1997; Kruger et al., 1998; Singleton et al., 2003; Chartier-Harlin et al., 2004; Farrer et al., 2004; Zarranz et al., 2004). Subsequently, antibodies against the α-synuclein protein were found to robustly stain Lewy bodies in the substantia nigra in familial parkinsonism and sporadic PD (Spillantini et al., 1997), demonstrating that the α-synuclein protein is a major component of the Lewy body. Furthermore, common genetic variability in the α-synuclein gene promoter has been implicated in sporadic PD (Pals et al., 2004; Mueller et al., 2005).

Autosomal recessive mutations in three genes—*parkin/PARK2, Dj-1/PARK6,* and *PINK1/PARK7*—are linked to early-onset parkinsonism (age less than 45 years at onset) (Kitada et al., 1998; Bonifati et al., 2003; Valente et al., 2004). At least five other genetic disorders have phenotypic overlap with PD (❷ *Table 4-3*). These diseases should be thought of separately from the *PARK* loci, as they rarely present clinically with parkinsonism only.

◻ **Table 4-3**
Genetic diseases with parkinsonism as part of the clinical spectrum

Disease	Chromosome	Gene	Clinical phenotype	Reference
SCA2	12q23-24	*ATXN2*	Ataxia, parkinsonism	Gwinn-Hardy et al. (2000)
SCA3	14q32	*ATXN3*	Ataxia, parkinsonism	Gwinn-Hardy et al. (2001)
FTDP-17	17q21-22	*MAPT, PGRN*	FTD, PD, PSP, CBD, ALS	Hutton et al. (1998), Baker et al. (2006)
XDP (DYT3)	Xp13.1	*Unknown*	Dystonia-parkinsonism	Nolte et al. (2003)
RDP (DYT12)	19q13	*ATP1A3*	Dystonia-parkinsonism	de Carvalho Aguiar et al. (2004)

Note: ALS, amyotrophic lateral sclerosis; ATP1A3, Na+/K+ ATPase alpha 3 polypeptide; CBD, corticobasal degeneration; FTD, frontotemporal dementia; FTDP-17, frontotemporal dementia with parkinsonism linked to chromosome 17; PD, Parkinson's disease; PSP, progressive supranuclear palsy; RDP, Rapid-onset dystonia-parkinsonism; SCA, spinocerebellar ataxia; XDP, X-linked dystonia parkinsonism

Despite the discovery of genetic defects in familial parkinsonism, the role of genetics in sporadic late-onset PD has remained controversial. This controversy has been caused by the clinical and the neuropathologic differences between sporadic PD and the hereditary forms of parkinsonism. However, this view is changing after recent identification of several pathogenic mutations in the leucine-rich repeat kinase 2 (*LRRK2*) gene associated with both familial parkinsonism and sporadic PD (Paisan-Ruiz et al., 2004; Zimprich et al., 2004a; Gilks et al., 2005; Kachergus et al., 2005). This seminal finding opens new avenues for investigating the mechanisms of brain dysfunction and degeneration. The discovery also raises new questions, as both clinical phenotype and neuropathologic findings in *LRRK2* vary, both within and between families. Here, we review the latest insights into *LRRK2*-associated parkinsonism.

2 Familial Parkinsonism Linked to the *PARK8* Locus

In 2002, Funayama and colleagues performed a genome-wide linkage analysis of a Japanese family with autosomal dominant parkinsonism (Funayama et al., 2002). In this family, also known as the Sagamihara kindred, members presented with clinical features that may not be distinguished from sporadic late-onset PD (Hasegawa and Kowa, 1997). The clinical symptoms responded well to levodopa, and mean age at symptom onset was 51 years. Neuropathologic examinations in four members of the kindred showed pure nigral degeneration without any identified Lewy bodies.

A parametric 2-point linkage analysis in this family generated a highly significant logarithm of odds (LOD) score of 4.32 at the marker D12S345. Haplotype analysis of markers on chromosome 12 shared by the affected family members defined the disease-associated haplotype to a relatively large 13.6-cm region located at 12p11-q13 (Funayama et al., 2002). This haplotype was shared not only by all affected family members, but also by some unaffected individuals, indicating that disease penetrance in the family was incomplete. The chromosome 12 locus differed from previously reported regions linked to familial parkinsonism and was assigned the symbol *PARK8*.

After identification of the *PARK8* locus, linkage to this region was confirmed in a study of autosomal dominant parkinsonism in 21 families originating from Europe and North America (Zimprich et al., 2004b). Based on analysis of the two kindreds with the highest LOD scores in this study (Family A and Family D), the most likely disease gene location was delineated to a 3.2-cm region on chromosome 12q12. A second study of four Basque families also found evidence for linkage of autosomal dominant PD to the *PARK8* locus, with a maximum 2-point LOD score of 3.21 (Paisan-Ruiz et al., 2005b). When the results of these studies were combined, they provided evidence that the *PARK8* locus is responsible for a subset of families with autosomal dominant parkinsonism and suggested that the locus may be relatively common and occur in patients from different populations.

The existence of a gene within the *PARK8* locus associated with familial parkinsonism was finally established when both the groups identified mutations in the novel *LRRK2* gene (Paisan-Ruiz et al., 2004; Zimprich et al., 2004a).

3 Structure, Function, and Molecular Genetics of *LRRK2*

The *LRRK2* gene is located close to the centromere on the long arm of chromosome 12, and the gene was not studied until the identification of pathogenic mutations in parkinsonian kindreds. To establish the complete cDNA sequence, the *LRRK2* gene was amplified from human brain cDNA using overlapping primers predicted by homology searches. The gene spans a genomic region of 144 kb, with a total of 51 exons encoding a 2,527 amino acid protein (❷ *Table 4-1*) (Zimprich et al., 2004a). Sequences of partial cDNA clones deposited in public libraries indicate alternative splicing of several exons of the gene, but this has not been studied in detail.

Using Northern blots and real-time reverse transcriptase–polymerase chain reaction methods, expression analyses have shown that the *LRRK2* gene is expressed at low levels throughout the adult human brain, with a slightly higher expression in the putamen and substantia nigra than in other brain regions. Of other tissues examined, the gene expression is highest in lungs (Paisan-Ruiz et al., 2004; Zimprich et al., 2004a).

Studies on lrrk2 protein function have not yet been published. However, in silico predictions and homology searches of similar proteins in other species indicate that lrrk2 is a member of the recently defined ROCO family (Bosgraaf and Van Haastert, 2003). In humans, mice, and rats, members of the ROCO protein family have five conserved functional domains (❷ *Figure 4-1*) (Bosgraaf and Van Haastert, 2003). These multidomain proteins have been found in species ranging from mammals to metazoans, and exhibit various functions including tumor suppression.

The lrrk2 protein has a large amino terminus ending with leucine-rich repeats, consisting of 12 strands of a 22–28 amino acid motif presented in a tandem array. The Roc (for Ras of complex proteins) domain contains a GTPase-like domain with homology to all four members of the GTPase superfamily. GTPases are small proteins that regulate a wide array of cellular processes, such as signaling, differentiation, and growth through binding and hydrolysis of guanosine triphosphate (GTP) (Bosgraaf and Van Haastert, 2003). Our alignment and molecular modeling studies have suggested that the protein region containing the R1441 codon within this domain, similar to a RabSF motif, may play a role in protein interactions and cellular localization (unpublished data). RabSF regions are located on the surface of Rab GTPases and are generally thought to be involved in molecular interactions and protein binding (Stenmark and Olkkonen, 2001; Ali et al., 2004). In addition, this motif has also been implicated in membrane targeting of some Rab GTPases to specific organelles (Ali et al., 2004).

All ROCO proteins contain a novel COR (C-terminal of Roc) domain, which is about 300–400 amino acids long. The function of this domain is currently unknown.

A kinase domain with a catalytic core common to serine, threonine, and tyrosine protein kinases is always present in this protein family. The kinase domain belongs to the MAPKKK subfamily of kinases. The active site is located in a cleft between an N-terminal and a C-terminal lobe and is covered by an activation

◘ Figure 4-1

Chromosome 12, the structure of the *LRRK2* gene, and the lrrk2 protein. (a) The *PARK8* locus is located on human chromosome 12q12. (b) The *LRRK2* gene has 51 exons; the exonic localization of mutations with proven pathogenicity is noted across the line. The *LRRK2* sequence and amino acid notation are based on National Center for Biotechnology Information sequence accession AY792511. (c) Pathogenic mutations are located within the predicted functional domains of the lrrk2 protein, which consists of 2,527 amino acids. COR, C-terminal of Roc; LRR, leucine-rich repeat; MAPKKK, mitogen-activated protein kinase kinase kinase; ROC, Ras in complex proteins; WD40, WD40 repeats

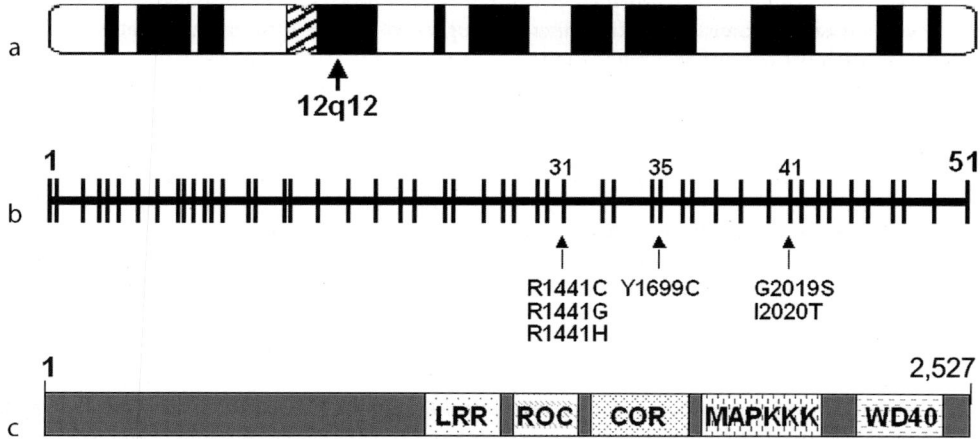

segment in its inactive form. The activation segment is a region of the kinase domain that undergoes crucial structural changes necessary to allow access to peptide substrates and also to orient key catalytic amino acids within the cleft of the kinase (Huse and Kuriyan, 2002). In other kinases, oncogenic mutations in residues within the activation segment of the kinase domain have an activating effect (Davies et al., 2002). We have therefore postulated that mutations in this region might have an activating effect on the kinase activity of *LRRK2* (Kachergus et al., 2005; Toft et al., 2005a). A mutation causing "gain of function" of the resulting protein would also be compatible with the dominant mode of disease transmission observed in the families. The pathogenic mutations in this region introduce serine and threonine residues, which may be potential targets for phosphorylation, altering substrate specificity.

There is a WD40 repeat domain at the carboxylate terminus. WD40 domains have previously been implicated in signal transduction, pre-mRNA processing, and cytoskeleton assembly (Smith et al., 1999).

4   *LRRK2* Mutations and Parkinsonism

Several mutations have been identified in the *LRRK2* gene, both in families with autosomal dominant parkinsonism and in individuals with seemingly sporadic disease. Segregation analyses within families provide statistical evidence for the pathogenicity of some of the published mutations (R1441C, R1441G, Y1699C, G2019S, and I2020T), whereas other mutations have been identified only in small families or single individuals and must be considered putative pathogenic variants.

Mutations have been identified in several of the functional domains of the lrrk2 protein (❷ *Figure 4-1*). Most of the patients described so far have presented a clinical phenotype with parkinsonism as the core feature, but other symptoms do also occur. Neuropathologic examinations have demonstrated nigral cell loss in all reported cases, but surprisingly, this has been associated with both α-synuclein and tau inclusions, and some patients have not presented any inclusion bodies (Funayama et al., 2002; Wszolek et al., 2004; Zimprich et al., 2004a; Gilks et al., 2005; Ross et al., 2005a). Each mutation and the clinical and pathologic features associated with it are discussed here in detail.

4.1 Pathogenic Mutations

Two putative pathogenic mutations have been identified in the leucine-rich repeat domain. In one family, an affected sib pair carried a mutation in exon 25 of the *LRRK2* gene causing an I1122V amino acid change (Zimprich et al., 2004a). In the same study, a second mutation in this domain (L1114L), possibly affecting splicing, was also found. Neither of the mutations was found in a large number of control subjects in the original study. In a second study, the L1114L mutation was, however, identified in both a patient and in an unaffected sibling of a patient not carrying this mutation, questioning the pathogenic relevance of this variant (Farrer et al., 2005).

In the Roc domain, three different mutations causing a substitution of the arginine at amino acid position 1441 in the lrrk2 protein have been described (R1441C, R1441G, and R1441H) (Paisan-Ruiz et al., 2004; Zimprich et al., 2004a; Zabetian et al., 2005; Mata et al., 2005a). Two other variants, an I1371V substitution (Paisan-Ruiz et al., 2005a) and an intronic mutation possibly affecting splicing (IVS31+3A>G) (Zabetian et al., 2005), have also been found in this part of the gene. However, both mutations have been identified in only one patient, and in the absence of segregation data, the pathogenicity of these mutations remains unclear.

The COR domain harbors the majority of coding variability, including amino acid substitutions, some of which might be expected to have an effect on protein structure (R1514Q, P1542S, V1598E, and R1725S) (Mata et al., 2005a). The relationship between these variants and disease is, however, not yet clear. The COR domain is also the location of the pathogenic Y1699C substitution segregating with disease in the large German-Canadian kindred (Family A) (Zimprich et al., 2004a) and in a large British kindred (Family PL) (Paisan-Ruiz et al., 2004).

The most common mutation in the *LRRK2* gene, G2019S, is located in the MAPKKK domain (DiFonzo et al., 2005; Hernandez et al., 2005; Kachergus et al., 2005). An I2020T mutation in this same region segregates with disease in a German family (Zimprich et al., 2004a) and in the first family linked to the *PARK8* locus, the Sagamihara kindred (Funayama et al., 2005). Both mutations cause substitutions of highly conserved amino acids within the regulatory activation segment of the kinase domain.

4.2 Mutation Frequencies

The G2019S substitution appears to be a more common cause of PD than other lrrk2 substitutions, and carriers originate from a large number of European and North American populations (DiFonzo et al., 2005; Gilks et al., 2005; Kachergus et al., 2005; Mata et al., 2005b; Nichols et al., 2005). In our study, although a family history was not always apparent, all mutation carriers shared a small ancestral haplotype of 145–154 kb. This indicates that the mutation arose in an ancient common founder (Kachergus et al., 2005). It also indicates that this specific mutation might be frequent and spread throughout different populations. The *LRRK2* G2019S substitution has recently also been found in patients from North Africa (Lesage et al., 2005), but was absent in a series of 1,000 ethnic Chinese PD patients from Asia (Tan et al., 2005).

In our series of 248 affected probands from families with autosomal dominant parkinsonism, 7 (2.8%) were identified as carriers of the *LRRK2* G2019S substitution (Kachergus et al., 2005). Two other studies found this mutation in 5% and 6.6% of families with autosomal dominant disease (DiFonzo et al., 2005; Nichols et al., 2005). Deng and colleagues found a total mutation frequency of 1.2% in a series of 326 PD patients, and 3 (2%) of 150 patients with familial PD carried the mutant allele. Each study used different criteria for autosomal dominant or familial parkinsonism, which probably to some extent explains the different mutation frequencies identified.

The G2019S mutation is also found in patients with sporadic PD, because the penetrance of *LRRK2* mutations is incomplete and depends on age. Several series of patients have been screened for the presence of this variant. Patients examined originated from different white populations and have been diagnosed as having idiopathic or sporadic PD; published mutation frequencies have varied between 0.5% and 1% (Deng et al., 2005; Farrer et al., 2005; Gilks et al., 2005; Kachergus et al., 2005; Zabetian et al., 2005). Hence, this mutation is the most common known genetic cause of so-called sporadic PD.

To date, few studies have reported the frequency of *LRRK2* mutations other than G2019S. The R1441G mutation was common (8%) in a series of patients from the Basque population (Paisan-Ruiz et al., 2004), and this mutation has also been found in patients from other regions of Spain (Mata et al., 2005b). R1441G and other mutations seem far less frequent in other populations. The R1441C mutation has also been identified in families from different populations, indicating that this mutation might be relatively frequent (Zimprich et al., 2004a).

In a community-based study of 786 patients from the United States, only 1 patient carried a mutation other than G2019S (Farrer et al., 2005). Zabetian and colleagues sequenced *LRRK2* exons 31, 35, and 41 in 371 consecutively recruited patients with PD and found mutations other than G2019S in 2 individuals (0.5%) (Zabetian et al., 2005). In a study of 435 Norwegian PD patients, only the G2019S mutation was identified (Aasly et al., 2005). No mutations were found in 188 Japanese patients examined for the presence of the 6 known mutations, including the I2020T substitution found in the Japanese family originally identified as having the *PARK8* gene (Funayama et al., 2005). Recently, heterozygosity for the *LRRK2* Gly2385Arg variant was found to be significantly more frequent among PD patients than controls in a population of Chinese ethnicity (Di Fonzo et al., 2006). This suggests that Gly2385Arg is a functionally relevant variant acting as a common risk factor for sporadic PD.

4.3 Clinical Features

Clinically, most individuals with *LRRK2* mutations have presented symptoms compatible with typical late-onset PD (❷ *Table 4-4*). In the Sagamihara kindred (with the I2020T mutation), affected members had asymmetric parkinsonism with a favorable response to dopaminergic treatment, and none of the individuals presented with any atypical symptoms (Hasegawa and Kowa, 1997; Funayama et al., 2002). Symptoms typical of sporadic PD responding to levodopa treatment were also found in the German family with the I2020T mutation (Zimprich et al., 2004a).

Slowly progressive parkinsonism with tremor as the presenting and initially predominant symptom was reported in R1441G-associated disease (Paisan-Ruiz et al., 2005b). These patients had a good response to dopaminergic treatment and developed motor complications typical of PD after 6–8 years of treatment. No cognitive decline was observed even after long disease duration, but delusional and paranoid hallucinations occurred in one affected individual. Writer's cramp and foot dystonia were present in a member of one family (Paisan-Ruiz et al., 2005b). In a second report of this mutation, the clinical features were consistent with typical late-onset PD (Mata et al., 2005b).

Resting tremor was not as prominent in Family D (Western Nebraska), a large family with an R1441C substitution affecting the same codon as the variant found in the Basque and Spanish families. The most common initial presentation was bradykinesia (60%) and unilateral resting hand tremor (40%). Response to levodopa therapy has been excellent, and motor complications have developed in half the patients receiving treatment. The phenotype of this family is indistinguishable from typical late-onset PD, except in one family member who also developed supranuclear gaze palsy, but remained responsive to levodopa until death (Wszolek et al., 1995, 2004).

We have observed other atypical symptoms in a second family, Family A (German-Canadian), carrying the Y1699C substitution. This family is characterized by a parkinsonian syndrome responsive to dopaminergic treatment, with subsequent development of motor complications. However, one family member had clinical amyotrophy characterized by muscle weakness, atrophy, and fasciculations. Two other mutation carriers in this family presented with dementia (Wszolek et al., 1997; Zimprich et al., 2004a).

More than 80 patients with the G2019S mutation have been reported so far, and few of these patients had any atypical symptoms. Dementia, autonomic dysfunction, and other neurologic signs do not seem to be more frequent than in other patients with PD (Aasly et al., 2005; DiFonzo et al., 2005; Gosal et al., 2005). The majority of patients have asymmetric onset of symptoms and good response to dopaminergic treatment, and develop levodopa-induced motor complications (Aasly et al., 2005; Nichols et al., 2005). One study found less severe clinical symptoms in G2019S carriers, despite an increased disease duration,

◘ Table 4-4

Comparison of different mutations within the *LRRK2* gene[a]

Domain and mutation (exon)	LRR		Roc		COR	MAPKKK	
	L1114L	I1122V	R1441C	R1441G	Y1699C	G2019S	I2020T
Family, no.	1	1	2	4	1	7	1
Affected individuals, no.	3	3	29	35	15	10	7
Mean (range) age at symptomatic onset, y	56 (52–60)	51 (49–53)	63 (48–78)	61 (50–79)	53 (36–65)	57 (43–70)	54 (48–59)
Mean (range) disease duration, y	23 (21–26)	18 (17–19)	14 (4–26)	NA	13 (5–18)	13.9 (5–25)	20 (12–27)
Predominant initial sign	RT	B	RT/B	RT	RT	RT/B	B
Parkinsonism							
Resting tremor	+	+	+	+	+	+	+
Bradykinesia	+	+	+	+	+	+	+
Rigidity	+	+	+	+	+	+	+
Postural or gait instability	+	+	+	+	+	+	+
Asymmetry of parkinsonism	NA	NA	+	+	+	+	NA
Response to levodopa	+	+	+	+	+	+	+
Levodopa-induced dyskinesia	–	–	+	+	+	+	–
Other clinical features	–	–	–[b]	Dystonia, delirium	Dementia, amyotrophy, dystonia	Dystonia, dementia, depression, RLS	–
Clinical phenotype	PD	PD	PD	PD	PD-plus syndrome	PD	PD

Note: B, Bradykinesia; COR, domain C-terminal of Roc; LRR, leucine-rich repeat; MAPKKK, mitogen-activated protein kinase kinase kinase; NA, not available; PD, Parkinson's disease; Roc, Ras in complex protein; RT, resting tremor; RLS, restless legs syndrome

[a]Data from Zimprich et al., 2004a; Aasly et al., 2005; Paisan-Ruiz et al., 2005b

[b]One case with supranuclear palsy responsive to levodopa therapy

compared with other patients with familial parkinsonism (Nichols et al., 2005). This indicates a slowed disease progression in these patients. Of interest, levodopa-responsive foot dystonia has been the initial symptom in several patients (Aasly et al., 2005; Gilks et al., 2005). Further studies are needed to assess the frequency of dystonia, but dystonia might be more frequent in *LRRK2*-associated parkinsonism. We have observed restless legs syndrome (RLS) in some patients with the G2019S mutation (Zbigniew K. Wszolek, personal observation, 2005).

As with PD in general, age is a risk factor for *LRRK2*-associated parkinsonism. The age of onset is variable, ranging from the fourth to the ninth decade, with the average age of onset between 55 and 65 years in the different families and studies (Zimprich et al., 2004a; Nichols et al., 2005; Paisan-Ruiz et al., 2005b).

Penetrance of *LRRK2* mutations depends on age, and varies among mutations and populations. In the Sagamihara kindred, segregation analysis of the disease-associated haplotype indicated penetrance of about 65% (Funayama et al., 2002). The oldest age at onset in families with the R1441G substitution was 80 years (Paisan-Ruiz et al., 2004).

We calculated the penetrance of the G2019S mutation and found that it increases in a close to linear fashion from 17% at age 50 years to 85% by age 70 years (Kachergus et al., 2005). Age at onset in this study was variable, both within and between different families, suggesting that other susceptibility factors, environmental or genetic, might influence the phenotype. Because the penetrance of *LRRK2* mutations depends on age, mutations are also found in patients with a negative family history for PD. This has important implications for genetic screening and counseling of PD patients.

4.4 Neuropathology

Neuropathologic findings have so far been reported for four *LRRK2* mutations (❍ *Figure 4-2*). In contrast to the relatively homogenous clinical presentation of most patients with *LRRK2*-associated parkinsonism, these examinations have shown strikingly diverse and pleomorphic findings. Neuronal loss and gliosis

❑ **Figure 4-2**
(a) Lewy bodies in substantia nigra in G2019S (hematoxylin-eosin stain); (b) pleomorphic Lewy bodies in Family D (α-synuclein stain); (c) glial and neuronal tau lesions in Family D (tau immunostain); and (d) ubiquitin neuronal inclusions in Family A (ubiquitin stain)

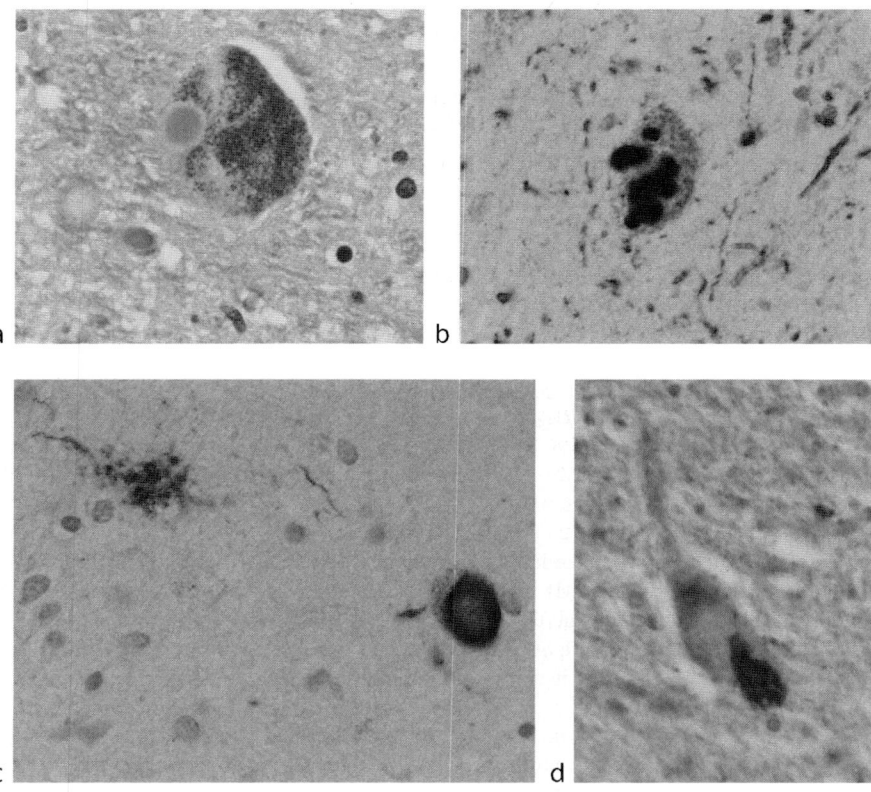

in the substantia nigra are found in all cases. However, various intracellular inclusions have been demonstrated.

In the Sagamihara kindred (I2020T), brain autopsy has been reported in four cases, which were all diagnosed as "pure nigral degeneration." The examinations showed mild to moderate nigral degeneration without any Lewy bodies or other coexisting pathology. No other pathologic changes were observed in other nuclei of the basal ganglia, in the cerebral cortex, or in the cerebellum (Hasegawa and Kowa, 1997; Funayama et al., 2002).

Four members of Family D (Western Nebraska) (R1441C) have come to autopsy, all presenting with neuronal loss and gliosis of the substantia nigra. However, variable α-synuclein and tau pathology were demonstrated in the affected individuals, all carrying the same *LRRK2* mutation. Lewy body pathology was found in two cases; in one case, Lewy bodies were restricted to brainstem nuclei, whereas the pathology was more widespread in the second patient. One family member had a tauopathy with neurofibrillary tangles and neuropil threads, qualitatively similar to the pathologic findings characteristic of PSP. The fourth family member showed ubiquitin-immunoreactive neuronal inclusions, without Lewy bodies or tau-related pathology (Wszolek et al., 2004; Zimprich et al., 2004a).

A number of cytoplasmic and nuclear inclusions were found in two members of Family A (German-Canadian) (Y1699C). The nuclear inclusions were similar to Marinesco bodies, although the neuronal cytoplasmic inclusions appear to be novel and are unclassified. Other patients from Family A had pathologic evidence of minimal anterior horn cell degeneration, indicating mild motor neuron disease, and one individual met neuropathologic criteria for Alzheimer's disease (AD) (Wszolek et al., 1997; Zimprich et al., 2004a). Coexisting Alzheimer-type pathology has been identified in several other *LRRK2* mutation carriers.

Two reports of autopsy findings for the more common LRRK2 G2019S mutation have been published. Histopathologic examination of three patients showed nigral cell loss with typical Lewy bodies. In two of the three patients, Lewy bodies were present in the limbic cortices, whereas the last case showed signs of pathologic aging, with diffuse senile plaques and occasional neurofibrillary tangles (Gilks et al., 2005). Reviewing two brain banks, Mayo Clinic in Jacksonville, Florida, and the National Parkinson Foundation/ University of Miami Brain Endowment Bank in Miami, Florida, we have recently identified eight further individuals with the G2019S mutation. Neuropathologic examination revealed brainstem-type Lewy body disease in four, transitional Lewy body disease in three, and diffuse Lewy body disease in one. In all respects, the Lewy body disease was typical in these cases, with prominent involvement in the brainstem monoaminergic nuclei, basal forebrain, and limbic cortex (Ross et al., 2005a).

5   *LRRK2* and Other Neurodegenerative Disorders

The majority of affected individuals with the *LRRK2* mutation present clinically with typical late-onset PD. However, some individuals carrying *LRRK2* mutations have exhibited cognitive dysfunction. Two patients in Family A (German-Canadian) (Y1699C) presented with dementia. Although none of these patients have come to autopsy, the third member of this family demonstrated concurrent AD pathology (Wszolek et al., 1997). One member of Family D (Western Nebraska) exhibited tau pathology and neurofibrillary tangles, the hallmark of PSP (Wszolek et al., 2004). Neuropathologic findings in *LRRK2*-affected families have therefore been pleomorphic and demonstrated various intracellular inclusions (Zimprich et al., 2004a). This indicates that the lrrk2 protein might be implicated in the aggregation of misfolded proteins in several neurodegenerative disorders, including AD and PSP.

5.1 Alzheimer's Disease

AD is the most common cause of dementia in elderly people and is considered a multifactorial disorder. A locus on chromosome 12 conferring susceptibility to late-onset AD was initially reported by Pericak-Vance et al. (1997). Additional evidence for a locus on this chromosome has been reported in subsequent linkage studies (Myers and Goate, 2001). In a fine-mapping study of this locus, evidence for linkage was greatest in families in which at least one affected individual had a neuropathologic diagnosis of a different form of dementia, namely, dementia with Lewy bodies (Scott et al., 2000). The linked region is large, containing a considerable number of genes. The *LRRK2* gene is located within this locus and is positioned underneath the peak in the study. *LRRK2* is therefore a plausible candidate gene for both late-onset AD and dementia with Lewy bodies because of the physical position and the coexisting Alzheimer-type pathology identified in several individuals with *LRRK2* mutations.

We studied the presence of seven known pathogenic mutations previously reported in the *LRRK2* gene in a series of 242 patients diagnosed clinically with dementia, the majority of whom received a diagnosis of AD. No mutation carriers were identified (Toft et al., 2005b). In a screening of 654 cases with pathologically confirmed AD, only one mutation carrier was found to have the G2019S substitution, and this finding may have been coincidental (Ross et al., 2005a). Nevertheless, this indicates that the examined *LRRK2* mutations are not common in AD.

5.2 Progressive Supranuclear Palsy

The occurrence of *LRRK2* mutations has also been examined in a series of 244 cases diagnosed as having PSP on the basis of pathologic findings. None of the individuals carried the studied amino acid substitutions, which included G2019S, I2020T, and the pleomorphic pathology-associated R1441C/G/H (Ross et al., 2005b).

These studies suggest that *LRRK2* mutations might not be frequent in neurodegenerative disorders other than familial parkinsonism and sporadic PD. However, several coding variants within the *LRRK2* gene are known, and these, or polymorphisms affecting gene expression and alternative splicing of protein isoforms, might associate with disease. The possibility of variants within this gene conferring susceptibility to AD, PSP, and other neurodegenerative disorders have not yet been studied.

6   Conclusions

Mutations in the *LRRK2* gene have been identified in numerous cases of both familial parkinsonism and sporadic PD. Further studies are needed to more accurately determine the frequency of the different mutations, and the clinical and pathologic features associated with *LRRK2* mutations. Nevertheless, we have gained important insights into PD pathogenesis from the studies that have been published so far. First, a single *LRRK2* G2019S mutation is established as the most common known genetic cause of PD. Second, although the majority of *LRRK2*-associated cases are clinically indistinguishable from typical sporadic PD and the majority of postmortem cases have Lewy body disease, *LRRK2* mutations may cause variable clinical and pathologic phenotypes. Thus, a single pathogenesis does not always result in a single pathologic end point. Finally, the variable clinical and pathologic phenotype and the wide range of age at disease onset indicate the presence of other modifiers, genetic or environmental.

The function of the lrrk2 protein is currently unknown, but the presence of a kinase domain indicates that the protein is involved in a signaling or second-messenger phosphorylation cascade. Lrrk2 kinase activity might directly or indirectly interact with proteins such as α-synuclein, tau, and ubiquitin and thereby promote the accumulation and deposition of these proteins into Lewy bodies and neurofibrillary tangles and thus drive neurodegeneration. This theory is supported by recent findings suggesting that α-synuclein phosphorylation may be a critical event in neurotoxicity (Chen and Feany, 2005). Alternatively, *LRRK2* mutations might act as primers for neurodegeneration, bringing neurons closer to dysfunction but without being sufficient to cause disease on their own (Singleton, 2005). In this model, the lrrk2 protein probably does not have a direct role in the metabolism of α-synuclein or tau.

Future studies will elucidate the complete role of the *LRRK2* gene in neurodegeneration. New insights are needed to pave the way not only to a better understanding of PD pathogenesis but also to the development of new therapies aimed at disease prevention.

Acknowledgments

Mayo Clinic in Jacksonville is a Morris K. Udall Parkinson's Disease Research Center of Excellence supported by National Institute of Neurological Disorders and Stroke grant P50 NS40256. M.T. has been supported by the Research Council of Norway and the Parkinson Disease Foundation. The authors thank patients and family members participating in the studies.

References

Aasly JO, Toft M, Fernandez-Mata I, Kachergus J, Hulihan M, et al. 2005. Clinical features of *LRRK2*-associated Parkinson's disease in Central Norway. Ann Neurol 57: 762-765.

Ali BR, Wasmeier C, Lamoreux L, Strom M, Seabra MC. 2004. Multiple regions contribute to membrane targeting of Rab GTPases. J Cell Sci 117: 6401-6412. Epub 2004 Nov 23.

Baker M, Litvan I, Houlden H, Adamson J, Dickson D, et al. 1999. Association of an extended haplotype in the tau gene with progressive supranuclear palsy. Hum Mol Genet 8: 711-715.

Bonifati V, Rizzu P, van Baren MJ, Schaap O, Breedveld GJ, et al. 2003. Mutations in the *DJ-1* gene associated with autosomal recessive early-onset parkinsonism. Science 299: 256-259. Epub 2002 Nov 21.

Bosgraaf L, Van Haastert PJ. 2003. Roc, a Ras/GTPase domain in complex proteins. Biochim Biophys Acta 1643: 5-10. Epub 2003 Sep 21.

Braak H, Del Tredici K, Rub U, de Vos RA, Jansen Steur EN, et al. 2003. Staging of brain pathology related to sporadic Parkinson's disease. Neurobiol Aging 24: 197-211. Epub 2002 Dec 10.

Chartier-Harlin MC, Kachergus J, Roumier C, Mouroux V, Douay X, et al. 2004. Alpha-Synuclein locus duplication as a cause of familial Parkinson's disease. Lancet 364: 1167-1169. Epub 2004 Sep 25.

Chen L, Feany MB. 2005. Alpha-synuclein phosphorylation controls neurotoxicity and inclusion formation in a Drosophila model of Parkinson disease. Nat Neurosci 8: 657-663. Epub 2005 Apr 17.

Davies H, Bignell GR, Cox C, Stephens P, Edkins S, et al. 2002. Mutations of the *BRAF* gene in human cancer. Nature 417: 949-954. Epub 2002 Jun 9.

de Carvalho Aguiar P, Sweadner KJ, Penniston JT, Zaremba J, Liu L, et al. 2004. Mutations in the Na+/K+ -ATPase *alpha3* gene ATP1A3 are associated with rapid-onset dystonia parkinsonism. Neuron 43: 169-175. Epub 2004 July 21.

de Rijk MC, Breteler MM, Graveland GA, Ott A, Grobbee DE, et al. 1995. Prevalence of Parkinson's disease in the elderly: The Rotterdam study. Neurology 45: 2143-2146.

Deng H, Le W, Guo Y, Hunter CB, Xie W, et al. 2005. Genetic and clinical identification of Parkinson's disease patients with *LRRK2* G2019S mutation [letter]. Ann Neurol 57: 933-934.

Di Fonzo A, Rohe CF, Ferreira J, Chien HF, Vacca L, et al. 2005. A frequent *LRRK2* gene mutation associated with autosomal dominant Parkinson's disease. Lancet 365: 412-415. Epub 2005 Jan 28.

Di Fonzo A, Wu-Chou YH, Lu CS, van Doeselaar M, Simons EJ, et al. 2006. A common missense variant in the *LRRK2* gene Gly2385Arg, associated with Parkinson's disease risk in Taiwan. Neurogenetics 7: 133-138.

Farrer M, Gwinn-Hardy K, Hutton M, Hardy J. 1999. The genetics of disorders with synuclein pathology and parkinsonism. Hum Mol Genet 8: 1901-1905.

Farrer M, Kachergus J, Forno L, Lincoln S, Wang DS, et al. 2004. Comparison of kindreds with parkinsonism and alpha-synuclein genomic multiplications. Ann Neurol 55: 174-179. Epub 2004 Jan 22.

Farrer M, Stone J, Mata IF, Lincoln S, Kachergus J, et al. 2005. *LRRK2* mutations in Parkinson disease. Neurology 65: 738-740.

Farrer M, Stone J, Mata IF, Lincoln S, Kachergus J, et al. 2005. *LRRK2* mutations in Parkinson disease. Neurology. Epub 2005 June 22.

Funayama M, Hasegawa K, Kowa H, Saito M, Tsuji S, et al. 2002. A new locus for Parkinson's disease (*PARK8*) maps to chromosome 12p11.2-q13.1. Ann Neurol 51: 296-301. Epub 2002 Feb 27.

Funayama M, Hasegawa K, Ohta E, Kawashima N, Komiyama M, et al. 2005. An *LRRK2* mutation as a cause for the parkinsonism in the original *PARK8* family. Ann Neurol 57: 918-921. Epub 2005 May 4.

Gasser T, Muller-Myhsok B, Wszolek ZK, Oehlmann R, Calne DB, et al. 1998. A susceptibility locus for Parkinson's disease maps to chromosome 2p13. Nat Genet 18: 262-265.

Gelb DJ, Oliver E, Gilman S. 1999. Diagnostic criteria for Parkinson disease. Arch Neurol 56: 33-39.

Gilks WP, Abou-Sleiman PM, Gandhi S, Jain S, Singleton A, et al. 2005. A common *LRRK2* mutation in idiopathic Parkinson's disease. Lancet 365: 415-416. Epub 2005 Jan 28.

Gosal D, Ross OA, Wiley J, Irvine GB, Johnston JA, et al. 2005. Clinical traits of *LRRK2*-associated Parkinson's disease in

Ireland: A link between familial and idiopathic PD. Parkinsonism Relat Disord 11: 349-352.

Gwinn-Hardy K, Chen JY, Liu HC, Liu TY, Boss M, et al. 2000. Spinocerebellar ataxia type 2 with parkinsonism in ethnic Chinese. Neurology 55: 800-805.

Gwinn-Hardy K, Singleton A, O'Suilleabhain P, Boss M, Nicholl D, et al. 2001. Spinocerebellar ataxia type 3 phenotypically resembling Parkinson disease in a black family. Arch Neurol 58: 296-299.

Hasegawa K, Kowa H. 1997. Autosomal dominant familial Parkinson disease: Older onset of age, and good response to levodopa therapy. Eur Neurol (38 Suppl) 1: 39-43.

Hernandez DG, Paisan-Ruiz C, McInerney-Leo A, Jain S, Meyer-Lindenberg A, et al. 2005. Clinical and positron emission tomography of Parkinson's disease caused by *LRRK2*. Ann Neurol 57: 453-456. Epub 2005 Feb 24.

Houlden H, Baker M, Morris HR, MacDonald N, Pickering-Brown S, et al. 2001. Corticobasal degeneration and progressive supranuclear palsy share a common tau haplotype. Neurology 56: 1702-1706.

Hughes AJ, Daniel SE, Kilford L, Lees AJ. 1992. Accuracy of clinical diagnosis of idiopathic Parkinson's disease: A clinico-pathological study of 100 cases. J Neurol Neurosurg Psychiatry 55: 181-184.

Huse M, Kuriyan J. 2002. The confirmational plasticity of protein kinases. Cell 109: 275-282. Epub 2002 May 17.

Hutton M, Lendon CL, Rizzu P, Baker M, Froelich S, et al. 1998. Association of missense and 5′-splice-site mutations in tau with the inherited dementia FTDP-17. Nature 393: 702-705.

Kachergus J, Mata IF, Hulihan M, Taylor JP, Lincoln S, et al. 2005. Identification of a novel *LRRK2* mutation linked to autosomal dominant parkinsonism: Evidence of a common founder across European populations. Am J Hum Genet 76: 672-680. Epub 2005 Feb 22.

Kitada T, Asakawa S, Hattori N, Matsumine H, Yamamura Y, et al. 1998. Mutations in the *Parkin* gene cause autosomal recessive juvenile parkinsonism. Nature 392: 605-608.

Kruger R, Kuhn W, Muller T, Woitalla D, Graeber M, et al. 1998. Ala30Pro mutation in the gene encoding alpha-synuclein in Parkinson's disease [letter]. Nat Genet 18: 106-108.

Lee VM, Goedert M, Trojanowski JQ. 2001. Neurodegenerative tauopathies. Annu Rev Neurosci 24: 1121-1159.

Leroy E, Boyer R, Auburger G, Leube B, Ulm G, et al. 1998. The ubiquitin pathway in Parkinson's disease [letter]. Nature 395: 451-452.

Lesage S, Ibanez P, Lohmann E, Agid Y, Durr A, et al. 2005. The G2019S *LRRK2* mutation in autosomal dominant European and North African Parkinson's disease is frequent and its penetrance is age-dependant: LBS.003. Neurology 64: 1826.

Mata IF, Kachergus JM, Taylor JP, Lincoln S, Aasly J, et al. 2005a. *LRRK2* pathogenic substitutions in Parkinson's disease. Neurogenetics 6: 171-177.

Mata IF, Taylor JP, Kachergus J, Hulihan M, Huerta C, et al. 2005b. *LRRK2* R1441G in Spanish patients with Parkinson's disease. Neurosci Lett 382: 309-311. Epub 2005 Apr 13.

Mueller JC, Fuchs J, Hofer A, Zimprich A, Lichtner P, et al. 2005. Multiple regions of alpha-synuclein are associated with Parkinson's disease. Ann Neurol 57: 535-541.

Myers AJ, Goate AM. 2001. The genetics of late-onset Alzheimer's disease. Curr Opin Neurol 14: 433-440.

Nichols WC, Pankratz N, Hernandez D, Paisan-Ruiz C, Jain S, et al. 2005. Genetic screening for a single common *LRRK2* mutation in familial Parkinson's disease. Lancet 365: 410-412. Epub 2005 Jan 18.

Nolte D, Niemann S, Muller U. 2003. Specific sequence changes in multiple transcript system DYT3 are associated with X-linked dystonia parkinsonism. Proc Natl Acad Sci USA 100: 10347-10352. Epub 2003 Aug 19.

Paisan-Ruiz C, Jain S, Evans EW, Gilks WP, Simon J, et al. 2004. Cloning of the gene containing mutations that cause *PARK8*-linked Parkinson's disease. Neuron 44: 595-600.

Paisan-Ruiz C, Lang AE, Kawarai T, Sato C, Salehi-Rad S, et al. 2005a. *LRRK2* gene in Parkinson disease: Mutation analysis and case control association study. Neurology 65: 696-700.

Paisan-Ruiz C, Saenz A, Lopez de Munain A, Martia I, Martinez Gil A, et al. 2005b. Familial Parkinson's disease: Clinical and genetic analysis of four Basque families. Ann Neurol 57: 365-372. Epub 2005 Feb 24.

Pals P, Lincoln S, Manning J, Heckman M, Skipper L, et al. 2004. Alpha-synuclein promoter confers susceptibility to Parkinson's disease. Ann Neurol 56: 591-595. Epub 2004 Sep 28.

Pericak-Vance MA, Bass MP, Yamaoka LH, Gaskell PC, Polymeropoulos MH, Lavedan C, Leroy E, Ide SE, Dehejia A, et al. 1997. Mutation in the alpha-synuclein gene identified in families with Parkinson's disease. Science 276: 2045-2047.

Ross OA, Toft M, Whittle AJ, Johnson JL, Papapetropoulos S, et al. 2005a. *LRRK2* and Lewy body disease. Ann Neurol 59: 388-393.

Ross OA, Whittle AJ, Cobb SA, Hulihan M, Lincoln SJ, et al. 2005b. *LRRK2* R1441 substitution and progressive supranuclear palsy. Neuropathol Appl Neurobiol. 32: 23-25.

Scott WK, Grubber JM, Conneally PM, Small GW, Hulette CM, et al. 2000. Fine mapping of the chromosome 12 late-onset Alzheimer disease locus: Potential genetic and phenotypic heterogeneity. Am J Hum Genet 66: 922-932. Epub 2000 Feb 24.

Singleton AB. 2005. Altered alpha-synuclein homeostasis causing Parkinson's disease: The potential roles of dardarin. Trends Neurosci 28: 416-421.

Singleton AB, Farrer M, Johnson J, Singleton A, Hague S, et al. 2003. Alpha-synuclein locus triplication causes Parkinson's disease. Science 302: 841.

Skipper L, Wilkes K, Toft M, Baker M, Lincoln S, et al. 2004. Linkage disequilibrium and association of *MAPT* H1 in Parkinson disease. Am J Hum Genet 75: 669-677. Epub 2004 Aug 3.

Smith TF, Gaitatzes C, Saxena K, Neer EJ. 1999. The WD repeat: A common architecture for diverse functions. Trends Biochem Sci 24: 181-185.

Spillantini MG, Schmidt ML, Lee VM, Trojanowski JQ, Jakes R, et al. 1997. Alpha-synuclein in lewy bodies [letter]. Nature 388: 839-840.

Stenmark H, Olkkonen VM. 2001. The Rab GTPase family. Genome Biol 2: 3007.

Tan EK, Shen H, Tan LC, Farrer M, Yew K, et al. 2005. The G2019S *LRRK2* mutation is uncommon in an Asian cohort of Parkinson's disease patients. Neurosci Lett 384: 327-329. Epub 2005 Jun 13.

Toft M, Mata IF, Kachergus JM, Ross OA, Farrer MJ. 2005a. *LRRK2* mutations and parkinsonism [letter]. Lancet 365: 1229-1230.

Toft M, Sando SB, Melquist S, Ross OA, White LR, et al. 2005b. *LRRK2* mutations are not common in Alzheimer's disease. Mech Ageing Dev 126: 1201-1205.

Valente EM, Abou-Sleiman PM, Caputo V, Muqit MM, Harvey K, et al. 2004. Hereditary early-onset Parkinson's disease caused by mutations in PINK1. Science 304: 1158-1160. Epub 2004 Apr 15.

Wszolek ZK, Pfeiffer B, Fulgham JR, Parisi JE, Thompson BM, et al. 1995. Western Nebraska family (Family D) with autosomal dominant parkinsonism. Neurology 45: 502-505.

Wszolek ZK, Pfeiffer RF, Tsuboi Y, Uitti RJ, McComb RD, et al. 2004. Autosomal dominant parkinsonism associated with variable synuclein and tau pathology. Neurology 62: 1619-1622.

Wszolek ZK, Vieregge P, Uitti RJ, Gasser T, Yasuhara O, et al. 1997. German-Canadian family (Family A) with parkinsonism, amyotrophy, and dementia: Longitudinal observations. Parkinsonism Relat Disord 3: 125-139.

Wszolek ZK. 2005. Personal Communication.

Zabetian CP, Samii A, Mosley AD, Roberts JW, Leis BC, et al. 2005. A clinic-based study of the *LRRK2* gene in Parkinson disease yields new mutations. Neurology 65: 741-744.

Zarranz JJ, Alegre J, Gomez-Esteban JC, Lezcano E, Ros R, et al. 2004. The new mutation, *E46K*, of alpha-synuclein causes Parkinson and lewy body dementia. Ann Neurol 55: 164-173. Epub 2003 Dec 30.

Zimprich A, Biskup S, Leitner P, Lichtner P, Farrer M, et al. 2004a. Mutations in *LRRK2* cause autosomal-dominant parkinsonism with pleomorphic pathology. Neuron 44: 601-607.

Zimprich A, Muller-Myhsok B, Farrer M, Leitner P, Sharma M, et al. 2004b. The *PARK8* locus in autosomal dominant parkinsonism: Confirmation of linkage and further delineation of the disease-containing interval. Am J Hum Genet 74: 11-19. Epub 2003 Dec 19.

5 Gene and Protein Expression Profiling in Parkinson's Disease: Quest for Neuroprotective Drugs

O. Weinreb · T. Amit · E. Grünblatt · P. Riederer · M. Youdim · S. Mandel

Abstract: In spite of the extensive studies performed on postmortem substantia nigra (SN) of Parkinson's disease (PD) brains, the etiology of the disease has not yet been established. The identification of single mutated genes linked to heritable forms of PD has further enlightened our understanding of its etio-pathophysiology in the nongenetic cases providing potential molecular pathways associated with those genes. There is a recognized consensus that in both the genetic and sporadic cases of PD there is a crucial implication of mitochondria and ubiquitin-proteasome system dysfunction that expresses itself with excess production of reactive oxygen species, protein misfolding, and aggregation into inclusion bodies. However, the precise identity of the pivotal genes involved in the neurotoxic cascade pathways leading to the death of the dopaminergic neurons in sporadic PD, which constitutes around 90% of the total cases of the disease, is still unknown. This chapter will review recent large-scale microarray gene expression profiling studies in human postmortem SN from sporadic PD brains and highlight gene candidates as putative molecular signatures for early diagnosis and future development of CNS "magic bullets" personalized drugs. We will discuss the application of transcriptomics and proteomics in the quest for neuroprotective drugs that might possess disease-modifying action.

List of Abbreviations: ADH5, alcohol dehydrogenase 5; BDNF, brain-derived neurotrophic factor; CDC42, cell division cycle 42; COX, cytochrome c oxidase; CSK, c-src tyrosine kinase; DAT, dopamine transporter; EGLN1, egl nine homolog 1; GADPH, glyceraldehyde 3-phosphate dehydrogenase; GAK, cyclin G associated kinase; GSH, glutathione; HIF, hypoxia-inducible factor; MAO, monoamine oxidase; MAP-2, microtubule associated protein; MARK-1, microtubule affinity regulating kinase; MPTP, N-methyl-4-phenyl-1,2,3,6-tetrahydropyridine; NIH, National Institutes of Health; NSF, N-ethyl malei-mide-sensitive factor; NINDS, National Institute of Neurological Disorders and Stroke; PHD2, proline hydroxylase-2; ROS, reactive oxygen species; SNpc, substantia nigra pars compacta; SYT1, synaptotagmin 1; UBE2J1, ubiquitin-conjugating enzyme E2; TPI, triosephosphate isomerase; UPS, ubiquitin-proteasome system; VEGF, vascular endothelial growth factor; VMAT2, vesicular monoamine member 2; VAV3, vav 3 oncogene

1 Introduction

Parkinson's disease (PD) is one of the most common neurological disorders, second only to Alzheimer's disease (AD), affecting approximately 2% of adults over the age of 60 years. It results mainly from the death of dopaminergic neurons in the substantia nigra pars compacta (SNpc) and consequential depletion of dopamine (DA). A number of mutations have been identified during the last decade causing rare familial forms of PD, some of them of early age-onset, but the majority of the cases are sporadic or "idiopathic." The cause of the idiopathic form of the disease, which constitutes more than 90% of total PD cases, is still unknown, but is considered to result from both environmental and genetic factors. The most widely accepted hypothesis suggests the involvement of a cascade/s of neurotoxic events acting in parallel and/or in a sequential chain leading eventually to the demise of the affected dopaminergic neurons, suggesting that the pathogenesis of PD is multifactorial. Among these events is the impairment in mitochondrial complex I activity (Schapira et al., 1990), failure of the ubiquitin-proteasome system (UPS) to adequately remove abnormal proteins (McNaught and Jenner, 2001) and a general environment of oxidative stress (OS) (Gotz et al., 1994; Jenner and Olanow, 1996; Youdim and Riederer, 1997). The presence of reactive oxygen species (ROS) would increase the amount of misfolded proteins and the demand for their disposition by the UPS, whose activity is however impaired in PD. This would result in the propagation of a vicious cycle which maybe self sustaining. Intimately related to this are studies showing a progressive accumulation of iron and ferritin in PD patients, specifically in the SNpc (Riederer et al., 1989; Gerlach et al., 1994). The brain is exquisitely susceptible to the devastating action of the Fenton and Haber-Weiss chemistry, catalyzed by free-chelatable iron (Fe II/III) leading to the formation of the aggressively reactive hydroxyl radical (Halliwell, 1992). Despite its short half-life, this radical can inflict prominent oxidation of cell membrane proteins and lipids. Furthermore, an increase in monoamine oxidase (MAO)-B activity within reactive microglia in PD brain tissues is considered to contribute to the formation of high levels of H_2O_2 as a by-product of amine turnover (Saura et al., 1994), which in turn cannot be efficiently removed because of low levels of reduced glutathione (GSH) in PD (Riederer et al., 1989; Di Monte et al., 1992). Low reduced glutathione levels and a shift in the FeII/FeIII ratio in favor of FeIII may significantly contribute to the particular susceptibility of the SNpc of PD patients and may constitute important factors underlying the etio-pathology of the disease.

The current concept regarding PD and other neurodegenerative disorders, considers them diseases of multiple etiological nature where several mechanisms are implicated in a cascade/s of events involving many biochemical and signaling pathways (Van der Schyf et al., 2006). An unresolved question however, is to determine which of these factors constitute the primary event, the sequence in which they act and where is the point of convergence or the final pathway by which the SNpc dopaminergic neurons die. It is at this point where high throughput transcriptomics and proteomics tools come to the help of neuroscientists to assist in dissecting signaling pathways and find networks that may shed light on the pathobiology of PD. This article will present primarily the most prominent findings from DNA microarray gene expression studies conducted in human postmortem SNpc from PD brains and discuss the implication of the molecular signatures found as future predictive biomarkers for early diagnosis and disease progression. A second illuminating topic will be the application of transcriptomics and proteomics in the quest for neuroprotective drugs that might possess disease-modifying action. These techniques are changing the view on the processes involved in progressive neurodegenerative disorders and their association with better and accurate target-directed drugs. Finally, recent data on molecular aspects of the aging brain will be reviewed since aging has long been recognized as a major risk factor for sporadic forms of neurodegenerative diseases.

2 Gene Expression Profiling in PD

Only very recently a number of groups have managed to produce gene expression profiling studies in human postmortem PD substantia nigra (SN), using high throughput gene-based platforms. Our group was the first to conduct a large-scale gene expression profiling study of postmortem SNpc of Parkinsonian brains and compared them to matched controls and unaffected brain regions (cerebellum and SN pars reticulate, SNr), employing Affymetrix HG-FOCUS oligonucleotide array (3,517 probe sets) (Grunblatt et al., 2004). Functional classification of Parkinsonian SN differential gene expression by the Gene Ontology Consortium (http://www.geneontology.org) revealed specific reductions in gene families such as signal transduction [i.e., phosphoinositide-3-kinase, catalytic, alpha polypeptide (PIK3CA), phospholipase C-like 1 (PLCL1), vascular endothelial growth factor (VEGF)] and protein handling [i.e., catalytic and regulatory subunits of the ubiquitin-proteasome system, S-phase kinase-associated protein 1A (p19A/SKP1A), and the heat shock 70 kDa protein 8, (HSPA8), coding for HSC70, a member of the HSP70 chaperone family]. SKP1 is an E3 ligase functioning with the Rbx family of RING proteins as a modular multiprotein *S*kp1, *C*ullin and a substrate-recognizing *F*-box protein (*SCF*) (Kamura et al., 1999). Thus, its progressive reduction during the course of PD and the decline in HSC-70, responsible for recognizing unfolded or aberrant proteins, may play a role in the accumulation of a wide spectrum of ubiquitinated protein aggregates in PD brains such as tyrosine hydroxylase (TH), synphilin-1, α-synuclein, and phosphorylated tau (Liani et al., 2004; Meredith et al., 2004; Zhang and Goodlett, 2004). Additional affected functional classes correspond to dopaminergic transmission/metabolism [i.e., cyclic AMP-regulated phosphoprotein (ARPP-21), solute carrier family 18 (vesicular monoamine member 2, VMAT2), alcohol dehydrogenase 5 (ADH5), and aldehyde dehydrogenase 1 family, member A1 (ALDH1A1)] and energy pathways/glycolysis (i.e., cytochrome c oxidase subunit VIa, phosphofructokinase, platelet 3-oxoacid CoA transferase).

The upregulated genes in PD clustered mainly in biological processes involving: (1) cell adhesion/cytoskeleton (i.e., selectin P ligand, parvin α, laminin β2, myomesin 1, filamin B); (2) cell cycle [(i.e., c-src tyrosine kinase (CSK), cyclin G associated kinase (GAK)]; (3) protein modification/phosphorylation; (4) protein metabolism, and (5) inflammation/stress [(i.e., CD22, Hsp40 B5 (DNAJB5)]. An important finding was the abnormal upregulation of the iron-oxygen sensor, egl nine homolog 1(EGLN1) gene, encoding the proline hydroxylase-2 (PHD2) enzyme (Epstein et al., 2001), which may exacerbate the OS status and promote iron-induced aggregation of α-synuclein (Ostrerova-Golts et al., 2000; Turnbull et al., 2001).

PHD2 is a member of the hypoxia-inducible factor (HIF) prolyl hydroxylases, a family of iron- and 2-oxoglutarate-dependent dioxygenases, which negatively regulate the stability of several proteins that have established roles in cell survival, proliferation, angiogenesis, glucose metabolism, and energy requirement for adaptation to hypoxia and oxidative stress processes. Thus, it is not surprising that low molecular weight HIF prolyl-4-hydroxylase inhibitors, which are iron chelators, have been proposed as novel neurological therapeutics for stroke as well as other diseases associated with oxidative stress (Siddiq et al., 2005).

Recently, a microarray study with PD brains (Miller et al., 2006) showed close similarity to those described above, with regard to biological processes or functional groups mainly affected in PD. The authors analyzed differential gene expression in the SN and striatum from postmortem PD brains and showed substantial downregulation of genes related mainly to dopaminergic phenotype [(i.e., dopamine transporter (DAT)], an aromatic amino acid decarboxylase (AADC)], vesicle trafficking, and synaptic transmission [i.e., syntaxin-1A, synaptogyrin 3 (SYNGR3), N-ethyl maleimide-sensitive factor (NSF)] and cytoskeleton maintenance [i.e., microtubule affinity regulating kinase (MARK-1)], microtubule associated protein (MAP-2), and dynein. Importantly, in the striatum, the most enriched functional class of differentially expressed genes was synaptic, sharing with the SN a decrease in a number of transcripts such as synapsins 1 and 2, SYNGR3 and 1, syntaxin 1A, and synaptotagmin 1 (SYT1). Thus, it is possible that either a premature dysregulation in a number of cytoskeleton components results in impairment of the bidirectional axonal transport between the SN and striatum, or that the synaptic changes precede the cytoskeleton dysfunction. In this context, it is interesting to note that all the gene expression profiling studies conducted so far in the SN of PD patients reported a dysfunction in vesicle trafficking and synaptic transmission functional group (see ❷ *Table 5-1* for a summary of all high throughput gene-based microarray studies conducted thus far in human late-stage Parkinsonian SN, highlighting major affected biological functional classes and candidate genes). More exciting is the identification of the same gene/s changes by independent laboratories, such as in the case of SYT1, which was found to be downregulated by Miller et al. (2006), Moran et al. (2006), Zhang et al. (2005), and Hauser et al. (2005) and of NSF, whose expression levels were found, by the first three out of the four groups aforementioned, to be decreased in the SN of PD patients. The fact that the same gene/s or intimately pathway-related genes have been identified by separate microarray examinations, employing different brain samples and experimental conditions, support the validity of the findings in the search for potential molecular biomarkers for the disease.

Zhang and colleagues (2005) were the first to conduct a multiple-area gene expression study in three different brain regions, the SN, putamen, and prefrontal cortex (BA9) of postmortem tissue from matched groups of PD or control subjects, ($n = 15$/group), using Affymetrix U133A GeneChip (22,000 probe sets). The authors focused their analysis on the most prominent gene alterations across the three brain areas aimed at identifying selected biological processes in PD. A comprehensive pathway analysis demonstrated that the most downregulated transcripts clustered in groups related to protein handling and degradation, such as various components of the proteasome, and energy pathways and mitochondrial electron transport chain [e.g., NADH dehydrogenase (ubiquinone) Fe-S protein 1, 75 kDa (NDUFS1) and COX11 homolog, cytochrome c oxidase assembly protein (COX11); see ❷ *Table 5-1* for prominent genes]. Support for the perturbation of the mitochondria-UPS network at the transcriptome level in PD was more recently provided by Duke et al. (2006). These authors examined gene expression in medial and lateral SN, as well as in the frontal cortex in brain tissue from 15 PD cases and 8 non-PD controls, using Affymetrix HG-U133 A & B array set (contains 33,000 well substantiated human gene sequences). A very significant correlation was demonstrated between the transcriptomic profiles of the mitochondria and UPS systems, both in disease and healthy brains, consistent with their driven hypothesis that this circuit is tightly connected. Moreover, an overall decrease in the expression of mitochondrial and UPS-associated transcripts [e.g., proteasome (prosome, macropain) subunit, alpha type, 6 (PSMA6); ubiquitin-conjugating enzyme E2, J1 (UBE2J1)] was found in both lateral and medial SN of PD when compared to control subjects. Also, the results of Hauser et al. (2005) with six human SN PD patients and five non-PD controls using Affymetrix U133A GeneChip identified significant alterations in the ubiquitination, chaperones, and mitochondrial gene pathways in the PD SN together with vesicle trafficking and nuclear-encoded genes. Collectively, the accumulated microarray data support the hypothesis that in PD there is a casual interconnection

◘ Table 5-1
Major affected biological functional classes and genes in Parkinsonian SN as summarized from high throughput gene-based microarray studies

Functional class	GenBank	Gene symbol	Gene title	Change	Reference
Cell adhesion/motility/cytoskeleton					
	NM 018650	MARK-1	Encode microtubule affinity regulating kinase	Down	Miller et al. (2006)
	NM 002374	MAP-2	Microtubule associated protein	Down	Miller et al. (2006)
	NM 004411	DNCI1	Dynein	Down	Miller et al. (2006)
	NM 004520.1	KIF2	Kinesin heavy chain member 2	Down	Grunblatt et al. (2004)
	BC004188.1	TUBB2	Tubulin, beta, 2	Down	Grunblatt et al. (2004)
	NM 002579.1	PALM	Paralemmin	Up	Grunblatt et al. (2004)
	NM 003803.1	MYOM1	Myomesin 1 (skelemin) 185 kDa	Up	Grunblatt et al. (2004)
	AL537457	NEFL	Neurofilament, light polypeptide	Down	Zhang et al. (2005), Moran et al. (2006)
	NM_014288.1	ITGB3BP	Integrin beta 3 binding protein (beta3-endonexin)	Up	Grunblatt et al. (2004)
	NM_000632.2	ITGAM	Integrin, alpha M (complement component receptor 3, alpha; also known as CD11b (p170), macrophage antigen alpha polypeptide)	Up	Grunblatt et al. (2004)
	M35543/ R37664	CDC42	Cell division cycle 42 (GTP binding protein, 25 kDa)	Down	Moran et al. (2006)
	NM006113	VAV3	Vav 3 oncogene	Down	Moran et al. (2006)
Protein handling/degradation					
	NM 004181	UCHL-1	Ubiquitin c-terminal hydrolase	Down	Moran et al. (2006) Miller et al. (2006)
	Not provided		Ubiquitin	Up	Miller et al. (2006)
	NM_002802.1	PSMC1	Proteasome (prosome,	Down	Duke et al. (2006)
	NM_006503.1	PSMC4	macropain) 26S subunit, ATPase 1/4	Down	Grunblatt et al. (2004)
	NM 002787.1	PSMA2	Proteasome (prosome,	Down	Grunblatt et al. (2004)
	NM 002790.1	PSMA3	macropain) subunit, alpha	Down	Grunblatt et al. (2004)
	NM 002788.1	PSMA5	type,2 ", type,3 ", type,5	Down	
	BC002979.1	PSMA6	Proteasome (prosome, macropain) subunit, alpha type, 6	Down	Duke et al. (2006)
	NM 005339.2	HIP2	Huntingtin interacting protein 2	Down	Grunblatt et al. (2004), Moran et al. (2006)
	NM 006930.1	SKP1A	S-phase kinase-associated protein 1A (p19A)	Down	Grunblatt et al. (2004)
	NM 016021.1	UBE2J1	Ubiquitin-conjugating enzyme E2, J1	Down	Duke et al. (2006)
Energy pathways: electron transport, glycolysis					
	NM 000188	HK1	Hexokinase	Down	Miller et al. (2006)
	NM 004373.1	COX6A1	Cytochrome c oxidase subunit VIa polypeptide 1	Down	Grunblatt et al. (2004)

◘ Table 5-1 (continued)

Functional class	GenBank	Gene symbol	Gene title	Change	Reference
	NM 002627.1	PFKP	Phosphofructokinase, platelet	Down	Grunblatt et al. (2004)
	NM 000158.1	GBE1	Glucan (1,4-alpha-), branching enzyme 1 (glycogen branching enzyme, Andersen disease, glycogen storage disease type IV)	Down	Grunblatt et al. (2004), Moran et al. (2006)
	AV727381	UQCRC2	Ubiquinol-cytochrome c reductase core protein II	Down	Grunblatt et al. (2004), Duke et al. (2006)
	AI348006	SDHA	Succinate dehydrogenase complex, subunit A	Down	Duke et al. (2006)
	NM 016243.1	NQO3A2	NAD(H) quinone oxidoreductase type 3	Down	Duke et al. (2006)
	NM 005006	NDUFS1	NADH dehydrogenase (ubiquinone) Fe-S protein 1, 75 kDa	Down	Zhang et al. (2005)
	BC005895	COX11	COX11 homolog, cytochrome c oxidase assembly protein (yeast)	Down	Zhang et al. (2005)
	NM_005917	MDH1	Malate dehydrogenase 1, NAD (soluble)	Down	Zhang et al. (2005)
	AF050641	NDUFA9	Mitochondrial	Down	Moran et al. (2006)
	NM 004542	NDUFA3	complex I subunit Mitochondrial complex III subunit	Down	Hauser et al. (2005)
Dopaminergic phenotype					
	S77154.1	NR4A2	Nuclear receptor subfamily 4, group A, member 2	Down	Grunblatt et al. (2004)
	NM 000689.1	ALDH1A1	Aldehyde dehydrogenase 1 family, member A1	Down	Grunblatt et al. (2004), Moran et al. (2006)
	NM 016300.1	ARPP-21	Cyclic AMP-regulated phosphoprotein, 21 kDa	Down	Grunblatt et al. (2004)
	AI269290	SLC18A2	Solute carrier family 18 (vesicular monoamine), member 2	Down	Grunblatt et al. (2004)
	NM 001044	SLC6A3	Dopamine transporter	Down	Miller et al. (2006)
	NM 000790	AADC	Aromatic amino acid decarboxylase	Down	Miller et al. (2006)
Response to stress					
	AK023253.1	DNAJB5	DnaJ (Hsp40) homolog, subfamily B, member 5	Up	Grunblatt et al. (2004),
	Not provided	DNAJB1 DNAJB6	DnaJ (Hsp40) homologue subfamily B, members 1 and 6	Up	Moran et al. (2006)
	Not provided	HERPUD1	Homocysteine-inducible, endoplasmic reticulum stress-inducible, ubiquitin-like domain member 1	Up	Moran et al. (2006)

◼ Table 5-1 (continued)

Functional class	GenBank	Gene symbol	Gene title	Change	Reference
	NM_001540	HSPB1	Heat shock 27 kDa protein 1	Up	Zhang et al. (2005), Moran et al. (2006)
	NM_005345	HSPA1A,	Heat shock protein A1A	Up	Hauser et al. (2005)
	NM 005346	HSPA1B	Heat shock protein A1B	Up	Hauser et al. (2005)
Cell transport/trafficking					
	AA890010	SEC22L1	SEC22 vesicle trafficking protein-like 1 (S. cerevisiae)	Down	Grunblatt et al. (2004), Zhang et al. (2005)
	NM 001860.1	SLC31A2	Solute carrier family 31 (copper transporters), member 2	Down	Grunblatt et al. (2004)
	NM 004731.1	SLC16A7	Solute carrier family 16 (monocarboxylic acid transporters), member 7	Down	Grunblatt et al. (2004)
	R15072	SLC16A14	Solute carrier family 16 (monocarboxylic acid transporters), member 14	Down	Moran et al. (2006)
Signal transduction					
	NM006218	PIK3CA	Phosphoinositide-3-kinase, catalytic, alpha polypeptide	Down	Grunblatt et al. (2004)
	NM_006226.1	PLCL1	Phospholipase C-like 1	Down	Grunblatt et al. (2004)
	AF022375.1	VEGF	Vascular endothelial growth factor	Down	Grunblatt et al. (2004)
	S77154.1	NR4A2	Nuclear receptor subfamily 4, group A, member 2	Down	Grunblatt et al. (2004), Moran et al. (2006)
	NM 004114	FGF13	Fibroblast growth factor 13	Down	Moran et al. (2006), Zhang et al. (2005)
Synaptic transmission					
	NM 004209	SYNGR3	Synaptogyrin	Down	Miller et al. (2006)
	NM 006178	NSF	N-ethyl maleimide-sensitive factor	Down	Miller et al. (2006), Zhang et al. (2005), Moran et al. (2006)
	NM 006950	SYN1	Synapsin 1	Down	Miller et al. (2006)
	NM 003178	SYN2	Synapsin 2	Down	Miller et al. (2006)
	NM 005639 AV731490 AV731490 AV731490	SYT1	Synaptotagmin 1	Down	Miller et al. (2006), Zhang et al. (2005), Moran et al. (2006), Hauser et al. (2005)
	NM004603	STX1A	Syntaxin 1A	Down	Miller et al. 2006),
	NM003165	STXBP1	Syntaxin binding protein 1	Down	Hauser et al. (2005)
	NM003081 L19760	SNAP25	Synaptosomal associated protein	Down	Moran et al. (2006), Zhang et al. (2005)

between failure of mitochondrial function, increase in ROS, and impairment in the activity of the UPS. The dysfunction in the crosstalk of these systems has been lately advocated as causative factors for the deposition of ubiquitinated proteins aggregates in the affected dopaminergic neurons of the SN (McNaught and Jenner, 2001; Zecca et al., 2004; Van der Schyf et al., 2006).

In a separate publication, but analyzing the same Affymetrix-derived data from Duke et al. (2006), Moran et al. (2006) reported that the most highly represented transcripts in PD fell into the biological classes of protein synthesis and breakdown, signaling, cell cycle, growth and development and transport, in agreement with previous studies (Grunblatt et al., 2004; Zhang et al., 2005). A number of priority sequences ($n = 25$) mapped to known PARK loci, representing candidates for as yet unidentified PD-causing genes, such as cell division cycle 42 (CDC42) and vav 3 oncogene (VAV3), which map to the chromosome 1p region and are involved in the integrin-mediated cell adhesion pathway. Interestingly, the cell adhesion functional class was the most highly represented group that was upregulated in the study of Grunblatt et al. (2004).

In the vast majority of the gene expression studies so far conducted in PD SN, there is a consistent elevation in heat-shock proteins/stress genes, such as the endoplasmic reticulum protein homocysteine-inducible, endoplasmic reticulum stress-inducible, ubiquitin-like domain member 1 (HERPUD1) and DnaJ (Hsp40) homologue subfamily B, member 1 (DNAJB1), considered top candidates by Moran et al. (2006). Additional selected genes are DnaJ (Hsp40) homolog, subfamily B, member 5 (DNAJB5) (Grunblatt), heat shock proteins A1A and A1B (HSPA1A, HSPA1B) (Hauser et al., 2005), and heat shock 27 kDa protein 1 (HSPB1), reported by both Zhang et al. (2005) and Moran et al. (2006).

It is apparent that although different gene platforms, tissue samples, and data analysis were employed, all of the aforementioned microarray studies in SN from PD subjects share in common the identification of same functional classes. Surprisingly, apoptosis-related genes constituted the smallest functional class in PD samples, questioning the relevance of programmed cell death in PD pathology, though they might have occurred at earlier stages thereby declining as the disease progresses.

3 Interpretation of Microarray Gene Profiling Results from Human PD Brains

While microarray technologies offer unprecedented opportunities to analyze global expression profiles of almost all transcripts in an organism, yet there are a number of issues that should be considered with caution before engaging in an experiment, from tissue selection to statistical analysis. To begin with, human brain tissue deserves an extra rigorous attention because of the innate problems to obtain suitable autopsy materials, with careful consideration of the disease pathology, autopsy time, age, gender, drug treatment, cause of death, and well-documented case history. Obviously, the information derived from postmortem PD patient brain tissue reflects late/end stage processes, questioning the relevance for earlier stages of the disease. In fact, optimal results would be obtained with presymptomatic and early diagnosed patient samples, but the probability to obtain such tissues is null. Another point that merits discussion is the fact that all the microarray expression studies conducted to date in human SN reflect changes in the whole tissue cell population and not in a homogeneous group of melanized surviving neurons. However, at the time of sample examination most of the melanin-containing dopaminergic neurons in the PD SNpc had died, so the transcriptional changes may reflect a combination of adaptation/compensation responses of the surviving cells or cell death processes. Thus, a matter of debate is whether the use of single dopaminergic neurons for transcriptomic profiling offers advantage over whole SN tissue, as the remaining cells may not necessarily express the initial gene changes caused by the neurotoxic environment to the original dopaminergic cell population. Furthermore, the single cell array technique requires sophisticated tools, as laser capture microdissection and round/s of RNA amplification because of the limited amount of relevant cells, thus increasing not only the cost but the inaccuracy of the results. To confront these obstacles there is a need for large data sets from postmortem brains obtained at different stages of the disease. A major help in the future will be the identification of valid biomarkers, which could be tested in peripheral tissues, such as blood or spinal fluid. This will allow monitoring individuals at risk before the onset of the disease and to apply a personalized drug therapy.

An important limitation of gene-based microarrays is their inability to monitor posttranscriptional and translational modifications or native protein alterations, thus providing a partial picture of the biochemical processes. Therefore, the integration of high throughput transcriptomic with parallel proteomic tools such as real time-quantitative RT-PCR and immunohistochemistry is crucial for a more precise global view of the interplay between biological processes and pathways, which is now considered the gold standard for research. A last issue, but by no means the less important, concerns the microarray data analysis. Usually, the "potential gene candidates" are selected on an arbitrary basis taking into account high p-values, gene clustering alignment into biological processes, or molecular functions and fold changes versus control non-PD subjects. The application of these personal criteria has a direct impact on the reproducibility and replication of the results from different laboratory data sets, since many relevant gene alterations in the disease samples could be simply overlooked because of the applied threshold cutoffs. This can be solved by applying identical data mining analysis methods and same statistical cutoffs to the diverse data sets from different sources, since each laboratory has different SN tissue preparations. The ideal program would be an international agreement to use exactly similar regions and specifically SNpc where melanin-containing dopamine neurons die. Such an initiative is currently being taken by the Edmond J. Safra Global Genetics Consortium on Gene Expression in PD sponsored by the Michael J Fox Foundation (http://www.michaeljfox. org/news/article.php?id = 114&sec = 1).

4 Gene and Protein Expression Profiling of the Neuroprotective, Anti-PD Drug, Rasagiline

Current therapies for PD ameliorate symptoms in the early phases of disease but become less effective over time. For this reason a major challenge is to prevent, or at least slow, the disease progression. With this in mind, the National Institute of Neurological Disorders and Stroke (NINDS), a part of the National Institutes of Health (NIH), has initiated a series of studies to evaluate potential neuroprotective agents in treating PD (Ravina et al., 2003). Using specific criteria for the evaluation of potential therapies, the authors found 12 compounds to be attractive candidates, including neurotrophic factors [e.g., brain-derived neurotrophic factor (BDNF)], bioenergetics (coenzyme Q10) and the novel anti-Parkinson drug rasagiline (Azilect; N-propargyl-1(R)-aminoindan), a second generation selective inhibitor of MAO-B (Youdim et al., 2001; Rascol et al., 2005). Rasagiline is a promising new drug for PD in light of lately reported benefits in PD patients with early illness (Group, 2004; Biglan et al., 2006). Recent multicenter double-blind monotherapy with rasagiline by the Parkinson Study Group (Group, 2004) and as adjunct therapy to L-DOPA (Rabey et al., 2000) have shown that rasagiline confers significant symptomatic improvement. Patients treated with rasagiline for 1 year showed less functional decline than patients whose treatment was delayed for 6 months, suggesting that the drug might possess a disease modifying property (Group, 2004; Blandini, 2005), and this advantage persisted six and a half years.[1]

Rasagiline has been shown to have a broad neuroprotective activity, which has been demonstrated in various models, both in vitro (Maruyama et al., 2001a, b; Akao et al., 2002) and in vivo, including the neurotoxins N-methyl-4-phenyl-1,2,3,6-tetrahydropyridine (MPTP)- and 6-hydroxydopamine (6-OHDA)-rodent models of PD (Heikkila et al., 1984; Tabakman et al., 2004; Blandini, 2005). In addition to its neuroprotective activity, the potential neurorescue/neuroregenerative action of rasagiline was initially demonstrated in a progressive apoptotic model of neuronal cell death induced by long-term serum deprivation (Bar-Am et al., 2005) and more recently, in post-MPTP-induced nigrostriatal dopamine neurodegeneration model in mice (Sagi et al., 2006). A continuous 10-day administration of rasagiline following MPTP lesion managed to restore the severe reduction in dopaminergic cell count, striatal DA content, and TH activity. Applying a parallel transcriptomic-proteomic study combined with a biology-based clustering method, Sagi et al. (2006) found that rasagiline caused induction of beta nerve growth factor (β-NGF) transcript with elevation of the tyrosine kinase receptor (Trk)-associated proteins including ShcC, SOS, AF6, Rin1, and Ras, in parallel with a specific increase in the Trk-downstream effecter PI3K proteins. Confirmatory immunohistochemical analysis indicated that this effect was associated with activation

---

[1] *The abstract entitled "Early rasagiline therapy shows long-term benefit for Parkinson's disease" was presented by the Parkinson Study Group in the 9th International Congress of Parkinson's disease and Movement Disorder, New Orleans, 2005.*

of the substrate of PI3K, Akt/PKB and phosphorylation/inactivation of glycogen synthase kinase-3β and Raf1, as depicted in ❷ *Figure 5-1*. This result supports and extends our previous and other laboratory studies, showing that the gene and protein expression of a number of Trk-ligands (BDNF and NGF) is induced by rasagiline and selegiline (Semkova et al., 1996; Tatton et al., 2002; Maruyama et al., 2004; Weinreb et al., 2004). These findings result demonstrated the essentiality of the activation of the Ras-PI3K–Akt survival pathway in rasagiline-mediated neurorescue effect. Collectively, a neuronal restoration of synaptic transmission by rasagiline in brains of PD animal model suggests a possible disease-modifying activity of rasagiline (Group, 2004). A much larger controlled clinical study is currently underway to prove this point.

❑ **Figure 5-1**

Proposed molecular mechanisms involved in the neurorescue action of rasagiline against MPTP in mice. Following MPTP administration, the fall in neurotrophic factor synthesis together with oxidative stress, may initiate the Rac-MKK4-JNK signaling pathway, leading to inactivation of basal Akt and activation of Forkhead translocation to the nucleus to induce Fasl expression and caspase-3-mediated apoptosis. The neurorescue effect of rasagiline against MPTP-induced neuronal death, can result from a specific induction of NGF and subsequent activation of Trk-A. The assembly of the Trk-related proteins by rasagiline may facilitate the recruitment of Ras activators which, in turn, activate H-Ras. H-Ras preferentially phosphorylates PI3K to activate Akt, resulting in phosphorylative inhibition of GSK-3β and inhibition of Fasl gene transcription. In addition, phosphorylation of Akt by rasagiline may inactivate the Raf-MAPK signaling. Red shape: upregulated by rasagiline; green shape: downregulated by rasagiline. For interpretation of color reference, the reader is referred to the web version of this article. Taken from (Sagi et al., 2006)

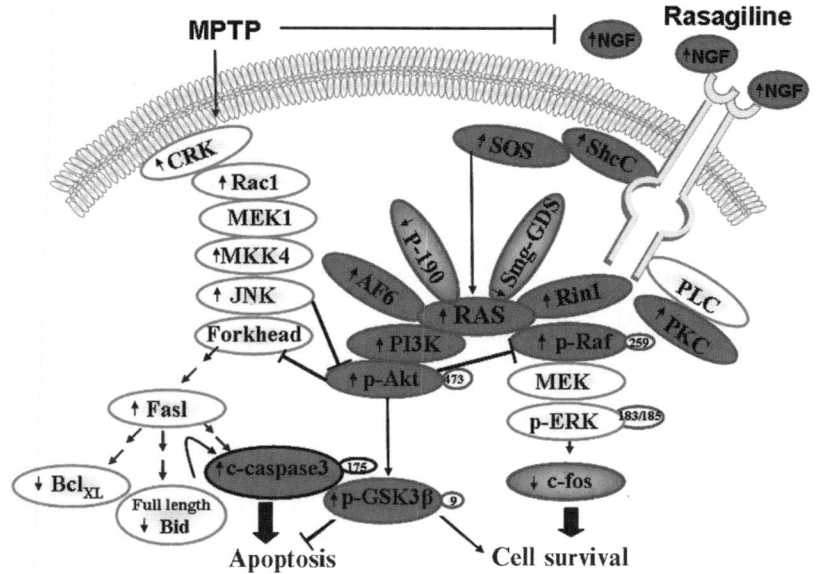

5 Transcriptomics and Proteomics in the Aging Brain

The aging process in the brain is not an outcome of a single gene or protein impairment, but rather a complex interaction of candidate genes and proteins that affect maintenance of the brain tissues (Rose et al., 2002). Currently, the human brain aging process is poorly understood, thus an important question is raised whether age-related neurodegenerative diseases, such as AD and PD, include components or by-products involved in aging. To date, there have been several in-vivo studies, in aged brain samples (Lee et al., 2000; Lukiw, 2004; Poon et al., 2004, 2005, 2006b; Weinreb et al., 2007), that succeeded in generating a list of genes

◻ Figure 5-2

A proposal diagram demonstrating the functional associations of selected proteins identified by proteomic analysis and significantly expressed in aged brains compared with young animals. Red, from (Poon et al., 2006b); blue, from (Poon et al., 2006a); black, from (Weinreb et al., 2007). For interpretation of color reference, the reader is referred to the web version of this article

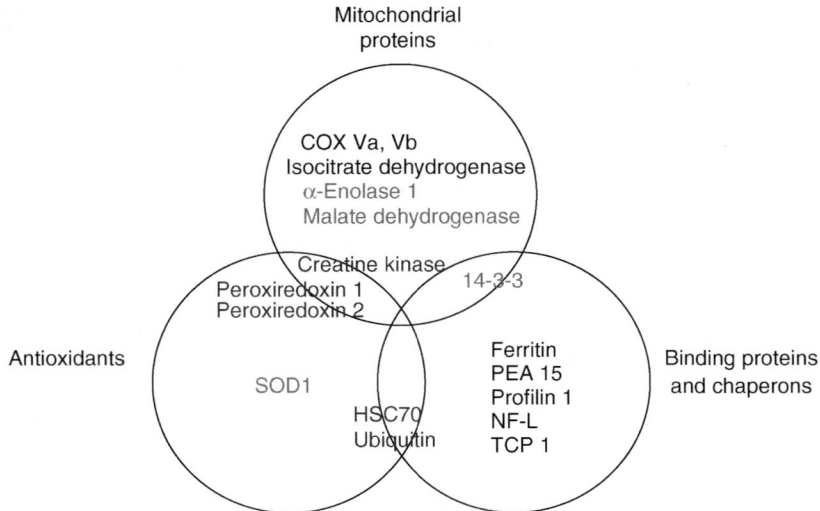

and proteins that are potentially involved in aging and age-related neurodegenerative diseases, stress-related signaling, binding, metabolism, and oxidation processes (❷ *Figure 5-2*).

The association of neurodegenerative diseases and age-related diseases with OS, as well as iron, has been discussed previously and supported by extensive studies (Jellinger et al., 1990; Olanow and Youdim, 1996; Youdim et al., 2004; Zecca et al., 2004). Increasing ROS in the mitochondria during aging has been suggested in causing DNA mutations and damaging the mitochondrial proteins (Lenaz et al., 2002). Taken together, alterations in proteins related to mitochondrial function and OS may reflect brain aging and underlie the higher risk for developing neurodegenerative diseases in older individuals (Chen et al., 2003). Our recent proteomic study in hippocampus tissues from old rats compared with young animals (Weinreb et al., 2007) demonstrated that the protein levels of two isoforms of the energy metabolic enzyme complex, cytochrome *c* oxidase (COX), subunits Va and Vb mitochondrial precursors, were decreased and the mRNA expression of subunit Va was downregulated as well, in old compared with young rats. COX subunits Va and Vb were shown to be involved in the transfer of electrons from cytochrome *c* to oxygen (Kadenbach et al., 2000). Thus, decreased levels of these subunits in aging and age-related diseases might affect the expression and activity of the other subunits and the entire complex of COX. In support, a previous immunohistochemical study demonstrated reduced levels of the nuclear coded subunit V of COX and the mitochondrial coded subunits II/III, as well as COX IV defects in the substantia nigra of old (70–90 years old) subjects (Itoh et al., 1996). Moreover, brains from AD patients showed 50–60% decreases in mRNA level of nuclear DNA-encoded subunit IV of COX (Chandrasekaran et al., 1997). Reduced gene and protein expression of the enzymes involved in energy metabolism, creatine kinase (BB isoform), α-enolase, triosephosphate isomerase (TPI), and glyceraldehyde 3-phosphate dehydrogenase (GADPH), in aging and age-related diseases (Castegna et al., 2002; Korolainen et al., 2005; Sultana et al., 2006), suggests impairments in mitochondria function and consequential ATP synthesis and excess of oxygen-derived species. It is well established that increased oxidative damage in aged brain is associated with dysfunction of the antioxidative defense mechanism (Reiter, 1995). Indeed, several studies (Chen et al., 2003; Jin et al., 2005; Miura et al., 2005; Poon et al., 2005; Weinreb et al., 2007) demonstrated reduced protein level of the mitochondrial antioxidant enzyme, peroxiredoxin 2, in aged human and animal brains. Peroxiredoxins are involved in the redox regulation of cellular signaling and differentiation and play a role in the antioxidant defense system, which eliminates peroxides generated during cellular metabolism (Rhee et al., 1994). It is likely that as a consequence of reduced activity of peroxiredoxin 2, which occurs during aging, the mitochondria will be subjected to enhanced OS.

The binding proteins and chaperones are another cluster of proteins demonstrated to be altered in aged animal brains (Poon et al., 2004, 2005). Profilin 1, an isoform of a family of actin-binding proteins (Witke et al., 2001), was increased in old brain rats (Weinreb et al., 2007). The induction of this protein may reflect an outcome of cytoskeletal changes occurring in aging and human PD brains [see ❷ *Table 5-1* and (Basso et al., 2004)]. Additionally, chaperonin containing TCP1, subunit 2 (beta), a cytosolic molecular chaperone assisting in the proper folding of actin, tubulin, and other cytosolic proteins (Yokota et al., 2001) was increased in the hippocampus of old rats (Weinreb et al., 2007). The abnormal accumulation of the major cytoskeletal constituent of neuronal cells, NF-L in the brain of old animals (Poon et al., 2004; Weinreb et al., 2006), supports previous data demonstrating NF-L intensification in the cytoplasm of neurons of sporadic ALS patients and other neurological diseases (Norgren et al., 2003; Lin et al., 2005). Neurofilament aggregation may be closely related to OS since oxidative exposure has been shown to promote its fibrilization or structure reorganization in vitro (Gelinas et al., 2000; Kim et al., 2004), suggesting that free radical activity might play an important role in the emergence of cytoplasmic inclusions leading to neuronal atrophy and death (Gelinas et al., 2000; Kim et al., 2004). Additionally, the induction of the iron regulation protein, ferritin heavy polypeptide 1, in aged hippocampal tissues (Weinreb et al., 2007) is in accordance with previous reports describing irregular enlargement of ferritin heavy and light peptides in different neuronal and extraneuronal tissues of aged, PD and AD human brains (Jellinger et al., 1993; Connor et al., 1995; Vidal et al., 2004). Hence, the iron homeostasis regulator gene, transferrin receptor (TfR) (Dassler et al., 2006) was completely diminished in the aged versus young hippocampus (Weinreb et al., 2007). The association of neurodegenerative diseases and age-related diseases with OS, as well as iron, has been discussed previously and supported by extensive studies (Jellinger et al., 1990; Olanow and Youdim, 1996; Zecca et al., 2004). Taken together, alterations in proteins related to mitochondrial function and OS may reflect brain aging and underlie the higher risk for developing neurodegenerative diseases in older individuals.

6 "All Roads Lead to Rome"

The gene–gene interactions and the pathogenic molecular mechanisms by which the mutated genes cause PD are not known, though, similar to sporadic disease, there is a degeneration of dopamine neurons and the patients respond well to L-DOPA. It can be assumed, therefore, that the final molecular intersections/pathways must be common to both sporadic and familial disease. As the saying goes "all roads lead to Rome" where Rome decided death or survival, implying a convergence of mechanisms responsible for DA neuron death in both classes of PD (❷ *Figure 5-3*). At present, genomic and proteomic analysis of familial PD have not been investigated. There is a compelling need to do so since it would illuminate whether the familial and sporadic diseases have a final common pathway. The working hypothesis is that subtle alterations in the kinetics of several affected proteins during the decades may have a cumulative effect underlying the slowly progressive neurodegeneration of the DA-containing neurons in PD. Thus, a progressive reduction/induction in a small number of genes may constitute a "convergence point" shared by both the hereditary and sporadic cases. Identifying this cluster is crucial for understanding the various pathways that gives rise to PD and developing effective neuroprotective and neurorescue drugs.

7 Conclusions and Future Directions

Although large-scale gene expression profiling studies conducted so far in human postmortem SN of PD patients are few in number, they have extended our current view on the molecular pathways underlying the

□ **Figure 5-3**

All roads lead to Rome. The cluster of gene expression changes demonstrated in sporadic Parkinsonian SNpc may very well represent the central core ("Rome" gene cascade) by which dopamine neurons degenerate in SNpc of sporadic, as well as familial PD. This is evidenced by the observation that there is homology between the sporadic and inherited disease, involving OS, iron dysregulation and a defect in the ubiquitin-proteasome system (UPS). However, to date no large-scale gene expression profile has been performed with any of the familial cases of PD, which could shed light on the crosstalk of genes participating in the cascade of neurodegeneration

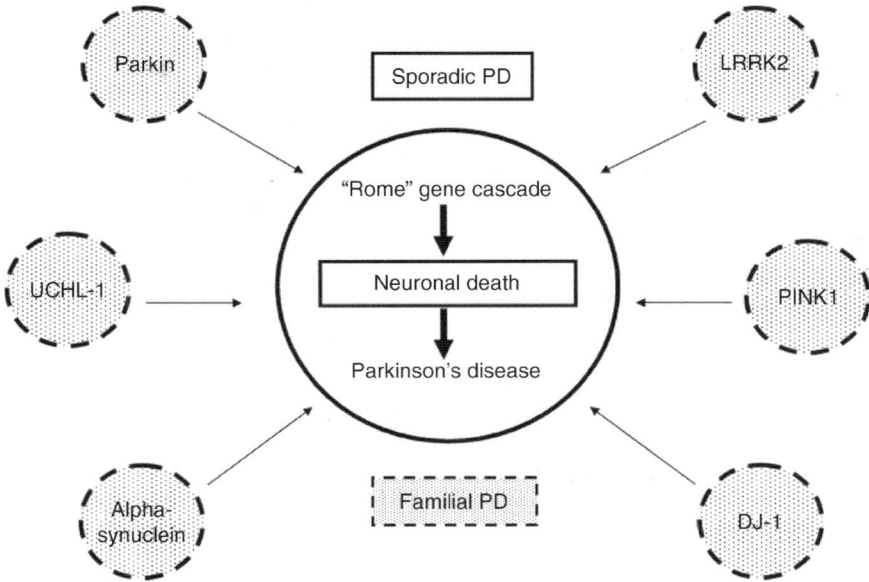

etiology and pathology of PD and have identified a selected number of candidate genes that may offer clues for potential gene intersections or crosstalks along the dopaminergic neurodegenerative cascade.

The signatures unveiled by these approaches could provide crucial information on diagnosis and development of surrogate markers for PD, reliable candidate genes as predictive early biomarkers to identify: (1) individuals at risk (susceptibility genes) before disease onset; (2) disease progression; (3) early pharmacological intervention; and (4) future development of CNS "magic bullets" targeted drugs.

References

Akao Y, Maruyama W, Shimizu S, Yi H, Nakagawa Y, et al. 2002. Mitochondrial permeability transition mediates apoptosis induced by *N*-methyl(R)salsolinol, an endogenous neurotoxin, and is inhibited by Bcl-2 and rasagiline, *N*-propargyl-1(R)-aminoindan. J Neurochem 82: 913-923.

Bar-Am O, Weinreb O, Amit T, Youdim MB. 2005. Regulation of Bcl-2 family proteins, neurotrophic factors, and APP processing in the neurorescue activity of propargylamine. FASEB J 19:1899-1901.

Basso M, Giraudo S, Corpillo D, Bergamasco B, Lopiano L, et al. 2004. Proteome analysis of human substantia nigra in Parkinson's disease. Proteomics 4: 3943-3952.

Biglan KM, Schwid S, Eberly S, Blindauer K, Fahn S, et al. 2006. Rasagiline improves quality of life in patients with early Parkinson's disease. Mov Disord 21: 616-623.

Blandini F. 2005. Neuroprotection by rasagiline: A new therapeutic approach to Parkinson's disease? CNS Drug Rev 11: 183-194.

Castegna A, Aksenov M, Aksenova M, Thongboonkerd V, Klein JB, et al. 2002. Proteomic identification of oxidatively modified proteins in Alzheimer's disease brain. Part I: Creatine kinase BB, glutamine synthase, and ubiquitin carboxy-terminal hydrolase L-1. Free Radic Biol Med 33: 562-571.

Chandrasekaran K, Hatanpaa K, Rapoport SI, Brady DR. 1997. Decreased expression of nuclear and mitochondrial DNA-encoded genes of oxidative phosphorylation in association neocortex in Alzheimer disease. Brain Res Mol Brain Res 44: 99-104.

Chen W, Ji J, Xu X, He S, Ru B. 2003. Proteomic comparison between human young and old brains by two-dimensional gel electrophoresis and identification of proteins. Int J Dev Neurosci 21: 209-216.

Connor JR, Snyder BS, Arosio P, Loeffler DA, LeWitt P. 1995. A quantitative analysis of isoferritins in select regions of aged, Parkinsonian, and Alzheimer's diseased brains. J Neurochem 65: 717-724.

Dassler K, Zydek M, Wandzik K, Kaup M, Fuchs H. 2006. Release of the soluble transferrin receptor is directly regulated by binding of its ligand ferritransferrin. J Biol Chem 281: 3297-3304.

Di Monte DA, Chan P, Sandy MS. 1992. Glutathione in Parkinson's disease: A link between oxidative stress and mitochondrial damage? Ann Neurol 32: S111-S115.

Duke DC, Moran LB, Kalaitzakis ME, Deprez M, Dexter DT, et al. 2006. Transcriptome analysis reveals link between proteasomal and mitochondrial pathways in Parkinson's disease. Neurogenetics 7: 139-148.

Epstein AC, Gleadle JM, McNeill LA, Hewitson KS, O'Rourke J, et al. 2001. C. elegans EGL-9 and mammalian homologs define a family of dioxygenases that regulate HIF by prolyl hydroxylation. Cell 107: 43-54.

Gelinas S, Chapados C, Beauregard M, Gosselin I, Martinoli MG. 2000. Effect of oxidative stress on stability and structure of neurofilament proteins. Biochem Cell Biol 78: 667-674.

Gerlach M, Ben-Shachar D, Riederer P, Youdim MBH. 1994. Altered brain metabolism of iron as a cause of neurodegenerative diseases? J Neurochem 793: 793-807.

Gotz ME, Kunig G, Riederer P, Youdim MB. 1994. Oxidative stress: Free radical production in neural degeneration. Pharmacol Ther 63: 37-122.

Group PS. 2004. A controlled, randomized, delayed-start study of rasagiline in early Parkinson disease. Arch Neurol 61: 561-566.

Grunblatt E, Mandel S, Jacob-Hirsch J, Zeligson S, Amariglo N, et al. 2004. Gene expression profiling of Parkinsonian substantia nigra pars compacta; alterations in ubiquitin-proteasome, heat shock protein, iron and oxidative stress regulated proteins, cell adhesion/cellular matrix and vesicle trafficking genes. J Neural Transm 111: 1543-1573.

Halliwell B. 1992. Reactive oxygen species and the central nervous system. J Neurochem 59: 1609-1623.

Hauser MA, Li YJ, Xu H, Noureddine MA, Shao YS, et al. 2005. Expression profiling of substantia nigra in Parkinson disease, progressive supranuclear palsy, and frontotemporal dementia with parkinsonism. Arch Neurol 62: 917-921.

Heikkila RE, Manzino L, Cabbat FS, Duvoisin RC. 1984. Protection against the dopaminergic neurotoxicity of 1-methyl-4-phenyl-1,2,5,6-tetrahydropyridine by monoamine oxidase inhibitors. Nature 311: 467-469.

Itoh K, Weis S, Mehraein P, Muller-Hocker J. 1996. Cytochrome c oxidase defects of the human substantia nigra in normal aging. Neurobiol Aging 17: 843-848.

Jellinger K, Paulus W, Grundke-Iqbal I, Riederer P, Youdim MBH. 1990. Brain iron and ferritin in Parkinson's and Alzheimer's diseases. J Neural Transm Park Dis Dement Sect 2: 327-340.

Jellinger KA, Kienzl E, Rumpelmaier G, Paulus W, Riederer P, et al. 1993. Iron and ferritin in substantia nigra in Parkinson's disease. Adv Neurol 60: 267-272.

Jenner P, Olanow CW. 1996. Oxidative stress and the pathogenesis of Parkinson's disease. Neurology 47: S161-S170.

Jin MH, Lee YH, Kim JM, Sun HN, Moon EY, et al. 2005. Characterization of neural cell types expressing peroxiredoxins in mouse brain. Neurosci Lett 381: 252-257.

Kadenbach B, Huttemann M, Arnold S, Lee I, Bender E. 2000. Mitochondrial energy metabolism is regulated via nuclear-coded subunits of cytochrome c oxidase. Free Radic Biol Med 29: 211-221.

Kamura T, Koepp DM, Conrad MN, Skowyra D, Moreland RJ, et al. 1999. Rbx1, a component of the VHL tumor suppressor complex and SCF ubiquitin ligase. Science 284: 657-661.

Kim NH, Jeong MS, Choi SY, Hoon Kang J. 2004. Oxidative modification of neurofilament-L by the Cu,Zn-superoxide dismutase and hydrogen peroxide system. Biochimie 86: 553-559.

Korolainen MA, Auriola S, Nyman TA, Alafuzoff I, Pirttila T. 2005. Proteomic analysis of glial fibrillary acidic protein in Alzheimer's disease and aging brain. Neurobiol Dis 20: 858-870.

Lee CK, Weindruch R, Prolla TA. 2000. Gene-expression profile of the ageing brain in mice. Nat Genet 25: 294-297.

Lenaz G, Bovina C, D'Aurelio M, Fato R, Formiggini G, et al. 2002. Role of mitochondria in oxidative stress and aging. Ann N Y Acad Sci 959: 199-213.

Liani E, Eyal A, Avraham E, Shemer R, Szargel R, et al. 2004. Ubiquitylation of synphilin-1 and alpha-synuclein by SIAH and its presence in cellular inclusions and lewy bodies imply a role in Parkinson's disease. Proc Natl Acad Sci USA 101: 5500-5505.

Lin H, Zhai J, Schlaepfer WW. 2005. RNA-binding protein is involved in aggregation of light neurofilament protein and is implicated in the pathogenesis of motor neuron degeneration. Hum Mol Genet 14: 3643-3659.

Lukiw WJ. 2004. Gene expression profiling in fetal, aged, and Alzheimer hippocampus: A continuum of stress-related signaling. Neurochem Res 29: 1287-1297.

Maruyama W, Youdim MBH, Naoi M. 2001a. Antiapoptotic properties of rasagiline, N-propargylamine-1(R)-aminoindan, and its optical (S)-isomer, TV1022. Ann N Y Acad of Sci 939: 320-329.

Maruyama W, Akao Y, Youdim MBH, Boulton AA, Davis BA, et al. 2001b. Transfection-enforced Bcl-2 overexpression and an anti-Parkinson drug, rasagiline, prevent nuclear accumulation of glyceraldehyde-3 phosphate dehydrogenase induced by an endogenous dopaminergic neurotoxin, N-methyl(R)salsolinol. J Neurochem 78: 727-735.

Maruyama W, Nitta A, Shamoto-Nagai M, Hirata Y, Akao Y, Youdim MBH, Furukawa S, Nabeshima T, Naoi M. 2004. N-Propargyl-1 (R)-aminoindan, rasagiline, increases glial cell line-derived neurotrophic factor (GDNF) in neuroblastoma SH-SY5Y cells through activation of NF-kappaB transcription factor. Neurochem Int 44: 393-400.

McNaught KS, Jenner P. 2001. Proteasomal function is impaired in substantia nigra in Parkinson's disease. Neurosci Lett 297: 191-194.

Meredith GE, Halliday GM, Totterdell S. 2004. A critical review of the development and importance of proteinaceous aggregates in animal models of Parkinson's disease: New insights into lewy body formation. Parkinsonism Relat Disord 10: 191-202.

Miller RM, Kiser GL, Kaysser-Kranich TM, Lockner RJ, Palaniappan C, et al. 2006. Robust dysregulation of gene expression in substantia nigra and striatum in Parkinson's disease. Neurobiol Dis 21: 305-313.

Miura Y, Kano M, Abe K, Urano S, Suzuki S, et al. 2005. Age-dependent variations of cell response to oxidative stress: Proteomic approach to protein expression and phosphorylation. Electrophoresis 26: 2786-2796.

Moran LB, Duke DC, Deprez M, Dexter DT, Pearce RK, et al. 2006. Whole genome expression profiling of the medial and lateral substantia nigra in Parkinson's disease. Neurogenetics 7: 1-11.

Norgren N, Rosengren L, Stigbrand T. 2003. Elevated neurofilament levels in neurological diseases. Brain Res 987: 25-31.

Olanow CW, Youdim MB. 1996. Iron and neurodegeneration: Prospects for neuroprotection. Neurodegeneration and Neuroprotection in Parkinson's Disease. Olanow CW, Jenner P, Youdim MB, editors. Academic Press: London pp. 55-69.

Ostrerova-Golts N, Petrucelli L, Hardy J, Lee JM, Farer M, et al. 2000. The A53T alpha-synuclein mutation increases iron-dependent aggregation and toxicity. J Neurosci 20: 6048-6054.

Poon HF, Calabrese V, Calvani M, Butterfield DA. 2006a. Proteomics analyses of specific protein oxidation and protein expression in aged rat brain and its modulation by ʟ-acetylcarnitine: Insights into the mechanisms of action of this proposed therapeutic agent for CNS disorders associated with oxidative stress. Antioxid Redox Signal 8: 381-394.

Poon HF, Vaishnav RA, Getchell TV, Getchell ML, Butterfield DA. 2006b. Quantitative proteomics analysis of differential protein expression and oxidative modification of specific proteins in the brains of old mice. Neurobiol Aging 27: 1010-1019.

Poon HF, Castegna A, Farr SA, Thongboonkerd V, Lynn BC, et al. 2004. Quantitative proteomics analysis of specific protein expression and oxidative modification in aged senescence-accelerated-prone 8 mice brain. Neuroscience 126: 915-926.

Poon HF, Farr SA, Banks WA, Pierce WM, Klein JB, et al. 2005. Proteomic identification of less oxidized brain proteins in aged senescence-accelerated mice following administration of antisense oligonucleotide directed at the Abeta region of amyloid precursor protein. Brain Res Mol Brain Res 138: 8-16.

Rabey JM, Sagi I, Huberman M, Melamed E, Korczyn A, et al. 2000. Rasagiline mesylate, a new MAO-B inhibitor for the treatment of Parkinson's disease: A double-blind study as adjunctive therapy to levodopa. Clin Neuropharmacol 23: 324-330.

Rascol O, Brooks DJ, Melamed E, Oertel W, Poewe W, et al. 2005. Rasagiline as an adjunct to levodopa in patients with Parkinson's disease and motor fluctuations (LARGO, lasting effect in adjunct therapy with rasagiline given once daily, study): A randomised, double-blind, parallel-group trial. Lancet 365: 947-954.

Ravina BM, Fagan SC, Hart RG, Hovinga CA, Murphy DD, et al. 2003. Neuroprotective agents for clinical trials in Parkinson's disease: A systematic assessment. Neurology 60: 1234-1240.

Reiter RJ. 1995. Oxidative processes and antioxidative defense mechanisms in the aging brain. FASEB J 9: 526-533.

Rhee SG, Kim KH, Chae HZ, Yim MB, Uchida K, et al. 1994. Antioxidant defense mechanisms: A new thiol-specific antioxidant enzyme. Ann N Y Acad Sci 738: 86-92.

Riederer P, Sofic E, Rausch WD, Schmidt B, Reynolds GP, et al. 1989. Transition metals, ferritin, glutathione, and ascorbic acid in Parkinsonian brains. J Neurochem 52: 515-520.

Rose MR, Mueller LD, Long AD. 2002. Pharmacology, genomics, and the evolutionary biology of ageing. Free Radic Res 36: 1293-1297.

Sagi Y, Mandel S, Amit T, Youdim MB. 2006. Activation of tyrosine kinase receptor signaling pathway by rasagiline facilitates neurorescue and restoration of nigrostriatal dopamine neurons in post-MPTP-induced Parkinsonism. Neurobiol Dis doi.10.1016/j.nbd.2006.07.020.

Saura J, Luque JM, Cesura AM, Da Prada M, Chan-Palay V, et al. 1994. Increased monoamine oxidase B activity in plaque-associated astrocytes of Alzheimer brains revealed by quantitative enzyme radioautography. Neuroscience 62: 15-30.

Schapira AH, Cooper JM, Dexter D, Clark JB, Jenner P, et al. 1990. Mitochondrial complex I deficiency in Parkinson's disease. J Neurochem 54: 823-827.

Semkova I, Wolz P, Schilling M, Krieglstein J. 1996. Selegiline enhances NGF synthesis and protects central nervous system neurons from excitotoxic and ischemic damage. Eur J Pharmacol 315: 19-30.

Siddiq A, Ayoub IA, Chavez JC, Aminova L, Shah S, et al. 2005. Hypoxia-inducible factor prolyl 4-hydroxylase inhibition. A target for neuroprotection in the central nervous system. J Biol Chem 280: 41732-41743.

Sultana R, Boyd-Kimball D, Poon HF, Cai J, Pierce WM, et al. 2006. Redox proteomics identification of oxidized proteins in Alzheimer's disease hippocampus and cerebellum: An approach to understand pathological and biochemical alterations in AD. Neurobiol Aging 27: 1564-1576.

Tabakman R, Lecht S, Lazarovici P. 2004. Neuroprotection by monoamine oxidase B inhibitors: A therapeutic strategy for Parkinson's disease? Bioessays 26: 80-90.

Tatton WG, Chalmers-Redman RM, Ju WJ, Mammen M, Carlile GW, Pong AW, Tatton NA. 2002. Propargylamines induce antiapoptotic new protein synthesis in serum- and nerve growth factor (NGF)-withdrawn, NGF-differentiated PC-12 cells. J Pharmacol Exp Ther 301: 753-764.

Turnbull S, Tabner BJ, El-Agnaf OM, Moore S, Davies Y, et al. 2001. Alpha-synuclein implicated in Parkinson's disease catalyses the formation of hydrogen peroxide in vitro. Free Radic Biol Med 30: 1163-1170.

Van der Schyf CJ, Geldenhuys WJ, Youdim MB. 2006. Multifunctional drugs with different CNS targets for neuropsychiatric disorders. J Neurochem 99: 1033-1048.

Vidal R, Ghetti B, Takao M, Brefel-Courbon C, Uro-Coste E, et al. 2004. Intracellular ferritin accumulation in neural and extraneural tissue characterizes a neurodegenerative disease associated with a mutation in the ferritin light polypeptide gene. J Neuropathol Exp Neurol 63: 363-380.

Weinreb O, Bar-Am O, Amit T, Chillag-Talmor O, Youdim MBH. 2004. Neuroprotection via pro-survival protein kinase C isoforms associated with Bcl-2 family members. FASEB J 18: 1471-1473.

Weinreb O, Drigues N, Sagi Y, Reznick AT, Amit T, Youdim MB. 2007. The application of proteomics and genomics to the study of age-related neurodegeneration and neuroprotection. Antioxid Redox Signal 9: 169-179.

Witke W, Sutherland JD, Sharpe A, Arai M, Kwiatkowski DJ. 2001. Profilin I is essential for cell survival and cell division in early mouse development. Proc Natl Acad Sci USA 98: 3832-3836.

Yokota S, Yanagi H, Yura T, Kubota H. 2001. Cytosolic chaperonin-containing t-complex polypeptide 1 changes the content of a particular subunit species concomitant with substrate binding and folding activities during the cell cycle. Eur J Biochem 268: 4664-4673.

Youdim MBH, Riederer P. 1997. Understanding Parkinson's disease. The smoking gun is still missing, but growing evidence suggests highly reactive substances called free radicals are central players in this common neurological disorder. Sci Am 276: 52-59.

Youdim MBH, Gross A, Finberg JPM. 2001. Rasagiline [N-propargyl-1R(+)-aminoindan], a selective and potent inhibitor of mitochondrial monoamine oxidase B. Br J Pharmacol 132: 500-506.

Youdim MBH, Stephenson G, Ben Shachar D. 2004. Ironing iron out in Parkinson's disease and other neurodegenerative diseases with iron chelators: A lesson from 6-hydroxydopamine and iron chelators, desferal and VK-28. Ann N Y Acad Sci 1012: 306-325.

Zecca L, Riederer P, Connor JR, Crichton RR. 2004. Iron, brain ageing and neurodegenerative disorders. Nat Rev Neurosci 5: 863-873.

Zhang J, Goodlett DR. 2004. Proteomic approach to studying Parkinson's disease. Mol Neurobiol 29: 271-288.

Zhang Y, James M, Middleton FA, Davis RL. 2005. Transcriptional analysis of multiple brain regions in Parkinson's disease supports the involvement of specific protein processing, energy metabolism, and signaling pathways, and suggests novel disease mechanisms. Am J Med Genet B Neuropsychiatr Genet 137: 5-16.

6 Neuroprotective Mechanisms: Oxidative Stress as a Target for Neuroprotective Therapies in Alzheimer's and Parkinson's Disease

A. Huber · A. Bürkle · G. Münch

Springer-Verlag Berlin Heidelberg 2007

Abstract: One of the major age-related damaging agents are reactive oxygen species (ROS). The brain is more vulnerable to oxidative stress than other organs as concomitant low activity and capacity of antioxidative protection systems allow for increased exposure of target molecules to ROS. Since neurons are postmitotic cells, they have to live with cellular damage accumulated over many decades. Increased levels of ROS (also termed "oxidative stress"), produced by normal mitochondrial activity, inflammation, and excess glutamate levels, is proposed to accelerate neurodegenerative processes characteristic for Alzheimer's disease (AD) and Parkinson's disease (PD). This chapter presents evidence for the importance of oxidative stress in the pathogenesis of these diseases and explains the nature of different types of ROS mediating neuronal damage. Furthermore, the potential beneficial effects of neuroprotective treatments, including antioxidants, energy supplements, and antiglutamatergic drugs, are discussed.

List of Abbreviations: AA, arachidonic acid; Aβ, β-amyloid peptide; AC, adenyl cyclase; AD, Alzheimer's disease; AGE, advanced glycation endproduct; AMPA, α-amino-3-hydroxy-5-methyl-4-isoxazolepropionic acid; APP, amyloid precursor protein; BB, blueberry; BGP, behavioral rating scale for geriatric patients; BH_4, tetrahydrobiopterin; cGMP, cyclic guanyl monophosphate; CHAP, Chicago Health and Aging Project; CIBIC-Plus, Clinician's Interview-Based Impression of Change Plus; CK, creatine kinase; CM, creatine monhohydrate; CR, creatine; DA, dopamine; DAG, diacylglycerol; DATATOP, deprenyl and tocopherol antioxidative therapy of parkinsonism; DHA, dehydroascorbic acid; DTT, dithiothreitol; DHLA, dihydro-lipoic acid; DPPH, 1,1-diphenyl-2-picryl-hydrazyl; EC, epicatechin; EGb, ginkgo biloba; EGC, ($-$)-epigal-locatechin; EGCG, ($-$)-epicatechin gallate; GC, guanylate cyclase; GSH, glutathione reduced; 4-HNE, 4-hydroxysynonenal; H_2O_2, hydrogen peroxide; HOCL, hypochlorous acid; HVA, homovanillic acid; IL-1β, interleukin 1β; IL-6, interleukin 6; IP_3, inositol-1,4,5,-trisphosphate; KA, kainic acid; LA, lipoic acid; LOOH, lipid hydroperoxidases; LP, lipoprotein; M-CSF, macrophage colony-stimulating factor; MDA, malondialdehyde; NAC, N-acetyl-cysteine; NADH, nicotine amide dinucleotide phosphate; NMDA, N-methyl-D-aspartate; MPTP, 1-methyl-4-phenyl-1,2,3,6-tetrahydro-pyridine; MtCK, mitochondrial creatine kinase; MTP, mitochondrial permeability transition pore; NO, nitric oxide; N_xO_y, nitrogen oxides; MMSE, mini-mental state examination; O_2^-, superoxide; OCl^-, hypochlorite; $\bullet$OH, hydroxyl radical; 6-OHDA, 6-hydroxy dopamine; $ONOO^-$, peroxyinitrite; PCP, phencyclidine; PCr, phosphocreatine; PD, Parkinson disease; PIP_2, L-3-phosphatidyl-inositol-4,5-bisphosphate; PKA, protein kinase A; PKC, protein kinase C; PLC, phospholipase C; RAGE, receptor for AGE; ROS, reactive oxygen species; SN, substantia nigra; SOD, superoxide dismutase; SPA, sweet potato anthocyanin; TH, tyrosine hydroxylase; TNF-α, tumor necrosis factor-α

1 Neurodegenerative Diseases: A Growing Burden in the Aging Population

Neurodegenerative diseases are initially characterized by subtle changes in the normal function of neurons, which is followed by overt neuronal dysfunction and cell death in later stage of the disease. Age is the single most important risk factor for the development of many neurodegenerative disorders including Parkinson's disease (PD) and Alzheimer's disease (AD). The number of people aged 80 and over will increase dramatically in the near future, and it is predicted that the proportion of elderly citizens above the age of 60 will double between now and the year 2025 (Hernandez, 1992). Patients with neurodegenerative diseases not only suffer emotionally and physically but also represent a significant financial and emotional burden for caregivers, society, and community. Thus, it is becoming increasingly important to find the cause(s) of neurodegenerative diseases and, based on these findings, develop and introduce novel therapies that prevent such diseases by providing neuroprotection.

2 Neurodegenerative Disorders

2.1 Alzheimer's Disease

2.1.1 Symptoms

AD is a progressive neurodegenerative brain disorder that gradually destroys a person's memory and ability to learn, make judgments, communicate with the social environment, and carry out daily activities. In the course of the disease, short-term memory is affected first, caused by neuronal dysfunction and cell death in the hippocampus and amygdala. As the disease progresses further, neurons also die in other cortical regions of the brain. At that stage, sufferers often experience dramatic changes in personality and behavior such as anxiety, suspiciousness, or agitation, as well as delusions or hallucinations.

2.1.2 Epidemiology

AD prevalence in the different age groups is 1% (65–69 years), 3% (70–74 years), 6% (75–79 years), 12% (80–84 years), and 25% (85 and over) (Yesavage et al., 2002). As the society ages, the number of AD patients will increase by 27% by 2020, 70% by 2030, and nearly 300% by 2050, unless science finds a way to slow the progression of the disease or prevent it. AD advances at widely different rates, and therefore the duration of the illness varies between 3 and 20 years.

2.1.3 Neuropathology and Biochemistry

AD is characterized by two characteristic lesions, amyloid plaques and neurofibrillary tangles, which are present in high numbers in the gray matter of affected brain areas.

Neurofibrillary tangles are intracellular deposits formed by hyperphosphorylated and extensively cross-linked tau protein. Tau is a microtubule associated protein that regulates a variety of properties of neuronal microtubules, especially their stability and orientation. In AD, however, tau is hyperphosphorylated and forms fibrillar inclusions. Presumably, this lead to neuronal dysfunction by disturbing cytoskeletal functions of neurons resulting in impaired axonal transport processes (Braak and Braak, 1995).

Senile plaques are the second characteristic hallmark in AD. These extracellular protein deposits are mainly composed of β-amyloid peptides (Aβs), which form β-sheeted fibrils and become insoluble. Aβ is derived from the β-amyloid precursor protein (APP), an integral membrane protein that is processed by β- and γ-secretases. In a variety of cell culture models, Aβ was shown to cause toxicity to neurons by various mechanism, many of which involving oxidative stress. For example, 4-hydroxynonenal and malondialdehyde (MDA), both markers of lipid peroxidation, were found in the hippocampus of patients with AD (Butterfield et al., 2002). Markers of protein oxidation, such as protein nitration, are also increased in the hippocampus and neocortex of individuals with AD (Sayre et al., 1997). The fact that Aβ binds strongly to metal ions, like copper and iron, and catalyzes the formation of the toxic hydroxyl radical (•OH) from hydrogen peroxide (H_2O_2) strongly suggests that it may be causally involved in oxidative stress in neurons.

Amyloid plaques, which can be present in the brain for decades, are a target for modification with advanced glycation endproducts (AGEs), i.e., sugar or carbonyl compounds attaching to lysine and arginine residues of long-lived proteins. AGEs have been found not only in amyloids but also in other proteins with slow turnover such as eye lens crystalline or cartilage and skin collagen. In AD, Aβ and AGE are both present in senile plaques and are resistant to degradation by macrophages due to extensive cross-linking. Aβ and AGE are able to chemoattract and activate microglial cells that surround these senile plaques (Dickson et al., 1996). AGEs and Aβ bind to the same receptor, i.e., the receptor for AGE (RAGE), and activate signal pathways involved in the induction of oxidative and nitrosative stress and proinflammatory effectors (Dukic-Stefanovic et al., 2003).

Inflammation, as evidenced by the activation of microglia and astroglia, is another hallmark of AD. Inflammation, particularly the induction of superoxide production (oxidative burst), is an important source of oxidative stress in AD patients. The inflammatory process occurs mainly around the amyloid plaques and is characterized by proinflammatory substances released from activated microglia (Halliday et al., 2000). Reactive oxygen species (ROS) are the most prominent molecules in the inflammatory process, along with prostaglandins, interleukin 1β (IL-1β), interleukin 6 (IL-6), macrophage colony-stimulating factor (M-CSF), and tumor necrosis factor-α (TNF-α)(Griffin et al., 1995; Meda et al., 1995; Pachter, 1997).

2.2 Parkinson's Disease

2.2.1 Epidemiology and Symptoms

Parkinson's disease (PD) is a progressive, degenerative disease of unknown etiology characterized by rhythmic tremor of the limbs, stooped posture, slowness of voluntary movements, and masklike facial expression. Idiopathic PD is the most common form of parkinsonism, which designates a group of movement disorders that have similar features and symptoms. Parkinson's disease is called "idiopathic" because its cause is unknown, in contrast to other forms of parkinsonism, where a cause has been established or is suspected. The disease is quite frequent—about four people in every thousand have PD. Symptoms of PD may appear at any age, but the average age of onset is 60.

2.2.2 Neuropathology and Biochemistry

Parkinson's results from the degeneration of dopaminergic neurons in the *substantia nigra* (SN) that develop inclusions called Lewy bodies. These deposits are composed of aggregated α-synuclein and ubiquitin. Ubiquitin is covalently attached to damaged or aged proteins in order to mark them for degradation in the proteasome. α-Synuclein is a synaptic protein of unknown function. When dopamine (DA) levels are depleted, the neuronal populations in the basal ganglia, to which the dopaminergic neurons project, become unable to coordinate and control the primary motor system, resulting in loss of prompt and smooth movement of limbs and trunk. Parkinson's disease patients have lost 80% or more of their DA-producing cells before symptoms appear, highlighting the great spare capacity normal human beings possess.

There is plenty of evidence that oxidative stress is occurring in PD. Partial deficiency of mitochondrial complex I leads to enhanced production of ROS and therefore to an inhibition of complex I (Tretter et al., 2004). Damaged neurons show reduced mitochondrial cytochrome *c*-oxidase activity, high concentrations of iron as well as ROS release (Bains and Shaw, 1997). An increase in lipid peroxidation products (Dexter et al., 1994), reactive protein carbonyls (Alam et al., 1997), and DNA oxidation provide further evidence for oxidative stress in PD.

Inflammation may also occur in this neurodegenerative disorder since reactive microglia and activated complement components are found in affected brain regions (McGeer and McGeer, 2004).

3 Mechanisms of Neurodegeneration

3.1 Oxidative Stress

3.1.1 Inflammation and Glia-Derived Oxidative Burst

One important source of ROS is the oxidative burst of activated macrophages and microglia, which release superoxide via the enzymatic activity of NADPH-oxidase. Under normal circumstances, the human body is able to scavenge these radicals but oxidant (particularly superoxide) production is dramatically increased in some inflammatory situations. Such inflammation is mediated through resident brain microglia or immigrated peripheral monocytes and macrophages. Their normal function is to destroy bacteria or virus-infected cells with an oxidative burst of $O_2^{\bullet-}$ and its dismutation product H_2O_2, as well as hypochlorous acid (HOCl) and nitric oxide (NO). These oxidants are generated by four different enzymes, i.e., NADPH oxidase, superoxide dismutase (SOD), myeloperoxidase, and nitric oxide synthase (NOS).

NADPH oxidase is inactive in resting phagocytes, but exposure to various stimuli (i.e., immune complexes, complement 5α, β-amyloid, AGEs) leads to its activation, i.e., catalysis of the conversion of molecular oxygen to superoxide.

$$2O_2 + NADPH \rightarrow 2O_2^- + NADP^+ + H^+$$

The oxidation of halide ions to hypohalous acid at the expense of H_2O_2 is carried out by the enzyme myeloperoxidase (Hampton et al., 1998):

$$Cl^- + H_2O_2 \rightarrow OCl^- + H_2O$$

Hypochlorite can then further react with hypochlorite, ammonia, or amines, the precursors of chloramines, a group of microbicidal oxidized halogens (Thomas, 1979).

3.1.2 Superoxide Production by Mitochondria

In the human body, 20% of the total oxygen consumption is needed for the brain, which is quite intriguing given that the brain takes only 2% space of the total tissue volume (Raichle and Gusnard, 2002). About 2% of the oxygen consumed in neuronal mitochondria is incompletely reduced to generate superoxide anions by electron leakage from the electron transport chain (Markesbery and Carney, 1999). These superoxide anions are usually converted by the enzyme SOD to H_2O_2 and then further converted to water by catalase or glutathione peroxidase (Fridovich, 1989).

Neurons in human brain are highly vulnerable to oxidative stress not only because of high oxygen consumption (and therefore high superoxide production) but also due to poor antioxidative mechanisms, particularly catalase activity, in comparison to other organs in the human body (Coyle and Puttfarcken, 1993). The cell membranes in the brain are especially vulnerable to attack by hydroxyl radicals as the double bonds of polyunsaturated fatty acids are easily oxidized.

There are many different species of oxygen free radicals, including superoxide, H_2O_2, and the hydroxyl radical, which are generated by reduction of molecular oxygen in three steps, and their generation will be described in detail below.

$$O_2 \xrightarrow{e^-} O_2^{\bullet-} \xrightarrow{SOD} H_2O_2 \xrightarrow[-Fe^{3+}]{Fe^{2+}} \bullet OH + OH^-$$

3.2 Reactive Oxygen Species

3.2.1 Superoxide

Superoxide ($O_2^{\bullet-}$) is a precursor for hydroxyl radicals; it is necessary for the formation of H_2O_2 and also acts as a reducing agent of transition metals. Most of the superoxide produced in human brain is generated by the electron transport chain of the mitochondria by incomplete reduction of molecular oxygen.

$$O_2 \xrightarrow{e^-} O_2^{\bullet-}$$

An increase in $O_2^{\bullet-}$ production also occurs when Ca^{2+} concentrations increase, as is the case during N-methyl-D-aspartate (NMDA) receptor overstimulation. Lipid metabolism is a second minor pathway for $O_2^{\bullet-}$ formation. When Ca^{2+} enters neurons, phospholipase A2 is activated and arachidonic acid (AA) is formed (Dumuis et al., 1988). Cyclooxygenase then converts AA to prostaglandins, with $O_2^{\bullet-}$ generated as a by-product (Needleman et al., 1986). $O_2^{\bullet-}$ formed within the neurons does not appear to cross plasma membrane (Patel et al., 1996) and is therefore only found in the intracellular environment following damage to the neuronal membrane. Superoxide is more reactive in organic solutions than in aqueous ones and can therefore damage hydrophobic membranes. Although $O_2^{\bullet-}$ on its own may have little significance in terms of toxicity, its conversion to peroxynitrite and hydroxyl radicals greatly increases neurotoxicity.

3.2.2 Hydrogen Peroxide

H_2O_2 is not a free radical per se, but belongs to the category of ROS. It is not particularly reactive, but it is the main source of hydroxyl radicals in the presence of transition metal ions (Cheeseman and Slater, 1993). It is also involved in the production of hypochlorous acid (HOCl) by neutrophils during oxidative burst (Mutze et al., 2003). H_2O_2 is generated in the dismutation reaction during which two molecules of superoxide are converted into one molecule of H_2O_2 plus molecular oxygen:

$$2O_2^- + 2H^+ \rightarrow H_2 + H_2O_2$$

O_2, the substrate, reacts with itself to give an oxidized product (i.e., oxygen) and a reduced product (i.e., H_2O_2). The enzyme responsible for this reaction is superoxide dismutase (SOD), which exists as a copper–zinc enzyme present in the cytoplasm and manganese isoform present in the mitochondria.

3.2.3 Hydroxyl Radical

The reactions of the hydroxyl radical, which is the most damaging free radical by far, can have devastating effects in any tissue type. The hydroxyl radical is a third generation species of radical which is derived from H_2O_2. H_2O_2 in turn, is derived from the superoxide radical through the action of SOD, and then reduced to hydroxyl radicals in the presence of transition metals such iron or copper. It reacts with any molecule including macromolecules such as DNA, membrane lipids, proteins, and carbohydrates. In DNA, the hydroxyl radical can induce strand breaks as well as chemical changes in the deoxyribose and in the purine and pyrimidine bases. Damaged proteins, many of them crucial enzymes in neurons, lose their efficiency and cellular function deteriorates. Protein oxidation mediated by hydroxyl radicals has been proposed as an explanation for the functional deficits associated with aging in many tissues and brain aging and neurodegeneration in particular.

The conversion of H_2O_2 to the hydroxyl radical depends on the presence of transition metals such as copper or iron. There is evidence that these metals accumulate in aged brain, and they may therefore contribute to neurodegenerative disorders. Some brain regions and some types of brain cells, for example oligodendrocytes, show above-average concentrations of iron (Smith et al., 1999), while their antioxidative protection systems are rather weak. Being transition metals, iron can exist in the oxidized (+III) or reduced (+II) forms and likewise copper as (+II) or (+I), which makes them important cofactors for many enzymes. Free transition metals may, however, generate ROS via the Fenton reaction. Here, Fe^{2+} is able to transfer one electron to H_2O_2, thus producing the hydroxy radical:

$$Fe^{2+} + H_2O_2 \rightarrow Fe^{3+} + \bullet OH + OH^{\bullet -}$$

The Haber-Weiss reaction also can form $\bullet OH$ in an interaction between $O_2^{\bullet -}$ and H_2O_2 in the presence of Fe^{2+} or Fe^{3+} (Markesbery, 1997). The body has developed chaperones to prevent such metal-dependent hydroxyl radical generation. Free copper is generally bound to cytochrome c oxidase, Cu/Zn SOD, and ceruloplasmin, while iron is mainly bound to ferritin. This prevents the inadvertent release of the metals and limits their availability to catalyze the Haber-Weiss reaction.

3.2.4 Nitric Oxide

Another radical, which has been proposed to be involved in neurodegeneration, is NO, created by various NO-synthetases. Ca^{2+} activated neuronal nitric oxide synthases (nNOS, bNOS, cNOS, TypeI) catalyze the oxidation of arginine to citrulline and NO, utilizing molecular oxygen:

$$Arginine + O_2 + NADPH \rightarrow NO + citrulline + NADP^+$$

In response to increased intracellular Ca^{2+}, nNOS interacts with Ca^{2+}/calmodulin (CaM), forms a complex with the cofactor tetrahydrobiopterin (BH_4), and translocates from the plasma membrane to the cytoplasm. The dephosphorylation of nNOS by calcineurin initates the production of NO. A further source of NO is the inducible NO isoform, which is upregulated in glia in response to proinflammatory stimuli present in AD and PD brains.

NO then has the capacity to activate several pathways, predominantly cyclic guanyl monophosphate (cGMP)-regulated signaling pathways, via the activation of guanylate cyclase (GC). NO undergoes a further reaction with $O_2^{\bullet-}$ radicals to form peroxynitrite ($ONOO^-$), one of the most potent oxidants in biological systems:

$$NO\bullet + O_2^- \rightarrow ONOO^-$$

Peroxynitrite causes nitration of tyrosine, phenylalanine, and tryptophan in proteins as well as the oxidation of lipids, DNA, and proteins. Lipid peroxidation is a significant problem for neurons as the destruction of lipidic cell membranes means that ionic gradients can no longer be maintained. Enhanced release of glutamate from presynaptic terminals further exacerbates this damage. The half-life of peroxynitrite is very short, and it decomposes in a liquid environment into further reactive oxygen- and nitrogen species, i.e., hydroxyl radicals and various nitrogen oxides (N_xO_y).

4   Overstimulation of Glutamate Receptors (Excitotoxicity)

Overstimulation of glutamate receptors, called excitotoxicity, is a further mechanism of neuronal cell death in neurodegeneration (Albin and Greenamyre, 1992). Glutamate is the major excitatory neurotransmitter in the mammalian CNS and is present in millimolar concentrations in the gray matter. Once released, it can act on ionotropic and metatropic receptors, two general receptors for glutamate. Metabotropic receptors are linked to G-proteins, which activate phospholipase C (PLC) or adenyl cyclase (AC). Activated PLC catalyzes the cleavage of L-3-phosphatidyl-inositol-4,5-bisphosphate (PIP_2) into inositol–1,4,5, trisphosphate (IP_3) and diacylglycerol (DAG). IP_3 is responsible for intracellular release of Ca^{2+} from the endoplasmic reticulum. Release is involved in the activation of protein kinase C (PKC) in neurons, which then phosphorylates cytosolic proteins (Weiss et al., 1989; Nicoletti et al., 1990).

Ionotropic receptors form ion channels and are subdivided into three further forms according to their respective agonists, i.e., kainic acid (KA), NMDA and α-amino-3-hydroxy-5-methyl-4-isoxazolepropionic acid (AMPA). KA and AMPA receptors are thought to subserve primarily rapid excitatory neurotransmission in the CNS and are mainly permeable to Na^+ and K^+ but not to Ca^{2+}. In contrast, NMDA receptors are permeable to Ca^{2+} and respond more slowly to glutamate. It is, therefore, believed that they will not function as mediators of rapid synaptic transmission but mediate the neurotoxic effects of glutamate in the adult brain (reviewed in Greenamyre and Porter, 1994).

A characteristic feature of the NMDA receptor is the so-called magnesium block, which must be released in a voltage-dependent manner for activation. It could be shown that glutamate becomes more neurotoxic when intracellular energy levels are reduced (Novelli et al., 1988). Since ATP levels are decreased in AD, neurons are more sensitive to the neurotoxic action of glutamate, because ATP production and functional Na^+, K^+-ATPases are necessary to generate a resting potential to maintain the voltage-dependent Mg^{2+} block of the NMDA receptor channel. Relief of the Mg^{2+} block enables the excitatory amino acids to act persistently at the NMDA receptor, resulting in its opening and subsequent neuronal damage.

Activation of NMDA receptors increases cytoplasmatic calcium levels and subverts of cytoplasmic Ca^{2+} homeostasis. The increased intracellular Ca^{2+} concentration leads to enhanced metabolic stress on mitochondria, resulting in excessive oxidative phosphorylation and production of ROS, e.g., via the activation of Ca^{2+}-dependent nitric oxide synthases (Toledo-Pereyra et al., 2004). Ca^{2+} toxicity is also mediated by activation of proteolytic enzymes like calpains, which are able to degrade essential proteins, or endonucleases that degrade DNA. $O_2^{\bullet-}$ and H_2O_2 are produced by xanthine oxidase, which arises from proteolytic cleavage of xanthine dehydrogenase by calcium-dependent enzymes to. Moreover, Ca^{2+} activates PLA_2 catalyzing the production of AA, which in turn is transformed by cyclooxygenases to prostaglandins, with $O_2^{\bullet-}$ formed as a by-product.

In summary, excitotoxicity also appears to involve oxidative stress, which makes an antioxidant therapy a promising alternative to NMDA receptor antagonists.

5   Neuroprotection

5.1   Introduction to the Principles of Neuroprotection

As discussed above, the role of oxidative stress in neurodegenerative disorders, such as PD and AD, is well documented. It should, therefore, be possible to provide neuroprotection by manipulating the redox state of the cells, e.g., by increasing the antioxidant capacity of the cells or by decreasing the pro-oxidative conditions in cells in the human brain. There are several therapeutic approaches for providing such neuroprotection, including supplementation of the patient with antioxidants, energy enhancing or anti-glutamatergic drugs. Antioxidants are used because of their ability to scavenge ROS and thus preventing damage to neurons. Since many plant-derived antioxidants are present in certain foods in different combinations, we will sometimes describe the positive effects of the foods rather than those of the isolated ingredients. In addition, energy supplementation is supposed to neuronal resistance to glutamate and ROS induced damages and initiate repair processes. Antiglutamatergic drugs will decrease pathophysiological calcium influx and calcium-related injury, including mitochondrial dysfunction. In the subsequent sections, a variety of natural and synthetic neuroprotective drugs or foods supplements are described, and evidence for their beneficial effects in cell culture and animal models of AD and PD as well as clinical evidence for their efficacy will be discussed.

5.2   Antioxidants

Antioxidants are low molecular weight substances which either block the production of free radicals or neutralize them by absorbing the unpaired electron, thereby turning into a radical that is less reactive. Endogenous antioxidant systems include antioxidative enzymes (e.g., glutathione peroxidase and catalase) and thiol based scavengers (e.g., glutathione). However, therapeutic manipulation of antioxidative enzymes is at a very early stage, since their expression is genetically determined and therefore difficult to increase by synthetic drugs without changing overall expression profiles. By contrast, exogenous antioxidants, such as vitamins C and E, have long been supplied with food in order to boost antioxidative defenses in age-related diseases. More and more foods, particularly those containing plant-derived antioxidants, are now offered by commercial vendors. Even in the absence of double-blind placebo controlled clinical trials, various antioxidants or antioxidant-rich plant extracts are therefore used as a therapeutic option in PD and AD.

5.2.1 Vitamin C

Occurrence   Vitamin C (ascorbate) is synthesized in the liver of most mammals from glucose. Guinea pigs, higher primates, and humans, however, have lost the ability to synthesize the vitamin and are relying on dietary supplementation (Rose, 1988). Vitamin C is found in high concentrations in many fruits and vegetables, including citrus fruits, kiwis, and tomatoes. The recommended daily allowance of vitamin C is 75 mg.

Function and Chemical Reactions   Vitamin C is a hydrophilic vitamin and can exist as a partially oxidized ascorbyl free radical or as dehydroascorbic acid (DHA) in the presence of mild oxidants, depending on how many electrons are lost. The latter form is very short lived (with a half-life of about 6 min), and it is reduced back to the monoanion form (e.g., by vitamin E), which is the most common form of ascorbate at physiological pH (Winkler, 1987). Hydrolysis of the lactone ring in DHA leads to the irreversible conversion to 2,3-diketo-1-gulonic acid (❷ *Figure 6-1*).

□ **Figure 6-1**

Oxidation of vitamin C (ascorbic acid) via dehydroascorbic acid as intermediate

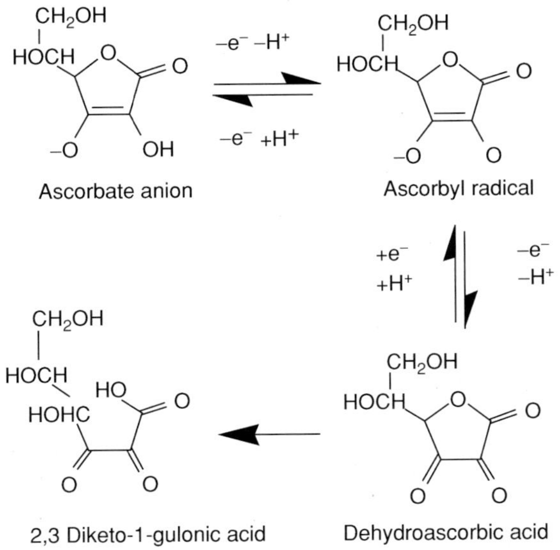

Ascorbate anion   Ascorbyl radical

2,3 Diketo-1-gulonic acid   Dehydroascorbic acid

Besides its functions in the synthesis of the amino acid hydroxyproline, as major building block of collagen in connective tissue and bones, and the absorption of iron, ascorbate plays a role as an antioxidant, protecting cells against oxidative stress, particularly in the cytosolic components of the cell, due to its reducing potential of its carbon–carbon double bond (❷ *Figure 6-1*). It mediates its antioxidant activity in the cytosol by donating two hydrogens and electrons to ROS like peroxyl and hydroxyl radicals, superoxide, singlet oxygen, and peroxynitrite (Sies, 1989). When associated with the plasma membrane, however, it is a powerful reductant of the α-tocopherol (vitamin E) radical.

Neuroprotective Effects in Cell Culture and Animal Models Since neurons have a tenfold higher oxidative metabolism than glial cells, vitamin C is an important antioxidant molecule in this cell type, and there is evidence that it also scavenges glutamate-generated ROS (Ciani et al., 1996). Using human cortical neurons, it could been shown that ascorbic acid inhibits apoptosis induced by H_2O_2 hydrogen peroxide, TNF-α, DA, or Aβ. This occurs through preventing the loss of mitochondrial membrane potential and DNA fragmentation (Medina et al., 2002). These findings are concurrent with the results from Lockhart et al. (1994), who have shown that vitamin C can attenuate the neurotoxic effect of β-amyloid in rat hippocampal cultures.

In PD, an increase in oxidative stress is observed by treatment of patients with levodopa, which is metabolized by monoaminooxidases in a radical-generating reaction. In an appropriate cell culture mode, levodopa proved toxic for the human neuroblastoma cell line NB69.

Tocopherol lacks significant preventive effect on levodopa toxicity, but ascorbic acid in millimolar concentrations prevents levodopa toxicity and quinone formation (Pardo et al., 1993). In a similar study, the role of ascorbic acid on DA oxidation-mediated cytotoxicity was studied using a different cell system, the neuronal cell line PC12. DA cytotoxicity was slightly attenuated by ascorbic acid, whereas the cytotoxicity of 6-hydroxy dopamine (6-OHDA), a DA oxidation product, was actually potentiated (Choi et al., 2000).

Neuroprotective Effects in Alzheimer's and Parkinson Patients Vitamin C, among other antioxidants, is often recommended for PD and AD patients. Epidemiological studies, however, yielded conflicting results. In the Rotterdam study, 5,395 participants, who did not display any signs or symptoms of AD, were analyzed in respect of the future onset of the disease. It was demonstrated that persons with a higher intake of vitamin C had a reduced risk of developing AD, compared to the ones with lower intake. An intake of 133 mg/day compared to 95 mg/day of vitamin C led to a 34% decrease in the development of AD (Engelhart et al., 2002). In contrast, the Chicago Health and Aging Project (CHAP), in which 3,888 healthy persons took part (of whom 815 persons, with the average age of 72 years, completed the study), found no relation between vitamin C intake and the time of onset of AD.

Plasma vitamin C in PD patients are significant lower than in controls, which may be evidence that oxidative stress is involved in PD (Rebec et al., 2003). However, no convincing positive epidemiological studies or clinical trials with vitamin C alone in PD patients have been reported so far.

5.2.2 Vitamin E

Occurrence   Vitamin E (α-tocopherol) is known as an essential nutrient since 1968. It is only produced in plants, particularly in wheat germ, plant oils, nuts, and soybeans. The daily requirement is reported as 30 mg/day, but this varies between ages and countries.

Function and Chemical Reactions   Vitamin E is a term for a group of tocopherols and tocotrienols of which α-tocopherol is the most active one. It is lipid-soluble and consists of a chromane ring with an isoprenoid side chain (❷ *Figure 6-2*).

◻ Figure 6-2
Chemical structure of vitamin E (tocopherol)

Since it is hydrophobic and concentrated in cell membranes, its main mechanism is the prevention of lipid peroxidation and other radical-driven oxidative events involving fatty acids (Tappel, 1972; Esterbauer et al., 1991). Vitamin E is regenerated by vitamin C, as was shown in both liposomal membrane systems and in homogeneous solutions (Niki, 1987).

Neuroprotective Effects in Cell Culture and Animal Models   It has been demonstrated that vitamin E has indirect neuroprotective effects through the suppression of microglial activation (Li et al., 2001). LPS-activated microglia showed attenuated expression of inflammatory mediators, such as IL-1α, TNF-α, or NO, when cocultured with vitamin E. Therefore, vitamin E has a therapeutic use in AD, since neurotoxic microglial–neuronal interactions are implicated in the pathogenesis in this type of dementia.

Further relationship between AD and vitamin E is the prevention of protein oxidation and ROS production induced by β-amyloid. It has been shown that this does not occur by inhibiting the Aβ fibril formation but rather by scavenging the Aβ associated free radicals (Yatin et al., 2000).

In a PD cell culture model, DA triggered apoptosis in PC12 cells, only the thiol-containing compounds, reduced glutathione (GSH), N-acetyl-cysteine (NAC), and dithiothreitol (DTT) were markedly protective, while vitamin E had no effect. The thiol antioxidants and vitamin C but not vitamin E prevented DA autooxidation and production of DA melanin (Offen et al., 1997).

Neuroprotective Effects in Alzheimer and Parkinson Patients   The effect of vitamin E in AD patients was investigated in the Rotterdam study. A daily intake of 15.5 mg of vitamin E compared to an intake less than 10.5 mg/day decreases the risk of developing AD to 43% (Engelhart et al., 2002). This finding was further confirmed in the CHAP, where a vitamin E intake of more than 11.4 IU/day resulted in a 70% reduced risk for AD development compared to an intake of less than 6.21 IU/day. The positive effect of vitamin E was not only significant but also dose dependent and seemed to have an increased protective role when taken over a longer period of time (Morris et al., 2002). In a further study, the combination of vitamin E and C in AD patients was evaluated. Use of vitamin E and C (ascorbic acid) supplements in combination was associated with reduced AD prevalence (adjusted odds ratio, 0.22; 95% confidence interval, 0.05–0.60) and incidence (adjusted hazard ratio, 0.36; 95% confidence interval, 0.09–0.99). A trend toward lower AD risk was also evident in users of vitamin E and multivitamins containing vitamin C, but no evidence of a protective effect with use of vitamin E or vitamin C supplements alone, with multivitamins alone, or with vitamin B–complex supplements was observed.

In an epidemiology study, a total of 371 incident PD cases were ascertained in the Nurses' Health Study, which comprised 76,890 women who were followed for 14 years, and the Health Professionals Follow-Up Study, which comprised 47,331 men who were followed for 12 years. In this study, the risk of PD was significantly reduced among men and women with high intake of dietary vitamin E. Since the beneficial effects were only seen with vitamin E from foods (not from synthetic sources), the authors suggest that other constituents of foods rich in vitamin E may be protective (Zhang et al., 2002).

In contrast, the community-based Rotterdam Study in the Netherlands collected data of 5,342 independently living individuals without dementia between 55 and 95 years of age, including 31 participants with PD. In this study, vitamin E was protective, and the association of the relative risk with vitamin E intake was dose dependent (de Rijk et al., 1997).

Vitamin E (and deprenyl) was tested by the Parkinson Study Group in the "Deprenyl and Tocopherol Antioxidative Therapy of Parkinsonism" (DATATOP) trial. However, deprenyl delayed the onset of disability associated with early, otherwise untreated Parkinson's disease, however, in contrast to the expectation of the authors, α-tocopherol proved to be ineffective in the DATATOP study (Miklya et al., 2003).

5.2.3 Lipoic Acid

Occurrence   Lipoic acid (LA) was first identified as a component of several human enzyme systems involved in the conversion of carbohydrates and fats into energy such as pyruvate dehydrogenase, α-ketoglutarate dehydrogenase. LA was not classified as a vitamin because some small amounts can be synthesized by the human body. However, lipoic as a therapeutic drug comes from chemically synthesized sources since amounts of 600 mg cannot be taken up with the normal diet.

Function and Chemical Reactions   Most antioxidants are either water or fat soluble. A few, such as LA, are both. This dual nature allows LA to function in both fatty and aqueous environment, an ability that is the reason why LA is often termed "universal antioxidant." LA provides therapeutic benefits in numerous oxidative stress related conditions and diseases including diabetes and heart disease. LA and its reduced form, dihydrolipoic acid (DHLA), show the ability to directly quench a variety of ROS (singlet oxygen, superoxides, peroxyl and hydroxyl radicals, hypochlorite, and peroxynitrite), inhibit reactive oxygen generators and spare and regenerate other antioxidants. LA can also regenerate a variety of antioxidants including glutathione, vitamins C and E, and the mitochondrial antioxidant coenzyme Q10 (Atwood et al., 2004).

Neuroprotective Effects in Cell Culture and Animal Models   LA can protect cortical neurons against cell death induced by Aβ and H_2O_2. It could be shown that this occurs through the Akt signaling pathway (Zhang et al., 2001). In a further study, the effects of LA and its reduced form, DHLA, in neuron cultures treated with amyloid β-peptide (Aβ 25–35), were studied. Pretreatment of dissociated primary hippocampal cultures with LA significantly protected against Aβ toxicity. In contrast, concomitant treatment of cultures with LA and Fe/H$_2$O$_2$ significantly potentiated the toxicity, most likely because LA has too be reduced to DHLA before exerting an antioxidant effect (Lovell et al., 2003).

Using PC12 cells as a cell model in PD, it could been shown that pretreatment with R-LA prevents depletion of the thiol compound GSH, impairment of mitochondrial complex I and ROS generation (Bharat et al., 2002).

Neuroprotective Effects in Alzheimer and Parkinson Patients Oxidative stress and energy depletion are characteristic biochemical hallmarks of AD, thus antioxidants with positive effects on glucose metabolism, such as LA, should exert positive effects in these patients. Therefore, 600-mg LA was given daily to patients with AD and related dementias (receiving a standard treatment with acetylcholinesterase inhibitors) in an open study over an observation period of, on average, 337±80 days. The treatment led to a stabilization of cognitive functions in the study group, demonstrated by constant scores in two neuropsychological tests (mini-mental state examination: MMSE and AD assessment scale, cognitive subscale: ADAScog) (Hager et al., 2001). Despite the encouraging data from cell culture models, no positive study with LA in PD patients has been reported so far.

5.3 Polyphenols

Polyphenols are secondary plant compounds and can be subdivided into different compound classes depending on the substituents attached to the three ring systems. All polyphenols share a common structure with two phenolic ring structures which are capable of scavenging free radicals. Four classes of polyphenols are particularly interesting as antioxidant drugs in human nutrition: flavanols, flavonols, and anthocyanidins (❍ *Figure 6-3*). Polyphenols have developed in plants in response to stress, including protection against ultraviolet light.

❑ Figure 6-3

Chemical structures of polyphenols including three classes of flavonoids (flavonols, flavanols, and anthocyanidins) and trihydroxystilben

Flavanols

Flavanols

Anthocyanidins

Trihydroxystilbenes

Plants often contain several different polyphenols, and it is difficult to judge which of them is the most potent antioxidant or neuroprotective compound in a plant derived food or beverage. Therefore, the next chapters can only describe the neuroprotective property of these food or plant preparations rather than been able to attribute these effects to a single polyphenol alone.

5.3.1 Flavanols (Catechins, Epicatechins)

Occurrence   Catechins are flavanols of which epicatechin (EC), (-)-epigallocatechin (EGC), and (-)-epicatechin gallate (EGCG) are the main ones present in fresh tea leaves (Higdon and Frei, 2003). Daily intakes in humans are up to 100 mg/day. Sources of flavanols are wine, apples, chocolate, and black and green tea. One cup (237 ml) of green tea contains typically 30–130 mg of EGCG while black tea contains only 0–70 mg of EGCG (❯ *Figure 6-4*).

◻ **Figure 6-4**
Chemical structures of catechins (a type of flavanol), main antioxidant ingredients of green tea

(–)–Epicatechin

(–)–Epigallocatechin

(–)–Epicatechin gallate

(–)–Epigallocatechin gallate

Function and Chemical Reactions   It is documented that catechins scavenge free radicals in vitro systems (Rice-Evans, 1999). Catechins are able to scavenge •NO in vitro (Haenen and Bast, 1999) as well ONOO⁻ (Pannala et al., 1997), even if the mechanisms of scavenging radicals is still unknown. Singlet oxygen ($^1O_2$), $O_2^{\bullet-}$, •OH, and peroxyl radicals •OOH are further radicals that are scavenged by catechins (Guo et al., 1999; Nanjo et al., 1999; Zhao et al., 2001).

Their superb antioxidant activity is because of a low reduction potential compared to vitamin E, but not to vitamin C. Therefore, EGCG and EGC are superior electron donors, which is concurrent with a better antioxidant activity.

Neuroprotective Effects in Cell Culture and Animal Models   Catechins in green tea are also capable of decreasing lipid peroxidation. Using pulse radiolysis, Bors and Michel (1999) have found that catechins rather than flavonols and flavones are the antioxidative compounds in tea. Catechins increase the activity of glutathione peroxidase and glutathione reductase, while decreasing the levels of lipid hydroperoxidases (LOOH), 4-hydroxysynonenal (4-HNE), and MDA (Skrzydlewska et al., 2002). The peroxidase activity of lipoxygenase and cyclooxygenase is one mechanism to increase oxidative stress, which can be inhibited by flavonoids and phenolic antioxidants in vitro (Laughton et al., 1991). Both enzymes increase oxidative stress because they take part in the synthesis of leukotrienes and prostaglandins, two molecules involved in inflammation.

Xanthine oxidase is another enzyme inhibited by catechins. This enzyme reduces O_2 to $O_2^{\bullet-}$ and H_2O_2 during the oxidation of xanthine and hypoxanthine to uric acid (Higdon and Frei, 2003), catechins however inhibit the production of these ROS. But besides the inhibition of enzymes, catechins are also able to induce some enzymes including phase II enzymes for detoxification or antioxidant enzymes such as glutathione peroxidase (GPx) (Khan et al., 1992), catalase, and superoxidase dismutase (SOD) (Agarwal et al., 1993).

EGCG has been shown to exert protective effects against $A\beta$-induced neurotoxicity and regulate secretory processing of nonamyloidogenic APP. In a cell culture model of AD, EGCG was shown to be able to protect PC12 cells against $A\beta$ toxicity in a dose-dependent manner. In addition, EGCG enhances (approximately sixfold) the release of the nonamyloidogenic and neurotrophic form of the amyloid precursor protein (sAPPα) into the conditioned media of human SH-SY5Y and PC12 neuroblastoma cells (Levites et al., 2003).

A further function of catechins in tea is the inhibition of upregulation of enzymes that propagate inflammation resulting in cellular oxidative stress. For example, it has been demonstrated that both green and black tea inhibit LPS-induced iNOS expression in macrophages by inhibiting the transcription factor for iNOS, which is NF-κB on its DNA binding activity and through phosphorylation of its inhibitor IκB (Lin and Lin, 1997; Lin et al., 1999).

The purpose of a further study was to investigate potential neuroprotective effects of tea extracts and possible signal pathway involved in a neuronal cell model of Parkinson's disease. 6-OHDA activated the iron-dependent inflammatory redox sensitive nuclear factor-κB in rat pheochromocytoma (PC12) and human neuroblastoma (NB) SH-SY5Y cells, respectively. Immunofluorescence and electromobility shift assays showed increased nuclear translocation and binding activity of NF-κB after exposure to 6-OHDA in NB SH-SY5Y cells, with a concomitant disappearance from the cytoplasm. Introduction of green tea extract before 6-OHDA inhibited both NF-κ nuclear translocation and binding activity induced by this toxin in NB SH-SY5Y cells. The authors conclude that neuroprotection was attributed to the potent antioxidant and iron chelating actions of the polyphenolic constituents of tea extracts, preventing nuclear translocation and activation of NF-κB (Levites et al., 2002).

In an animal model of Parkinson's disease (PD), the effect of green tea toxicity of 1-methyl-4-phenyl-1,2,3,6-tetrahydropyridine (MPTP) was investigated in mice. Both tea and the oral administration of EGCG prevented the loss of tyrosine hydroxylase (TH)-positive cells in the SN and of TH activity in the striatum. These treatments also preserved striatal levels of DA and its metabolites, 3,4-dihydroxyphenylacetic acid and homovanillic acid (HVA). Also tea plus MPTP and EGCG plus MPTP treatments decreased expressions of neuronal NO synthase (nNOS) at the similar levels of EGCG treatment group (Choi et al., 2002).

Neuroprotective effects in Alzheimer and Parkinson Patients   In a study in the northern health region of England, risk factors for presenile AD (onset blow 65 years of age) were analyzed in 109 patients. In this study, which was originally aimed to assess the impact of aluminum of dementia, no significant relationship between AD and exposure to tea and antacids was found (Forster et al., 1995).

A protective role of tea consumption for PD has been shown only in Asian countries, where green tea is a major beverage. A study in Hong Kong in which 215 PD patients and 313 controls took part found that a regular tea drinking is protective against PD (Chan et al., 1998). In a study undertaken in Singapore, the relationship between coffee and tea drinking, cigarette smoking, and other environmental factors and risk of PD among 300 PD and 500 population (ethnic Chinese) was analyzed. The authors calculated that three cups/day of tea for 10 years would lead to 28% risk reduction of PD (Tan et al., 2003). However, a study in France showed the opposite. One hundred and forty patients and 240 hospital controls showed an increased risk by 90% to develop this dementia (Preux et al., 2000). It is also reported from a study in Spain that there was no significant association between tea consumption and the risk of PD (Morano et al., 1994).

5.3.2 Flavonols (Quercetin, Kaempferol)

Occurrence   Flavonols including quercetin and kaempferol are characterized by a keto group at the 4-position in the heterocyclic C ring. They are present in red wine, onions, apples, tea and broccoli, and in the leaves of the Gingko biloba tree. In a normal European diet, daily flavonol intakes of 3–64 mg have been reported. The Ginkgo tree has evolved strategies to live for up to 1,000 years (Bilia, 2002). Because of the high antioxidant content of its leaves, the pharmaceutical industry has developed the standardized leaf extract EGb 761 to treat vascular and neurodegenerative diseases. This extract consists of 24% flavone glycosides (also called ginkgosides, mainly quercetin, kaempferol, and isorhamnetin glycosides) and 6% lactone terpenes, consisting of ginkgolides A, B, C, and J as well as bilobalide. The amount of ginkgolic acids is less than 0.0005% because it has been described that 100 μM of it causes allergic symptoms and neuronal cell death (Ahlemeyer et al., 2001; Chatterjee et al., 2003).

Neuroprotective Effects in Cell Culture and Animal Models   It has been shown that quercetin or structurally similar flavonoids are potent neuroprotective antioxidants in red wine or related fruit derived foods. It is thought that the presence of 2,3 unsaturation together with an oxo function at position 4 in the ring C, coplanarity of the molecule and a 3′,4′-dihydroxy catechol structure in the B ring are responsible for the antioxidant scavenger activity (Dajas et al., 2003). The standardized leaf extract EGb 761 is mainly used for the treatment of vascular diseases and for neurodegenerative dementia. The antioxidant effects are mediated by the ginkgosides (Bedir et al., 2002); however, terpenes are also thought to contribute via their anti-inflammatory effects (Kidd, 1999). It has been demonstrated in rats that the protein level and activity of antioxidant enzymes is increased by EGb 761, and it is therefore believed that this drug exerts its effect by stabilization of the redox status as by radical scavenging (Colak et al., 1998; Bridi et al., 2001). EGb 761 is also able to scavenge NO (Marcocci et al., 1994). Since increased nNOS and iNOS expression is found in AD, EGb 761 has been proposed as therapeutic treatment for this type of dementia conditions. Oxidative damage caused by superoxide leakage from mitochondrial respiration can be decreased by EGb 761, most likely by increasing the messenger RNA (mRNA) and protein level of subunit I of mitochondrial nicotine amide dinucleotide phosphate (NADH) dehydrogenase.

There is also plenty of evidence for neuroprotective activities of Gingko biloba extracts in AD specific disease models. Aβ, the main compound of senile plaques in AD, originates from a larger APP, the APP. This protein can then either be cleaved by β- and γ-secretases, as described the neuropathology and biochemistry section of AD, or by α-secretases that release a large and soluble extracellular domain of APP, the so-called αAPPs, that represents the nontoxic form of Aβ. Using hippocampal slices cultures, it could be shown that EGb761 increases αAPPs release (Colciaghi et al., 2004).

EGb761 has also been shown to decrease ROS in a neuroblastoma cell line that stably expresses an AD-associated double mutation, which exhibits both increased secretion and intracellular accumulation of Aβ when stimulated. This effect could also been shown in vivo using *Caenorhabditis elegans* that constitutively expresses human Aβ. Both, the cell line and *C. elegans* showed a rise in levels of H_2O_2 compared with wild-type controls, before stimulating (Smith and Luo, 2003).

The effect of the flavonol kaempferol on Aβ-induced toxicity in PC12 neuroblastoma and T47D human breast cancer cells was investigated, and a protective effect of kaempferol (comparable to that observed with estradiol) was observed. Since the effects of the weak estrogen receptor agonists α-estradiol and kaempferol were found to be similar to the effects of the strong estrogen receptor agonist β-estradiol, the authors suggested a mode of action independent from the nuclear estrogen receptor such as radical scavenging (Roth et al., 1999). A beneficial effect of extracts of Ginkgo biloba leaves (EGb) was observed in the Parkinson's disease (PD) model induced by MPTP. MPTP was microinjected into SN of rats to induce a behavior change of rotation. EGb treatment decreased the duration and frequency of the rotation of rats as well as the level of the lipid peroxidation product MDA. In addition, the decrease of DA levels was attenuated, which indicates increased viability of dopaminergic neurons (Yang et al., 2001). Similar positive effects were observed in the 6-OHDA model, where neurotoxicity in the nigrostriatal dopaminergic system of the rat brain was attenuated by the EGb extract (Kim et al., 2004).

Neuroprotective Effects in Alzheimer and Parkinson Patients   A study with 216 patients having mild-to-moderate degenerative dementia of AD or multi-infarct dementia was carried out. Patients were treated with 240 mg of EGb 761 or placebo for 24 weeks. Using memory functions, independence, coping skills, and other parameters as the basis of measures, it was concluded that EGb 761 has clinical efficacy in AD patients (Kanowski et al., 1996). In another study, the potential association between the use EGb761 and AD was investigated. A case-control study was nested in a cohort of 1,462 community-dwelling elderly woman aged over 75 years. Sixty-nine women with AD were compared with 345 women whose cognitive function remained normal. Analysis revealed that fewer women who developed AD had been prescribed EGb761 for at least 2 years, but to establish a causal relationship, prospective studies have to be carried out to confirm these findings (Andrieu et al., 2003).

5.3.3 Anthocyanins (Cyanidine and Others)

Occurrence   Anthocyanins are natural red and blue colors of plants and fruits, including various berries (strawberries, blueberries, raspberries, black currants) and red grapes (❷ *Figure 6-3*). The average daily intake in humans is about 200 mg/day.

Function and Chemical Reactions   Anthocyanins also belong to the class of polyphenols and consist of three carbon rings with two aromatic (A and B) and one O-heterocyclic ring, which is positively charged (❷ *Figure 6-3*).

Neuroprotective Effects in Cell Culture and Animal Models   The effects of purple sweet potato anthocyanin (SPA) on lipid peroxidation, 1,1-diphenyl-2-picryl-hydrazyl (DPPH) radicals and cognitive deficits were examined. SPA was shown to exhibit DPPH radical scavenging activities and to effectively inhibit lipid peroxidation initiated by Fe^{2+} and ascorbic acid in rat brain homogenates. Furthermore, SPA markedly enhanced cognitive performance, assessed by passive avoidance test in ethanol-treated mice. These results demonstrate that anthocyanin prepared from purple sweet potato exhibits memory enhancing effects, which may be associated with its antioxidant properties (Cho et al., 2003). In another study, Fischer 344 rats received dietary supplementation with fruit or vegetable extracts high in anthocyanins (e.g., blueberry (BB) or spinach, respectively). This diet decreased the age-related vulnerability to oxidative stress in these rats as assessed in vivo by examining reductions in neuronal signaling and behavioral deficits and in vitro via H_2O_2-induced decrements in striatal synaptosomal calcium buffering. Examinations have also revealed that BB supplementations are effective in antagonizing other age-related changes in brain and behavior, as well as decreasing indices of inflammation and oxidative stress in muscles. The authors conclude that anthocyanins show the most efficacy in penetrating the cell membrane and in providing antioxidant protection (Galli et al., 2002). A diet rich in blueberries was also shown to prevent behavioral deficits in an AD model, the APP overexpressing transgenic mouse. The authors believe that the protective effect is due to the enhancement of redox-sensitive memory-associated neuronal signaling and alterations in neutral sphingomyelin-specific PLC activity (Forbes et al., 2003).

5.3.4 Hydroxystilbenes (Resveratrol)

Occurrence   Resveratrol (3,4', 5-trihydroxystilbene) is natural polyphenol, which differs in structure from the flavonoids having only two (aromatic) ring systems (❷ *Figure 6-3*). It is synthesized by plants in response to fungal infection and helps the plant to survive due to its strong antifungal properties. In human beverages, high amounts of resveratrol are found in red wine—highlighting an interesting fact that fungi indirectly increase the beneficial effect of a food or beverage (❷ *Figure 6-4*). Because it is released from the skin of grapes, the fermentation time is a major factor in its effectiveness. White wine (which contains very low levels of anthocyanins) has also a lower content of resveratrol compared to red wine due to its shorter maceration time.

Function and Chemical Reactions   The effects of resveratrol include inhibition of lipid peroxidation, chelation of copper, free-radical scavenging, and anti-inflammatory activity (reviewed) (Fremont, 2000).

Neuroprotective Effects in Cell Culture and Animal Models   Resveratrol also exerts its antioxidative action by enhancing the intracellular free-radical scavenger glutathione. Reveratol was able to protect SH-SY5Y neuroblastoma cells from Aβ and H_2O_2-induced cell death and was also able to increase the level of reduced glutathione (Savaskan et al., 2003). In a further study, the authors hypothesized that lipoproteins (LP) enhance the Aβ-mediated toxicity to neurons and that uptake of oxidized LP by neuron leads to an acceleration of intracellular oxidative pathways and deteriorate cell death. Using PC12 cells, it could be shown that the combination of Aβ and oxidized LP leads to an enhanced cell death. However, this effect could be ameliorated by resveratrol, showing that it acts as a potent neuroprotective antioxidant (Sun et al., 2001). Furthermore, the combination of *trans*-resveratrol and vitamin C and/or E were more effective in protecting cells than was any of these three antioxidants alone (Chanvitayapongs et al., 1997).

Neuroprotective Effects in Alzheimer and Parkinson Patients   Moderated consumption of red wine—despite a high-fat intake—is considered to be the cause of a low incidence of cardiovascular disease in France, the so-called "French paradox."

In line with this type of studies, a prospective community study in the Bordeaux area in France has demonstrated that the consumption of red wine significantly decreases the risk of AD. 3,777 seniors (65 years and older) were observed in respect of their alcohol consumption and onset of dementia. In moderate drinkers who had 250–500 ml of red wine per day, red wine seemed to have a strong favorable effect since the odds ratio was 0.28 compared to nondrinkers (Orgogozo et al., 1997). In a similar study, the Washington Heights Inwood-Columbia Aging Project, the intake of up to three daily servings of wine was associated with a lower risk of AD (hazard ratio = 0.55) but intake of liquor, beer, and total alcohol was not associated with a lower risk of AD (Luchsinger et al., 2004).

5.4   Energy Supplementation

There is evidence that bioenergetic dysfunction plays a role in many neuropathological disorders. The mitochondrial respiratory chain and oxidative phosphorylation system are responsible for the production of ATP and defects in energy production are therefore thought to be a cause for neurodegeneration. The link pathways between energy depletion and neurodegenerative diseases are various, including free-radical generation, impaired calcium buffering, and the mitochondrial permeability transition, all leading to cell death. When neutralization of radicals has failed, energy supplementation is another opportunity to ensure that there is enough ATP for survival and for the repair of the damage occurring in the cell. Therefore, the application of bioenergetic drugs, including creatine and pyruvate, is a further option for neuroprotective treatment.

5.4.1 Creatine

Occurrence   Creatine (CR) can be produced endogenously by various organs, such as pancreas, liver, and kidney using arginine, glycine, and methionine (❷ *Figure 6-5*), but can also be taken up by ingestion of

☐ Figure 6-5

Chemical structures of the proenergetic compounds creatine and pyruvate

Creatine   Pyruvate

meat products (Juhn and Tarnopolsky, 1998) or as a supplement in the form of creatine monohydrate (CM). Latter is commercially available as a water soluble powder, in capsule or chewable. The daily recommended dose of creatine is about 5–10 g (Tarnopolsky and Beal, 2001).

Function and Chemical Reactions A healthy brain contains approximately 4.5-mM phosphocreatine (PCr) (Bottomley et al., 1992), which serves as an energy buffer in muscles and brain through transferring the phosphoryl group to ADP, catalyzed by cytosolic creatine kinase (CK).

Neuroprotective Effects in Cell Culture and Animal Models It has been documented that creatine concentrations are decreased in AD (Pettegrew et al., 1994). CR can also function together with PCr as an energy buffer between the cytosol and mitochondria. Mitochondrial creative kinase (MtCK) is the mitochondrial isoform of CK and exists in an octameric and dimeric form. MtCK in its octameric form inhibits the opening of the mitochondrial permeability transition pore (MTP) (Brdiczka et al., 1998), but dimerizes in the presence of radicals, allowing the MTP to open and impair mitochondrial respiration leading to cell death (O'Gorman et al., 1997). It has been shown that creatine attenuate these negative effects (Pulido et al., 1998). Furthermore, creatine administration has been found to protect against glutamate and amyloid toxicity in rat hippocampal neurons in vitro (Brewer and Wallimann, 2000).

Neuroprotective Effects in Alzheimer and Parkinson Patients It has been documented that creatine concentrations are decreased in AD (Pettegrew et al., 1994). However, no data of clinical trials for creatine supplementation for AD or PD have been published so far.

5.4.2 Pyruvate

Function and Chemical Reactions A common feature of AD and PD is an impairment of glucose metabolism (Beal, 1992). The resulting decrease in pyruvate (❯ *Figure 6-5*), which is an intermediate in glycolyis, is thought to cause a deficit of energy, which could be ameliorated by supplementation from exogenous sources. Pyruvate acts also as an antioxidant. Pyruvate and other α-ketoacids can react nonenzymatically with H_2O_2 through a reaction in which carbon dioxide is liberated, and the α-ketoacid is converted into the corresponding carboxylic acid (Desagher et al., 1997).

$$R - RCCOOH + H_2O_2 \rightarrow RCOOH + CO_2 + H_2O$$

Pyruvate is shuttled by the H^+-monocarbonylate cotransporter between the intracellular and extracellular space and can therefore scavenge H_2O_2 inside and outside the cell (Poole and Halestrap, 1993; Garcia et al., 1994). However, the neuroprotective effects of pyruvate are decreased in high concentration since this leads to intracellular acidification, mainly generated by the H^+ cotransport across the plasma membrane (Nedergaard and Goldman, 1993; Areosa and Sherriff, 2003) and to a minor extent, from undissociated pyruvic acid (Bakker and van Dam, 1974).

Neuroprotective Effects in Cell Culture and Animal Models Mechanisms responsible for the neuroprotective effects of pyruvate are controversial. It was suggested that the neuroprotective effect of pyruvate is more likely due to its antioxidant effect rather through its improvement of energy metabolism because the neuroprotection could be reproduced with other α-ketoacids (which are able to react with H_2O_2) but not with lactate, another neuronal energy substrate (Desagher et al., 1997). However, it was also shown that the neuroprotective effect of pyruvate is not related to its ability to undergo a decarboxylation process by counteracting the deep decrease in the neuronal ATP content, indicating the neurons are protected by rescuing the cellular energy charge (Maus et al., 1999). It was demonstrated that increased glutamate release, such as that in excitotoxity, stimulates glycolysis in astrocytes and that the endproduct pyruvate may contribute to neuronal protection (Pellerin and Magistretti, 1994). In addition, glial cells release more of the energy substrate lactate in the presence of glutamate. The release of glutamate from nerve terminals may therefore cause a negative feedback, increasing neurotoxicity (Tsacopoulos and Magistretti, 1996).

Neuroprotective Effects in Alzheimer and Parkinson Patients Pyruvate administration for therapeutic use may provide benefit for patients because unlike exogenous catalase, pyruvate and other α-ketoacids can cross the blood–brain barrier (Oldendorf, 1973; Garcia et al., 1994). However, until now clinical trials with pyruvate in AD or PD patients are still missing.

5.5 NMDA Antagonists

5.5.1 Function and Chemical Reactions

Since excitotoxicity plays a role in AD as described above, several NMDA antagonists have been developed for the treatment of neurodegenerative disorders. Synthetic drugs like memantine, MK-801, or dizocilpine have made into different stages of clinical development, also natural components like kynurenic acid or quinolinic acid. However, the most promising NMDA receptor antagonist so far is memantine (❯ *Figure 6-6*). Memantine acts as a noncompetitive (channel-blocking) antagonist. It inhibits pathopysiological Ca^{2+} influx without blocking physiological activation of the NMDA receptor, which is important for learning and memory. Although the mode of action is not clarified, yet it seems that memantine is combined and released with the ion channel depending on electric potential in the same way as the magnesium ion (Kato, 2004). The main advantage of memantine is its low-moderate affinity in comparison to phencyclidine

□ **Figure 6-6**
Chemical structure of the *N*-methyl-ᴅ-aspartate NMDA-receptor antagonist memantine, a 1-aminoadamantan derivative

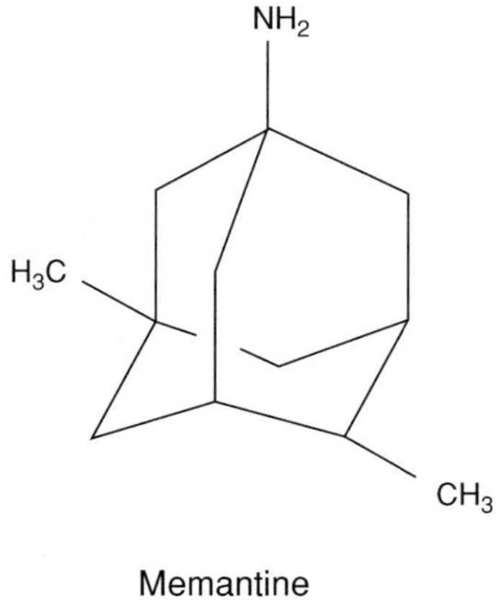

Memantine

(PCP) and MK-801, which are high-affinity NMDA receptors antagonists and are therefore not practical due to adverse side effects such as schizophrenic symptoms (Medvedev et al., 2004). Kynurenic acid is a natural metabolite of tryptophan and can antagonize non-NMDA as well as NMDA receptor activation (Stone, 2001).

5.5.2 Neuroprotective Effects in Cell Culture and Animal Models

In an AD animal model, Sprague-Dawley rats received memantine and β-amyloid 1–40 [Aβ(1–40)]. Aβ(1–40) injections into hippocampus led to neuronal loss in the CA1 subfield, evidence of widespread apoptosis, and astrocytic and microglial activation and hypertrophy. Memantine treated animals had significant reductions in the amount of neuronal degeneration, pyknotic nuclei and glial fibrillary acidic protein (GFAP) immunostaining as compared with vehicle treated animals. These data suggest that memantine, at therapeutically relevant concentrations, can protect against neuronal degeneration induced by Aβ (Miguel-Hidalgo et al., 2002).

Memantine was able to abolish toxic effects of glutamate in a rat model with parkinsonian syndromes. At a concentration as low as 5-mg/kg body weight, memantine significantly prevented the development of oligokinesea and muscular ricidity caused by application of the neurotoxin MPTP (Kucheryanu and Kryzhanovskii, 2000).

5.5.3 Neuroprotective Effects in Alzheimer and Parkinson Patients

After successful clinical trials, memantine is now used as a therapeutic drug for AD in the EU and USA. In one of the clinical trials, 166 AD patients were treated with 10 mg/day of memantine. It was shown that there was an improvement in the care dependence subscore on the Behaviorial Rating Scale for Geriatric Patients (BGP) of 3.1 points compared to 1.1 points in the placebo group (Areosa and Sherriff, 2003). These data were consistent with another study, in which Reisberg and coworker analyzed 252 patients with moderate-to-severe AD, who were treated with 20 mg of memantine daily for 28 weeks (Reisberg et al., 2003). According to the result of the Clinician's Interview-Based Impression of Change Plus (CIBIC-Plus), the change from baseline was significantly better with memantine compared to placebo. Two more studies reported statistically significant benefits of memantine. In a US-28-week placebo-controlled clinical trial, memantine was determined for clinical efficacy and safety for people with AD. 20 mg/day of memantine caused a positive effect on cognition, mood, and behavior, however clinical detectable changes could not be observed (Areosa and Sherriff, 2003). In a further European trial, memantine was investigated in severely demented nursing home patients suffering from AD. Memantine-treated patients showed a functional and global improvement, reduced care dependency, compared to placebo-treated persons (Winblad and Poritis, 1999).

Positive effects of NMDA receptor antagonists, such as memantine, have also been reported for PD patients. However, rather than being neuroprotective (which can only be demonstrated in long-term studies), these drugs are rather beneficial in PD patients by alleviating characteristic PD symptoms such as dyskinesia, suggesting an effect on glutamatergic signal transduction. In an open-fashion conducted study, 5 out of 14 PD patients improved their main Parkinson features, while 6 further patients of this group improved their "off" episodes (Rabey et al., 1992). In a further case report, memantine was used with three cognitively impaired, dyskinetic parkinsonian patients, and two seemed to benefit from this medication regarding their dyskinesia (Lokk, 2004).

Acknowledgments

We thank Heiner Körner and Peter Riederer for their helpful discussion. This study was supported by a JCU Career Development Grant, the Alzheimer Forschungs Initiative e.V and the DFG (Mu 1011–13).

References

Ahlemeyer B, Selke D, Schaper C, Klumpp S, Krieglstein J. 2001. Ginkgolic acids induce neuronal death and activate protein phosphatase type-2C. Eur J Pharmacol 430: 1-7.

Alam ZI, Daniel SE, Lees AJ, Marsden DC, Jenner P, et al. 1997. A generalised increase in protein carbonyls in the brain in Parkinson's but not incidental Lewy body disease. J Neurochem 69: 1326-1329.

Albin RL, Greenamyre JT. 1992. Alternative excitotoxic hypotheses. Neurology 42: 733-738.

Andrieu S, Gillette S, Amouyal K, Nourhashemi F, Reynish E, et al. 2003. Association of Alzheimer's disease onset with ginkgo biloba and other symptomatic cognitive treatments in a population of women aged 75 years and older from the EPIDOS study. J Gerontol A Biol Sci Med Sci 58: 372-377.

Areosa SA, Sherriff F. 2003. Memantine for dementia. Cochrane Database Syst Rev 3: CD003154.

Atwood CS, Perry G, Zeng H, Kato Y, Jones WD, et al. 2004. Copper mediates dityrosine cross-linking of Alzheimer's amyloid-beta. Biochemistry 43: 560-568.

Bains JS, Shaw CA. 1997. Neurodegenerative disorders in humans: the role of glutathione in oxidative stress-mediated neuronal death. Brain Res Brain Res Rev 25: 335-358.

Bakker EP, van Dam K. 1974. The movement of monocarboxylic acids across phospholipid membranes: evidence for an exchange diffusion between pyruvate and other monocarboxylate ions. Biochim Biophys Acta 339: 285-289.

Beal MF. 1992. Does impairment of energy metabolism result in excitotoxic neuronal death in neurodegenerative illnesses? Ann Neurol 31: 119-130.

Bedir E, Tatli II, Khan RA, Zhao J, Takamatsu S, et al. 2002. Biologically active secondary metabolites from Ginkgo biloba. J Agric Food Chem 50: 3150-3155.

Bharat S, Cochran BC, Hsu M, Liu J, Ames BN, et al. 2002. Pre-treatment with R-lipoic acid alleviates the effects of GSH depletion in PC12 cells: implications for Parkinson's disease therapy. Neurotoxicology 23: 479-486.

Bilia AR. 2002. Ginkgo biloba L. Fitoterapia 73: 276-279.

Bors W, Michel C. 1999. Antioxidant capacity of flavanols and gallate esters: pulse radiolysis studies. Free Radic Biol Med 27: 1413-1426.

Bottomley PA, Cousins JP, Pendrey DL, Wagle WA, Hardy CJ, et al. 1992. Alzheimer dementia: quantification of energy metabolism and mobile phosphoesters with P-31 NMR spectroscopy. Radiology 183: 695-699.

Braak H, Braak E. 1995. Staging of Alzheimer's disease-related neurofibrillary changes. Neurobiol Aging 16: 271-278; discussion 78–84.

Brdiczka D, Beutner G, Ruck A, Dolder M, Wallimann T. 1998. The molecular structure of mitochondrial contact sites. Their role in regulation of energy metabolism and permeability transition. Biofactors 8: 235-242.

Brewer GJ, Wallimann TW. 2000. Protective effect of the energy precursor creatine against toxicity of glutamate and beta-amyloid in rat hippocampal neurons. J Neurochem 74: 1968-1978.

Bridi R, Crossetti FP, Steffen VM, Henriques AT. 2001. The antioxidant activity of standardized extract of Ginkgo biloba (EGb 761) in rats. Phytother Res 15: 449-451.

Butterfield DA, Griffin S, Münch G, Pasinetti GM. 2002. Amyloid beta-peptide and amyloid pathology are central to the oxidative stress and inflammatory cascades under which Alzheimer's disease brain exists. J Alzheimers Dis 4: 193-201.

Chan DK, Woo J, Ho SC, Pang CP, Law LK, et al. 1998. Genetic and environmental risk factors for Parkinson's disease in a Chinese population. J Neurol Neurosurg Psychiatry 65: 781-784.

Chanvitayapongs S, Draczynska-Lusiak B, Sun AY. 1997. Amelioration of oxidative stress by antioxidants and resveratrol in PC12 cells. Neuroreport 8: 1499-1502.

Chatterjee SS, Kondratskaya EL, Krishtal OA. 2003. Structure-activity studies with Ginkgo biloba extract constituents as receptor-gated chloride channel blockers and modulators. Pharmacopsychiatry 36(Suppl. 1): S68-S77.

Cheeseman KH, Slater TF. 1993. An introduction to free radical biochemistry. Br Med Bull 49: 481-493.

Cho J, Kang JS, Long PH, Jing J, Back Y, et al. 2003. Antioxidant and memory enhancing effects of purple sweet potato anthocyanin and cordyceps mushroom extract. Arch Pharm Res 26: 821-825.

Choi HY, Song JH, Park DK, Ross GM. 2000. The effects of ascorbic acid on dopamine-induced death of PC12 cells are dependent on exposure kinetics. Neurosci Lett 296: 81-84.

Choi JY, Park CS, Kim DJ, Cho MH, Jin BK, et al. 2002. Prevention of nitric oxide-mediated 1-methyl-4-phenyl-1,2,3,6-tetrahydropyridine-induced Parkinson's disease in mice by tea phenolic epigallocatechin 3-gallate. Neurotoxicology 23: 367-374.

Ciani E, Groneng L, Voltattorni M, Rolseth V, Contestabile A, et al. 1996. Inhibition of free radical production or free radical scavenging protects from the excitotoxic cell death mediated by glutamate in cultures of cerebellar granule neurons. Brain Res 728: 1-6.

Colak O, Sahin A, Alatas O, Inal M, Yasar B, et al. 1998. The effect of Ginkgo biloba on the activity of catalase and lipid peroxidation in experimental strangulation ileus. Int J Clin Lab Res 28: 69-71.

Colciaghi F, Borroni B, Zimmermann M, Bellone C, Longhi A, et al. 2004. Amyloid precursor protein metabolism is regulated toward alpha-secretase pathway by Ginkgo biloba extracts. Neurobiol Dis 16: 454-460.

Coyle JT, Puttfarcken P. 1993. Oxidative stress, glutamate, and neurodegenerative disorders. Science 262: 689-695.

Dajas F, Rivera-Megret F, Blasina F, Arredondo F, Abin-Carriquiry JA, et al. 2003. Neuroprotection by flavonoids. Braz J Med Biol Res 36: 1613-1620.

de Rijk MC, Breteler MM, den Breeijen JH, Launer LJ, Grobbee DE, et al. 1997. Dietary antioxidants and Parkinson disease. The Rotterdam Study. Arch Neurol 54: 762-765.

Desagher S, Glowinski J, Premont J. 1997. Pyruvate protects neurons against hydrogen peroxide-induced toxicity. J Neurosci 17: 9060-9067.

Dexter DT, Holley AE, Flitter WD, Slater TF, Wells FR, et al. 1994. Increased levels of lipid hydroperoxides in the parkinsonian substantia nigra: an HPLC and ESR study. Mov Disord 9: 92-97.

Dickson DW, Sinicropi S, Yen SH, Ko LW, Mattiace LA, et al. 1996. Glycation and microglial reaction in lesions of Alzheimer's disease. Neurobiol Aging 17: 733-743.

Dukic-Stefanovic S, Gasic-Milenkovic J, Deuther-Conrad W, Münch G. 2003. Signal transduction pathways in mouse microglia N-11 cells activated by advanced glycation endproducts (AGEs). J Neurochem 87: 2609-2615.

Dumuis A, Sebben M, Haynes L, Pin JP, Bockaert J. 1988. NMDA receptors activate the arachidonic acid cascade system in striatal neurons. Nature 336: 68-70.

Engelhart MJ, Geerlings MI, Ruitenberg A, van Swieten JC, Hofman A, et al. 2002. Dietary intake of antioxidants and risk of Alzheimer disease. JAMA 287: 3223-3229.

Esterbauer H, Dieber-Rotheneder M, Striegl G, Waeg G. 1991. Role of vitamin E in preventing the oxidation of low-density lipoprotein. Am J Clin Nutr 53:314S-321S.

Forbes JM, Thallas V, Thomas MC, Founds HW, Burns WC, et al. 2003. The breakdown of preexisting advanced glycation end products is associated with reduced renal fibrosis in experimental diabetes. FASEB J 17: 1762-1764.

Forster DP, Newens AJ, Kay DW, Edwardson JA. 1995. Risk factors in clinically diagnosed presenile dementia of the Alzheimer type: a case-control study in northern England. J Epidemiol Commun Health 49: 253-258.

Fremont L. 2000. Biological effects of resveratrol. Life Sci 66: 663-673.

Fridovich I. 1989. Superoxide dismutases. An adaptation to a paramagnetic gas. J Biol Chem 264: 7761-7764.

Galli RL, Shukitt-Hale B, Youdim KA, Joseph JA. 2002. Fruit polyphenolics and brain aging: nutritional interventions targeting age-related neuronal and behavioral deficits. Ann N Y Acad Sci 959: 128-132.

Garcia CK, Goldstein JL, Pathak RK, Anderson RG, Brown MS. 1994. Molecular characterization of a membrane transporter for lactate, pyruvate, and other monocarboxy-lates: implications for the Cori cycle. Cell 76: 865-873.

Greenamyre JT, Porter RH. 1994. Anatomy and physiology of glutamate in the CNS. Neurology 44: S7-S13.

Griffin WS, Sheng JG, Roberts GW, Mrak RE. 1995. Interleukin-1 expression in different plaque types in Alzheimer's disease: significance in plaque evolution. J Neuropathol Exp Neurol 54: 276-281.

Guo Q, Zhao B, Shen S, Hou J, Hu J, et al. 1999. ESR study on the structure-antioxidant activity relationship of tea catechins and their epimers. Biochim Biophys Acta 1427: 13-23.

Haenen GR, Bast A. 1999. Nitric oxide radical scavenging of flavonoids. Methods Enzymol 301: 490-503.

Hager K, Marahrens A, Kenklies M, Riederer P, Münch G. 2001. Alpha-lipoic acid as a new treatment option for Alzheimer type dementia. Arch Gerontol Geriatr 32: 275-282.

Halliday G, Robinson SR, Shepherd C, Kril J. 2000. Alzheimer's disease and inflammation: a review of cellular and therapeutic mechanisms. Clin Exp Pharmacol Physiol 27: 1-8.

Hampton MB, Kettle AJ, Winterbourn CC. 1998. Inside the neutrophil phagosome: oxidants, myeloperoxidase, and bacterial killing. Blood 92: 3007-3017.

Hernandez R. 1992. Demography of ageing. Bold 2: 8-12.

Higdon JV, Frei B. 2003. Tea catechins and polyphenols: health effects, metabolism, and antioxidant functions. Crit Rev Food Sci Nutr 43: 89-143.

Juhn MS, Tarnopolsky M. 1998. Oral creatine supplementation and athletic performance: a critical review. Clin J Sport Med 8: 286-297.

Kanowski S, Herrmann WM, Stephan K, Wierich W, Horr R. 1996. Proof of efficacy of the ginkgo biloba special extract EGb 761 in outpatients suffering from mild to moderate primary degenerative dementia of the Alzheimer type or multi-infarct dementia. Pharmacopsychiatry 29: 47-56.

Kato T. 2004. Memantine: a therapeutic drug for Alzheimer's disease and the comparison with MK-801. Nippon Yakurigaku Zasshi 124: 145-151.

Khan SG, Katiyar SK, Agarwal R, Mukhtar H. 1992. Enhancement of antioxidant and phase II enzymes by oral feeding of green tea polyphenols in drinking water to SKH-1 hairless mice: possible role in cancer chemoprevention. Cancer Res 52: 4050-4052.

Kidd PM. 1999. A review of nutrients and botanicals in the integrative management of cognitive dysfunction. Altern Med Rev 4: 144-161.

Kim MS, Lee JI, Lee WY, Kim SE. 2004. Neuroprotective effect of Ginkgo biloba L. extract in a rat model of Parkinson's disease. Phytother Res 18: 663-666.

Kucheryanu VG, Kryzhanovskii GN. 2000. Effect of glutamate and antagonists of N-methyl-D-aspartate receptors on experimental parkinsonian syndrome in rats. Bull Exp Biol Med 130: 629-632.

Laughton MJ, Evans PJ, Moroney MA, Hoult JR, Halliwell B. 1991. Inhibition of mammalian 5-lipoxygenase and cyclo-oxygenase by flavonoids and phenolic dietary additives. Relationship to antioxidant activity and to iron ion-reducing ability. Biochem Pharmacol 42: 1673-1681.

Levites Y, Amit T, Youdim MB, Mandel S. 2002. Involvement of protein kinase C activation and cell survival/ cell cycle genes in green tea polyphenol (–)-epigallocatechin 3-gallate neuroprotective action. J Biol Chem 277: 30574-30580.

Levites Y, Amit T, Mandel S, Youdim MB. 2003. Neuroprotection and neurorescue against Abeta toxicity and PKC-dependent release of nonamyloidogenic soluble precursor protein by green tea polyphenol (-)-epigallocatechin-3-gallate. FASEB J 17: 952-954.

Li Y, Liu L, Barger SW, Mrak RE, Griffin WS. 2001. Vitamin E suppression of microglial activation is neuroprotective. J Neurosci Res 66: 163-170.

Lin YL, Lin JK. 1997. (-)-Epigallocatechin-3-gallate blocks the induction of nitric oxide synthase by down-regulating lipopolysaccharide-induced activity of transcription factor nuclear factor-kappaB. Mol Pharmacol 52: 465-472.

Lin YL, Tsai SH, Lin-Shiau SY, Ho CT, Lin JK. 1999. Theaflavin-3,3'-digallate from black tea blocks the nitric oxide synthase by down-regulating the activation of NF-kappaB in macrophages. Eur J Pharmacol 367: 379-388.

Lockhart BP, Benicourt C, Junien JL, Privat A. 1994. Inhibitors of free radical formation fail to attenuate direct beta-amyloid25–35 peptide-mediated neurotoxicity in rat hippocampal cultures. J Neurosci Res 39: 494-505.

Lokk J. 2004. Memantine can relieve certain symptoms in Parkinson disease. Improvement achieved in two out of three described cases with dyskinesia and cognitive failure. Lakartidningen 101: 2003-2006.

Lovell MA, Xie C, Xiong S, Markesbery WR. 2003. Protection against amyloid beta peptide and iron/hydrogen peroxide toxicity by alpha lipoic acid. J Alzheimers Dis 5: 229-239.

Luchsinger JA, Tang MX, Siddiqui M, Shea S, Mayeux R. 2004. Alcohol intake and risk of dementia. J Am Geriatr Soc 52: 540-546.

Marcocci L, Packer L, Droy-Lefaix MT, Sekaki A, Gardes-Albert M. 1994. Antioxidant action of Ginkgo biloba extract EGb 761. Methods Enzymol 234: 462-475.

Markesbery WR. 1997. Oxidative stress hypothesis in Alzheimer's disease. Free Radic Biol Med 23: 134-147.

Markesbery WR, Carney JM. 1999. Oxidative alterations in Alzheimer's disease. Brain Pathol 9: 133-146.

Maus M, Marin P, Israel M, Glowinski J, Premont J. 1999. Pyruvate and lactate protect striatal neurons against N-methyl-D-aspartate-induced neurotoxicity. Eur J Neurosci 11: 3215-3224.

McGeer PL, McGeer EG. 2004. Inflammation and the degenerative diseases of aging. Ann N Y Acad Sci 1035: 104-116.

Meda L, Cassatella MA, Szendrei GI, Otvos L Jr, Baron P, et al. 1995. Activation of microglial cells by beta-amyloid protein and interferon-gamma. Nature 374: 647-650.

Medina S, Martinez M, Hernanz A. 2002. Antioxidants inhibit the human cortical neuron apoptosis induced by hydrogen peroxide, tumor necrosis factor alpha, dopamine and beta-amyloid peptide 1–42. Free Radic Res 36: 1179-1184.

Medvedev IO, Malyshkin AA, Belozertseva IV, Sukhotina IA, Sevostianova NY, et al. 2004. Effects of low-affinity NMDA receptor channel blockers in two rat models of chronic pain. Neuropharmacology 47: 175-183.

Miguel-Hidalgo JJ, Alvarez XA, Cacabelos R, Quack G. 2002. Neuroprotection by memantine against neurodegeneration induced by beta-amyloid(1–40). Brain Res 958: 210-221.

Miklya I, Knoll B, Knoll J. 2003. A pharmacological analysis elucidating why, in contrast to (-)-deprenyl (selegiline), alpha-tocopherol was ineffective in the DATATOP study. Life Sci 72: 2641-2648.

Morano A, Jimenez-Jimenez FJ, Molina JA, Antolin MA. 1994. Risk-factors for Parkinson's disease: case-control study in the province of Caceres, Spain. Acta Neurol Scand 89: 164-170.

Morris MC, Evans DA, Bienias JL, Tangney CC, Bennett DA, et al. 2002. Dietary intake of antioxidant nutrients and the risk of incident Alzheimer disease in a biracial community study. JAMA 287: 3230-3237.

Mutze S, Hebling U, Stremmel W, Wang J, Arnhold J, et al. 2003. Myeloperoxidase-derived hypochlorous acid antagonizes the oxidative stress-mediated activation of iron regulatory protein 1. J Biol Chem 278: 40542-40549.

Nanjo F, Mori M, Goto K, Hara Y. 1999. Radical scavenging activity of tea catechins and their related compounds. Biosci Biotechnol Biochem 63: 1621-1623.

Nedergaard M, Goldman SA. 1993. Carrier-mediated transport of lactic acid in cultured neurons and astrocytes. Am J Physiol 265: R282-R289.

Needleman P, Turk J, Jakschik BA, Morrison AR, Lefkowith JB. 1986. Arachidonic acid metabolism. Annu Rev Biochem 55: 69-102.

Nicoletti F, Magri G, Ingrao F, Bruno V, Catania MV, et al. 1990. Excitatory amino acids stimulate inositol phospholipid hydrolysis and reduce proliferation in cultured astrocytes. J Neurochem 54: 771-777.

Niki E. 1987. Interaction of ascorbate and alpha-tocopherol. Ann N Y Acad Sci 498: 186-199.

Novelli A, Reilly JA, Lysko PG, Henneberry RC. 1988. Glutamate becomes neurotoxic via the N-methyl-D-aspartate receptor when intracellular energy levels are reduced. Brain Res 451: 205-212.

Offen D, Ziv I, Panet H, Wasserman L, Stein R, et al. 1997. Dopamine-induced apoptosis is inhibited in PC12 cells expressing Bcl-2. Cell Mol Neurobiol 17: 289-304.

O'Gorman E, Beutner G, Dolder M, Koretsky AP, Brdiczka D, et al. 1997. The role of creatine kinase in inhibition of mitochondrial permeability transition. FEBS Lett 414: 253-257.

Oldendorf WH. 1973. Carrier-mediated blood-brain barrier transport of short-chain monocarboxylic organic acids. Am J Physiol 224: 1450-1453.

Orgogozo JM, Dartigues JF, Lafont S, Letenneur L, Commenges D, et al. 1997. Wine consumption and dementia in the elderly: a prospective community study in the Bordeaux area. Rev Neurol (Paris) 153: 185-192.

Pachter JS. 1997. Inflammatory mechanisms in Alzheimer disease: the role of beta-amyloid/glial interactions. Mol Psychiatry 2: 91-95.

Pannala AS, Rice-Evans CA, Halliwell B, Singh S. 1997. Inhibition of peroxynitrite-mediated tyrosine nitration by catechin polyphenols. Biochem Biophys Res Commun 232: 164-168.

Pardo B, Mena MA, Fahn S, Garcia de Yebenes J. 1993. Ascorbic acid protects against levodopa-induced neurotoxicity on a catecholamine-rich human neuroblastoma cell line. Mov Disord 8: 278-284.

Patel M, Day BJ, Crapo JD, Fridovich I, McNamara JO. 1996. Requirement for superoxide in excitotoxic cell death. Neuron 16: 345-355.

Pellerin L, Magistretti PJ. 1994. Glutamate uptake into astrocytes stimulates aerobic glycolysis: a mechanism coupling neuronal activity to glucose utilization. Proc Natl Acad Sci USA 91: 10625-10629.

Pettegrew JW, Panchalingam K, Klunk WE, McClure RJ, Muenz LR. 1994. Alterations of cerebral metabolism in probable Alzheimer's disease: a preliminary study. Neurobiol Aging 15: 117-132.

Poole RC, Halestrap AP. 1993. Transport of lactate and other monocarboxylates across mammalian plasma membranes. Am J Physiol 264: C761-C782.

Preux PM, Condet A, Anglade C, Druet-Cabanac M, Debrock C, et al. 2000. Parkinson's disease and environmental factors. Matched case-control study in the Limousin region, France. Neuroepidemiology 19: 333-337.

Pulido SM, Passaquin AC, Leijendekker WJ, Challet C, Wallimann T, Ruegg UT. 1998. Creatine supplementation improves intracellular Ca^{2+} handling amd survival in mdx skeletal muscle cells. FEBS Lett 439: 357-362.

Rabey JM, Nissipeanu P, Korczyn AD. 1992. Efficacy of memantine, an NMDA receptor antagonist, in the treatment of Parkinson's disease. J Neural Transm Park Dis Dement Sect 4: 277-282.

Raichle ME, Gusnard DA. 2002. Appraising the brain's energy budget. Proc Natl Acad Sci USA 99: 10237-10239.

Rebec GV, Barton SJ, Marseilles AM, Collins K. 2003. Ascorbate treatment attenuates the Huntington behavioral phenotype in mice. Neuroreport 14: 1263-1265.

Reisberg B, Doody R, Stoffler A, Schmitt F, Ferris S, et al. 2003. Memantine in moderate-to-severe Alzheimer's disease. N Engl J Med 348: 1333-1341.

Rice-Evans C. 1999. Implications of the mechanisms of action of tea polyphenols as antioxidants in vitro for chemoprevention in humans. Proc Soc Exp Biol Med 220: 262-266.

Rose RC. 1988. Transport of ascorbic acid and other watersoluble vitamins. Biochim Biophys Acta 947: 335-366.

Roth A, Schaffner W, Hertel C. 1999. Phytoestrogen kaempferol (3,4',5,7-tetrahydroxyflavone) protects PC12 and T47D cells from beta-amyloid-induced toxicity. J Neurosci Res 57: 399-404.

Savaskan E, Olivieri G, Meier F, Seifritz E, Wirz-Justice A, et al. 2003. Red wine ingredient resveratrol protects from beta-amyloid neurotoxicity. Gerontology 49: 380-383.

Sayre LM, Zelasko DA, Harris PL, Perry G, Salomon RG, et al. 1997. 4-Hydroxynonenal-derived advanced lipid peroxidation end products are increased in Alzheimer's disease. J Neurochem 68: 2092-2097.

Sies H. 1989. Relationship between free radicals and vitamins: an overview. Int J Vitam Nutr Res 30(Suppl.): 215-223.

Skrzydlewska E, Ostrowska J, Farbiszewski R, Michalak K. 2002. Protective effect of green tea against lipid peroxidation in the rat liver, blood serum and the brain. Phytomedicine 9: 232-238.

Smith JV, Luo Y. 2003. Elevation of oxidative free radicals in Alzheimer's disease models can be attenuated by Ginkgo biloba extract EGb 761. J Alzheimers Dis 5: 287-300.

Smith KJ, Kapoor R, Felts PA. 1999. Demyelination: the role of reactive oxygen and nitrogen species. Brain Pathol 9: 69-92.

Stone TW. 2001. Kynurenic acid antagonists and kynurenine pathway inhibitors. Expert Opin Investig Drugs 10: 633-645.

Sun WG, Liao HL, Huang ZS. 2001. Intervention effect of Chinese herbal medicine on changes of immune system in senile dementia. Zhongguo Zhong Xi Yi Jie He Za Zhi 21: 716-718.

Tan EK, Tan C, Fook-Chong SM, Lum SY, Chai A, et al. 2003. Dose-dependent protective effect of coffee, tea, and smoking in Parkinson's disease: a study in ethnic Chinese. J Neurol Sci 216: 163-167.

Tappel AL. 1972. Vitamin E and free radical peroxidation of lipids. Ann N Y Acad Sci 203: 12-28.

Tarnopolsky MA, Beal MF. 2001. Potential for creatine and other therapies targeting cellular energy dysfunction in neurological disorders. Ann Neurol 49: 561-574.

Thomas EL. 1979. Myeloperoxidase, hydrogen peroxide, chloride antimicrobial system: nitrogen-chlorine derivatives of bacterial components in bactericidal action against *Escherichia coli*. Infect Immun 23: 522-531.

Toledo-Pereyra LH, Lopez-Neblina F, Toledo AH. 2004. Reactive oxygen species and molecular biology of ischemia/reperfusion. Ann Transplant 9: 81-83.

Tretter L, Sipos I, Adam-Vizi V. 2004. Initiation of neuronal damage by complex I deficiency and oxidative stress in Parkinson's disease. Neurochem Res 29: 569-577.

Tsacopoulos M, Magistretti PJ. 1996. Metabolic coupling between glia and neurons. J Neurosci 16: 877-885.

Weiss S, Ellis J, Hendley DD, Lenox RH. 1989. Translocation and activation of protein kinase C in striatal neurons in primary culture: relationship to phorbol dibutyrate actions on the inositol phosphate generating system and neurotransmitter release. J Neurochem 52: 530-536.

Winblad B, Poritis N. 1999. Memantine in severe dementia: results of the 9M-Best Study (Benefit and efficacy in severely demented patients during treatment with memantine). Int J Geriatr Psychiatry 14: 135-146.

Winkler BS. 1987. In vitro oxidation of ascorbic acid and its prevention by GSH. Biochim Biophys Acta 925: 258-264.

Yang SF, Wu Q, Sun AS, Huang XN, Shi JS. 2001. Protective effect and mechanism of Ginkgo biloba leaf extracts for Parkinson disease induced by 1-methyl-4-phenyl-1,2,3,6-tetrahydropyridine. Acta Pharmacol Sin 22: 1089-1093.

Yatin SM, Varadarajan S, Butterfield DA. 2000. Vitamin E prevents Alzheimer's amyloid beta-peptide (1–42)-induced neuronal protein oxidation and reactive oxygen species production. J Alzheimers Dis 2: 123-131.

Yesavage JA, O'Hara R, Kraemer H, Noda A, Taylor JL, et al. 2002. Modeling the prevalence and incidence of Alzheimer's disease and mild cognitive impairment. J Psychiatr Res 36: 281-286.

Zhang L, Xing GQ, Barker JL, Chang Y, Maric D, et al. 2001. Alpha-lipoic acid protects rat cortical neurons against cell death induced by amyloid and hydrogen peroxide through the Akt signalling pathway. Neurosci Lett 312: 125-128.

Zhang SM, Hernan MA, Chen H, Spiegelman D, Willett WC, et al. 2002. Intakes of vitamins E and C, carotenoids, vitamin supplements, and PD risk. Neurology 59: 1161-1169.

Zhao B, Guo Q, Xin W. 2001. Free radical scavenging by green tea polyphenols. Methods Enzymol 335: 217-231.

7   Drug Treatment of Motor Symptoms in Parkinson's Disease

T. Müller

Abstract: Insidious onset of mild, unspecific, vegetative, psychopathological, cognitive, and perceptive disturbances with a resulting change in personal behavior often precede the initial intermittent onset of motor symptoms in patients with Parkinson's disease (PD). Currently available diagnostic tools and concepts suggest this appearance of motor symptoms as the criterion for diagnosis of PD. Motor symptom control can be achieved by dopaminergic substitution, which should deliver a dopamine-substituting drug in a continuous fashion to the brain, This enables imitation of physiologic conditions of dopaminergic neurotransmission. This delays onset of motor complications which mostly appear after a certain period probably due to progression of the disease with loss of compensatory mechanisms, which reduce alternating levels of dopamine-substituting compounds in the synaptic cleft and at striatal postsynaptic dopaminergic uptake sites. These long-term motor complications are fluctuations of movement and are mostly associated with levodopa owing to the short half-life of the drug. They considerably burden quality of life in PD patients and their caregivers. These complications follow the so-called honeymoon period of well-tolerated levodopa administration and are looked upon as one essential clinical marker of dopaminergic chronic neurodegeneration and progression in PD patients. Dopamine agonists delay onset of these motor complications because of their longer half-life in plasma and their affinity to striatal postsynaptic dopaminergic uptake sites. However, their efficacy and tolerability differ in the individual patient and is generally reduced compared with levodopa. Dopaminergic substitution with dopamine agonists or MAO-B inhibitors instead of levodopa also prevent levodopa-associated homocysteine elevation which may contribute to progression of PD and increase the risk for onset of nonmotor features like dementia and depression in the long run. These nonmotor PD symptoms predominantly result from nondopaminergic neurodegeneration and additionally considerably reduce quality of life, in particular in advanced stages of the disease.

List of Abbreviations: COMT, catechol-O-methyltransferase; DDI, decarboxylase inhibitor; DEC, dihydroergocryptin; DIRD, DA-induced respiratory disorders; GABA, gamma-aminobutyric acid; MAO-B, Monoaminooxidase-B; MAO, monoaminooxidase; NMDA, N-methyl-D-aspartate; PD, Parkinson's disease; UPDRS, Unified Parkinson's Disease Rating Scale; 3-OMD, 3-O-methyldopa; 5-HT, serotonine

1 Introduction

James Parkinson's original 1817 essay "shaking palsy" described slow deterioration of motor function as one of the cardinal symptoms of Parkinson's disease (PD) with the term akinesia. Moreover, he mentioned the onset of rigidity and tremor beside an array of furthermore secondary clinical signs, which are not directly related to the dysfunction of the motor system (Parkinson, 1817). The slowly preceding loss of nigral dopaminergic presynaptic neurons in PD is mainly responsible for the occurrence of akinesia, rigidity, and to a lesser extent tremor, all of which result from a reduction of about 70–80% striatal dopamine in combination with a less distinct altered transmission of further neurotransmitters, that is, noradrenaline, serotonine, acetylcholine, and adenosine (Bernheimer et al., 1961; Riederer and Wuketich, 1976; Braak et al., 2003). These neurochemical findings of decreased dopamine levels in predominant nigrostriatal pathways of brains of PD patients led to the concept of dopaminergic substitution, which was a milestone in the treatment of PD, whereas the less pronounced reduction of further neurotransmitters inaugurated the nondopaminergic treatment strategies (Bernheimer et al., 1961; Hornykiewicz, 1998; Jellinger, 1999; Foley et al., 2000; Braak et al., 2003).

1.1 Relevant Structures for Nigrostriatal Dopaminergic Neurotransmission

There are two families of dopamine (D) receptors, named D_1 and D_2. The D_1 receptor family consists of D_1-(D_{1A}) and the D_5-(D_{1B}) receptors, both of which increase the synthesis of cAMP and activate the dopamine-sensitive adenylatcyclase. D_1 receptors are predominantly located in striatal, hippocampal, and olfactory neurons, whereas D_5-uptake sites occur in the hippocampal and hypothalamic regions. D_5 receptors are more sensitive to dopamine application in the micromolar range compared with D_1-uptake sites. D_2 receptor family stimulation, which corresponds to dopamine application in the nanomolar range, causes antiparkinsonian efficacy with predominant improvement of motor symptoms. Several subtypes are identified: D_{2A}, D_{2B}, D_3, and D_4 receptors. D_2-uptake sites mostly appear in striatal neurons, whereas D_3 and D_4 receptors predominantly occur in the limbic system (Stanwood et al., 2000; Gerlach et al., 2003). The presynaptic dopaminergic reuptake sites are at least six to ten times more sensitive to DA compared with postsynaptic dopaminergic receptors (Skirboll et al., 1979). They mediate a certain inhibitory feedback regulation. Chronic dopaminergic substitution may diminish this autoreceptor function and induce some slowly evolving postsynaptic pharmacodynamic change in dopamine receptor function and interaction, but this loss of presynaptic dopaminergic autoreceptor function is also a typical feature of the progression of PD (Müller et al., 1997; Ekesbo et al., 1999).

1.2 Aims of Dopaminergic Substitution

Dopaminergic substitution in PD focuses on a careful and slow titration of a combination of dopaminergic compounds to provide an optimum improvement of motor symptoms and quality of life without reducing the PD patients' compliance due to onset of adverse effects. The efficacy of dopaminergic drugs depends on their half-life in plasma and their receptor affinity to postsynaptic dopaminergic reuptake sites in the nigrostriatal system. It is essential that these drugs are delivered to these receptors in continuous fashion to imitate physiological conditions of dopaminergic neurotransmission (Chase et al., 1998). Moreover, slowing of progression of PD by applied compounds due to their putative neuroprotective and neuroroegenerative properties, which were convincingly only demonstrated in preclinical research, plays an important role.

2 Levodopa (LD)

Levodopa (LD) is the blood–brain barrier trespassing precursor of the physiologic neurotransmitter dopamine. Then LD is converted to dopamine by the enzyme tyrosine hydroxylase within the brain predominantly in presynaptic dopaminergic neurons. Initially, LD was administered in dosages up to 9 g/day, which caused side effects, that is, nausea, due to the fast peripheral degradation, since LD has a short plasma half-life (Cotzias et al., 1969). The introduction of the additional administration of an aromatic amino acid decarboxylase inhibitor (DDI) markedly reduced the peripheral LD degradation, lengthened the half-life up to 60–90 min, and as a consequence reduced the necessary daily administered LD dosage generally below 1000 mg for improvement of motor symptoms in PD patients. Two different DDI are available, benserazide is administered with levodopa on 1:4 basis, carbidopa on 1:10 basis (Kent et al., 1990). Nowadays, each standard LD formulation is applied with a peripherally acting DDI. Therefore, LD is predominantly converted to its metabolite 3-O-methyldopa (3-OMD) by the ubiquitous enzyme catechol-O-methyltransferase (COMT) in blood, peripheral tissues, and in nigrostriatal neurons. The next development was the introduction of COMT inhibitors as adjuncts to LD/DDI applications, which further prolong the half-life of LD and therefore convincingly improve the clinical efficacy on motor symptoms. Pharmacokinetic and placebo-controlled trials with the peripherally acting entacapone and the additionally centrally working tolcapone proved this therapeutic principle. They showed that the antiparkinsonian clinical effect of LD was further enhanced by its substantial extension of the plasma elimination half-life and its subsequently increased brain delivery (Baas et al., 1997; Müller et al., 2000b; Poewe et al., 2002; Gershanik et al., 2003; Olanow et al., 2004). This principle of LD administration was improved with the launch of the compound Stalevo, which combines LD, carbidopa, and entacapone in one tablet. The clinically more efficacious also centrally acting COMT inhibitor, tolcapone, caused severe toxic side effects in the liver. Moreover, there was a controversy whether this compound may accelerate neurodegeneration. Therefore, it was temporarily withdrawn nearly all over the world for several years. Recently, it was again launched; however, its prescription demands a strict control of liver enzyme activity on a regular basis (Kuhn et al., 1998; Russ et al., 1999; Storch et al., 2000; Müller et al., 2000b; Gerlach et al., 2001). More rapidly absorbed LD preparations, that is, water-soluble forms of LD/benserazide, are also often used in clinical practice for treatment of early morning akinesia or severe sometimes suddenly appearing akinetic states, the so-called unpredictable off-periods. A further rapid LD preparation, the levodopa ethylester, did not survive in clinical trials due to a missing significant reduction of these off-states. Retarded release of LD/DDI compounds is also used despite their decreased clinical efficacy of up to 30% compared with the normal LD/DDI preparations. They are often administered for the last LD intake before sleep at night to improve disturbed sleep behavior and early morning akinesia (Block et al., 1997; Koller et al., 1999). To date there exist no trials that compare the clinical efficacy of these delayed release formulations with the combination of LD/DDI and entacapone or tolcapone in this indication.

2.1   Clinical Efficacy of LD

To date LD/DDI is the most efficacious and best initially tolerated dopaminergic antiparkinsonian compound, which was again convincingly demonstrated in the ELLDOPA trial (Fahn et al., 2004). All idiopathic PD patients respond to LD application with a distinct improvement of predominant akinesia and rigidity and to a lesser extent tremor. This also serves as a specific diagnostic criterion (Hughes et al., 1992; Müller et al., 2003a). A missing response may indicate a gastrointestinal absorption problem or makes the diagnosis of idiopathic PD uncertain. Efficacy of LD depends on its gastrointestinal absorption and plasma appearance, which increases with progression of PD and inversely correlates with body weight (Murata et al., 1996; Müller et al., 2000a; Muhlack et al., 2004; Goetze et al., 2005).

Drug-induced increased gastrointestinal motility, reduction of concomitant application of anticholinergics, and controversially discussed reduction of food consisting of big neutral amino acids and proteins are conditions that may support the gastrointestinal and blood–brain barrier trespassing the pharmacokinetic behavior of LD and thus its efficacy (Kurlan et al., 1988; Ogawa, 2000; Goetze et al., 2005).

2.2   Side Effects of LD

LD is well tolerated after intake. Onset of nausea, orthostatic syndrome, or dizziness may occur within a fast titration regime up to efficient dosages, which depends on the severity of motor symptoms. The clinical relevant disadvantages of this agent are long-term complications and metabolic changes.

2.3   Long-Term Motor Complications Dependent on LD Intake

Initial motor complications are end-of-dose phenomena or so-called predictable fluctuations of movement. They appear with increased onset of more severe motor symptoms in a time-dependent manner from the last LD intake within 5 to 10 years after initiation of LD treatment. In this stage, more frequent dosing intervals, addition of COMT inhibitors, and/or use of retarded LD preparations may reduce these complications to a certain extent for a certain interval, but this recurrent, predictable loss of efficacy of LD also indicates the end of the so-called "honeymoon period" of LD administration (Obeso et al., 2000; Ogawa, 2000; Olanow and Stocchi, 2004).

2.4   Unpredictable Long-Term Motor Fluctuations Independent of LD Intake

The motor fluctuations become more and more intense and unpredictable, sometimes in combination with onset of involuntary movements or even dystonic cramps. This hyperkinetic behavior, termed dyskinesia, appears despite increase in dosages and more frequent dosing of LD/DDI in the later course of PD. Subtypes of dyskinesias are peak-dose dyskinesia, associated with the maximum of the LD plasma level, whereas biphasic dyskinesia appears with the increasing and/or decreasing LD plasma level. Off-dose dystonia often appears early in the morning before the first LD intake (Jenner, 2000, 2004; Rascol, 2000). It is characterized by spasms in the extremities, which shortly disappear after intake of a water-soluble LD compound or subcutaneous injection of apomorphine.

2.5 Treatment of Fluctuations Depends on the Modes of LD Delivery to the Brain

There are several hypotheses on the origin of these LD-associated motor complications. Loss of presynaptic dopaminergic autoreceptor function with resulting unphysiologic high synaptic dopamine concentrations and consecutive altered postsynaptic dopamine receptor activation and further downstream intracellular changes currently represents one of the most widely accepted ones. Thus, a continuous nigrostriatal postsynaptic dopaminergic receptor stimulation may prevent the onset of motor complications (Durif, 1999; Chase and Oh, 2000a). Therefore, stabilizing LD plasma levels via further inhibition of LD metabolism with COMT inhibitors as adjuncts to LD/DDI may be useful from the theoretical point of view, since this may provide a more continuous brain delivery of LD. Accordingly, placebo-controlled trials with the COMT inhibitors entacapone and tolcapone as adjuncts to an existing LD/DDI long-term application showed a certain benefit, in particular in predictable end-of-dose fluctuations in clinical trials (Olanow and Obeso, 2000). However, there is no convincing trial yet, which compares the efficacy of further LD/DDI titration with early adjustment versus additional COMT inhibitor application from the beginning in terms of delay of onset of motor complications; but this trial is on the way. Moreover, addition of COMT inhibitors often induced dyskinesia in clinical trials, since the addition of COMT inhibitors increases the amount of levodopa, which is delivered to the brain (Gerlach et al., 2004). Therefore, there is a need to titrate or adjust the oral LD/DDI intake, when a COMT inhibitor is additionally introduced. Before the launch of COMT inhibitors for the treatment of PD, various types of slow-release LD formulations were looked on as an alternative, preventive approach to motor fluctuations. However, the correspondingly performed clinical trials failed, since they did not last long enough and did not examine the onset of motor fluctuations enough in detail.

2.6 Metabolic Changes in Long-Term Application of LD

These motor side effects of long-term LD therapy initiated a long debate on the pro's and con's of the putative neurotoxicity of LD with the performance of preclinical studies in various cell culture and animal models and numerous reviews, which mainly focus on progression of PD, but not on the putative atherosclerosis-promoting effects of LD (Fahn, 1996). However, occurrence of increased hazard ratios for both ischemic heart and cerebrovascular disease is known in LD/DDI-treated PD patients (Ben Shlomo and Marmot, 1995). One did not consider that O-methylation of levodopa to 3-OMD via the COMT requires Mg^{2+} as cofactor and S-adenosylmethionine as methyl donor, which is associated with conversion of S-adenosylmethionine to S-adenosylhomocysteine and subsequently homocysteine in the long term. Homocysteine induces substantial impairment of endothelial function and is looked on as an independent risk factor for atherosclerotic disease. Accordingly, elevated homocysteine levels appeared in treated PD patients compared with matched controls and significantly correlated to daily levodopa dosage (Blandini et al., 2001; Kuhn et al., 2001; Müller et al., 2001). Treated but not previously untreated PD subjects showed augmented plasma concentrations of homocysteine. Nevertheless, a putative influence of other antiparkinsonian drugs beneath levodopa/DDI on plasma homocysteine was a critical issue of all trials. However, the demonstrated association between homocysteine and 3-OMD plasma levels provided further evidence for a certain impact of levodopa metabolism on homocysteine concentrations. Since the plasma half-life of 3-OMD is about 15 h in contrast to levodopa with $\sim$1h, there was no relation between homocysteine and levodopa, respectively, and 3-OMD and levodopa (Müller et al., 2002). Homocysteine induces substantial impairment of endothelial function and subsequent atherosclerosis. Atherosclerotic disease of striatal cerebral vessels hypothetically results in subsequent onset of increased susceptibility to impaired mitochondrial energy metabolism, oxidative stress, and basal ganglia circuit dysfunction, all of which represent typical pathophysiologic features of PD (Chambers et al., 1998). Moreover, exposure of the endothelium to homocysteine induces release of nitric oxide, a further excitotoxic compound under suspicion for the contribution of the ensuing neuronal degeneration in PD (Schlussel et al., 1995; Lipton et al., 1997; Beal, 1998; Gerlach et al., 1999) Thus long-term application of LD/DDI with its consecutive impact on homocysteine metabolism may promote vascular disease, polyneuropathy, and hypothetical progression of PD (Miller et al., 2003; Nakaso et al., 2003; Müller et al., 2003e, 2004a). Further, predisposing factors for this homocysteine-mediated toxicity are certain genetic mutations of homocysteine metaboliz-ing and catalyzing enzymes and/or nutrition. From this point of view, monitoring and subsequent therapeutic decrease in homocysteine levels may possibly prevent and/or delay the onset of vascular disease in PD. The relationship shown between total plasma homocysteine levels and brain atrophy in healthy elderly individuals supports this hypothesis. Putative therapeutic approaches for reduction of homocyste-ine levels are additional folic acid supplementation, since, that is, folic acid and cobalamine catalyze and enhance metabolism of homocysteine to methionine, or, hypothetically, application of peripherally acting COMT inhibitors as adjunct to LD/DDI treatment. On the one hand, COMT inhibitors increase the peripheral bioavailability of LD, but on the other hand the combination of LD/DDI with COMT inhibitors reduces O-methylation of LD and thus decreases homocysteine levels (Miller et al., 1997, 2002; Müller et al., 2001, 2003e, 2004a; Seshadri et al., 2002; Nakaso et al., 2003; Nissinen et al., 2005).

3   Dopamine Agonists (DA)

Ergot and non-ergot derivative DA are used in the treatment of PD, since DA delay the onset of levodopa-associated motor complications, which limit the quality of life to a considerable extent in patients in later stages of PD (Przuntek et al., 1996; Rascol et al., 2000). Moreover, controversially discussed results of long-term trials with additional performance of functional brain imaging techniques of the nigrostriatal dopaminergic system indicate that progression of PD is slower with DA long-term application than with levodopa monotherapy (Marek et al., 2002; Morrish, 2002, 2003; Whone et al., 2003; Przuntek et al., 2004).

The emetic DA apomorphine stimulates D_1- and D_2-uptake sites, shows a very close structural resemblance to dopamine in terms of its affinity to postsynaptic dopaminergic structures, trespasses the blood–brain barrier very well, and has a short half-life of 10 min. Apomorphine is predominantly subcuta-neously administered in dosages between 2 and 5 mg with a pen inject system for testing the motor response to dopaminergic stimulation or for the treatment of severe off-phases. Oral, nasal, and transdermal forms are currently under investigation, because there is a pronounced first pass effect of this sedative drug after gastrointestinal resorption. Since apomorphine may cause severe nausea, a concomitant therapy with domperidone eases tolerability. Apomorphine may also continuously be administered via a pump system with daily dosages between 20 and 50 mg with considerable anitparkinsonian benefit; however, only for a limited time period, since granuloma due to local inflammation may appear at the injection site. Induction of psychosis is rare due to the sedative component of this compound (Factor, 2004; Tyne et al., 2004; Katzenschlager et al., 2005). Bromocriptine shows a strong D_2 receptor agonistic and a postulated weak effect at the D_1-uptake site. Therefore, one assumes that endogenous dopamine improves its clinical efficacy. There are numerous trials on this compound since the early 1970s, demonstrating its symptomatic benefit on motor symptoms. Moreover, long-term trials showed that LD substitution to an at least 30% extent by bromocriptine postpones LD-associated motor complications and reduced mortality from cardiovascular disease. Treatment with bromocriptine alone or in combination with LD is beneficial; however, 90% of study participants with initial monotherapy with bromocriptine were on a combination regimen with LD due to the onset of orthostatic syndrome, nausea, dizziness, and insufficient clinical efficacy in the treatment of motor symptoms within several years (Przuntek et al., 1992, 1996; Hely et al., 1999). Cabergoline predominantly stimulates D_2-uptake sites. This agent has a very long plasma half-life of about 65 h, since it is not metabolized in the liver and has a low binding potential to plasma proteins. Therefore PD patients, who are still engaged in their professional life, prefer this drug and take it once daily mostly in the morning. Combination of cabergoline with LD reduces off-periods and delays the onset of dyskinesia, whereas the side effects are the same compared with other ergolines (Hutton et al., 1993; Marsden, 1998; Curran and Perry, 2004). Alpha-dihydroergocryptin (DEC) has a plasma half-life of 16 h with a distinct first-pass effect and hepatic metabolism. There are various clinical trials demonstrating its symptomatic benefit versus placebo or bromocriptine, and lisuride. These studies report and emphasize a low incidence of side effects, such as orthostatic syndrome or onset of psychosis; however, predominantly dosages of DEC were lower than the one used in daily clinical practice, which normally ranges between 60 and 120 mg. The 8-β-ergoline lisuride with a plasma half-life of 1–3 h shows a higher affinity to the D_2 receptor family ($D_2 > D_3 > D_4$) compared with bromocriptine and pergolide and binds to postsynaptic D_1-uptake sites. There are many studies on the symptomatic benefit of this compound even in monotherapy; long-term trials showed reduction of mortality and delayed onset of motor complications. Lisuride is administered several times daily in clinical practice sometimes even in combination with each LD intake. Some authors postulate an increased risk of psychosis due to the high binding potential of this compound to dopaminergic and serotonergic receptors (Battistin et al., 1999; Deleu et al., 2002; Stocchi et al., 2002; Woitalla et al., 2004). Pergolide stimulates D_1, D_2, and D_3 receptors and has a fast 1–2 h lasting gastrointestinal absorption and a plasma half-life of about 27 h. The PELMO-PET study failed to demonstrate a reduced progression of PD compared with LD application alone due to the use of different PET scanners in the course of the trial; however, there was a significantly reduced appearance of dyskinesia in the pergolide-treated patients. The not globally available ergoline D_2 receptor agonist, piribedil, shares the previously mentioned properties of other DA in terms of symptomatic benefit on motor symptoms and long-term effects. The non-ergot derivative, pramipexole, has a plasma half-life of about 8 h and predominantly binds to the subtypes of the D_2 receptor family, in particular to D_3-uptake sites; nearly 90% of the compound is not metabolized in the periphery. There are numerous trials on the symptomatic dopaminergic benefit of this agent. The CALM-PD(-CIT) trial showed a lower incidence of dyskinesia and reduced progression of PD compared with levodopa after 5 years. Ropinirol has a similar receptor-binding profile to dopaminergic receptors like pramipexole, but a shorter plasma half-life of 6 h. Hepatic metabolism via the isoenzyme 1 A2 of the cytochrome P450 system is an essential step of its pharmacokinetic behavior. Numerous trials demonstrated its symptomatic benefit (Matheson and Spencer, 2000; Hubble, 2002; Müller, 2002; Rascol et al., 2002; Inzelberg et al., 2003). There was an initial association of occurrence of sudden so-called "sleep attacks" for both non-ergoline compounds, which called patients' and prescribing physicians' attention in a negative manner. Then there were more and more reports on this side effect induced by direct and indirect dopamine substituting agents in PD patients. Now the whole issue is looked on as class effect of all dopaminergic drugs (Olanow et al., 2000); additional intake of modafinil provides symptomatic relief (Högl et al., 2002).

3.1 Tolerability and Efficacy of DA

There are numerous study results on the efficacy of each compound in monotherapy and/or combination with LD versus placebo and/or levodopa alone. Long-term studies comparing DA with LD monotherapy correspondingly show the phenomenon that reduction of scores of applied PD rating scales is less pronounced with DA compared with levodopa monotherapy (Holloway et al., 2004). This may indicate a limited antiparkinsonian effect and/or tolerability of DA, which prevented optimum titration in particular in the initial phase of the trial. However, one may also postulate a certain antidepressant efficacy of DA, which improves the coping of PD patients with their insufficient motor situation and their reduced quality of life (Hubble, 2002; Holloway et al., 2004; Moller et al., 2005). To date, the performed head-to-head trials between various DA predominantly suffer from insufficient designs in terms of application of equivalent DA dosages. Therefore, a postulated superiority of one DA over the other should be interpreted very cautiously. All DA show a limited tolerability in the initiation period compared with LD. The titration of DA should be performed in a slow and cautious manner, since the tolerability is individually different and depends on the putative, additional temporary intake of domperidone against nausea and midodrine due to the onset of an orthostatic syndrome. Moreover, PD patients often complain of loss of appetite, sleepiness, and/or edema, all of which influence the compliance. However, the availability of various DA enables the therapeutic approach of switching from one DA to another to test the optimum tolerability and sometimes to provide a further therapeutic benefit, for instance, on motor fluctuations and/or response. Switching between DA is often performed from 1 day to the other even in an outpatient setting. The switch over from bromocriptine to pergolide is best studied. It renewed benefit in patients, whose efficacy from bromocriptine waned or vice versa (Müller, 2002). There are also now some new results on the transdermal delivery of DA, which demonstrate efficacy on motor symptoms and postulate a continuous delivery of the drug to the brain, which may contribute to a delayed onset of motor complications in the long term. However, one must also consider the affinity of the administered dopaminergic compound to the postsynaptic dopaminergic receptor. Therefore, lisuride with its distinct higher affinity to postsynaptic dopaminergic striatal terminals than other dopamine agonists was effective in the treatment of motor complications despite its short plasma half-life after oral intake, which is circumvented by the transdermal delivery of lisuride (Rinne, 1999; Müller, 2002; Woitalla et al., 2004). Trials with patches containing rotigotine or piribedil also showed a decrease in off-periods and symptomatic efficacy against motor symptoms (Montastruc et al., 1999; Metman et al., 2001). However, there may be a certain risk of local allergic skin reactions, which may appear after a certain interval in predisposed patients. This suggests a delayed allergic immune reaction triggered with a still unknown long-lasting immune reaction cascade.

3.2   Long-Term Side Effects of DA

Long-term DA-associated complications are rare, but DA-induced respiratory disorders (DIRD) represent the most serious and sometimes life-threatening condition with delayed appearance and diagnosis due to their insidious onset, since symptoms often emerge after several years of well-tolerated DA treatment. There is an observed increased onset of fibrotic reactions during long-term application of bromocriptine and pergolide in particular in PD patients with former exposure to asbestos according to the literature. However, this incidence for bromocriptine and pergolide is biased by their partial availability as generic drug, their global distribution and use, the applied mean dosage, and the time point of launch in different countries (Müller and Fritze, 2003b; Van Camp et al., 2004; Tintner et al., 2005). Moreover, considering all these issues, the risk for onset of DIRD seems to be reduced in particular in the case of pramipexole and ropinirole with no published report available to date. The mechanism of onset of DIRD is poorly understood. One suggested risk factor is the hypothesis of an idiosyncratic immune response with the drug acting as a hapten, since pathological studies of drug-induced fibrotic reactions showed only minimal inflammatory infiltrates, occasionally with mononuclear and/or eosinophilic cells or vasculitis, but without evidence of complement or immune complex deposition. One further possible mechanism could be associated with an altered function of 5-HT transmission, since most of these ergot agents have certain, at least partially, very specific various 5-HT receptor affinities, that is on the 5-HT$_2$ receptor with its consecutive stimulating impact on the regulation of the key mediator of fibrosis, the transforming growth factor-β 1, which can direct both proliferative and fibrotic signals in various mesothelial cell types (Müller and Fritze, 2003b; Horowski et al., 2004).

4   Further Compounds for Amelioration of Motor Symptoms in with Parkinson's Disease (PD) Patients

4.1   Amantadine

Better knowledge of the neural mechanisms underlying dyskinesias, which were initially shown in MPTP-primed monkeys, led to the hypothesis that increased synaptic efficacy of N-methyl-D-aspartate (NMDA) receptors expressed on basal ganglia neurons could play a role in the pathophysiology of LD-induced dyskinesias. Selective antagonists of NMDA receptors diminished choreic LD-induced dyskinesias. Since specific and selective NMDA antagonists are not currently available for treatment of PD patients, marketed drugs with antiglutamatergic properties, such as amantadine, were then used in open and controlled phase II and III studies to decrease levodopa-induced dyskinesias in PD patients without concomitant worsening of symptoms of PD. These first promising results suggest that adjunctive application of amantadine to LD improves dyskinesias and supports the view that hyperfunction of NMDA receptors contributes to the pathogenesis of LD-induced dyskinesias (Uitti et al., 1996; Verhagen et al., 1998a, b; Crosby et al., 2003; Thomas et al., 2004). These studies on a relatively low number of subjects and clinical observations led to the resurgence of amantadine in the treatment of PD (Verhagen et al., 1998a, b; Rascol, 2000). Longer controlled studies of amantadine, which assess if this effect lasts for intervals corresponding to the real therapeutic management of PD, are not available. Nevertheless, particularly intravenous application of amantadine sulfate, a drug available in Germany, induces a rapid improvement of bradykinesia, rigidity, and tremor in treated PD patients (Uitti et al., 1996; Müller et al., 2003c).

4.2 Anticholinergics

Anticholinergics are one of the oldest drugs that are used in the treatment of PD. Atropa belladonna tinctures were applied for treatment of salivation in PD patients in 1887. This also improved tremor. Thus at the beginning of twentieth century, belladonna alkaloids represented a therapeutic option for the treatment of extrapyramidal disorders. Later increasing evidence showed (i) a significant cholinergic, muscarinergic, input into most nuclei of the basal ganglia from the pedunculopontine nucleus and (ii) that interneurons within the striatum are largely muscarinergic. Nicotinic receptors exist within the striatum, most notably on the terminals of the nigrostriatal pathway. Nowadays, the manipulation of cholinergic function within the basal ganglia is an established approach to treat PD, which has been discussed by many reviews (Hughes et al., 1971; Ransmayr et al., 1992; Katzenschlager et al., 2003). Anticholinergic drugs acting on muscarinic receptors are used to treat mild symptoms of PD, particularly tremor. Currently applied anticholinergic compounds are nonselective for the muscarinic receptor subtypes within the basal ganglia. Their clinical use suffer from the occurrence of peripheral and central side effects like dryness of the mouth, obstipation, miction problems, tachyarrhythmia, delirium, and dementia (Cooper et al., 1993; Pondal et al., 1996; Bosboom et al., 2003).

4.3 Adenosine A2a Receptor Antagonists

Adenosine A2a receptors are highly localized to cholinergic interneurons and to the cell bodies of the strio-GPe indirect output pathway. Through such a selective localization, adenosine A2a receptors can influence both striatal gamma-aminobutyric acid (GABA) and acetylcholine release (Zeng et al., 2000). Indeed, the adenosine A2a antagonist KW6002 produces motor activation in rodent models of PD. In MPTP-treated primates, KW6002 induces a modest increase in locomotor activity while reversing the motor disabilities exhibited by these animals. Importantly, when KW6002 is administered in an acute or long-term manner to levodopa-primed MPTP-treated primates, it does not provoke a dyskinetic response (Zeng et al., 2000). Clinicians initially observed in treated fluctuating PD patients an increase of "on"-time after addition of the nonselective adenosine antagonist theophylline, which was confirmed with further more selective adenosine A2a receptor antagonists in clinical trials in the meantime (Bibbiani et al., 2003; Hauser et al., 2003; Jenner, 2003; Agnati et al., 2004).

4.4 Budipine

Budipine is a compound with NMDA antagonistic and anticholinergic properties combined with a positive impact on neurotransmission of 5-HT, norepinephrine, and dopamine. This drug licensed only in Germany supports the onset of an acquired long QT syndrome; its clinical use is limited by a demand for ECG controls on a regular basis by the regulatory health authorities. Previous clinical budipine trials demonstrated the therapeutic efficacy of budipine on motor symptoms in PD patients on an insufficient antiparkinsonian drug regime. These studies in previously treated PD patients pointed out the efficacy of budipine on tremor probably due to its anticholinergic profile. Moreover, budipine provided an additional therapeutic benefit in PD patients on a stable, prior optimum titrated dopaminergic drug regime with levodopa, bromocriptine, and optionally selegiline. Budipine significantly reduced the UPDRS scores for bradykinesia and rigidity, but not for tremor in previously untreated PD patients (Przuntek and Müller, 1999; Przuntek et al., 2002; Müller et al., 2004b).

4.5 Monoaminooxidase-B Inhibitors

Inhibitors of monoaminooxidase-B (MAO-B) have been used in the drug treatment of PD for more than 30 years particularly due to the assumption that these compounds delay progression PD in addition to their symptomatic effect in the treatment of motor symptoms (Birkmayer et al., 1985; Riederer et al., 2004).

4.5.1 Selegiline

Selegiline is metabolized via the CYP450 system, in particular CYP1A2 and CYP3A4. Its main metabolites are N-desmethyl selegiline and mainly amphetamine derivatives (Heinonen and Lammintausta, 1991). N-Desmethyl selegiline has neuroprotective and antiapoptotic properties in vitro, whereas amphetamine-like compounds showed neurotoxic effects in cultures of dopaminergic neurons. A retrospective analysis of mortality data of PD patients revealed an increased life expectancy in PD patients with a combined levodopa/selegiline therapeutic regimen in comparison with the ones who were treated with levodopa alone (Riederer et al., 2004). This was later confirmed by analysis of long-term data of prospective long-term trials with selegiline, but did not turn out statistically significant. However, one assumed that the amphetamine derivatives of selegiline may contribute to the onset of cardiovascular, psychiatric, and neurological side effects. This leads to the development and launch of zydis selegiline®/Xilopar®, a formulation of selegiline, which circumvents the first-pass effect in the liver and thus decreases metabolism to amphetamine derivatives due to rapid absorption in the mouth. The clinical efficacy of 1.25 mg zydis selegiline®/Xilopar® has been shown to be equivalent to oral application of 10 mg selegiline, that is, according to outcomes of a 12 weeks lasting trial, in which this compound also reduced the off-time in fluctuating PD patients (Seager, 1998; Clarke et al., 2003a, b; Waters et al., 2004).

4.5.2 Rasagiline

A further MAO-B inhibitor is rasagiline (R(+)-N-propargyl-1-aminoindane mesylate), which inhibits 3- to 15-fold more potently than selegiline the metabolism of monoamines, that is, dopamine and 5-HT. In the TEMPO-trial, rasagiline significantly reduced the UPDRS total score (mean difference to placebo: 4.2 points) in PD patients after 6 months. In particular, tremor and bradykinesia improved in comparison with placebo administration in this 6-month trial. The LARGO- and the PRESTO-study confirmed the therapeutic efficacy of rasagiline in more advanced PD patients, who experienced a prior optimum titration of their levodopa therapy. Rasagiline significantly reduced the severity of motor complications and duration of off-times in both studies (Parkinson Study Group, 2004, 2005; Stern et al., 2004; Rascol et al., 2005; Clarke, 2005).

5 Neuroprotection of the Dopaminergic System

Slowing of progression of PD is a very important goal of treatment, since results of the neuroregenerative transplantation and curative growth factor trials were very disappointing and future stem cell therapy is far from clinical testing (Harrower and Barker, 2004). The essential problem of all these more regenerative therapeutic approaches is their specific focus on the dopaminergic system in PD and the still missing ability of regulation of dopamine release in the nigrostriatal system in a physiologic manner. Therefore, the transplantation trials in idiopathic PD patients failed due to the onset of dyskinesia and cognitive disturbances, which hypothetically result from dopamine overflow after a certain interval (Freed et al., 2001; Gill et al., 2003; Olanow et al., 2003; Clarke, 2004b). Thus long-term performance of neuroprotection still remains as the one essential option. Many trials investigated the issue of neuroprotection with an outstandingly high scientific interest and financial support both in clinical and basic research, in particular, in the case of PD. Their more or less disappointing results do not only suffer from a relative late diagnosis of PD with respect to the preexisting degeneration of dopaminergic terminals at the level of onset of motor symptoms, but also from the relative short duration of the trials. Moreover, the interpretation of the study results are confounded by an overlap with the symptomatic effect of the investigated compounds or placebo and/or too short wash-out periods, which do not allow a consistent interpretation of the found effect in terms of modification of the disease progress (Parkinson Study Group, 1993; Sano et al., 1997; Foley and Riederer, 1999; Marek et al., 2002; Whone et al., 2003; Fahn et al., 2004; Müller et al., 2004b). Therefore, the promising, at least partially controversially discussed, clinical relevant positive results on neuroprotective and/or life span increasing properties in clinical trials are still controversial and do not reflect the neuroprotective potential, which was assumed after positive outcomes in cell culture and animal models of PD with the tested compounds. Furthermore, the applied biological surrogate marker, that is, the functional brain imaging techniques with various radiotracers, focused only on the dopaminergic system, but they did not consider that neurodegeneration in PD also affects other neurotransmitters, that is, 5-HT, norepinephrine, acetylcholine, with the corresponding onset of nonmotor features of PD (Foley and Riederer, 1999; Wolters et al., 2000; Braak et al., 2003; Przuntek et al., 2004). This drawback also additionally concerns the research in cell culture models with its specific focus on certain types of neuronal cells and in animal models with their exposure to selective toxins of the dopaminergic system.

6   Treatment Recommendations in the Real World

From the clinical point of view, the onset of in earlier stages predictable and then later in the course of PD unforeseen motor fluctuations, hyperkinesias, and dyskinesias led to the still controversially debated attitude to postpone LD application as long as possible and to reduce the need for LD especially in PD patients with a prospective long remaining lifetime with necessary treatment of PD. However, emerging costs of health care systems with their resulting budget problems for practitioners and/or specialists in neurology do not support this treatment strategy, since a high-dose dopamine agonist therapy is much more expensive than monotherapy with LD preparations nearly all over the world (Clarke, 2002, 2004a; Müller et al., 2004c). Moreover, treatment options of initial side effects of dopamine agonist application, for instance nausea or dizziness, only exist in some countries due to the availability of domperidone, a peripheral dopamine receptor blocker. Therefore, a putative initial monotherapy with dopamine agonists with a slow and low beginning suffers from compliance problems and additionally the patients' need for more contact, motivation, educational work, and cooperation. In contrast initial antiparkinsonian drug treatment with LD preparations is more efficacious, enables a faster titration followed by an earlier therapeutic success with respect to motor symptoms, and shows less and then mainly severe short-term side effects in the beginning. Nevertheless, appearance of LD-associated long-term complications stimulated research on the causes of these phenomena and the search for therapeutic alternatives for the treatment of the parkinsonian dopaminergic striatal deficit, since these changes of the therapeutic response to levodopa and dopamine agonists represent one of the major causes of disability in PD patients of more advanced stages of PD.

7   The Principle of Continuous Stimulation and Clinical Practice

In order to prevent or delay the onset of motor complications in the long run, application of predominantly postsynaptic acting dopamine agonists with long half-life and high receptor affinity may be the best option. Nowadays, it is generally accepted that dopamine-containing neurons comprising the nigrostriatal system characteristically manifest slow (about 4–5 Hz), single-spike activity, which is occasionally interrupted by short bursts of faster (usually in the 15–20 Hz range) spiking in response to salient visual or auditory stimuli

(Grace and Bunney, 1984a, b). Since postsynaptic receptor stimulation is roughly proportional to impulse activity at the presynaptic terminal, intrasynaptic dopamine concentrations normally remain fairly constant. Accordingly, the most physiological approach to dopamine replacement in PD is to maintain stable normal intrasynaptic levels of the transmitter amine. However with disease progression, this goal becomes progressively less attainable. As nigral dopaminergic neurons degenerate, the amount of exogenous levodopa entering striatal terminals diminishes. Moreover, due to increasing presynaptic dopaminergic degeneration with resulting less availability of dopaminergic autoreceptors, loss of dopaminergic autoreceptor function occurs in advanced stages of PD (Müller et al., 1997; Ekesbo et al., 1999). The main purpose of this presynaptic inhibitory feedback regulation, particularly in the dorsal putamen, is to maintain congruity within the dopaminergic system in response to antiparkinsonian medication. This reduced inhibitory feedback regulation appears with the progression of nigrostriatal degeneration even in previously untreated PD patients and/or due to chronic dopamine agonist treatment. Instead, increasing amounts of LD are taken up and converted to dopamine in other decarboxylase-containing cells, especially serotonergic neurons (Ng et al., 1972; Wachtel and Abercrombie, 1994; Chase et al., 2000b). In the absence of appropriate mechanisms for storing or regulating the release of dopamine, the newly synthesized amine leaks into the extracellular compartment and diffuses into contact with nearby dopamine receptors. Under such circumstances, intrasynaptic dopamine concentrations reflect the wide swings in cerebral LD levels that occur with standard precursor dosing regimens. LD therapy in patients with advanced PD thus results, at best, in only episodic restoration of physiological dopamine levels. Finally, this ends up in a vicious circle. Administration of short-acting dopamine agonists with low receptor affinity or of LD in patients with advanced disease mainly stimulates postsynaptic dopaminergic receptors at subthreshold levels; interrupted soon after each dose when it briefly rises into the physiological range. This resulting pulsatile stimulation of postsynaptic dopaminergic receptors mainly contributes to the appearance and intensity of predominant and in the beginning predictable motor complications. Postmortem determinations of striatal dopamine concentrations and measurements of spinal fluid homovanillic acid levels suggest that transmitter levels in PD patients receiving standard LD therapy ordinarily peak well above the physiological range, presumably at levels approximating to those achieved in in vitro models exposed to high-intensity stimulation (Grace and Bunney, 1984a; Hornykiewicz, 1998). With a sufficient loss of dopamine terminals, dopaminergic transmission thus tends to be compromised whether or not dopaminomimetic treatment is initiated. Clinically, the whole system with its nonphysiological pattern of dopaminergic stimulation becomes more sensitive to the appearance of loss of drug efficacy, for example, off-periods or onset of psychosis, hyperkinesia, and dyskinesia as symptoms of hyperstimulation of postsynaptic dopaminergic receptors. Thus (i) preponderantly performed combination therapy with LD with its various adjunctive enzyme blockers of metabolization and dopamine agonists, (ii) storage and subsequent uncontrolled release of dopamine as "false" transmitter from serotonergic neurons (Ng et al., 1972; Wachtel and Abercrombie, 1994), and (iii) peripheral resorption problems complicate the treatment of PD patients, since clinicians are more and more unable to predict the time point of occurrence of these side effects of dopaminergic substitution therapy in relation to intake of dopaminergic drugs in particular levodopa. Moreover, a further clinical problem in advanced stages of PD with concomitant temporary onset of biphasic hyperkinesia is to differentiate whether the dopaminergic system is too much stimulated, which results in peak-dose dyskinesia, dystonia, or akinesia, or too less substituted, which leads to similar symptoms, classified as "off"-akinesia, -dystonia, or -akinesia. In this regard, it is noteworthy to discuss results of clinical phase II and phase III trials on the efficacy of an antiparkinsonian compound with selected and well-defined participating PD patients. Sometimes study results sound convincing, but after launch of the compound with widespread use of the drug – clinical practice shows limited efficacy and severe side effects, that is, hepatotoxicity in the case of tolcapone.

8 Novel Therapy Approaches

There also now some new promising results of oral drugs, that is, adenosine A2 antagonists, which modify or modulate function of presynaptic and postsynaptic nigrostriatal dopaminergic neurons. However, it is still rather unsecure whether all these drugs will survive in clinical practice in the long term after taking the first step of approval by the FDA or the EMEA, since there are nearly no data on the impact of these drugs on the efficacy of improvement of quality of life of PD patients (Hutton et al., 2001). However, health insurance systems increasingly demand for these kinds of data to perform reimbursement. Moreover, the availability of cheaper generic drugs influences the acceptance and survival of more developed drug. A typical future example may be rasagiline, which still cannot prove its distinct superiority over its now generic precursor selegiline in terms of clinical efficacy that is measured with a rating or a quality of life instrument. This still necessary acceptance by financing health insurances also concerns treatment strategies, which modify the delivery of the dopaminergic compound to provide more continuous plasma concentrations, which represents a suitable accepted therapeutic approach for improvement of motor fluctuations. Such a novel promising therapeutic alternative is the transdermal drug delivery of dopamine-substituting compounds, which enable more constant plasma levels resulting in an amelioration of intensity and frequency of on–off phenomena, but their superiority over the more probable cheaper oral application of dopamine agonists still need to be proven in the long term with respect to prevention of motor complications within a head-to-head study approach.

9   Pharmacoeconomic and Social Aspects

These issues of reimbursement also concern future neuroprotective treatment strategies. To overcome the increase in restrictions in view of rising health economical problems in various healthcare systems, trials on progression of PD with DA should also focus on the efficacy of the tested compound in preventing and/or delaying patients' referral to nursing homes and/or their need for ambulatory aids (Clarke, 2004a, b). Moreover, they should report on reduction of concomitant therapy with drugs, which treat or impair, that is, psychopathological features, endogenously related to the underlying disease process itself, and/or exogenously caused by necessary symptomatic drugs, that is, psychosis induced by dopamine-substituting agents, in the long run. Positive results on these kinds of long-term complications, which nowadays potentially increase the total treatment costs of chronic degenerative diseases, will help the pharmaceutical companies in their approval by regulatory authorities and in the marketing process of their compound and thus will bring medical progress to PD patients. Nowadays treatment of PD patients in the "real world" with general practitioners and neurologists predominantly starts with LD, since titration with dopamine agonists doubles or even triples daily medication costs. Moreover, PD patients are mostly only referred to movement disorder specialists or PD clinics once motor complications and more severe conditions have arisen (Dodel et al., 1998a, b; Müller et al., 2003d, 2004c; Clarke, 2004b).

10   Treatment Strategies in PD

The neuroprotective treatment strategies of neurodegeneration suffer from diagnosis that is too late in the course of disease onset. An effective neuroprotective therapeutic intervention has to start far more earlier. Therefore, there is a need for easy to handle, cheap diagnostic tools, which may be used in screening procedures for the relative unspecific, subtle clinical signs of the premotor syndrome in PD or related clinical similar symptoms, that is, mild cognitive impairment in Alzheimer's disease. Moreover, one should consider that diagnosis of idiopathic PD according to the current criteria only reflects a Parkinson syndrome, but not the various types of the disease with its individual different course, presentation of symptoms, and response to treatment (Korczyn, 1999; Przuntek and Müller, 1999; Przuntek et al., 2004). The start of therapeutic intervention in terms of substitution of an affected neurotransmitter system, that is, dopamine in PD, should consider strategies that only modulate or support the physiologic neurotransmission process with drugs like amantadine, selegiline, and so on. The need for a direct dopaminergic substitution to achieve a better control of motor symptoms should consider strategies that deliver precursors of dopamine or dopamine agonists to the brain in a more constant fashion. This enables a less alternating stimulation of postsynaptic dopaminergic receptors, but these strategies should also demand for the further regulating role of presynaptic autoreceptors within physiologic ranges to imitate physiologic conditions for the remaining healthy or less affected neurons and glial cells. This will hopefully delay the onset of dyskinesias or unpredictable on–off fluctuations, but will not prevent them in patients with a not yet explored predisposition for these kinds of motor complications. With the very severe onset of these motor complications, the PD patient enters the stage of deep brain stimulation, which extensively reduces the dosages of dopaminergic drug substitution and provides a better quality of life. But deep brain stimulation may also cause psychosocial problems due to change of personal behavior in the long run and its cost effectiveness has still to be proven for financing health care systems in the long run. Deep brain stimulation only improves motor symptoms probably by an electric stimulation of the endogenous, subthalamic dopamine synthesis (Meissner et al., 2002, 2003). Taking into account all the existing hardware problems, one should consider a further still to be developed approach: The direct infusion of the dopaminergic compound, that is, dopamine, LD, or apomorphine in the affected brain region. The technology exists and needs to be simplified and modified only to a certain extent in particular for long-term use (de Yebenes et al., 1988; Gill et al., 2003), but this procedure will allow a more sensitive tuning of motor behavior according to the patients' demands and thus will better imitate central physiologic dopaminergic neurotransmission (Gill et al., 2003; Nutt et al., 2003). The present infusion systems, which administer dopaminergic drugs in a more continuous fashion, showed clinically relevant effects in the treatment of motor complications. Their subcutaneous, intravenous, or gastrointestinal application modes may cause local inflammatory reactions, that is, subcutaneous noodles, at the administration site and have the potential to cause hardware problems in the long run. Moreover, these systems are expensive at present and increase caregiver burden (Nilsson et al., 2001; Stocchi et al., 2002; Tyne et al., 2004). Future research on treatment options in PD patients will also focus on nonmotor features of the PD; however one may suppose that there is no commercial interest on the issue whether early long-term folic acid supplementation will lower the incidence of cognitive deficits, depression, or even psychosis in PD patients once they enter the stage with a need for LD intake with associated homocysteine increase as risk factor for onset of these non motor features of PD.

References

Agnati LF, Leo G, Vergoni AV, Martinez E, Hockemeyer J, et al. 2004. Neuroprotective effect of L-DOPA co-administered with the adenosine A2A receptor agonist CGS 21680 in an animal model of Parkinson's disease. Brain Res Bull 64: 155-164.

Baas H, Beiske AG, Ghika J, Jackson M, Oertel WH, et al. 1997. Catechol-O-methyltransferase inhibition with tolcapone reduces the "wearing off" phenomenon and levodopa requirements in fluctuating parkinsonian patients. J Neurol Neurosurg Psychiatry 63: 421-428.

Battistin L, Bardin PG, Ferro-Milone F, Ravenna C, Toso V, et al. 1999. Alpha-dihydroergocryptine in Parkinson's disease: A multicentre randomized double blind parallel group study. Acta Neurol Scand 99: 36-42.

Beal MF. 1998. Excitotoxicity and nitric oxide in Parkinson's disease pathogenesis. Ann Neurol 44: S110-S114.

Ben Shlomo Y, Marmot MG. 1995. Survival and cause of death in a cohort of patients with parkinsonism: Possible clues to aetiology? J Neurol Neurosurg Psychiatry 58: 293-299.

Bernheimer H, Birkmayer W, Hornykiewicz O. 1961. [Distribution of 5-hydroxytryptamine (serotonin) in the human brain and its behavior in patients with Parkinson's syndrome]. Klin Wochenschr 39: 1056-1059.

Bibbiani F, Oh JD, Petzer JP, Castagnoli N, Chen JF, Jr, et al. 2003. A2A antagonist prevents dopamine agonist-induced motor complications in animal models of Parkinson's disease. Exp Neurol 184: 285-294.

Birkmayer W, Knoll J, Riederer P, Youdim MB, Hars V, et al. 1985. Increased life expectancy resulting from addition of L-deprenyl to Madopar treatment in Parkinson's disease: A longterm study. J Neural Transm 64: 113-127.

Blandini F, Fancellu R, Martignoni E, Mangiagalli A, Pacchetti C, et al. 2001. Plasma homocysteine and L-dopa metabolism in patients with Parkinson disease. Clin Chem 47: 1102-1104.

Block G, Liss C, Reines S, Irr J, Nibbelink D. 1997. Comparison of immediate-release and controlled release carbidopa/levodopa in Parkinson's disease. A multicenter 5-year study. The CR First Study Group. Eur Neurol 37: 23-27.

Bonuccelli U, Colzi A, Del Dotto P. 2002. Pergolide in the treatment of patients with early and advanced Parkinson's disease. Clin Neuropharmacol 25: 1-10.

Bosboom JL, Stoffers D, Wolters EC. 2003. The role of acetylcholine and dopamine in dementia and psychosis in Parkinson's disease. J Neural Transm Suppl: 185-195.

Braak H, Rub U, Gai WP, Del Tredici K. 2003. Idiopathic Parkinson's disease: Possible routes by which vulnerable neuronal types may be subject to neuroinvasion by an unknown pathogen. J Neural Transm 110: 517-536.

Chambers JC, McGregor A, Jean-Marie J, Kooner JS. 1998. Acute hyperhomocysteinaemia and endothelial dysfunction [letter]. Lancet 351: 36-37.

Chase TN, Oh JD. 2000a. Striatal mechanisms and pathogenesis of parkinsonian signs and motor complications. Ann Neurol 47: S122-S129.

Chase TN, Oh JD, Blanchet PJ. 1998. Neostriatal mechanisms in Parkinson's disease. Neurology 51: S30-S35.

Chase TN, Oh JD, Konitsiotis S. 2000b. Antiparkinsonian and antidyskinetic activity of drugs targeting central glutamatergic mechanisms. J Neurol 247 (Suppl. 2): II36-II42.

Clarke CE. 2002. Medical management of Parkinson's disease. J Neurol Neurosurg Psychiatry 72 (Suppl. 1): I22-I27.

Clarke CE. 2004a. A "cure" for Parkinson's disease: Can neuroprotection be proven with current trial designs? Mov Disord 19: 491-498.

Clarke CE. 2004b. Neuroprotection and pharmacotherapy for motor symptoms in Parkinson's disease. Lancet Neurol 3: 466-474.

Clarke CE. 2005. Rasagiline for motor complications in Parkinson's disease. Lancet 365: 914-916.

Clarke A, Brewer F, Johnson ES, Mallard N, Hartig F, et al. 2003a. A new formulation of selegiline: Improved bioavailability and selectivity for MAO-B inhibition. J Neural Transm 110: 1241-1255.

Clarke A, Johnson ES, Mallard N, Corn TH, Johnston A, et al. 2003b. A new low-dose formulation of selegiline: Clinical efficacy, patient preference and selectivity for MAO-B inhibition. J Neural Transm 110: 1257-1271.

Cooper JA, Sagar HJ, Sullivan EV. 1993. Short-term memory and temporal ordering in early Parkinson's disease: Effects of disease chronicity and medication. Neuropsychologia 31: 933-949.

Cotzias GC, Papavasiliou PS, Gellene R. 1969. Modification of Parkinsonism – chronic treatment with L-dopa. N Engl J Med 280: 337-345.

Crosby NJ, Deane KH, Clarke CE. 2003. Amantadine for dyskinesia in Parkinson's disease. Cochrane Database Syst Rev CD003467.

Curran MP, Perry CM. 2004. Cabergoline: A review of its use in the treatment of Parkinson's disease. Drugs 64: 2125-2141.

de Yebenes JG, Fahn S, Jackson-Lewis V, Jorge P, Mena MA, et al. 1988. Continuous intracerebroventricular infusion of dopamine and dopamine agonists through a totally implanted drug delivery system in animal models of Parkinson's disease. J Neural Transm Suppl 27: 141-160.

Deleu D, Northway MG, Hanssens Y. 2002. Clinical pharmacokinetic and pharmacodynamic properties of drugs used

in the treatment of Parkinson's disease. Clin Pharmacokinet 41: 261-309.

Dodel RC, Eggert KM, Singer MS, Eichhorn TE, Pogarell O, et al. 1998a. Costs of drug treatment in Parkinson's disease. Mov Disord 13: 249-254.

Dodel RC, Singer M, Kohne-Volland R, Szucs T, Rathay B, et al. 1998b. The economic impact of Parkinson's disease. An estimation based on a 3-month prospective analysis. Pharmacoeconomics 14: 299-312.

Durif F. 1999. Treating and preventing levodopa-induced dyskinesias: Current and future strategies. Drugs Aging 14: 337-345.

Ekesbo A, Rydin E, Torstenson R, Sydow O, Laengstrom B, et al. 1999. Dopamine autoreceptor function is lost in advanced Parkinson's disease. Neurology 52: 120-125.

Factor SA. 2004. Literature review: Intermittent subcutaneous apomorphine therapy in Parkinson's disease. Neurology 62: S12-S17.

Fahn S. 1996. Controversies in the therapy of Parkinson's disease. Adv Neurol 69: 477-486.

Fahn S, Oakes D, Shoulson I, Kieburtz K, Rudolph A, et al. 2004. Levodopa and the progression of Parkinson's disease. N Engl J Med 351: 2498-2508.

Foley P, Mizuno Y, Nagatsu T, Sano A, Youdin MBH, et al. 2000. The L-DOPA story – an early Japanese contribution. Parkinsonism Relat Disord 6: 1.

Foley P, Riederer P. 1999. Pathogenesis and preclinical course of Parkinson's disease. J Neural Transm Suppl 56: 31-74.

Freed CR, Greene PE, Breeze RE, Tsai WY, DuMouchel W, et al. 2001. Transplantation of embryonic dopamine neurons for severe Parkinson's disease. N Engl J Med 344: 710-719.

Gerlach M, Blum-Degen D, Lan J, Riederer P. 1999. Nitric oxide in the pathogenesis of Parkinson's disease. Adv Neurol 80: 239-245.

Gerlach M, Double K, Arzberger T, Leblhuber F, Tatschner T, et al. 2003. Dopamine receptor agonists in current clinical use: Comparative dopamine receptor binding profiles defined in the human striatum. J Neural Transm 110: 1119-1127.

Gerlach M, van den BM, Blaha C, Bremen D, Riederer P. 2004. Entacapone increases and prolongs the central effects of L-DOPA in the 6-hydroxydopamine-lesioned rat. Naunyn Schmiedebergs Arch Pharmacol 370: 388-394.

Gerlach M, Xiao AY, Kuhn W, Lehnfeld R, Waldmeier P, et al. 2001. The central catechol-O-methyltransferase inhibitor tolcapone increases striatal hydroxyl radical production in L-DOPA/carbidopa treated rats. J Neural Transm 108: 189-204.

Gershanik O, Emre M, Bernhard G, Sauer D. 2003. Efficacy and safety of levodopa with entacapone in Parkinson's disease patients suboptimally controlled with levodopa alone, in daily clinical practice: An international, multicentre, open-label study. Prog Neuropsychopharmacol Biol Psychiatry 27: 963-971.

Gill SS, Patel NK, Hotton GR, O'Sullivan K, McCarter R, et al. 2003. Direct brain infusion of glial cell line-derived neurotrophic factor in Parkinson disease. Nat Med 9: 589-595.

Goetze O, Wieczorek J, Müeller T, Przuntek H, Schmidt WE, et al. 2005. Impaired gastric emptying of a solid test meal in patients with Parkinson's disease using 13C-sodium octanoate breath test. Neurosci Lett 375: 170-173.

Grace AA, Bunney BS. 1984a. The control of firing pattern in nigral dopamine neurons: Burst firing. J Neurosci 4: 2877-2890.

Grace AA, Bunney BS. 1984b. The control of firing pattern in nigral dopamine neurons: Single spike firing. J Neurosci 4: 2866-2876.

Harrower TP, Barker RA. 2004. Is there a future for neural transplantation? BioDrugs 18: 141-153.

Hauser RA, Hubble JP, Truong DD. 2003. Randomized trial of the adenosine A(2A) receptor antagonist istradefylline in advanced PD. Neurology 61: 297-303.

Heinonen EH, Lammintausta R. 1991. A review of the pharmacology of selegiline. Acta Neurol Scand Suppl 136: 44-59.

Hely MA, Morris JG, Traficante R, Reid WG, O'Sullivan DJ, et al. 1999. The sydney multicentre study of Parkinson's disease: Progression and mortality at 10 years. J Neurol Neurosurg Psychiatry 67: 300-307.

Högl B, Saletu M, Brandauer E, Glatzl S, Frauscher B, et al. 2002. Modafinil for the treatment of daytime sleepiness in Parkinson's disease: A double-blind, randomized, crossover, placebo-controlled polygraphic trial. Sleep 25: 905-909.

Holloway RG, Shoulson I, Fahn S, Kieburtz K, Lang A, et al. 2004. Pramipexole vs levodopa as initial treatment for Parkinson disease: A 4-year randomized controlled trial. Arch Neurol 61: 1044-1053.

Hornykiewicz O. 1998. Biochemical aspects of Parkinson's disease. Neurology 51: S2-S9.

Horowski R, Jahnichen S, Pertz HH. 2004. Fibrotic valvular heart disease is not related to chemical class but to biological function: 5-HT2B receptor activation plays crucial role. Mov Disord 19: 1523-1524.

Hubble JP. 2002. Long-term studies of dopamine agonists. Neurology 58: S42-S50.

Hughes AJ, Daniel SE, Kilford L, Lees AJ. 1992. Accuracy of clinical diagnosis of idiopathic Parkinson's disease: A clinico-pathological study of 100 cases. J Neurol Neurosurg Psychiatry 55: 181-184.

Hughes RC, Polgar JG, Weightman D, Walton JN. 1971. Levodopa in Parkinsonism: The effects of withdrawal of anticholinergic drugs. Br Med J 2: 487-491.

Hutton JT, Metman LV, Chase TN, Juncos JL, Koller WC, et al. 2001. Transdermal dopaminergic D(2) receptor agonist therapy in Parkinson's disease with N-0923 TDS: A double-blind, placebo-controlled study. Mov Disord 16: 459-463.

Hutton JT, Morris JL, Brewer MA. 1993. Controlled study of the antiparkinsonian activity and tolerability of cabergoline. Neurology 43: 613-616.

Inzelberg R, Schechtman E, Nisipeanu P. 2003. Cabergoline, pramipexole and ropinirole used as monotherapy in early Parkinson's disease: An evidence-based comparison. Drugs Aging 20: 847-855.

Jellinger KA. 1999. Post mortem studies in Parkinson's disease – is it possible to detect brain areas for specific symptoms? J Neural Transm Suppl 56: 1-29.

Jenner P. 2000. Pathophysiology and biochemistry of dyskinesia: Clues for the development of non-dopaminergic treatments. J Neurol 247 (Suppl. 2): II43-II50.

Jenner P. 2003. A2A antagonists as novel non-dopaminergic therapy for motor dysfunction in PD. Neurology 61: S32-S38.

Jenner P. 2004. Avoidance of dyskinesia: Preclinical evidence for continuous dopaminergic stimulation. Neurology 62: S47-S55.

Katzenschlager R, Hughes A, Evans A, Manson AJ, Hoffman M, et al. 2005. Continuous subcutaneous apomorphine therapy improves dyskinesias in Parkinson's disease: A prospective study using single-dose challenges. Mov Disord 20: 151-157.

Katzenschlager R, Sampaio C, Costa J, Lees A. 2003. Anticholinergics for symptomatic management of Parkinson's disease. Cochrane Database Syst Rev CD003735.

Kent AP, Stern GM, Webster RA. 1990. The effect of benserazide on the peripheral and central distribution and metabolism of levodopa after acute and chronic administration in the rat. Br J Pharmacol 100: 743-748.

Koller WC, Hutton JT, Tolosa E, Capilldeo R. 1999. Immediate-release and controlled-release carbidopa/levodopa in PD: A 5-year randomized multicenter study. Carbidopa/Levodopa Study Group. Neurology 53: 1012-1019.

Korczyn AD. 1999. Parkinson's disease: One disease entity or many? J Neural Transm Suppl 56: 107-111.

Kuhn W, Hummel T, Woitalla D, Müller T. 2001. Plasma homocysteine and MTHFR C677T genotype in levodopa-treated patients with PD. Neurology 56: 281-282.

Kuhn W, Woitalla D, Gerlach M, Russ H, Müller T. 1998. Tolcapone and neurotoxicity in Parkinson's disease. Lancet 352: 1313-1314.

Kurlan R, Rothfield KP, Woodward WR, Nutt JG, Miller C, et al. 1988. Erratic gastric emptying of levodopa may cause "random" fluctuations of parkinsonian mobility. Neurology 38: 419-421.

Lipton SA, Kim WK, Choi YB, Kumar S, D'Emilia DM, et al. 1997. Neurotoxicity associated with dual actions of homocysteine at the N-methyl-D-aspartate receptor. Proc Natl Acad Sci USA 94: 5923-5928.

Marek K, Seibyl J, Shoulson I, Holloway R, Kieburtz K, et al. 2002. Dopamine transporter brain imaging to assess the effects of pramipexole vs levodopa on Parkinson disease progression. JAMA 287: 1653-1661.

Marsden CD. 1998. Clinical experience with cabergoline in patients with advanced Parkinson's disease treated with levodopa. Drugs 55 (Suppl. 1): 17-22.

Matheson AJ, Spencer CM. 2000. Ropinirole: A review of its use in the management of Parkinson's disease. Drugs 60: 115-137.

Meissner W, Harnack D, Paul G, Reum T, Sohr R, et al. 2002. Deep brain stimulation of subthalamic neurons increases striatal dopamine metabolism and induces contralateral circling in freely moving 6-hydroxydopamine-lesioned rats. Neurosci Lett 328: 105-108.

Meissner W, Harnack D, Reese R, Paul G, Reum T, et al. 2003. High-frequency stimulation of the subthalamic nucleus enhances striatal dopamine release and metabolism in rats. J Neurochem 85: 601-609.

Metman LV, Gillespie M, Farmer C, Bibbiani F, Konitsiotis S, et al. 2001. Continuous transdermal dopaminergic stimulation in advanced Parkinson's disease. Clin Neuropharmacol 24: 163-169.

Miller JW, Green R, Mungas DM, Reed BR, Jagust WJ. 2002. Homocysteine, vitamin B6, and vascular disease in AD patients. Neurology 58: 1471-1475.

Miller JW, Selhub J, Nadeau MR, Thomas CA, Feldman RG, et al. 2003. Effect of L-dopa on plasma homocysteine in PD patients: Relationship to B-vitamin status. Neurology 60: 1125-1129.

Miller JW, Shukitt-Hale B, Villalobos-Molina R, Nadeau MR, Selhub J, et al. 1997. Effect of L-Dopa and the catechol-O-methyltransferase inhibitor Ro 41-0960 on sulfur amino acid metabolites in rats. Clin Neuropharmacol 20: 55-66.

Möller JC, Oertel WH, Koster J, Pezzoli G, Provinciali L. 2005. Long-term efficacy and safety of pramipexole in advanced Parkinson's disease: Results from a European multicenter trial. Mov Disord 20: 602-610.

Montastruc JL, Ziegler M, Rascol O, Malbezin M. 1999. A randomized, double-blind study of a skin patch of a dopaminergic agonist, piribedil, in Parkinson's disease. Mov Disord 14: 336-341.

Morrish P. 2002. Is it time to abandon functional imaging in the study of neuroprotection? Mov Disord 17: 229-232.

Morrish PK. 2003. REAL and CALM: What have we learned? Mov Disord 18: 839-840.

Muhlack S, Woitalla D, Welnic J, Twiehaus S, Przuntek H, et al. 2004. Chronic levodopa intake increases levodopa plasma bioavailability in patients with Parkinson's disease. Neurosci Lett 363: 284-287.

Müller T. 2002. Dopaminergic substitution in Parkinson's disease. Expert Opin Pharmacother 3: 1393-1403.

Müller T, Benz S, Börnke C, Russ H, Przuntek H. 2003a. Repeated rating improves value of diagnostic dopaminergic challenge tests in Parkinson's disease. J Neural Transm 110: 603-609.

Müller T, Eising EG, Reiners C, Przuntek H, Jacob M, et al. 1997. 2-[123I]-iodolisuride SPET visualizes dopaminergic loss in de-novo parkinsonian patients: Is it a marker of striatal pre-synaptic degeneration? Nucl Med Commun 18: 1115-1121.

Müller T, Fritze J. 2003b. Fibrosis associated with dopamine agonist therapy in Parkinson's disease. Clinical Neuropharmacology 26: 109-111.

Müller T, Kuhn W, Przuntek H. 2004b. Efficacy of budipine and placebo in untreated patients with Parkinson's disease. J Neural Transm. 112: 1015-1023.

Müller T, Kuhn W, Schulte T, Przuntek H. 2003c. Intravenous amantadine sulphate application improves the performance of complex but not simple motor tasks in patients with Parkinson's disease. Neurosci Lett 339: 25-28.

Müller T, Renger K, Kuhn W. 2004a. Levodopa associated homocysteine increase and sural axonal neurodegeneration. Arch Neurol 61: 657-660.

Müller T, Voss B, Hellwig K, Josef SF, Schulte T, et al. 2004c. Treatment benefit and daily drug costs associated with treating Parkinson's disease in a Parkinson's disease clinic. CNS Drugs 18: 105-111.

Müller T, Voss B, Hellwig K, Przuntek H. 2003d. Treatment benefit correlates with increase of daily drug costs in Parkinson's disease clinics. NeuroRehabilitation 18: 271-275.

Müller T, Woitalla D, Fowler B, Kuhn W. 2002. 3-OMD and homocysteine plasma levels in parkinsonian patients. J Neural Transm 109: 175-179.

Müller T, Woitalla D, Hauptmann B, Fowler B, Kuhn W. 2001. Decrease of methionine and S-adenosylmethionine and increase of homocysteine in treated patients with Parkinson's disease. Neurosci Lett 308: 54-56.

Müller T, Woitalla D, Kuhn W. 2003e. Benefit of folic acid supplementation in parkinsonian patients treated with levodopa. J Neurol Neurosurg Psychiatry 74: 549.

Müller T, Woitalla D, Saft C, Kuhn W. 2000a. Levodopa in plasma correlates with body weight of parkinsonian patients. Parkinsonism Relat Disord 6: 171-173.

Müller T, Woitalla D, Schulz D, Peters S, Kuhn W, et al. 2000b. Tolcapone increases maximum concentration of levodopa. J Neural Transm 107: 113-119.

Murata M, Mizusawa H, Yamanouchi H, Kanazawa I. 1996. Chronic levodopa therapy enhances dopa absorption: Contribution to wearing-off. J Neural Transm 103: 1177-1185.

Nakaso K, Yasui K, Kowa H, Kusumi M, Ueda K, et al. 2003. Hypertrophy of IMC of carotid artery in Parkinson's disease is associated with L-DOPA, homocysteine, and MTHFR genotype. J Neurol Sci 207: 19-23.

Ng LK, Chase TN, Colburn RW, Kopin IJ. 1972. L-dopa in Parkinsonism. A possible mechanism of action. Neurology 22: 688-696.

Nilsson D, Nyholm D, Aquilonius SM. 2001. Duodenal levodopa infusion in Parkinson's disease–long-term experience. Acta Neurol Scand 104: 343-348.

Nissinen E, Nissinen H, Larjonmaa H, Vaananen A, Helkamaa T, 2005. The COMT inhibitor, entacapone, reduces levodopa-induced elevations in plasma homocysteine in healthy adult rats. J Neural Transm 112: 1213-1221.

Nutt JG, Burchiel KJ, Comella CL, Jankovic J, Lang AE, et al. 2003. Randomized, double-blind trial of glial cell line-derived neurotrophic factor (GDNF) in PD. Neurology 60: 69-73.

Obeso JA, Olanow CW, Nutt JG. 2000. Levodopa motor complications in Parkinson's disease. Trends Neurosci 23: S2-S7.

Ogawa N. 2000. Factors affecting levodopa effects in Parkinson's disease. Acta Med Okayama 54: 95-101.

Olanow CW, Goetz CG, Kordower JH, Stoessl AJ, Sossi V, et al. 2003. A double-blind controlled trial of bilateral fetal nigral transplantation in Parkinson's disease. Ann Neurol 54: 403-414.

Olanow CW, Kieburtz K, Stern M, Watts R, Langston JW, et al. 2004. Double-blind, placebo-controlled study of entacapone in levodopa-treated patients with stable Parkinson disease. Arch Neurol 61: 1563-1568.

Olanow CW, Obeso JA. 2000. Pulsatile stimulation of dopamine receptors and levodopa-induced motor complications in Parkinson's disease: Implications for the early use of COMT inhibitors. Neurology 55: S72-S77.

Olanow CW, Schapira AH, Roth T. 2000. Falling asleep at the wheel: Motor vehicle mishaps in people taking pramipexole and ropinirole [letter; comment]. Neurology 54: 274-277.

Olanow CW, Stocchi F. 2004. COMT inhibitors in Parkinson's disease: Can they prevent and/or reverse levodopa-induced motor complications? Neurology 62: S72-S81.

Parkinson Study Group 1993. Effects of tocopherol and deprenyl on the progression of disability in early Parkinson's disease. The Parkinson Study Group. N Engl J Med 328: 176-183.

Parkinson Study Group 2004. A controlled, randomized, delayed-start study of rasagiline in early Parkinson disease. Arch Neurol 61: 561-566.

Parkinson Study Group 2005. A randomized placebo-controlled trial of rasagiline in levodopa-treated patients with Parkinson disease and motor fluctuations: The PRESTO study. Arch Neurol 62: 241-248.

Parkinson J. 1817. An essay on the shaking palsy. London: Sherwood Neely & Jones.

Poewe WH, Deuschl G, Gordin A, Kultalahti ER, Leinonen M. 2002. Efficacy and safety of entacapone in Parkinson's disease patients with suboptimal levodopa response: A 6-month randomized placebo-controlled double-blind study in Germany and Austria (Celomen study). Acta Neurol Scand 105: 245-255.

Pondal M, Del Ser T, Bermejo F. 1996. Anticholinergic therapy and dementia in patients with Parkinson's disease. J Neurol 243: 543-546.

Przuntek H, Bittkau S, Bliesath H, Büttner U, Fuchs G, et al. 2002. Budipine provides additional benefit in patients with Parkinson disease receiving a stable optimum dopaminergic drug regimen. Arch Neurol 59: 803-806.

Przuntek H, Müller T. 1999. Clinical efficacy of budipine in Parkinson's disease. J Neural Transm Suppl 56: 75-82.

Przuntek H, Müller T, Riederer P. 2004. Diagnostic staging of Parkinson's disease: Conceptual aspects. J Neural Transm 111: 201-216.

Przuntek H, Welzel D, Blumner E, Danielczyk W, Letzel H, et al. 1992. Bromocriptine lessens the incidence of mortality in L-dopa-treated parkinsonian patients: Prado-study discontinued. Eur J Clin Pharmacol 43: 357-363.

Przuntek H, Welzel D, Gerlach M, Blumner E, Danielczyk W, et al. 1996. Early institution of bromocriptine in Parkinson's disease inhibits the emergence of levodopa-associated motor side effects. Long-term results of the PRADO study. J Neural Transm Gen Sect 103: 699-715.

Ransmayr G, Kunig G, Gerstenbrand F. 1992. Modern therapy of Parkinson's disease. J Neural Transm Suppl 38: 129-140.

Rascol O. 2000. The pharmacological therapeutic management of levodopa-induced dyskinesias in patients with Parkinson's disease. J Neurol 247 (Suppl. 2): II51-II57.

Rascol O, Brooks DJ, Korczyn AD, De Deyn PP, Clarke CE, et al. 2000. A five-year study of the incidence of dyskinesia in patients with early Parkinson's disease who were treated with ropinirole or levodopa. 056 Study Group. N Engl J Med 342: 1484-1491.

Rascol O, Brooks DJ, Melamed E, Oertel W, Poewe W, et al. 2005. Rasagiline as an adjunct to levodopa in patients with Parkinson's disease and motor fluctuations (LARGO, Lasting effect in Adjunct therapy with Rasagiline Given Once daily, study): A randomised, double-blind, parallel-group trial. Lancet 365: 947-954.

Rascol O, Goetz C, Koller W, Poewe W, Sampaio C. 2002. Treatment interventions for Parkinson's disease: An evidence based assessment. Lancet 359: 1589-1598.

Riederer P, Lachenmayer L, Laux G. 2004. Clinical applications of MAO-inhibitors. Curr Med Chem 11: 2033-2043.

Riederer P, Wuketich S. 1976. Time course of nigrostriatal degeneration in parkinson's disease. A detailed study of influential factors in human brain amine analysis. J Neural Transm 38: 277-301.

Rinne UK. 1999. [Combination therapy with lisuride and L-dopa in the early stages of Parkinson's disease decreases and delays the development of motor fluctuations. Long-term study over 10 years in comparison with L-dopa monotherapy]. Nervenarzt 70 (Suppl. 1): S19-S25.

Russ H, Müller T, Woitalla D, Rahbar A, Hahn J, et al. 1999. Detection of tolcapone in the cerebrospinal fluid of parkinsonian subjects. Naunyn Schmiedebergs Arch Pharmacol 360: 719-720.

Sano M, Ernesto C, Thomas RG, Klauber MR, Schafer K, et al. 1997. A controlled trial of selegiline, alpha-tocopherol, or both as treatment for Alzheimer's disease. The Alzheimer's Disease Cooperative Study. N Engl J Med 336: 1216-1222.

Schlussel E, Preibisch G, Putter S, Elstner EF. 1995. Homocysteine-induced oxidative damage: Mechanisms and possible roles in neurodegenerative and atherogenic processes. Z Naturforsch [C.] 50: 699-707.

Seager H. 1998. Drug-delivery products and the Zydis fast-dissolving dosage form. J Pharm Pharmacol 50: 375-382.

Seshadri S, Beiser A, Selhub J, Jacques PF, Rosenberg IH, et al. 2002. Plasma homocysteine as a risk factor for dementia and Alzheimer's disease. N Engl J Med 346: 476-483.

Skirboll LR, Grace AA, Bunney BS. 1979. Dopamine auto- and postsynaptic receptors: Electrophysiological evidence for differential sensitivity to dopamine agonists. Science 206: 80-82.

Stanwood GD, Lucki I, McGonigle P. 2000. Differential regulation of dopamine D2 and D3 receptors by chronic drug treatments. J Pharmacol Exp Ther 295: 1232-1240.

Stern MB, Marek KL, Friedman J, Hauser RA, LeWitt PA, et al. 2004. Double-blind, randomized, controlled trial of rasagiline as monotherapy in early Parkinson's disease patients. Mov Disord 19: 916-923.

Stocchi F, Ruggieri S, Vacca L, Olanow CW. 2002. Prospective randomized trial of lisuride infusion versus oral levodopa in patients with Parkinson's disease. Brain 125: 2058-2066.

Storch A, Blessing H, Bareiss M, Jankowski S, Ling ZD, et al. 2000. Catechol-O-methyltransferase inhibition attenuates levodopa toxicity in mesencephalic dopamine neurons. Mol Pharmacol 57: 589-594.

Thomas A, Iacono D, Luciano AL, Armellino K, Di Iorio A, et al. 2004. Duration of amantadine benefit on dyskinesia of severe Parkinson's disease. J Neurol Neurosurg Psychiatry 75: 141-143.

Tintner R, Manian P, Gauthier P, Jankovic J. 2005. Pleuropulmonary fibrosis after long-term treatment with the dopamine agonist pergolide for Parkinson disease. Arch Neurol 62: 1290-1295.

Tyne HL, Parsons J, Sinnott A, Fox SH, Fletcher NA, et al. 2004. A 10 year retrospective audit of long-term apomorphine use in Parkinson's disease. J Neurol 251: 1370-1374.

Uitti RJ, Rajput AH, Ahlskog JE, Offord KP, Schroeder DR, et al. 1996. Amantadine treatment is an independent predictor of improved survival in Parkinson's disease. Neurology 46: 1551-1556.

Van Camp G, Flamez A, Cosyns B, Weytjens C, Muyldermans L, et al. 2004. Treatment of Parkinson's disease with pergolide and relation to restrictive valvular heart disease. Lancet 363: 1179-1183.

Verhagen ML, Del Dotto P, Blanchet PJ, van den MP, Chase TN. 1998a. Blockade of glutamatergic transmission as treatment for dyskinesias and motor fluctuations in Parkinson's disease. Amino Acids 14: 75-82.

Verhagen ML, Del Dotto P, van den MP, Fang J, Mouradian MM, et al. 1998b. Amantadine as treatment for dyskinesias and motor fluctuations in Parkinson's disease. Neurology 50: 1323-1326.

Wachtel SR, Abercrombie ED. 1994. L-3,4-dihydroxyphenylalanine-induced dopamine release in the striatum of intact and 6-hydroxydopamine-treated rats: Differential effects of monoamine oxidase A and B inhibitors. J Neurochem 63: 108-117.

Waters CH, Sethi KD, Hauser RA, Molho E, Bertoni JM. 2004. Zydis selegiline reduces off time in Parkinson's disease patients with motor fluctuations: A 3-month, randomized, placebo-controlled study. Mov Disord 19: 426-432.

Whone AL, Watts RL, Stoessl AJ, Davis M, Reske S, et al. 2003. Slower progression of Parkinson's disease with ropinirole versus levodopa: The REAL-PET study. Ann Neurol 54: 93-101.

Woitalla D, Müller T, Benz S, Horowski R, Przuntek H. 2004. Transdermal lisuride delivery in the treatment of Parkinson's disease. J Neural Transm Suppl: 89-95.

Wolters EC, Francot C, Bergmans P, Winogrodzka A, Booij J, et al. 2000. Preclinical (premotor) Parkinson's disease. J Neurol 247 (Suppl 2): II103-II109.

Zeng BY, Pearce RK, Mac Kenzie GM, Jenner P. 2000. Alterations in preproenkephalin and adenosine-2a receptor mRNA, but not preprotachykinin mRNA correlate with occurrence of dyskinesia in normal monkeys chronically treated with L-DOPA. Eur J Neurosci 12: 1096-1104.

8 Progressive Supranuclear Palsy and Corticobasal Degeneration

F. Geser · S. W. Scholz · G. K. Wenning

Abstract: Progressive supranuclear palsy (PSP) and corticobasal degeneration (CBD) are commonly referred to as "atypical parkinsonian syndromes" or "frontotemporal dementias." Clinically, they are characterized by an akinetic-rigid syndrome and cognitive impairment combined with a variety of so-called "plus features." On pathological grounds, their hallmarks are certain neurodegenerative lesion patterns coupled with tau protein-positive intracellular protein aggregates in both neuronal and glial cells bodies or their processes. The pathogenesis of these disorders is not fully elucidated yet. Both genetic and environmental factors may be contributory. Here we review briefly the clinical presentation, comment on the neuropathology, and discuss in more detail the current state of pathogenetic concepts of these disorders with a special focus on recent genetic findings.

List of Abbreviations: AD, Alzheimer's disease; CBD, corticobasal degeneration; FTDP-17, frontotemporal dementia with parkinsonism linked to chromosome 17; 3R-tau, 3 repeat tau; 4R-tau, 4 repeat tau; PD, Parkinson's disease; PiD, Pick's disease; PSP, progressive supranuclear palsy; TIQs, tetrahydroisoquinolines

1   Clinical Presentation

In 1964, John Steele, Clifford Richardson, and Jerzy Olszewski were the first to describe progressive supranuclear palsy (PSP), nowadays also known as Steele–Richardson–Olszewski syndrome (Steele et al., 1964). More recently, epidemiological studies have shown that PSP actually is a frequent cause of atypical parkinsonism (for review see Burn and Lees (2002) and Litvan (2003)). In fact, incidence and age-adjusted prevalence rates were shown to be as high as 5.3 (in the population aged 50–99 years) and 6.4, respectively (Bower et al., 1997; Schrag et al., 1999; Poewe and Wenning, 2003). Mean age at disease onset is around 63 years and mean survival close to 6 years (Litvan, 1996b; Poewe and Wenning, 2003). Salient clinical features were reported to comprise poorly levodopa-responsive parkinsonism, supranuclear ophthalmoplegia (vertical before horizontal gaze palsy), postural instability with recurrent falls early on, and dementia (Burn and Lees, 2002; Poewe and Wenning, 2003). Further clinical features include slowing of vertical saccades, a staring facial expression, neck rigidity, erect posture, pseudobulbar palsy with dysphagia and dysarthria, palilalia or palilogia, and pseudobulbar crying or laughing (Poewe and Wenning, 2003). The fixed facial expression may reflect both a very low blink rate and frontalis muscle overactivity (Burn and Lees, 2002; Poewe and Wenning, 2003; Litvan, 2005a). The National Institute of Neurologic Disorders and Stroke and The Society for PSP, Inc. diagnostic criteria for PSP have been established as the "gold standard" for the clinical diagnosis of PSP since their introduction more than 10 years ago (Litvan et al., 1996a; Poewe and Wenning, 2003). Despite the fact that these criteria define supranuclear vertical gaze palsy as obligatory feature for a diagnosis of probable PSP, this sign is rather nonspecific and may also occur in corticobasal degeneration (CBD) and a variety of other disease of the central nervous system such as arteriosclerotic pseudo-parkinsonism, dementia with Lewy bodies, Creutzfeldt–Jakob disease, Whipple's disease, or Huntington's disease (Poewe and Wenning, 2003; Litvan, 2005a). Sometimes, in early disease stages, PSP patients may not unequivocally show postural instability or ophthalmoplegia and may be still responsive to levodopa and might be, therefore, difficult to distinguish from Parkinson's disease (PD). In this context, it is interesting to note that the classic clinical description of PSP including supranuclear gaze palsy, early falls, and dementia (see above) may not adequately describe all of the PSP patients. In fact, in 103 consecutive cases of pathologically confirmed PSP, Williams and colleagues (2005) identified two clinical phenotypes by factor analysis, which they have named Richardson's syndrome and PSP-parkinsonism. Cases of Richardson's syndrome made up 54% of all cases and were characterized by the early onset of postural instability and falls, supranuclear vertical gaze palsy, and cognitive dysfunction. The PSP-parkinsonism group, accounting for 32% of cases, was characterized by asymmetric onset, tremor, a moderate initial therapeutic response to levodopa and was frequently confused with PD. Cases that could not be separated according to these criteria constituted 14%. Taken these data together, PSP-parkinsonism seems to represent a second discrete clinical phenotype accounting for about one-third of patients and needs to be clinically distinguished from both classical PSP and PD. Despite any attempts of symptomatic therapy, PSP is a relentlessly progressive disorder. In late-disease stages, PSP patients become wheelchair or bed bound and their voice is reduced to a distinctive slurred growl or groan (Burn and Lees, 2002; Poewe and Wenning, 2003). Dressing and eating is not possible anymore and profound difficulties in swallowing bear the danger of food aspiration with consequent aspiration pneumonia. Eventually, the patient ends in a state of immobility with akinesia, rigidity, or dystonia, and palliative care is very important (Poewe and Wenning, 2003).

CBD is an orphan disease and its true incidence or prevalence rates are not clearly established yet (Mahapatra et al., 2004). Nonetheless, according to a study conducted in Japan, the prevalence of CBD has been estimated as 1 for every 2.5 PSP patients; this corresponds to a CBD prevalence rate of 1.7 (Morimatsu and Negoro, 2002). Notwithstanding this, clinical and pathological studies usually suggest this disorder does occur less frequently (Hughes et al., 2002; Riley and Moro-de-Casillas, 2005). CBD typically manifests around the seventh decade of life (Kumar et al., 2002). In 1968, Rebeiz and colleagues reported on "corticodentatonigral degeneration with neuronal achromasia" (Rebeiz et al., 1968; Wakabayashi and Takahashi, 2004). The terms "achromasia" refers to the finding of swollen neurons devoid of Nissl substance (Mahapatra et al., 2004). Since this seminal description, it took about two more decades to re-awake interest in this disease (Gibb et al., 1989; Watts et al., 1989; Riley et al., 1990; Poewe and Wenning, 2003). The actual term "corticobasal degeneration" was coined in 1989 by Gibb and colleagues. The clinical features may vary and presenting features include progressive asymmetrical akineto-rigid parkinsonism and apraxia, progressive aphasia, impaired ocular movements, and cognitive impairment (Wakabayashi and Takahashi, 2004). According to Poewe and Wenning (2003), other movement disorders include, most commonly, dystonia and myoclonus coupled with the "alien limb" phenomena and cortical sensory loss. A substantial proportion of patients eventually become depressed and show frontal-lobe type behavioral changes comprising apathy or disinhibition, impulsiveness, and irritability (Rinne et al., 1994; Wenning et al., 1998; Cummings and Litvan, 2000; Poewe and Wenning, 2003). As Mahapatra and colleagues (2004) pointed out, cognitive dysfunction was assumed for a long time to be only rarely encountered in CBD (Rebeiz et al., 1968; Rinne et al., 1994). However, it is becoming increasingly clear that cognitive impairment actually is a frequent symptom and may even be the presenting feature of this disorder (Wenning et al., 1998; Grimes et al., 1999; Graham et al., 2003; Lang, 2003). Interestingly, recently, a large clinicopathological study was published showing that a substantial proportion of cases that presented clinically with frontotemporal dementia were classified as tauopathies on pathological grounds (Forman et al., 2006); and among the tauopathies, CBD was the most common neuropathological diagnosis similar to an earlier report (Kertesz et al., 2005; Forman et al., 2006). Moreover, it was shown that the pattern of clinical features considered characteristic of CBD is associated with heterogeneous pathologies including PSP, Alzheimer's disease (AD), and Pick's disease (PiD) (Boeve et al., 1999). These results challenge the idea of CBD as a distinct clinicopathologic entity as suggested by the early reports. In keeping with this, PSP and CBD may share many clinical features (Bergeron et al., 1997; Boeve et al., 2003; Josephs and Dickson, 2003). However, in contrast to PSP, the parkinsonian syndrome in CBD is typically appendicular rather than axial (Riley and Moro-de-Casillas, 2005). Several sets of clinical diagnostic criteria have been already proposed for CBD (Riley et al., 1990; Maraganore et al., 1992; Lang et al., 1994; Riley and Lang, 2000; Kumar et al., 2002; Morimatsu and Negoro, 2002; Boeve et al., 2003; Riley and Moro-de-Casillas, 2005). They may facilitate the recognition of CBD in clinical practice and advance certain kinds of clinical research. According to Poewe and Wenning (2003), the motor disorders associated with CBD usually remain clearly asymmetric, eventually spreading from the affected arm to the ipsilateral leg, and progress relentlessly until death. Mean survival is about 7 years (Kumar et al., 2002). Bilateral parkinsonian features (including tremor, rigidity, and bradykinesia) as well as frontal lobe syndrome at first neurological evaluation predict shorter survival (Wenning et al., 1998).

2 Neuropathology

The frontotemporal dementias PSP and CBD are tauopathies characterized by a degeneration of various brain areas associated with an accumulation of insoluble, filamentous, abnormally hyperphosphorylated, tau-protein in both neurons and glial cells.

Although macroscopic examination of a given PSP case might be unrevealing, common findings are atrophy or destruction of several subcortical structures including brainstem (such as the superior colliculus, periaqueductal grey matter, pretectal areas, substantia nigra, midbrain and pontine reticular formation) as well as basal ganglia (such as the subthalamic nucleus and pallidum) (Jellinger and Blancher, 1992; Dickson, 1999; Cordato et al., 2000; Halliday et al., 2000; Burn and Lees, 2002). PSP is histopathologically characterized by tau-positive globose-type neurofibrillary tangles and neuropil threads. And the tufted astrocyte represents the signature lesion of this disorder (Williams, 2006). This pathology is found in, especially, several subcortical structures as described above. Deeper cortical layers, in particular the motor and premotor cortices, might also be involved, albeit to a lesser degree (Hauw et al., 1994; Toloso et al., 2002). In the cerebellum, grumose degeneration is frequently encountered and consists of degeneration of terminal axons of Purkinje cells in the dentate nucleus (Cruz-Sanchez et al., 1992). Coiled bodies are comma-shaped oligodendroglial inclusions that occur both in PSP and CBD (Wakabayashi and Takahashi, 1996; Arima et al., 1997; Tawana and Ramsden, 2001; Burn and Lees, 2002; Williams, 2006).

The macroscopic appearance of CBD is characterized by focal cortical asymmetric parietofrontal (or frontotemporal) atrophy (Armstrong, 2000; Dickson et al., 2002, Mahapatra et al., 2004; Wakabayashi and Takahashi, 2004; Litvan, 2005b). There is inconsistently a depigmentation of the substantia nigra and the midbrain may be atrophic as a whole (Dickson, 1999). As Poewe and Wenning (2003) pointed out, CBD affects the nigrostriatal system and a variety of other subcortical structures including varying degrees of cell loss coupled with gliosis in the thalamus, subthalamic nucleus, globus pallidus, red nucleus, and scattered changes in other brainstem nuclei do also occur. The dentate nucleus in the cerebellum might be affected, too. Actually, the lateral two-thirds of the substantia nigra have been reported to exhibit neuronal drop out and gliosis (Rebeiz et al., 1968; Gibb et al., 1989). In the affected cortical areas, there is also loss of neurons accompanied by gliosis, and there might be spongiosis of superficial cortical areas in addition (Kumar et al., 2002). CBD is further characterized by achromatic balloon-shaped or "ballooned neurons" and some small neurofibrillary tangles or many pretangles (Uchihara et al., 1998; Oyanagi et al., 2001; Ikeda et al., 2002; Wakabayashi and Takahashi, 2004). The cytoplasm of these ballooned neurons stain positively for phosphorylated neurofilaments or αB-crystalline and sometimes for ubiquitin and tau (Gibb et al., 1989; Dickson, 1999; Smith et al., 1992; Dickson et al., 2000; Kumar et al., 2002, Mahapatra et al., 2004). Further, these cells show a lack of apparent Nissl substance (Dickson, 1999) different to central chromatolysis (Rebeiz et al., 1968; Case Records of the Massachusetts General Hospital, 1993; Mahapatra et al., 2004). Although ballooned neurons occur in a high density in CBD, they are not specific for this disease and may be also be encountered in other neurodegenerative diseases such as PSP, albeit in a lesser frequency (Dickson, 1999; Mahapatra et al., 2004, Wakabayashi and Takahashi, 2004). Marked and widespread thread-like lesions in both the grey and white matter is a main feature of CBD. Astrocytic plaques are frequently encountered in CBD and coiled bodies may also be seen (see above) (Ikeda et al., 1994; Feany and Dickson, 1996; Wakabayashi and Takahashi, 2004; Litvan, 2005b).

Criteria for the postmortem diagnosis of CBD have been developed and validated a couple of years ago (Dickson et al., 2002). According to these neuropathological criteria, core features include neuronal loss in focal cortical, nigral, and striatal areas associated with both neuronal and glial pathology (especially astrocytic plaques and threads) in both white and grey matter. The ballooned neurons are only incorporated as supportive features into these criteria. Hence, a diagnosis of CBD can be established in a given case with scarce or no detectable ballooned neurons.

Despite this attempt to establish a set of diagnostic criteria featuring the minimal pathological hallmarks required for a diagnosis of CBD, this disorder may show considerable neuropathological overlap with other frontotemporal dementias, in particular PSP (for review see Wakabayashi and Takahashi, 2004). This is exemplified by the substantia nigra that is affected in both of these disorders. Further, the dentate nucleus in the cerebellum is degenerated in both of these diseases, too. However, this feature is more frequently observed in PSP compared with CBD. In PSP, the subthalamic nucleus is more affected than the striatum; this contrasts to CBD (Wakabayashi and Takahashi, 1996; Tsuchiya et al., 1997; Wakabayashi and Takahashi, 2004). Balloon-shaped neurons are of higher diagnostic value for CBD compared with PSP (Mackenzie and Hudson, 1995; Higuchi et al., 1995; Togo and Dickson, 2002; Wakabayashi and Takahashi, 2004). Coiled bodies contribute little to the differential diagnosis of these diseases. In contrast, high numbers of astrocytic plaques (i.e., an accumulation of the tau protein in distal cellular processes) or tufted astrocytes (i.e., an accumulation of tau in the entire length of the processes) may be associated with CBD and PSP, respectively, and this may help in differentiating these two diseases (Feany and Dickson, 1995; Feany et al., 1996; Komori et al., 1998; Komori, 1999; Wakabayashi and Takahashi, 2004; Litvan, 2005b).

The ultrastructural differences between PSP and CBD have been summarized by Dickson (1999) as follows: The neurofibrillary tangles in PSP contain mostly straight filaments (Powell et al., 1974; Bugiani et al., 1979; Yagishita et al., 1979). In CBD, the filaments show a paired helical appearance (Dickson, 1999). AD cases mostly show paired helical filaments, too. However, they may exhibit further distinct ultrastructural characteristics that may allow differentiation from those occurring in CBD (for more details see Dickson, 1999). More recently, the physical parameters of the core of tau-paired helical filaments were studied and described in detail by von Bergen and colleagues (2006).

3   Pathogenesis

Substantial clinical and histopathological overlap between different tauopathies has been reported (for a detailed neuropathological review see Armstrong et al., 2005). Furthermore, the phenotypes of patients classified within the same disorder may show considerable variations. As outlined above, PSP and CBD may present with complex clinicopathological heterogeneity that represents a challenge for a proper diagnostic work-up. This further emphasizes the pathophysiological complexity of tauopathies. As with all syndromes in the group "frontotemporal dementias," the etiology of PSP and CBD is not fully understood yet. However, a multifactorial genesis including both endogenous (genetic) and exogenous (environmental) influences or mechanism eventually leading to protein aggregation and death of both neuronal and glial cells might be assumed.

3.1 Environmental Factors

The pathogenetic role of environmental factors in PSP has been reviewed in detail elsewhere (Burn and Lees, 2002; Litvan, 2003, 2005a), and some of these ideas are thus summarized: Actually, there are single case-studies reporting on the environmental exposure to organic solvents in PSP (McCrank and Rabheru, 1989; Pezzoli et al., 1989; McCrank, 1990; Pezzoli et al., 1990; Tetrud et al., 1994; Litvan, 2003, 2005a). However, taken together the small number of case-control studies conducted so far, no environmental factor (such as pesticide exposure, nonsmoking, and head trauma) has been clearly established to cause PSP (Davis et al., 1988; Golbe et al., 1996; Dickson, 1999; Vanacore et al., 2000; Vanacore et al., 2001; Litvan, 2005a). Interestingly, Vanacore and colleagues (2000) reported that different groups of neurodegenerative disease seem to be associated with different smoking habits insofar as the inverse association with smoking found previously in PD is shared by multiple system atrophy but not by PSP. For CBD, there are no comparable case-control studies available (Dickson, 1999). As Burn and Lees (2002) pointed out, the consuming of tropical plants or fruits and herbal teas in Guadeloupe (French West Indies) have been linked to a parkinsonian syndrome clinically and pathologically reminiscent to PSP; stopping their consumption might have stabilized or even improved some symptoms (Caparros-Lefebvre and Elbaz, 1999; Burn and Lees, 2002; Caparros-Lefebvre et al., 2002). Tetrahydroisoquinolines (TIQs) have been found in the tropical fruits and teas. They are also present in varying amounts in Western food (e.g., milk, eggs, cocoa, or bananas) (Makino et al., 1988; Niwa et al., 1989b; Litvan, 2005a). Chronic exposure to TIQs might lead to their accumulation in the brain (Makino et al., 1988; Litvan, 2003) and maybe eventually cause atypical parkinsonism (Caparros-Lefebvre and Elbaz, 1999). And indeed, Niwa and colleagues (1989a) found TIQs in parkinsonian brains. As Litvan pointed out (Litvan, 2003, 2005a), TIQs might have a neurotoxic potential and its injection into mice, rats, or primates generates a parkinsonian movement disorder (Makino et al., 1990; Yoshida et al., 1990; Kikuchi et al., 1991; Tasaki et al., 1991; Kotake et al., 1995; 1996; Kawai et al., 1998; Friedrich, 1999; Lannuzel et al., 2000; Soto-Otero et al., 2001). Taken together, the concept of an exogenous toxic substance emerging from the clinically and pathologically similar cases on

Guadeloupe in the pathogenesis of tauopathies such as PSP or CBD remains enigmatic. Further, link between an environmental compound and the actual tau-pathology in the brain is not known, and, for sure, endogenous or genetic factors may be contributory and are therefore reviewed in the following section.

3.2 Tau Pathology

As outlined above, varying degrees and kinds of tau pathology are present in CBD and PSP as well as in other frontotemporal dementias including AD, Lewy body variant of AD, PiD, frontotemporal dementia with parkinsonism linked to chromosome 17 (FTDP-17), argyrophilic grain disease, and tangle-predominant senile dementia (Forman et al., 2006). These diseases are therefore collectively subsumed as tauopathies. In keeping with this, the tau gene and its gene products may merit close scrutiny. Basically, tau is known to be a microtubule-associated protein with essential functions in microtubule assembly, stabilization, and axonal transport (Lee et al., 2001). In the normal brain, tau is soluble and binds reversibly to microtubules. In contrast, many tauopathies are characterized by an alteration of the tau protein, i.e., usually loss or reduction of its affinity for microtubules, resistance to proteolysis, and insolubility. Further, in normal neurons, tau is predominantly expressed in axons, whereas in tauopathies it is redistributed to the cell body and dendrites (Burn and Lees, 2002).

The molecular biology of normal and abnormal tau is reviewed in detail by Lee et al. (2001) and some of the main ideas are summarized here: Human tau proteins are encoded by a single gene that is located on chromosome 17q21 and that consists of 16 exons. Alternative splicing of exons 2, 3, and 10 in the adult human brain results in six different isoforms; their length ranges between 352 and 441 amino acids (Goedert et al., 1989; Lee et al., 2001) (❷ Figure 8-1). They differ by the presence or absence of 29- or 58-amino acid inserts located in the N-terminus and a 31-amino acid repeat located in the carboxy-terminal half of the protein. The latter is a transcript of exon 10. Its inclusion into the final protein product gives rise to 3 isoforms containing 4 repeats each (4R-tau), whereas its exclusion leads to the production of

◘ Figure 8-1

Schematic representation of the human tau gene and the six central nervous system (CNS) tau isoforms generated by alternative mRNA splicing. The human tau gene contains 16 exons, including exon (E)0, which is part of the promoter. Alternative splicing of E2, E3, and E10 (grey boxes) produces the six tau isoforms. E6 and E8 (stippled boxes) are not transcribed in the human CNS. E4a (striped box), which is also not transcribed in the human CNS, is expressed in the peripheral nervous system, leading to the larger tau isoforms, termed big tau (see text). The *black bars* depict the 18-amino acid microtubule-binding repeats and are designated R1 to R4. The relative sizes of the exons and introns are not drawn to scale. Reprinted, with permission, from the *Annual Review of Neuroscience*, Volume 24 (2001), by Annual Reviews (www.annualreviews.org)

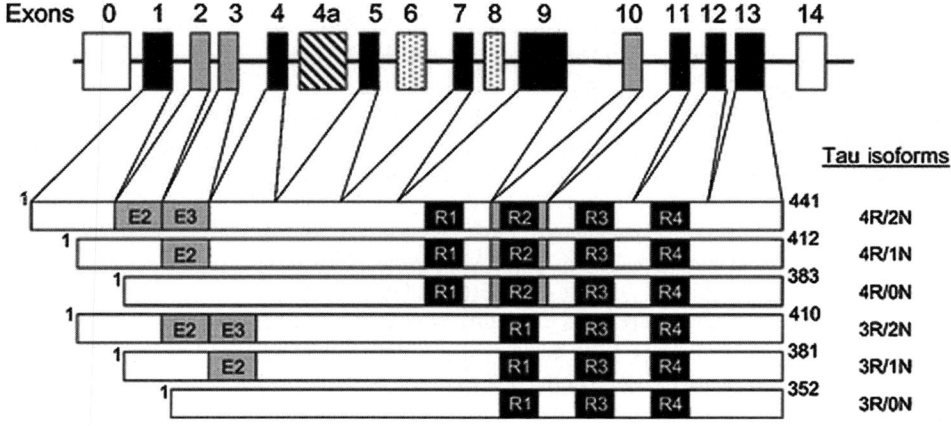

3 isoforms, each containing 3 repeats (3R-tau). There seems to be a developmental regulation of the alternative splicing of tau (Lee et al., 2001). In fact, whereas all six isoforms are expressed in the postnatal period of the human brain, only the shortest tau isoform (3R/0N) appears in the fetal brain (Goedert et al., 1989). In the normal adult brain tissue, the ratio of 3R-tau to 4R-tau is nearly 1 with a slight preponderance of 3R-tau (Flament et al., 1991). Interestingly, the term "big tau" refers to the expression of a higher M_r tau as the consequence of the inclusion of E4a in the amino terminal half of the protein (Georgieff et al., 1991; Couchie et al., 1992; Goedert et al., 1992; Lee et al., 2001).

The tau isoforms in CBD and PSP contain predominantly four repeat sequences (Buee and Delacourte, 1999). This contrasts to AD (with both 3R and 4R tau isoforms) and PiD (with a predominance of the 3R isoforms). As the microtubule-binding domains are localized within these tandem repeat sequences, the number of repeat sequences affects the affinity and binding capacity of tau for microtubules (Lee et al., 2001). Interestingly, tau mutations discovered in some families with FTDP-17 also lead to accumulation of abnormal tau with predominant 4R-tau. These pathogenic mutations are localized in exon 10 and in the proximity of the 5′-splice site, downstream of exon 10 (Hutton, 2000). They affect splice regulatory sequences in exon 10 or disrupt a stem–loop structure formed in pre-mRNA that results in overexpression of 4R-tau (Jiang et al., 2000; Varani et al., 2000). It follows that a correct splicing of exon 10 of the tau gene and consequently the ratio of resulting 4R-tau to 3R-tau isoforms might be essential for preventing neurodegeneration. Furthermore, heterogenous nuclear ribonucleoproteins, which are part of the spliceosome, and serine-rich proteins are involved in splice-site selection. Transfection experiments demonstrated that several serine-rich proteins have the capability to promote the exclusion of exon 10 (Kondo et al., 2000; Jiang et al., 2003). It is tempting to wonder if mutations in these proteins may cause altered splicing of exon 10. Besides a genetic cause for the disruption of the stem–loop structure or alternative splicing regulation, toxic substances should also be considered (Golbe, 2000; Burn and Lees, 2002). Because of this striking similarity of PSP and CBD with some FTDP-17 cases, a common pathogenetic mechanism with respect to abnormal tau isoform expression might be implicated. Interestingly, there is rather new evidence that the isoform composition of insoluble tangle-tau isolated from the basal pons differed significantly between the two clinical phenotypes of PSP, i.e., Richardson's syndrome and PSP-parkinsonism (see ❯ Sect. 1) (Williams et al., 2005). In the former, the mean 4R/3R-tau ratio was 2.84 and in the latter 1.63. This different tau isoform deposition suggests that these two clinical syndromes may ultimately prove to represent discrete nosological entities even on a biochemical level.

It is known for some time that the biochemical profile of detergent-insoluble tau from brain material of typical PSP and CBD cases is similar revealing a characteristic doublet of 64 and 68 kDa in Western blotting (Flament et al., 1991; Ksiezak-Reding et al., 1994; Vermersch et al., 1994; Dickson, 1999). The notion that the isoforms found in pathological tau accumulation may differ between CBD and PSP insofar as inclusions associated with CBD might lack exon 3 sequences, is debatable (Lee et al., 2001). Howsoever, similar material from AD brains shows three major tau-immunoreactive bands of 60, 64, and 68 kDa (Greenberg and Davis, 1990; Dickson, 1999; Lee et al., 2001). Interestingly, the triplet pattern associated with AD may be shared by the entorhinal cortex of PSP (Dickson, 1999) or clinical "atypical PSP" patients (Morris et al., 2002) and cases of the Parkinson–dementia complex of Guam (Hof, 1994; Burn and Lees, 2002). This finding provides a further hint toward a common pathogenetic mechanism of these tauopathies. Most likely, abnormal hyperphosphorylation of tau may play an important role in this context (Lee et al., 2001).

Other posttranslational processes may also contribute to the pathogenesis of neurofibrillary changes in AD (for review see Gong et al., 2005) or different tauopathies. In keeping with this assumption, in vitro experiments have shown that the formation of insoluble tau filament can be provoked by glycation and transglutamination (Miller and Johnson, 1995; Goedert et al., 1996; Perez et al., 1996). And, in fact, advanced glycation end-products have been immunohistochemically localized in neurofibrillary tangles of AD tissue or in intracellular granules of hippocampal pyramidal neurons of AD, PSP, PiD, and Parkinsonism–dementia complex of Guam cases (Sasaki et al., 1998; Burn and Lees, 2002). Further, Zemaitaitis and colleagues (2000) showed significantly higher levels of transglutaminase-induced ε-(γ-glutamyl)-lysine cross-linking of paired helical filament tau in globus pallidus and pons regions of PSP cases compared with barely detectable cross-links in controls. As transglutaminases may be involved in the formation and/or stabilization of neuro-fibrillary tangles in selectively vulnerable brain regions in PSP cases, they may represent potential targets for therapeutic intervention (Zemaitaitis et al., 2003). There is also evidence suggesting that nitrative injury is directly linked to the formation of tau inclusions (Horiguchi et al., 2003). However, the direct link between the various posttranslational modifications and the genetic background remains to be established.

In 1997, Conrad and colleagues identified a dinucleotide repeat polymorphism in intron 9 of the tau gene (Conrad et al., 1997). Subjects with the homozygous tau A0 alleles were excessively represented in the PSP compared with an age-matched healthy control group. Similar results were found in CBD cases by Di Maria and colleagues in 2000. Hence, it was suggested that CBD and PSP may share a common genetic background. Further studies have shown that this polymorphism is inherited as part of two extended haplotypes (H1 and H2) that cover a 900-kb stretch on the long arm of chromosome 17, containing several genes including microtubule-associated protein tau (Baker et al., 1999; Stefansson et al., 2005). H1 and H2 haplotypes differ from each other in both sequence and in terms of orientation. H2 haplotype is only found in Caucasian populations and recent evidence suggests that it is derived from *Homo neandertha-lensis* and entered the *Homo sapiens* population during the coexistence of these species in Europe (Hardy et al., 2005). The two haplotypes show no evidence of recombination because of an inversion polymor-phism (Stefansson et al., 2005). H1, the more common haplotype, is over-represented in patients with PSP (Litvan et al., 2001) and CBD (Houlden et al., 2001) (❷ *Table 8-1*). There is some preliminary evidence suggesting greater variability in the disease phenotype in PSP cases that are heterozygous (H1/H2). However, the H1 haplotype dosage does not affect age at onset, symptom severity, or survival in PSP (Litvan et al., 2001). It was suggested that a 238 bp sequence flanking exon 10 in the H1 haplotype might affect exon 10 splicing, thereby increasing the relative proportion of 4R-tau (Baker et al., 1999). Of special interest in the tau locus is the saitohin gene, which is located within intron 9 of the microtubule associated protein tau gene. In fact, an association of the nonsilent Q7R polymorphism of this gene and PSP has been shown by de Silva and colleagues (2003). PSP patients who had the Q/Q genotype constituted 95.9%. By contrast, 59.7% of normal controls showed the Q/Q genotype. The Q allele, which is represented in the H1 haplotype, is also the most common haplotype in controls. Saitohin might play an essential role in the tau expression regulation, and the Q7R polymorphism could interfere with this function, causing a predisposition to the pathogenesis of PSP and CBD (de Silva et al., 2003). However, taking together the studies published so far, the causal link between the H1 haplotype and the increased risk for PSP and CBD remains enigmatic. It has been proposed that the H1 haplotype might directly or indirectly affect the pathway that leads to aggregation of α-synuclein into the filaments found in PD (Spillanitini and Goedert, 2001). Although a concept of a pathogenetic link between synucleinopathies and tauopathies would follow, no predisposing effect of α-synuclein or synphilin on the development of PSP has been reported so far (Morris et al., 2000; Burn and Lees, 2002); moreover, a study on different loci, which had been previously reported in PD, did not reveal a role of these polymorphisms in a variety of neurodegenerative disease including PSP (Nicholl et al., 1999).

◻ **Table 8-1**

Tau haplotype and genotype frequency in PSP, CBD, and a control group

	PSP	CBD	Controls
Haplotype			
H1	93.7	92.1	76.6
H2	6.3	7.9	23.4
Genotype			
H1/H1	87.5	84.2	59.6
H1/H2	12.5	15.8	33.6
H2/H2	0	0	6.8

Adapted from Houlden H, Baker M, Morris HR, MacDonald N, Pickering-Brown S, et al. (2001). Corticobasal degeneration and progressive supranuclear palsy share a common tau haplotype. Neurology 56: 1702-1706 (with permission) and Litvan I (2003). Update on epidemiological aspects of progressive supranuclear palsy. Movement Disorders 18(Suppl 6): S43-S50 (with permission). Values are expressed as percentages

PSP, progressive supranuclear palsy; CBD, corticobasal degeneration

It is known that the clinical manifestations of PSP and CBD can be very similar to those seen in several FTDP-17 tau mutations (CBD: Bugiani et al., 1999; Delisle et al., 1999; PSP: Stanford et al., 2000; Spillantini et al., 2000; Pastor et al., 2001; Poorkaj et al., 2002; Morris et al., 2003). Notwithstanding this, a sequence analysis of the tau gene in both familial and sporadic PSP cases has been negative (Morris et al., 2002) confirming the commonly held notion of PSP and CBD as sporadic disorders (Burn and Lees, 2002). Nonetheless, rare familial cases have been described. Apart from the notion that this scarcity may reflect a real phenomenon, it may be related to a lack of recognition owing to the variable phenotypic expression or death of gene carriers occurring before manifestation of any clinical symptoms (de Yebenese et al., 1995). In fact, Rojo and colleagues (1999) described 12 families in which at least eight families had two affected generations. The absence of consanguinity suggested autosomal dominant transmission with incomplete penetrance. Further, de Yebenese and colleagues (1995) described one family with PSP and discussed another six already published families with neurodegenerative disorders associated with pathological findings consistent with PSP in at least one member. The pattern of inheritance was reported to be compatible with autosomal dominant transmission, although other patterns of transmission could not be excluded. More recently, Ros and colleagues (2005) presented evidence that PSP is linked to a 3.4 cM region on chromosome 1q31.1 in a large Spanish family with an autosomal dominant mode of inheritance. Further research will be necessary to narrow down this region of linkage and hopefully reveal the pathogenetic changes in this locus. For sure, scrutinizing these rare familial cases may eventually prove helpful in identifying different factors contributing to the complex disease expression.

3.3 Oxidative Stress

During the last years, several studies on free-radical-induced oxidative stress in various neurodegenerative diseases have emerged (Toloso et al., 2002). A recent study pointed out that oxidative stress response seems to be likely—at least—one out of several mechanisms contributing to the pathology of FTDP-17 (Schweitzer et al., 2006). For sure, membrane lipid peroxidation as well as mitochondrial dysfunction may be one factor—among others—in the pathogenesis of PSP (Toloso et al., 2002; Litvan, 2005a). In fact, it has been suggested that that mitochondrial DNA aberration occurs in PSP, causes electron transport chain pathology, and can produce oxidative stress (Swerdlow et al., 2000). Further, an increased labeling for heme-oxygenase 1, a marker for oxidative injury, in the pathological signature lesions both of CBD and PiD was shown, suggesting that oxidative stress may play an important role in lesion formation (Castellani et al., 1995). More recently, it was proposed that the phosphorylation of tau in the lesion of PSP and PiD is a direct consequence of the oxidative stress-induced activation of mitogen-activated protein kinases (Hartzler et al., 2002). However, whether and how mitochondrial dysfunction is of toxic or genetic origin or how it might exactly relate to tau metabolism and to which degree it is necessary to lead to specific neurodegeneration is unknown.

3.4 Neuroinflammation

Inflammatory processes are commonly thought to play a role in the pathogenesis of several neurodegenerative diseases (Litvan, 2003, 2005a). In accordance with this assumption, microglia and complement activation has been shown to occur in tauopathies such as AD or PSP (Uchihara et al., 1994; Yamada et al., 1994; Drache et al., 1997; Lippa et al., 1998; Ishizawa and Dickson, 2001; Litvan, 2005a). The pattern of microglial activation seems to differ between PSP and CBD in a system-specific and neuroanatomic-specific manner and correlates directly with the tau burden in most brain areas as shown by in vivo imaging (Gerhard et al., 2001, 2004). But still, the extent to which inflammation contributes to the pathogenesis of tauopathies is unclear. Further research has to be carried out to investigate the implication of inflammation in tau-linked neurodegeneration, and the disease-modifying potential of antiinflammatory agents remains to be established.

4   Conclusion

Although much progress was achieved in the knowledge about the tauopathies during the last decades, we are now far from understanding the exact pathogenetic mechanism. Further, as both PSP and CBD share many features on a clinical, pathological, and neurobiological level with other tauopathies, it is very questionable if they really represent separate nosological entities. Further hints into the pathogenesis of these sporadic diseases may arise from scrutinizing their genetically well-characterized counterparts, although direct extrapolations from the rare genetic cases to the much more common sporadic disease should be performed with caution. Future pathogenetic concepts should synoptically appreciate an interplay of both endogenous and exogenous factors in order to identify the best targets for the development of new therapeutic strategies for these devastating disorders.

References

Arima K, Nakamura M, Sunohara N, Ogawa M, Anno M, et al. 1997. Ultrastructural characterization of the tau-immunoreactive tubules in the oligodendroglial perikarya and their inner loop processes in progressive supranuclear palsy. Acta Neuropathol (Berl) 93(6): 558-566.

Armstrong RA, Cairns NJ, Lantos PL. 2000. A quantitative study of the pathological lesions in the neocortex and hippocampus of twelve patients with corticobasal degeneration. Exp Neurol 163(2): 348-356.

Armstrong R, Lantos PL, Cairns NJ. 2005. Overlap between neurodegenerative disorders. Neuropathol 25: 111-124.

Baker M, Litvan I, Houlden H, Adamson J, Dickson D, et al. 1999. Association of an extended haplotype in the tau gene with progressive supranuclear palsy. Hum Mol Genet 8: 711-715.

Bergeron C, Pollanen MS, Weyer L, Lang AE. 1997. Cortical degeneration in progressive supranuclear palsy. A comparison with cortical-basal ganglionic degeneration. J Neuropathol Exp Neurol 56(6): 726-734.

Boeve BF, Maraganore DM, Parisi JE, Ahlskog JE, Graff-Radford N, et al. 1999. Pathologic heterogeneity in clinically diagnosed corticobasal degeneration. Neurology 53(4): 795-800.

Boeve BF, Lang AE, Litvan I. 2003. Corticobasal degeneration and its relationship to progressive supranuclear palsy and frontotemporal dementia. Ann Neurol 54 (Suppl. 5): S15-S19.

Bower J, Maraganore D, McDonnell S, Rocca W. 1997. Incidence of progressive supranuclear palsy and multiple system atrophy in Olmsted County, Minnesota, 1976 to 1990. Neurology 49: 1284-1288.

Buee L, Delacourte A. 1999. Comparative biochemistry of tau in progressive supranuclear palsy, corticobasal degeneration, FTDP-17 and Pick's disease. Brain Pathol 9: 681-693.

Bugiani O, Murrell JR, Giaccone G, Hasegawa M, Ghigo G, et al. 1999. Frontotemporal dementia and corticobasal degeneration in a family with a P301S mutation in tau. J Neuropathol 58: 667-677.

Bugiani O, Mancardi GL, Brusa A, Ederli A. 1979. The fine structure of subcortical neurofibrillary tangles in progressive supranuclear palsy. Acta Neuropathol 45: 147-152.

Burn DJ, Lees AJ. 2002. Progressive supranuclear palsy: Where are we now? Lancet Neurol 1(6): 359-369.

Caparros-Lefebvre D, Elbaz A. 1999. Possible relation of atypical parkinsonism in the French West Indies with consumption of tropical plants: A case-control study. Caribbean Parkinsonism Study Group. Lancet 354(9175): 281-286.

Caparros-Lefebvre D, Sergeant N, Lees A, Camuzat A, Daniel S, et al. 2002. Guadeloupean parkinsonism: A cluster of progressive supranuclear palsy-like tauopathy. Brain 125 (Pt 4): 801-811.

Case Records of the Massachusetts General Hospital 1993. Case 38–1985: Case of corticonigral degeneration with neuronal achromasia. N Eng J Med 329: 1560–1567.

Castellani R, Smith MA, Richey PL, Kalaria R, Gambetti P, et al. 1995. Evidence for oxidative stress in Pick disease and corticobasal degeneration. Brain Res 696(1–2): 268-271.

Conrad C, Andreadis A, Trojanowski JQ, Dickson DW, Kang D, et al. 1997. Genetic evidence for the involvement of τ in progressive supranuclear palsy. Ann Neurol 41: 277-281.

Cordato NJ, Halliday GM, Harding AJ, Hely MA, Morris JG. 2000. Regional brain atrophy in progressive supranuclear palsy and Lewy body disease. Ann Neurol 47(6): 718-728.

Couchie D, Mavilia C, Georgieff IS, Liem RK, Shelanski ML, et al. 1992. Primary structure of high molecular weight tau present in the peripheral nervous system. Proc Natl Acad Sci USA 89(10): 4378-4381.

Cummings JL, Litvan I. 2000. Neuropsychiatric aspects of corticobasal degeneration. "Corticobasal Degeneration". Advances in Neurology, Vol. 82. Litvan I, Goetz CG, Lang AE, editors. Philadelphia: Lippincott Williams & Wilkins; pp. 147-152.

Cruz-Sanchez FF, Rossi ML, Cardozo A, Picardo A, Tolosa E. 1992. Immunohistological study of grumose degeneration of the dentate nucleus in progressive supranuclear palsy. J Neurol Sci 110(1–2): 228-231.

Davis PH, Golbe LI, Duvoisin RC, Schoenberg BS. 1988. Risk factors for progressive supranuclear palsy. Neurology 38(10): 1546-1552.

Delisle MB, Murrell JR, Richardson R, Trofatter JA, Rascol O, et al. 1999. A mutation at codon 279 (N279K) in exon 10 of the tau gene causes a tauopathy with dementia and progressive supranuclear palsy. Acta Neuropathol 98: 62-77.

de Silva R, Hope A, Pittman A, Weale ME, Morris HR, et al. 2003. Strong association of the Saitohin gene Q7 variant with progressive supranuclear palsy. Neurology 61: 407-409.

de Yebenese JG, Sarasa JL, Daniel SE, Lees AL. 1995. Familial progressive supranuclear palsy: Description of a pedigree and review of the literature. Brain 118: 1095-1103.

Dickson DW. 1999. Neuropathologic differentiation of progressive supranuclear palsy and corticobasal degeneration. J Neurol 246: II6-15.

Dickson DW, Bergeron C, Chin SS, Duyckaerts C, Horoupian D, et al. 2002. Office of rare diseases neuropathologic criteria for corticobasal degeneration. J Neuropathol Exp Neurol 61(11): 935-946.

Dickson DW, Liu WK, Ksiezak-Reding H, Yen SH. 2000. Neuropathologic and molecular considerations. Adv Neurol 82: 9-27.

di Maria E, Tabaton M, Vigo T, Abbruzzese G, Bellone E, et al. 2000. Corticobasal degeneration shares a common genetic background with progressive supranuclear palsy. Ann Neurol 47(3): 374-377.

Drache B, Diehl GE, Beyreuther K, Perlmutter LS, Konig G. 1997. Bcl-xl-specific antibody labels activated microglia associated with Alzheimer's disease and other pathological states. J Neurosci Res 47(1): 98-108.

Feany MB, Dickson DW. 1995. Widespread cytoskeletal pathology characterizes corticobasal degeneration. Am J Pathol 146: 1388-1396.

Feany MB, Dickson DW. 1996. Neurodegenerative disorders with extensive tau pathology: A comparative study and review. Ann Neurol 40(2): 139-148.

Feany MB, Mattiace LA, Dickson DW. 1996. Neuropathologic overlap of progressive supranuclear palsy, Pick's disease and corticobasal degeneration. J Neuropathol Exp Neurol 55: 53-67.

Flament S, Delacourte A, Verny M, Hauw JJ, Javoy-Agid F. 1991. Abnormal tau proteins in progressive supranuclear palsy: Similarities and differences with the neurofibrillary degeneration of the Alzheimer type. Acta Neuropathol 81: 591-596.

Forman MS, Farmer J, Johnson JK, Clark CM, Arnold SE, et al. 2006. Frontotemporal dementia: Clinicopathological correlations. Ann Neurol 59(6): 952-962.

Friedrich MJ. 1999. Pesticide study aids Parkinson research. JAMA 282: 2200.

Georgieff IS, Liem RK, Mellado W, Nunez J, Shelanski ML. 1991. High molecular weight tau: Preferential localization in the peripheral nervous system. J Cell Sci 100(1): 55-60.

Gerhard A, Banati RB, Cagnin A, Brooks DJ. 2001. In vivo imaging of activated microglia with [^{11}C]PK11195 positron emission tomography (PET) in idiopathic and atypical Parkinson's disease. Neurology 56: 270.

Gerhard A, Watts J, Trender-Gerhard I, Turkheimer F, Banati RB, et al. 2004. In vivo imaging of microglial activation with [^{11}C](R)-PK11195 PET in corticobasal degeneration. Mov Disord 19(10): 1221-1226.

Gibb WRG, Luthert PJ, Marsden CD. 1989. Corticobasal degeneration. Brain 112: 1171-1192.

Goedert M, Jakes R, Spillantini MG, Hasegawa M, Smith MJ, et al. (1996) Assembly of microtubule-associated protein tau into Alzheimer-like filaments induced by sulfated glycosaminoglycans. Nature 282: 550-553.

Goedert M, Spillantini MG, Crowther RA. 1992. Cloning of a big tau microtubule-associated protein characteristic of the peripheral nervous system. Proc Natl Acad Sci USA 89(5): 1983-1987.

Goedert M, Spillantini MG, Jakes R, Rutherford D, Crowther RA. 1989. Multiple isoforms of human microtubule-associated protein tau: Sequences and localization in neurofibrillary tangles in Alzheimer's disease. Neuron 3: 519-526.

Golbe LI. 2000. Progressive supranuclear palsy in the molecular age. Lancet 356(9233): 870-871.

Golbe LI, Rubin RS, Cody RP, Belsh JM, Duvoisin RC, et al. 1996. Follow-up study of risk factors in progressive supranuclear palsy. Neurology 47(1): 148-154.

Gong CX, Liu F, Grundke-Iqbal I, Iqbal K. 2005. Post-translational modifications of tau protein in Alzheimer's disease. J Neural Transm 112(6): 813-838.

Graham NL, Bak TH, Hodges JR. 2003. Corticobasal degeneration as a cognitive disorder. Mov Disord 18(11): 1224-1232.

Greenberg SG, Davis P. 1990. A preparation of Alzheimer paired helical filaments that displays distinct tau proteins by polyacrylamide gel electrophoresis. Proc Natl Acad Sci USA 87(15): 5827-5831.

Grimes DA, Lang AE, Bergeron CB. 1999. Dementia as the most common presentation of cortical-basal ganglionic degeneration. Neurology 53(9): 1969-1974.

Halliday GM, Hardman CD, Cordato NJ, Hely MA, Morris JG. 2000. A role for the substantia nigra pars reticulata in the gaze palsy of progressive supranuclear palsy. Brain 123 (Pt 4): 724-732.

Hardy J, Pittman A, Myers A, Gwinn-Hardy K, Fung HC, 2005. Evidence suggesting that *Homo neanderthalensis* contributed the H2 MAPT haplotype to *Homo sapiens.* 32(4): 582–585.

Hartzler AW, Zhu X, Siedlak SL, Castellani RJ, Avila J, et al. 2002. The p38 pathway is activated in Pick disease and progressive supranuclear palsy: A mechanistic link between mitogenic pathways, oxidative stress, and tau. Neurobiol Aging 23(5): 855-859.

Hauw JJ, Daniel SE, Dickson D, Horoupian DS, Jellinger K, et al. 1994. Preliminary NINDS neuropathologic criteria for Steele-Richardson-Olszewski syndrome (progressive supranuclear palsy). Neurology 44(11): 2015-2019.

Higuchi Y, Iwaki T, Tateishi J. 1995. Neurodegeneration in the limbic and paralimbic system in progressive supranuclear palsy. Neuropathol Appl Neurobiol 21: 246-254.

Hof PR, Nimchinsky EA, Buee-Scherrer V, Buee L, Nasrallah J, et al. 1994. Amyotrophic lateral sclerosis/parkinsonism-dementia complex of Guam: Quantitative neuropathology, immunohistochemical analysis of neuronal vulnerability, and comparison with related neurodegenerative disorders. Acta Neuropathol 88(5): 397-404.

Horiguchi T, Uryu K, Giasson BI, Ischiropoulos H, Light Foot R, et al. 2003. Nitration of tau protein is linked to neurodegeneration in tauopathies. Am J Pathol 163(3): 1021-1031.

Houlden H, Baker M, Morris HR, Mac Donald N, Pickering-Brown S, et al. 2001. Corticobasal degeneration and progressive supranuclear palsy share a common tau haplotype. Neurology 56: 1702-1706.

Hughes AJ, Daniel SE, Ben-Shlomo Y, Lees AJ. 2002. The accuracy of diagnosis of parkinsonian syndromes in a specialist movement disorder service. Brain 125(Pt 4): 861-870.

Hutton M. 2000. "Missing" tau mutation identified. Ann Neurol 47: 417-418.

Ikeda K, Akiyama H, Arai T, Tsuchiya K. 2002. Pick-body-like inclusions in corticobasal degeneration differ from Pick bodies in Pick's disease. Acta Neuropathol (Berl) 103: 115-118.

Ikeda K, Akiyama H, Haga C, Kondo H, Arima K, et al. 1994. Argyrophilic thread-like structure in corticobasal degeneration and supranuclear palsy. Neurosci Lett 174: 157-159.

Ishizawa K, Dickson DW. 2001. Microglial activation parallels system degeneration in progressive supranuclear palsy and corticobasal degeneration. J Neuropathol Exp Neurol 60: 647-657.

Jellinger KA, Blancher C. 1992. Neuropathology. Progressive Supranuclear Palsy: Clinical and Research Approaches. Litvan I, Agid Y, editors. Oxford: Oxford University Press; pp. 44-88.

Jiang Z, Cote J, Kwon JM, Goate AM, Wu JY. 2000. Aberrant splicing of tau pre-mRNA caused by intronic mutations associated with the inherited dementia frontotemporal dementia with Parkinsonism linked to chromosome 17. Mol Cell Biol 20: 4036-4048.

Jiang Y, Tang H, Havlioglu N, Zhang X, Stamm S, et al. 2003. Mutations in tau gene exon 10 associated with FTDP-17 alter the activity of an exonic splicing enhancer to interact with Tra2β. J Biol Chem 278: 18997-19007.

Josephs KA, Dickson DW. 2003. Diagnostic accuracy of progressive supranuclear palsy in the Society for Progressive Supranuclear Palsy brain bank. Mov Disord 18(9): 1018-1026.

Kawai H, Makino Y, Hirobe M, Ohta S. 1998. Novel endogenous 1,2,3,4-tetrahydroisoquinoline derivative: Uptake by dopamine transporter and activity to induce parkinsonism. J Neurochem 70: 745-751.

Kertesz A, McMonagle P, Blair M, Davidson W, Munoz DG. 2005. The evolution and pathology of frontotemporal dementia. 128(Pt 9): 1996-2005.

Kikuchi K, Nagatsu Y, Makino Y, Mashino T, Ohta S, et al. 1991. Metabolism and penetration through blood-brain barrier of parkinsonism-related compounds 1,2,3,4-tetrahydroisoquinoline and 1-methyl-1,2,3,4-tetrahydroisoquinoline. Drug Metab Dispos 19(1): 257-262.

Komori T. 1999. Tau-positive glial inclusions in progressive supranuclear palsy, corticobasal degeneration and Pick's disease. Brain Pathol 9: 663-679.

Komori T, Arai N, Oda M, Nakayama H, Mori H, et al. 1998. Astrocytic plaques and tufts of abnormal fibers do not coexist in corticobasal degeneration and progressive supranuclear palsy. Acta Neuropathol (Berl) 96(4): 401-408.

Kondo S, Yamamoto N, Murakami T, Okumura M, Mayeda A, et al. 2000. Tra2β, SF2/ASF and SRp30c modulate the function of an exonic splicing enhancer in exon 10 of tau pre-mRNA. Genes Cells 9: 121-130.

Kotake Y, Tasaki Y, Yea. Makino 1995. 1-Benzyl-1,2,3,4-tetrahydroisoquinoline as a parkinsonism-inducing agent: A novel endogenous amine in mouse brain and parkinsonian CSF. J Neurochem 65: 2633-2638.

Kotake T, Yoshida M, Ogawa M, Tasaki Y, Hirobe M, et al. 1996. Chronic administration of 1-benzyl-1,2,3,4-tetrahydroisoquinoline, an endogenous amine in the brain, induces parkinsonism in a primate. Neurosci Lett 217: 69-71.

Ksiezak-Reding H, Morgan K, Mattiace LA, Davies P, Liu W-K, et al. 1994. Ultrastructure and biochemical composition of paired helical filaments in corticobasal degeneration. Am J Pathol 145: 1496-1508.

Kumar R, Bergeron C, Lang AE. 2002. Corticobasal degeneration. Parkinson's Disease and Movement Disorders, Jankovic J, Tolosa E, editors. 4th edition. Philadelphia: Lippincott Williams & Wilkins; pp. 185-198.

Lang AE. 2003. Corticobasal degeneration: Selected developments. Mov Disord 18(Suppl. 6): S51-S56.

Lang A, Riley D, Bergeron C. 1994. Cortical-basal degeneration. Neurodegenerative Diseases. Calne D, editor. Philadelphia: Saunders; pp. 877-894.

Lannuzel A, Michel PP, Abaul MJ, Caparros-Lefebvre D, Ruberg M. 2000. Neurotoxic effects of alkaloids from *Annona muricata* (soursop) on dopaminergic neurons: Potential role in etiology of atypical parkinsonism in French West Indies. Mov Disord 13(Suppl. 3): 28.

Lee V, Goedert M, Tojanowski JQ. 2001. Neurodegenerative tauopathies. Annu Rev Neurosci 24: 1121-1159.

Lippa CF, Flanders KC, Kim ES, Croul S. 1998. TGF-β receptors-I and -II immunoexpression in Alzheimer's disease: A comparison with aging and progressive supranuclear palsy. Neurobiol Aging 19(6): 527-533.

Litvan I. 2003. Update on epidemiological aspects of progressive supranuclear palsy. Mov Disord 18 (Suppl. 6): S43-S50.

Litvan I. 2005a. Progressive supranuclear palsy. Scientific Basis for Treatment of Parkinson's disease. Galvez-Jimenez N, editor. London, NewYork: Taylor & Francis; pp. 267-278.

Litvan I. 2005b. Progressive supranuclear palsy and corticobasal degeneration. Animal Models of Movement Disorders. LeDoux M, editor. Amsterdam: Elsevier; pp. 505-514.

Litvan I, Baker M, Hutton M. 2001. Tau genotype: No effect on onset, symptom severity, or survival in progressive supranuclear palsy. Neurology 57: 138-140.

Litvan I, Agid Y, Calne D, Campbell G, Dubois B, et al. 1996a. Clinical research criteria for the diagnosis of progressive supranuclear palsy (Steele-Richardson-Olszewski syndrome): Report of the NINDS-SPSP international workshop. Neurology 47(1): 1-9.

Litvan I, Mangone CA, McKee A, Verny M, Parsa A, et al. 1996b. Natural history of progressive supranuclear palsy (Steele-Richardson-Olszewski syndrome) and clinical predictors of survival: A clinicopathological study. J Neurol Neurosurg Psychiat 60: 615-620.

Mackenzie IRA, Hudson LP. 1995. Achromatic neurons in the cortex of progressive supranuclear palsy. Acta Neuropathol (Berl) 90: 615-619.

Mahapatra RK, Edwards MJ, Schott JM, Bhatia KP. 2004. Corticobasal degeneration. Lancet Neurol 3(12): 736-743.

Makino Y, Ohta S, Tachikawa O, Hirobe M. 1988. Presence of tetrahydroisoquinoline and 1-methyl-tetrahydro-isoquinoline in foods: Compounds related to Parkinson's disease. Life Sci 43: 373-378.

Makino Y, Ohta S, Tasaki Y, Tachikawa O, Kashiwasake M, et al. 1990. A novel and neurotoxic tetrahydroisoquinoline derivative in vivo: Formation of 1,3-dimethyl-1,2,3,4-tetrahydroisoquinoline, a condensation product of amphetamines, in brains of rats under chronic ethanol treatment. J Neurochem 55: 963-969.

Maraganore D, Ahlskog J, Petersen R. 1992. Progressive asymmetric rigidity with apraxia: A distinct clinical entity. Mov Disord 7(Suppl. 1): 80.

McCrank E. 1990. PSP risk factors. Neurology 40(10): 1637.

McCrank E, Rabheru K. 1989. Four cases of progressive supranuclear palsy in patients exposed to organic solvents. Can J Psychiatry 34(9): 934-936.

Miller ML, Johnson JVW. 1995. Transglutaminase crosslinking of the τ protein. J Neurochem 65: 1760-1770.

Morimatsu M, Negoro K. 2002. Provisional diagnostic criteria of corticobasal degeneration (CBD) and the survey of patients with CBD in Japan. Rinsho Shinkeigaku 42(11): 1150-1153.

Morris HR, Vaughan JR, Datta SR, Bandopadhyay R, Rohan De Silva HA, et al. 2000. Multiple system atrophy/progressive supranuclear palsy: α-Synuclein, synphilin, tau, and APOE. Neurology 55(12): 1918-1920.

Morris HR, Gibb G, Katzenschlager R, Wood NW, Hanger DP, et al. 2002. Pathological, clinical and genetic heterogeneity in progressive supranuclear palsy. Brain 125: 969-975.

Morris HR, Osaki Y, Holton J, Lees AJ, Wood NW, et al. 2003. Tau exon 10+16 mutation FTDP-17 presenting clinically as sporadic young onset PSP. Neurology 61: 102-104.

Nicholl DJ, Bennett P, Hiller L, Bonifati V, Vanacore N, et al. 1999. A study of five candidate genes in Parkinson's disease and related neurodegenerative disorders. European Study Group on Atypical Parkinsonism. Neurology 53(7): 1415-1421.

Niwa T, Takeda N, Sasaoka T, Kaneda T, et al. 1989a. Detection of tetrahydroisoquinoline, in parkinsonian brain as an endogenous amine by use of gas chromatography-mass spectrometry. J Chromatogr 491: 397-403.

Niwa T, Yoshizumi H, Aea. Tatematsu 1989b. Presence of tetrahydroisoquinoline, a parkinsonism-related compound, in foods. J Chromatogr 493: 347-352.

Oyanagi K, Tsuchiya K, Yamazaki M, Ikeda K. 2001. Substantia nigra in progressive supranuclear palsy, corticobasal degeneration, and parkinsonism-dementia complex of Guam: Specific pathological features. J Neuropathol Exp Neurol 60: 393-402.

Pastor P, Pastor E, Carnero C, Vela R, Garcia T, et al. 2001. Familial atypical progressive supranuclear palsy associated with homozygosity for del N296 mutation in the tau gene. Ann Neurol 49: 263-267.

Perez M, Valpuesta JM, Medine M, Montejo de Garcini E, Avila J. 1996. Polymerization of tau into filaments in the presence of heparin: The minimal sequence requirement for tau-tau interactions. J Neurochem 67: 1183-1190.

Pezzoli G, Ricciardi S, Masotto C, Mariani CB, Carenzi A. 1990. n-Hexane induces parkinsonism in rodents. Brain Res 531(1–2): 355-357.

Pezzoli G, Barbieri S, Ferrante C, Zecchinelli A, Foa V. 1989. Parkinsonism due to n-Hexane exposure. Lancet 2(8667): 874.

Poewe W, Wenning GK. 2003. Atypical parkinsonism. Neurological Disorders: Course and Treatment. Brandt T, Caplan LR, Dichans J, Diener HC, Kennard C, editors. Amsterdam: Elsevier; pp. 1081-1098.

Poorkaj P, Muma NA, Zhukareva V, Cochran EJ, Shannon KM, et al. 2002. An R5L mutation in a subject with a progressive supranuclear palsy phenotype. Ann Neurol 52: 511-516.

Powell HC, London GW, Lampert PW. 1974. Neurofibrillary tangles in progressive supranuclear palsy: Electron microscopic observations. J Neuropathol Exp Neurol 33: 98-106.

Rebeiz JJ, Kolodny EH, Richardson EP Jr. 1968. Corticodentatonigral degeneration with neuronal achromasia. Arch Neurol 18(1): 20-33.

Riley DE, Lang AE, Lewis A, Resch L, Ashby P, et al. 1990. Cortical-basal ganglionic degeneration. Neurology 40(8): 1203-1212.

Riley DE, Lang AE. 2000. Clinical diagnostic criteria. Adv Neurol 82: 29-34.

Riley DE, Moro-de-Casillas ML. 2005. Cortico-basal degeneration. Scientific Basis for the Treatment of Parkinson's Disease. Galvez-Jimenez N, editor. London, NewYork: Taylor & Francis; pp. 279-289.

Rinne JO, Lee MS, Thompson PD, Marsden CD. 1994. Corticobasal degeneration. A clinical study of 36 cases. Brain 117 (Pt 5): 1183-1196.

Rojo A, Pernaute RS, Fontan A, Ruiz PG, Honnorat J, et al. 1999. Clinical genetics of familial progressive supranuclear palsy. Brain 122: 1233-1245.

Ros R, Garre PG, Hirano M, Tai YF, Ampuero I, et al. 2005. Genetic linkage of autosomal dominant progressive supranuclear palsy to 1q31.1. Ann Neurol 57: 634-641.

Sasaki N, Fukatsu R, Tsuzuki K, Hayashi Y, Yoshida T, et al. 1998. Advanced glycation end products in Alzheimer's disease and other neurodegenerative diseases. Am J Pathol 153(4): 1149-1155.

Schrag A, Ben Shlomo Y, Quinn NP. 1999. Prevalence of progressive supranuclear palsy and multiple system atrophy: A cross-sectional study. Lancet 354(9192): 1771-1775.

Schweitzer K, Decker E, Zhu L, Miller RE, Mirra SS, et al. 2006. Aberrantly regulated proteins in frontotemporal dementia. Biochem Biophys Res Commun 348(2): 465-472.

Smith TW, Lippa CF, de Girolami U. 1992. Immunocytochemical study of ballooned neurons in cortical degeneration with neuronal achromasia. Clin Neuropathol 11(1): 28-35.

Soto-Otero R, Mendez-Alvarez E, Sanchez-Sellero I, Cruz-Landeira A, Lopez-Rivadulla Lamas M. 2001. Reduction of rat brain levels of the endogenous dopaminergic pro-neurotoxins 1,2,3,4-tetrahydroisoquinoline and 1,2,3,4-tetrahydro-β-carboline by cigarette smoke. Neurosci Lett 298: 187-190.

Spillantini MG, Goedert M. 2001. Tau and Parkinson disease. JAMA 286: 2324-2326.

Spillantini MG, Yoshida H, Rizzini C, Lantos PL, Khan N, et al. 2000. A novel mutation (N296N) in familial dementia with swollen achromatic neurons and corticobasal inclusion bodies. Ann Neurol 48: 939–943.

Stanford PM, Halliday GM, Brooks WS, Kwok JB, Store CE, et al. 2000. Progressive supranuclear palsy pathology caused by a novel silent mutation in exon 10 of the tau gene: Expansion of the disease phenotype caused by tau gene mutations. Brain 123: 880-893.

Steele JC, Richardson JC, Olszewski J. 1964. Progressive supranuclear palsy. A heterogeneous degeneration involving the brain stem, basal ganglia and cerebellum with vertical gaze and pseudobulbar palsy, nuchal dystonia and dementia. Arch Neurol 10: 333-359.

Stefansson H, Helgason A, Thorleifsson G, Steinthorsdottir V, Masson G, et al. 2005. A common inversion under selection in Europeans. Nat Genet 37(2): 129-137.

Swerdlow RH, Golbe LI, Parks JK, Cassarino DS, Binder DR, et al. 2000. Mitochondrial dysfunction in cybrid lines expressing mitochondrial genes from patients with progressive supranuclear palsy. J Neurochem 75: 1681-1685.

Tasaki Y, Makino Y, Ohta S, Hirobe M. 1991. 1-Methyl-1,2,3,4-tetrahydroisoquioline, decreasing in 1-methyl-4-phenyl-1,2,3,6-tetrahydropyridine-treated mouse, prevents parkinsonism-like behavior abnormalities. J Neurochem 57: 1940-1943.

Tawana K, Ramsden DB. 2001. Progressive supranuclear palsy. Mol Pathol 54(6): 427-434.

Tetrud JW, Langston JW, Irwin I, Snow B. 1994. Parkinsonism caused by petroleum waste ingestion. Neurology 44(6): 1051-1054.

Togo T, Dickson D. 2002. Ballooned neurons in PSP are usually due to concurrent argyrophilic grain disease. Acta Neuropathol (Berl) 104: 5356.

Toloso E, Valldeoriola F, Pastor P. 2002. Progressive supranuclear palsy. Parkinson's Disease and Movement Disorders. Jankovic J, Tolosa E, editors. 4th edition. Philadelphia: Lippincott Williams & Wilkins; pp. 152-169.

Tsuchiya K, Uchihara T, Oda T, Arima K, Ikeda K, et al. 1997. Basal ganglia lesions in corticobasal degeneration differ from those in Pick's disease and progressive supranuclear palsy: A topographic neuropathological study of six autopsy cases. Neuropathology 17: 208-216.

Uchihara T, Mitani K, Mori H, Kondo H, Yamada M, et al. 1994. Abnormal cytoskeletal pathology peculiar to corticobasal degeneration is different from that of Alzheimer's disease or progressive supranuclear palsy. Acta Neuropathol (Berl) 88(4): 379-383.

Uchihara T, Mizusawa H, Tsuchiya K, Kondo H, Oda T, et al. 1998. Discrepancy between tau immunoreactivity and argyrophilia by the Bodian method in neocortical neurons of corticobasal degeneration. Acta Neuropathol 96: 553-557.

Vanacore N, Bonifati V, Colosimo C, Fabbrini G, De Michele G, et al. 2001. Epidemiology of progressive supranuclear palsy. ESGAP Consortium. European Study Group on Atypical Parkinsonisms. Neurol Sci 22(1): 101-103.

Vanacore N, Bonifati V, Fabbrini G, Colosimo C, Marconi R, et al. 2000. Smoking habits in multiple system atrophy and progressive supranuclear palsy. European Study Group on Atypical Parkinsonisms. Neurology 54(1): 114-119.

Varani L, Spillantini MG, Goedert M, Varani G. 2000. Structural basis for recognition of the RNA major groove in the tau exon 10 splicing regulatory element by aminoglycoside antibiotics. Nucleic Acids Res 28(3): 710-719.

Vermersch P, Robitaille Y, Berneir L, Wattez A, Gauvreau D, et al. 1994. Biochemical mapping of neurofibrillary degeneration in a case of progressive supranuclear palsy: Evidence for general cortical involvement. Acta Neuropathol 87: 572-577.

von Bergen M, Barghorn S, Muller SA, Pickhardt M, Biernat J, et al. 2006. The core of tau-paired helical filaments studied by scanning transmission electron microscopy and limited proteolysis. Biochemistry 45(20): 6446-6457.

Wakabayashi K, Takahashi H. 1996. Similarities and differences among progressive supranuclear palsy, corticobasal degeneration and Pick's disease. Neuropathology 16: 262-268.

Wakabayashi K, Takahashi H. 2004. Pathological heterogeneity in progressive supranuclear palsy and corticobasal degeneration. Neuropathology 24(1): 79-86.

Watts RL, Mirra SS, Young RR. 1989. Corticobasal ganglionic degeneration (CBDG) with neuronal achromasia: Clinical-pathological study of two cases. Neurology 39: 140.

Wenning GK, Litvan I, Jankovic J, Granata R, Mangone CA, et al. 1998. Natural history and survival of 14 patients with corticobasal degeneration confirmed at postmortem examination. J Neurol Neurosurg Psychiatry 64(2): 184-189.

Williams DR. 2006. Tauopathies: Classification and clinical update on neurodegenerative diseases associated with microtubule-associated protein tau. Intern Med J 36(10): 652-660.

Williams DR, de Silva R, Paviour DC, Pittman A, Watt HC, et al. 2005. Characteristics of two distinct phenotypes in pathologically proven progressive supranuclear palsy: Richardson's syndrome and PSP-parkinsonism. Brain 128(6): 1247-1258.

Yagishita S, Itoh Y, Amano N, Nakano T, Saitoh A. 1979. Ultrastructure of neurofibrillary tangles in progressive supranuclear palsy. Acta Neuropathol 48: 27-30.

Yamada T, Moroo I, Koguchi Y, Asahina M, Hirayama K. 1994. Increased concentration of C4d complement protein in the cerebrospinal fluids in progressive supranuclear pals. Acta Neurol Scand 89: 42-46.

Yoshida M, Niwa T, Nagatsu T. 1990. Parkinsonism in monkeys produced by chronic administration of an endogenous substance of the brain, tetrahydroisoquinoline: The behavioral and biochemical changes. Neurosci Lett 119: 109-113.

Zemaitaitis MO, Lee JM, Troncoso JC, Muma NA. 2000. Transglutaminase-induced cross-linking of tau proteins in progressive supranuclear palsy. J Neuropathol Exp Neurol 59(11): 983-989.

Zemaitaitis MO, Kim SY, Halverson RA, Troncoso JC, Lee JM, et al. 2003. Transglutaminase activity, protein, and mRNA expression are increased in progressive supranuclear palsy. J Neuropathol Exp Neurol 62(2): 173-184.

9 Central Tremors

T. Glöckler · H. Reichmann

Abstract: The most common types of central tremor are essential tremor (ET) and parkinsonian tremor (PT), which may produce substantial physical and psychological disabilities. The separation between these pathological and physiological tremors in normal subjects may be difficult and could be a challenge. This article summarizes the progress in the field of central tremors and describes the pathophysiology, characteristic clinical findings, diagnostic procedures, and therapeutic opportunities for the various types of tremor.

List of Abbreviations: DATScan, Dopamine Transporter Imaging; EMG, surface electomyography; ET, Essential Tremor; HMSN, Hereditary Motor-Sensory Neuropathy; MS, Multiple Sclerosis; PD, Parkinson's disease; PET, Positron emission tomography with 18F-dopa; PT, Parkinsonian Tremor; SPECT, Single Photon Emission Computerised Tomography; VIM, Ventralis intermedius nucleus of thalamus

1   Introduction

Tremor is a rhythmic, involuntary oscillatory movement of a body part with consistent rate, amplitude, and pattern (Deuschl et al., 1998; Gillespie and Tremor, 1991). It is classified into resting tremor without any voluntary activity and action tremor during voluntary activity. It is produced by alternating or irregular but synchronous contractions of agonist and antagonist muscles.

Tremor is a disseminated symptom of all motor activity. The causes of tremor are heterogeneous. Essential tremor (ET) and the tremor associated with Parkinson's disease (PD), parkinsonian tremor (PT), are the most common types of tremor encountered in clinical practice. Other types of tremor include physiological and cerebellar intention (Gillespie and Tremor, 1991).

Action tremor is divided into postural tremor, when a position is voluntarily maintained against gravity, and kinetic tremor, which is provoked by voluntary movements. Kinetic tremor may present as simple kinetic tremor, intention tremor, or action correlated tremor.

Different frequencies and activation patterns determine the various types of pathological tremor. Clinical distinguishing features are examination and activation conditions (at rest, with outstretched arms, during arm movements), frequency (❷ *Table 9-1*) and topic distribution (distal or proximal). With regard to frequency we distinguish low (<4 Hz), middle (4–7 Hz), and high frequency (>7 Hz) tremor (Deuschl and Lauk, 1995).

❏ Table 9-1

Types of tremor with typical frequencies

Type of tremor	Typical frequency in the upper limb (Hz)	Activation pattern
Cerebellar intention	2, 5–5	i (p)
Holmes	2–5	i>p>r
Parkinson's disease (rest)	3–6	r>p, r>p, r=p
(postural)	4–8	p
Essential	4–11	p/k/a/(r)
Dystonic	4–7	p/k
Neuropathic	3–8	p
Primary writing	5–7	a
physiological	8–12	p
Orthostatic	13–18	a

a action tremor, i intention tremor, k kinetic tremor, p postural tremor, r resting tremor

2   Basic Principles of Pathogenesis

There are three basic principles (Hallett, 1998) of pathogenesis:

The first principle is mechanical oscillation of the hand–muscle system comparable with physiological tremor.

The second mechanism might be upregulation of reflexes of the central nervous system (Deuschl and Lauk, 1995) maintaining abnormal oscillations like in exaggerated physiological tremor or psychogenic tremor.

Finally, central oscillators may play an important role in creating motor output of a certain frequency as seen in ET and PT.

A number of structures seem to be involved in the tremor-generating oscillatory neuronal network: First, the olivocerebellar system appears to play a pivotal role. This is supported by animal experiments on harmaline-induced tremor which is considered as an animal model of ET (Llinàs and Volkind, 1973; Lamarre, 1984), as well as by functional imaging studies in humans demonstrating increased glucose metabolism in the medulla oblongata (Hallett and Dubinsky, 1993) and increased blood flow in the cerebellum (Jenkins et al., 1993). Second, the basal ganglia are directly involved in the central oscillatory mechanisms and are components of several larger cortico-subcortical pathways including the thalamus. Thalamic single neuron activity recorded during stereotactic operations in patients with ET is strongly correlated with forearm electromyography signals (Hua et al., 1998; Lopez delVal and Santos 2003). The presence of this tremor correlated activity at the site where a lesion abolishes ET (Jankovic, 2000). The cells responsible for tremor frequency in the ventral thalamic nuclei do not show significant differences in proportion and nuclear location in ET and PD. Twenty-seven percent of the nerve cells in the nucleus ventralis intermedius (VIM) and 25% in the nuclei ventralis oralis posterior et anterior (Kobayashi et al., 2003) are tremor generating cells. Finally, the cerebral cortex seems to be a part of the tremor-generating neuronal network. Coherence analysis of simultaneous electroencephalography and surface electromyography (EMG) recordings in patients with unilaterally activated ET revealed tremor-correlated activity in the contralateral sensorimotor cortex (Hellwig et al., 2001).

In ET, the central oscillators in the right and left brain are not entirely independent of each other. They may dynamically synchronize, presumably by interhemispheric coupling via the corpus callosum (Hellwig et al., 2003).

3 Types of Tremor

3.1 Exaggerated Physiological Tremor

This tremor displays a frequency between 8 and 12 Hz similar to physiological tremor, is visible, mainly postural and reversible. Physiological tremor is a mechanical tremor with a small amplitude, normal posture, and a central oscillator. Its function is to keep the motor system in an oscillating state in order to gate the timing of motor events. Exaggerated physiologic tremor is a mechanical reflex tremor (Hallett, 1998). It has no associated neurological symptoms and is caused by metabolic disturbances, alcohol withdrawal, or lithium intoxication.

3.2 Essential Tremor

ET is the most common type of tremor with a quoted prevalence of 300–415 per 100,000 persons (Brin and Koller, 1998; Louis et al., 1998a). ET is generated by central oscillators (Llinàs and Volkind, 1973; Hallett and Dubinsky, 1993; Jenkins et al., 1993; Hua et al., 1998), which are strongly influenced by sensory input (Hallett, 1998). ET occurs with a frequency between 4 and 11 Hz as a visible bilateral mono-symptomatic postural, kinetic or action tremor of the upper limbs with slow progression (Pahwa and Lyons, 2003). Cerebellar signs such as intention tremor may be present, but other neurological abnormalities, do not occur (Deuschl et al., 1998). It may also affect the head, voice, trunk, and rarely the legs. Some patients may have a rest component which usually occurs later in the course of the disease (Rajput et al., 1993) Data of some clinical trials suggest that the kinetic tremor component of ET is more severe than the postural tremor component (Brennan et al., 2002). ET is sometimes asymmetrical (Louis et al., 1998b) and may rarely even be unilateral.

The consensus criteria of the Tremor Investigation Group aid in the diagnosis of ET: the diagnosis of ET is "definite" if the tremor exists for at least 5 years, it is only "probable" if the tremor exists for 3 years and "possible" if the tremor exists for less than 1 year (Jankovic, 2000; Lorenz et al., 2004). Drug related, endocrine or other symptomatic causes should be excluded.

In more than 50% of cases the tremor responds well to alcohol and shows a variable response to beta-blockers. A positive family history is seen in around 70% of patients (usually with an autosomal-dominant trait) (Jankovic, 2000), 30–40% are sporadic cases. The marked clinical variability of ET noted in some families is probably the result of genetic heterogeneity. Investigators found different gene loci for familial ET on chromosome 3q13 (Icelandic ET families) and 2p22-p25 (Czech-American families) (Jankovic, 2000). This suggests that additional gene loci may be identified in the near future.

ET may start at any age with peaks in the second and sixth decade (Jankovic, 2000) Patients with early onset (<30 years) of ET symptoms have considerably more upper limb involvement and are more likely to have associated dystonia. They show better improvement by alcohol intake than those with later onset (>40 years) of ET symptoms.

Early in the disease course an olfactory deficit is present in patients with ET (Louis and Jurewicz, 2003), but it is often milder than in patients with PD. Later in the disease course and at higher ages the tremor amplitude increases and the tremor frequency diminishes.

The diagnosis of ET and its differentiation from other types of tremor is often difficult.

Subtle extrapyramidal signs (reduced facial expression, reduced arm swing or stooped posture) are often found in the elderly. When these signs occur in patients with ET, they may reduce diagnostic certainty. A subgroup of ET patients (nearly 20%) may develop additional PD (relatively often patients with a positive family history) (Geraghty et al., 1985; Jankovic, 1989). Other investigators believe that the coexistence of the two disorders simply represents a random co-occurrence of two common diseases (Pahwa and Koller, 1993). It is also possible that some ET patients with isolated rest tremor may not have early PD (Louis and Jurewicz, 2003).

The clinical features of the tremor sometimes do not permit to distinguish ET from PD. In these cases, functional imaging with dopamine transporter ligands (such as DATScan and SPECT or ^{18}F-dopa and PET) are very helpful (Marshall and Grosset, 2003) In ET, both imaging techniques show normal striatal binding or uptake of the radioligand.

Patients with ET appear to have a higher frequency of deafness (Ondo and Jankovic, 1997) and a mild form of postural instability (Henderson et al., 1996) which may suggest a cochleovestibular disorder or cerebellar dysfunction. Furthermore, ET-like tremor occurs in genetic diseases, such as Kennedy's disease, also known as X-linked recessive spinal and bulbar muscular atrophy, which is caused by an expansion of CAG repeats in the androgen receptor gene on the X chromosome. ET-like tremor has also been described in patients with hereditary myoclonus and in patients with hereditary sensorimotor neuropathy (Jankovic, 2000).

3.3 Tremor in Parkinson's Disease

The exact anatomical basis of PT is unknown. The oscillator is probably situated in the upper part of the brainstem or thalamus (Deuschl and Lauk, 1995; Kobayashi et al., 2003). PD-tremor is often located in one or both hands and forearms and less frequently affects the feet, jaw, lips, and tongue. Three types of PT are distinguished:

Type I is the classical 4–6 Hz resting tremor (Louis et al., 2001) ("pill rolling" tremor) with a combination of flexion–extension of the fingers and adduction–abduction of the thumb (Henderson et al., 1994) (characterized by alternating EMG activity between antagonistic muscles) or a resting tremor and a postural tremor with the same frequency.

Type II is a resting tremor and a postural tremor with different frequencies.

Type III is an isolated postural tremor, which is rarely observed in PD. It shows a synchronous pattern of EMG activity (Henderson et al. 1994).

In contrast to ET, where the postural tremor is seen immediately when the patients outstretch their arms, postural tremor in patients with PD only develops with delay of several seconds after the limb assumed a horizontal position (Jankovic, 2000). This so-called "reemergent tremor" is actually considered to be a form of rest tremor (Jankovic et al., 1999) and appears to correlate better with motor disability than the typical rest tremor, which correlates mainly with the social handicap and is often exacerbated by walking. The absence of such a delay does not exclude the diagnosis of PD (Marshall and Grosset, 2003). Sometimes we find monosymptomatic PD with pure or predominant resting tremor and without brady-kinesia, rigidity, or postural instability. According to current clinical diagnostic criteria, PD cannot be diagnosed without bradykinesia, but in these patients the other symptoms of PD occur later in the course of the disease. Another group of PD patients presents with resting tremor as the main feature in concordance with mild bradykinesia, rigidity, or both (tremor dominant PD). These patients have a better prognosis with slower progression compared to patients with akinetic-rigid PD.

Patients with early PD (Hoehn and Yahr stage I) have reduced striatal binding in both dopamine transporter and ^{18}F-dopa imaging.

Optimal classification of PT and ET patients may be obtained by analyzing tremor frequency and two selected amplitude parameters from the resting limb (Burne et al., 2002).

The most important differential diagnostic criteria for ET and PT are summarized in ❯ *Table 9-2*.

◻ Table 9-2
Differential diagnostic criteria in ET and PT

Criteria	Essential tremor	Parkinsonian tremor
Positive family history	+++	(+)
Head localization	+++	(+)
Voice localization	+++	–
Leg localization	+	+++
Response to alcohol	++	–
Tremor at rest	+	++
^{18}F-Dopa-PET (striatal binding)	Normal	–
DAT scan	Normal	–

3.4 Orthostatic Tremor

Orthostatic tremor was first described by Heilmann in 1984.

This relatively rapid tremor of the thighs, sometimes known as "shaky legs syndrome," is characterized by a frequency of 13 and 18 Hz when standing (McManis and Sharbrough, 1993; Mastain et al., 1998; Sander et al., 1998). Electromyographic studies showed high frequency rhythmic discharges in weight-bearing muscles. Middle-aged or elderly patients describe a feeling of restlessness or unsteadiness (McManis and Sharbrough, 1993; Mastain et al., 1998) in the legs and a fear of falling when standing, rarely a tremor. The tremor is attenuated by walking and abolished immediately by sitting. This type of tremor is not a special subtype of ET but a tremor with central oscillators (Hallett, 1998), and it is likely to originate in so far unidentified brainstem nuclei (Deuschl et al. 1998). This generator seems to receive a modulating input from the motor cortex (Spiegel et al., 2004).

3.5 Primary Writing Tremor (Task-Specific Tremor)

Primary writing tremor is a symptomatic focal tremor with a frequency between 5 and 7 Hz (Modugno et al., 2002). The patients develop tremor in the upper extremity, primarily when they attempt to write or draw or play a musical instrument (Elble et al., 1990).

The pathophysiology of primary writing tremor is still unknown. It has been classified as a focal form of ET and as a tremolous form of writer's cramp. Some authors studied cortical and spinal excitability, which is normal in patients with primary writing tremor compared to patients with ET and writer's cramp. They could show that the pathophysiology of primary writing tremor is different from that of writer's cramp and partially also of ET (Modugno et al., 2002).

3.6 Dystonic Tremor

The frequency of dystonic tremor which occurs in a body part affected by focal or generalized dystonia lies between 4 and 7 Hz. It presents as postural (Jankovic, 2000) (similar to ET) or kinetic tremor and is irregular in amplitude and frequency. Dystonic head tremor, which is an example of dystonic tremor, can be suppressed by a tick maneuvr (geste antagoniste). The head tremor may precede the spasmodic torticollis by several years. Dystonic tremor may also originate from the basal ganglia (Deuschl et al. 1998).

3.7 Cerebellar Tremor

Cerebellar tremor is a symptomatic tremor with a frequency between 2.5 and 5 Hz. The typical clinical manifestation is that of an intention tremor (amplitude increases during visually guided movements toward a target at the end of movement) mainly a postural tremor as a cerebellar outflow tremor in patients with multiple sclerosis (MS). It is a three per second body tremor in anterior or posterior direction. Sometimes it occurs as titubation, with slow frequency oscillations of the head and trunk.

Cerebellar tremor is partly caused by a disturbance of the cerebellar feedforward control of voluntary movements (Deuschl and Bergman, 2002).

3.8 Holmes Tremor (Rubral, Thalamic, Peduncular, or Midbrain Tremor)

This symptomatic tremor occurs after a lesion in the brainstem, midbrain, or cerebellum. It is a very slow (2–5 Hz) resting and intention tremor that often results in a large amplitude, pronation/supination oscillation of the forearm. It is mainly an intention tremor, but also a postural or resting tremor (Deuschl et al., 1998). Usually a delay of at least 4 weeks to 2 years between the brain lesion and the outset of tremor is observed. In addition, other symptoms of midbrain or cerebellar damage are frequently detected. Frequent causes are MS, olivo-ponto-cerebellar atrophies, brainstem or cerebellar ischemias (Benedikt Syndrome), tumors of the brainstem, midbrain or thalamus, and PD.

3.9 Neuropathic Tremor

This tremor correlates directly with the reduction in conduction velocity of the peripheral nerves and occurs in diseases like hereditary motor-sensory neuropathy (HSMN), demyelinating neuropathy and IgM-paraproteinaemias or Roussy Levy-Syndrom. There is no unique picture: the tremor is irregular or rhythmic, proximal or distal, and mainly postural with a frequency between 3 and 8 Hz. Neuropathic tremor is believed to be caused by an abnormal function of reflex pathways (delays in peripheral loops) (Hallett, 1998) and by intoxication or drugs (Deuschl et al., 1998).

3.10 Psychogenic Tremor

This type of tremor has various clinical presentations. Usually the patients show sudden onset of tremor, no progression, spontaneous remission, decrease in amplitude, and variation in frequency. Typical findings are unusual combinations of tremors, decreased intensity during mental tasks, and increased intensity with concentration (Gironell et al., 1997). Often a noticeable muscular tension of the extremities is encountered. In addition, patients have a previous history of somatization and of additional and unrelated neurological signs (Kim et al., 1999). To differentiate psychogenic tremor from tremor in PD dopamine transporter imaging and also accelerometry is helpful. Patients with psychogenic tremor show larger changes of tremor frequency and a higher intraindividual variability of repetitive motor tasks than patients with ET and PT (Zeuner et al., 2003).

3.11 Wilson's Tremor

The most common initial symptom and the most constant sign of Wilson's disease is tremor (Moringlane, 2000). Patients with Wilson's disease may show a resting tremor of parkinsonian type, usually mixed with action tremor (batswing tremor) and other extrapyramidal signs (Topaloglu et al., 1990). The tremor involves upper and lower limbs, head, and trunk and intermittently worsens dramatically for several minutes.

4 Diagnostic Procedures to Differentiate Tremor Forms

The most important procedure to differentiate the various types of tremor is to take a detailed history with particular emphasis on age and region of onset, sequence of tremor spread, response to alcohol, family history, and medication followed by clinical examination. Findings like akinesia or rigidity in PD or a family history of tremor are helpful to distinguish the various types of tremor. Tremor rating scale, writing test and standardized tremor examination, laboratory tests (❯ *Table 9-3*), and neurophysiological investigations, such as accelerometry and EMG, complete the diagnostic work up.

◻ Table 9-3
Laboratory tests in tremor

Laboratory parameters	Diseases with tremor
T3, T4, TRH	Hyperthyroidism
Electrolytes	Metabolic disturbances
Serum electrophoresis	Paraproteinaemia
GGT, ALAT, ASAT	Alcohol withdrawal, hepatocellular degeneration
Blood glucose	Hypoglycemia
Cortisol	Adrenocortical dysfunction
Parathyroid hormone	Hypo-/hyperparathyreoidism
Catecholamines	Pheochromocytoma
Coeruloplasmin, copper	Wilson's disease
As, Bi, Br, Hg, Pb, ethanol	Intoxications

Excellent methods are *accelerometry and surface EMG* (Burne et al., 2002) combined with Fast-Fourier spectral analysis while holding a weight, which allows separation of mechanical reflex tremor and tremor arising from central oscillators. Bipolar surface EMG electrodes are placed over forearm flexors and extensors, and accelerometer sensors are placed over the dorsum of the hand (hand tremor) and over the forearm (elbow tremor) 10 cm proximal to the wrist.

Each patient should also receive a cranial CT or MRI.

Dopamine transporter imaging (DAT imaging) which reflects the integrity of presynaptic dopaminergic neurons within the striatum and *Positron emission tomography (PET)* might also be of help in differentiating the various types of tremor. In PD, we find reduced striatal binding using DAT imaging, also in Wilson's tremor and orthostatic tremor (Marshall and Grosset, 2003). In PD, we also find a reduced ^{18}F-dopa uptake (PET) in the putamen. In ET, the ^{18}F-dopa uptake rate (Marshall and Grosset, 2003) (PET) and striatal binding (DAT imaging) are both normal, which also applies to Holmes tremor, dystonic tremor, cerebellar tremor, and neuropathic tremor.

5 Therapy

Multiple pharmacological treatments have been studied and recommended for the various types of tremor. The Tremor Investigation Study Group has published recommendations for the drug treatment of tremor. They recommend to start with the lowest possible effective drug dosage to prevent side effects and then to increase the dose to the highest possible and tolerable dosage. If the symptoms do not improve or even worsen or if intolerable side effects occur, we reduce the dosage and change to another recommended drug. Mildly affected patients may not need any treatment at all, and the potential benefits must always be judged against the possible side effects. It is important for the physician to realize the variety of symptomatic treatments available. Besides the traditional and often effective oral medications, new therapeutic options, such as botulinum toxin and deep brain stimulation, may also reduce tremor and improve the quality of life for patients with ET and PT.

5.1 β-Blockers

These drugs are effective in ET, PT, MS (proximal or resting tremor), and sometimes also in neuropathic or posttraumatic tremor.

Propranolol is one of the first-line drugs for the therapy of ET (Uitti, 1998; Wasielewski et al., 1998; Lambert and Waters, 1999; Gunall et al., 2000; Pahwa and Lyons, 2003; Crosby et al., 2003) although it is not tolerated well by all patients. It is unlikely that patients will respond miraculously to another medication if their response to poropranolol is minimal. Some subtypes of ET, however, such as the kinetic predominant type, may respond better to other medications (benzodiazepines). It is recommended to start with a very low dosage of 12.5 mg and to increase slowly to 240 mg/day.

5.2 Primidon

Since the 1970s and 1980s, the therapeutic efficacy of β-blockers and primidon has been proven for most patients with ET (Uitti, 1998; Wasielewski et al., 1998; Jankovic, 2000), orthostatic tremor, and postural tremor in PD, also for some patients with writing/task-specific tremor, neuropathic tremor, and posttraumatic tremor. It is recommended to start with 62.5 mg and to increase to a maximum of 500 mg/day. In a 1-year follow-up study, low doses of primidone (250 mg/day) were equally or even more effective than high doses (750 mg/day) (Serrano-Duenas, 2003).

5.3 Benzodiazepines

Clonazepam is effective in ET and ET with a predominant kinetic component (Uitti, 1998; Jankovic, 2000; Pahwa and Lyons, 2003) as well as in orthostatic tremor (Jankovic, 2000). In ET with a predominant postural component alprazolam, a carbonic anhydrase inhibitor (Uitti, 1998; Gunall et al., 2000), is effective. It can be used as an alternative agent in elderly patients with ET who cannot tolerate primidone or propranolol. The mean effective daily dose of alprazolam is 0.75 mg.

Sometimes Benzodiazepines can also be used in orthostatic tremor (Clonazepam) (Jankovic, 2000), dystonic tremor, MS, and occasionally in neuropathic or rest tremor.

5.4 Anticholinergics

Anticholinergics are effective drugs in PT (Jankovic, 2000), ET, primary writing tremor, dystonic tremor, and Holmes tremor. Due to their side effects (mouth dryness, hypohidrosis, accommodation disturbances, constipation, frequency of urination, deterioration of glaucoma), they should be only chosen if the patients did not respond to the drugs discussed above. Following dosages are recommended:

Biperiden 1–12 mg/day
Bornaprin 3–60 mg/day
Metixen 7.5–60 mg/day
Trihexiphenidyl 1–10 mg/day

5.5 Gabapentin

This drug is efficacious in ET (Ondo and Jankovic, 1997; Pahwa et al., 1998; Gironell et al., 1999; Pahwa and Lyons, 2003), occasionally in PT, orthostatic tremor (Jankovic, 2000; Lopez del Val and Santos, 2003), MS (Lopez del Val and Santos, 2003), or primary writing tremor. It is recommended to start with 300 mg/day and gradually increase to 1,800 mg/day. In some clinical trials, the response was similar with high (3,600 mg) and low doses. No significant predictors for response to gabapentin could be observed in patients with ET. They demonstrated only a response in some cases (Ondo et al., 2000), other trials showed significantly reduced tremor power (Pahwa et al., 1998; Gironell et al., 1999).

5.6 Clozapin

This atypical neuroleptic drug is effective in ET and postural PT in patients without any response to commonly used drugs (Factor and Friedman, 1997; Ceravolo et al., 1999). The lowest single dose should be 6.25 mg/day increasing up to 75 (250) mg/day. Weekly blood counts are necessary during the first 18 weeks. A significant reduction of tremor has been reported with chronic clozapine treatment in some trials so that clozapine should be considered before deep brain stimulation (Ceravolo et al., 1999).

5.7 Botulinumtoxine

Injections with botulinumtoxine ("no-no"-type) are a further option in patients with ET and head tremor (Heilman, 1984; Lambert and Waters, 1999; Pahwa and Lyons, 2003), whereas its effect is variable in hand tremor (Lambert, 1999). Dystonic head tremor is reported to be responsive (Deuschl et al., 2002) as well as the positional and postural abnormality of torticollis. In head tremor, injections are given into both splenei muscles and the sternocleidomastoid muscle, particularly if there is a "yes-yes"-component of the tremor. Recommended doses are between 800 and 1,000 mouse units (Dysport).

5.8 Deep Brain Stimulation

Although medication can improve tremor in some patients, approximately 50% of patients with ET have drug-resistant symptoms (Lyons, 2004). The revival of stereotactic surgery has led to further improvement for those patients affected by severe ET, PT (Volkmann, 2004), and dystonic tremor. Better surgical techniques and the introduction of deep brain stimulation have transformed these procedures into safer and more widely applied tools for the treatment of severe tremor. Deep brain stimulation should be considered because it carries the lowest risk out of all available surgical procedures, and it should be given preference to thalamotomy (Lambert, 1999; Pahwa and Lyons, 2003). The preferred targets are the VIM in ET and tremor-dominant PT (older PD-patients), and VIM and/or globus pallidus (Gpi) in dystonic tremor. Since akinesia and rigidity may develop at later stages of PD and then necessitate drug treatment with all the associate risks of long-term side effects, subthalamic stimulation has replaced thalamic stimulation for younger patients with tremor-dominant PD (Deuschl et al., 2002). Typically, improvement of tremor occurs within days to months. A variety of studies of thalamic deep brain stimulation have shown that it is efficacious in ET and PT, often with secondary improvement of voice and head tremor (Pahwa and Lyons, 2003; Sydow et al., 2003; Lyons, 2004). A satisfyable tremor reduction is found on average in 50–80%

of ET patients and can be maintained for more than 6 years (Sydow et al. 2003). Adverse effects related to stimulation are usually mild and can be managed by changing the stimulation parameters. Surgical procedures and follow-up should be concentrated in relatively few centers, which will then acquire a high degree of expertise. With careful patient selection, DBS can also provide significant functional benefit in the management of MS tremor (Moringlane, 2000; Berk et al., 2002; Nandi and Aziz, 2004).

References

Berk C, Carr J, Sinden M, Maritzke J, Honey CR. 2002. Thalamic deep brain stimulation for the treatment of tremor due to multiple sclerosis: A prospective study of tremor and quality of life. J Neurosurg 97: 815-820.

Brennan KC, Jurewicz EC, Ford B, Pullman SL, Louis ED. 2002. Is essential tremor predominantly a kinetic or a postural tremor? A clinical and electrophysiological study. Mov Disord 17 (2): 313-316.

Brin MF, Koller W. 1998. Epidemiology and genetics of essential tremor. Mov Disord 13: 55-63.

Burne JA, Hayes MW, Fung VS, Yiannikas C, Bojevac D. 2002. The contribution of tremor studies to diagnosis of parkinsonian and essential tremor: A statistical evaluation. J Clin Neurosci 9 (3): 237-242.

Ceravolo R, Salvetti S, Piccini P, Lucetti C, Gambaccini G, et al. 1999. Acute and chronic effects of clozapine in essential tremor. Mov Disord 14 (3): 468-472.

Crosby NJ, Deane KH, Clarke CE. 2003. Beta-blocker therapy for tremor in Parkinson's disease. Cochrane Database Syst Rev (1): CD003361.

Deuschl G, Lauk M. 1995. Tremor classification and tremor time series analysis. Chaos 5 (1): 48-51.

Deuschl G, Bain P, Brin M. 1998. Consensus statement of Movement Disorder Society on Tremor ad hoc Scientific Committee. Mov Disord 13 (Suppl.): 2-23.

Deuschl G, Bergman H. 2002. Pathophysiology of nonparkinsonian tremors. Mov Disord 17 (Suppl. 3): 41–48.

Deuschl G, Fogel W, Hahne M, Kupsch A, Müller D, et al. 2002. Deep brain stimulation for Parkinson's disease. J Neurol 249 (Suppl. 3): III/36-III/39.

Elble RJ, Moody C, Higgins C. 1990. Primary writing tremor. A form of focal dystonia? Mov Disord 5 (2): 118-126.

Factor SA, Friedman JH. 1997. The emerging role of clozapine in the treatment of movement disorders. Mov Disord 12 (24): 483-496.

Geraghty JJ, Jankovic J, Zetusky WJ. 1985. Association between essential tremor and Parkinson's disease. Ann Neurol 17: 329-333.

Gillespie MM, Tremor J. 1991. J Neurosci Nurs 23 (3): 170-174.

Gironell A, Lopez-Villegas D, Barbanoj M, Kulisevsky J. 1997. Psychogenic tremor: Clinical, electrophysiologic and psychopathologic assessment. Neurologia 12 (7): 293-299.

Gironell A, Kulisevsky J, Barbanoj M, Lopez-Vilegas D, Hernandez G, et al. 1999. A randomised placebo-controlled comparative trial of gabapentin and propranolol in essential tremor. Arch Neurol 56 (4): 475-480.

Gunall DI, Afsar N, Bekiroglu N, Aktan S. 2000. New alternative agents in essential tremor therapy: Double-blind placebp-controlled study of alprazolam and azetazolamide. Neurol Sci 21 (5): 315-317.

Hallett M. 1998. Overview of human tremor physiology. Mov Disord 13 (Suppl. 3): 43-48.

Hallett M, Dubinsky RM. 1993. Glucose metabolism in the brains of patients with essential tremor. J Neurol Sci 114: 45-48.

Heilman KM. 1984. Orthostatic tremor. Arch Neurol 1: 880-881.

Hellwig B, Häussler S, Schelter B, Lauk M, Guschlbauer B, et al. 2001. Tremor-correlated cortical activity in essential tremor. Lancet 357: 519-523.

Hellwig B, Schelter B, Guschlbauer B, Timmer J, Lucking CH. 2003. Dynamic synchronisation of central oscillatiors in essential tremor. Clin Neurophysiol 4 (8): 1462-1467.

Henderson E, Overby A, Jankovic J. 1996. Postural control in essential tremor. Neurology 46: A273.

Henderson JM, Yia nnikas C, Morris JG, Einstein R, Jackson D, et al. 1994. Postural tremor of Parkinson's disease. Clin Neuropharmacol 17 (3): 277-285.

Hua SE, Lenz FA, Zirh TA, Dougherty PM. 1998. Thalmic neuronal activity correlated with essential tremor. J Neurol Neurosurg Psychiatry 64: 273-276.

Jankovic J. 1989. Essential tremor and Parkinson's disease. Ann Neurol 25: 211-212.

Jankovic J. 2000. Essential tremor clinical characteristics. Neurology 54 (11 Suppl. 4): S21-S25.

Jankovic J, Shwartz KS, Ondo W. 1999. Reemergent tremor of Parkinson's disease. J Neurol Neurosurg Psychiatry 67: 646-650.

Jenkins IH, Bain PG, Colebatch JG, Thompson PD, Findley LJ, et al. 1993. A positron emission tomography study of essential tremor: Evidence for overactivity of cerebellar connections. Ann Neurol 34: 82-90.

Kim YJ, Pakian AS, Lang AE. 1999. Historical and clinical features of psychogenic tremor: A review of two cases. Can J Neurol Sci 26: 190-195.

Kobayashi K, Katayama Y, Kasai M, Oshima H, Fukaya C, et al. 2003. Localization of thalamic cells with tremor-frequency activity in Parkinson's disease and essential tremor. Acta Neurochir 87 (Suppl.): 137-139.

Lamarre Y. 1984. Animal models of physiologic, essential and parkinsonian-like tremors. Movement Disorders: Tremor. Findley LJ, Capildeo R, editors. London: Macmillan Press; pp. 183-194.

Lambert D, Waters CH. 1999. Essential tremor. Curr Treat Options Neurol 1 (1): 6-13.

Llinàs R, Volkind RA. 1973. The olivo-cerebellar system: Functional properties as revealed by harmaline-induced tremor. Exp Brain Res 18: 69-87.

Lopez del Val LJ, Santos S. 2003. Gabapentin in the treatment of tremor. Rev Neurol 36 (4): 32-36.

Lorenz D, Frederiksen H, Moises H, Kopper F, Deuschl G, et al. 2004. High concordance for essential tremor in mono-zygotic twins of old age. Neurology 62 (2): 208-211.

Louis ED, Jurewicz EC. 2003. Olfaction in essential tremor patients with and without isolated rest tremor. Mov Disord 18 (11): 1387-1389.

Louis ED, Ottman R, Hauser WA. 1998a. How common is the most common adult movement disorder? Estimates of the prevalence of essential tremor throughout the world. Mov Disord 13 (1): 5-10.

Louis ED, Wendt KJ, Pullmann SL, Ford B. 1998b. Is essential tremor symmetric? Observational data from a community-based study of essential tremor. Arch Neurol 55 (12): 1553-1559.

Louis ED, Levy G, Cote LJ, Mejia H, Fahn S, 2001. Clinical correlates of action tremor in Parkinson's disease. Arch Neurol 58: 1630-1634.

Lyons KE, Pahwa R. 2004. Deep brain stimulation and essential tremor. J Clin Neurophysiol 21 (1): 2-5.

Marshall V, Grosset DG. 2003. Role of dopamine transporter imaging in the diagnosis of a typical tremor disorders. Mov Disord 8 (Suppl. 7): S22-S27.

Mastain B, Cassim F, Guieu JD, Destee A. 1998. "Primary" orthostatic tremor. 10 clinical electrophysiologic observations. Rev Neurol (Paris) 154 (4): 322-329.

McManis PG, Sharbrough FW. 1993. Orthostatic tremor: Clinical and electrophysiologic characteristics. Muscle Nerve 16 (11): 125-160.

Modugno N, Nakamura Y, Bestmann S, Curra A, Beradelli A, et al. 2002. Neurophysiological investigations in patients with primary writing tremor. Mov Disord 17 (2): 313-316.

Moringlane JR. 2000. Chronische Elektrostimulation zur Behandlung des Intentionstremors und der Ataxie der Multiplen Sklerose. Nervenheilkunde 6: 2-5.

Nandi D, Aziz TZ. 2004. Deep brain stimulation in the management of neuropathic pain and multiple sclerosis tremor. J Clin Neurophysiol 21 (1): 31-34.

Ondo W, Jankovic J. 1997. Hearing loss in essential tremor. Ann Neurol 42: 449.

Ondo W, Hunter C, Vuong KD, Schwartz K, Jankovic J. 2000. Gabapentin for essential tremor: A multiple-dose, double-blind, placebo-controlled trial. Mov Disord 15 (4): 678-682.

Pahwa R, Koller WC. 1993. Is there a relationship between Parkinson's disease and essential tremor? Clin Neuropharmacol 16: 30-35.

Pahwa R, Lyons KE. 2003. Essential tremor: Differential diagnosis and current therapy. Am J Med 115 (2): 134-142.

Pahwa R, Lyons K, Hubble JP, Busenbark K, Rienerth JD, et al. 1998. Double-blind controlled trial of gabapentin in essential tremor. Mov Disord 13 (3): 465-467.

Rajput AH, Jamieson H, Hirsh S, Quavaiski A. 1975. Relative efficacy of alcohol and propranolol in action tremor. Can J Neurol Sci 2: 31-35.

Rajput AH, Rozdilsky B, Ang L, Rajput A. 1993. Significance of parkinsonian manifestation in essential tremors. Can J Neurol Sci 20: 114-117.

Sander HW, Masdeu JC, Tavoulareas G, Walters A, Zimmermann T, et al. 1998. Orthostatic tremor: An electrophysiological analysis. Mov Disord 13 (4): 735-738.

Schrag A, Munchau A, Bhatia KP, Quinn NP, Marsden CD. 2000. Essential tremor: An overdiagnosed condition? J Neurol 247 (12): 955-959.

Serrano-Duenas M. 2003. Use of primidone in low doses (250 mg/day) versus high doses (750 mg/day) in the management of essential tremor. Double-blind comparative study with one-year follow up. Parkinsonism Relat Disord 10 (1): 29-33.

Spiegel J, Fuss G, Krick C, Dillmann U. 2004. Impact of different types on orthostatic tremor. Clin Neurophysiol 115 (3): 569-575.

Sydow O, Thobois S, Alesch F, Speelman JD. 2003. Multi-centre European of thalamic stimulation in essential tremor: A six-year follow-up. J Neurol Neurosurg Psychiatry 74 (10): 1387-1391.

Topaloglu H, Gucuyener K, Orkun C, Renda Y. 1990. Tremor of tongue and dysarthria as the sole manifestation of Wilson's disease. Clin Neurol Neurosurg 92 (3): 295-296.

Uitti RJ. 1998. Medical treatment of essential tremor and Parkinson's disease. Geriatrics 53 (5): 46-48, 53-57.

Volkmann J. 2004. Deep brain stimulation for the treatment of Parkinson's disease. J Clin Neurophysiol 21 (1): 6-17.

Wasielewski PG, Burns JM, Koller WC. 1998. Pharmacologic treatment of tremor. Mov Disord 13 (Suppl. 3): 90-100.

Zeuner KE, Shoge RO, Goldstein SR, Dambrosia JM, Hallett M. 2003. Accelerometry to distinguish psychgenic from essential or parkinsonian tremor. Neurology 61 (4): 548-550.

10 Molecular Mechanisms of Methamphetamine-Induced Neurotoxicity: Insights Obtained Through cDNA Array Analyses

S. Jayanthi · M. T. McCoy · J. L. Cadet

Abstract: Methamphetamine (METH) is an abused psychostimulant that can cause psychotic symptoms, cognitive and psychomotor deficits, stroke, and death. METH-induced neurotoxicity includes neural cell death which occurs via processes that resemble apoptosis. We have used cDNA array technology to investigate transcriptional responses that occur in the rodent brain after a toxic dose of the drug. These studies showed that the drug can cause an early pattern of gene induction which is characterized by changes in transcription factors. A more delayed response included changes in the expression of genes involved in DNA repair, regulation of trophic factors, and in cell death pathways. Interestingly, regulation of some of these transcription factors has been shown to be involved in controlling the expression of Fas ligand (FasL) which can trigger apoptosis in a number of models of cell death. Thus, these results indicate that METH can cause some of its degenerative effects, in part, via activation of a receptor-mediated cell death pathway consequent to multiple transcription factor-mediated FasL up-regulation. When taken together, these observations indicate that METH-induced toxicity results from the balance of pro-death/protective mechanisms in the rodent brain.

List of Abbreviations: AICD, activity-induced cell death; AP-1, activator protein-1; AP-site, apurinic/apyrimidinic site; BER, base excision repair; DA, dopamine; DAT, dopamine transporter; FGF, fibroblast growth factor; IGF, insulin growth factor; MMR, mismatch repair; NER, nucleotide excision repair; OGG1, 8-oxo-guanine glycosylase-1; ROS, reactive oxygen species; TGF, transforming growth factor; TH, tyrosine hydroxylase

1 Introduction

Methamphetamine (METH, Speed) is widely abused because of its powerful stimulant effects (McCann et al., 1998; Sekine et al., 2001; Volkow et al., 2001a; Farrell et al., 2002). The number of METH abusers has increased worldwide and recent data from monitoring the future (MTF) survey in the U.S. indicate that 6.2% of high school seniors had reported lifetime use of METH in 2004 (NIDA report; see Johnston et al., 2006). In addition to its stimulant effects, METH can cause psychotic, cognitive, and psychomotor impairments in humans (Buffenstein et al., 1999; London et al., 2004; Dore and Sweeting 2006). Humans who abused METH also show neurodegenerative effects in their brains. These abnormalities are reflected by significant decrements in the density of DA transporters (Volkow et al., 2001b; Sekine et al., 2003) and serotonin transporters (Sekine et al., 2006). Long-term damage to presynaptic dopaminergic and serotonergic terminals are also observed in rodents (Ricaurte et al., 1980; Wagner et al., 1980; Harvey et al., 2000). In addition to the changes in monoaminergic terminals, degeneration of cell bodies both in vitro (Cadet et al., 1997; Stumm et al., 1999; Deng et al., 2002) and in vivo (Eisch et al., 1998; Deng and Cadet, 1999; Deng and Cadet, 2000; O'Dell and Marshall, 2000) are also observed. The cell-body loss has been shown to occur via a process that resembles apoptosis (Cadet et al., 1997; Stumm et al., 1999; Deng and Cadet, 2000; Lotharius et al., 2005).

METH-induced neurotoxic effects are thought to be secondary to oxidative stress (Fumagalli et al., 1999; LaVoie and Hastings, 1999; Deng and Cadet, 2000; Lotharius et al., 2005). The generation of reactive oxygen species (ROS) occur secondary to METH-induced changes in DA turnover (for review see Cadet and Brannock, 1998) and through the release of glutamate (Sonsalla et al., 1989; Staszewski and Yamamoto, 2006). Nitric oxide appears to also play a significant role in the long-term behavioral and toxic effects of the drug (Deng and Cadet, 1999; Itzhak et al., 2000; Ali et al., 2005). Transcription factors have also emerged as important factors in METH toxicity. These transcription factors include the activator protein-1 (AP-1) families of transcription factors (Cadet et al., 2001; Jayanthi et al., 2002; Krasnova et al., 2002; Salzmann et al., 2003) and p53 (Hirata and Cadet, 1997; Imam et al., 2001). A number of approaches had suggested that c-fos, the heterodimeric partner of c-jun in AP-1 complexes, is activated by toxic METH doses (Thiriet et al., 2001). When taken together, the accumulated evidence had thus hinted at the possibility that METH-induced neurodegeneration might involve complex cellular and molecular mechanisms. Because the single-gene approach was thought to be somewhat restrictive, we opted to approach the METH toxicity dilemma by using an approach that might provide a more general view of the involved mechanisms. The cDNA array approach held the promise of identifying patterns of changes in the expression of several genes that might be affected during the course of METH toxicity.

Alterations in gene expression patterns are essential for the maintenance of normal brain function. In most cases, they can also provide windows to the pathobiology of diseased states. Gene expression profiles can help to choose candidates for future hypothesis-driven studies into the molecular mechanisms that control the transitions between normal function and cell death. These studies can also help to identify targets for pharmacotherapeutic interventions.

2 Involvement of AP-1-Related Transcription Factors in METH-Induced Neurotoxicity

Our first data using cDNA array analysis gene expression was obtained using the mouse Atlas cDNA Expression Array (total of 588 genes; Clontech, Mountain View, CA). The frontal cortex of METH-treated and saline-treated mice euthanized at 2, 4, and 16 h after a single dose of METH (40 mg/kg) or saline, respectively, were used (see ❷ *Figure 10-1* and ❷ *Figure 10-2*). The analysis provided an extensive temporal profile of the effects of METH on gene expression in the mouse brain (Cadet et al., 2001). We used the criteria of threefold changes in gene expression at all the three time points (2, 4, and 16 h) (see ❷ *Figure 10-1*) after the METH injections in order to select genes for further analyses. This approach resulted in a total of 91 genes that were either upregulated or downregulated. These genes were then organized by hierarchical clustering (see ❷ *Figure 10-2a*). The cluster analysis is based on the algorithm generated by GeneSpring (version 4.0.2, Agilent Technologies, Palo Alto, CA). GeneSpring software arranges the genes according to their expression profiles in such a way that genes with similar patterns of expression across all of the time points are clustered adjacent to each other. Because genes with similar patterns of expression in responses to METH are grouped together by common branches of the dendrogram (see also Eisen et al., 1998, for a more extensive discussion), we were able to further divide the large cluster into seven subclusters (see the graphical representation; (❷ *Figure 10-2b–2h*). To make sense of the large data sets, color coding is used. The primary coding scheme in GeneSpring is to map relative expression levels to colors; high expression levels are colored red and low expression levels are colored blue.

These observations suggest that METH administration can have profound effects on the molecular machinery of the brain. For example, there were significant changes in the expression of transcription factors, changes which occur within the first 2 h after injection of METH. These include upregulation of c-jun, c-fos, jun B, as well as jun D (Cadet et al., 2001), members of the family of transcription factors named AP-1 (Shaulian and Karin, 2002). These genes are usually induced by manipulations that cause neuronal injury (Pennypacker, 1997). These changes are probably related to METH-induced generation of ROS. ROS such as hydroxyl and superoxide radicals can induce the expression of many genes via their regulation of AP-1 family of transcription factors (Dalton et al., 1999). These changes were further confirmed by our subsequent analyses in which we used WT and c-fos$^{+/-}$ mice in order to further evaluate the role of c-fos in protective or protoxic mechanisms in response to METH injections. A previous study from this laboratory suggested that c-fos might play protective roles (Deng et al., 1999) but the mechanisms involved needed to be clarified. This study made use of c-fos knockout mice to study METH toxicity. These knockout mice showed METH-induced decreases in dopamine (DA) uptake sites, dopamine transporter (DAT) protein, tyrosine hydroxylase (TH) activity, and increases in the number of apoptotic cells which were of greater magnitude than that observed in WT mice (Deng et al., 1999).

For studying the molecular basis of c-fos induced neuroprotection, we used the Atlas Mouse Cancer 1.2 Arrays (Clontech), which contain a total of 1176 cDNA segments spotted on a nylon membrane. We used two time points (2 and 12 h) after the last dose of saline or METH. ❷ *Table 10-1* shows the magnitude of the effects of METH on the expression of AP-1 family genes. There were significant METH-induced decreases in c-fos, FRA1, and fosB expression in the striatum of the c-fos$^{+/-}$ mice (❷ *Table 10-1*). ❷ *Table 10-2* provides a list of the genes whose responses to METH were similar to those of c-jun, c-fos, fosB, and FRA1. ❷ *Figure 10-3* shows the cluster analysis of the expression profiles for these 46 genes. Genes that show similar expression profile to FRA1 and fosB include genes that participate in cell–cell and cell–matrix interactions (see also Cadet et al., 2002).

3 Role of DNA Repair Genes in METH-Induced Toxicity

The experiments, previously described, also identified other genes of interest (Cadet et al., 2002). These include: ATM, NBS1, XPG, PMS1, polB, MSH3, and XRCC5 (Ku80). Some genes showed higher basal level expression in either WT or fos$^{+/-}$ mice. Genes that were upregulated in the WT mice belong to classes of cell matrix, adhesion proteins, and transcription factors. Genes that were upregulated in fos$^{+/-}$ mice belong to classes of growth factors and cytokines. METH administration induced several genes that belong to classes of growth factors, cytokines, and DNA repair processes in WT mice (Cadet et al., 2002). Of interest among these genes are the genes for neurotrophic growth factors, EGF and FGF10, because their expression levels were decreased in fos$^{+/-}$ mice after METH administration. These observations suggest that these trophic factors might participate in protective mechanisms that are induced in METH-mediated induction of c-fos. It is also interesting to note that levels of genes involved in protective mechanisms and those of cell matrix and cell adhesion, which showed higher basal expression in WT mice, were negatively impacted after METH treatment.

It is of interest to relate the data on DNA repair genes to the reports that c-fos proteins might be involved in defense against UV radiation and other DNA-damaging agents (Dosch and Kaina, 1996; Tanos et al., 2005). Specifically, c-fos$^{-/-}$ fibroblasts were found to be highly sensitive to the cytotoxic effects of UV radiation (Haas and Kaina, 1995; Lackinger et al., 2001) and cadmium which causes its DNA damage via oxidative mechanisms (Matsuoka et al., 2000). Of related interest, even though c-fos$^{-/-}$ cells could perform DNA repair at low levels of radiation, they were unable to do so at higher doses of UV exposure (Haas and Kaina, 1995). These results are consistent with an earlier report that ribozyme-induced cleavage of c-fos mRNA caused reduction in the synthesis of enzymes involved in DNA synthesis and repair (Scanlon et al., 1991). The microarray data are also consistent with the report that c-fos might be involved in repairing oxidative DNA damage (Matsuoka et al., 2000).

Oxidative stress can cause single- and double-strand breaks (Li and Trush, 1993). This type of damage can be repaired via base excision repair (BER), nucleotide excision repair (NER), mismatch repair (MMR), and DNA damage reversal (Gros et al., 2002; Slupphaug et al., 2003; Barzilai and Yamamoto, 2004; Valko et al., 2004). Apurinic/apyrimidinic site (AP-site) and single-strand breaks, which are strongly promutagenic, are repaired by the BER pathway (Fortini et al., 2003). 8-oxo-G is excised from DNA by the 8-oxo-guanine glycosylase-1 (OGG1) protein leaving an AP-site, which is then acted upon by the AP-endonuclease-1 (APEX/APE1) enzyme (Fung and Demple, 2005). BER also involves polymerase B (*POLB*), ligase-3 (*LIG3*), exonuclease-1 (*EXO1*), and proliferating cell nuclear antigen (*PCNA*) genes (Barnes and Lindahl, 2004). DNA ligase I participates in a second BER pathway, which is carried out by a multiprotein complex in which DNA ligase I (LIG1) interacts directly with DNA polymerase β (Tomkinson and Mackey, 1998). Our observations that METH treatment can cause upregulation of APEX, PolB, and LIG1 suggest that these changes might be compensatory increases aimed at counteracting METH-mediated ROS-induced DNA damage through the BER pathway. This proposition is also supported with a previous report that oxidative stress can induce the expression of APE/Ref-1, which was able to provide protection against toxic levels of ROS such as hydrogen peroxide (Ramana et al., 1998).

It is also of interest that transcript levels for genes involved in MMR and NER such as ATM, MSH3, XPG, and PMS1 are also upregulated in WT mice but not in the c-fos$^{+/-}$ knockout animals. These

◘ Figure 10-1

Methamphetamine (METH) causes time-dependent changes in gene expression. (a) through (c) show scatter plot of correlations between two duplicate spots on the Mouse Atlas cDNA Expression Array (588 genes). All genes (*grey squares*) selected had a minimum expression value of 0.3. They also met the threefold criteria cutoffs represented by the *diagonal lines*. (d) through (f) show scatter plot of correlations between expression values obtained from saline- and METH-treated mice at 2, 4, and 16 h. In these scatter plots (d–f), the abscissa is the average of the two spots for saline and the ordinates are average of the two spots for METH. Deviations from the center within the area defined by the *diagonal lines* indicate changes in METH from saline, with *circles* indicating upregulated and *triangles* showing downregulated genes. The cluster analysis for these genes are shown in ❷ *Figure 10-2*

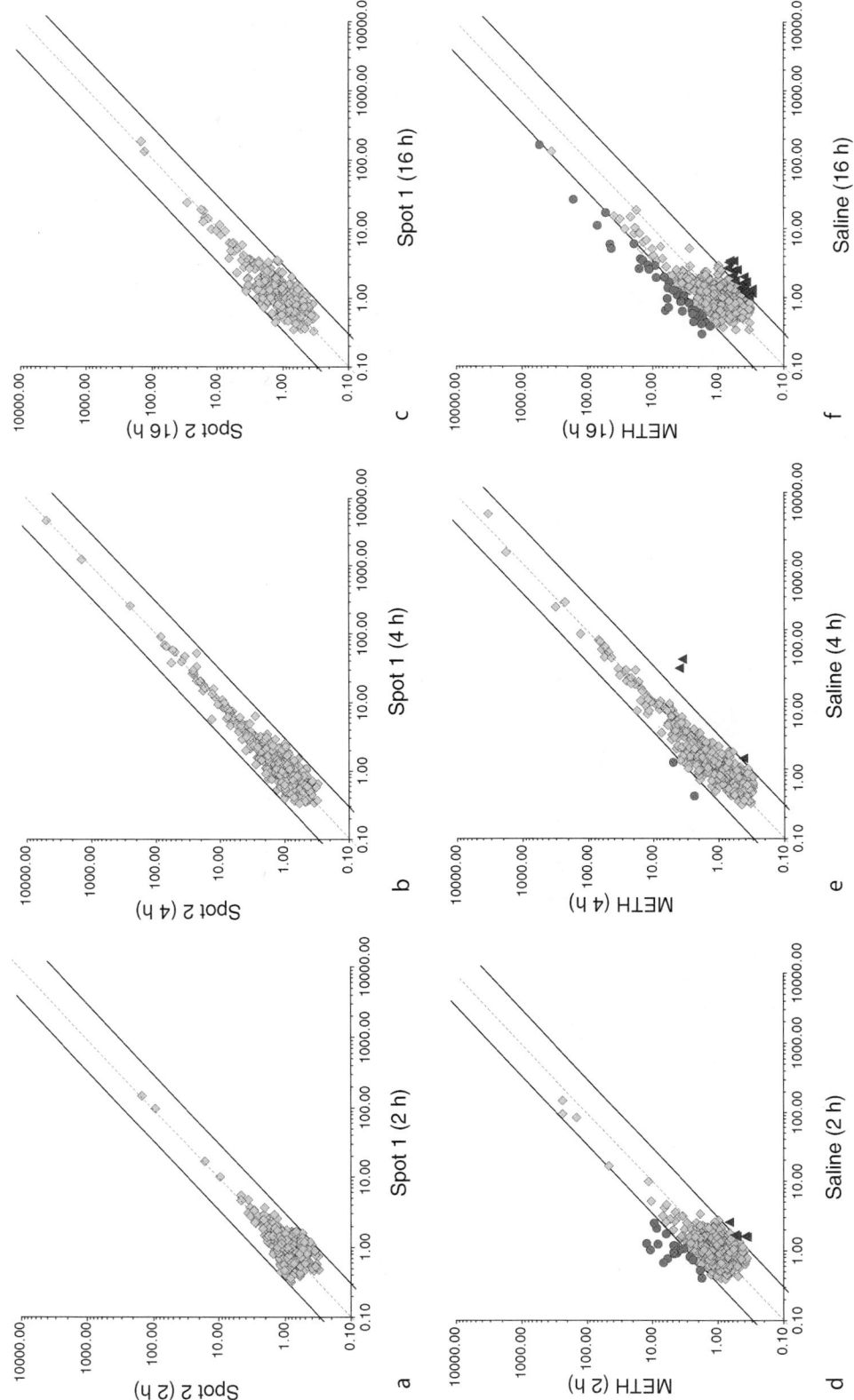

☐ Figure 10-2

(a) Cluster analysis showing the effects of methamphetamine (METH) at 2, 4, and 16 h after administration of a single dose (40 mg/kg) of the drug. 91 genes whose transcript levels were affected by METH were selected based on criteria described in the text. These genes were clustered according to the similarity of their expression profiles using GeneSpring. The expression of each gene is shown as a *horizontal strip* whereas all the genes for each time point are shown as *columns*. The dendrogram on the left of the cluster shows the relatedness of the genes in the cluster. The letters (I to VII) on the right side of the larger cluster identify subclusters. The color scale at the bottom represents the relative expression levels for the selected genes at the various time points. The subclusters are shown in (b–h)

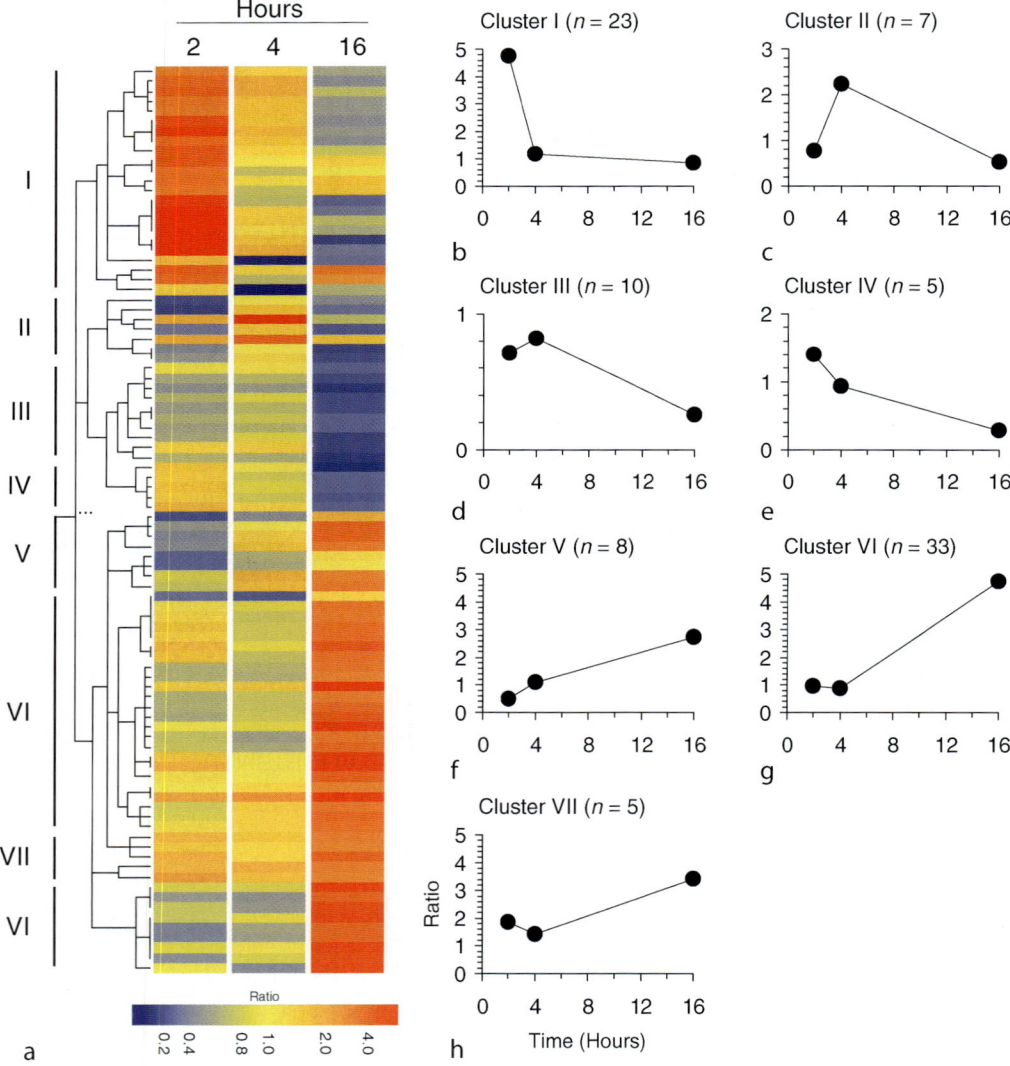

observations suggest that METH-induced oxidative DNA damage can be repaired not only by BER but also by NER and MMR. NER is indeed known to also participate in the repair of ROS-induced DNA damage (Hutsell and Sancar, 2005). ATM and Nbs1 activation suggests the possibility of the early occurrence of DNA double-strand breaks after injections of toxic doses of METH (Varon et al., 1998; Andegeko et al., 2001).

◻ Table 10-1
METH effects on AP-1-related genes in c-fos$^{+/-}$ mice

GenBank accession no.	Description	WT 2 h METH/WT saline	WT 12 h METH/WT saline	Fos$^{+/-}$ 2 h METH/Fos$^{+/-}$ saline	Fos$^{+/-}$ 12 h METH/Fos$^{+/-}$ saline	WT saline/Fos$^{+/-}$ saline	WT 2 h METH/Fos$^{+/-}$ 2 h METH	WT 12 h METH/Fos$^{+/-}$ 12 h METH
	AP-1 Genes							
V00727	c-fos protooncogene (FOS)	-2.19	-2.24	1.91	1.06	3.87	-1.08	1.63
X14897	FBJ osteosarcoma oncogene B (FOSB)	-2.21	-3.13	-1.11	-1.60	2.05	1.03	1.05
AF017128	fos-like antigen 1 (FRA1)	-2.55	-5.30	-1.04	-1.25	3.59	1.47	-1.18
X83971	fos-related antigen 2 (FRA2)	1.15	-1.36	1.69	1.45	1.03	-1.43	-1.92
X12761	Jun oncogene (JUN)	1.47	2.23	1.23	1.12	-1.12	1.07	1.77
J03236	junB protooncogene (JUNB)	1.32	1.57	1.05	-1.41	-1.69	-1.34	1.30
J05205	junD1 protooncogene (JUND)	-1.50	-1.22	-1.11	-1.45	-1.14	-1.55	1.04

This table was generated from comparisons of METH-induced changes in gene expression in the two genotypes. The values represent average ratios obtained from the two independent experiments. Positive values represent increases while negative numbers show decreases in gene expression

▫ Table 10-2

Classification of genes similar to c-jun, c-fos, fosB, and FRA1 in the mouse striatum after METH treatment

GenBank accession no.	Description	WT 2 h METH/WT saline	WT 12 h METH/WT saline	$Fos^{+/-}$ 2 h METH /$Fos^{-/-}$ saline	$Fos^{+/-}$ 12 h METH /$Fos^{-/-}$ saline	WT saline / $Fos^{-/-}$ saline	WT 2 h METH/$Fos^{-/-}$ 2h METH	WT 12 h METH/$Fos^{-/-}$ 12 h METH
	c-jun-like							
X02594	Proliferin-related protein (PLFR)	2.03	3.03	1.03	-1.23	-2.04	-1.03	1.84
AF019048	Tumor necrosis factor superfamily member 11 (TNFSF11)	2.01	3.20	1.22	1.11	-1.70	-1.03	1.70
X06381	Leukemia inhibitory factor (LIF)	1.73	2.20	1.07	1.10	-1.07	1.50	1.86
M23376	CD3 antigen ε (CD3E)	1.49	2.04	1.17	-1.19	-1.36	-1.06	1.79
X12761	c-jun	1.47	2.23	1.23	1.12	-1.12	1.07	1.77
AF092840	Nibrin (NBN)	1.41	2.74	1.24	1.33	-1.42	-1.25	1.45
X62622	Tissue inhibitor of metalloproteinase 2 (TIMP2)	1.36	2.42	1.40	1.27	-1.02	-1.05	1.87
Z22866	Myomesin 1 (MYOM1)	1.19	2.49	1.16	1.35	-1.02	1.01	1.82
X99018	Uracil-DNA glycosylase (UNG)	1.14	2.13	1.31	1.27	1.09	-1.05	1.83
U28724	Postmeiotic segregation increased 2 (PMS2)	1.02	2.21	1.27	1.23	-1.01	-1.26	1.78
	FRA1-like							
L10656	v-abl Abelson murine leukemia oncogene 1 (ABL1)	-1.97	-5.78	-1.17	-1.51	4.09	2.42	1.07
S65038	Gli oncogene; zinc finger transcription factor (GLI)	-2.01	-6.03	-1.71	-1.26	2.91	2.47	-1.65
AF013107	A disintegrin and metalloproteinase domain 7 (ADAM7)	-2.01	-3.26	1.04	-1.48	2.78	1.33	1.26
M64429	B-raf protooncogene (B-raf)	-2.30	-4.06	-1.44	-1.59	2.71	1.70	1.06
X12616	Protooncogene tyrosine protein kinase fes/fps (FES)	-2.32	-6.43	1.04	1.08	4.97	2.07	-1.40
X97818	Semaphorin G precursor (SEMA5B)	-2.35	-18.58	1.02	-1.51	5.74	2.40	-2.14

Accession	Gene						
M84487	Vascular cell adhesion molecule 1 precursor (VCAM1)	-2.53	-7.37	-1.08	1.06	2.83	1.21	-2.77
AF017128	fos-like antigen 1 (FRA1)	-2.55	-5.30	-1.04	-1.25	3.59	1.47	-1.18
S36676	Macrophage galactose N-acetyl-galactosamine-specific lectin (MGL)	-2.77	-5.43	1.19	1.12	3.94	1.19	-1.54
Y00769	Integrin β1 (ITGB1)	-2.81	-5.67	-1.34	-1.32	3.74	1.78	-1.15
U48420	Glutathione-S-transferase, θ2 (GSTT2)	-3.20	-4.42	-1.08	1.02	2.93	-1.02	-1.53
U22058	A disintegrin and metalloproteinase domain 4 (ADAM4)	-3.26	-10.49	-1.08	-1.00	4.88	1.62	-2.14
U07634	Eph receptor A2 (EPHA2)	-3.55	-7.43	-1.62	-1.11	3.17	1.45	-2.10
U22059	A disintegrin and metalloproteinase domain 5 (ADAM5)	-3.58	-7.63	-1.07	-1.21	4.00	1.20	-1.57
fosB-like							
U12147	Laminin α2 (LAMA2)	-1.48	-4.22	-1.16	-2.06	1.61	1.26	-1.27
U16242	A disintegrin and metalloprotease domain 2 (ADAM2)	-1.78	-4.12	-1.36	-1.35	1.74	1.33	-1.75
M13071	raf-related oncogene (Araf)	-1.94	-3.20	-1.01	-1.02	2.29	1.20	-1.37
X74335	Desmoglein 1 (DSG1)	-1.98	-5.10	1.03	-1.85	1.93	-1.06	-1.43
M13018	Cysteine-rich intestinal protein (CRIP)	-2.20	-2.48	-1.14	-1.32	1.51	-1.27	-1.25
X14897	FBJ osteosarcoma oncogene B (FOSB)	-2.21	-3.13	-1.11	-1.60	2.05	1.03	1.05
M68513	eph receptor A3 (EPHA3)	-2.23	-3.47	-1.12	-1.59	1.46	-1.36	-1.49
U87948	Epithelial membrane protein 3 (EMP3)	-2.34	-4.68	1.22	-1.02	2.33	-1.23	-1.96
X91144	P-selectin glycoprotein ligand 1 (SELP1)	-2.38	-3.29	1.01	1.04	1.82	-1.32	-1.89
U83903	Tumor necrosis factor induced protein 6 (TNFIP6)	-2.64	-2.65	1.21	1.03	2.11	-1.51	-1.30

continued

■ Table 10-2 (continued)

GenBank accession no.	Description	WT 2 h METH/WT saline	WT 12 h METH/WT saline	Fos$^{+/-}$ 2 h METH/Fos$^{+/-}$ saline	Fos$^{+/-}$ 12 h METH/Fos$^{+/-}$ saline	WT saline / Fos$^{+/-}$ saline	WT 2 h METH/Fos$^{+/-}$ 2h METH	WT 12 h METH/Fos$^{+/-}$ 12 h METH
	c-fos-like							
U83509	Angiopoietin (AGPT)	−1.51	−2.53	−1.21	1.00	2.40	1.93	−1.06
AF030311	Killer cell lectin-like receptor, subfamily D, member 1 (KLRD1)	−1.70	−1.57	1.14	−1.06	2.13	1.10	1.44
AF114753	Receptor for MCF virus 1 (RMC1)	−1.73	−1.64	1.51	1.27	2.72	1.04	1.30
J05287	Lysosomal membrane glycoprotein 2 (LAMP2)	−1.87	−1.96	−1.06	−1.25	1.86	1.06	1.18
V00727	c-fos	−2.19	−2.24	1.91	1.06	3.87	−1.08	1.63
M31314	High-affinity IgG Fc receptor I (FCGR)	−2.36	−2.12	−1.50	−1.07	2.22	1.41	1.12
M12848	Myeloblastosis protooncogene (MYB)	−2.42	−1.96	1.58	1.78	3.26	−1.18	−1.07
X52191	c-fgr protooncogene (FGR)	−2.59	−3.15	2.11	1.81	4.41	−1.24	−1.29
U58992	Mothers against decapentaplegic homolog 1 (MADH1)	−2.75	−2.21	−1.14	−1.03	2.57	1.06	1.20
U89484	Agouti-related protein (AGRP)	−3.75	−2.83	−1.73	−1.73	2.33	1.08	1.43

☐ Figure 10-3

Cluster analysis showing the expression profiles for a total of 46 genes whose RNA levels changed in a fashion similar to the four listed AP-1 genes after methamphetamine (METH) treatment. Genes were selected using GeneSpring based on the similarity of their expression profiles to those of c-jun, c-fos, fosB, or FRA1. These genes were then clustered based on their temporal expression profiles using GeneSpring. Each *row* represents a gene while each *column* represents a specific genotype and the corresponding treatment group. The dendrogram on the left depicts the degree of relatedness of the genes in the cluster, with the shortest branches showing the highest degree of correlation. The pseudocolor scale is provided to represent relative expression of genes, with *red* showing high and *blue* showing low transcript levels

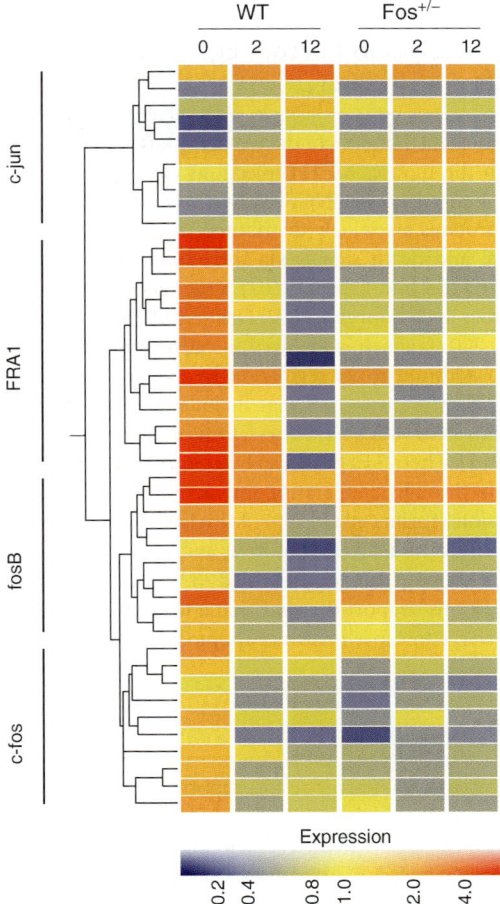

Both MSH3 and PMS1 whose transcript levels are unregulated by METH play important roles in MMR (Prolla et al., 1994; Saparbaev et al., 1996).

4 Bcl-2-Related Genes and METH Toxicity

Another interesting family of genes that showed differential regulation after METH administration was the Bcl-2 family genes (see ❷ *Figure 10-1* and ❷ *Table* in Cadet et al., 2001). These changes occur late in the course after METH injections (Cadet et al., 2001; Jayanthi et al., 2001). RT-PCR analysis and Western blot experiments confirmed the changes observed in the array analyses, with the proapoptotic genes, BAX and

BID, showing significant upregulation and the antideath genes, Bcl-2 and Bcl-X_L, showing prolonged down-regulation. Other group of researchers also showed that METH treatment can downregulate Bcl-2 in the striatum in mice (Imam et al., 2001; He et al., 2004).

Our results suggest that METH acts to promote a shift in the prodeath/antideath balance with the ultimate result being the death of the cells within which this shift has occurred (Jayanthi et al., 2001). This suggestion is in accordance with our previous report that overexpression of Bcl-2 can protect against METH-induced apoptosis in vitro (Cadet et al., 1997). Further examination of these issues indicates that the mitochondrial death pathway interacts with pathological events in the endoplasmic reticulum to cause METH-induced damage in the brain (Jayanthi et al., 2004).

5 Involvement of Transcription Factors Egr and Fas Ligand in METH-Induced Neuronal Apoptosis

In order to expand these studies further, we conducted array analyses of the toxic effects of METH in rats (Jayanthi et al., 2005). As in the mouse, METH injections caused increased expression of several genes including members of the Egr, AP1, and NGFI-B/Nur77 subfamilies of transcription factors in rats (Jayanthi et al., 2005). Egr proteins are widely distributed zing finger TFs that are involved in mediating development, differentiation, and death signals via their influence on the expression of various genes (Beckmann and Wilce, 1997; Thiel and Cibelli, 2002). One of these genes is Fas ligand (FasL, TNFSF6) (Li-Weber et al., 1999; Li-Weber and Krammer, 2002; Droin et al., 2003), a member of the TNF superfamily of cytokines (Locksley et al., 2001), which is involved in causing apoptosis in various models of neuronal injury (Qiu et al., 2002). In addition to its regulation by Egr proteins, Fas ligand (FasL) expression is influenced by an array of transcription factors that include AP-1 and Nur77 factors (Toth et al., 2001; Li-Weber and Krammer, 2002), all of which were also significantly induced by the injection of the toxic dose of METH (Jayanthi et al., 2005). These TFs have also been implicated in the regulation of activity-induced cell death (AICD) of T cells through cooperative interactions with the FasL promoter with resulting upregulation of FasL expression (Latinis et al., 1997; Mittelstadt and Ashwell, 1999; Macian et al., 2000). We thus used quantitative PCR to measure FasL mRNA and found significant increases in its expression. Further experiments revealed cleavage of caspase-8 and -3, which participate in the Fas death pathway (Nagata, 1999). These observations support the view that a toxic dose of METH is able to recruit multiple molecular pathways that converge to cause neuronal apoptosis in the rat brain.

6 General Discussion

Our data are similar to the recent cDNA array results reported using Atlas Rat 1.2 Array (Clontech) after METH exposure by Asanuma and coworkers (2004). They are also consistent with the observations of Thomas and coworkers (2004) who used Affymetrix oligonucleotide arrays (12,488 genes). Our studies and those of these investigators have revealed a very complex response of the brain to METH. These responses include profound changes in inflammatory/immune response elements, receptor/signal transduction components, and ion channel/transport proteins (Thomas et al., 2004). Interestingly, protoxic cascades appear to be counterbalanced by corresponding reactions in trophic factors and antioxidant enzymes, which are also activated after METH administration (Jayanthi et al., 1998; Cadet et al., 2001). This includes the activation of insulin growth factor (IGF), transforming growth factor (TGF), fibroblast growth factor (FGF), NF-κB, and AKT, all of which have been shown to exert significant antideath activities (Suzuki, 2003).

7 Summary

In conclusion, the purpose of this chapter was to provide a brief review of the insight our group has obtained into the molecular bases of METH toxicity by using the cDNA approach. These data support the view that oxidative mechanisms, transcription factors, and cell death-related genes are involved in METH-induced neurodegeneration. Future studies will help to target these mechanisms for therapeutic interventions.

References

Ali SF, Imam SZ, Itzhak Y. 2005. Role of peroxynitrite in methamphetamine-induced dopaminergic neurodegeneration and neuroprotection by antioxidants and selective NOS inhibitors. Ann N Y Acad Sci 1053: 97-98.

Andegeko Y, et al. 2001. Nuclear retention of ATM at sites of DNA double strand breaks. J Biol Chem 276: 38224-38230.

Asanuma M, et al. 2004. Specific gene expression and possible involvement of inflammation in methamphetamine-induced neurotoxicity. Ann N Y Acad Sci 1025: 69-75.

Barnes DE, Lindahl T. 2004. Repair and genetic consequences of endogenous DNA base damage in mammalian cells. Annu Rev Genet 38: 445-476.

Barzilai A, Yamamoto K. 2004. DNA damage responses to oxidative stress. DNA Repair (Amst) 3: 1109-1115.

Beckmann AM, Wilce PA. 1997. Egr transcription factors in the nervous system. Neurochem Int 31: 477-510; discussion 517-476.

Buffenstein A, Heaster J, Ko P. 1999. Chronic psychotic illness from methamphetamine. Am J Psychiatry 156: 662.

Cadet JL, Brannock C. 1998. Free radicals and the pathobiology of brain dopamine systems. Neurochem Int 32: 117-131.

Cadet JL, et al. 2001. Temporal profiling of methamphetamine-induced changes in gene expression in the mouse brain: Evidence from cDNA array. Synapse 41: 40-48.

Cadet JL, McCoy MT, Ladenheim B. 2002. Distinct gene expression signatures in the striata of wild-type and heterozygous c-fos knockout mice following methamphetamine administration: Evidence from cDNA array analyses. Synapse 44: 211-226.

Cadet JL, Ordonez SV, Ordonez JV. 1997. Methamphetamine induces apoptosis in immortalized neural cells: Protection by the proto-oncogene, bcl-2. Synapse 25: 176-184.

Dalton TP, Shertzer HG, Puga A. 1999. Regulation of gene expression by reactive oxygen. Annu Rev Pharmacol Toxicol 39: 67-101.

Deng X, Cadet JL. 1999. Methamphetamine administration causes overexpression of nNOS in the mouse striatum. Brain Res 851: 254-257.

Deng X, Cadet JL. 2000. Methamphetamine-induced apoptosis is attenuated in the striata of copper-zinc superoxide dismutase transgenic mice. Brain Res Mol Brain Res 83: 121-124.

Deng X, et al. 2002. Methamphetamine induces apoptosis in an immortalized rat striatal cell line by activating the mitochondrial cell death pathway. Neuropharmacology 42: 837-845.

Deng X, Ladenheim B, Tsao LI, Cadet JL. 1999. Null mutation of c-fos causes exacerbation of methamphetamine-induced neurotoxicity. J Neurosci 19: 10107-10115.

Dore G, Sweeting M. 2006. Drug-induced psychosis associated with crystalline methamphetamine. Australas Psychiatry 14: 86-89.

Dosch J, Kaina B. 1996. Induction of c-fos, c-jun, junB and junD mRNA and AP-1 by alkylating mutagens in cells deficient and proficient for the DNA repair protein O6-methylguanine-DNA methyltransferase (MGMT) and its relationship to cell death, mutation induction and chromosomal instability. Oncogene 13: 1927-1935.

Droin NM, Pinkoski MJ, Dejardin E, Green DR. 2003. Egr family members regulate nonlymphoid expression of Fas ligand, TRAIL, and tumor necrosis factor during immune responses. Mol Cell Biol 23: 7638-7647.

Eisch AJ, Schmued LC, Marshall JF. 1998. Characterizing cortical neuron injury with Fluoro-Jade labeling after a neurotoxic regimen of methamphetamine. Synapse 30: 329-333.

Eisen MB, Spellman PT, Brown PO, Botstein D. 1998. Cluster analysis and display of genome-wide expression patterns. Proc Natl Acad Sci USA 95: 14863-14868.

Farrell M, Marsden J, Ali R, Ling W. 2002. Methamphetamine: Drug use and psychoses becomes a major public health issue in the Asia Pacific region. Addiction 97: 771-772.

Fortini P, et al. 2003. 8-Oxoguanine DNA damage: At the crossroad of alternative repair pathways. Mutat Res 531: 127-139.

Fumagalli F, et al. 1999. Increased methamphetamine neurotoxicity in heterozygous vesicular monoamine transporter 2 knock-out mice. J Neurosci 19: 2424-2431.

Fung H, Demple B. 2005. A vital role for Ape1/Ref1 protein in repairing spontaneous DNA damage in human cells. Mol Cell 17: 463-470.

Gros L, Saparbaev MK, Laval J. 2002. Enzymology of the repair of free radicals-induced DNA damage. Oncogene 21: 8905-8925.

Haas S, Kaina B. 1995. c-Fos is involved in the cellular defence against the genotoxic effect of UV radiation. Carcinogenesis 16: 985-991.

Harvey DC, Lacan G, Tanious SP, Melega WP. 2000. Recovery from methamphetamine induced long-term nigrostriatal dopaminergic deficits without substantia nigra cell loss. Brain Res 871: 259-270.

He J, et al. 2004. Neuroprotective effects of olanzapine on methamphetamine-induced neurotoxicity are associated with an inhibition of hyperthermia and prevention of Bcl-2 decrease in rats. Brain Res 1018: 186-192.

Hirata H, Cadet JL. 1997. p53-knockout mice are protected against the long-term effects of methamphetamine on dopaminergic terminals and cell bodies. J Neurochem 69: 780-790.

Hutsell SQ, Sancar A. 2005. Nucleotide excision repair, oxidative damage, DNA sequence polymorphisms, and cancer treatment. Clin Cancer Res 11: 1355-1357.

Imam SZ, et al. 2001. Methamphetamine-induced alteration in striatal p53 and bcl-2 expressions in mice. Brain Res Mol Brain Res 91: 174-178.

Itzhak Y, Martin JL, Ail SF. 2000. nNOS inhibitors attenuate methamphetamine-induced dopaminergic neurotoxicity but not hyperthermia in mice. Neuroreport 11: 2943-2946.

Jayanthi S, et al. 2001. Methamphetamine causes differential regulation of pro-death and anti-death Bcl-2 genes in the mouse neocortex. FASEB J 15: 1745-1752.

Jayanthi S, et al. 2004. Methamphetamine induces neuronal apoptosis via cross-talks between endoplasmic reticulum and mitochondria-dependent death cascades. FASEB J 18: 238-251.

Jayanthi S, et al. 2005. Calcineurin/NFAT-induced up-regulation of the Fas ligand/Fas death pathway is involved in methamphetamine-induced neuronal apoptosis. Proc Natl Acad Sci USA 102: 868-873.

Jayanthi S, Ladenheim B, Cadet JL. 1998. Methamphetamine-induced changes in antioxidant enzymes and lipid peroxidation in copper/zinc-superoxide dismutase transgenic mice. Ann N Y Acad Sci 844: 92-102.

Jayanthi S, McCoy MT, Ladenheim B, Cadet JL. 2002. Methamphetamine causes coordinate regulation of SRC, cas, crk, and the jun N-terminal kinase-jun pathway. Mol Pharmacol 61: 1124-1131.

Johnston L, O'Malley PM, Bachman JG, Schulenberg J. 2006. Monitoring the future national results on adolescent drug use: Overview of key findings, 2005. Bethesda, MD: (NIH Publication No. 06-5882).

Krasnova IN, McCoy MT, Ladenheim B, Cadet JL. 2002. cDNA array analysis of gene expression profiles in the striata of wild-type and Cu/Zn superoxide dismutase transgenic mice treated with neurotoxic doses of amphetamine. FASEB J 16: 1379-1388.

Lackinger D, Eichhorn U, Kaina B. 2001. Effect of ultraviolet light, methyl methanesulfonate and ionizing radiation on the genotoxic response and apoptosis of mouse fibroblasts lacking c-Fos, p53 or both. Mutagenesis 16: 233-241.

Latinis KM, Norian LA, Eliason SL, Koretzky GA. 1997. Two NFAT transcription factor binding sites participate in the regulation of CD95 (Fas) ligand expression in activated human T cells. J Biol Chem 272: 31427-31434.

La Voie MJ, Hastings TG. 1999. Dopamine quinone formation and protein modification associated with the striatal neurotoxicity of methamphetamine: Evidence against a role for extracellular dopamine. J Neurosci 19: 1484-1491.

Li Y, Trush MA. 1993. DNA damage resulting from the oxidation of hydroquinone by copper: Role for a Cu(II)/Cu(I) redox cycle and reactive oxygen generation. Carcinogenesis 14: 1303-1311.

Li-Weber M, Krammer PH. 2002. The death of a T-cell: Expression of the CD95 ligand. Cell Death Differ 9: 101-103.

Li-Weber M, Laur O, Krammer PH. 1999. Novel Egr/NF-AT composite sites mediate activation of the CD95 (APO-1/Fas) ligand promoter in response to T cell stimulation. Eur J Immunol 29: 3017-3027.

Locksley RM, Killeen N, Lenardo MJ. 2001. The TNF and TNF receptor superfamilies: Integrating mammalian biology. Cell 104: 487-501.

London ED, et al. 2004. Mood disturbances and regional cerebral metabolic abnormalities in recently abstinent methamphetamine abusers. Arch Gen Psychiatry 61: 73-84.

Lotharius J, et al. 2005. Progressive degeneration of human mesencephalic neuron-derived cells triggered by dopamine-dependent oxidative stress is dependent on the mixed-lineage kinase pathway. J Neurosci 25: 6329-6342.

Macian F, Garcia-Rodriguez C, Rao A. 2000. Gene expression elicited by NFAT in the presence or absence of cooperative recruitment of Fos and Jun. EMBO J 19: 4783-4795.

Matsuoka M, Wispriyono B, Igisu H. 2000. Increased cytotoxicity of cadmium in fibroblasts lacking c-fos. Biochem Pharmacol 59: 1573-1576.

McCann UD, et al. 1998. Reduced striatal dopamine transporter density in abstinent methamphetamine and methcathinone users: Evidence from positron emission tomography studies with [^{11}C]WIN-35,428. J Neurosci 18: 8417-8422.

Mittelstadt PR, Ashwell JD. 1999. Role of Egr-2 in up-regulation of Fas ligand in normal T cells and aberrant double-negative lpr and gld T cells. J Biol Chem 274: 3222-3227.

Nagata S. 1999. Fas ligand-induced apoptosis. Annu Rev Genet 33: 29-55.

O'Dell SJ, Marshall JF. 2000. Repeated administration of methamphetamine damages cells in the somatosensory cortex: Overlap with cytochrome oxidase-rich barrels. Synapse 37: 32-37.

Pennypacker K. 1997. Transcription factors in brain injury. Histol Histopathol 12: 1125-1133.

Prolla TA, et al. 1994. MLH1, PMS1, and MSH2 interactions during the initiation of DNA mismatch repair in yeast. Science 265: 1091-1093.

Qiu J, et al. 2002. Upregulation of the Fas receptor death-inducing signaling complex after traumatic brain injury in mice and humans. J Neurosci 22: 3504-3511.

Ramana CV, Boldogh I, Izumi T, Mitra S. 1998. Activation of apurinic/apyrimidinic endonuclease in human cells by reactive oxygen species and its correlation with their adaptive response to genotoxicity of free radicals. Proc Natl Acad Sci USA 95: 5061-5066.

Ricaurte GA, Schuster CR, Seiden LS. 1980. Long-term effects of repeated methylamphetamine administration on dopamine and serotonin neurons in the rat brain: A regional study. Brain Res 193: 153-163.

Salzmann J, et al. 2003. Importance of ERK activation in behavioral and biochemical effects induced by MDMA in mice. Br J Pharmacol 140: 831-838.

Saparbaev M, Prakash L, Prakash S. 1996. Requirement of mismatch repair genes MSH2 and MSH3 in the RAD1–RAD10 pathway of mitotic recombination in Saccharomyces cerevisiae. Genetics 142: 727-736.

Scanlon KJ, et al. 1991. Ribozyme-mediated cleavage of c-fos mRNA reduces gene expression of DNA synthesis enzymes and metallothionein. Proc Natl Acad Sci USA 88: 10591-10595.

Sekine Y, et al. 2001. Methamphetamine-related psychiatric symptoms and reduced brain dopamine transporters studied with PET. Am J Psychiatry 158: 1206-1214.

Sekine Y, et al. 2003. Association of dopamine transporter loss in the orbitofrontal and dorsolateral prefrontal cortices with methamphetamine-related psychiatric symptoms. Am J Psychiatry 160: 1699-1701.

Sekine Y, et al. 2006. Brain serotonin transporter density and aggression in abstinent methamphetamine abusers. Arch Gen Psychiatry 63: 90-100.

Shaulian E, Karin M. 2002. AP-1 as a regulator of cell life and death. Nat Cell Biol 4: E131-E136.

Slupphaug G, Kavli B, Krokan HE. 2003. The interacting pathways for prevention and repair of oxidative DNA damage. Mutat Res 531: 231-251.

Sonsalla PK, Nicklas WJ, Heikkila RE. 1989. Role for excitatory amino acids in methamphetamine-induced nigrostriatal dopaminergic toxicity. Science 243: 398-400.

Staszewski RD, Yamamoto BK. 2006. Methamphetamine-induced spectrin proteolysis in the rat striatum. J Neurochem 96: 1267-1276.

Stumm G, et al. 1999. Amphetamines induce apoptosis and regulation of bcl-x splice variants in neocortical neurons. FASEB J 13: 1065-1072.

Suzuki YJ. 2003. Growth factor signaling for cardioprotection against oxidative stress-induced apoptosis. Antioxid Redox Signal 5: 741-749.

Tanos T, et al. 2005. Phosphorylation of c-Fos by members of the p38 MAPK family. Role in the AP-1 response to UV light. J Biol Chem 280: 18842-18852.

Thiel G, Cibelli G. 2002. Regulation of life and death by the zinc finger transcription factor Egr-1. J Cell Physiol 193: 287-292.

Thiriet N, Zwiller J, Ali SF. 2001. Induction of the immediate early genes egr-1 and c-fos by methamphetamine in mouse brain. Brain Res 919: 31-40.

Thomas DM, Francescutti-Verbeem DM, Liu X, Kuhn DM. 2004. Identification of differentially regulated transcripts in mouse striatum following methamphetamine treatment—an oligonucleotide microarray approach. J Neurochem 88: 380-393.

Tomkinson AE, Mackey ZB. 1998. Structure and function of mammalian DNA ligases. Mutat Res 407: 1-9.

Toth R, et al. 2001. Activation-induced apoptosis and cell surface expression of Fas (CD95) ligand are reciprocally regulated by retinoic acid receptor α and γ and involve nur77 in T cells. Eur J Immunol 31: 1382-1391.

Valko M, et al. 2004. Role of oxygen radicals in DNA damage and cancer incidence. Mol Cell Biochem 266: 37-56.

Varon R, et al. 1998. Nibrin, a novel DNA double-strand break repair protein, is mutated in Nijmegen breakage syndrome. Cell 93: 467-476.

Volkow ND, et al. 2001a. Loss of dopamine transporters in methamphetamine abusers recovers with protracted abstinence. J Neurosci 21: 9414-9418.

Volkow ND, et al. 2001b. Association of dopamine transporter reduction with psychomotor impairment in methamphetamine abusers. Am J Psychiatry 158: 377-382.

Wagner GC, et al. 1980. Long-lasting depletions of striatal dopamine and loss of dopamine uptake sites following repeated administration of methamphetamine. Brain Res 181: 151-160.

11 Huntington's Disease: Unraveling the Pathophysiological Cascade Behind a Singular Gene Defect

C. M. Kosinski · B. Landwehrmeyer · A. Ludolph

Abstract: Huntington's disease (HD), an autosomal dominantly inherited disease, belongs to a group of neurodegenerative disorders, which share the same genetic defect, i.e., a CAG triplet expansion that is translated into polyglutamine chains within different proteins. Although HD is caused by a singular gene defect, this disease shows a high variability in age of onset and in its clinical manifestations, which include characteristic motor, cognitive, and psychiatric symptoms. This variability can only partly be explained by changes in CAG length. While some other rare neurodegenerative diseases share some of the clinical phenotype with HD, some perplexing phenocopies of HD called "HD-like diseases" have been identified recently, which are related to different gene defects. The key step in the molecular pathology is a conformational change in the expanded polyglutamine above a certain length threshold, which leads to formation of β-strands and polyglutamine aggregation. As a histopathological hallmark, intranuclear huntingtin-positive aggregates have been identified, making HD another neurodegenerative disease, with the presence of insoluble protein aggregates, like Alzheimer's or Parkinson's disease. Furthermore, key steps in molecular pathology include early changes in transcriptional regulation. Numerous transgenic mouse, fly, and in vitro cellular models help us to elucidate disease mechanism and evaluate therapeutic strategies. This allows more specific therapeutic interventions in the near future, in this model disease, for hereditary neurodegenerative disorders.

List of Abbreviations: APOE, apolipoprotein E; BDNF, brain-derived neurotrophic factor; CBP, CREB-binding protein; DRPLA, dentato-rubro-pallido-luysian atrophy; HARP, hypoprebetalipoproteinemia, acanthocytosis, retinitis pigmentosa, and pallidal degeneration; HD, Huntington's disease; Hdh, mouse HD gene homologue; IT15, interesting transcript 15, name of the HD gene on chromosome 4; NAA, N-acetylaspartate; NMDA, N-methyl-D-aspartate; PEG, percutaneous endoscopic gastrostomy; SAHA, suberoylanilide hydroxamic acid; SCA, spinocerebellar atrophy; UTR, untranslated region

1   History

The description by George Huntington in 1872 of the disease that has subsequently borne his name is remarkable for its clarity and comprehensiveness (Huntington, 1872). It was not the first description of the disorder (see, for instance, Waters, 1842; Lund, 1860), but it stands out as the first full delineation of the condition as a specific disease entity, quite separate from other forms of chorea. All the cardinal features of Huntington's disease (HD) are recognized in this description: the adult onset, the progressive course and eventually fatal outcome, the choreic movements combined with mental impairment, even the pattern of inheritance; George Huntington's role as a family doctor, following his father and grandfather to give a 78-year total period of observation, gave him an unique perspective to appreciate the hereditary nature and the clinical features of HD. The recognition of juvenile HD (Hoffmann, 1888) led to the realization that chorea was not the only motor disorder in HD, and that some patients show a predominance of rigidity and hypokinesia (Westphal variant [Westphal, 1883]), usually in young adults or children, that occur in the same families as cases with chorea as the presenting feature. It was not earlier than the 1920s that there was general agreement that the neuropathological abnormalities in HD were primarily degenerative and atrophic, despite Jelgersma's description in 1908 (Jelgersma, 1908) reporting generalized shrinkage of the HD brain and atrophy of the caudate nucleus (reduced to one-third of its original volume).

2   Genetics

2.1   HD: A CAG Triplet Disease

Huntington's disease displays an autosomal dominant pattern of inheritance: equal sex incidence, equal transmission by both sexes, a 50% proportion of affected offspring born to an affected parent, and lack of transmission by unaffected family members are all features that are evident from any large HD pedigree. Ten years after the mapping of the HD gene to chromosome 4, a gene in the critical region, named "interesting transcript 15" (IT15), proved to contain a CAG repeat in its first exon, which was expanded in HD patients (Huntington's et al., 1993).

The CAG expansion appears to vary. Expansions of more than 36 CAG repeats are associated with the HD phenotype. There is, however, a narrow region of overlap between the normal and the pathological range, even though few individuals are represented in this borderline region, and therefore, there is no absolute separation between the normal and the pathological range—a conclusion of practical importance when the information is being used diagnostically.

2.2 CAG Length–Phenotype Correlation

The CAG repeat expansion is unstable on meiotic transmission; one of the key consequences of genetic instability in HD is that different individuals carrying the mutation will have different degrees of expansion in the CAG repeat sequence. A strong inverse correlation between repeat number and age at onset was demonstrated (❷ *Figure 11-1*) (Andrew et al., 1993; Duyao et al., 1993; MacMillan et al., 1993; Snell et al., 1993; Zuhlke et al., 1993; Claes et al., 1995). The CAG repeat length appears to account for 47%–73% of the variance in age of onset in studies of HD populations (Ranen et al., 1995; Brinkman et al., 1997; Rosenblatt et al., 2001). However, it is important to note the considerable variation for each repeat number. For example, with a repeat number of 41 ($n = 98$), onset ranged from 35 to 75 years. The strength of the correlation was greatest for the largest repeats frequently associated with juvenile onset of disease and was weakest in the lower part of the abnormal range, corresponding to the vast majority of adult-onset HD cases.

In contrast to the clear relationship between CAG repeat number and age of disease onset, there is little evidence for a similar correlation with rate of disease progression (Ashizawa et al., 1994; Kieburtz et al., 1994; Brandt et al., 1996). In conclusion, the number of repeats in an HD mutation carrier is strongly correlated with age at onset of disease, especially for large repeat expansions; it is also correlated with

❑ Figure 11-1

Negative correlation between the length of the (CAG)$_n$ repeat in the HD allele and age of disease onset (Image from Cleas et al. 1995. Arch Neurol 113: 749–753)

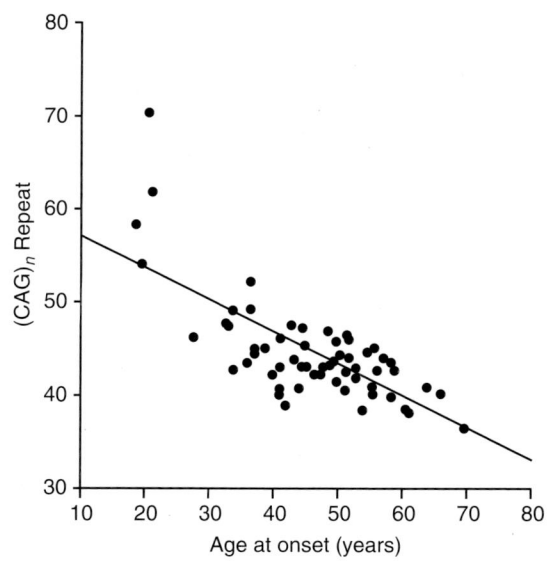

severity of brain pathology (Penney et al., 1997), but not with the development of psychiatric features or rate of decline.

2.3 Anticipation and De Novo Mutations

Most juvenile and early-onset cases of HD are paternally transmitted; in addition, there appeared to be "anticipation," i.e., progressively earlier onset in successive generations in male-transmitted cases. Until 1993, these observations were unexplained. Next, two-generation data showed clearly that the anticipation seen in individual families was closely paralleled by an increased number of CAG repeats (Andrew et al., 1993; Duyao et al., 1993; Huntington's et al., 1993). In a series of 254 parent–child pairs with HD (Kremer et al., 1995), expansion was seen in 70% of meioses. Although small increases in repeat number (up to seven repeats) were seen in transmission by both sexes (47.1% of males and 37.8% of females), large increases (more than seven repeats) were seen almost exclusively in the offsprings of males (21.0% versus 0.7% in the offsprings of females) (Trottier et al., 1994; Ranen et al., 1995). Analysis of single sperm (Leeflang et al., 1995; Chong et al., 1997) gives a clear picture of the distribution of sperm with different repeat numbers generated at meiosis in males, confirming a high frequency (>90%) of increased repeat number for alleles in the pathological range. For an intermediate allele carrier (definition of "intermediate allele," see above), this figure was still over 50%, with 8% in the clinically significant range. Large series of sporadic cases showed the HD mutation in most clinically typical cases (MacMillan et al., 1993; Davis et al., 1994; Mandich et al., 1996). When parents of such patients were studied, it became clear that while the clinical phenotype of HD might arise "de novo," the molecular defect did not, with one parent invariably showing an expanded allele usually in the 30–38 repeat range, but not associated with evidence of clinical disease even at an advanced age (Goldberg et al., 1993; Myers et al., 1993). In all six cases in the series investigated by Goldberg et al., where parental origin could be determined, the father was the one with the intermediate allele. The recognition of intermediate allele carriers has led to an assessment of penetrance of the HD phenotype in this borderline repeat range.

2.4 Mechanism of CAG Instability

One of the questions that arose from a study of HD and other diseases caused by trinucleotide repeat expansions is the nature of the molecular mechanism by which those expansions occurred in the first place; the same mechanism is likely involved in the generation of intermediate alleles and in their expansion into the HD range as well. DNA undergoes a number of processes during which size changes in repeats could possibly occur. These include replication, recombination, transcription, and repair. The best evidence for the mechanism of instability is that the HD transgenic R6/1 models when crossed with $Msh2^{-/-}$ mice have no expansion of the HD mutation in germ cells (Kovtun and McMurray, 2001); this almost certainly reflects a loss of gap repair function rather than DNA strand slippage.

2.5 Genetic Modifiers

Three genetic modifiers of the HD gene have been reported, each probably accounting for only a small proportion of the variance in the age of onset of HD. A noncoding TAA repeat polymorphism in the 3'-UTR of the kainate receptor (GluR6) gene was reported to account for varying amounts of the residual age at onset variation in HD (Pickering et al., 1993; Rubinsztein et al., 1997). APOE genotype had a small but significant effect with the ε2ε3 genotype associated with an earlier onset in males than females (Kehoe et al., 1999). APOE genotype is the only robust genetic risk factor associated with late onset AD (Panas et al., 1999), where it does have an influence on age at onset. Yet another genetic variation detected as influencing age of onset in HD is a length polymorphism in a polyglutamine tract encoded in the CA150 gene, a transcription factor of unknown function, which interacts directly with huntingtin (Holbert et al., 2001). A recent 10-cM density genomewide scan in 629 affected sibling pairs provides suggestive evidence for linkage at 4p16, 6p21-23, and 6q24-26 (Li et al., 2003).

3   HD: The Clinical Phenotype

Clinically HD is characterized by a triad of symptoms and signs: (1) a prominent movement disorder (in adult onset patients' chorea), (2) behavioral and emotional alterations, and (3) a cognitive decline.

3.1   Onset of Symptoms and Signs

A consensus exists among clinicians that a clinical diagnosis of HD should be made with certainty only in the presence of a motor disorder. Thus, defining the clinical onset of HD as the onset of the motor disorder yields a more or less reproducible way to conduct, for example, onset age surveys or genotype–phenotype correlation studies. However, it should be realized that the onset of motor abnormalities is neither a sensitive measure of disease onset nor specific and reliable, since even well-trained motor raters may overinterpret and overdiagnose mild motor abnormalities resulting in poor interobserver agreement in assessing the presence of motor signs (de Boo et al., 1998). Prior to motor manifestations, mood disorders and cognitive deficits disclosed by neuropsychological testing may be present, accompanied or followed by nonspecific, minor motor abnormalities (e.g., general restlessness, abnormal eye movements or impaired optokinetic nystagmus, hyperreflexia, impaired finger tapping or rapid alternating hand movements, etc.) (Penney et al., 1990), which eventually evolve into a recognizable extrapyramidal syndrome (Kirkwood et al., 2000).

3.2   The Classic Phenotype

On neurological examination, HD is characterized by both extrapyramidal motor abnormalities and an impairment of voluntary movements affecting gait, speech, and swallowing. *Chorea* is the major motor sign of HD, hence, the designation "Huntington's chorea." Choreic movements are excessive movements: involuntary, abrupt, irregularly timed, and randomly distributed typically though with some distal (finger and toes) accentuation. Choreic movements are continuously present during waking hours and worsen during stress. Mild and intermittent choreic movements often go unnoticed by patients, while observers tend to describe them as "nervousness." Severity may vary from restlessness, fidgeting movements of the hands, unstable, dance-like gait to a continuous flow of exhausting and severely disabling, violent movements (Young et al., 1986).

Bradykinesia and *rigidity*, best known as core features in Parkinson's disease (PD), appear gradually and often dominate the final stages of HD (Young et al., 1986; Penney et al., 1990). Early in the illness, bradykinesia alone may contribute to impairment in voluntary motor performance; some patients display a significant decrease in overall daytime motor activity, suggestive of hypokinesia, or paucity of movements (van Vugt et al., 1996). The use of neuroleptic drugs, intended to suppress choreic movements, may aggravate any existing bradykinesia and rigidity.

Dystonia, characterized by slow abnormal twisting movements and abnormal posturing with increased muscle tone, may become a prominent feature toward the later stages of HD (Young et al., 1986). In a recent study of patients who attended an HD specialty clinic, the prevalence of dystonia of any severity was found to be 95%, while in 7 of 42 patients (17%), the dystonia was severe and constant although most patients did not complain and appeared hardly bothered (Louis et al., 1999). Dystonia is particularly prominent in juvenile onset patients and in patients with late-stage disease. The dystonia-predominant adult-onset form of the disease was recently reported to constitute 11.8% of an adult population seen at an HD specialty clinic (Louis et al., 2000). *Oculomotor disturbances* are among the earliest signs and are present in the vast majority of affected patients (Schubotz et al., 1976; Lasker and Zee, 1997). Saccadic eye movements are primarily disturbed with inability to suppress reflexive glances to suddenly appearing novel stimuli, and delayed initiation of voluntary saccades (Lasker and Zee, 1997). Later in the disease, slowing of saccades may be seen in up to 75% of symptomatic individuals, especially in early-onset cases, affecting vertical more than horizontal movements (Oepen et al., 1981; Beenen et al., 1986; Lasker et al., 1988; Lasker and Zee, 1997). Impaired pursuit with saccadic intrusions, impairment of gaze fixation due to distractibility, slowing of optokinetic nystagmus, and inability to suppress blinking during saccades may also occur (Oepen et al., 1981; Beenen et al., 1986; Penney et al., 1990; Tian et al., 1991; Lasker and Zee, 1997). *Voluntary motor function* is impaired early (Folstein et al., 1983; Hefter et al., 1987; Thompson et al., 1988). Patients and their families describe clumsiness in common daily activities. Disturbances in motor speed, fine motor control, and gait correlate with disease progression and appear to be better measures of duration of illness than chorea (Folstein et al., 1983). Mechanisms proposed to underlie the slowing of voluntary motor activities include impaired internal motor cueing (Thompson et al., 1988; Phillips et al., 1996; Curra et al., 2000), impairment in constructing and refining internal representations of movement (Georgiou et al., 1997), or a reduced ability to process relevant afferent input (Schwarz et al., 2001). *Gait* disturbances ultimately result in severe disability. Patients experience frequent falls and are ultimately confined to a wheelchair. Neuroleptic treatment, which suppresses choreic movements, does not improve the gait disturbance (Koller and Trimble, 1985).

3.3   Late Disease Stage Symptoms

The final years of most HD patients are dominated by loss of independence, severe restrictions in functioning, and dependence on others for activities of daily living. During this phase, most patients live in nursing homes.

Speech is often severely dysarthric and may impair communication to a great extent. *Swallowing* is equally impaired, which requires very careful assistance by the nursing staff during meals, or alternatively, feeding through a nasogastric tube or a percutaneous endoscopic gastrostomy (PEG) tube. Patients lose independent gait and spend their days in a chair, a wheelchair, or in bed. An increase in muscle tone which often develops in late stage of HD may result in secondary joint contractures, while immobility increases the risk of pressure sores. Reflexes tend to be brisk but hyporeflexia is also often seen. However, in patients in whom weight loss is prominent, it may be hard to judge whether lower motor neuron pathology with muscle atrophy is present, or just plain malnourishment.

Weight loss and striking emaciation are features of late-stage HD (Kremer and Roos, 1992). This weight loss may occur in conjunction with an adequate diet or even increased carbohydrate intake (Sanberg et al., 1981; Farrer, 1985; Morales et al., 1989). Sedentary energy expenditure is higher than in controls in proportion to the severity of the movement disorder. Total free-living energy expenditure is not higher, however, because patients with HD appear to engage in less voluntary physical activity (Pratley et al., 2000). Intriguingly, a relationship has been found between weight at initial examination and rate of progress of the disease (Myers et al., 1991). The leading causes of death in persons with HD are pneumonia (33%), and reportedly, heart disease (24%) (Haines and Conneally, 1986; Lanska et al., 1988). Pneumonia occurs five times more commonly in HD than in controls and is likely to be secondary to the significant dysphagia, which results in choking and aspiration pneumonia.

3.4   Nonmotor Signs and Symptoms

Although the motor symptoms are most immediately evident, there is little doubt that it is the nonmotor symptoms that have great impact on patients' daily lives and contribute prominently to patients' loss of independence. People with HD have specific and characteristic cognitive difficulties, with other aspects of cognitive function remaining well preserved. *Cognitive deficits* lie particularly in the realm of executive functions, which include the ability to plan, organize, and monitor behavior, to show mental flexibility, and to switch from one way of responding to another (mental set shifting) (Butters et al., 1986; Bylsma et al., 1991). In addition, procedural memory and psychomotor skills are impaired. Primary tools of cognition, such as language, are relatively preserved. In their daily lives, people with HD often exhibit poor planning and judgment. They may appear impulsive and show an absence of forethought, their actions being governed by immediate rather than long-term considerations. People with HD have difficulty coping with multiple tasks simultaneously, suggesting difficulty in the allocation and switching of attention (Brandt et al., 1984). Moreover, different domains of cognitive function do not decline in a uniform fashion (Bamford et al., 1995; Snowden et al., 2001). Therapeutic trials that seek to slow the natural course of disease will need to evaluate performance over years rather than months in order to demonstrate efficacy.

Psychiatric changes are characterized by major depressions or schizophrenia and are most distressing for patients and carers alike. Many changes are amenable to symptomatic treatment so that early detection is vital. Behavioral problems in HD are, however, complex and need to be assessed individually to improve patient care.

4   Differential Diagnosis

4.1   HD-Like Diseases

It seems there is a group of rare cases that may or may not show atypical symptoms, which are not due to an expansion in the HD gene (Vuillaume et al., 2000). Some of these disorders are rare autosomal dominant diseases, which look like HD, and some of these have now genes associated with them. An HD-like disease in a single pedigree was mapped to chromosome 20p12 (Moore et al., 2001), the site of the prion protein (PrP) gene, and found to segregate with a 192-nucleotide insertion in the PrP gene, encoding eight octapeptide repeats. This disease was called HD-like 1 (HDL1). Margolis et al. (2001) reported an HD-like disease (HDL2) with very similar symptoms and pathology to HD, almost certainly associated with an expanded polyglutamine tract, since an antibody which detects only expanded polyglutamine-containing inclusions found such inclusions in postmortem brain from a member of this family. A consanguineous family affected by an autosomal recessive, progressive neurodegenerative HD-like disorder was recently described in Saudi Arabia (Kambouris et al., 2000). The disease manifests at approximately 3–4 years and is characterized by both pyramidal and extrapyramidal abnormalities, including chorea, dystonia, ataxia, gait instability, spasticity, seizures, mutism, and intellectual impairment. Linkage with an LOD score of 3.03 was initially achieved with a marker at 4p15.3 (Kambouris et al., 2000). Finally, a late-onset basal ganglia disease first detected in families in the north of England was associated with an insertion in the ferritin light-chain gene and is called neuroferritinopathy (Curtis et al., 2001).

4.2   Other Hereditary Neurodegenerative Diseases with Chorea

Related polyglutamine diseases like DRPLA, SCA3, and SCA17 may mimic HD (Quinn and Schrag, 1998; Bauer et al., 2004). In addition, neurological conditions associated with acanthocytes may resemble HD (Hardie et al., 1991; Danek et al., 2001).

Choreoacanthocytosis, or neuroacanthocytosis, is one of a heterogeneous group of neurological disorders associated with irregular spiny erythrocytes, or acanthocytes, which can be detected in a peripheral blood smear. Among the neurological disorders with acanthocytes, two groups may be distinguished. The first may additionally be characterized by low-serum betalipoproteins and includes Bassen–Kornzweig disease (abetalipoproteinemia: a recessive disease with vitamin E dependent polyneuropathy and ataxia); a form of hypobetalipoproteinemia (Mars et al., 1969); and the HARP syndrome—hypoprebetalipoproteinemia, acanthocytosis, retinitis pigmentosa, and pallidal degeneration (Higgins et al., 1992). A second group comprises those with normal serum betalipoproteins and, apart from a few patients with

Hallervorden–Spatz–like disease, includes patients with neuromuscular and basal ganglia disease with or without the McLeod blood group phenotype. This latter group is known as choreoacanthocytosis, neuroacanthocytosis, familial amyotrophic chorea with acanthocytosis, or the Levine–Critchley syndrome (Critchley et al., 1967).

Choreoacanthocytosis is a slowly progressive neurodegenerative disorder that affects the basal ganglia, peripheral nerves, and muscle (Hardie et al., 1991). The clinical feature that distinguishes neuroacantho-cytosis from HD is the presence of neuromuscular abnormalities, consisting of hypo- or areflexia of tendon reflexes and distal amyotrophy. Serum CK activity is elevated in the majority of cases, while nerve conduction studies reveal reduced sensory action potentials in about 50% of cases (Hardie et al., 1991).

4.3   Senile Chorea

Senile chorea or chorea of late onset in the elderly person without a family history of other affected individuals may clinically resemble HD or any of the aforementioned disorders (Shinotoh et al., 1994; Garcia Ruiz et al., 1997). In fact senile chorea of unknown origin and late-onset HD may be clinically indistinguishable, requiring CAG repeat length assessment to produce a diagnosis (Garcia Ruiz et al., 1997). The cause of non-HD senile chorea is a clinical diagnostic challenge. All of the noninherited causes that are discussed in the previous paragraph should be considered.

5   Neuropathology

5.1   Gross Pathology

Macroscopic inspection of the brain of advanced HD patients at autopsy shows that HD brains weigh about 10–20% less than brains of age-matched controls. Reduction in size occurs in the cerebral hemispheres, the diencephalon, the cerebellum, and also the brainstem and the spinal cord (Forno and Norville, 1979). The most striking neuropathological feature is the shrunken appearance of the neostriatum with gross atrophy of the caudate nucleus and putamen with the caudate nucleus reduced to a thin rim of tissue. Reduction in the size of the caudate is accompanied by secondary enlargement of the lateral ventricles.

5.2   Microscopic Pathology

At the microscopic level, the atrophied neostriatum shows marked neuronal loss and astrogliosis which has been the topic of many quantitative analyses (Bruyn, 1979; Roos et al., 1985; Vonsattel et al., 1985; Myers et al., 1988; Heinsen et al., 1994). Striatal pathology likely underlies involuntary movements (chorea and dystonia), disordered planning, impulsive behaviors, and diminished emotional control as well as some other symptoms of HD (Crossman, 1987; Reiner et al., 1988; Albin et al., 1989). The extent of gross and microscopic striatal pathology provides a basis for dividing the severity of HD pathology into five grades (0–4) of increasing pathologic severity, which correlate with clinical progression (Vonsattel et al., 1985; Myers et al., 1988). Grade 0 cases have a strong clinical and familial history suggesting HD but no detectable histological neuropathology at autopsy. In grade 1 cases, neuropathological changes can be detected microscopically with as much as 50% depletion of striatal neurons but without visible gross atrophy. In more severe grades (2–3), gross atrophy, neuronal depletion, and gliosis are progressively more pro-nounced, and pallidal pathology becomes evident. In the most severe grade (4), more than 90% of striatal neurons are lost, and microscopic studies predominantly reveal the remaining astrocytes. There is a dorsal to ventral, anterior to posterior, and medial to lateral progression of neuronal death with the dorsomedial striatum affected earliest and relative sparing of the ventral striatum and nucleus accumbens (Bots et al., 1981; Roos et al., 1985; Vonsattel et al., 1985).

6   Molecular Pathology

6.1   HD Protein (Huntingtin) and the HD Mutation

The HD gene contains 67 exons (Huntington's et al., 1993) and encodes a 350-kDa protein named huntingtin. The CAG repeat is located within the first exon and is translated into a polyglutamine tract. Huntingtin is ubiquitously expressed not only throughout the brain but also in many other tissues like muscle, liver, and lymphocytes, and its expression pattern cannot account for the regional selectivity of neurodegeneration (Li et al., 1993; Landwehrmeyer et al., 1995; Schilling et al., 1995; Sharp et al., 1995). It is of note that when the gene was cloned it had no identifying homologies in the databases and to a very large extent this remains true today.

Since the cloning of the HD gene, several other genes with expanded CAG repeats have been isolated (for review, see Cummings et al., 2000). All lead to neurodegeneration, have similar sizes of disease-causing repeats, and contain expanded glutamine tracts in their cognate proteins. Whereas neurodegeneration occurs in different brain regions in all these CAG repeat disorders, another similarity between these diseases is that the CAG repeat is always located within the coding region of the protein so that the disease-causing proteins share a region with an expanded polyglutamine repeat.

6.2   Polyglutamine Aggregation

It was Max Perutz who first hypothesized that instability in the chemical properties of polyglutamines might account for a conformational change above a certain threshold length of the polyglutamine stretch, which might in turn be related to disease occurrence:

▶   Poly-L-glutamines form pleated sheets of β-strands held together by hydrogen bonds between their amides. Incorporation of glutamine repeats into a small protein of known structure made it associate irreversibly into oligomers. That association took place during the folding of the protein molecules and led to their becoming firmly interlocked by either strand- or domain-swapping. Thermodynamic considerations suggest that elongation of glutamine repeats beyond a certain length may lead to a phase change from random coils to hydrogen-bonded hairpins. (Perutz, 1996)

A length-dependent oligomerization of polyglutamines into amyloid-like aggregates was demonstrated in vitro and in vivo shortly afterward (Scherzinger et al., 1997). Interestingly, the polyglutamine length above which oligomerization takes place is exactly in the range between 30 and 40 repeats, which in HD and other CAG triplet disorders delineates a normal from a disease gene. For the first time a plausible explanation for this threshold effect could be given.

Further pending questions were whether such polyglutamine oligomers are in fact present in HD brains and how these may possibly cause neurodegeneration. In fact, amyloid-like aggregates had already been described by Roizin et al. (1979) predominantly in neuronal nuclei when electron microscopy studies were applied to examine brain biopsies from HD patients. Without the knowledge about the HD gene defect and a possible role of *polyglutamine aggregates*, however, the significance of this finding could not be acknowledged at that time.

The first clear evidence for the development of amyloid-like aggregates in HD, which are formed from a polyglutamine-containing fragment of the HD protein, came from the development of HD transgenic mice. Gill Bates' group was the first who developed a transgenic mouse which expressed only exon 1 of the human HD gene with a massively expanded polyglutamine (around 150 repeats) (Mangiarini et al., 1996). Aggregates in these mice were ubiquitinated and could be stained with antibodies against N-terminal huntingtin. Because of their predominant localization in nuclei and absence in glial cells, these aggregates were named "neuronal nuclear inclusions" (NII). Meanwhile, polyglutamine-containing amyloid-like aggregates, which are highly ubiquitinated have also been described in postmortem HD brains, where they were not only found in neuronal nuclei but also in dystrophic neurites (DiFiglia et al., 1997).

6.3   Key Steps in Molecular Pathology

While the presence of polylgutamine-containing ubiquitinated aggregates is widely accepted as a key feature of CAG triplet disorders (Davies et al., 1997), it is still disputed what their exact pathophysiological role is and how important their presence is for the process of neuronal degeneration.

6.3.1 Huntingtin Cleavage

From more detailed analyses of NII in HD, it is known that these aggregates do not contain the entire huntingtin protein but only an N-terminal fragment, including the polyglutamine chain. In a number of HD models cleavage of the polyglutamine-containing fragment from the rest of the protein can be induced in a length-dependent manner by activation of proteinases such as caspase-1 (Wellington et al., 1997). Cleavage seems to be a precondition for the translocation of the polyglutamine-containing fragment from the usual cytoplasm localization of the normal huntingtin protein into the nucleus (❷ *Figure 11-2*). In turn, blockade of this cleavage process by caspase inhibition can reduce disease progression in HD models.

6.3.2 Nuclear Localization of Polyglutamines

These experiments not only demonstrate that cleavage of huntingtin is a pathophysiological key step but they also emphasize the importance of nuclear localization of expanded polyglutamine oligomers for disease development (❷ *Figure 11-2*). This fact has been supported also by Saudou et al. (1998) who showed very elegantly in genetically modified cellular models of HD that nuclear localization of the expanded poly-glutamine is necessary to induce cell death. This notion is also supported by studies in which normal cell cultures were exposed to chemically synthesized polyglutamine peptides, which were readily taken up by the cells into the cytoplasm. Those polyglutamines became only toxic to the cells when they also contained a nuclear localization signal and did thus enter the nucleus (Yang et al., 2002).

6.3.3 Huntingtin Interacting Proteins

To further elucidate the function of huntingtin in how the expanded polyglutamines may cause cell death, yeast two hybrid systems and immunoprecipitation assays were developed to screen for interacting protein partners and abnormal protein interactions. Within this screen a number of transcription factors (Sp1, TAFII130, N-CoR, mSin3, p53, CREB-binding protein, Gln-Ala repeat transcriptional activator CA150) were identified, which interact with mutant huntingtin in a polyglutamine length-dependent manner, i.e., increasing aggregation with expansion of the polyglutamine (Steffan et al., 2000; Holbert et al., 2001; Dunah et al., 2002; Li et al., 2002; Boutell et al., 2004). This is not very surprising since it is well known that many transcription factors contain polyglutamine-rich regions themselves. Also a number of transcription factors like CREB-binding protein (CBP) and p53 were identified as components of neuronal nuclear inclusions in HD (Steffan et al., 2000; Nucifora et al., 2001). Cell nuclei were depleted of these transcription factors not only in various disease models but also in HD brain tissue, and compensation for CBP loss through CBP over-expression could rescue cells from polyglutamine-induced neuronal toxicity in vitro (Steffan et al., 2000; Nucifora et al., 2001).

□ Figure 11-2

Key steps in HD molecular pathology and possible therapeutic interventions. HT, huntingtin protein; PolyQ$_n$, polyglutamines; HDAC, histone deacetylase. In former therapeutic trials, drugs were used which mainly modified late-stage events which are summarized as the "common final pathway of neurodegenerative diseases." Future therapeutic interventions will focus on rather HD-specific and early molecular events using HDAC inhibitors, inhibitors of aggregate formation, or caspase inhibitors

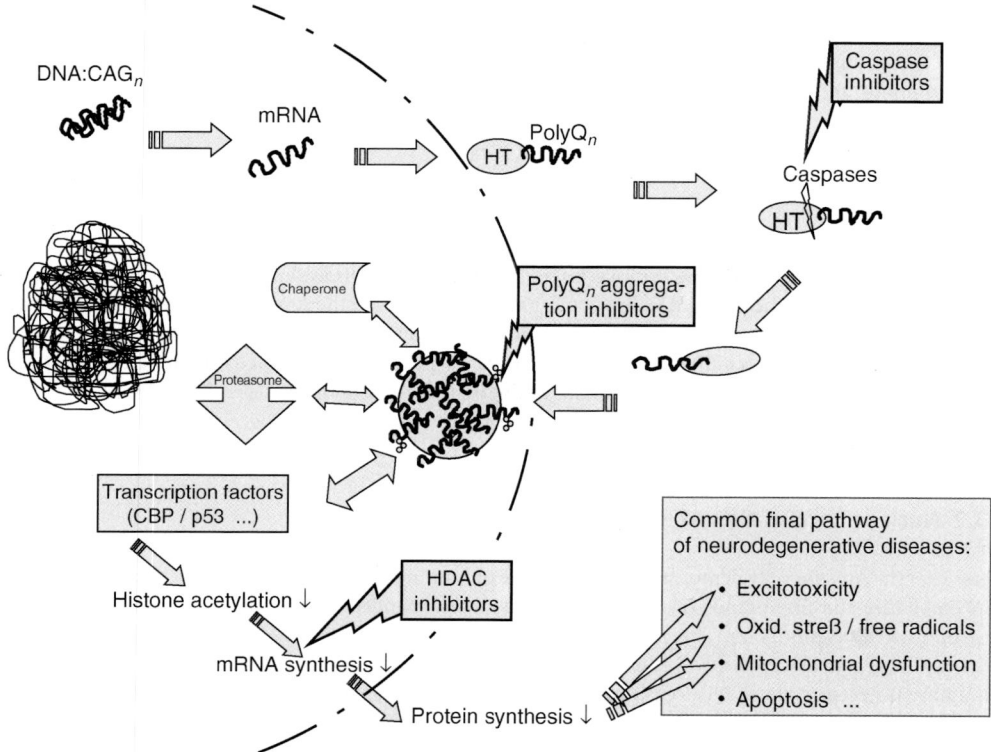

6.3.4 Transcriptional Dysregulation

The expected consequence of nuclear transcription factor inactivation through interaction with polyglutamines from mutant huntingtin protein is a change in the transcriptional activity of the affected cell (❷ *Figure 11-2*). In fact, early and very specific changes in the mRNA expression pattern were first demonstrated in HD transgenic mouse models for various neurotransmitter receptors (Cha et al., 1998, 1999). Meanwhile, this concept has been largely expanded through mRNA expression profiling applying microchip array techniques. A characteristic pattern of reduced mRNA expression was found for encoding components of neurotransmitter systems, proteins involved in calcium homeostasis, and also the retinoid signaling pathways. Similar changes were observed in different HD transgenic mouse models, which are known to be critical to neuronal function. Longitudinal studies also revealed that mRNA changes develop in parallel with pathologic and behavioral deterioration in these animals (Luthi-Carter et al., 2000, 2002; Chan et al., 2002).

Interestingly, CBP-dependent transcriptional dysregulation can be partly compensated by treatment with histone deacetylase inhibitors, like suberoylanilide hydroxamic acid (SAHA) or sodium butyrate, which have been recently developed for cancer therapy. Treatment with these drugs in animal models of HD nicely validates the hypothesis of transcriptional dysregulation in that treated animals survive longer and show attenuated behavioral and morphological abnormalities (Steffan et al., 2001; Hockly et al., 2003; Ferrante et al., 2004).

Thus, transcriptional dysregulation due to interaction of misfolded mutant polyglutamines in cell nuclei with transcription factors is now thought to be a key feature of HD pathology (Cha, 2000). As a consequence of transcriptional dysregulation, neuronal dysfunction may occur in parallel at many different cell sites, which are all in itself essential for cell survival. A number of well-described examples which may thus be all understood as rather late events in the cascade of molecular events are glutamate-related excitotoxicity, mitochondrial dysfunction, and deficits in energy metabolism, as well as induction of the apoptotic signaling pathway (Beal, 1995; Koroshetz et al., 1997; Browne et al., 1999; Schapira, 1999; Cepeda et al., 2001; Ross, 2002; Behrens et al., 2002; Panov et al., 2002; Zeron et al., 2002; Friedlander, 2003; Hickey and Chesselet, 2003).

6.4   Excitotoxicity

Glutamate-mediated excitotoxicity has been postulated to play a central role in the pathogenesis of HD since intrastriatal injections of glutamate agonists, particularly of those acting at the N-methyl-D-aspartate (NMDA) receptor, reproduce neuropathological features of HD (Beal et al., 1991). More recent studies in vitro and in a transgenic mouse model of HD expressing full-length mutant huntingtin postulate an interaction of mutant huntingtin protein with NMDA receptors, i.e., sensitization of neurons for glutamate-induced calcium influx and NMDA receptor–mediated cell death (Levine et al., 1999; Zeron et al., 2001; Zeron et al., 2002). As a further contributing factor to induce glutamate-mediated excitotoxicity, it has been demonstrated recently that the glial glutamate transporters which are mainly responsible for the clearance of glutamate from the synaptic cleft are also decreased in HD transgenic mice (Lievens et al., 2001; Behrens et al., 2002). Corresponding to these experimental data, studies in HD patients using MRI spectroscopy demonstrated increased levels of striatal glutamate/glutamine and lactate suggesting that glutamatergic function and abnormalities in energy metabolism may in fact contribute to the pathology of HD (Koroshetz et al., 1997).

6.5   Mitochondrial Dysfunction and Oxidative Stress

As a secondary consequence of transcriptional dysregulation, defects in energy metabolism may contribute significantly to neurodegeneration. In accordance with this hypothesis experimental data demonstrate that mitochondrial toxins can mimic the HD characteristic i.e., selective striatal damage in animals. HD transgenic mice also develop marked decreases in N-acetylaspartate (NAA) concentrations before neuronal loss (Jenkins et al., 2000) and have an increased vulnerability toward mitochondrial toxin (Bogdanov et al., 1998).

Studies on postmortem HD tissue describe a reduction in mitochondrial electron transport enzymes (Jenkins et al., 1993; Browne et al., 1997; Koroshetz et al., 1997). Studies in HD patients also show increased susceptibility of mitochondria to depolarization in lymphoblasts and fibroblasts (Sawa et al., 1999). In addition, lactate levels are elevated in the cerebral cortex and basal ganglia of HD patients, and there is reduced phosphocreatine/inorganic phosphate in resting muscle of HD patients.

Further evidence for a significant contribution of deficits in energy metabolism to HD pathology comes from a number of experimental therapeutic studies in HD transgenic mice, which demonstrate neuroprotective effects with substances that improve neuronal energy metabolism like coenzyme Q10, dichloroacetate, or creatine. (Andreassen et al., 2001; Ferrante et al., 2002; Dedeoglu et al., 2003).

6.6   Activation of Apoptotic Pathways

Neurodegeneration via apoptosis, i.e., activation of cell death programs, is thought to play a crucial role in a number of neurologic diseases. In chronic neurodegenerative diseases, it is rather the activation of caspase-mediated apoptotic pathways that may be relevant in mediating cell dysfunction and death

(Friedlander, 2003). Caspases are the major executioners in the apoptotic program and are cysteine-dependent, aspartate-specific proteases.

In HD transgenic mice, one of the earliest events is transcriptional upregulation of caspase-1 (Ona et al., 1999). As the disease progresses, caspase-3 is transcriptionally upregulated and the protein is activated (Chen et al., 2000). Activation of caspase-8 and caspase-9 and release of cytochrome *c* have also been demonstrated (Sanchez et al., 1999).

Several of the findings in mouse models of HD have also been demonstrated in human striatal brain tissue, including activation of caspases-1, -3, -8, and -9 and release of cytochrome *c* (Sanchez et al., 1999, 2003).

Caspases may not only kill neurons through activation of the apoptotic cascade. It is also important to understand that huntingtin is a substrate for caspases (Wellington et al., 1997). In fact, huntingtin cleavage through caspases is required for translocation of the polyglutamine fragment into the nucleus where the fragments build aggregates and interfere with transcriptional regulation. As the disease progresses, increased caspase-mediated cleavage of huntingtin increases the generation of huntingtin fragments and depletes wild-type huntingtin (Ona et al., 1999). The degree of caspase activity may thus be the key step that decides the speed in which the molecular pathology of HD progresses.

The efficacy of caspase inhibition to modify disease progression in HD has been well demonstrated in HD transgenic mice, which were either genetically modified through cross breeding with dominant negative caspase-1 mutants or through treatment with minocycline, a tetracycline which inhibits caspase-1 and caspase-3 expression (Ona et al., 1999; Chen et al., 2000).

6.7 Gain of Function Versus Loss of Function Theories

As an autosomal dominant disease HD is thought to be a gain-of-function disease. There is, however, accumulating evidence for a loss of function in HD due to loss of wild-type huntingtin, which seems to have a number of beneficial effects in neurons.

Wild-type huntingtin is a cytoplasmic protein involved in vesicle trafficking in the secretory and endocytic pathways (DiFiglia et al., 1995). It is essential during development (MacDonald et al., 1996), and it is important for neuronal survival in the adult. Wild-type huntingtin is also antiapoptotic in neurons (Rigamonti et al., 2000) and reduces the toxicity of mutant huntingtin in vivo (Leavitt et al., 2001).

These neuroprotective effects of wild-type huntingtin are most likely due to its important role in the delivery of brain-derived neurotrophic factor (BDNF) from cortical neurons to the striatum (Zuccato et al., 2001, 2003). BDNF is essential for the survival and differentiation of striatal neurons and protects striatal neurons from excitotoxic neurodegeneration. Wild-type huntingtin positively modulates the transcription of the *BDNF* gene. BDNF gene transcription and protein production are lost in HD transgenic mice, resulting in downregulation of cortically derived BDNF in the striatum (Zuccato et al., 2001). Thus, selective vulnerability of striatal neurons in HD may partly result from the loss of activity of the wild-type huntingtin in cortical neurons crucial for the functioning of these cells.

7 Disease Models

With the knowledge about the gene defect in HD, various genetic models to study disease pathophysiology have been developed. It is very important to understand that all of these models have proven to be superior to study certain disease aspects. Only the parallel evaluation of various disease aspects in these different models led to an emerging conclusive picture of the HD disease mechanism.

7.1 Transgenic Cells, Yeasts, Worms, and Flies

These models include in vitro models of either transient or stable transfected cell lines in which the HD gene defect induces increased sensitivity to various forms of cell stress and early apoptotic cell death. These models are very useful in studying short-term consequences of the gene defect on cell metabolism (Saudou et al., 1998; Kim et al., 1999). Yeast models in two-hybrid systems turned out to be very helpful for screens of proteins that interact with huntingtin and in studying the effect of the HD mutation on protein interactions (Holbert et al., 2001).

Transfer of the HD gene defect was also achieved on *Caenorhabdites elegans* worms and *Drosophila* fruit flies. These organisms have a very simple nervous system; the entire genome has been identified, and transfer of the HD gene defect leads to neurodegeneration. These models have proven extremely helpful in identifying genetic modifiers of the disease (Sipione and Cattaneo, 2001).

7.2 Transgenic Mice

Of great importance for HD research in the last decade was the development of genetic mouse models. The obvious advantage of mouse models is to study the consequences of the gene defect in a mammalian brain also allowing the complex intercellular interactions, which might contribute to the disease. HD mouse models vary in a number of aspects which include length of the CAG repeat, length of the huntingtin protein expressed together with the repeat, expression rate of the mutant protein depending on the promotor used, and last but not least, the differences in the genetic background due to various strain lines used. All available HD mouse models mimic only certain aspects of the human disease.

7.2.1 Genotype–Phenotype Correlation

Imitation of the HD mutation in mice is reached most closely in so-called *knockin mice*, in which an expanded CAG repeat with a repeat length in the typical range of human HD has been introduced into the mouse HD gene homolog (Hdh). These mice develop no neurological phenotype, have a normal life span (around 2 years), and show almost no signs of striatal neuronal degeneration (Wheeler et al., 1999). In contrast, mice with a massively increased CAG repeat length (110–150 CAG repeats) expressed together with only a small part (exon 1) of the human HD gene develop a progressive neurological phenotype and die after a short period of time (3–4 months). Although some important aspects of the morphological characteristics can be seen in these mice (intranuclear aggregates), there are only mild signs of striatal neurodegeneration and astrocytosis even in late disease stages (Mangiarini et al., 1996). In between these groups are some transgenic and knockin mice with intermediate characteristics (Reddy et al., 1998). ❷ *Figure 11-3* gives a list of some of the HD mouse models with phenotype–genotype correlation.

In conclusion, the imitation of a human hereditary disease, which starts usually not before age 35, has some obvious limitations in mice with a life expectancy of not more than 2 years. To induce a disease onset within this relatively short life span, some exaggeration of human disease conditions are necessary. Disease development largely depends on length of CAG expansion, length of the huntingtin fragment, and mutant protein load (i.e., expression level/promotor).

7.2.2 Behavioral Abnormalities in HD Mice

The behavioral characteristics of HD transgenic mice have been most extensively tested in the R6/2 mouse model since these mice develop the most obvious phenotype (Mangiarini et al., 1996). The earliest abnormality of motor behavior is spontaneous hyperactivity most closely related to hyperkinetic state in human disease. Later during disease, mice become clearly hypoactive similar to late stages of human HD in which parkinsonian and dystonic features often occur (Lüesse et al., 2001). With 6–8 weeks, mice start to develop a progressive loss of motor coordination skills (Carter et al., 1999). Lifting of these mice at the tail induces a dystonic behavior of the hind limbs, which was called "clasping." Mice also show early cognitive deficits in spatial orientation and memory (Lione et al., 1999). Thus, key features of HD motor and cognitive abnormalities can be demonstrated in HD mouse models.

□ **Figure 11-3**

Inverse genotype–phenotype correlation in HD mouse models. Knockin mouse models, which resemble human HD mutation most perfectly in mice, develop almost no phenotype in a mouse life span. Neurologic disease can be accelerated in mice by three factors: increase of CAG length, increased cellular load with mutated protein by using highly active promoters, and expression of the polyglutamine together with only small portions of the huntingtin protein

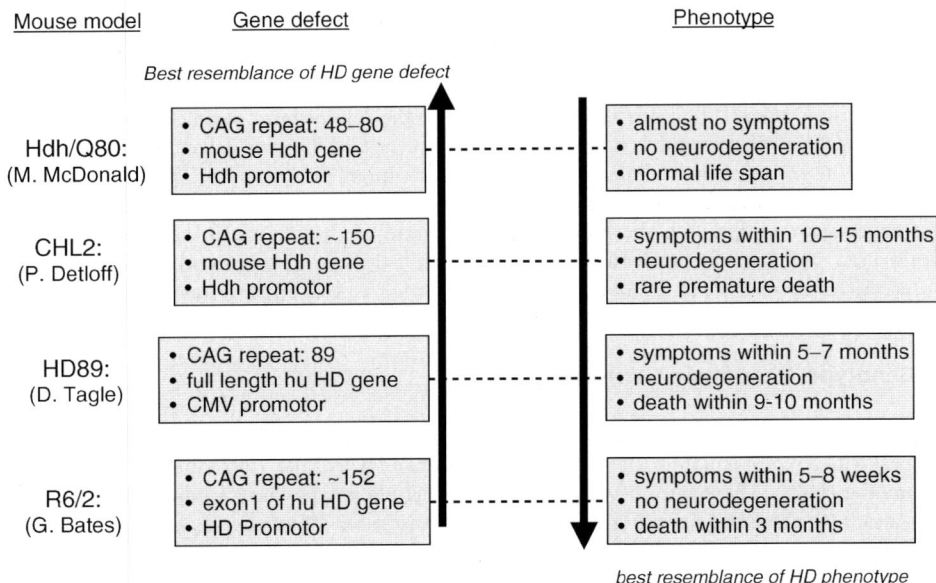

7.2.3 Morphological Characteristics of HD Mice

As a morphological key feature, all mouse models show *ubiquitinated intranuclear inclusions* to various extents (Davies et al., 1998). Classical neuropathological techniques, however, fail to detect signs of cell loss, neurodegeneration, and astrogliosis in most HD mouse models. This means that the mice phenotype cannot be explained by neuronal degeneration and that neuronal dysfunction instead can precede neurodegeneration for a long period in these mice. Other techniques turned out to be extremely helpful in demonstrating signs of neuronal dysfunction in various HD mouse models. With receptor autoradiography it is possible to demonstrate specific changes in *neurotransmitter receptor expression*, including dopamine D_1 and D_2 receptors, adenosine A_{2a} receptors, metabotropic glutamate receptors, and others even at early disease stages (Cha et al., 1998, 1999). Microchip array techniques demonstrated complex changes in mRNA expression pattern (Luthi-Carter et al., 2000, 2002; Chan et al., 2002). These changes not only replicate many findings known from human HD but also occur and go in parallel very closely with the progressive neurological phenotype of these mice.

8   Experimental Therapeutics for HD

More than 10 years after the gene defect in HD has been identified, there is still a number of unresolved problems in our understanding of the molecular disease mechanisms. Examples are the exact function of the normal huntingtin protein and the regional selectivity of neurodegeneration. But it should be emphasized that our current understanding offers entirely new opportunities for therapeutic approaches. A number of these approaches have already been successfully tested in various HD models. Especially, HD transgenic mice have been used systematically to evaluate potential therapeutics.

Most promising for a therapeutic intervention in HD patients are obviously results with drugs which are already approved by the FDA for diseases or which do not need approval since they are understood as natural food supplements. Examples are inhibition of caspase-1 and -3 activity with minocycline (Chen et al., 2000), inhibition of nuclear polyglutamine aggregation with the disaccharide trehalose (Tanaka et al., 2004), compensation for loss of CBP-dependent transcriptional activity through treatment with the histone deacetylase inhibitors SAHA (suberoylanilide hydroxamic acid) or sodium butyrate (Steffan et al., 2001; Ferrante et al., 2004; Hockly et al., 2003), and improvement of mitochondrial energy metabolism through treatment with creatine (Dedeoglu et al., 2003; Ferrante et al., 2000).

References

Albin RL, Young AB, Penney JB. 1989. The functional anatomy of basal ganglia disorders. Trends Neurosci 12: 366-375.

Andreassen OA, Ferrante RJ, Huang HM, Dedeoglu A, Park L, et al. 2001. Dichloroacetate exerts therapeutic effects in transgenic mouse models of Huntington's disease. Ann Neurol 50: 112-117.

Andrew SE, Goldberg YP, Kremer B, Telenius H, Theilmann J, et al. 1993. The relationship between trinucleotide (CAG) repeat length and clinical features of Huntington's disease. Nat Genet 4: 398-403.

Ashizawa T, Wong LJ, Richards CS, Caskey CT, Jankovic J. 1994. CAG repeat size and clinical presentation in Huntington's disease. Neurology 44: 1137-1143.

Bamford KA, Caine ED, Kido DK, Cox C, Shoulson I. 1995. A prospective evaluation of cognitive decline in early Huntington's disease: Functional and radiographic correlates. Neurology 45: 1867-1873.

Bauer P, Laccone F, Rolfs A, Wullner U, Bosch S, et al. 2004. Trinucleotide repeat expansion in SCA17/TBP in white patients with Huntington's disease-like phenotype. J Med Genet 41: 230-232.

Beal MF. 1995. Aging, energy, and oxidative stress in neurodegenerative diseases. Ann Neurol 38: 357-366.

Beal MF, Ferrante RJ, Swartz KJ, Kowall NW. 1991. Chronic quinolinic acid lesions in rats closely resemble Huntington's disease. J Neurosci 11: 1649-1659.

Beenen N, Buttner U, Lange HW. 1986. The diagnostic value of eye movement recordings in patients with Huntington's disease and their offspring. Electroencephalogr Clin Neurophysiol 63: 119-127.

Behrens PF, Franz P, Woodman B, Lindenberg KS, Landwehrmeyer GB. 2002. Impaired glutamate transport and glutamate-glutamine cycling: Downstream effects of the Huntington mutation. Brain 125: 1908-1922.

Bogdanov MB, Ferrante RJ, Kuemmerle S, Klivenyi P, Beal MF. 1998. Increased vulnerability to 3-nitropropionic acid in an animal model of Huntington's disease. J Neurochem 71: 2644.

Boutell JM, Thomas P, Neal JW, Weston VJ, Duce J, et al. 1999. Aberrant interactions of transcriptional repressor proteins with the Huntington's disease gene product, huntingtin. Hum Mol Genet 8(9)1647-1655.

Brandt J, Bylsma F, Gross R, Stine O, Ranen N, et al. 1996. Trinucleotide repeat length and clinical progression in Huntington's disease. Neurology 46: 531-535.

Brandt J, Strauss ME, Larus J, Jensen B, Folstein SE, et al. 1984. Clinical correlates of dementia and disability in Huntington's disease. J Clin Neuropsychol 6: 401-412.

Brinkman RR, Mezei MM, Theilmann J, Almqvist E, Hayde MR. 1997. The likelihood of being affected with Huntington disease by a particular age, for a specific CAG size. Am J Hum Genet 60: 1202-1210.

Browne SE, Bowling AC, Mac Garvey U, Baik MJ, Berger SC, et al. 1997. Oxidative damage and metabolic dysfunction in Huntington's disease: Selective vulnerability of the basal ganglia. Ann Neurol 41: 653.

Browne SE, Ferrante RJ, Beal MF. 1999. Oxidative stress in Huntington's disease. Brain Pathol 9: 147-163.

Bruyn GW. 1979. Huntington's chorea. Tijdschr Ziekenverpl 32: 101-105.

Butters N, Wolfe J, Granholm E, Martone M. 1986. An assessment of verbal recall, recognition and fluency abilities in patients with Huntington's disease. Cortex 22: 11-12.

Bylsma FW, Rebok GW, Brandt J. 1991. Long-term retention of implicit learning in Huntington's disease. Neuropsychologia 29: 1213-1221.

Carter RJ, Lione LA, Humby T, Mangiarini L, Mahal A, et al. 1999. Characterization of progressive motor deficits in mice transgenic for the human Huntington's disease mutation. J Neurosci 19: 3248-3257.

Cepeda C, Ariano MA, Calvert CR, Flores-Hernandez J, Chandler SH, et al. 2001. NMDA receptor function in mouse models of Huntington disease. J Neurosci Res 66: 525-539.

Cha JH. 2000. Transcriptional dysregulation in Huntington's disease. Trends Neurosci 23: 387-392.

Cha JH, Frey AS, Alsdorf SA, Kerner JA, Kosinski CM, et al. 1999. Altered neurotransmitter receptor expression in transgenic mouse models of Huntington's disease. Philos Trans R Soc Lond B Biol Sci 354: 981-989.

Cha JH, Kosinski CM, Kerner JA, Alsdorf SA, Mangiarini L, et al. 1998. Altered brain neurotransmitter receptors in transgenic mice expressing a portion of an abnormal human huntington disease gene. Proc Natl Acad Sci USA 95(11): 6480-6485.

Chan EY, Luthi-Carter R, Strand A, Solano SM, Hanson SA, et al. 2002. Increased huntingtin protein length reduces the number of polyglutamine-induced gene expression changes in mouse models of Huntington's disease. Hum Mol Genet 11: 1939-1951.

Chen M, Ona VO, Li M, Ferrante RJ, Fin, KB, et al. 2000. Minocycline inhibits caspase-1 and caspase-3 expression and delays mortality in a transgenic mouse model of huntington disease. Nat Med 6: 797-801.

Chong SS, Almqvist E, Telenius H, La Tray L, Nichol K, et al. 1997. Contribution of DNA sequence and CAG size to mutation frequencies of intermediate alleles for Huntington disease: Evidence from single sperm analyses. Hum Mol Genet 6: 301-309.

Claes S, Van Zand K, Legius E, Dom R, Malfroid M, et al. 1995. Correlations between triplet repeat expansion and clinical features in Huntington's disease. Arch Neurol 52: 749-753.

Critchley EM, Clark, DB, Wikler A. 1967. An adult form of acanthocytosis. Trans Am Neurol Assoc 92: 132-137.

Crossman AR. 1987. Primate models of dyskinesia: The experimental approach to the study of basal ganglia-related involuntary movement disorders. Neuroscience 21: 1-40.

Cummings CJ, Zoghbi HY. 2000. Trinucleotide repeats: Mechanisms and pathophysiology. Annu Rev Genomics Hum Genet 1: 281-328.

Curra A, Agostino R, Galizia P, Fittipaldi F, Manfredi M, et al. 2000. Sub-movement cueing and motor sequence execution in patients with Huntington's disease. Clin Neurophysiol 111: 1184-1190.

Curtis AR, Fey C, Morris CM, Bindoff LA, Ince PG, et al. 2001. Mutation in the gene encoding ferritin light polypeptide causes dominant adult-onset basal ganglia disease. Nat Genet 28: 350-354.

Danek A, Tison F, Rubio J, Oechsner M, Kalckreuth W, et al. 2001. The chorea of McLeod syndrome. Mov Disord 16: 882-889.

Davies S, Beardsall K, Turmaine M, Di Figlia M, Aronin N, et al. 1998. Are neuronal intranuclear inclusions the common neuropathology of triplet-repeat disorders with polyglutamine-repeat expansions? Lancet 351: 131-133.

Davies S, Turmaine M, Cozens B, Di Figlia M, Sharp A, et al. 1997. Formation of neuronal intranuclear inclusions (NII) underlies the neurological dysfunction in mice transgenic for the HD mutation. Cell 90: 537-548.

Davis MB, Bateman D, Quinn NP, Marsden CD, Harding, AE. 1994. Mutation analysis in patients with possible but apparently sporadic Huntington's disease. Lancet 344: 714-717.

de Boo G, Tibben A, Hermans J, Maat A, Roos RA. 1998. Subtle involuntary movements are not reliable indicators of incipient Huntington's disease. Mov Disord 13: 96-99.

Dedeoglu A, Kubilus JK, Yang L, Ferrante KL, Hersch SM, et al. 2003. Creatine therapy provides neuroprotection after onset of clinical symptoms in Huntington's disease transgenic mice. J Neurochem 85: 1359-1367.

DiFiglia M, Sapp E, Chase K, Davies S, Bates G, et al. 1997. Aggregation of huntingtin in neuronal intranuclear inclusions and dystrophic neurites in brain. Science 277: 1990-1993.

DiFiglia M, Sapp E, Chase K, Schwarz C, Meloni A, et al. 1995. Huntingtin is a cytoplasmic protein associated with vesicles in human and rat brain neurons. Neuron 14: 1075-1081.

Dunah AW, Jeong H, Griffin A, Kim YM, Standaert DG, et al. 2002. Sp1 and TAFII130 transcriptional activity disrupted in early Huntington's disease. Science 296: 2238-2243.

Duyao M, Ambrose C, Myers R. 1993. Trinucleotide repeat length instability and age of onset in Huntington's disease. Nature Genet 4: 387-392.

Farrer LA. 1985. Diabetes mellitus in Huntington disease. Clin Genet 27(1): 62-67.

Ferrante RJ, Andreassen OA, Dedeoglu A, Ferrante KL, Jenkins BG, et al. 2002. Therapeutic effects of coenzyme Q10 and remacemide in transgenic mouse models of Huntington's disease. J Neurosci 22: 1592-1599.

Ferrante RJ, Andreassen OA, Jenkins BG, Dedeoglu A, Kuemmerle S, et al. 2000. Neuroprotective effects of creatine in a transgenic mouse model of Huntington's disease. J Neurosci 20: 4389-4397.

Ferrante RJ, Kubilusz JK, Lee J, yu H, eesen A, et al. 2004. Histone deacetylase inhibition by sodium butyrate chemotherapy ameliorates the neurodegenerative phenotype in Huntington's disease mice. J Neurosci 23: 9418-9427.

Folstein SE, Jensen B, Leigh RJ, Folstein MF. 1983. The measurement of abnormal movement: Methods developed for Huntington's disease. Neurobehav Toxicol Teratol 5: 605-609.

Forno LS, Norville RL. 1979. Ultrastructure of the neostriatum in Huntington's and Parkinson's disease. Adv Neurol 23: 123-139.

Friedlander RM. 2003. Apoptosis and caspases in neurodegenerative diseases. N Engl J Med 348: 1365-1375.

Garcia Ruiz PJ, Gomez-Tortosa E, del Barrio A, Benitez J, Morales B, et al. 1997. Senile chorea: A multicenter prospective study. Acta Neurol Scand 95: 180-183.

Georgiou N, Phillips JG, Bradshaw JL, Cunnington R, Chiu E. 1997. Impairments of movement kinematics in patients with Huntington's disease: A comparison with and without a concurrent task. Mov Disord 12: 386-396.

Goldberg YP, Andrew SE, Theilmann J, Kremer B, Squitieri F, et al. 1993. Familial predisposition to recurrent mutations causing Huntington's disease: Genetic risk to sibs of sporadic cases. J Med Genet 30: 987-990.

Haines JL, Conneally PM. 1986. Causes of death in Huntington disease as reported on death certificates. Genet Epidemiol 3: 417-423.

Hardie RJ, Pullon HW, Harding AE, Owen JS, Pires M, et al. 1991. Neuroacanthocytosis. A clinical, haematological and pathological study of 19 cases. Brain 114 (Pt. 1A):13-49.

Hefter H, Homberg V, Lange HW, Freund HJ. 1987. Impairment of rapid movement in Huntington's disease. Brain 110: 585-612.

Heinsen H, Strik M, Bauer M, Luther K, Ulmar G, et al. 1994. Cortical and striatal neurone number in Huntington's disease. Acta Neuropathol 88: 320-333.

Hickey MA, Chesselet MF. 2003. Apoptosis in Huntington's disease. Prog Neuropsychopharmacol Biol Psychiatry 27: 255-265.

Higgins JJ, Patterson MC, Papadopoulos NM, Brady RO, Pentchev PG, et al. 1992. Hypoprebetalipoproteinemia, acanthocytosis, retinitis pigmentosa, and pallidal degeneration (HARP syndrome). Neurology 42: 194-198.

Hockly E, Richon VM, Woodman B, Smith DL, Zhou X, et al. 2003. Suberoylanilide hydroxamic acid, a histone deacetylase inhibitor, ameliorates motor deficits in a mouse model of Huntington's disease. Proc Natl Acad Sci USA 100: 2041-2046.

Hoffmann J. 1888. Über Chorea chronica progressiva (Huntingtonsche Chorea, Chorea hereditaria). Virchows Archiv für pathologische Anatomie 111: 513-548.

Holbert S, Denghien I, Kiechle T, Rosenblatt A, Wellington C, et al. 2001. The Gln-Ala repeat transcriptional activator CA150 interacts with huntingtin: Neuropathologic and genetic evidence for a role in Huntington's disease pathogenesis. Proc Natl Acad Sci USA 98: 1811-1816.

Huntington G. 1872. On chorea. Med Surg Rep 26: 317-321.

Huntington's Disease Collaborative Research Group. 1993. A novel gene containing a trinucleotide repeat that is expanded and unstable on Huntington's disease chromosomes. Cell 72: 971–983.

Jelgersma. 1908. Die anatomischen Veränderungen bei Paralysis agitans und chronischer Chorea. Verhandlungen der Gesellschaft deutscher Naturforscher und Ärzte 2: 383-388.

Jenkins B, Koroshetz W, Beal MF, Rosen B. 1993. Evidence for an energy metabolism defect in Huntington's disease using localized proton spectroscopy. Neurology 43: 2689-2695.

Jenkins BG, Klivenyi P, Kustermann E, Andreassen OA, Ferrante RJ, et al. 2000. Non-linear decrease over time in N-acetylaspartate levels in the absence of neuronal loss and increases in glutamine and glucose in transgenic Huntington's disease mice. J Neurochem 74: 2119.

Kambouris M, Bohlega S, Al Tahan A, Meyer BF. 2000. Localization of the gene for a novel autosomal recessive neurodegenerative Huntington-like disorder to 4p15.3. Am J Hum Genet 66: 445-452.

Kehoe P, Krawczak M, Harper PS, Owen MJ, Jones AL. 1999. Age of onset in Huntington disease: Sex specific influence of apolipoprotein E genotype and normal CAG repeat length. J Med Genet 36: 108-111.

Kieburtz K, Mac Donald M, Shih C, Feigin A, Steinberg K, et al. 1994. Trinucleotide repeat length and progression of illness in Huntington's disease. J Med Genet 31: 872-874.

Kim M, Lee HS, La Foret G, McIntyre C, Martin EJ, et al. 1999. Mutant huntingtin expression in clonal striatal cells: Dissociation of inclusion formation and neuronal survival by caspase inhibition. J Neurosci 19: 964-973.

Kirkwood SC, Siemers E, Hodes ME, Conneally PM, Christian JC, et al. 2000. Subtle changes among presymptomatic carriers of the Huntington's disease gene. J Neurol Neurosurg Psychiatry 69: 773-779.

Koller WC, Trimble J. 1985. The gait abnormality of Huntington's disease. Neurology 35: 1450-1454.

Koroshetz WJ, Jenkins BG, Rosen BR, Beal MF. 1997. Energy metabolism defects in Huntington's disease and effects of coenzyme Q10. Ann Neurol 41: 160-165.

Kovtun IV, McMurray CT. 2001. Trinucleotide expansion in haploid germ cells by gap repair. Nat Genet 27: 407-411.

Kremer B, Almqvist E, Theilmann J, Spence N, Telenius H, et al. 1995. Sex-dependent mechanisms for expansions and contractions of the CAG repeat on affected Huntington disease chromosomes. Am J Hum Genet 57: 343-350.

Kremer HP, Roos RA. 1992. Weight loss in Huntington's disease. Arch Neurol 49: 349.

Landwehrmeyer GB, McNeil SM, Dure LS, Ge P, Aizawa H, et al. 1995. Huntington's disease gene: Regional and cellular expression in brain of normal and affected individuals. Ann Neurol 37: 218-230.

Lanska DJ, Lanska MJ, Lavine L, Schoenberg BS. 1988. Conditions associated with Huntington's disease at death. A case-control study. Arch Neurol 45: 878-880.

Lasker AG, Zee DS. 1997. Ocular motor abnormalities in Huntington's disease. Vision Res 37: 3639-3645.

Lasker AG, Zee DS, Hain TC, Folstein SE, Singer HS. 1988. Saccades in Huntington's disease: Slowing and dysmetria. Neurology 38: 427-431.

Leavitt BR, Guttman JA, Hodgson JG, Kimel GH, Singaraja R, et al. 2001. Wild-type huntingtin reduces the cellular toxicity of mutant huntingtin in vivo. Am J Hum Genet 68: 313-324.

Leeflang EP, Zhang L, Tavare S, Hubert R, Srinidhi J, et al. 1995. Single sperm analysis of the trinucleotide repeats in the Huntington's disease gene: Quantification of the mutation frequency spectrum. Hum Mol Genet 4: 1519-1526.

Levine MS, Klapstein GJ, Koppel A, Gruen E, Cepeda C, et al. 1999. Enhanced sensitivity to N-methyl-D-aspartate receptor activation in transgenic and knock-in mouse models of Huntington's disease. J Neurosci Res 58: 515-532.

Li JL, Hayden MR, Almqvist EW, Brinkman RR, Durr A, et al. 2003. A genome scan for modifiers of age at onset in Huntington disease: The HD MAPS study. Am J Hum Genet 73: 682-687.

Li SH, Cheng AL, Zhou H, Lam S, Rao M, et al. 2002. Interaction of Huntington disease protein with transcriptional activator. Sp1 Mol Cell Biol 22: 1277-1287.

Li SH, Schilling G, Young W3, Li XJ, Margolis RL, et al. 1993. Huntington's disease gene (IT15) is widely expressed in human and rat tissues. Neuron 11: 985-993.

Lievens JC, Woodman B, Mahal A, Spasic-Boscovic O, Samuel D, et al. 2001. Impaired glutamate uptake in the R6 Huntington's disease transgenic mice. Neurobiol Dis 8: 807-821.

Lione LA, Carter RJ, Hunt MJ, Bates GP, Morton AJ, et al. 1999. Selective discrimination learning impairments in mice expressing the human Huntington's disease mutation. J Neurosci 19: 10428-10427.

Louis ED, Anderson KE, Moskowitz C, Thorne DZ, Marder K. 2000. Dystonia-predominant adult-onset Huntington disease: Association between motor phenotype and age of onset in adults. Arch Neurol 57: 1326-1330.

Louis ED, Marder K, Moskowitz C, Greene P. 1999. Arm elevation in Huntington's disease: Dystonia or levitation? Mov Disord 14: 1035-1038.

Lüesse H, Schiefer J, Sprünken A, Puls C, Block F, et al. 2001. Evaluation of R6/2 transgenic mice for therapeutic studies in Huntington's disease: Behavioral testing and impact of diabetes mellitus. Behav Brain Res 126: 185-195.

Lund JC. 1860. Chorea St Vitus Dance in Saetersdalen, p. 137, quoted by Orbeck, editor. Report of Health and Medicine and Medical Conditions in Norway in 1860.

Luthi-Carter R, Hanson SA, Strand AD, Bergstrom DA, Chun W, et al. 2002. Dysregulation of gene expression in the R6/2 model of polyglutamine disease: Parallel changes in muscle and brain. Hum Mol Genet 11: 1911-1926.

Luthi-Carter R, Strand A, Peters NL, Solano SM, Hollingsworth ZR, et al. 2000. Decreased expression of striatal signaling genes in a mouse model of Huntington's disease. Hum Mol Genet 9: 1259-1271.

MacDonald M, Duyao M, Calzonetti T, Auerbach A, Ryan A, et al. 1996. Targeted inactivation of the mouse Huntington's disease gene homolog Hdh. Cold Spring Harb Symp Quant Biol 61: 627-638.

MacMillan JC, Snell RG, Tyler A, Houlihan GD, Fenton I, et al. 1993. Molecular analysis and clinical correlations of the Huntington's disease mutation. Lancet 342: 954-958.

Mandich P, Di Maria E, Bellone E, Ajmar F, Abbruzzese G. 1996. Molecular analysis of the IT15 gene in patients with apparently "sporadic" Huntington's disease. Eur Neurol 36: 348-352.

Mangiarini L, Sathasivam K, Seller M, Cozens B, Harper A, et al. 1996. Exon 1 of the HD gene with an expanded CAG repeat is sufficient to cause a progressive neurological phenotype in transgenic mice. Cell 87: 493-506.

Margolis RL, O'Hearn E, Rosenblatt A, Willour V, Holmes SE, et al. 2001. A disorder similar to Huntington's disease is associated with a novel CAG repeat expansion. Ann Neurol 50: 373-380.

Mars H, Lewis LA, Robertson AL Jr, Butkus A, Williams GH Jr. 1969. Familial hypo-beta-lipoproteinemia: A genetic disorder of lipid metabolism with nervous system involvement. Am J Med 46: 886-900.

Moore RC, Xiang F, Monaghan J, Han D, Zhang Z, et al. 2001. Huntington disease phenocopy is a familial prion disease. Am J Hum Genet 69: 1385-1388.

Morales LM, Estevez J, Suarez H, Villalobos R, Chacin dB, et al. 1989. Nutritional evaluation of Huntington disease patients. Am J Clin Nutr 50: 145-150.

Myers RH, Mac Donald ME, Koroshetz WJ, Duyao MP, Ambrose CM, et al. 1993. De novo expansion of a (CAG)n repeat in sporadic Huntington's disease. Nat Genet 5: 168-173.

Myers RH, Sax DS, Koroshetz WJ, Mastromauro C, Cupples LA, et al. 1991. Factors associated with slow progression in Huntington's disease. Arch Neurol 48: 800-804.

Myers RH, Vonsattel JP, Stevens TJ, Cupples LA, Richardson P, et al. 1988. Clinical and neuropathologic assessment of severity in Huntington's disease. Neurology 38: 341-347.

Nucifora FC Jr, Sasaki M, Peters MF, Huang H, Cooper JK, et al. 2001. Interference by huntingtin and atrophin-1 with cbp-mediated transcription leading to cellular toxicity. Science 291: 2423-2428.

Oepen G, Clarenbach P, Thoden U. 1981. Disturbance of eye movements in Huntington's chorea. Arch Psychiatr Nervenkr 229: 205-213.

Ona VO, Li M, Vonsattel JP, Andrews LJ, Khan SQ, et al. 1999. Inhibition of caspase-1 slows disease progression in a mouse model of Huntington's disease. Nature 399: 263-267.

Panas M, Avramopoulos D, Karadima G, Petersen MB, Vassilopoulos D. 1999. Apolipoprotein E and presenilin-1 genotypes in Huntington's disease. J Neurol 246: 574-577.

Panov AV, Gutekunst CA, Leavitt B, Hayden MR, Burke JR, et al. 2002. Early mitochondrial calcium defects in Huntington's disease are a direct effect of polyglutamines. Nat Neurosci 5: 731-736.

Penney JB Jr, Vonsattel JP, Mac Donald ME, Gusella JF, Myers RH. 1997. CAG repeat number governs the development rate of pathology in Huntington's disease. Ann Neurol 41: 689-692.

Penney JB Jr, Young AB, Shoulson I, Starosta-Rubenstein S, Snodgrass SR, et al. 1990. Huntington's disease in Venezuela: 7 years of follow-up on symptomatic and asymptomatic individuals. Mov Disord 5: 93-99.

Perutz M. 1996. Glutamine repeats and inherited neurodegenerative diseases: Molecular aspects. Curr Opin Struct Biol 6: 848-858.

Phillips JG, Bradshaw JL, Chiu E, Teasdale N, Iansek R, et al. 1996. Bradykinesia and movement precision in Huntington's disease. Neuropsychologia 34: 1241-1245.

Pickering DS, Thomsen C, Suzdak PD, Fletcher EJ, Robitaille R, et al. 1993. A comparison of two alternatively spliced forms of a metabotropic glutamate receptor coupled to phosphoinositide turnover. J Neurochem 61: 8-92.

Pratley RE, Salbe AD, Ravussin E, Caviness JN. 2000. Higher sedentary energy expenditure in patients with Huntington's disease. Ann Neurol 47: 64-70.

Quinn N, Schrag A. 1998. Huntington's disease and other choreas. J Neurol 245: 709-716.

Ranen N, Stine O, Abbot M, Sherr M, Codori A, et al. 1995. Anticipation and instability of IT-15 (CAG)n repeats in parent-offspring pairs with Huntington disease. Am J Hum Genet 57: 593-602.

Reddy PH, Williams M, Charles V, Garrett L, Pike-Buchanan L, et al. 1998. Behavioural abnormalities and selective neuronal loss in HD transgenic mice expressing mutated full-length HD cDNA. Nat Genet 20: 198-202.

Reiner A, Albin RL, Anderson KD, D'Amato CJ, Penney JB, et al. 1988. Differential loss of striatal projection neurons in Huntington disease. Proc Natl Acad Sci USA 85: 5733-5737.

Rigamonti D, Bauer JH, De Fraja C, Conti L, Sipione S, et al. 2000. Wild-type huntingtin protects from apoptosis upstream of caspase-3. J Neurosci 20: 3705-3713.

Roizin L, Stellar S, Liu J. 1979. Neuronal nuclear-cytoplasmic changes in Huntingtons chorea: Electron microscope investigations. Advance in Neurology, Huntington's Disease. Wexler N, Barbeau A, editors. New York: Raven Press; pp. 195-122.

Roos RA, Pruyt JF, de Vries J, Bots GT. 1985. Neuronal distribution in the putamen in Huntington's disease. J Neurol Neurosurg Psychiatry 48: 422-425.

Rosenblatt A, Brinkman RR, Liang KY, Almqvist EW, Margolis RL, et al. 2001. Familial influence on age of onset among siblings with Huntington disease. Am J Med Genet 105: 399-403.

Ross CA. 2002. Polyglutamine pathogenesis: Emergence of unifying mechanisms for Huntington's disase and related disorders. Neuron 35: 819-822.

Rubinsztein DC, Leggo J, Chiano M, Dodge A, Norbury G, et al. 1997. Genotypes at the GluR6 kainate receptor locus are associated with variation in the age of onset of Huntington disease. Proc Natl Acad Sci USA 94: 3872-3876.

Sanberg PR, Fibiger HC, Mark RF. 1981. Body weight and dietary factors in Huntington's disease patients compared with matched controls. Med J Aust 1: 407-409.

Sanchez I, Mahlke C, Yuan J. 2003. Pivotal role of oligomerization in expanded polyglutamine neurodegenerative disorders. Nature 421: 373-379.

Sanchez I, Xu CJ, Juo P, Kakazika A, Blenis J, et al. 1999. Caspase-8 is required for cell death induced by expanded polyglutamine repeats. Neuron 22: 633.

Saudou F, Finkbeiner S, Devys D, Greenberg M. 1998. Huntingtin acts in the nucleus to induce apoptosis but death does not correlate with the formation of intranuclear inclusions. Cell 95: 55-66.

Sawa A, Wiegand GW, Cooper J, Margolis RL, Sharp AH, et al. 1999. Increased apoptosis of Huntington disease lymphoblasts associated with repeat length dependent mitochondrial depolarization. Nat Med 5: 1194-1198.

Schapira AH. 1999. Mitochondrial involvement in Parkinson's, Huntington's disease, hereditäry spastic paraplegia and Friedreich's ataxia. Biochem Biophys Acta 1410: 99-102.

Scherzinger E, Lurz R, Turmaine M, Mangiarini L, Hollenbach B, et al. 1997. Huntingtin-encoded polyglutamine expansions form amyloid-like protein aggregates in vitro and in vivo. Cell 90: 549-558.

Schilling G, Sharp AH, Loev SJ, Wagster MV, Li SH, et al. 1995. Expression of the Huntington's disease (IT15) protein product in HD patients. Hum Mol Genet 4: 1365-1371.

Schubotz R, Hausmann L, Kaffarnik H, Zehner J, Oepen H. 1976. Fatty acid patterns and glucose tolerance in Huntington's chorea. Res Exp Med (Berl) 167(3): 203-215.

Schwarz M, Fellows SJ, Schaffrath C, Noth J. 2001. Deficits in sensorimotor control during precise hand movements in Huntington's disease. Clin Neurophysiol 112: 95-106.

Sharp AH, Loev SJ, Schilling G, Li SH, Li XJ, et al. 1995. Widespread expression of Huntington's disease gene (IT15) protein product. Neuron 14: 1065-1074.

Shinotoh H, Calne DB, Snow B, Hayward M, Kremer B, et al. 1994. Normal CAG repeat length in the Huntington's disease gene in senile chorea. Neurology 44: 2183-2184.

Sipione S, Cattaneo E. 2001. Modeling Huntington's disease in cells, flies, and mice. Mol Neurobiol 23: 21-51.

Snell RG, MacMillan JC, Cheadle JP, Fenton I, Lazarou LP, et al. 1993. Relationship between trinucleotide repeat expansion and phenotypic variation in Huntington's disease. Nat Genet 4: 393-397.

Snowden J, Craufurd D, Griffiths H, Thompson J, Neary D. 2001. Longitudinal evaluation of cognitive disorder in Huntington's disease. J Int Neuropsychol Soc 7: 33-44.

Steffan JS, Bodai L, Pallos J, Poelman M, McCampbell A, et al. 2001. Histone deacetylase inhibitors arrest polyglutamine-dependent neurodegeneration in Drosophila. Nature 413: 739-743.

Steffan JS, Kazantsev A, Spasic-Boskovic O, Greenwald M, Zhu YZ, et al. 2000. The Huntington's disease protein interacts with p53 and CREB-binding protein and represses transcription. Proc Natl Acad Sci USA 97: 676-678.

Tanaka M, Machida Y, Niu S, Ikeda T, Jana NR, et al. 2004. Trehalose alleviates polyglutamine-mediated pathology in a mouse model of Huntington disease. Nat Med 10: 148-154.

Thompson PD, Berardelli A, Rothwell JC, Day BL, Dick JP, et al. 1988. The coexistence of bradykinesia and chorea in Huntington's disease and its implications for theories of basal ganglia control of movement. Brain 111: 223-244.

Tian JR, Zee DS, Lasker AG, Folstein SE. 1991. Saccades in Huntington's disease: Edictive tracking and interaction between release of fixation and initiation of saccades. Neurology 41: 875-881.

Trottier Y, Biancalana V, Mandel JL. 1994. Instability of CAG repeats in Huntington's disease: Relation to parental transmission and age of onset. J Med Genet 31: 377-382.

van Vugt JP, van Hilten BJ, Roos RA. 1996. Hypokinesia in Huntington's disease. Mov Disord 11: 384-388.

Vonsattel JP, Myers RH, Stevens TJ, Ferrante RJ, Bird ED, et al. 1985. Neuropathological classification of Huntington's disease. J Neuropathol Exp Neurol 44: 559-577.

Vuillaume I, Meynieu P, Schraen-Maschke S, Destee A, Sablonniere B. 2000. Absence of unidentified CAG repeat expansion in patients with Huntington's disease-like phenotype. J Neurol Neurosurg Psychiatry 68: 672-675.

Waters CO. 1842. *Practice in Medicine*. Dunglison R, editor. Philadelphia: Lee and Blanchard; p. 312.

Wellington CL, Brinkman RR, Kusky J, Hayden MR. 1997. Toward understanding the molecular pathology of Huntington's disease. Brain Pathol 7: 979-1002.

Westphal C. 1883. Über eine dem Bilde der cerebrospinalen grauen Degeneration ähnlichen Erkrankung des centralen Nervensystems ohne anatomischen Befund, nebst einigen Bemerkungen über paradoxe Contractionen. Arch Psychiatr Nervenkr 14: 187-194.

Wheeler VC, Auerbach W, White JK, Srinidhi J, Auerbach A, et al. 1999. Length-dependent gametic CAG repeat instability in the Huntington's disease knock-in mouse. Hum Mol Genet 8: 115-122.

Yang W, Dunlap JR, Andrews RB, Wetzel R. 2002. Aggregated polyglutamine peptides delivered to nuclei are toxic to mammalian cells. Hum Mol Genet 11: 2905-2917.

Young AB, Shoulson I, Penney JB, Starosta-Rubinstein S, Gomez F, et al. 1986. Huntington's disease in Venezuela neurologic features and functional decline. Neurology 36: 244-249.

Zeron MM, Chen N, Moshaver A, Ting-Chun Lee A, Wellington CL, et al. 2001. Mutant huntingtin enhances excitotoxic cell death. Mol Cell Neurosci 17: 41-53.

Zeron MM, Hansson O, Chen N, Wellington CL, Leavitt BR, et al. 2002. Increased sensitivity to N-methyl-D-aspartate receptor-mediated excitotoxicity in a mouse model of Huntington's disease. Neuron 33: 849-860.

Zuccato C, Ciammola A, Rigamonti D, Leavitt BR, Goffredo D, et al. 2001. Loss of huntingtin-mediated BDNF gene transcription in Huntington's disease. Science 293: 493-498.

Zuccato C, Tartari M, Crotti A, Goffredo D, Valenza M, et al. 2003. Huntingtin interacts with REST/NRSF to modulate the transcription of NRSE-controlled neuronal genes. Nat Genet 35: 76-83.

Zuhlke C, Riess O, Schroder K, Siedlaczck I, Epplen JT, et al. 1993. Expansion of the (CAG)n repeat causing Huntington's disease in 352 patients of German origin. Hum Mol Genet 2: 1467-1469.

12 Dementia: The Significance of Cerebral Metabolic Disturbances in Alzheimer's Disease. Relation to Parkinson's Disease

S. Hoyer · L. Frölich

Abstract: Alzheimer's disease (AD) and Parkinson's disease (PD) are the predominant neurodegenerative disorders of late life. Both disorders do not form nosological entities. A minority of all AD cases is caused by mutations on chromosomes 1, 14 and 21, a minority of all PD cases by mutations on chromosomes 1, 2, 4, 6 and 12, whereas the majority of all both AD and PD cases is age-related and sporadic in origin. In sporadic AD, the desensitization of the neural insulin receptor similar to non-insulin dependent diabetes mellitus may be the core abnormality inducing a cascade-like process of disturbances in the insulinergic, acetyl-cholinergic and glutamatergic systems. Cellular glucose metabolism and its derivatives acetylcholine, cholesterol and ATP are most compromised probably inducing abnormalities in the metabolism of the amyloid precursor protein APP and of tau-protein resulting in the formation of both amyloidiogenic derivatives and hyperphosphorylated tau-protein. As a result, mental deficits in cognition, learning and memory predominate the clinical feature.

Although mental deficits are also found in the course of PD due to an involvement of the acetyl-cholinergic system, this disorder starts with movement disturbances due to structural and metabolic abnormalities of nigrostriatal dopaminergic system. Thus, clear differences become obvious between sporadic AO and sporadic PD in their pathophysiologies.

List of Abbreviations: ACTH, adrenocorticotrophic hormone; AD, Alzheimer's disease; AGE, advanced glycation endproduct; APOE, apolipoprotein E; APP, amyloid precursor protein; ATP, adenosine tri-phosphate; βA4, β-amyloid 4; CA, cornu ammonis; CDK, cyclin-dependent kinase; CNS, central nervous system; CRF, corticotropic releasing factor; CSF, cerebrospinal fluid; DNA, desoxyribonucleic acid; ER, endoplasmic reticulum; GA, Golgi apparatus; GABA, γ-amino butyric acid; GRP, glucose-regulated protein; GSK, glycogen synthase kinase; HPA, hypothalamic-pituitary-adrenal; HSP, heat shock protein; IDE, insulin degrading enzyme; IGF, insulin-like growth factor; IRS, insulin receptor substrate; MAP, mitogen-activated protein; mRNA, messenger ribonucleic acid; NEK, nuclear factor kappa; NMDA, N-methyl-D-aspartate; PD, Parkinson's disease; PDH, pyruvate dehydrogenase; PI, phosphatidylinositol; PS, presenilin; RAGE, receptor of advanced glycation endproduct; SAPK/JNK, stress-activated protein kinase/c-Jun N-terminal kinase; TCAC, tricarboxylic acid cycle; UDP, uridine diphosphate

1   Introduction

Dementia is the most frequent mental disability in the elderly. Among all different forms of dementia, Alzheimer's disease (AD) is its most common form. Nosologically, AD is not a single disorder, although the clinical phenotype of differently caused AD is fairly uniform. Evidence has been provided that a very small proportion of 404 families worldwide (by August 2006) of all Alzheimer cases is caused by missense mutations in the presenilin (PS) gene 1 on chromosome 14 (315 families ~ 78%), in the presenilin gene 2 on chromosome 1 (18 families ~ 4%), and in the amyloid precursor protein (APP) gene on chromosome 21 (71 families ~ 18%) (http://molgen-www.uia.ac.be/ADMutations//), leading to autosomal dominant fa-milial AD with an early onset. Because these different causative mutations can be explained to lead to a common pathophysiology, these mutations form the scientific basis of the amyloid cascade hypothesis of AD. This explains well the genetically induced excess formation of the APP derivative βA4 that aggregates to form amyloid deposited extracellularly as senile plaques, and somewhat less well the intraneural formation and accumulation of hyperphosphorylated tau-protein as neurofibrillary tangles (Hardy and Selkoe, 2002). In contrast, the great majority of patients suffering from AD (approximately 25 million patients worldwide) are sporadic in origin and of late onset. Morphologically, sporadic AD as compared to familial AD is characterized by a similar but less accentuated extracellular deposition of senile plaques and an intraneural accumulation of hyperphosphorylated tau-protein as neurofibrillary tangles. A common morphological feature of both AD forms is synapse pathology and cell loss.

However, excess formation of βA4 has not been proven to be necessary for the generation and development of sporadic AD (Joseph et al., 2001). Thus, the amyloid cascade hypothesis may not apply to sporadic AD. Instead, there is increasing evidence that a cerebrometabolic deficiency is the proximate cause of sporadic AD (Blass et al., 2002). In this context, susceptibility genes may contribute to the generation of sporadic AD. Best known are allelic abnormalities of the apolipoprotein E (APOE) gene on chromosome 19, which are responsible for both anticipated onset and increase in severity of both inherited and sporadic AD. In synergy with the allelic variations of APOE, the polymorphism of the tumor necrosis factor (TNF)-α gene has been found to increase the risk of sporadic AD (McCusker et al., 2001). Also in sporadic AD, single nucleotide polymorphism has been found in the gene coding for 11β-hydroxysteroid dehydrogenase I, which is associated with a sixfold-increased risk for sporadic AD (de Quervain et al., 2004). 11β-Hydroxysteroid dehydrogenase has been demonstrated to be expressed in the brain, predominantly localized within neurons and less in glial cells, and accentuated in the hippocampus and cerebral cortex (Sakai et al., 1992). This enzyme acts as a dehydrogenase/reductase, catalyzing the interconversion of active glucocorticoids and inert 11-keto forms (dehydrogenase) and regenerating active glucocorticoids from inert 11-keto forms (reductase) (Duax et al., 1996; Jellinck et al., 1999). The reaction direction of 11β-hydroxysteroid dehydrogenase I as an oxireductase is determined by hexose-6-phosphate dehydrogenase (Atanasov et al., 2004), which cooperates with 11β-hydroxysteroid dehydrogenase in the lumen of the endoplasmic reticulum (ER), whereby glucose-6-phosphate stimulated the reductase activity of 11β-hydroxysteroid dehydrogenase I (Banhegyi et al., 2004). In the brain, 11β-hydroxysteroid dehydrogenase I acts predominantly as a reductase, thus amplifying the glucocorticoid action (Seckl and Walker, 2001). The expression of 11β-hydroxysteroid dehydrogenase I mRNA in the brain is distinct from that of 11β-hydroxysteroid dehydrogenase II found in discrete brain areas (Roland et al., 1995).

Besides, the APOE gene and 11β-hydroxysteroid dehydrogenase I gene are other potential candidate susceptibility genes for sporadic AD whose contributory significance is speculative as yet (Rocchi et al., 2003).

Such a genetic predisposition together with risk factors, in the case of sporadic AD, the major risk factor aging (Hoyer, 1995, 2000; Ott et al., 1995) may then cause the disease (Holness et al., 2000). In this chapter, it will be discussed which metabolic abnormalities are associated with both early-onset AD and sporadic AD and in which way these abnormalities may contribute to the generation of the morphological hallmarks synaptic damage, senile plaques, neurofibrillary tangles, and cell loss in the brains of both AD types. Furthermore, the common pathophysiology of sporadic AD and Parkinson's disease (PD) is discussed.

2   Cerebral Metabolism in the Normal Adult Brain

2.1   Glucose and Energy Metabolism

Glucose has been found to be the major nutrient of the adult mature, healthy brain and to be the source of functionally important metabolites such as acetyl-CoA and ATP (Gibbs et al., 1942; Gottstein et al., 1963; Hoyer, 1970, 1992; Siesjö, 1978; Erecinska and Silver, 1989). Various mechanisms contribute to the uptake of glucose from the arterial blood into the brain and to the regulation of glucose metabolism in the brain. Glucose enters the brain by crossing the luminal endothelial layer of the blood–brain barrier via the 55-kDa glucose transport protein-1. The transport from the endothelial cell into the abluminal extracellular space is obviously mediated by astrocytes (Farrell and Pardridge, 1991; Maher et al., 1993; Fillenz and Lowry, 1998). The glucose transport across the blood–brain barrier into the extracellular space of the brain is obviously insulin dependent (Bachelard, 1971; Hertz et al., 1981; Kahn, 1985). The mRNA of glucose transport protein-1 has been found to be regulated by neuropeptides in all probability also by insulin (Pardridge et al., 1985, 1990; Boado, 1995). The glucose transport protein-3 mediates the glucose transport from the extracellular space into neurons (Harr et al., 1995) and the glucose transport protein-5 into glial cells (Payne et al., 1997). Some neuron populations in the basal frontal brain, cerebral cortex, hippocampus, and cerebellum were found to express the insulin-sensitive glucose transport protein-4 besides glucose transport protein-3 (Apelt et al., 1999).

The glycolytic breakdown of glucose in the nerve cell is controlled by the allosteric enzymes hexokinase and pyruvate kinase working in a concerted way under the influence of phosphofructokinase (Newsholm and Start, 1973). The oxidation of the glycolytically formed pyruvate starts by means of the preeminent multienzyme complex pyruvate dehydrogenase (PDH) (Garland and Randle, 1964) yielding the energy-rich compound acetyl-CoA that is used (1) for further oxidation in the tricarboxylic acid cycle (TCAC) to ATP (more than 95% of acetyl-CoA), (2) for the formation of the neurotransmitter acetylcholine (1%–2% of acetyl-CoA) (Gibson et al., 1975), and (3) for the formation of cholesterol in the 3-hydroxy-2-methylglutaryl-CoA cycle (Michikawa and Yanagisawa, 1999). Cholesterol is the main sterol in membranes, and it also serves as the basic compound from which neurosteroids derive (Rupprecht and Holsboer, 1999). Glia-derived cholesterol has been demonstrated to promote synaptogenesis in nervous tissue (Mauch et al., 2001).

PDH activity is reduced by its product acetyl-CoA, for example, when fatty acids or ketone bodies are used for oxidation instead of glucose (Garland et al., 1964). A small pool of free fatty acids exists in the brain (Westerberg et al., 1987). In the brain, free fatty acids can be metabolized via acetyl-CoA by beta-oxidation (Singh et al., 1989). The activity of the enzyme choline acetyltransferase, which catalyzes acetylcholine formation from acetyl-CoA and choline, is closely linked functionally to the PDH complex (Perry et al., 1980). Acetylcholine acts (1) as an excitatory neurotransmitter for learning and memory capacities (Drachman et al., 1980) and (2) as a regulator for regional cerebral blood flow (Sato and Sato, 1995).

From acetyl-CoA, glucose-derived carbon is rapidly transferred into amino acids via the TCAC and the γ-aminobutyric acid (GABA) shunt (Sacks, 1957, 1965). Glutamate, glutamine, aspartate, and GABA are formed most abundantly (Wong and Tyce, 1983).

Dehydrogenating multienzymes complexes working in the TCAC provide redox equivalents that are oxidized in the respiratory chain to yield ATP. Energy-rich compounds, such as ATP and UDP, represent the driving force for nearly all cellular and molecular work. Some out of many others are listed here.

In general:
- Folding of proteins (Braakman et al., 1992; Gething and Sambrook, 1992)
- Sorting of proteins (Rothman and Wieland, 1996)
- Transport of proteins (Rothman and Orci, 1992; Rothman, 1996)

In particular:
- Function of the ER/Golgi apparatus (GA)
- 1. Maintenance of a pH of 6 (Seksek et al., 1995; Demaurex et al., 1998)
- 2. Transport of membrane proteins (Dorner et al., 1990; Verde et al., 1995)
- 3. Glycosylation of proteins (Wiegandt, 1982; Roth et al., 1985; Hirschberg and Snider, 1987; for review Roth, 1987; Lannert et al., 1998)

- Neurotransmission at the synapse, being the site of the highest energy utilization out of all cellular compartments (Sokoloff, 1981; Kadekaro et al., 1985; Kauppinen and Nicholls, 1986; Keller et al., 1997)
- Maintenance of Na^+, K^+ transmembranous flux (Erecinska and Silver, 1989)

2.2 Insulin and Insulin Receptor Signaling

2.2.1 Insulin Expression, Insulin Receptor Distribution

Substantial evidence has been gathered in support of the presence of both insulin and insulin receptors in the brain. The main source of brain insulin is the pancreas. Insulin is known to cross the blood–brain barrier by a saturable transport mechanism. The transporter is unevenly distributed throughout the brain with the olfactory bulb having the fastest transport rate of any brain region (Banks, 2004). A smaller proportion of insulin is produced in the brain itself (for review Plata-Salaman, 1991). Glucose transport protein 2 mRNA has been found in a limited number of brain nuclei and the olfactory bulb (Leloup et al., 1994). In pancreatic β cells, glucose transport protein 2, coupled with glucokinase, participates in the glucose-sensing mechanism of insulin release. In all likelihood, this transport protein has the same effect in the brain, where it is expressed in different brain regions at a low level (Brant et al., 1993).

Insulin gene expression and insulin synthesis have been demonstrated in both immature (Schechter et al., 1992, 1996) and mature mammalian neuronal cells. Insulin mRNA was found to be distributed in a highly specific pattern, with the highest density in pyramidal cells of the hippocampus and high densities in medial prefrontal cortex, the entorhinal cortex, perirhinal cortex, thalamus, and the granule layer of the olfactory bulb. Neither insulin mRNA nor synthesis of the hormone was observed in glia cells (Devaskar et al., 1994). Glucose stimulates the release of insulin from the brain synaptosomes (Santos et al., 1999). There is recent clear evidence that the transfer of insulin across the brain occurs rapidly from the olfactory nerve to cerebrospinal fluid (CSF) (Born et al., 2002).

Insulin receptors have been demonstrated to be dispersed throughout the brain following also a highly specific pattern, with the highest density detected in olfactory bulb, hypothalamus, cerebral cortex, and hippocampus (van Houten et al., 1979a; Baskin et al., 1986; Hill et al., 1986; Werther et al., 1987; Unger et al., 1989). Nerve terminals show enriched densities of insulin receptors (van Houten et al., 1979b; Abbott et al., 1999), which bind insulin in a highly specific and rapid manner (Raizada et al., 1988). Two different types of insulin receptors have been found in the adult mammalian brain: (1) a peripheral type detected in lower density on glia cells (α-subunit 130 kDa, β-subunit 95 kDA), which is downregulated by insulin, and (2) a neuron-specific brain type with high concentrations on neurons (α-subunit 118 kDa, β-subunit 91 kDA), which is not downregulated by insulin (Zahniser et al., 1984; Lowe et al., 1986; for review Adamo et al., 1989). It has been shown that the location of phosphotyrosine-containing proteins corresponds to the distribution of the insulin receptor (Moss et al., 1990). It has also been established that the insulin receptor substrate (IRS)-1 colocalizes with these phosphotyrosines (Baskin et al., 1993). IRS-1 and the insulin receptor are found to be coexpressed in discrete neuron populations in rat hippocampus and olfactory bulb whereas their proteins showed the highest densities in the synaptic neuropile (Baskin et al., 1994).

2.2.2 Insulin Degradation

Insulin is degraded by the insulin-degrading enzyme (IDE) (Kurochkin and Goto, 1994), the gene of which maps on chromosome 10. It is a cytosolic metalloendoprotease that is expressed differently in different tissues (Duckworth, 1988). IDE mRNA increases from the first postnatal week to adulthood, when levels are high in several organs (Kuo et al., 1993). IDE degrades different groups of substrates. Insulin, transforming growth factor α, atrial natriuretic peptide, and insulin-like growth factor (IGF)-2 are high-affinitiy substrates with $K_m \sim 0.1$ μM. In contrast, substrates such as glucagon, epidermal growth factor, IGF-I, β-endorphin, and β-amyloid analogs show a lower affinity at $K_m > 2$ μM (Kurochkin and Goto, 1994; Perez et al., 2000; Farris et al., 2003). That is, with respect to insulin and the AD-related β-amyloid protein Aβ, the degradation capacity of IDE is different and depends on the intracellular metabolic state. In the normal human brain, IDE is the main soluble β-amyloid-degrading enzyme at neutral pH (Kurochkin and Goto, 1994), but the greatest β-amyloid-degrading capacity occurs between pH 4 and 5. This indicates that IDE may act as an "amyloidase," preventing the accumulation of amyloidogenic derivatives (McDermott and Gibson, 1997), the extracellular level of which is regulated by neuronal IDE (Vekrellis et al., 2000). In a physiological rat cortical cell system, IDE was shown to eliminate the neurotoxic effects of both β-amyloid protein-(1-40) and β-amyloid protein-(1-42) and to prevent the deposition Aβ-(1-40) onto a synthetic amyloid (Mukherjee et al., 2000). Excess insulin inhibits the degradation of both Aβ-(1-40) and Aβ-(1-42) nearly completely, and both Aβ derivatives inhibit insulin degradation in a dose-dependent manner (Perez et al., 2000).

In functional studies with IDE-knockout mice, the IDE(−/−) mice showed a greater than 50% decrease in β-amyloid degradation, and a 1.6-fold [β-amyloid (1-40)] increase and 1.4-fold [β-amyloid (1-42)] increase, associated with a sixfold increase in the level of the γ-secretase-generated C-terminal fragment of APP, and a similar deficit in insulin degradation in liver. Glucose intolerance and hyperinsulinemia were also found (Farris et al., 2003; Miller et al., 2003). These studies confirm that IDE is definitely one of four or five proteases that regulate the level of Aβ protein throughout the brain and most likely the most important protease.

2.2.3 Insulin Receptor Regulation

The major molecular structure and most of the biochemical properties of the neuronal insulin receptor have been demonstrated to be indistinguishable from those in nonnervous tissues (Heidenreich et al., 1983). Binding of insulin to the extracellular α-subunit of its receptor induces autophosphorylation of the intracellular β-subunit by phosphorylation of the receptor's intrinsic tyrosine residues for activation. The receptor's phosphorylation state is regulated by the action of both phosphotyrosine phosphatases (dephosphorylation) (Goldstein, 1993) and serine/threonine kinases (Häring, 1991; for review Avruch, 1998).

Insulin receptor is desensitized by glucocorticoids that have been shown to inhibit its phosphorylation of the tyrosine residues (Giorgino et al., 1993). β-Adrenergic catecholamines decrease both the activity of the receptor's tyrosine kinase by cAMP-mediated phosphorylation of the serine/threonine residues (Häring et al., 1986) and the insulin-induced tyrosine phosphorylation of the IRS-1 and -2 (Klein et al., 1999). The cytokine TNF-α has been demonstrated to decrease both the insulin-stimulated insulin receptor autophosphorylation (Hotamisligil et al., 1994) and the phosphorylation of the IRS-1 and -2 (Hotamisligil et al., 1994; Peraldi et al., 1996; Valverde et al., 1998). Additionally, TNF-α was demonstrated to cause the phosphorylation of serine/threonine residues of IRS-1 that, in turn, induces insulin resistance by means of an inhibition of insulin receptor tyrosine kinase activity (Hotamisligil et al., 1996). Thus, several different mechanisms help to regulate the function and activity of the insulin receptor. The most prominent substrate of the insulin receptor tyrosine kinase is the IRS-1, that transfers the insulin signal to a wide spectrum of cellular and molecular mechanisms (for review White and Kahn, 1994). Dexamethasone downregulates IRS-1 (Turnbow et al., 1994).

Recently, first evidence has been provided that insulin and insulin receptor are functionally linked to cognition in general and spatial memory in particular by upregulation of the insulin mRNA in the hippocampus and increased accumulation of insulin receptor protein in hippocampal synaptic membranes (Zhao et al., 1999; for review Park, 2001; Zhao et al., 2004).

2.3 The Action of Insulin on Metabolic Pathways in the Brain with Relevance to Neurodegeneration Diseases

2.3.1 Glucose and Energy Metabolism

Acute stimulation of the cerebral insulin receptor was achieved through a single intracerebroventricular injection of insulin. This procedure led to a dose-dependent stimulation of the glycolytic key enzymes hexokinase and phosphofructokinase in the cerebral cortex (Hoyer et al., 1993). Also, acute stimulatory effects of the hormone in the brain have been demonstrated for PDH (Rinaudo et al., 1987) and choline acetyltransferase (Kyriakis et al., 1987). These data may indicate that both glycolytic flux and pyruvate oxidation in the brain are stimulated by insulin paralleling the hormone's effect in nonnervous tissue. Short-term (1 day) or long-term (7 and 21 days) intracerebroventricular infusion of insulin have been found to exhibit a discrete anabolic effect on energy metabolism in the hippocampus as can be concluded from an 11% increase in the concentration of creatine phosphate, the storage form of ATP (Henneberg and Hoyer, 1994). Of great functional importance and closely related to energy metabolism is the regulation of the Na^+/K^+-ATPase by insulin (Sweeney and Klip, 1998).

2.3.2 Glycogen Synthase Kinase-3 Cascade

The insulin/insulin receptor-mediated signal transduction controls the activity of several enzymes in a cascade-like manner (Avruch, 1998). Phosphatidylinositol-3 kinase is insulin regulated and activates protein kinase B (Alessi and Cohen, 1998; Vanhaesebroeck and Alessi, 2000). This latter enzyme regulates the activity of the glycogen synthase kinases (GSK)-3α and -3β (Cross et al., 1995, 1997). The insulin-stimulated signal decreased the activity of GSK-3α by about 50% (Ramakrishna and Benjamin, 1998) and GSK-3β by about 70%, obviously by phosphorylation of the serine 9 residue through the S6 kinase. The inhibition of both enzymes was reversed by protein phosphatase 2A dephosphorylating the GSKs at serine 9 (Welsh and Proud, 1993; Sutherland and Cohen, 1994).

GSK-3α is inhibited by phosphorylation at serine 21 and GSK-3β by phosphorylation at serine 9 (Shaw et al., 1997; Frame and Cohen, 2001; for review Cohen and Frame, 2001). GSK-3 is constitutively phosphorylated at tyrosine 279 (GSK-3α) and at tyrosine 216 (GSK-3β), which is associated with an activation. Dephosphorylation of these tyrosine residues induces the inactivation of this enzyme (Hughes et al., 1993; Bhat et al., 2000). Interestingly, the inhibition of protein phosphatase-1 being involved to a lesser extent in the regulation of tau-protein phosphorylation inhibited up to about 80% of GSK-3 activity (Bennecib et al., 2000). Independently from insulin, cAMP-dependent protein kinase A exerts the same effect on both GSK-3α and GSK-3β, as does the insulin-stimulated protein kinase B (Fang et al., 2000;

Li et al., 2000). GSK-3 has been turned out to play a key role in numerous cell functions, which holds true also for the metabolism of the APP and the regulation of tau-protein phosphorylation.

In general, the diminished activity of phosphatidylinositol-3 kinase causes a decrease in the activity of protein kinase B (Alessi and Cohen, 1998; Vanhaesebroeck and Alessi, 2000). As a consequence, the activities of both GSK-3α and GSK-3β may be disinhibited by reduced phosphorylation at serine 21 (GSK-3α) and at serine 9 (GSK-3β). In postmortem sporadic Alzheimer brain, reduced levels of phosphatidylinositol-3 kinase and both phosphorylated protein kinase B and GSK-3 were found. The unchanged level of total GSK-3 may indicate its reduced phosphorylation (Steen et al., 2005) pointing to an increased capacity for protein phosphorylation.

2.3.3 GSK and Presenilin 1

Presenilin (PS) 1 is a polytopic protein residing in the ER. It is enriched in neurons and interacts with γ-catenin found in the synaptic adherens junction (Uchida et al., 1996; Zhou et al., 1997). In the NH_2- and COOH-terminal fragments of PS are components of a ~250 kDa complex containing β-catenin (Yu et al., 1998). The latter has been shown to be regulated by GSK-3β and to be associated with PS 1 (Takashima et al., 1998; Kirschenbaum et al., 2001a). It might be of functional significance that GSK-3β can exclusively influence the level of the COOH-terminal fragment by phosporylation of serine 397 in the PS 1 loop domain (Kirschenbaum et al., 2001b) since the NH_2- and COOH-terminal fragments represent the functional PS 1 structure.

2.3.4 Cyclin-Dependent Kinase 5 Cascade

Cyclin-dependent kinase 5 (CDK5) is a prolin-directed protein kinase with close homology to the cell cycle kinases CDK1 and CDK2. p35 and p39, two neuronal specific proteins, activate CDK5. This enzyme plays a pivotal role in the development of the nervous system and may also be involved in neurodegeneration. Transfection experiments have shown that CDK5 plays an important role in axonal extension (Nikolic et al., 1996; Pigino et al., 1997; Connell-Crowley et al., 2000). CDK5 has also been shown to phosphorylate and regulate the activity of calcium channels and NMDA glutamate receptors (Li et al., 2001; Tomizawa et al., 2002) pointing to its role in synaptic plasticity. Through its interaction with p25 cleavage and p35 production, CDK5 can be regulated in several neurotoxic conditions (Kusakawa et al., 2000; Lee et al., 2000; Nath et al., 2000).

CDK5 is a major enzyme for phosphorylating tau in the brain (Imahori and Uchida, 1997). Several experiments have provided evidence that p35 cleavage and p25 production are the crucial events leading to alteration in substrate specificity of CDK5 and thus contribute to tau hyperphosphorylation (Ahlijanian et al., 2000; van den Haute et al., 2000; Hashiguchi et al., 2002; Cruz et al., 2003; Noble et al., 2003). CDK5 has also been shown to phosphorylate APP on threonine 668 (Iijima et al., 2000), which may have a direct impact on APP processing and amyloid pathology (Tsai et al., 2004).

Thus, both GSK-3 and CDK5 are important enzymes involved in the phosphorylation of tau and possibly APP, both of them appear to be changed in sporadic AD and implicated in feedback loops in the pathogenesis of AD.

2.3.5 The Insulin Signal Cascade and Amyloid Precursor Protein

The cytoplasmic domain of recombinant APP has been demonstrated to be phosphorylated by GSK-3β (Aplin et al., 1996). GSK-3α has been shown to regulate the production of the APP derivatives, β-amyloid peptides (Phiel et al., 2003), which are generated intracellularly in the ER/intermediate compartment of the GA (Cook et al., 1997; Hartmann et al., 1997; Wild-Bode et al., 1997; Xu et al., 1997; Greenfield et al., 1999). Also, in the ER/GA, APP undergoes posttranslational modification: O-GlcNAcylation of the cytoplasmic domain of APP (Abeijon and Hirschberg, 1988; Capasso et al., 1988; Griffith and Schmitz, 1995; Griffith et al., 1995) and glycosylation (Haass et al., 1995). Additionally, a minor fraction of the N-glycosylated form of APP complexes with the presenilins (Xia et al., 1997). The work of both ER and GA is highly ATP dependent. ATP ensures the pH of 6 necessary for normal functioning, the transport, and the secretion of proteins (Dorner et al., 1990; Seksek et al., 1995; Verde et al., 1995).

The intracellular accumulation of both β-amlyoid (1-40) and β-amlyoid (1-42) is reduced by accelerating the APP/β-amlyoid transport from the *trans*-Golgi network to the plasma membrane assisted by mitogen-activated protein kinase signaling. The promotion of APP secretion from the intracellular to the extracellular space and the inhibition of its degradation by IDE is mediated by insulin and the tyrosine kinase activity of the insulin receptor (Gasparini et al., 2001). The same holds true for APP release into the extracellular space, which is dependent on the activation of phosphatidylinositol-3 kinase (Solano et al., 2000). Further evidence had been provided for the release of APP derivatives into the extracellular space mediated by the activation of the acetylcholinergic system in general (Rossner et al., 1997), and for its muscarinic m1 and m3 receptors (Nitsch et al., 1992a).

APP potentiates the activity of neurotrophic factors (Wallace et al., 1997a) via the insulin-signaling pathway (Wallace et al., 1997b). Both derivatives of APP, β-amlyoid (1-40) and β-amlyoid (1-42), reduce the binding of insulin to its receptor by decreasing the affinity of insulin binding, resulting in a reduced receptor autophosphorylation (Xie et al., 2002).

2.3.6 The Insulin Signal Cascade and Tau-Protein

The tau-protein belongs to a family of microtubule-associated proteins that stimulate the generation and stabilization of microtubules. Tau-protein is present in all neuronal cell types. The extent of its phosphorylation is different in different locations with tau-protein in the distal axon and in the white matter being less heavily phosphorylated. Most of the phosphorylation sites are concentrated at serine and threonine residues in two regions, 181–235 and 396–422 (Watanabe et al., 1993; for review, Lovestone and Reynolds, 1997). The phosphorylation and dephosphorylation of tau-protein are regulated by several protein kinases and by protein phosphatases. Among the tau-phosphorylating protein kinases are the ATP-dependent PK^{erk36} and PK^{erk40} (Röder and Ingram, 1991), protein kinase-1-glycogen synthase kinase-β (Ishiguro et al., 1992, 1993), and the protein kinase FA/GSK-3α (Mandelkow et al., 1992), which all work in an insulin-dependent manner (see above) (Cross et al., 1995, 1997; Hong and Lee, 1997; Lesort et al., 1999).

Dephosphorylation of tau-protein at serine/threonine residues is performed by a family of protein phosphatases classified as types 1, 2A, and 2B (Gong et al., 1994a, b; Wang et al., 1995; for review Lovestone and Reynolds, 1997). Protein phosphatase 2A has been demonstrated to be differently regulated during the cell cycle: high activity during the S phase, intermediate during G1, and low in G2- and M-phases (Sontag et al., 1995). Insulin has been found to have a dual effect on tau-protein phosphorylation. After a short period of insulin treatment (1 min), tau-protein phosphorylation and GSK-3β activity are increased (Lesort et al., 1999). In contrast, prolonged exposure to insulin induced a downregulation of the GSK-3β activity and, thus, decreased phosphorylation of tau-protein (Cross et al., 1997; Hong and Lee, 1997).

As with APP, tau-protein is posttranslationally modified by O-linked *N*-acetylglucosamine (O-GlcNA) cylation. This process is found exclusively on nuclear and cytosolic proteins and is as abundant as phosphorylation (for review Hart, 1997; Hayes and Hart, 1998; Vosseller et al., 2002). O-GlcNAcylation and phosphorylation are mutually exclusive (Kelly et al., 1993; Comer and Hart, 2001). Insulin–glucose have been found to trigger O-GlcNAcylation (Yki-Jarvinen et al., 1998; Buse et al., 2002). Additionally, insulin and glucose have been shown to be important regulators of the expression of the DNA repair enzyme xeroderma pigmentosum complementation group D (Merkel et al., 2003).

2.3.7 Sulfonylurea Receptors and Their Relation to Insulin

While glucose is the central substrate for the intracellular oxidative metabolism in all neurons (see above), a subpopulation of neurons regulates their firing rate as a function of glucose availability (Ashford et al., 1990; Levin and Dunn-Meynell, 1997). Such glucose-responsive neurons are functionally linked to sulfonylurea receptors and ATP-sensitive K^+ channels (Inagaki et al., 1996). The density of the sulfonylurea receptors seems to be species specific with differences notably in the hippocampus, globus pallidus, and substantia nigra (Mourre et al., 1989; Treherne and Ashford, 1991; Zini et al., 1993). Different kinds of binding sites were found to reside on neurons: low-affinity sites primarily on cell bodies (Dunn-Meynell et al., 1997) and high-affinity sites primarily on axon terminals (Lee et al., 1996). Binding to the sulfonylurea receptor blocks the ATP-regulated K^+ channel causing an increase of intracellular K^+, cell depolarization, and firing (Bernardi et al., 1988; Gopalakrishan et al., 1993). Endosulfine has been found to be the endogenous ligand for the sulfonylurea receptor in the brain. Endosulfine is equally recognized by sulfonylurea receptors of the brain and by those receptors of insulin-secreting β cells, the latter of which are stimulated to release insulin (Virsolvy-Vergine et al., 1992). Otherwise, insulin itself activates ATP-sensitive K^+ channels in hypothalamic neurons (Spanswick et al., 2000). However, it is as yet not known whether insulin is released (as from pancreatic β cells) from neurons after ligand binding to the sulfonylurea receptor.

2.4 Hypoglycemia as a Model for Metabolic Disturbances in Neurodegenerative Diseases

2.4.1 Alterations in Brain Glucose Metabolism

A fall of the physiological nutrient glucose in the arterial blood caused a decrease in cerebral glucose utilization from 50% to 67% of normal whereas cerebral blood flow and oxygen consumption did not change (Eisenberg and Seltzer, 1962; Gottstein and Held, 1967). Thus, the same disproportion between oxygen and glucose becomes obvious as in sporadic AD when arterial glucose concentration is normal (see above). In hypoglycemia, the energy state in cerebral cortex remained unchanged unless the decrease in blood glucose concentration was below 1 μmol/g (Lewis et al., 1974).

There is clear evidence that, apart from glucose, arterial ketone bodies are taken up by the brain to support cerebral energy production when arterial glucose concentration had fallen during a fast lasting several hours and ketone bodies had increased likewise (Owen et al., 1967; Gottstein et al., 1971; Hawkins et al., 1971; Ruderman et al., 1974). It should be emphasized that ketone bodies are not taken up by the brain in normoglycemia or acute arterial hypoglycemia. In acute and short-lasting arterial hypoglycemia, the amino acid concentration in cerebral cortex changed in that glutamate decreased, and both aspartate and ammonia increased indicating the utilization of glutamate instead of glucose (Norberg and Siesjö, 1976). This metabolic pattern resembles the one occurring in early-onset and late-onset AD brain (Procter et al., 1988; Hoyer and Nitsch, 1989; Hoyer et al., 1990). The activity of the TCAC is low except succinate that increased markedly (Norberg and Siesjö, 1976). It may be assumed that an anaplerotic reaction is set into motion as is the case in cerebral ischemia: the phylogenetically developed metabolism is shifted to a phylogenetically primitive one (Folbergrova et al., 1974; Hoyer and Krier, 1986).

2.4.2 Alterations in Metabolism of Amyloid Precursor Protein and Tau-Protein

The metabolism of the APP and tau-protein was studied in experimental acute glucose deprivation. Astroglial cells deprived of 95% of its glucose increased the APP expression time-dependent with a peak at 24-h glucose deprivation (Shi et al., 1997). In hippocampal brain slices, glucose deprivation caused an excessive release of APP (El Tamer et al., 1998). Also, clear evidence has been provided that starvation up to 72 h induces hypoglycemia and hypoinsulinemia, but hypercortisolemia and an increase of β-hydroxybutyric acid in the arterial blood (Ahima et al., 1996). In the brain, glucose concentration decreases in many areas by around 30% such as cortex, hippocampus, and hypothalamus (Garriga and Cusso, 1992). Arterial hypoglycemia accompanied by a reduction of blood insulin concentration by 85% caused enhanced phosphorylation of tau-protein at serine residues 199, 396, 413, and 422, and at threonine residue 205, most pronounced in the hippocampus, and lesser in cerebral cortex. The starvation-induced tau-hyperphosphorlyation was reversible within 1 day by refeeding (Yanagisawa et al., 1999). GSK-3β has been found to be one out of more candidates for tau-protein phosphorylation at serine 9 (see above). However during starvation, GSK-3β activity was decreased most likely due to phosphorylation at serine 9 (Imahori and Uchida, 1997). This reduced activity may be assisted by a fall in the activity of protein phosphatase 2A that normally activates GSK-3β by its dephosphorylation at serine 9 (Sutherland and Cohen, 1994). However, during starvation the activity of protein phosphatase 2A is reduced (Planel et al., 2001), consequently leading to a diminished dephosphorylation capacity of GSK-3β (Bennecib et al., 2000).

Alterations in glucose metabolism were found to be associated with hypothermia in experimental mice. Hypothermia was demonstrated to be a major component of tau-hyperphosphorylation and inhibition of both tau-kinases and tau-phosphatases, particularly protein phosphatase 2A (Planel et al., 2004). As detailed above, tau-protein is posttranslationally modified by O-linked *N*-acetylglucosamine cylation (for review Hart, 1997). This process is triggered by insulin–glucose. Starved mice showing low glucose uptake/ metabolism in the brain produced a decrease in O-GlycNAcylation accompanied by tau-protein hyperphosphorylation at the majority of the phosphorylation sites (Liu et al., 2004). These data may indicate that the metabolism of both APP and tau-protein are regulated by the insulin-controlled glucose metabolism.

3 The Effects of Aging on Cerebral Metabolism

Both, PD and sporadic AD have been demonstrated to be age-related neurodegenerative disorders. Numerically, the latter condition is the predominating brain disorder in old age. The prevalence of sporadic AD increases from 0.5% at the age of 60 years to nearly 50% at 85 years and older (Evans et al., 1989; Ott et al., 1995). These studies show an exponential increase of sporadic AD from 65 to about 85 years. However, thereafter, the prevalence curve flattens and approaches to around 75% cognitive impairment in centenarians (Baltes et al., 1999; Kliegel et al., 2004). The finding that obviously not more than 50% of people aged 85 years are affected by cognitive impairment, whereas 50% are not, and that these numbers change to 75% and 25%, respectively, in centenarians point to additional factors besides aging to generate sporadic AD.

Numerous studies have clearly shown multiple minor changes of biological parameters during the aging process of the human brain up to the seventh to the eighth decade of life. Beyond this threshold, more marked changes are found at the circulatory, cellular, and molecular levels (for review Hoyer, 1986, 1995; Hoyer and Plaschke, 2004). Out of the multitude of age-related changes, some of them should be emphasized in the context of this chapter.

3.1 Glucose and Energy Metabolism

The age-related variations in the regulation system of neuronal glucose metabolism may cause a reduction of the glycolytic glucose breakdown from young adulthood to senescence, whereas only slight variations have been found in the oxidative processes of the tricarboxylic acid cycle and the respiratory chain with aging (Hoyer, 1995; for review Hoyer and Plaschke, 2004). As consequences of the reduced capacity of the glycolytic glucose breakdown, pyruvate production decreased causing (1) increased lactate formation that changes both the cytoplasmic redox potential and intracellular pH in terms of an acidic shift (Hoyer and Krier, 1986; Roberts and Sick, 1996) and (2) the formation of acetyl-CoA, leading to the reduction of both the synthesis of acetylcholine to 65% in senescence and the acetylcholine release to 25% compared to young adulthood (Gibson et al., 1981; Bowen, 1984) associated with a decrease in the number of muscarinic receptors by 22% (Lippa et al., 1981). These age-related changes may indicate a diminished function of the cholinergic system (Bigl et al., 1987). Moreover, the reduced formation of acetyl-CoA may reduce the formation of both cholesterol and neurosteroids (Rupprecht and Holsboer, 1999), which may decrease the brain function in general. The decreased function of the cholinergic system may lead to a shift in the balance between acetylcholinergic and noradrenergic innervation of the microvasculature and of basic biochemical processes in terms of an increase in the sympathetic tone based on an increase of the noradrenaline concentration in cerebral cortex with aging and prolongation of the release of noradrenaline after stress (Ida et al., 1982; Harik and McCracken, 1986; Perego et al., 1993).

Another close functional relationship exists between the acetylcholinergic and glutamatergic system, which is most widely distributed in the brain (Walaas and Fonnum, 1980; Davies et al., 1984; Cotman et al., 1987; Gasic and Heinemann, 1991). With aging, the neuronal concentration of glutamate is reduced (Mullamy et al., 1996). Otherwise, glutamate binding increased progressively to hippocampal-binding sites but not because of receptor affinity for glutamate (Baudry et al., 1981). In this respect, differences between the acetylcholinergic and glutamatergic systems become obvious.

When glucose consumption and energy formation of the aging brain are considered, it has to be noted that in physically and mentally healthy senescent people, a 23% reduction in glucose consumption was found without any change in cerebral oxygen utilization compared with healthy younger people (for review Dastur, 1985; Hoyer, 1986). When ATP formation is calculated on the basis of these data, a slight decrease by around 7% becomes obvious (Hoyer, 1992). This also holds true for experimental animals: the availability of energy was found to be reduced by 5% in parietotemporal cerebral cortex of 104-week-old rats compared to 52-week-old animals under resting conditions, but decreased by 15% in very old (130 weeks of age) rats in general (Hoyer, 1985; Dutschke et al., 1994). However, when rats were discriminated in behaviorally well and poorly performing animals, some differences became obvious. In good performers, creatine phosphate and total energy pool fell by 9%, in poor performers, ATP, creatine phosphate, and total energy pool decreased by 9% too, in both parietotemporal cerebral cortex and hippocampus under resting conditions from 52 to 104 weeks of age. Long-term mental activation increased ATP formation by around 12% in both good and poor performers in partietotemporal cerebral cortex with age. However, whereas energy turnover nearly doubled in good performers in both parietotemporal cerebral cortex and hippocampus, changes of energy turnover could not be found in cerebral cortex and to a small extent only in hippocampus in poor performers (Hoyer et al., 2004). This deficit in energy utilization under stressful conditions in old age may be assumed to be due to the perturbed activity of the ATP/ADP translocase, hindering the extrusion of ATP from the mitochondria into the cytoplasm (Nohl and Krämer, 1980; Yan and Sohal, 1998). Thus, an imbalance between energy formation and the capacity to use it becomes obvious in old age in conditions prone to cognitive impairment.

ATP-consuming nonmitochondrial ATPases in synaptic plasma membranes and synaptic vesicles are assumed to play a fundamental role in the regulation of presynaptic nerve-ending homeostasis and postsynaptic activities (Schwartz et al., 1975; Kometiani et al., 1984). In hippocampal synaptic plasma membranes, no age-related changes were found in Na^+, K^+-ATPase activity, but specific Mg^{2+}-ATPase decreased with aging (Villa et al., 2002) pointing to a varied modulation of the turnover of neurotransmitters (Wyllie and Gilbert, 1979).

During aging, the brain may be compromised more frequently than in adulthood by external abnormal conditions such as arterial hypotension, arterial hypoglycemia, and arterial hypoxemia. These additional stressful events may be assumed to aggravate the changes normally occurring in cerebral oxidative/energy metabolism and related metabolism. In profound arterial hypoxemia of short duration only, changes in cerebral glucose metabolism were more severe in aged than in adult animals, pointing to insufficient compensation mechanisms during aging (Degrell et al., 1993) and to increased vulnerability to anoxia (Roberts and Chih, 1995). Recovery of the energy pool in cerebral cortex after arterial hypoglycemia was found to be more markedly compromised in aged than in adult animals, as was the restoration of the cerebral amino acid pool. As a marker of the severity of the damage, the concentration of ammonia more than doubled in aged brains when compared to adult ones (Benzi et al., 1984). Cerebral ischemia is metabolically characterized by changes of adaptation in glycolytic glucose breakdown, an anaplerotic reaction in the TCAC, and a more pronounced fall and loss in the adenosine nucleotide pool in aged as compared to adult animals (Hoyer and Krier, 1986). In the postischemic recirculation period, the delayed decrease in energy-rich phosphates was more severe in aged than in young adult animals and was more pronounced in hippocampus as compared to cerebral cortex (Hoyer and Betz, 1988).

3.2   Insulin and Insulin Receptors

The neuronal insulin signal transduction system has been found to undergo reduction with aging. The concentration of insulin, density of insulin receptors, and the activity of the tyrosine kinase have been shown to fall beyond the age of 60 years (Frölich et al., 1998). In cerebral cortex of rodents, an age-related reduction of insulin receptor number and insulin receptor mRNA has been demonstrated (Zaia and Piantanelli, 2000). On glucose stimulation, the insulin concentration in dialysate of the hypothalamus of aged rats was half the level of young animals (Gerozissis et al., 2001).

3.3 Glucocorticoids and HPA-Axis

As has been pointed out above, cortisol is an important inhibitor of the function of the insulin receptor. Clear evidence exists that a higher basal cortisol concentration becomes effective with aging (Oxenkrug et al., 1983; Halbreich et al., 1984; Lupien et al., 1994). The increased cortisol concentration in plasma has been found to be mirrored in the CSF of healthy elderly people who showed enhanced CSF cortisol levels as compared to young subjects (Swaab et al., 1994).

In parallel, the number of corticosteroid receptors mainly those ones of type I nearly exclusively located on neurons of limbic structures has been demonstrated to be selectively reduced with aging in the hippocampus (Reul et al., 1988; Lorens et al., 1990). In contrast, no change or even an increase was found in type II receptors that are widely distributed throughout the brain and are mainly activated under stress conditions (Reul et al., 1991). Glucocorticoids have been assumed to be toxic to hippocampal neurons (Sapolsky et al., 1986; Chrousos and Gold, 1992), which may result in functional/structural changes in the hippocampus and may, thus, lead to a reduction of its major inhibitory control on cortisol secretion (Sapolsky et al., 1983, 1991). This decrease in the inhibitory function of the hippocampus may be assumed to disinhibit the HPA-axis (Cizza et al., 1994) resulting in an increased and prolonged response to stress (Sapolsky et al., 1986; de Kloet, 1991; Cizza et al., 1995; Sapolsky, 1999). In parallel, the basal concentration of ACTH has been demonstrated to be increased (Reul et al., 1991; Hauger et al., 1994). Social conditions have been found to affect both cortisol and ACTH plasma levels in that the response to CRF was demonstrated to be higher in younger than in old monkeys (Gust et al., 2000). From these variations, it may be deduced that the basal tone of the HPA-axis increases with aging leading to hypercortisolemia in a cascade-like manner that finally may compromise the function of the neuronal insulin receptor by a dysregulation of its phosphorylation of the tyrosine residues (Giorgino et al., 1993). Thus, a marked imbalance between insulin function and cortisol effects exists in the aging brain.

3.4 Neuroinflammation

As pointed out above, both the HPA-axis and the sympathetic system were found to be activated with aging. Glucocorticoids increase the expression of genes in astrocytes and microglia with aging (Nichols, 1999). In such activated cells, the production of inflammatory markers is increased (Eikelenboom et al., 1994). In general, in plasma of healthy elderly, the cytokines interleukin-6, interleukin-1β, and TNF-α were (most consistently) found to be increased (Fagiolo et al., 1993; Maes et al., 1999). Both interleukin-1 and TNF-α cause an activation of nuclear factor-kappa B (NF-κB) (Lieb et al., 1996), which has been proposed to play a pivotal role in survival or death of neurons in a stimulus- and activation kinetic-dependent manner (Lipton, 1997; Chiarugi, 2002).

Independent from the aging process, an increase of noradrenaline was associated with an induction of NF-κB-dependent gene expression (Bierhaus et al., 2003) and an increase in interleukin-6 in CSF (Stover et al., 2003). From the above data, it may tentatively be concluded that there is an increased inflammatory capacity with aging induced by both glucocorticoids and noradrenaline.

3.5 Free Radicals and Mixed Function Oxidation

One out of other hypotheses of aging is based on the increase of free radicals or reactive oxygen species (Leibovitz and Siegel, 1980; Harman, 1981) due to an imbalance between an increased formation and reduced detoxification capacity. Fe^{2+} contributes to the generation of the highly toxic hydroxyl radical •OH in two different reactions (1) the Fenton-reaction and (2) the Haber–Weiss reaction. It is assumed that, during aging, the normally stable redox state of iron as Fe^{3+} is transferred into the soluble Fe^{2+}-chelate. As a consequence, the increasingly formed superoxide radical H_2O_2 is available for the Fenton reaction and Haber–Weiss reaction so that high levels of the hydroxyl radical •OH are formed. Superoxide radicals were first found on plasma membranes in the brain leading to lipid peroxidation and damage of membrane fluidity (Sawada et al., 1992).

Mixed function oxidation, also termed metal-catalized oxidation, is related to glucose/energy metabolism and possesses a great damaging potency. Continuous intracellular protein turnover includes oxidative inactivation by various enzymatic and nonenzymatic mixed function oxidation processes mediated by free radicals and preceding proteolysis (Levine et al., 1981; Stadtman and Oliver, 1991; Stadtman, 1992). Thus, mixed function oxidation, i.e., the amount of oxidized proteins has been demonstrated to increase in the brain cortex during normal aging (Carney et al., 1991; Smith et al., 1991). The oxidation of amino acids to carbonyl derivatives also mediated by the formation of free radicals is synergistically affected by nucleoside phosphates and triphosphates and by bicarbonate (Stadtman and Berlett, 1991). The degradation of proteins has been found to be ATP dependent. Thus, the aged brain may be characterized by a disproportion between enhanced formation of oxidized, damaged proteins and the cellular capacity to degrade them.

3.6   Advanced Glycation End Products

In long-lasting metabolic processes, glucose and other sugars are able to glycolyse proteins irreversibly by nonenzymatic reactions to finally generate advanced glycation end products (AGEs) (Vlassara et al., 1994). These posttranslational modifications were found to be prominent during aging (Monnier and Cerami, 1981). AGEs show a high affinity to its receptor (RAGE). Receptor binding is followed by an expression of both growth factors and transcription factors as well as by changes of membrane permeability (Vlassara et al., 1985, 1994). RAGEs have been demonstrated to be present in the CNS (Brett et al., 1993). Interestingly, βA4 has been shown to bind to RAGE (Yan et al., 1996). With aging, AGEs have been found to accumulate in the perikaryon of large neurons in the hippocampus and the dentate gyrus, and in pyramidal cells of the cortical layers III, V, and VI (Li et al., 1995). The cell layers have been demonstrated to be particularly prone to degeneration in sporadic Alzheimer's disease.

3.7   Cell Cycle Activity in the Brain and its Relevance to Neurodegeneration

There is recent evidence that insulin may induce mitogenesis throughout two independent pathways acting in synergy: PI-3 kinase/p70 S6 kinase and p43/p44-MAP kinase (Virkamäki et al., 1999), accompanied by an increase in the proportion of proliferating cells in S and G2/M phases of the cell cycle (Conejo and Lorenzo, 2001). Lacking insulin as well as the inhibition of both the PI-3 kinase and the MAP kinase activities have been demonstrated to result in a reduced incorporation of (^{3}H) thymidine, initiation of apoptosis obviously due to NF-κB, and of growth arrest in the G0/G1 phases of the cell cycle with a high proportion (90%) of quiescent cells (Conejo and Lorenzo, 2001; Conejo et al., 2001). An abnormality in the insulin signal transduction cascade by a deficit of the IRS-1 has been found to result in the inability of cells to increase DNA synthesis and to enter into the S/G2/M phases of the cell cycle (Valverde et al., 2001). The clear age-associated fall of both insulin concentration and tyrosine kinase activity of the insulin receptor (Frölich et al., 1998) may be assumed to reduce the action of IRS-1, thus inhibiting the S/G2/M phases and stimulating the G0/G1 phases of the cell cycle (Hoyer, 2000a).

In AD, a hypothesis has been proposed that entry of postmitotic neurons into the cell cycle, followed by an uncoordinated progression through the cycle, leads to degeneration of neurons in diseases such as AD. Disturbances of neuronal plasticity, i.e., processes driven by neuronal activity, in AD might cause neurons to enter the cell cycle (Arendt, 2003). Additionally, neurons under metabolic stress, e.g., due to the functional consequences induced by impaired insulin-signaling mechanisms may be more vulnerable to entering the cell cycle. Furthermore, APP interacts with the cell cycle protein APP-binding protein 1 (APP-BP1), and activation of APP-BP 1 can cause neurons to enter the cell cycle (Chen et al., 2004).

3.8 Gene Expression Profile

With respect to gene expression with aging, there is now first evidence that the aging process of the brain is accompanied by variations in the gene expression profile. Some genes were switched off whereas other ones are switched on or upregulated eventually in a region-specific manner (Salehi et al., 1996; Wu and Lee, 1997; Hung et al., 2000). The balance between genes expressing proteins working protectively and those accelerating atrophy may be assumed to be disturbed (Whittemore et al., 1986; Parhad et al., 1995). More specific studies showed that the gene expression of some ATPases and of proteins working in synaptic transmission was reduced two- to tenfold in the cerebral cortex and the hypothalamus (Jiang et al., 2001), which may have impacts on transmembrane ion fluxes as, e.g., in the ER and the GA (Seksek et al., 1995; Verde et al., 1995). In the hippocampus, age-associated changes in gene expression were described to be related to proteins working in energy metabolism and in cell cycle regulation (Cho et al., 2002). However, exposure to an enriched environment was found to oppositely regulate gene expression as compared to aging at rest (Jiang et al., 2001). Another novel aspect has been reported recently. A stressful event originated to offsprings by maternal separation at postnatal day 9 was found to reduce the expression of the brain-derived growth factor and NMDA receptor subunits in rat hippocampus at postnatal day 72 (Roceri et al., 2002). These data may indicate that perinatal stress to the fetus/offspring may influence or even determine the quality of the gene expression profile with aging (Holness et al., 2000).

3.9 Synopsis: Aging as the Main Risk Factor for Neurodegeneration

With aging, a multitude of inherent variations in fundamental metabolic processes mainly in glucose/energy metabolism and its control are set into motion at the cellular, molecular, and genetic levels, which result from functional imbalances of regulative systems such as:

– Insulin action (reduced) versus cortisol action (increased, i.e., increased tone of the HPA-axis)
– Energy production (reduced) versus energy turnover (increased)
– Acetylcholine action (reduced) versus noradrenaline action (increased), indicating an increased sympathetic tone
– Formation of reactive oxygen species and oxidized proteins (increased) versus capacity of their degradation (reduced)
– Shift in the gene expression profile from anabolic site (reduced) to catabolic site (increased)
– Inappropriate increase of inflammation processes

All these variations/shifts may indicate an uncoupling of synchronization that has been demonstrated to exist in biological systems (Mirollo and Strogatz, 1990). This model may correspond to the increase in entropy, which is an elemental, inherent principle of chemical and biological processes (Hess, 1983, 1990; Prigoine, 1989). In the physical sciences, the term criticality is used to describe a self-organized metalabile steady state (metalabile equilibrium in entropy). Smaller additional internal or external events, even one that is ineffective in itself, may change biological and/or biophysical properties of the aging brain. Such events may shift a system in a stepwise manner from supercriticality to criticality to subcriticality/catastrophic reaction (Bak et al., 1988; Held et al., 1990), i.e., a disease in medical terms.

4 Alzheimer's Disease

Although different in origin, the hereditary early-onset type of AD and the sporadic late-onset type of AD bear some common features of metabolic abnormalities in cerebral glucose/energy metabolism and in its control. Similarities and differences are discussed below.

4.1   The Hereditary Early-Onset Type of Alzheimer's Disease

Former studies on the brain blood flow and oxidative metabolism clearly demonstrated a drastic reduction by around 50% in cerebral glucose utilization whereas both blood flow and oxygen consumption of the brain were found to be unchanged (Hoyer et al., 1988, 1991). In this presenile type of Alzheimer's disease, the reduction of glucose utilization was regionally accentuated in frontoparietotemporal cortical association areas (Mielke et al., 1992). However, there is evidence that this neuroglucopenia is substituted, at least transiently, by the utilization of the brain-derived glucoplastic amino acids, in particular glutamate (Hoyer and Nitsch, 1989).

The latter results in the intraneural formation of both aspartate and neurotoxic ammonia from the aspartate aminotransferase reaction (Hoyer et al., 1990). The expression of the aspartate aminotransferase gene is glucose dependent and is induced by neuroglucopenia (Plee-Gautier et al., 1998). The intraneuronal accumulation of both ammonia and aspartate may be assumed to contribute to cellular and molecular damage: inhibition of mitochondrial dehydrogenases such as α-ketoglutarate dehydrogenase, isocitrate dehydrogenase, and malate dehydrogenase (Lai and Cooper, 1991; for review Hoyer, 1996).

The decrease of cerebral glucose consumption by around 50% would expect a deficit in energy formation when ATP formation is calculated on the basis of the utilization rates of both oxygen and glucose. However, since oxygen not needed for the oxidation of glucose may be used to oxidize endogenous brain substrates such as amino acids (see above) and in all probability fatty acids from membrane phospholipids too, no significant differences in the brain ATP concentration could be found between early-onset AD and controls (Hoyer, 1992).

The genetic abnormalities associated with hereditary early-onset Alzheimer's disease are all characterized by the permanent generation of β-amyloid (1-40) and β-amyloid (1-42) in particular, beginning early in life (Selkoe 1999; Hardy and Selkoe, 2002). Both these derivatives of the APP reduce the binding of insulin to its receptor and receptor autophosphorylation (Xie et al., 2002). The disruption of autophosphorylation by ATP may result in a decrease/lack of receptor tyrosine kinase activity and thus in a failure of postreceptor effects exerted via IRS-1 (Chou et al., 1987). This dysfunction of the insulin signal transduction cascade may cause the drastic fall in cerebral glucose utilization detailed above.

With respect to glycolytic glucose breakdown, mRNA for the brain phosphofructokinase was found to be decreased in β-amyloid plaque-associated neurons in aged transgenic mice overexpressing the Swedish mutation of human APP. In contrast, in reactive astrocytes, mRNA of phosphofructokinase was upregulated resulting in an overall undiminished enzyme activity in the brain cortex and hippocampus. These data may indicate a further impairment of cerebral glucose metabolism as a consequence of long-lasting β-amyloid burden (Bigl et al., 2003), the toxicity of which may be exacerbated by inhibition of glycolysis (Arias et al., 2002). The effect of β-amyloid may be strengthened because of the decreased capacity of the β-amyloid-degrading enzyme neprilysin (Apelt et al., 2003). This zinc metalloproteinase has been found to be the major β-amyloid-degrading enzyme (Iwata et al., 2000; Carson and Turner, 2002). Although upregulated in reactive astrocytes, it may not hinder β-amyloid deposition.

Reactive gliosis may be assumed to be a rather uniform response to neuronal damage. It may, thus, be of pathophysiological significance that β-amyloid induced the glial expression of both pro- and anti-inflammatory cytokines in the brain cortex of mice carrying the Swedish APP mutation (Apelt and Schliebs, 2001). As was pointed out above, inflammatory cytokines involving nuclear factor-κB have been demonstrated to play a pivotal role in apoptotic cell processes.

Besides mutations in the APP gene, the PS genes are genetically modified in early-onset Alzheimer's disease. As was detailed above, GSK complexes with PS, which is involved in both the controlled intracellular metabolism of APP and the controlled phosphorylation of tau-protein. Although not shown as yet, the mutated PS may be assumed to vary the function of the GSK in terms of a mismetabolism of APP generating more β-amyloid and of the hyperphosphorylation of tau-protein. Thus, the mutated genes or its derivatives may cause diverse cellular/molecular abnormalities of different biological systems of pivotal functional significance in early-onset AD. These abnormalities form an important part of the amyloid cascade hypothesis as the scientific basis for early-onset Alzheimer's disease (Hardy and Selkoe, 2002).

4.2   The Sporadic Late-Onset Type of Alzheimer's Disease

4.2.1 Gene Expression Profile

Recent findings clearly demonstrated that insulin gene expression in the hippocampus and hypothalamus was reduced in postmortem AD brains relative to controls. IGF-1 mRNA was found to be five- to sixfold lower in AD hypothalamus and frontal cortex. Insulin receptor mRNA was decreased in cortex, hippocampus, and hypothalamus in the AD brain, as was IFG-1 receptor mRNA in hippocampus and hypothalamus (Steen et al., 2005).

More evidence has been provided for the gene expression profile in general in sporadic AD. Postmortem studies on the human brain tissue revealed upregulation of 31 genes of chronic inflammation, cell adhesion, cell proliferation, and protein synthesis. Conversely, downregulation was found in 87 genes of signal transduction, energy metabolism, stress response, and synaptic vesicle synthesis, and function. Genes associated with early-onset AD, such as APP and the presenilins, were not altered (Loring et al., 2001).

Downregulation of genes was strongest in relation to signal transduction and energy metabolism. About 50% of downregulated genes belong to these categories. In this context, it may be of significance that the GSK gene (see below) and the protein phosphatase 2A catalytic subunit alpha-gene (Gong et al., 2000) are among the downregulated signal transduction genes, and the mitochondrial ADP/ATP translocator gene (Nohl and Krämer, 1980; Adrian et al., 1986) is among the downregulated genes in energy metabolism. It may be also of significance that signal transduction at synapses is the site of the highest energy utilization (Sokoloff, 1981; Kadekaro et al., 1985; Kauppinen and Nicholls, 1986; Erecinska and Silver, 1989; Keller et al., 1997), i.e., a negative synergistic effect may functionally be expected. Most of the clinical symptoms summarized as dementia may be due to the downregulation of genes being active in signal transduction and energy metabolism.

The causation of the downregulation of these categories of genes is unknown as yet. As was detailed above, the contributing factor aging may reduce expression of genes working in synaptic transmission and energy metabolism (Jiang et al., 2001; Cho et al., 2002). Experimentally induced inhibition of the neuronal insulin receptor by intracerebroventricular application of the diabetogenic substance streptozotocin (Nitsch and Hoyer, 1991; Lannert and Hoyer, 1998) resulted in concerted alterations of gene expression in cerebral cortex, striatum, and cerebellum. Downregulation as well as upregulation of gene expression were observed. Downregulated gene categories were growth factors such as insulin-like growth factor receptor and nerve growth factor. Upregulation was present in genes such as potassium channels, GABA receptors, glutamate transporter, glial-derived neurotrophic factor, brain-derived growth factor, and integrin-α-M, pointing to repair/rescue attempts after damage (Grünblatt et al., 2004a; Grünblatt et al., in preparation).

4.2.2 Glucose and Energy Metabolism

Brain Blood Flow and Metabolic Rate of Glucose   Subjects suffering from memory impairment without general cognitive decline (Mini-Mental State Examination 25), which progressed later to probable AD showed a 22% reduction of the cerebral glucose metabolic rate in the posterior cingulate cortex (Minoshima et al., 1997). In longitudinal studies of regional cerebral glucose utilization, early and severe abnormalities were found paralleling the worsening of clinical dementia symptoms. A decline of cerebral glucose utilization was noted in the whole brain but was most pronounced in parietotemporal and frontal association cortices sparing relatively the primary sensorimotor cortex and primary visual cortex (Mielke et al., 1994; Minoshima et al., 1997). The predictive power of changes in the cerebral metabolic rate of glucose was greater than changes in psychometric measures in mild cognitive impairment and conversion in AD (Arnaiz et al., 2001).

The more dementia progresses, the lesser prominent were the local abnormalities in glucose utilization, especially in severe dementia (Foster et al., 1984; Duara et al., 1986; Mielke et al., 1992; Herholz et al., 2002). As in early-onset AD, a disproportion between the cerebral utilization rates of oxygen and glucose became obvious. Cerebral oxygen utilization is less severely diminished than glucose utilization in sporadic late-onset AD. The same holds true for cerebral blood flow (Hoyer et al., 1991; Fukuyama et al., 1994).

Interestingly, this hypometabolism in cerebral cortex is particularly pronounced in structures with both high glucose demands and insulin sensitivity (for review Henneberg and Hoyer, 1995).

Cellular Glucose Metabolism   The diminished cerebral glucose utilization in Alzheimer brain may be mediated by reduced capacities of key enzymes working in glycolytic glucose breakdown and in both the formation and oxidation of acetyl-CoA. There is a fall in enzyme activity of phosphofructokinase in neurons but an increase of both expression and activity in astrocytes (Bigl et al., 1996, 1999, 2000). The capacity of the key dehydrogenating enzyme complexes PDH (Perry et al., 1980; Sorbi et al., 1983) and α-ketoglutarate dehydrogenase (Mastrogiacomo et al., 1993) was found to be reduced.

The reduced PDH activity results in a decreased level of acetyl-CoA, which may have consequences for the synthesis of (1) acetylcholine, (2) cholesterol, and (3) for the formation of ATP.

1.   Decreased acetyl-CoA, together with the diminished activity of choline acetyltransferase, reduces the synthesis of acetylcholine in the presynaptic neuron (Sims et al., 1983a). It is, however, discussed that the cholinergic deficit occurs only late in AD and that in earliest stages even an upregulation of the cholinergic activity takes place in the brain (for review Frölich, 2002) (see also below).
2.   The diminished availability of acetyl-CoA causes a reduced level of cholesterol in the brain tissue membranes and postmortem CSF of patients suffering from AD (Svennerholm and Gottfries, 1994; Mulder et al., 1998; Eckert et al., 2000). Any major damage of membrane integrity through a loss of membrane constituents, such as, e.g., cholesterol, is incompatible with normal cellular function. Severe membrane damage generally leads to cell death (Michikawa and Yanigasawa, 1999; for review Klein, 2000). It is, as yet, unknown whether the synthesis of neurosteroids changes in the Alzheimer brain.
3.   A decisive pathophysiological consequence of the markedly perturbed cerebral glucose metabolism inclusive of the diminished generation of acetyl-CoA and, additionally, a generalized depression of activity of all electron transport chain complexes, in particular cytochrome c oxidase (Parker et al., 1994), is the decrease in ATP production from glucose by around 50% in the beginning of sporadic AD. The oxidative utilization of substrates other than glucose restores ATP formation to 80% of normal (see below), but thereafter ATP levels decrease throughout the course of the disease (Hoyer, 1992). A fall in ATP formation in the sporadic AD brain has also been demonstrated by other investigators (Sims et al., 1983b; Brown et al., 1989). This energy deficit may compromise ATP-dependent processes in a hierarchical manner (Buttgereit and Brand, 1995), including cellular and molecular mechanisms in particular at synapses (Arias et al., 2002) and in the ER and GA (see above). A depletion of cellular ATP prevents the dissociation of chaperone/ protein complexes and, thus, blocks secretion of these proteins (Dorner et al., 1990). Misfolded or malfolded protein complexes may accumulate in the ER and are indicative of "endoplasmic reticulum stress" (Kaufman, 1999; Paschen and Frandsen, 2001). In the Alzheimer brain, abnormal processing of multiple proteins has been demonstrated (Zhang et al., 1989).

Molecular chaperones belong to a highly effective defense system against different kinds of stresses at the cellular level. Chaperones appear to be active in highly specialized cellular compartments such as the synapse and the ER (for review Ohtsuka and Suzuki, 2000). With respect to the latter, glucose-regulated proteins (GRP) are induced in the presence of malfolded proteins (Munro and Pelham, 1986; Kozutsumi et al., 1988; Bole et al., 1989). Besides the GRP78, heat shock protein (HSP)72 located in the cytoplasm may play a significant role (Welch and Suhan, 1985).

In sporadic AD brains, chaperones were found to be expressed differently (Yoo et al., 2001). HSP72, normally expressed at low levels in the brain, increased markedly in sporadic Alzheimer's disease, and was found to be localized in neuritic plaques and neurofibrillary tangles. GRP78 also increased in relatively spared, surviving neurons but was not present in neuritic plaques, neurofibrillary tangles, or damaged neurons (Hamos et al., 1991). The activated expression of both HSP72 in damaged neurons and GRP78 in surviving neurons may be indicative of a response to metabolic abnormalities characteristic for sporadic AD, in particular to disturbances in oxidative glucose/energy metabolism. In this context, the binding of GRP78 to the APP and the reduction of the secretion of both β-amyloid (1-40) and (1-42) (see below) may be of significance (Yang et al., 1998).

Morphologically, the ER/GA shows a clear diminution in sporadic AD compared to normal status, which is assumed to be a consequence of metabolic abnormalities (for review Salehi and Swaab, 1999).

4.2.3 Glucose-Related Metabolism

The above-mentioned imbalance between (less) diminished cerebral oxygen consumption and (more markedly) reduced glucose consumption may indicate that substrates other than glucose are oxidized to form energy. In all probability, glucoplastic amino acids are, at least transiently, utilized as they are in hereditary early-onset AD (see above). In early/middle stages of sporadic AD, an increase of aspartic acid was found in cerebral cortex associated with a reduction in glutamic acid concentration. In the late stage of the disease, no elevation of aspartic acid was found, and the glutamic acid content was reduced by much less compared to early stage disease (Procter et al., 1988). However, there is clear evidence that fatty acids from membranes may be used to compensate for the glucose deficit for longer. The major phospholipids phosphatidylcholine and phosphatidylethanolamine were found to be reduced by 15%–20% whereas the phospholipids catabolites glycerophosphocholine and glycerophosphoethanolamine increased by 60%–80% (Blusztajn et al., 1990; Nitsch et al., 1992b; Pettegrew et al., 1995). This process may be facilitated by ATP depletion favoring the degradation of membrane phospholipids (Sun et al., 1993).

4.2.4 Insulin and Insulin Receptor Signaling

In contrast to the hypoglycemia-induced metabolic alterations in brain glucose metabolism, the disturbances in cerebral glucose utilization found in patients suffering from AD are accompanied by normoglycemia although a slight but significant fall in arterial glucose concentration was observed in AD. However, the latter variation may be of insignificant functional meaning. Likewise, the plasma insulin concentration was enhanced, and insulin-mediated glucose disposal was diminished (Bucht et al., 1983; Craft et al., 1998, 1999). Interestingly, a decreased peripheral glucose regulation was associated with decreased general cognitive performance, memory deficits short of dementia, and atrophy of the hippocampus (Convit et al., 2003). The general alteration in carbohydrate metabolism in sporadic AD becomes also obvious by an increase of the IGF-1 and -2 in serum and of IGF-2 and insulin-like growth factor binding protein in CSF (Tham et al., 1993).

The abnormalities in neuronal glucose metabolism and in glucose-related metabolism are assumed to be caused by a disturbance in the control of glucose utilization at the level of neuronal insulin signal transduction. Although insulin concentration and the activity of insulin receptor tyrosine kinase are diminished in brain tissue, and CSF insulin concentration is also decreased in AD (Craft et al., 1998), these changes are not significantly different from age-matched healthy adults older than 60 years. Thus, these abnormalities may be age-related rather than disease-related. However, insulin receptor density is upregulated in AD (Frölich et al., 1998), indicating an impairment of the insulin signal transduction cascade similar to that seen in non-insulin-dependent diabetes mellitus. So, the hypothesis was forwarded that sporadic AD is the brain equivalent of non-insulin-dependent diabetes mellitus (Hoyer, 1998).

The dysfunction of the neuronal insulin receptor may be assumed to have drastic impacts on systems regulated by the neuronal insulin signal transduction cascade (see above). Besides enzymes working in neuronal glucose metabolism found to be reduced (see above), the preeminent metabolic system phosphatidylinositol-3 kinase/protein kinase B/GSK-3-cascade may be changed in a deleterious way: the neuronal insulin receptor desensitization may result in a reduced activation of phosphatidylinositol-3 kinase, which may have different effects on the metabolism of the APP and of the tau-protein (see below).

In general, the diminished activity of phosphatidylinositol-3 kinase causes a decrease in the activity of protein kinase B (Alessi and Cohen, 1998; Vanhaesebroeck and Alessi, 2000). As a consequence, the activities of both GSK-3α and GSK-3β may be disinhibited by reduced phosphorylation at serine 21 (GSK-3α) and at serine 9 (GSK-3β). The function of the GSKs may also be compromised by PS 1 and 2, which complex with GSKs and whose mRNAs have been shown to be downregulated in the sporadic AD brain. However, an upregulation of PS 2-gene mRNA was found in astrocytes adjacent to senile plaques (McMillan et al., 1996, 2000; Takami et al., 1997). In vitro inhibition of energy metabolism downregulated the PS 2-gene expression through a modification of its promoter activity (Ghidoni et al., 2003).

Mechanisms of Desensitization of the Neuronal Insulin Receptor As was pointed out above, with aging, the basal tone of the HPA-axis increases leading to hypercortisolemia in a cascade-like manner. Likewise, noradrenaline increases indicating an increased sympathetic tone. These age-related variations have been recognized to jeopardize the function of the neuronal insulin receptor. In sporadic AD single nucleotide polymorphism in the 11β-hydroxysteroid dehydrogenase I gene regulating cortisol turnover increased the risk for sporadic AD sixfold (de Quervain et al., 2004). After a 1-min cold pressure test, both cortisol and noradrenaline increased in sporadic AD patients indicating an increased HPA-axis responsiveness hyperactivity and an increased sympathetic tone significantly higher than in age-matched controls (Pascualy et al., 2000). With respect to cortisol response, same holds true when low dose of ACTH was applied (O'Brien et al., 1996). Also, under resting conditions, basal cortisol concentrations in the circulating blood were found to be higher in sporadic AD compared with age-matched controls (Davis et al., 1986; Belanoff et al., 2001), and cortisol was independently associated to dementia (de Bruin et al., 2002). The enhanced concentration of cortisol is mirrored by its increase in CSF in sporadic AD. This increase in CSF cortisol concentration in sporadic AD is much higher than in healthy and middle-age/aged adults (Swaab et al., 1994; Peskind et al., 2001).

With respect to noradrenaline/adrenalin, increased plasma adrenaline concentration was found in sporadic AD compared with young and old healthy subjects. Both resting CSF noradrenalin/adrenalin were also higher in sporadic AD than in older or young subjects, and increased with dementia severity in sporadic AD (Peskind et al., 1995, 1998). The increase of noradrenaline in the CSF in sporadic AD may be due to an increased activity of surviving locus ceruleus neurons, which is assumed to act as a compensatory mechanism because of the disease-related loss of neurons in the locus ceruleus (Hoogendijk et al., 1999). The increase in noradrenaline in the sporadic AD brain may cause an elevation of the cAMP concentration as was found in CSF of sporadic AD patients, indicating that the cAMP messenger system is upregulated (Martinez et al., 1999). Cyclic AMP has been shown to be involved in the α-secretase cleavage of APP and the secretion of its derivative APPs. This process is inhibited by cAMP participating in the intracellular accumulation of APPs (Efthimiopoulos et al., 1996).

Another molecule having been shown to inhibit the function of the insulin receptor and its substrates is the TNF-α (Feinstein et al., 1993; Rui et al., 2001). Normal aging was accompanied by a marked increase in serum TNF-α whereas no changes were detected in patients suffering from sporadic AD (Maes et al., 1999). In contrast, TNF-α was found to be 25-fold higher in CSF of sporadic AD compared with controls. However, the role of TNF-α has not been clearly defined in pathophysiological terms since in vitro studies indicated neuroprotective effects of this molecule (Tarkowski et al., 1999).

The data presented may give rise to the assumption that the age-associated increases of both cortisol and noradrenaline are exacerbated by an as yet unknown "disease-related factor" that contributes markedly to the desensitization of the neuronal insulin receptor in sporadic AD.

4.2.5 Tau-Protein Hyperphosphorylation

In the context discussed here, tau-protein hyperphosphorylation may start with the disinhibition of GSK-3β (Hanger et al., 1992; Cross et al., 1995, 1997; Hong and Lee, 1997; Lesort et al., 1999). However, tau hyperphosphorylation has been found to be a complex mechanism. Besides the phosphorylating capacity of the dephoshorylated GSK-3α and GSK-3β enzyme complexes, the dephosphorylating capacity of the phosphatase 2A would have to be considered. As mentioned above, starvation-induced hypoglycemia causes hyperphosphorylation of tau-protein. Surprisingly, plausible tau kinases such as GSK-3β were inhibited, but the activity of phosphatase 2A decreased under hypoglycemic conditions (Planel et al., 2001). In brain slices treated with okadaic acid plus calyculin A protein phosphatase 2A was inhibited by 100% but GSK-3 by 80% only, leading to hyperphosphorylation of tau-protein at serines 198, 199, 202, 396, 404, and 422. This may point to a hierarchic order between the GSKs and protein phosphatase 2A in that the latter enzyme is more susceptible to damage such as disturbances in glucose metabolism in the human brain.

Besides phosphorylation by the dephosphorylated, activated GSKs, the stress-activated protein kinase/ c-Jun N-terminal kinase (SAPK/JNK) phosphorylates tau-protein at threonine 205 and at serine 422, which are highly phosphorylated in the sporadic AD brain (Reynolds et al., 1997; Ferrer et al., 2001). Recently, evidence has been provided that the AMPA-induced induction of SAPK/JNK is inhibited by insulin possibly by activating protein kinase B and blocking SAPK/JNK (Kim and Han, 2005).

Protein phosphatase 2A exists as heterotrimer containing scaffolding, regulatory, and catalytic subunits. The latter undergoes carboxyl methylation catalized by a specific methyltransferase (Lee and Stock, 1993) abundantly found in neurons. In the sporadic AD brain, a loss of protein phosphatase 2A methyltransferase was found in hippocampal areas paralleling the severity of tau pathology but not of amyloid plaque burden (Sontag et al., 2004a). The activity of of protein phosphatase 2A was markedly diminished in the sporadic Alzheimer's disease brain also correlating with tau pathology but not with amyloid burden (Sontag et al., 2004b). At the mRNA level, protein phosphatase 2A expression was significantly reduced in hippocampus of sporadic Alzheimer's disease brains (Vogelsberg-Ragaglia et al., 2001; see also Loring et al., 2001).

Apart from AD brain, it has been shown that the protein phosphatase 2A mRNA expression was increased whereas the protein expression and enzymatic function were defective in Alzheimer's disease fibroblasts (Zhao et al., 2003). The latter finding is in agreement with the reduced activity of protein phosphatase 2A in the AD brain (Sontag et al., 2004b). Once formed, hyperphosphorylated tau-protein is the core of neurofibrillary tangles, which assemble in paired helical filaments stored in neurons. The generation of these abnormal proteins in the sporadic Alzheimer brain may be assumed to be largely a consequence of glucose mismetabolism in the sporadic Alzheimer brain.

4.2.6 Expression of Amyloid Precursor Protein

Damage to the cholinergic system plays a pivotal role in sporadic AD and in the molecular response of the APP. To start, cholinergic neurons have been shown to be selectively vulnerable to alterations such as aging (for review Bowen, 1984), and alterations in the choline metabolism have been considered to be a basis for the selective vulnerability of presynaptic cholinergic neurons (Wurtman, 1992). In the sporadic AD brain, the synthesis of acetylcholine in the presynaptic neuron is markedly reduced (Sims et al., 1983). It is noteworthy that the degeneration of the cholinergic system correlates with the progression of mental disturbances in patients with AD (Baskin et al., 1999).

In animal studies, a rapid, and in most cases long-lasting (up to 2 months) increase of APP mRNA in cholinergic projection areas, such as hippocampus and cerebral cortex, was achieved by cholinergic antagonists and cholinotoxic lesions accompanied by an increased synthesis of APP (Wallace et al., 1993; Beach et al., 1996; Bigl and Schliebs, 1998; Leanza, 1998; Lin et al., 1998). Selective immunolesions of the basal forebrain cholinergic projections induced an increase of APP in hippocampus and cerebral cortex 8 months postlesion (Ramirez et al., 2001). These findings clearly indicated that the enhanced expression of the APP is not a primary event but subsequent to preceding metabolic lesions.

The expression of APP mRNA and APP protein in the AD brain has been intensely studied. However, the findings are inconclusive and contradictory. Reductions as well as increases of APP mRNA have been reported (Johnson et al., 1988; Palmert et al., 1988; see also Loring et al., 2001). Same holds true for APP protein: no differences between control and sporadic AD brain, and increased APP (Arai et al., 1991; Nordstedt et al., 1991). A recent study on APP, PS 1, and rab 3a, a specific synaptic vesicle protein, showed significantly reduced levels of the 3 proteins in association cortex and in hippocampus of sporadic AD brains (Davidsson et al., 2001). Independent of whether APP protein concentration is altered in the sporadic Alzheimer's disease brain, abnormalities in the phosphatidylinositol-3 kinase/protein kinase B/ GSK-3-cascade exert different effects on the metabolism of the APP. The diminished/inhibited activity of phosphatidylinositol-3 kinase alters the metabolism of the APP in that the release of both secreted APP and β-amyloid from the cell is hampered, and full-length APP accumulates in the cell (Petanceska and Gandy, 1999), i.e., the phosphatidylinositol-3 kinase-stimulated release of both secreted APP and β-amyloid (Solano et al., 2000) is compromised. GSK-3α may be disinhibited by reduced phosphorylation at serine 21 leading to mismetabolism of the APP and the formation of β-amyloid (1-40) and (1-42) (Phiel et al., 2003). The intracellular accumulation of both β-amyloid (1-40) and (1-42) is reduced by accelerating the APP/β-amyloid transport from the *trans*-Golgi network to the plasma membrane assisted by mitogen-activated protein kinase signaling. This process is regulated by insulin and insulin receptor tyrosine kinase

(Gasparini et al., 2001). If, as in sporadic AD, the function of the neuronal insulin/insulin receptor signal transduction is perturbed, both secreted APP and β-amyloid may be expected to be retained in the cell, in particular, in the ER that is assumed to be a core event in the pathophysiology of sporadic AD (Greenfield et al., 1999; Hartmann, 1999; Wilson et al., 1999). Surprisingly, inhibition of the protein phosphatases 1 and 2A reduced β-amyloid production by 50%–80% in in vitro studies indicating that phosphorylation of the APP may inhibit the generation of β-amyloid (Buxbaum et al., 1993) and may facilitate the generation of secreted APP (Gillespie et al., 1992). Also, the release of derivatives of the APP secreted APP and β-amyloid from the cell is stimulated by the acetylcholinergic m1 and m3 receptors (Nitsch et al., 1992). Reduced availability of acetylcholine, and reduced phosphorylation of APP, e.g., due to an ATP-deficit may increase the formation of β-amyloid and, thus, the accumulation of both secreted APP and β-amyloid in the cell. Such an accumulation may contribute to "endoplasmic reticulum stress," and may damage cellular function (Kaufman, 1999; Paschen and Frandsen, 2001; see also above).

Within the cell, β-amyloid may be polymerized into insoluble aggregates to be stored in lysosomes (Yang et al., 1999) accumulating with time (Skovronsky et al., 1998). In the sporadic Alzheimer brain, evidence for β-amyloid accumulation in neurons was provided (Gouras et al., 2000). β-Amyloid (1-42) first accumulates in the perikaryon of pyramidal cells presumably in lysosomes. This process was found to be accompanied by lysis of the β-amyloid-burdened neurons that may be the source of (extracellular) neuritic plaques (D'Andrea et al., 2001). Lysed cells may activate microglial cells that can trigger astrocystes. Both cell types can secrete compounds toxic to surrounding neurons (D'Andrea et al., 2004). These assumed neurotoxic factors may contribute to the vicious cascade of cell damage and cell death besides other factors such as hypometabolism, synaptic dysfunction, or hyperphosphorylation of tau-protein.

4.2.7 Degradation of Amyloid Precursor Protein and β-Amyloid

Amyloid Precursor Protein   Physiological metabolism of the APP starts with the cleavage of this holoprotein by α-secretase between lysine 687 and leucine 688 yielding the soluble secreted APP and the C-terminal fragment, which is proteolytically cleaved at valine 711/isoleucine 713 by γ-secretase. No amyloidogenic derivatives are formed by these to pathways except a small amount of the amyloidogenic β-amyloid generated by APP-cleavage at methionine 671 by β-secretase. This cleavage results in the formation of a fragment with a molecular weight of 12,000, which is proteolytically degraded by γ-secretase in the ER/intermediate compartment [β-amyloid (1-42)] and in the GA [β-amyloid (1-40)] (Cook et al., 1997; Hartmann et al., 1997) in picomolar/nanomolar concentrations found in plasma and CSF (Haass et al., 1992; Seubert et al., 1992). Normally, the interaction among these different secretases seems to be well balanced.

Although in sporadic AD brain, APP mRNA in all probability is reduced in brain cortex (Johnson et al., 1988; Davidsson et al., 2001), different factors may be assumed to contribute to the enhanced formation and intracellular accumulation of β-amyloid peptides in the ER/GA. The diminution in size of the latter may indicate a reduced metabolic activity (for review Salehi and Swaabm, 1999; Dubelaar et al., 2004). The reduced ATP availability may damage the function of the ER/GA, leading to a slowing down/inhibition of the passage in these compartments (Dorner et al., 1990; Seksek et al., 1995; Verde et al., 1995; Wild-Bode et al., 1997; Demaurex et al., 1998). Both, the reduced activity of the cholinergic system (Nitsch et al., 1992; Rossner et al., 1997) and of the insulin signal transduction system (Solano et al., 2000; Gasparini et al., 2001) may retain APP and its derivatives in the cell. Also, intracellular APP increased, and both extracellular secreted APP and β-amyloid decrease due to a moderate energy deficit. However, severe abnormalities in energy formation/utilization were accompanied by a drastic fall in the intracellular APP by at leas 50%, paralleled by an accelerating reduction in the extracellular concentrations of both secreted APP and β-amyloid (Hoyer et al., 2005). The abnormal cleavage of APP may be favored by the decreased activity of α-secretase (Lannfelt et al., 1995) and by the 2.7-fold increase of β-secretase expression (Holsinger et al., 2002), indicating the imbalance of secretases in the sporadic AD brain.

The β-amyloid peptide fragment 25–25 has been demonstrated to be neurotoxic particularly during inhibition of both glycolytic glucose breakdown and mitochondrial ATP production. Synapses may be particularly sensitive to metabolic dysfunction pointing to the interaction of the latter and β-amyloid accumulation (Arias et al., 2002). Also, the intracellular accumulation of β-amyloid (1-42) has been found to damage neurons by induction of apoptosis (Kienlen-Campard et al., 2002).

β-Amyloid Two different enzymes have been demonstrated to degrade β-amyloid: (1) IDE and (2) neprilysin. The zinc metalloprotease IDE has a preferential affinity for insulin because of its K_m of ~0.1 μM whereas the K_m for β-amyloid analogs is >2 μM, i.e., the presence of insulin may not favor IDE-mediated degradation of β-amyloid (see above; Craft et al., 2000). There is some evidence that variants of IDE are associated with sporadic Alzheimer's disease within APOE-negative patients (Edland et al., 2003). In the sporadic AD brain, the β-amyloid-degrading capacity of IDE is about 50% of that of control brains, but degradation of the (reduced) insulin decrease by about 30% only (Perez et al., 2000). Decreased IDE mRNA and IDE activity have been found in the hippocampus of sporadic AD brain. This effect was more pronounced in patients carrying an APOE-4-allele compared to patients without the APOE-4-allele (Cook et al., 2003). However, in neurons adjacent to neuritic plaques, IDE is upregulated (Bernstein et al., 1999). The different changes in the degradation capacity of IDE for β-amyloid and insulin in the sporadic AD brain may explain the increased formation and acculumation of β-amyloid but not the diminished insulin concentration found in postmortem sporadic AD brain and CSF (Craft et al., 1998; Frölich et al., 1998). Although insulin inhibits the degradation of both β-amyloid (1-40) and (1-42), the diminished insulin concentration as was found in the sporadic Alzheimer's disease brain may not be assumed to favor the degradation of β-amyloid by IDE. In contrast, IDE activity is reduced in all probability due to alterations in gene expression and due to the varied (pathological) metabolic state of cell.

The zinc metalloprotease neprilysin has been shown also to degrade β-amyloid (Iwata et al., 2001; Carson and Turner, 2002). It is a 94-kDa plasma membrane glycoprotein and is expressed in the brain in a cell- and tissue-specific manner in the subfields CA1–3 of the hippocampus in the stratum pyramidale and stratum lacunosum-moleculare, in the molecular layer of the dentate gyrus and in layer II/III of the brain cortex (Li et al., 1995; Fukami et al., 2002) besides the caudate nucleus, putamen, olfactory tubercle, and substantia nigra (Waksman et al., 1986). Neprilysin was found to be present at presynaptic terminals and axons. With age, brain cortical neprilysin mRNA decreased significantly, but, by contrast, neprilysin protein did not and was unchanged during aging. In transgenic Tg 2576 mice, a local upregulation of neprilysin was found in dystrophic neuritis and in reactive astrocytes adjacent to β-amyloid deposits (Apelt et al., 2003). However, in the sporadic AD brain, both neprilysin mRNA and neprilysin protein have been found to be downregulated by approximately 50% as compared with age-matched controls, also in high plaques areas (Akiyama et al., 2001; Yasojima et al., 2001a, b).

In sporadic AD, no increased expression of APP, except in the very early period of the disease, could be detected (Loring et al., 2001; Iwata et al., 2002; see also above). However, several abnormal metabolic processes at the cellular and molecular levels modify APP metabolism and generate β-amyloid in higher concentration than normal. Additionally, the capacity to degrade β-amyloid is downregulated so that the brain is overflown with this derivative.

4.2.8 Neuroinflammation

Alzheimer himself described glial cells with enlarged nuclei and enhanced formation of fibrillary glial cells particularly in deeper layers of the cerebral cortex in dementia occurring late in life (Alzheimer, 1911). Hypertrophic astrocytes, in conjunction with proliferating microglia, are the general response to different kinds of neuronal stress such as aging or severe injury in disease conditions (Kreutzberg, 1996). The morphological and metabolic changes of these glial cells are termed reactive gliosis (Schipper, 1996; Cotrina and Nedergaard, 2002). A laminar gliosis of the layers II, III, and V has been reported in cerebral association cortex of sporadic AD (Beach et al., 1989). Reactive fibrillary astrocytes along with frequently paired astrocytic nuclei have been found in early degeneration preceding neuronal dropout (Brun and Englund, 1981).

As with aging, systemic alterations in plasma concentrations of acute-phase proteins, such as increased levels of interleukin-6 and TNF-α and reduced albumin concentration, have been found (Maes et al., 1999; Schmidt et al., 2002) in sporadic AD, and in elderly subjects at risk of sporadic AD. In the sporadic AD brain, the activation of microglia involves its transformation into phagocytotic cells that are capable to ingest cell debris and also to release cytokines such as interleukin-1, interleukin-6, and TNF-α (Dickson and Rogers, 1992; Dickson et al., 1993). Astrocytes are assumed to play a role in the initial phase of neuroinflammation as they can be stimulated by interleukin-1 and -6 (Das and Potter, 1995) and are capable to secrete the complement proteins C4 and C9 (Walker et al., 1998).

The causation of reactive gliosis in the Alzheimer brain may be assumed to be different between hereditary AD and sporadic AD. In hereditary AD, the permanent and over time increasing production of β-amyloid is capable to activate microglia without the assistance of costimulatory molecules (Lue et al., 2001; Walker et al., 2001; Münch et al., 2001). In contrast, in sporadic AD, the initial stimulatory effect of β-amyloid is lacking. As with aging, elevated concentrations of cortisol and noradrenaline (Hoogendijk et al., 1999; de Quervain et al., 2004) are characteristic for sporadic AD. Cortisol increases the expression of genes in astrocytes and microglia (Nichols, 1999), and once activated, these cells start to increase the production of inflammatory compounds (Eikelenboom et al., 1994). Also noradrenaline caused an increase of interleukin-6 in CSF (Stover et al., 2003). These data may indicate that inflammatory mechanisms are differently involved in the early steps of the pathological cascades of both hereditary and sporadic AD (Eikelenboom and van Gool, 2004).

The excessive neuroinflammation occurring in each of the two pathological conditions contributes to neurodegeneration, which may be assumed to be the result of a concerted action (Aisen and Davis, 1994; Akiyama et al., 2000) in which energy failure, transmission failure, imbalances in ion exchange, and shifts in redox states participate to name only a few out of many other metabolic candidates.

5   Sporadic Alzheimer's Disease Versus Parkinson's Disease Dementia-Related Abnormalities

5.1   Morphological Abnormalities

Clear evidence was provided that the severity of neuropathological changes in sporadic DA, such as cell loss, neurofibrillary tangles, and neuritic plaques, was maximal in the hippocampus, the entorhinal cortex, and cortical association areas. A cell layer-specific accumulation of neurofibrillary tangles and neuritic plaques was found in that neurofibrillary tangles were mainly observed in cerebral cortical layers III and V preferentially in pyramidal cells. Neuritic plaques occurred in all layers throughout the cerebral cortex but with predominance in layers II and III (Pearson et al., 1985). In entorhinal cortex, layers II, III, and IV revealed severe cell loss, and in layers II and III, neurofibrillary tangles accumulated. Neuritic plaques were abundantly found in the middle portion of the stratum moleculare of the dentate gyrus (Hyman et al., 1984, 1986). In hippocampus, a marked cell loss and large number of neurofibrillary tangles were seen in the subiculum and the CA1 subfield affecting pyramidal cells, whereas cells in the adjacent CA3 subfield were rarely affected. Neuritic plaques were mainly found in the dorsal half of the stratum pyramidale of the subiculum and in the stratum radiatum of the adjacent CA1 subfield. In this area, cell loss is extensive, neurofibrillary tangles are found to be enriched in pyramidal cells, and neuritic plaques in the pyramidal cell layer (Hyman et al., 1984, 1986). Interestingly, both densities of insulin receptor and N-methyl-D-aspartate (NMDA) receptor are very high, and APP mRNA is abundantly available. The distribution of neuritic plaques failed to correspond to cholinergic, noradrenergic, and serotonergic afferents to the stratum moleculare of the dentate gyrus and the CA1 hippocampal subfield (Crain and Burger, 1988). In these brain areas mentioned, glutamatergic and acetylcholinergic pyramidal cells are predominantly affected in sporadic AD.

In contrast in the CA3 subfield, the densities of both insulin receptor and NMDA receptor were found to be lower than in CA1 subfield, and cell loss, and the formation of neuritic plaques and neurofibrillary tangles are rather rare or less pronounced than in the CA1 subfield (for review Hoyer, 1988).

In contrast to sporadic AD, in PD, clinical symptoms do not appear until 50% of nigral neurons and 80% of striatal dopamine are lost (Marsden, 1990). The substantia nigra is an anatomically heterogeneous nucleus with regional variations in striatal projections and distribution of histochemical markers. Hassler (1938) was the first to describe neuronal loss being most severe in its caudal and ventrolateral parts. These findings were confirmed in a series of subsequent investigations (Jellinger and Danielczyk, 1968; Andrews et al., 1970; Sharpe et al., 1973; Borit et al., 1975). In more detail, cell loss was greatest in the lateral ventral tier (71%) and the dorsal tier (56%) (Fearnley and Lees, 1991). Dopaminergic neurons appear to be particularly vulnerable and affected to degenerate in PD when they contain melatonin and/or Lewy bodies (Gibb and Lees, 1988; Hirsch et al., 1988).

5.2 Gene Expression

As with AD, PD is not a nosological entity. A minority of all PD cases is caused by mutations on chromosomes 1, 2, 4, 6, and 12, whereas the majority is age-related and sporadic in origin. Interestingly, early-onset cases carrying mutations on chromosomes 1 and 6 are autosomal recessive, whereas cases with later onset carrying mutations on chromosomes 2, 4, and 12 are autosomal dominant (Dekker et al., 2003). A recent study on the expression of genes in substantia nigra pars compacta of sporadic PD patients revealed 68 downregulated and 69 upregulated genes. Downregulation comprised gene classes for signal transduction, protein degradation, dopaminergic transmission/metabolism, ion transport, protein modification/phosphorylation, and energy pathway/glycolysis. Upregulation was found in gene classes of cell adhesion/cytoskeleton, extracellular matrix components, cell cycle, protein modification/phosphorylation, protein, metabolism, transcription, and inflammation/stress (Grünblatt et al., 2004b).

The comparison of the gene expression profiles between sporadic AD (Loring et al., 2001; see above) and sporadic PD shows a large overlap of gene classes being either upregulated or downregulated. Although a detailed analysis is presently not possible, the involvement of different single genes out of the same gene classes, e.g., energy metabolism and signal transduction may indicate a common pathological feature for the induction of cell damage and probably of dementia, too, in the two age-related neurodegenerative disorders.

5.3 Neurotransmitter Changes

In sporadic AD, neuropathological abnormalities were found most in the acetylcholinergic and noradrenergic systems followed by the serotonergic system and the dopaminergic system. The changes in the latter system are of minor quantity only (Moll et al., 1990). There is a strong correlation between the degree of morphological degeneration and the decrease in the concentrations of neurotransmitters: acetylcholinergic > noradrenergic = serotonergic > dopaminergic (Gsell et al., 1996, 2004).

Although James Parkinson himself stated that "... the senses and intellects being uninjured" (Parkinson, 1817), there is no doubt that dementia can be associated with PD. The deficit of dopamine may not explain the occurrence of a dementia in terms of a decline of cognitive and memory capacities. However, in patients suffering from PD plus dementia, more marked dropouts of neurons in the basal nucleus of Meynert were found than in PD only (Jellinger, 1987). It, thus, becomes obvious that the pathological condition PD/dementia involves obviously multiple neurotransmitter systems, at least the dopaminergic and the acetylcholinergic systems. However, the pathophysiology is not fully understood as yet. Due to the obvious differences in pathomorphology and pathobiochemistry between sporadic AD and sporadic PD, differences in the quality of dementia symptoms and its time course may be expected. Sporadic AD starts with the cardinal failure of increasingly developing reductions in mental capacities such as learning, memory, and cognition (cortical dementia). Dementia associated with sporadic PD develops later in the course of existing movements disturbance and is characterized by a progressive dysexecutive syndrome (Bosboom et al., 2004), also classified as subcortical dementia (Albert et al., 1974; Turner et al., 2002), with dementia rates between 17% and 62% (Reid et al., 1996).

5.4 Neurodegeneration

The mode of damage and destruction of dopaminergic neurons is characterized by an excessive iron accumulation in the substantia nigra (Riederer et al., 1989; Youdim et al., 1993; Götz et al., 1994). It has been shown that the intracellular iron concentration is regulated by two proteins, IRP1 and IRP2, and ubiquitation. If IRP2 and/or ubiquitation fails, iron accumulates in the cell (Ciechanover and Brundin,

2003; Youdim, 2003). Age-related changes, such as shifts in the cellular redox-state to acidosis (see above), may transfer iron from Fe^{3+} and Fe^{2+} (Rehncrona et al., 1989). The increasingly formed superoxide radical H_2O_2 is available for the Fenton and the Haber–Weiss reactions so that high concentrations of the hydroxyl radical $\bullet OH$ are formed, the redox potential of which is $+2$ V. Oxidative stress, also resulting from mitochondrial dysfunction and oxidation of dopamine (Götz et al., 1994; Jenner, 1998), along with iron has been shown to cause the formation of toxic aggregates from α-synuclein that is considered to contribute to the formation of Lewy bodies (Ebadi et al., 2001; Turnbull et al., 2001). This cell damage induced predominantly by oxidative stress is accompanied by reactive gliosis (Schipper, 1996) and inflammation accelerating and exacerbating the course of the disease (Herrera et al., 2005). As far as hereditary PD is concerned, endoplasmic reticulum dysfunction acts as a pivotal pathological mechanism (Imai et al., 2000; Paschen and Frandsen, 2001).

It, thus, becomes clear that the mechanisms of neurodegeneration in age-associated sporadic AD and age-associated sporadic PD start differently: sporadic AD starts by a transmission failure of insulin signal transduction in cerebral cortex and hippocampus causing energy failure, acetylcholinergic neurotransmission failure, neuroinflammation, and free radical formation in a cascade-like manner; sporadic PD starts in the substantia nigra by excessive iron accumulation with subsequent free radical formation. Neurodegeneration spreads more or less over the whole brain.

In so far, Parkinson's early notion of ". . . senses and intellects being uninjured" can be largely confirmed by actual studies. The dementias associated with sporadic AD and sporadic PD are caused differently in regional and biochemical terms generating a different clinical symptomatology (Pillon et al., 1986; Mindham et al., 1992).

6   Synopsis

Each of the neurodegenerative disorders, AD and PD, do not form a nosogical entity. Hereditary types are different from sporadic types whereby the former represent minorities only; in the case of hereditary AD a few hundred families worldwide. The majority of patients suffer from sporadic AD and sporadic PD which both are age-associated pathological conditions.

In AD, glutamatergic and acetylcholinergic pyramidal neurons in different cerebral cortical areas and the hippocampus degenerate. PD starts in nigrostriatal brain area in which dopaminergic neurons degenerate.

A failure in the neuronal insulin signal transduction may cause a cascade-like pathological process in sporadic AD in which cell degeneration and cell loss, hyperphosphorylated tau-protein and neurofibrillary tangles, and increased β-amyloid and neuritic plaques develop. In sporadic Parkinson's disease, an excessive iron accumulation in dopaminergic is the central pathological event followed by enhanced formation of both free radicals and neuroinflammation.

Sporadic AD starts with deficits in mental capacities such as cognition, learning, and memory. Movement disturbances can be found in advanced disease stages. Sporadic PD starts with movement disturbances without failure in mental capacities. In later advanced stages, subcortical dementia can develop being clinically different from dementia of Alzheimer type.

References

Abbott MA, Wells DG, Fallon JR. 1999. The insulin receptor tyrosine kinase substrate p 58/53 and the insulin receptor are components of CNS synapses. J Neurosci 19: 7300-7308.

Abeijon C, Hirschberg CB. 1988. Intrinsic membrane glycoproteins with cytosol-orientied sugards in the endoplasmic reticulum. Proc Natl Acad Sci USA 85: 1010-1014.

Adamo M, Raizada MK, Le Roith D. 1989. Insulin and insulin-like growth factor receptors in the nervous system. Mol Neurobiol 3: 71-100.

Adrian GS, McCammon MT, Montgomery DL, Douglas MG. 1986. Sequences required for delivery and localization of the ADP/ATP translocator to the mitochrondial inner membrane. Mol Cell Biol 6: 626-634.

Ahima RS, Prabakaran D, Mantzoros C, Qu D, Lowell B, et al. 1996. Role of leptin in the neuroendocrine response to fasting. Nature 382: 250-252.

Ahlijanian MK, Barrezueta NX, Williams RD, Jakowski A, Kowsz KP, et al. 2000. Hyperphosphorylated tau and neurofilament and cytoskeletal disruptions in mice over-expressing human p25, an activator of cdk5. Proc Natl Acad Sci USA 97: 2910-2915.

Aisen P, Davis K. 1994. Inflammatory mechanisms in Alzheimer's disease: Implications for therapy. Am J Psychiatry 151: 1105-1113.

Akiyama H, Barger S, Barnum S, Bradt B, Bauer J, et al. 2000. Inflammation and Alzheimer's disease. Neurobiol Aging 21: 393-421.

Akiyama H, Kondo H, Ikeda K, Kato M, McGeer PL. 2001. Immunohistochemical localization of neprilysin in the human cerebral cortex: Inverse association with vulnerability to amyloid β-protein (Aβ) deposition. Brain Res 902: 277-281.

Albert ML, Feldman RG, Willis AL. 1974. The "subcortical dementia" of progressive supranuclear palsy. J Neurol Neurosurg Psychiatry 37: 121-130.

Alessi DR, Cohen P. 1998. Mechanism of activation and function of protein kinase B. Curr Opin Genet Dev 8: 55-62.

Alzheimer A. 1911. Über eigenartige Krankheitsfälle des späteren Alters. Ztschr. ges. Neurol Psychiatry 4: 356-385.

Andrews JM, Terry RD, Spataro J. 1970. Striatonigral degeneration. Clinical-pathological correlations and response to stereotaxic surgery. Arch Neurol 23: 319-239.

Apelt J, Ach K, Schliebs R. 2003. Aging-related down-regulation of neprilysin, a putative β-amyloid-degrading enzyme, in transgenic Tg2576 Alzheimer-like mouse brain is accompanied by an astroglial upregulation in the vicinity of β-amyloid plaques. Neurosci Lett 339: 182-186.

Apelt J, Mehlhorn G, Schliebs R. 1999. Insulin-sensitive GLUT 4 glucose transporters are localized with GLUT 3-expressing cells and demonstrate a chemically distinct neuron-specific localization in rat brain. J Neurosci Res 57: 693-705.

Apelt J, Schliebs R. 2001. β-Amyloid-induced glial expression of both pro- and anti-inflammatory cytokines in cerebral cortex of aged transgenic Tg2576 mice with Alzheimer plaque pathology. Brain Res 894: 21-30.

Aplin AE, Gibb GM, Jacobsen JS, Gallo JM, Anderton BH. 1996. In vitro phosphorylation of the cytoplasmic domain of the amyloid precursor protein by glycogen synthase kinase-3β. J Neurochem 67: 699-707.

Arai H, Lee VMY, Messinger ML, Greenberg BD, Lowery DE, et al. 1991. Expression of pattern of β-amyloid precursor protein (β-APP) in neural and nonneuronal human tissues from Alzheimer's disease and control subjects. Ann Neurol 30: 686-693.

Arendt T. 2003. Synaptic plasticity and cell cycle activation in neurons are alternative effector pathways: The 'Dr. Jekyll and Mr. Hyde concept' of Alzheimer's disease or the yin and yang of neuroplasticity. Prog Neurobiol 71: 83-248.

Arias C, Montiel T, Quiroz-Baez R, Massieu L. 2002. β-Amyloid neurotoxicity is exacerbated during glycolysis

inhibition and mitochondria impairment in the rat hippocampus in vivo and in isolated nerve terminals: Implications for Alzheimer's disease. Exp Neurol 176: 163-174.

Arnaiz E, Jelic V, Almkvist O, Wahlund LO, Winblad B, et al. 2001. Impaired cerebral glucose metabolism and cognitive functioning predict deterioration in mild cognitive impairment. Neuroreport 12: 851-855.

Ashford ML, Boden PR, Treherne JM. 1990. Glucose-induced excitation of hypothalamic neurons is mediated by ATP-sensitive K^+ channels. Pflügers Arch 415: 479-483.

Atanasov AG, Nashev LG, Schweizer RAS, Frick C, Odermatt A. 2004. Hexose-6 phosphate dehydrogenase determines the reaction direction of 11β-hydroxysteroid dehydrogenase type 1 as an oxoreductase. FEBS Lett 571: 129-133.

Avruch J. 1998. Insulin signal transduction through protein kinase cascades. Mol Cell Biochem 182: 31-48.

Bachelard HS. 1971. Specific and kinetic properties of monosaccharide uptake into guinea pig cerebral cortex in vitro. J Neurochem 13: 213-222.

Bak P, Tang C, Wiesenfeld K. 1988. Self-organized criticality. Physic Rev A 38: 365-374.

Baltes PB, Mayer KU, Helmchen H, Steinhagen-Thiessen E. 1999. The Berlin Aging Study (BASE): Sample, Design, and Overviews of Measures. Baltes PB, Mayer KU, editors. Cambridge: Cambridge University Press; pp. 15-55.

Banhegyi G, Benedetti A, Fulceri R, Senesi S. 2004. Cooperativity between 11β-hydroxysteroid dehydrogenase type 1 and hexose-6-phosphate dehydrogenase in the lumen of the endoplasmic reticulum. J Biol Chem 279: 27017-27021.

Banks WA. 2004. The source of cerebral insulin. Eur J Pharmacol 490: 5-12.

Baskin DG, Brewitt B, Davidson DA, Crop E, Paquette T, et al. 1986. Quantitative autoradiographic evidence for insulin receptors in the choroid plexus of the rat brain. Diabetes 35: 246-249.

Baskin DG, Schwartz MW, Sipols AJ, D'Alessio DA, Goldstein BJ, et al. 1994. Insulin receptor substrate-1 (IRS-1) expression in rat brain. Endocrinology 134: 1952-1955.

Baskin DG, Sipols AJ, Schwartz MW, White MF. 1993. Immunocytochemical detection of insulin receptor substrate-1 (IRS-1) in rat brain: Colocalization with phosphotyrosine. Regul Pept 48: 257-266.

Baskin DS, Browning JL, Pirozzollo FJ, Korporaal S, Baskin JA, et al. 1999. Brain choline acetyltransferase and mental function in Alzheimer disease. Arch Neurol 56: 1221-1223.

Baudry M, Arst DS, Lynch G. 1981. Increased (^{3}H) glutamate receptor binding in aged rats. Brain Res 223: 195-198.

Beach TG, Walker DG, Cynader MS, Hughes LH. 1996. Increased β-amyloid precursor protein mRNA in the rat cerebral cortex and hippocampus after chronic systemic atropine treatment. Neurosci Lett 210: 13-16.

Beach TG, Walker R, McGeer EG. 1989. Patterns of gliosis in Alzheimer's disease and aging cerebrum. Glia 2: 420-436.

Belanoff JK, Gross K, Yager A, Schatzberg AF. 2001. Corticosteroids and cognition. J Psychiatric Res 35: 127-145.

Bennecib M, Gong CX, Grundke-Iqbal I, Iqbal K. 2000. Role of protein phosphatase-2A and -1 in the regulation of GSK-3, cdk 5 and cdc 2 and the phosphorylation of tau in rat forebrain. FEBS Lett 485: 87-93.

Benzi G, Pastoris O, Villa RF, Giuffrida-Stella AM. 1984. Effect of aging on cerebral cortex energy metabolism in hypoglycemia and posthypoglycemic recovery. Neurobiol Aging 5: 457-463.

Bernardi H, Fosset M, Lazdunski M. 1988. Characterization, purification, and affinity labeling of the brain (^{3}H) glibenclamide-binding protein, a putative neuronal ATP-regulated K^+ channel. Proc Natl Acad Sci USA 85: 9816-9820.

Bernstein HG, Ansorge S, Riederer P, Reiser FM, Frölich L, et al. 1999. Insulin-degrading enzyme in Alzheimer's disease brain: Prominent localization in neurons and senile plaques. Neurosci Lett 263: 161-164.

Bhat RV, Shanley J, Correll MP, Fieles WE, Keith RA, et al. 2000. Regulation and localization of tyrosine[216] phosphorylation of glycogen synthase kinase-3β in cellular and animal models of neuronal degeneration. Proc Natl Acad Sci USA 97: 11074-11079.

Bierhaus A, Wolf J, Andrassy M, Rohleder N, Humpert PM, et al. 2003. A mechanism converting psychosocial stress into mononuclear cell activation. Proc Natl Acad Sci USA 100: 1920-1925.

Bigl M, Apelt J, Eschrich K, Schliebs R. 2003. Cortical glucose metabolism is altered in aged transgenic Tg 2576 mice that demonstrate Alzheimer plaque pathology. J Neural Transm 110: 77-94.

Bigl M, Beck M, Bleyl AD, Bigl V, Eschrich K. 2000. Altered prosphofructokinase mRNA levels but unchanged isoenzyme pattern in brains from patients with Alzheimer's diseae. Mol Brain Res 76: 411-414.

Bigl M, Bleyl AD, Zedlick D, Arendt T, Bigl V, et al. 1996. Changes of activity and isoenzyme pattern of phosphofructokinase in the brains of patients with Alzheimer's disease. J Neurochem 67: 1164-1171.

Bigl M, Brückner MK, Arendt T, Bigl V, Eschrich K. 1999. Activities of key glycolytic enzymes in the brains of patients with Alzheimer's disease. J Neural Transm 106: 499-511.

Bigl V, Arendt T, Fischer S, Werner M, Arendt A. 1987. The cholinergic system in aging. Gerontology 33: 172-180.

Bigl V, Schliebs R. 1998. Simulation of cortical cholinergic deficits: A novel experimental approach to study pathogenetic aspects of Alzheimer's disease. J Neural Transm (Suppl. 54): 237–247.

Blass JP, Gibson GE, Hoyer S. 2002. The role of the metabolic lesion in Alzheimer's disease. J Alzheimer Dis 4: 225-232.

Blusztajn JK, Lopez Gonzales-Coviella I, Logue M, Growdon JH, Wurtman RJ. 1990. Levels of phospholipid catabolic intermediates, glycerophosphocholine and glycerophosphoethanolamine, are elevated in brains of Alzheimer's disease but not of Down's syndrome patients. Brain Res 536: 240-244.

Boado RJ. 1995. Brain-derived peptides regulate the steady state levels and increase stability of the blood-brain barrier GLUT-1 glucose transporter mRNA. Neurosci Lett 197: 179-182.

Bole DG, Dowin R, Doriaux M, Jamieson JD. 1989. Immunocytochemical localization of BiP to the rough endoplasmic reticulum: Evidence for protein sorting by selective retention. J Histochem Cytochem 37: 1817-1823.

Borit A, Rubinstein LJ, Urich H. 1975. The striatonigral degenerations: Putaminal pigments and nosology. Brain 98: 101-112.

Born J, Lange T, Kern W, McGregor GP, Bickel U, et al. 2002. Sniffing neuropeptides: A transnasal approach to the human brain. Nat Neurosci 5: 514-516.

Bosboom JLW, Stoffers D, Wolters EC. 2004. Cognitive dysfunction and dementia in Parkinson's disease. J Neural Transm 111: 1303-1315.

Bowen DM. 1984. Cellular aging: Selective vulnerability of cholinergic neurons in human brain. Monogr Dev Biol 17: 42-59.

Braakman I, Helenius J, Helenius A. 1992. Role of ATP and disulfide bonds during protein folding in the endoplasmic reticulum. Nature 356: 260-261.

Brant AM, Jess TJ, Milligan G, Brown CM, Gould GW. 1993. Immunological analysis of glucose transporters expressed in different regions of the rat brain and central nervous system. Biochem Biophys Res Commun 192: 1297-1302.

Brett J, Schmidt AM, Yan SD, Zou YD, Weidman E, et al. 1993. Survey of the distribution of an newly characterized receptor for advanced glycation end products in tissues. Am J Pathol 143: 1699-1712.

Brown GG, Levine SR, Gorell JM, Pettegrew JW, Gdowski JE, et al. 1989. In vitro ^{31}P-NMR profiles of Alzheimer disease and multiple subcortical infarct dementia. Neurology 39: 1423-1427.

Brun A, Englund E. 1981. Regional pattern of degeneration in Alzheimer's disease: Neuronal loss and histopathological grading. Histopathology 5: 549-564.

Bucht G, Adolfsson R, Lithner F, Winblad B. 1983. Changes in blood glucose and insulin secretion in patients with senile dementia of Alzheimer type. Acta Med Scand 213: 387-392.

Buse MG, Robinson KA, Marshall BA, Hresko RC, Mueckler MM. 2002. Enhanced O-GlcNAc protein modification is associated with insulin resistance in GLUT1-overexpressing muscles. Am J Physiol Endocrinol Metab 283: E241-E250.

Buttgereit F, Brand MD. 1995. A hierarchy of ATP-consuming processes in mammalian cells. Biochem J 312: 163-167.

Buxbaum JD, Koo EH, Greengard P. 1993. Protein phosphorylation inhibits production of Alzheimer amyloid β/A4 peptide. Proc Natl Acad Sci USA 90: 9195-9198.

Capasso JM, Abeijon C, Hirschberg CB. 1988. An intrinsic membrane glycoprotein of the Golgi apparatus with O-linked N-acetylglucosamine facing the cytosol. J Biol Chem 263: 19778-19782.

Carney JM, Starke-Reed PE, Oliver CN, Landum RW, Cheng MS, et al. 1991. Reversal and -age-related increase in brain protein oxidation, decrease in enzyme activity, and loss in temporal and spatial memory by chronic administration of the spin-trapping compound N-tert-butyl-alpha-phenylnitrone. Proc Natl Acad Sci USA 88: 3633-3636.

Carson JA, Turner AJ. 2002. Beta-amyloid catabolism: Roles for neprilysin (NP) and other metallopeptidases? J Neurochem 81: 1-8.

Chen YZ. 2004. APP induces neuronal apoptosis through APP-BP1-mediated downregulation of beta-catenin. Apoptosis 9: 415-422.

Chiarugi A. 2002. Characterization of the molecular events following impairment of NF-kB-driven transcription in neurons. Mol Brain Res 109: 179-188.

Cho KS, Choi J, Ha CM, Son YJ, Choi WS, et al. 2002. Comparison of gene expression in old versus young rat hippocampus by cDNA array. Neuro Report 13: 285-289.

Chou CK, Dull TJ, Russel DS, Gherzi R, Lebwohl D, et al. 1987. Human insulin receptors mutated at the ATP-binding site lack protein tyrosine activity and fail to mediate postreceptor effects of insulin. J Biol Chem 262: 1842-1847.

Chrousos GP, Gold PW. 1992. The concept of stress and stress system disorders. J Am Med Assoc 267: 1244-1252.

Ciechanover A, Brundin P. 2003. The ubiquitoin proteasome system in neurodegenerative diseases: Sometomes the chicken, sometimes the egg. Neuron 40: 427-446.

Cizza G, Calogero AE, Brady LS, Bagdy G, Bergamini E, et al. 1994. Male Fischer 344/rats show a progressiv impairmaint of the hypothalamic-pituitary-adrenal axis with advancing age. Endocrinology 134: 1611-1620.

Cizza G, Gold PW, Chrousos GP. 1995. Aging is associated in the 344/N Fischer rat with decreased stress responsitivity of central and peripheral catecholaminergic systems and impairment of the hypothalamic-pituitary-adrenal axis. Ann N Y Acad Sci 771: 491-511.

Cohen P, Frame S. 2001. The renaissance of GSK. Mol Cell Biol 2: 769-776.

Comer FI, Hart GW. 2001. Reciprocity between 0-GlcNac and 0-phosphate on the carboxyl terminal domain of RNA polymerase II. Biochemistry 40: 7845-7852.

Conejo R, Lorenzo M. 2001. Insulin signalling leading to proliferation, survival and membrane ruffling in C2 C12 myoblasts. J Cell Physiol 187: 96-108.

Conejo R, Valverde AM, Benito M, Lorenzo M. 2001. Insulin produces myogenesis in C2 C12 myoblasts by induction of NF-kappa B and downregulation of AP-1 activities. J Cell Physiol 186: 82-94.

Connell-Crowley L, Le Gall M, Vo DJ, Giniger E. 2000. The cyclin dependent kinase Cdk5 controls multiple aspects of axon patterning in vivo. Curr Biol 10: 599-602.

Convit A, Wolf OT, Tarshish C, de Leon MJ. 2003. Reduced glucose tolerance is associated with poor memory performance and hippocampal atrophy among normal elderly. Proc Natl Acad Sci USA 100: 2019-2022.

Cook DG, Forman FS, Sung IC, Leight S, Kolson DI, et al. 1997. Alzheimer's Aβ (1–42) is generated in the endoplasmic reticulum/intermediate compartment of NT2N cells. Nat Med 3: 1021-1023.

Cook DG, Leverenz JB, McMillan PJ, Kulstad JJ, Ericksen S, et al. 2003. Reduced hippocampal insulin-degrading enzyme in late-onset Alzheimer's disease is associated with the apolipoprotein E-Σ4 allele. Am J Pathol 162: 313-319.

Cotman CW, Monaghan DT, Ottersen OP, Storm-Mathisen H. 1987. Anatomical organization of excitatory amino acid receptors and their pathways. Trends Neurosci 10: 273-280.

Cotrina ML, Nedergaard M. 2002. Astrocytes in the aging brain. J Neurosci Res 67: 1-10.

Craft S, Asthana S, Schellenberg G, Cherrier M, Baker LD, et al. 1999. Insulin metabolism in Alzheimer'disease differs according to apolipoprotein E genotype and gender. Neuroendocrinology 70: 146-152.

Craft S, Peskind E, Asthana S, Watson GS, Baker LD, et al. 2000. Effects of insulin on cerebrospinal fluid levels of Aβ 42. Neurobiol Aging 21: S272 (1245).

Craft S, Peskind E, Schwartz MW, Schellenberg GD, Raskind M, et al. 1998. Cerebrospinal fluid and plasma insulin levels in Alzheimer's disease: Relationship to severity of dementia and apolipoprotein E genotype. Neurology 50: 164-168.

Crain BJ, Burger PC. 1988. The laminar distribution of neuritic plaques in the fascia dentate of patients with Alzheimer's disease. Acta Neuropathol 76: 86-93.

Cross DA, Watt PW, Shaw M, Kaay von der J, Downes CP, et al. 1997. Insulin activates protein kinase B, inhibits glycogen synthase kinase-3 and activatives glycogen synthase by rapamycin-sensitive pathways in skeletal muscle and adipose tissue. FEBS Lett 406: 211-215.

Cross DAE, Alessi DR, Cohen P, Andjelkovich M, Hemmings BA. 1995. Inhibition of glycogen synthase kinase-3 by insulin mediated protein kinase. Nature 378: 785-789.

Cruz JC, Tseng HC, Goldman JA, Shih H, Tsai LH. 2003. Aberrant cdk5 activation by p25 triggers pathological events leading to neurodegeneration and neurofibrillary tangles. Neuron 40: 471-483.

D'Andrea MR, Cole GM, Ard MD. 2004. The microglial phagocytic role with specific plaque types in the Alzheimer disease brain. Neurobiol Aging 25: 675-683.

D'Andrea MR, Nagele RG, Wang HY, Peterson PA, Lee DHS. 2001. Evidence that neurones accumulating amyloid can undergo lysis to form amyloid plaques in Alzheimer's disease. Histopathology 38: 120-134.

Das S, Potter H. 1995. Expression of the Alzheimer amyloid-promoting factor antichymotrypsin is induced in human astrocytes by IL-1. Neuron 14: 447-456.

Dastur DK. 1985. Cerebral blood flow and metabolism in normal human aging, pathological aging, and senile dementia. J Cereb Blood Flow Metab 5: 1-9.

Davidsson P, Bogdanovic N, Lannfelt L, Blennow K. 2001. Reduced expression of amyloid precursor protein, presenilin-1 and rab 3a in cortical brain regions in Alzheimer's disease. Dement Geriatr Cogn Disord 12: 243-250.

Davies SG, McBean JG, Roberts PJ. 1984. A glutamatergic innervation of the nucleus basalis/substantia innominata. Neurosci Lett 45: 105-110.

Davis KL, Davis BM, Greenwald BS, Mohs RC, Mathé AA, et al. 1986. Cortisol and Alzheimer's disease, I: Basal studies. Am J Psychiatry 143: 300-305.

de Bruin VMS, Vieira MCM, Rocha MNM, Viana GSP. 2002. Cortisol and dehydroepiandrosterone sulfate plasma levels and their relationship to aging, cognitive function, and dementia. Brain Cogn 50: 316-323.

de Kloet ER. 1991. Brain corticosteroid receptor balance and homeostatic control. Front Neuroendocrinol 12: 95-165.

de Quervain DJF, Poirier R, Wollmer MA, Grimaldi LME, Tsolaki M, et al. 2004. Glucocorticoid-related genetic susceptibility for Alzheimer's disease. Human Mol Genet 13: 47-52.

Degrell I, Krier C, Hoyer S. 1993. Neuropathology and Neuropharmacology. Aging, Vol 21. Cervos-Navarro J, Sarkander MI, editors. New York; Raven; pp. 289-300.

Dekker MC, Bonifati V, van Duijn CM. 2003. Parkinson's disease: Piecing together a genetic jigsaw. Brain 126: 1722-1733.

Demaurex N, Furuya W, D'Souza S, Bonifacino JS, Grinstein S. 1998. Mechanism of acidification of the trans-Golgi network (TGN). In situ measurements of pH using retrieval of TGN 38 and furin from the cell surface. J Biol Chem 273: 2044-2051.

Devaskar SU, Giddings SJ, Rajakumar PA, Carnaghi LR, Menon RK, et al. 1994. Insulin gene expression and insulin synthesis in mammalian neuronal cells. J Biol Chem 269: 8445-8454.

Dickson D, Lee S, Mattiace L, Yen S, Brosnan C. 1993. Microglia and cytokines in neurological disease, with specific reference to AIDS and Alzheimer's disease. Glia 7: 75-83.

Dickson D, Rogers J. 1992. Neuroimmunology of Alzheimer's disease. Neurobiol Aging 15: 793-798.

Dorner AJ, Wasley LC, Kaufman RJ. 1990. Protein dissociation from GRP78 and secretion are blocked by depletion of cellular ATP levels. Proc Natl Acad Sci USA 87: 7429-7432.

Drachman DA, Noffsinger D, Sahakian BJ, Kurdziel S, Fleming P. 1980. Aging, memory and the cholinergic system: A study of dichotic listening. Neurobiol Aging 1: 39-43.

Duara R, Grady C, Haxby J, Sundaram S, Cutler NR, et al. 1986. Positron emission tomography in Alzheimer's disease. Neurology 36: 879-887.

Duax WL, Griffin JF, Ghosh D. 1996. The fascinating complexities of steroid-binding enzymes. Curr Opin Struct Biol 6: 813-823.

Dubelaar EJG, Verwer RWH, Hofman MA, van Heerikhuize JJ, Ravid R, et al. 2004. ApoE Σ4 genotype is accompanied by lower metabolic activity in nucleus basalis of Meynert neurons in Alzheimer patients and controls as indicated by the size of Golgi apparatus. J Neuropathol Exp Neurol 63: 159-169.

Duckworth WC. 1988. Insulin degradation: Mechanisms, products and significance. Endocr Rev 9: 319-345.

Dunn-Meynell AA, Routh VH, McArdle JJ, Levin BE. 1997. Low affinity sulfonylurea binding sites reside on neuronal cell bodies in the brain. Brain Res 745: 1-9.

Dutschke K, Nitsch RM, Hoyer S. 1994. Short-term mental activation accelerates the age-related decline of high-energy phosphates in rat cerebral cortex. Arch Gerontol Geriatr 19: 43-51.

Ebadi M, Govitrapong P, Sharma S, Murali-Krishan D, Shavali S, et al. 2001. Ubiquinone (coenzyme q10) and mitochondria in oxidative stress of Parkinson's disease. Biol Signals Recept 10: 224-253.

Eckert GP, Cairns NJ, Maras A, Gattaz WF, Müller WE. 2000. Cholesterol modulates the membrane-disordering effects of beta-amyloid peptides in the hippocampus: Specific changes in Alzheimer's disease. Demen Geriatr Cogn Disord 11: 181-186.

Edland SD, Wavrant-De Vriesé F, Compton D, Smith GE, Ivnik R, et al. 2003. Insulin degrading enzyme (IDE) genetic variants and risk of Alzheimer's disease: Evidence of effect modification by apolipoprotein E (APOE). Neurosci Lett 345: 21-24.

Efthimiopoulos S, Punj S, Manolopoulos V, Pangalos M, Wang GP, et al. 1996. Intracellular cyclic AMP inhibits constitutive and phorbol ester-stimulated secretory cleavage of amyloid precursor protein. J Neurochem 67: 872-875.

Eikelenboom P, van Gool WA. 2004. Neuroinflammatory perspectives on the two faces of Alzheimer's disease. J Neural Transm 111: 281-294.

Eikelenboom P, Zhan SS, van Gool WA, Eikelenboom P. 1994. Inflammatory mechanisms in Alzheimer's disease. Trends Pharmacol Sci 15: 147-150.

Eisenberg S, Seltzer HS. 1962. The cerebral metabolic effects of acutely induced hypoglycaemia in human subjects. Metabolism 11: 1162-1168.

El Tamer A, Raikoff K, Hanin I. 1998. Effect of glucose-deprivation on amyloid precursor protein (APP) release from hippocampal (HIP) slices of rat. Soc Neurosci Abstr 24.

Erecinska M, Silver IA. 1989. ATP and brain function. J Cereb Blood Flow Metab 9: 2-19.

Evans DA, Funkenstein H, Albert MS, Scherr PA, Cook NR, et al. 1989. Prevalence of Alzheimer's disease in a community population of older persons: Higher than previously reported. J Am Med Assoc 262: 2551-2556.

Fagiolo U, Cossarizza A, Scala E, Fanales Belasio E, Ortolani C, et al. 1993. Increased cytokine production in mononuclear cells of healthy elderly people. Eur J Immunol 23: 2375-2378.

Fang X, Yu SX, Lu Y, Bast RC Jr, Woodgett JR, et al. 2000. Phosphorylation and inactivation of glycogen synthase kinase 3 by protein kinase A. Proc Natl Acad Sci USA 97: 11960-11965.

Farrell CL, Pardridge WM. 1991. Blood-brain barrier glucose transporter is asymmetrically distributed on brain capillary endothelial luminal and ablumenal membranes: An electron microscopic immunogold study. Proc Natl Acad Sci USA 88: 5779-5783.

Farris W, Mansourian S, Chang Y, Lindsay L, Eckman EA, et al. 2003. Insulin-degrading enzyme regulates the levels of insulin, amlyoid-β protein, and the β-amyloid precursor protein intracellular domain in vivo. Proc Natl Acad Sci USA 100: 4162-4167.

Fearnley JM, Lees AJ. 1991. Aging and Parkinson's disease: Substantia nigra regional selectively. Brain 114: 2283-2301.

Feinstein R, Kanety H, Papa MZ, Lunenfeld B, Karasik A. 1993. Tumor necrosis factor-α suppresses insulin-induced tyrosine phosphorylation of insulin receptor and its substrates. J Biol Chem 268: 26055-26058.

Ferrer I, Blanco R, Carmona M, Puig B. 2001. Phosphorylated mitogen-activated protein kinase (MAPK/ERK-P), protein kinase of 38 kDa (p38-P), stress activated protein kinase (SAPK/JNK-P), and calcium/calmodulin-dependent kinase II (CaMkinase II) are differently expressed in tau deposits in neurons and glial cells in tauopathies. J Neural Transm 108: 1397-1415.

Fillenz M, Lowry JP. 1998. Studies of the source of glucose in the extracellular compartment of the rat brain. J Dev Neurosci 20: 365-368.

Folbergrova J, Ljunggren B, Siesjö BK. 1974. Influence of complete ischemia on glycolytic metabolites of complete

ischemia on glycolytic metabolites, citric and cycle intermediates, and associated amino acids in the rat cerebral cortex. Brain Res 80: 265-279.

Foster NL, Chase TN, Mansi K, Brooks R, Fedio P, et al. 1984. Cortical abnormalities in Alzheimer's disease. Ann Neurol 16: 649-654.

Frame S, Cohen P. 2001. GSK 3 takes centre stage more than 20 years after its discovery. Biochem J 359: 1-16.

Frölich L. 2002. The cholinergic pathology in Alzheimer's disease: Discrepancies between clinical experience and pathophysiological findings. J Neural Transm 109: 1003-1014.

Frölich L, Blum-Degen D, Bernstein HG, Engelsberger S, Humrich J, et al. 1998. Insulin and insulin receptors in the brain in aging and sporadic Alzheimer's disease. J Neural Transm 105: 423-438.

Fukami S, Watanabe K, Iwata N, Haraoka H, Lu B, et al. 2002. Aβ-degrading endopeptidase, neprilysin, in mouse brain: Synaptic and axonal localization inversely correlating with Aβ pathology. Neurosci Res 43: 39-56.

Fukuyama H, Ogawa M, Yamauchi H, Yamaguchi S, Kimura J, et al. 1994. Altered cerebral energy metabolism in Alzheimer's disease: A PET study. J Nucl Med 35: 1-6.

Garland PB, Newsholme EA, Randle PJ. 1964. Regulation of glucose uptake by muscle. 9. Effects of fatty acids and ketone bodies, and of alloxan-diabetes and starvation, on pyruvate metabolism and lactate/pyruvate and L-glycerol-3-phosphate/ dihydro- xyacetone phosphate concentration ratios in the heart and rat diaphragm muscles. Biochem J 93: 665-678.

Garland PB, Randle PJ. 1964. Control of pyruvate dehydrogenase in the perfused rat heart by the intracellular concentration of acetyl-coenzyme A. Biochem J 91: 76C-77C.

Garriga J, Cusso R. 1992. Effect of starvation on glycogen and glucose metabolism in different areas of the brain. Brain Res 591: 277-282.

Gasic GP, Heinemann S. 1991. Receptors coupled to ionic channels: The glutamate receptor family. Curr Opin Neurobiol 1: 20-26.

Gasparini L, Gouras GK, Wang R, Gross RS, Beal MF, et al. 2001. Stimulation of β-amyloid precursor protein trafficking by insulin reduces intraneural β-amyloid and requires mitogen-activated protein kinase signalling. J Neurosci 21: 2561-2570.

Gerozissis R, Rouch C, Lemiere S, Nicolaidis S, Orosco M. 2001. A potential role of central insulin in learning and memory related to feeding. Cell Mol Neurobiol 21: 389-401.

Gething MJ, Sambrook J. 1992. Protein folding in the cell. Nature 355: 33-45.

Ghidoni R, Gasparini K, Alberici A, Benussi L, Barbiero L, et al. 2003. Inhibition of energy metabolism downregulates the Alzheimer related presenilin 2 gene. J Neural Transm 110: 1029-1039.

Gibb WRG, Lees AJ. 1988. The relevance of the Lewy body to the pathogenesis of idiopathic Parkinson's disease. J Neurol Neurosurg Psychiatry 51: 745-752.

Gibbs EL, Lennox WG, Nims LF, Gibbs FA. 1942. Arterial and cerebral venous blood: Arterial venous differences in man. J Biol Chem 144: 325-332.

Gibson GE, Jope R, Blass JP. 1975. Decreased synthesis of acetylcholine accompanying impaired oxidation of pyruvic acid in rat brain minces. Biochem J 148: 17-23.

Gibson GE, Petersen C, Jand enden DJ. 1981. Brain acetylcholine synthesis declines with senescence. Science 213: 674-676.

Gillespie SL, Golde TE, Younkin SG. 1992. Secretory processing of the Alzheimer amyloid β/A4 protein precursor is increased by protein phosphorylation. Biochem Biophys Res Commun 187: 1285-1290.

Giorgino F, Almahfouz A, Goodyear LJ, Smith RJ. 1993. Glucocorticoid regulation of insulin receptor and subtrate IRS-1 tyrosine phosphorylation in rat skeletal muscle in vivo. J Clin Invest 91: 2020-2030.

Goldstein BJ. 1993. Regulation of insulin receptor signalling by protein-tyrosine dephosphorylation. Receptor 3: 1-15.

Gong CX, Grundke-Iqbal I, Iqbal K. 1994a. Dephosphorylation of Alzheimer's disease abnormally phosphorylated tau by protein phosphatase-2A. Neuroscience 61: 765-772.

Gong CX, Lidsky T, Wegiel J, Zuck L, Grundke-Iqbal I, et al. 2000. Phosphorylation of microtubule-associated protein tau is regulated by protein phosphatase 2A in mammalian brain: Implications for neurofibrillary degeneration in Alzheimer's disease. J Biol Chem 275: 5535-5544.

Gong CX, Singh TJ, Grundke-Iqbal I, Iqbal K. 1994b. Alzheimer's disease abnormally phosphorylated tau is dephosphorylated by phosphatase-2B (calcineurin). J Neurochem 62: 803-806.

Gopalakrishan M, Janis RA, Triggle DJ. 1993. ATP-sensitive K+ channels: Pharmacologic properties, regulation, and therapeutic potential. Drug Dev Res 28: 95-127.

Gottstein U, Bernsmeier A, Sedlmeyer I. 1963. Der Kohlenhydratstoffwechsel des menschlichen Gehirns. I. Untersuchungen mit substratspezifischen enzymatischen Methoden bei normaler Hirndurchblutung. Klin Wschr 41: 943-948.

Gottstein U, Held K. 1967. Insulinwirkung auf den menschlichen Hirnmetabolismus unter dem Einfluss intravenöser Infusionen von Glukose Glukagon und Glukose-Insulin. Klin Wschr 43: 965-975.

Gottstein U, Müller W, Berhoff W, Gärtner H, Held K. 1971. Zur Utilisation von nicht-veresterten Fettsäuren und Ketonkörper im Gehirn des Menschen. Klin Wschr 49: 406-411.

Götz ME, Kunig G, Riederer P, Youdim MB. 1994. Oxidative stress: Free radical production in neuronal degeneration. Pharmacol Ther 63: 37-122.

Gouras GK, Tsai J, Naslund J, Vincent B, Edgard M, et al. 2000. Intraneuronal βA42 accumulation in human brain. Am J Pathol 156: 15-20.

Greenfield JP, Tsai J, Gouras GK, Hai B, Thinakaran G, et al. 1999. Endoplasmic reticulum and trans-Golgi network generate distinct populations of Alzheimer β-amyloid peptides. Proc Natl Acad Sci USA 96: 742-747.

Griffith LS, Mathes M, Schmitz B. 1995. β-amyloid precursor protein is modified with O-linked N-acetylglucosamine. J Neurosci Res 41: 270-278.

Griffith LS, Schmitz B. 1995. O-linked N-acetylglucosamine is upregulated in Alzheimer brains. Biochem Biophys Res Commun 213: 423-431.

Grünblatt E, Hoyer S, Riederer P. Changes in gene expression in brain cortex and striatum of the streptozotocin rat model for sporadic Alzheimer's disease (in preparation).

Grünblatt E, Hoyer S, Riederer P. 2004a. Gene expression profile in streptozotocin rat model for sporadic Alzheimer's disease. J Neural Transm 111: 367-386.

Grünblatt E, Mandel S, Jacob-Hirsch J, Zeligson S, Amariglo N, et al. 2004b. Gene expression profiling of parkinsonian substantia nigra pars compacta; alterations in ubiquitin-proteaseome, heat shock protein, iron and oxidative stress regulated proteins, cell adhesion/cellular matrix and vesicle trafficking genes. J Neural Transm 111: 1543-1573.

Gsell W, Jungkunz G, Riederer P. 2004. Functional neurochemistry of Alzheimer's disease. Curr Pharm Design 10: 265-293.

Gsell W, Strein I, Riederer P. 1996. The neurochemistry of Alzheimer type, vascular type and mixed type dementia compared. J Neural Transm 47: 73-101.

Gust DA, Wilson ME, Stocker T, Conrad S, Plotsky PM, et al. 2000. Activity of the hypothalamic-pituitary-adrenal axis is altered by aging and exposure to social stress in female rhesus monkeys. J Clin Endocrinol Metab 85: 2256-2563.

Haass C, Lemere CA, Capell A, Citron M, Seubert P, et al. 1995. The Swedish mutation causes early-onset Alzheimer's disease bei β-secretase cleavage within the secretory pathway. Nat Med 1: 1291-1296.

Haass C, Schlossmacher M, Hung A, Vigo-Pelfrey C, Mellon A, et al. 1992. Amyloid β-peptide is produced by cultured cells during normal metabolism. Nature 359: 322-324.

Halbreich U, Asnis GM, Zumoff B, Nathan RS, Shindledecker R. 1984. Effect of age and sex on cortisol secretion in depressives and normals. Psychiatry Res 13: 221-229.

Hamos JE, Oblas B, Pulaski-Salo D, Welch WJ, Bole DG, et al. 1991. Expression of heat shock proteins in Alzheimer's disease. Neurology 41: 345-350.

Hanger DP, Hughes K, Woodgett JR, Brion JP, Anderton BH. 1992. Glycogen synthase kinase-3 induces Alzheimer's disease-like phosphorylation of tau: Generation of paired helical filament epitopes and neuronal localisation of the kinase. Neurosci Lett 147: 58-62.

Hardy J, Selkoe DJ. 2002. The amyloid hypothesis of Alzheimer's disease: Progress and problems on the road of therapeutics. Science 297: 353-356.

Harik SI, McCracken KA. 1986. Age-related increase in presynaptic noradrenergic markers of the rat cerebral cortex. Brain Res 381: 125-130.

Häring HU. 1991. The insulin receptor: Signalling mechanism and contribution to the pathogenesis of insulin resistance. Diabetologica 34: 848-861.

Häring HU, Kirsch D, Obermeier B, Ermel B, Machicao F. 1986. Decreased tyrosine kinase activity of insulin receptor isolated from rat adipocytes rendered insulin-resistant by catecholamine treatment in vitro. Biochem J 234: 59-66.

Harman D. 1981. The aging process. Proc Natl Acad Sci USA 78: 7124-7128.

Harr SD, Simonian NA, Hyman BT. 1995. Functional alterations in Alzheimer's disease: Decreased glucose transporter 3 immunoreactivity in the perforant pathway terminal zone. J Neuropathol Exp Neurol 54: 38-41.

Hart GW. 1997. Dynamic O-linked glycosylation of nuclear and cytoskeletal proteins. Annu Rev Biochem 66: 315-335.

Hartmann T. 1999. Intracellular biology of Alzheimer's disease amyloid beta peptides. Eur Arch Psychiatry Clin Neurosci 249: 291-298.

Hartmann T, Bieger SC, Brüh lB, Tienari PJ, Ida N, et al. 1997. Distinct sites of intracellular production for Alzheimer's disease Aβ 40/42 amyloid peptides. Nat Med 3: 1016-1020.

Hashiguchi M, Saito T, Hisanaga S, Hashiguchi T. 2002. Truncation of CDK5 activator p35 induces intensive phosphorylation of Ser202/Thr205 of human tau. J Biol Chem 277: 44525-44530.

Hassler R. 1938. Zur Pathologie der Paralysis agitans und des postenzephatischen Parkinsonismus. Z Psychiat Neurol 48: 387-476.

Hauger RL, Thrivikraman KV, Plotsky PM. 1994. Age-related alterations of hypothalamic-pituitary-adrenal axis function in male Fischer 344 rats. Endocrinology 134: 1528-1536.

Hawkins RA, Williamson DH, Krebs HA. 1971. Ketone-body utilization by adult and suckling rat brain in vivo. Biochem J 122: 13-18.

Hayes BK, Hart GW. 1998. Protein O-GlycNAcylation: Potential mechanisms for the regulation of protein function. Adv Exp Med Biol 435: 85-94.

Haute Van den C, Spittaels K, Van Dorpe J, Lasrado R, Vandezande K, et al. 2000. Coexpression of human cdk5 and its activator p35 with human protein tau in neurons in brain of triple transgenic mice. Neurobiol Dis 8: 32-44.

Heidenreich KA, Zahniser NR, Berhanu P, Brandenburg D, Olefsky JM. 1983. Structural differences between insulin receptors in the brain and peripheral target tissues. J Biol Chem 258: 8527-8530.

Held GA, Solina DH, Keane DT, Haag WJ, Horn PM, et al. 1990. Experimental study of critical mass-fluctuations in an evolving sandpile. Phys Rev Lett 69: 1120-1123.

Henneberg N, Hoyer S. 1994. Short-term or long-term intracerebroventricular (i.c.v.) infusion of insulin exhibits a discrete anabolic effect on cerebral energy metabolism in the rat. Neurosci Lett 175: 153-156.

Henneberg N, Hoyer S. 1995. Desensitiziation of the neuronal insulin receptor: A new approach in the etiopathogenesis of late-onset sporadic dementia of the Alzheimer type (SDAT)? Arch Gerontol Geriatr 21: 63-74.

Herholz K, Salmon E, Perani D, Baron DJ, Holthoff V, et al. 2002. Discrimination between Alzheimer dementia and controls by automated analysis of multicenter FDG PET. NeuroImage 17: 302-316.

Herrera A, Tomás-Camardiel M, Venero JL, Cano J, Machado A. 2005. Inflammatory process as a determinant factor for the degeneration of substantia nigra dopaminergic neurons. J Neural Transm 112: 111-120.

Hertz MM, Paulson OB, Barry DI, Christiansen JS, Svendsen PA. 1981. Insulin increases glucose transfer across the blood-brain barrier. J Clin Invest 67: 597-604.

Hess B. 1983. Non-equilibrium dynamics of biochemical processes. Hoppe-Seylers Z Physiol Chem 364: 1-20.

Hess B. 1990. Order and chaos in chemistry and biology. Fresenius Anal Chem 337: 459-468.

Hill JM, Lesniak MA, Pert CB, Roth J. 1986. Autoradiographic localization of insulin receptors in rat brain: Promince in olfactory and limbic areas. Neuroscience 17: 1127-1138.

Hirsch E, Graybiel AM, Agid YA. 1988. Melanized dopaminergic neurons are differentially susceptible to degeneration in Parkinson's disease. Nature 334: 345-348.

Hirschberg CB, Snider MD. 1987. Topography of glycosylation in the rough endoplasmic reticulum and Golgi apparatus. Ann Rev Biochem 56: 63-87.

Holness MJ, Langdown M, Sugden MC. 2000. Early-life programming of susceptibility to dysregulation of glucose metabolism and the development of type 2 diabetes mellitus. Biochem J 349: 657-665.

Holsinger RMD, McLean CA, Beyreuther K, Masters CL, Evin G. 2002. Increased expression of the amyloid precursor β-secretase in Alzheimer's disease. Ann Neurol 51: 783-786.

Hong MF, Lee VMY. 1997. Insulin and insulin-like growth factor-1 regulate tau phosyphorylation in cultured human neurons. J Biol Chem 272: 19547-19553.

Hoogendijk WJG, Feenstra MGP, Botterblom MHA, Gilhuis J, Sommer IEC, et al. 1999. Increased activity of surviving locus ceruleus neurons in Alzheimer's disease. Ann Neurol 45: 82-89.

Hotamisligil GS, Murray DL, Choy LN, Spiegelmann BM. 1994. Tumor necrosis factor α inhibits signalling from the insulin receptor. Proc Natl Acad Sci USA 91: 4854-4858.

Hotamisligil GS, Peraldi P, Budavari A, Ellis R, White MF, et al. 1996. IRS-1-mediated inhibition of insulin receptor tyrosine kinase activity in TNF-α- and obesity-induced insulin resistance. Science 271: 665-668.

Hoyer A, Bardenheuer JH, Martin E, Plaschke K. 2005. Amyloid precursor protein (APP) and its derivatives change after cellular energy depletion. An in vitro study. J Neural Transm 112: 239-253.

Hoyer S. 1970. Der Aminosäurenstoffwechsel des normalen menschlichen Gehirns. Klin Wschr 48: 1239-1243.

Hoyer S. 1985. The effect of age on glucose and energy metabolism in brain cortex of rats. Arch Gerontol Geriatr 4: 193-203.

Hoyer S. 1986. Senile dementia and Alzheimer's disease: Brain blood flow and metabolism. Prog Neuropsychopharmacol Biol Psychiatry 10: 447-478.

Hoyer S. 1988. Glucose and related brain metabolism in dementia of Alzheimer type and its morphological significance. Age 11: 158-166.

Hoyer S. 1992. Oxidative energy metabolism in Alzheimer brain. Studies in early-onset and late onset cases. Mol Chem Neuropathol 16: 207-224.

Hoyer S. 1995. Age-related changes in cerebral oxidative metabolism. Implications for drug therapy. Drugs Aging 6: 210-218.

Hoyer S. 1996. Oxidative metabolism deficiencies in brains of patients with Alzheimer's disease. Acta Neurol Scand Suppl 165: 18-24.

Hoyer S. 1998. Is sporadic Alzheimer disease the brain type of non-insulin dependent diabetes mellitus? A challenging hypothesis. J Neural Transm 105: 415-422.

Hoyer S. 2000a. The aging brain. Changes in the neuronal insulin/insulin receptor signal transduction cascade trigger late-onset sporadic Alzheimer disease (SAD). A mini-review. J Neural Transm 109: 991-1002.

Hoyer S. 2000b. Brain glucose and energy metabolism abnormalities in sporadic Alzheimer disease. Causes and consequences: An update. Exp Gerontol 35: 1363-1372.

Hoyer S. 2002. The brain insulin signal transduction system and sporadic (type II) Alzheimer disease: An update. J Neural Transm 109: 341-360.

Hoyer S. 2004a. Causes and consequences of disturbances of cerebral glucose metabolism in sporadic Alzheimer disease: Therapeutic implications. Adv Exp Med Biol 541: 135-152.

Hoyer S. 2004b. Glucose metabolism and insulin receptor signal transduction in Alzheimer disease. Eur J Pharmacol 490: 115-125.

Hoyer S, Betz K. 1988. Abnormalities in glucose and energy metabolism are more severe in the hippocampus than in the cerebral cortex in postischemic recovery in aged rats. Neurosci Lett 94: 167-172.

Hoyer S, Krier C. 1986. Ischemia and the aging brain. Studies on glucose and energy metabolism in rat cerebral cortex. Neurobiol Aging 7: 23-29.

Hoyer S, Lannert H, Latteier E, Meisel T. 2004. Relationship between cerebral energy metabolism in parietotemporal cortex and hippocampus and mental activity during aging in rats. J Neural Transm 111: 575-589.

Hoyer S, Nitsch R. 1989. Cerebral excess release of neurotransmitter amino acids subsequent to reduced cerebral glucose metabolism in early-onset dementia of Alzheimer type. J Neural Transm 75: 227-232.

Hoyer S, Nitsch R, Oesterreich K. 1990. Ammonia is endogenously generated in the brain in the presence of presumed and verified dementia of Alzheimer type. Neurosci Lett 117: 358-362.

Hoyer S, Nitsch R, Oesterreich K. 1991. Predominant abnormality in cerebral glucose utilization in late-onset dementia of the Alzheimer-type: A cross-sectional comparison against advanced late-onset dementia and incipient early-onset cases. J Neural Transm (PD-Sect) 3: 1-14.

Hoyer S, Oesterreich K, Wagner O. 1988. Glucose metabolism as the site of the primary abnormality in early-onset dementia of Alzheimer type. J Neurol 235: 143-148.

Hoyer S, Plaschke K. 2004. Brain damage and repair. From Molecular Research to Clinical Therapy. Herdegen T, Dilgado Garccia J, editors. Dordrecht, Boston, London: Kluwer; pp. 1-22.

Hoyer S, Prem L, Sorbi S, Amaducci L. 1993. Stimulation of glycolytic key enzymes in cerebral cortex by insulin. Neuroreport 4: 991-993.

Hughes K, Nikolakaki E, Plyte SE, Totty NF, Woodgett JR. 1993. Modulation of the glycogen synthase kinase-3 family by tyrosine phosphorylation. EMBO J 12: 803-808.

Hung HC, Tsai MJ, Wu HC, Lee EHY. 2000. Age-dependent increase in C7-1 gene expression in frontal rat cortex. Mol Brain Res 75: 330-336.

Hyman BT, van Hoesen GW, Damasio AR, Barnes CL. 1984. Alzheimer's disease: Cell-specific pathology isolates the hippocampal formation. Science 225: 1168-1170.

Hyman BT, van Hoesen GW, Kromer LJ, Damasio AR. 1986. Perforant pathway changes and the memory impairment of Alzheimer's disease. Ann Neurol 20: 472-481.

Ida Y, Tanaka M, Kohno Y, Nakagawa R, Iiomori K, et al. 1982. Effect of age and stress on regional noradrenaline metabolism in the rat brain. Neurobiol Aging 3: 233-236.

Iijima K, Ando K, Takeda S, Satoh Y, Seki T, et al. 2000. Neuron-specific phosphorylation of Alzheimer's beta-amyloid precursor protein by cyclindependent kinase 5. J Neurochem 75: 1085-1091.

Imahori K, Uchida T. 1997. Physiology and pathology of tau protein kinases in relation to Alzheimer's disease. J Biochem 121: 179-188.

Imai Y, Soda M, Takahashi R. 2000. Parkin suppresses unfolded protein stress-induced cell death through its E3 ubiquitation-protein ligase activity. J Biol Chem 275: 35661-35664.

Inagaki N, Gonor T, Clement JP, Wang CZ, Aquilar-Bryan L, et al. 1996. A family of sulfonylurea receptors determines the pharmacological properties of ATP-sensitive K^+ channels. Neuron 16: 1101-1107.

Ishiguro K, Shiratsuchi A, Sato S, Omori A, Arioka M, et al. 1993. Glycogen synthase kinase 3-beta is identical to tau protein kinase I generating several epitopes of paired helical filaments. FEBS Lett 325: 167-172.

Ishiguro K, Takamatsu M, Tomizawa K, Omori A, Takahashi M, et al. 1992. Tau protein kinase I converts normal tau protein into A68-like component of paired helical filaments. J Biol Chem 267: 10897-10901.

Iwata N, Tsubuki S, Takaki Y, Shirotani K, Lu B, et al. 2001. Metabolic regulation of brain Aβ by neprilysin. Science 292: 1550-1552.

Iwata N, Tsubuki S, Takaki Y, Watanabe K, Sekiguchi M, et al. 2000. Identification of the major Aβ 1–42-degrading catabolic pathway in brain parenchyma: Suppression leads to biochemical and pathological deposition. Nat Med 6: 143-150.

Jellinck PH, Pavlides C, Sakai RR, McEwen BS. 1999. 11β-hydroxysteroid dehydrogenase functions reversibly as an oxidoreductase in the rat hippocampus in vivo. J Steroid Biochem Mol Biol 71: 139-144.

Jellinger K. 1987. Movement Disorders, Vol. 2. Marsden CD, Fahn S, editors. London: Butterworth; pp. 124-165.

Jellinger K, Danielczyk W. 1968. Striato-nigrale degeneration. Acta Neuropathol 10: 242-257.

Jenner P, 1998. Oxidative mechanisms in nigral cell death in Parkinson's disease. Mov Disord 13 (Suppl.): 24-34.

Jiang CH, Tsien JZ, Schultz PG, Hu Y. 2001. The effects of aging on gene expression in the hypothalamus and cortex of mice. Proc Natl Acad Sci USA 98: 1930-1934.

Johnson SA, Pasinetti GM, May PC, Ponte PA, Cordell B, et al. 1988. Selective reduction of mRNA for the beta-amyloid precursor protein that lacks a Kunitz-type protease inhibitor motif in cortex from Alzheimer brains. Exp Neurol 102: 264-268.

Joseph J, Shukitt-Hale B, Denisova NA, Martin A, Perry G, et al. 2001. Copernicus revisited: Amyloid beta in Alzheimer's disease. Neurobiol Aging 22: 131-146.

Kadekaro M, Crane AM, Sokoloff L. 1985. Differential effects of electrical stimulation of sciatic nerve on metabolic activity in spinal cord and dorsal root ganglion in the rat. Proc Natl Acad Sci USA 82: 6010-6013.

Kahn CR. 1985. The molecular mechanism of insulin action. Ann Rev Med 36: 429-451.

Kaufman RJ. 1999. Stress signalling from the lumen of the endoplasmic reticulum: Coordination of gene transcriptional and translational controls. Genes Dev 13: 1211-1233.

Kauppinen RA, Nicholls DG. 1986. Failure to maintain glycolysis in anoxic nerve terminals. J Neurochem 47: 1864-1869.

Keller JN, Pang Z, Geddes JW, Begley JG, Germeyer A, et al. 1997. Impairment of glucose and glutamate transport and induction of mitochondrial oxidative stress and dysfunction in synaptosomes by amyloid β-peptide: Role of the lipid peroxidation product 4-hydroxynonenal. J Neurochem 69: 273-284.

Kelly WG, Dahmus ME, Hart GW. 1993. RNA polymerase II is a glycoprotein: Modification of the COOH-terminal domain by O-GlcNAc. J Biol Chem 268: 10416-10424.

Kienlen-Campard P, Miolet S, Tasiaux B, Octave JN. 2002. Intracellular amyloid-β 1–42, but not extracellular soluble amyloid-β peptides, induces neuronal apoptosis. J Biol Chem 277: 15666-15670.

Kim SJ, Han Y. 2005. Insulin inhibits AMPA-induced neuronal damage via stimulation of protein kinase B (Akt). J Neural Transm 112: 179-191.

Kirschenbaum F, Hsu SC, Cordell B, McCarthy JV. 2001a. Substitution of a glycogen synthase kinase-3 phosphorylation site in presenilin 1 separates presenilin function from β-catenin signalling. J Biol Chem 276: 7366-7375.

Kirschenbaum F, Hsu SC, Cordell B, McCarthy JV. 2001b. Glycogen synthase kinase-3β regulates presenilin 1 C-terminal fragment levels. J Biol Chem 276: 30701-30707.

Klein J. 2000. Membrane breakdown in acute and chronic neurodegeneration: Focus on choline-containing phospholipids. J Neural Transm 107: 1027-1063.

Klein J, Fasshauer M, Ito M, Lowell BB, Benito M, et al. 1999. β3-Adrenergic stimulation differentially inhibits insulin signalling and decreases insulin-induced glucose uptake in brown adipocytes. J Biol Chem 274: 34795-34802.

Kliegel M, Moor C, Rott C. 2004. Cognitive status and development in the oldest old: A longitudinal analysis from the Heidelberg Centenarian Study. Arch Gerontol Geriatr 39: 143-156.

Kometiani ZP, Tsakadze LG, Jariashvili TY. 1984. Functional significance of the effects of neurotransmitters in the Na^+, K^+-ATPase system. J Neurochem 42: 1246-1250.

Kozutsumi Y, Segal M, Normington K, Gething MJ, Sambrook J. 1988. The presence of malfolded proteins in the endoplasmic reticulum signals the induction of glucose regulated proteins. Nature 332: 462-464.

Kreutzberg W. 1996. Microglia: A sensor for pathological events in the CNS. Trends Neurosci 19: 312-318.

Kuo WL, Montag AG, Rosner MR. 1993. Insulin-degrading enzyme is differentially expressed and developmentally regulated in various rat tissues. Endocrinology 132: 604-611.

Kurochkin IV, Goto S. 1994. Alzheimer's beta-amyloid peptide specifically interacts with and is degraded by insulin degrading enzyme. FEBS Lett 345: 33-37.

Kusakawa G, Saito T, Onuki R, Ishiguro K, Kishimoto T, et al. 2000. Calpain-dependent proteolytic cleavage of the p35 cyclindependent kinase 5 activator to p25. J Biol Chem 275: 17166-17172.

Kyriakis JM, Hausman RE, Peterson SW. 1987. Insulin stimulates choline acetyltransferase activity in cultured embryonic chicken retina neurons. Proc Natl Acad Sci USA 84: 7463-7467.

Lai JCK, Cooper AJL. 1991. Neurotoxicity of ammonia and fatty acids: Differential inhibition of mitochondrial dehydrogenases by ammonia and fatty acids coenzyme A derivatives. Neurochem Res 16: 795-803.

Lannert H, Gorgas K, Meißner I, Wieland FT, Jeckel D. 1998. Functional organization of the Golgi apparatus in glycosphingolipid biosynthesis. J Biol Chem 273: 2939-2946.

Lannert H, Hoyer S. 1998. Intracerebroventricular administration of streptozotocin causes long-term diminutions in learning and memory abilities and in cerebral energy metabolism in adult rats. Behav Neurosci 112: 1199-1208.

Lannfelt L, Basun H, Wahlund LO, Rowe BA, Wagner SL. 1995. Decreased α-secretase-cleaved amyloid precursor protein as a diagnostic marker for Alzheimer's disease. Nat Med 1: 829-832.

Leanza G. 1998. Chronic elevation of amyloid precursor protein expression in the neocortex and hippocampus of rats with selective cholinergic lesions. Neurosci Lett 257: 53-56.

Lee J, Stock J. 1993. Protein phosphatase 2A catalytic subunit is methylesterified at its carboxyl terminus by a novel methyltransferase. J Biol Chem 268: 19192-19195.

Lee K, Dixon AK, Rowe ICM, Ashford MJK, Richardson PJ. 1996. The high-affinity sulfonylurea receptors regulates K_{ATP} channels in nerve terminal of the rat motor cortex. J Neurochem 66: 2562-2571.

Lee MS, Kwon YT, Li M, Peng J, Friedlander RM, et al. 2000. Neurotoxicity induces cleavage of p35 to p25 by calpain. Nature 405: 360-364.

Leibovitz BE, Siegel BV. 1980. Aspects of free radical reactions in biological systems: aging. J Gerontol 35: 45-56.

Leloup C, Arluison M, Lepetit N, Cartier N, Marfaing-Jallat P, et al. 1994. Glucose transporter 2 (GLUT2): Expression in specific brain nuclei. Brain Res 638: 221-226.

Lesort M, Jope RS, Johnson GVW. 1999. Insulin transiently increases tau phosphorylation: Involvement of glycogen synthase kinase-3β and Fyn tyrosine kinase. J Neurochem 72: 576-584.

Levin BE, Dunn-Meynell AA. 1997. In vivo and in vitro regulation of (^{3}H) glyburide binding to brain sulfonylurea receptors in obesity-prone and resistant rats by glucose. Brain Res 776: 146-153.

Levine RL, Oliver CN, Fulks RM, Stadtman ER. 1981. Turnover of bacterial glutamine synthetase: Oxidative inactivation precedes proteolysis. Proc Natl Acad Sci USA 78: 2120-2124.

Lewis LD, Ljunggren B, Ratcheson RA, Siesjö BK. 1974. Cerebral energy state in insulin-induced hypoglycaemia, related to blood glucose and to EEG. J Neurochem 23: 673-679.

Li BS, Sun MK, Zhang L, Takahashi S, Ma W, et al. 2001. Regulation of NMDA receptors by cyclin-dependent kinase-5. Proc Natl Acad Sci USA 98: 12742-12747.

Li C, Booze RM, Hersh LB. 1995a. Tissue-specific expression of rat neutral endopeptidase (neprilysin) mRNAs. J Biol Chem 270: 5723-5728.

Li JJ, Surini M, Catsicas S, Kawashima E, Bouras C. 1995b. Age-dependent accumulation of advanced glycation end products in human neurons. Neurobiol Aging 16: 69-76.

Li M, Wang X, Meintzer M, Laessig T, Birnbaum MJ, et al. 2000. Cyclic AMP promotes neuronal survival by phosphorylation of glycogen synthase kinase 3β. Mol Cell Biol 20: 9356-9363.

Lieb K, Fiebich BL, Schaller H, Berger M, Bauer J. 1996. Interleukin-1 beta and tumor necrosis factor-alpha induce expression of alpha 1-antichymotrypsin in human astrocytoma cells by activation of nuclear factor-kappa B. J Neurochem 67: 2039-2044.

Lin T, Le Blanc CJ, Deacon TW, Isascon O. 1998. Chronic cognitive deficits and amyloid precursor protein elevation after selective immunnotoxin lesions of the basal forebrain cholinergic system. Neuroreport 9: 547-552.

Lippa AS, Critchett DJ, Ehlert F, Yamamura HI, Enna SJ, et al. 1981. Age-related alterations in neurotransmitter receptors: An electrophysiological and biochemical analysis. Neurobiol Aging 2: 3-8.

Lipton SA. 1997. Janus faces of NFKB activity: Neurodestruction versus neuroprotection. Nat Med 3: 20-22.

Liu F, Iqbal K, Grundke-Iqbal I, Hart GW, Gong CX. 2004. O-GlycNacylation regulates phosphorylation of tau: A unique mechanism involved in Alzheimer's disease. Proc Natl Acad Sci USA 101: 19804-19809.

Lorens SA, Hata N, Handa RJ, van de Kar LD, Guschwan M, et al. 1990. Neurochemical, endocrine and immunological responses to stress in young and old Fischer 344 male rats. Neurobiol Aging 11: 139-150.

Loring JF, Wen X, Lee JM, Seilhamer J, Somogyi R. 2001. A gene expression profile of Alzheimer's disease. DNA Cell Biol 20: 683-695.

Lovestone S, Reynolds CH. 1997. The phosphorylation of tau: A critical stage in neurodevelopment and neurodegenerative processes. Neuroscience 78: 309-324.

Lowe WL Jr, Boyd FT, Clarke DW, Raizada MK, Hart C, et al. 1986. Development of brain insulin receptors: Structural and functional studies of insulin receptors from whole brain and primary cell cultures. Endocrinology 119: 25-35.

Lue L, Walker D, Rogers J. 2001. Modeling microglial activation in Alzheimer's disease with human postmortem microglial cultures. Neurobiol Aging 22: 945-956.

Lupien S, Lecors A, Lussier I, Schwartz G, Nair N, et al. 1994. Basal cortisol levels and cognitive deficits in human aging. J Neurosci 14: 2893-2903.

Maes M, De Vos N, Wauters A, Demedts P, Maurits VW, et al. 1999. Inflammatory markers in younger vs elderly normal volunteers and in patients with Alzheimer's disease. J Psychiatry Res 33: 397-405.

Maher F, Simpson IA, Vannucci SJ. 1993. Alterations in brain glucose transporter proteins GLUT1 and GLUT3, in streptozotocin diabetic rats. Adv Exp Med Biol 331: 9-12.

Mandelkow EM, Drewes G, Biernet J, Gustke N, van Lint J, et al. 1992. Glycogen synthase kinase-3 and the Alzheimer-like state of microtubule-associated protein tau. FEBS Lett 314: 215-221.

Marsden CD. 1990. Parkinson's disease. Lancet 335: 948-952.

Martinez M, Fernandez E, Frank A, Guaza C, de la Fuente M, et al. 1999. Increased cerebrospinal fluid cAMP levels in Alzheimer's disease. Brain Res 846: 265-267.

Mastrogiacomo F, Bergeron C, Kish SJ. 1993. Brain α-ketoglutarate dehydrogenase complex activity in Alzheimer's disease. J Neurochem 61: 2007-2014.

Mauch DH, Nägler J, Schumacher S, Göritz EC, Otto A, et al. 2001. CNS synaptogenesis promoted by glia-derived cholesterol. Science 294: 1354-1357.

McCusker SM, Curran MD, Dynan KB, McCullagh CD, Urquhart DD, et al. 2001. Association between polymorphism in regulatory region of gene encoding tumour necrosis factor α and risk of Alzheimer's disease and vascular dementia: A case-control study. Lancet 337: 436-439.

McDermott JR, Gibson AM. 1997. Degradation of Alzheimer's beta-amyloid protein by human and rat brain peptidases: Involvement of insulin-degrading enzyme. Neurochem Res 22: 49-56.

McMillan PJ, Leverenz JB, Dorsa DM. 2000. Specific downregulation of presenilin 2 gene expression is prominent during early stages of sporadic late-onset Alzheimer's disease. Brain Res Mol Brain Res 78: 138-145.

McMillan PJ, Leverenz JB, Poorkaj P, Schellenberg GC, Dorsa DM. 1996. Neuronal expression of STM2 mNRA in human brain is reduced in Alzheimer's disease. J Histochem Cytochem 44: 1215-1222.

Merkel P, Khoury N, Bertolotto C, Perfetti R. 2003. Insulin and glucose regulate the expression of the DNA repair enzyme XPD. Mol Cell Endocrinol 201: 75-85.

Michikawa M, Yanagisawa K. 1999. Inhibition of cholesterol production but not of nonsterol isoprenoid products induces neuronal cell death. J Neurochem 72: 2278-2285.

Mielke R, Herholz K, Grond M, Heiss WD. 1994. Clinical deterioration in probable Alzheimer's disease correlates with progressive metabolic impairment of association areas. Dementia 5: 36-41.

Mielke R, Herholz K, Grond M, Kessler J, Heiss WD. 1992. Differences of regional cerebral glucose metabolism between presenile and senile dementia of Alzheimer type. Neurobiol Aging 13: 93-98.

Miller BC, Eckman EA, Sambamurti K, Dobbs N, Chon KM, et al. 2003. Amyloid-β peptide levels in brain are inversely correlated with insulysin activity levels in vivo. Proc Natl Acad Sci USA 100: 6221-6226.

Mindham RH, Ahmed SW, Clough CG. 1992. A controlled study of dementia in Parkinson's disease. J Neurol Neurosurg Psychiatry 45: 969-974.

Minoshima S, Giordani B, Berent S, Frey KA, Foster NL, et al. 1997. Metabolic reduction in the posterior cingulate cortex in very early Alzheimer's disease. Ann Neurol 42: 85-94.

Mirollo RE, Strogatz SH. 1990. Synchronization of pulse-coupled biological oscillators. SIAM J Appl Math 50: 1645-1649.

Moll G, Gsell W, Wichart I, Jellinger K, Riederer P. 1990. Alzheimer's Disease: Epidemiology, Neuropathology, Neurochemistry, and Clinics. Maurer K, Riederer P, Beckmann H, editors. New York: Springer Wien; pp. 235-243.

Monnier VM, Cerami A. 1991. Nonenzymatic browning in vivo: Possible process for aging of long-lived proteins. Science 211: 491-493.

Moss AM, Unger JW, Moxley RT, Livingston JN. 1990. Location of phosphotyrosine-containing proteins by immunocytochemistry in the rat forebrain corresponds to the distribution of insulin receptor. Proc Natl Acad Sci USA 87: 4453-4457.

Mourre C, Ben Ari Y, Bernardi H, Fosset M, Lazdunski M. 1989. Antidiabetic sulfonylureas: Localization of binding sites in the brain and effects on the hyperpolarization induced by anoxia in hippocampal slices. Brain Res 486: 159-164.

Mukherjee A, Song E, Kihiko-Ehmann M, Goodman JP Jr, St Pyrek J, et al. 2000. Insulysin hydrolyzes amyloid β peptide to products that are neither neurotoxic nor deposit on amyloid plaques. J Neurosci 20: 8745-8749.

Mulder M, Ravid R, Swaab DF, de Kloet ER, Haasdijk ED, et al. 1998. Reduced levels of cholesterol phospholipids, and fatty acids in cerebrospinal fluid of Alzheimer disease patients are not related to apolipoprotein E4. Alzheimer Dis Assoc Disord 12: 198-203.

Mullamy P, Conolly S, Lynch MA. 1996. Ageing is associated with changes in glutamate release, protein tyrosine kinase and Ca^{2+}/ calmodulin-dependent protein kinase II in rat hippocampus. Eur J Pharmacol 309: 311-315.

Münch G, Apelt J, Kientsch-Engel R, Stahl P, Lüth H, et al. 2003. Advanced glycation endproducts and pro-inflammatory cytokines in transgenic Tg 2576 mice with amyloid plaque pathology. J Neurochem 86: 283-289.

Münch G, Schinzel R, Loske C, Wong A, Durany N, et al. 1998. Alzheimer's disease-synergistic effects of glucose deficit, oxidative stress and advanced glycation endproducts. J Neural Transm 105: 439-461.

Munro S, Pelham HR. 1986. An Hsp 70-like protein in the ER: identity with the 78 kd glucose-regulated protein and immunoglobulin heavy chain binding protein. Cell 46: 291-300.

Nath R, Davis M, Probert AW, Kupina NC, Ren X, et al. 2000. Processing of cdk5 activator p35 to its truncated form (p25) by calpain in acutely injured neuronal cells. Biochem Biophys Res Commun 274: 16-21.

Newsholm EA, Start C. 1973. Regulation in metabolism. Wiley; Chichester: pp. 88-145.

Nichols NR. 1999. Glial responses to steroids as markers of brain aging. J Neurobiol 40: 585-601.

Nikolic M, Dudek H, Kwon YT, Ramos YF, Tsai LH. 1996. The cdk5/ p35 kinase is essential for neurite outgrowth during neuronal differentiation. Genes Dev 10: 816-825.

Nitsch R, Hoyer S. 1991. Local action of the diabetogenic drug, streptozotocin, on glucose and energy metabolism in rat brain cortex. Neurosci Lett 128: 199-202.

Nitsch RM, Blusztajn JK, Pittas AG, Slack BE, Grodwon JH, et al. 1992b. Evidence for a membrane defect in Alzheimer's disease brain. Proc Natl Acad Sci USA 89: 1671-1675.

Nitsch RM, Slack BE, Wurtman RJ, Growdon J. 1992a. Release of Alzheimer amyloid precursor derivatives stimulated by activation of muscarinic acetylcholine receptors. Science 258: 304-307.

Noble W, Olm V, Takata K, Casey E, Mary O, et al. 2003. Cdk5 is a key factor in tau aggregation and tangle formation in vivo. Neuron 38: 555-565.

Nohl H, Krämer R. 1980. Molecular basis of age-dependent changes in the activity of adenine nucleotide translocase. Mech Ageing Dev 14: 137-144.

Norberg K, Siesjö BK. 1976. Oxidative metabolism of the cerebral cortex of the rat in insulin-induced hypoglycaemia. J Neurochem 26: 345-352.

Nordstedt C, Candy SE, Alafuzoff I, Caporaso GL, Iverfeldt K, et al. 1991. Alzheimer beta/A4 amyloid precursor protein in human brain: Aging-associated increases in holoprotein and in a proteolytic fragment. Proc Natl Acad Sci USA 88: 8910-8914.

O'Brien JT, Ames D, Schweitzer I, Mastwyk M, Colman P. 1996. Enhanced adrenal sensitivity to adrenocorticotrophic hormone (ACTH) is evident of HPA axis hyperactivity in Alzheimer's disease. Psychol Med 26: 7-14.

Ohtsuka K, Suzuki T. 2000. Roles of molecule chaperones in the nervous system. Brain Res Bull 53: 141-146.

Ott A, Breteler MMB, van Harskamp F, Claus JJ, Cammen van der TJM, et al. 1995. Prevalence of Alzheimer's disease and vascular dementia: Association with education. The Rotterdam study. Br J Med 310: 970-973.

Owen OE, Morgan AP, Kemp HG, Sullivan JM, Herrera MG, et al. 1967. Brain metabolism during fasting. J Clin Invest 46: 1589-1597.

Oxenkrug GF, Pomara N, McIntyre IM, Branconnier RJ, Stanley M, et al. 1983. Aging and cortisol resistance to suppression by dexamethasone: A positive correlation. Psychiatry Res 10: 125-130.

Palmert MR, Golde TE, Cohen ML, Kovacs DM, Tanzi RE, et al. 1988. Amyloid protein precursor messenger RNAs: Differential expression in Alzheimer's disease. Science 241: 1080-1084.

Pardridge WM, Boado RJ, Farrell CR. 1990. Brain-type glucose transporter (GLUT-1) is selectively localized to the blood-brain barrier. Studies with quantitative Western blotting and in situ hybridization. J Biol Chem 265: 18035-18040.

Pardridge WM, Eisenberg J, Yang J. 1985. Human blood-brain barrier insulin receptor. J Neurochem 44: 1771-1778.

Parhad IM, Scott JN, Cellars LA, Bains JS, Krekoski CA, et al. 1995. Axonal atrophy in aging is associated with a decline in neurofilament gene expression. J Neurosci Res 41: 355-366.

Park CR. 2001. Cognitive effects of insulin in the central nervous system. Neurosci Biohav Rev 25: 311-323.

Parker WD, Parks J, Filley CM, Kleinschmidt-De Masters BK. 1994. Electron transport chain defects in Alzheimer's disease brain. Neurology 44: 1090-1096.

Parkinson J. 1817. Essay on the Shaking Palsy. London: Whittingham & Bowland.

Paschen W, Frandsen A. 2001. Endoplasmic reticulum dysfunction-a common denominator for cell injury in acute and degenerative diseases of the brain? J Neurochem 79: 719-725.

Pascualy M, Petrie EC, Brodkin K, Peskind ER, Wilkinson W, et al. 2000. Hypothalamic pituitary adrenocortical and sympathetic nervous system responses to the cold pressure test in Alzheimer's disease. Biol Psychiatry 48: 247-254.

Payne J, Maher F, Simpson U, Mattice L, Davies P. 1997. Glucose transporter Glut 5 expression in microglial cells. Glia 21: 327-331.

Pearson RCA, Esiri MM, Hiorns RW, Wilcock GK, Powell TPS. 1985. Anatomical correlates of the distribution of the pathological changes in the neocortex in Alzheimer disease. Proc Natl Acad Sci USA 82: 4531-4534.

Peraldi Hotamisligil GS, Buurman WA, White MF, Spiegelman BM. 1996. Tumor necrosis factor (TNF)-α inhibits insulin signaling through stimulation of the p55TNF receptor and activation of sphingomyelinase. J Biol Chem 271: 13018-13022.

Perego C, Vetrugno CC, de Simoni MG, Algeri S. 1993. Aging prolongs the stress-induced release of noradrenaline in rat hippocampus. Neurosci Lett 157: 127-130.

Perez A, Morelli L, Cresto JC, Castano EM. 2000. Degradation of soluble amyloid β-peptides 1–40, 1–42, and the Dutch variant 1–40 Q by insulin degrading enzyme from Alzheimer disease and control brain. Neurochem Res 25: 247-255.

Perry EK, Perry RG, Tomlinson BE, Blessed G, Gibson PH. 1980. Coenzyme A-acetylating enzymes in Alzheimer's diseae: Possible cholinergic "compartment" of pyruvate dehydrogenase. Neurosci Lett 18: 105-110.

Peskind ER, Elrod R, Dobie DJ, Pascualy M, Petrie E, et al. 1998. Cerebrospinal fluid epinephrine in Alzheimer's disease and normal aging. Neuropsychopharmacology 19: 465-471.

Peskind ER, Wilkinson CW, Schellenberg EC, Raskind MA. 2001. Increased CSF cortisol in AD is a function of APOE genotype. Neurology 56: 1094-1098.

Peskind ER, Wingerson D, Murray S, Pascualy M, Dobie DJ, et al. 1995. Effects of Alzheimer's disease and normal aging on cerebrospinal fluid norepinephrine responses to yohimbine and clonidine. Arch Gen Psychiatry 52: 774-782.

Petanceska SS, Gandy S. 1999. The phosphatidylinositol 3-kinase inhibitor wortmannin alters the metabolism of the Alzheimer's amyloid precursor protein. J Neurochem 73: 2316-2320.

Pettegrew JW, Klunk WE, Kanal E, Panchalingam K, McClure RJ. 1995. Changes in brain membrane phospholipids and high-energy phosphate metabolism precede dementia. Neurobiol Aging 16: 973-975.

Phiel CJ, Wilson CA, Lee VMY, Klein PS. 2003. GSK-3α regulates production of Alzheimer's disease amyloid-β peptides. Nature 423: 435-439.

Pigino G, Paglini G, Ulloa L, Avila J, Caceres A. 1997. Analysis of the expression, distribution and function of cyclin dependent kinase 5 (cdk5) in developing cerebellar macroneurons. J Cell Sci 110 (Pt. 2): 257-270.

Pillon B, Dubois B, Lhermitte F, Agid Y. 1986. Heterogeneity of cognitive impairment in progressive supranuclear palsy Parkinson's disease, and Alzheimer's disease. Neurology 36: 1179-1185.

Planel E, Miyasaka I, Launey T, Chui DH, Tanermura K, et al. 2004. Alteration in glucose metabolism induce hypothermia leading to tau hyperphosphorylation through differential inhibition of kinase and phosphatase activities: implications for Alzheimer's disease. J Neurosci 24: 2401-2411.

Planel E, Yasutake K, Fujita SC, Ishiguro K. 2001. Inhibition of protein phosphatase 2A overrides tau protein kinase I/glycogen synthase kinase 3β and cyclin-dependent kinase 5 inhibition and results in tau hyperphosphorylation in the hippocampus of starved mice. J Biol Chem 276: 34298-34306.

Plata-Salaman CR. 1991. Insulin in the cerebrospinal fluid. Neurosci Biobehav Rev 15: 243-258.

Plee-Gautier E, Grimal H, Aggerbeck M, Barouki R, Forest C. 1998. Cytosolic aspartate aminotransferase gene is a member of the glucose-regulated protein gene family in adipocytes. Biochem J 329: 37-40.

Prigoine I. 1989. What is entropy? Naturwissenschaften 76: 1-8.

Procter AW, Palmer AM, Francis PT, Lowe SL, Neary D, et al. 1988. Evidence of glutamatergic denervation and possible abnormal metabolism in Alzheimer's disease. J Neurochem 50: 790-802.

Raizada ML, Shemer J, Judkins JH, Clarke DW, Masters BA, et al. 1988. Insulin receptors in the brain: Structural and physiological characterization. Neurochem Res 13: 297-303.

Ramakrishna S, Benjamin WB. 1998. Insulin action rapidly decreases multifunctional protein kinase activity in rat adipose tissue. J Biol Chem 263: 12677-12681.

Ramirez MJ, Ridley RM, Baker HF, Maclean CJ, Honer WG, et al. 2001. Chronic elevation of amyloid precursor protein in the neocortex or hippocampus of marmosets with selective cholinergic lesions. J Neural Transm 108: 809-826.

Rehncrona S, Hauge HN, Siesjö BK. 1989. Enhancement of iron-catalized free radical formation by acidosis in brain homogenates: Difference in effect by lactic acid and CO_2. J Cereb Blood Flow Metab 9: 65-70.

Reid WG, Hely MA, Morris JGL, Broe GA, Adena M, et al. 1996. A longitudinal study of Parkinson's disease: Clinical and neuropathological correlates of dementia. J Clin Neurosci 3: 327-333.

Reul JMHM, Rothuizen J, Dekloet ER. 1991. Age-related changes in the dog hypothalamic-pituitary-adrenocortical system: Neuroendocrine activity and corticosteroid receptors. J Steroid Biochem Mol Biol 40: 63-69.

Reul JMHM, Tonnaer JADM, de Kloet ER. 1988. Neurotropic ACTH analogue promotes plasticity of type I corticosteroid receptors in brain of senescent rats. Neurobiol Aging 9: 253-260.

Reynolds CH, Utton MA, Gibb GM, Yates A, Anderton BH. 1997. Stress-activated protein kinase/c-jun N-terminal kinase phosphorylates tau-protein. J Neurochem 68: 1736-1744.

Riederer P, Sofic E, Rausch WD, Schmidt B, Reynolds GP, et al. 1989. Transition metals, ferritin, glutathione, and ascorbic acid in parkinsonian brains. J Neurochem 52: 515-520.

Rinaudo MT, Curto M, Bruno R, Marino C, Rossetti V, et al. 1987. Evidence of an insulin generated pyruvate dehydrogenase stimulating factor in rat brain plasma membranes. Ital J Biochem 19: 909-913.

Roberts EK Jr, Chih CP. 1995. Age-related alterations in energy metabolism contribute to the increased vulnerability of the aging brain to anoxic damage. Brain Res 678: 83-90.

Roberts EL Jr, Sick TJ. 1996. Aging impairs regulation of intracellular pH in rat hippocampal slices. Brain Res 735: 339-342.

Rocchi A, Pellegrini S, Siciliano G, Murri K. 2003. Causative and susceptibility genes for Alzheimer's disease: A review. Brain Res Bull 61: 1-24.

Roceri M, Hendriks W, Racagni G, Ellenbroek BA, Riva MA. 2002. Early maternal deprivation reduces the expression of BDNF and NMDA receptor subunits in rat hippocampus. Mol Psychiatry 7: 609-616.

Röder HM, Ingram VM. 1991. Two novel kinases phosphorylate tau and the KSP site of heavy neurofilament subunits in high stoichiometric ratios. J Neurosci 11: 3325-3342.

Roland BL, Li KXZ, Funder JW. 1995. Hybridization histochemical localization of 11β-hydroxysteroid dehydrogenase type 2 in rat brain. Endocrinology 136: 4697-4700.

Rossner S, Weberham U, Yu J, Kirazov L, Schliebs R, et al. 1997. In vivo regulation of amyloid precursor protein secretion in rat neocortex by cholinergic activity. Eur J Neurosci 9: 2125-2134.

Roth J. 1987. Subcellular organization of glycosylation in mammalian cells. Biochim Biophys Acta 906: 405-436.

Roth J, Taatjes DJ, Lucocq JM, Weinstein J, Paulson JC. 1985. Demonstration of an extensive trans-tubular network continuous with the Golgi apparatus stack that may function in glycosylation. Cell 43: 287-295.

Rothman JE. 1996. The protein machinery of vesicle budding and fusion. Protein Sci 5: 185-194.

Rothman JE, Orci L. 1992. Molecular dissection of the secretory pathway. Nature 355: 409-415.

Rothman JE, Wieland FT. 1996. Protein sorting by transport vesicles. Science 272: 227-234.

Ruderman NB, Ross PS, Berger M, Goodman MN. 1974. Regulation of glucose and ketone body metabolism in brain of unaesthetized rats. Biochem J 138: 1-10.

Rui L, Aguirre V, Kim JK, Shulman GI, Lee A, et al. 2001. Insulin/IGF-1 and TNF-α stimulate phosphorylation of IRS-1 at inhibitory Ser[307] via distinct pathways. J Clin Invest 107: 181-189.

Rupprecht R, Holsboer F. 1999. Neuroactive steroids: Mechanism of action and neuropsychopharmacological perspectives. Trends Neurosci 22: 410-416.

Sacks W. 1957. Cerebral metabolism of isotopic glucose in normal human subjects. J Appl Physiol 10: 37-44.

Sakai RR, Lakshmi V, Monder C, McEwen BS. 1992. Immunocytochemical localization of 11 beta-hydroxysteroid dehydrogenase in hippocampus and other brain regions of the rat. J Neuroendocrinol 4: 101-106.

Salehi A, Swaab DF. 1999. Diminished neuroal metabolic activity in Alzheimer's diseae. J Neural Transm 106: 955-986.

Salehi M, Hodgkins MA, Merry BJ, Goyns MJ. 1996. Age-related changes in gene expression in the rat brain revealed by differential display. Experientia 52: 888-891.

Santos MS, Pereira EM, Carvaho AP. 1999. Stimulation of immunoreactive insulin release by glucose in rat brain synaptosomes. Neurochem Res 24: 33-36.

Sapolsky RM. 1999. Glucocorticoids, stress, and their adverse neurological effects: Relevance to aging. Exp Gerontol 34: 721-732.

Sapolsky RM, Krey LC, McEwen BS. 1983. The adrenocortical stress-response in the aged male rat: Impairment of recovery from stress. Exp Gerontol 18: 55-64.

Sapolsky RM, Krey LC, McEwen BS. 1986. The neuroendocrinology of stress and aging. The glucocorticoid cascade hypothesis. Endocrin Res 7: 284-300.

Sapolsky RM, Zola-Morgan S, Squire LR. 1991. Inhibition of glucocorticoid secretion by the hippocampal formation in the primate. J Neurosci 11: 3695-3704.

Sato A, Sato Y. 1995. Cholinergic neural regulation of regional cerebral blood flow. Alzheimer Dis Assoc Dis 9: 28-38.

Sawada M, Sester U, Calson JC. 1992. Superoxide radical formation and associated biochemical alterations in the plasma membrane of brain, heart, and liver during the lifetime of the rat. J Cell Biochem 58: 296-304.

Schechter R, Beju D, Gaffney T, Schaefer F, Whetsell L. 1996. Preproinsulin I and II mRNAs and insulin electron microscopic immuno-reaction are present within the fetal nervous system. Brain Res 736: 16-27.

Schechter R, Whitmire J, Holtzelaw L, George M, Devaskar SU. 1992. Developmental regulation of insulin in the mammalian central nervous system. Brain Res 582: 27-37.

Schipper HM. 1996. Astrocytes, brain aging and neurodegeneration. Neurobiol Aging 17: 467-480.

Schmidt R, Schmidt H, Curb JD, Masaki K, White LR, et al. 2002. Early inflammation and dementia: A 25-year follow-up of the Honolulu-asian aging study. Ann Neurol 52: 168-174.

Schwartz A, Lindenmayer GE, Allen JC. 1975. The sodium-potassium adenosine triphosphatase. Pharmacological, physiological and biochemical aspects. Pharmacol Rev 27: 3-134.

Seckl JR, Walker BR. 2001. Minireview: 11β-hydroxysteroid dehydrogenase type 1-a tissue-specific amplifier of glucocorticoid action. Endocrinology 142: 1371-1376.

Seksek O, Biwersi J, Verkman AS. 1995. Direct measurement of trans-Golgi pH on living cells and regulation by second messengers. J Biol Chem 270: 4967-4970.

Selkoe DJ. 1993. Physiological production of the β-amyloid protein and the mechanism of Alzheimer's disease. Trends Neurosci 16: 403-409.

Seubert P, Vigo-Pelfrey C, Esch F, Lee M, Dovey H, et al. 1992. Isolation and quantification of soluble Alzheimer's β-peptide from biological fluids. Nature 359: 325-327.

Sharpe JA, Rewcastle NB, Lloyd KG, Hornykiewicz O, Hill M, et al. 1973. Striatonigral degeneration: Response to levadopa therapy with pathological and neurochemical correlation. J Neural Sci 19: 275-286.

Shaw M, Cohen P, Alessi DR. 1997. Further evidence that the inhibition of glycogen synthase kinase-3β by IGF-1 is mediated by PDK1/PKB-induced phosphorylation of Ser-9 and not by dephosphorylation of Tyr-216. FEBS Lett 416: 307-311.

Shi J, Xiang Y, Simpkins JW. 1997. Hypoglycemia enhances the expression of mRNA encoding β-amylod precursor protein in rat primary cortical astroglial cells. Brain Res 772: 247-251.

Siesjö BK. 1978. Brain Energy Metabolism. Chichester; Wiley; pp. 1-28, 151–209.

Sims NR, Bowen DM, Allen SJ, Smith CCT, Neary D, et al. 1983a. Presynaptic cholinergic dysfunction in patients with dementia. J Neurochem 40: 503-509.

Sims NR, Bowen DM, Neary D, Davison AN. 1983b. Metabolic processes in Alzheimer's disease: Adenine nucleotide content and production of $^{14}CO_2$ from (U ^{14}C) glucose in vivo in human neocortex. J Neurochem 41: 1329-1334.

Singh H, Usher S, Poulos A. 1989. Mitchondrial and peroxisomal beta-oxidation of stearic and lignoceric acids by rat brain. J Neurochem 53: 1711-1718.

Skovronsky DM, Doms RW, Lee VMY. 1998. Detection of a novel intraneuronal pool of insoluble amyloid β protein that accumulates with time in culture. J Cell Biol 141: 1031-1039.

Smith CD, Carney JM, Starke-Reed PE, Oliver CN, Stadtman ER, et al. 1991. Excess brain protein oxidation and enzyme dysfunction in normal aging and in Alzheimer disease. Proc Natl Acad Sci USA 88: 10540-10543.

Smith MA, Taneda S, Richey PL, Miyata S, Yan SD, et al. 1994. Advanced Maillard reaction end products are associated with Alzheimer disease pathology. Proc Natl Acad Sci USA 91: 5710-5714.

Sokoloff L. 1981. Localization of functional activity in the central nervous system by measurement of glucose utilization with radioactive deoxyglucose. J Cereb Blood Flow Metab 1: 7-36.

Solano DC, Sironi M, Bonfini C, Solarte SB, Govoni S, et al. 2000. Insulin regulates soluble amyloid precursor protein release via phosphatidyl inositol 3 kinase-dependent pathway. FASEB J 14: 1015-1022.

Sontag E, Hladik C, Montgomery L, Luangpirom A, Mudrak I, et al. 2004a. Downregulation of protein phosphatase 2A carboxyl methylation and methyltransferase may contribute to Alzheimer disease pathogenesis. J Neuropathol Exp Neurol 63: 1080-1091.

Sontag E, Luangpirom A, Hladik C, Mudrack I, Ogris E, et al. 2004b. Altered expression levels of the protein phosphatase 2 A ABαC enzyme are associated with Alzheimer disease pathology. J Neuropathol Exp Neurol 63: 287-301.

Sontag E, Nunbhakdi-Craig V, Bloom GS, Mumby MC. 1995. A novel pool of protein phosphatase 2A is associated with microtubules and is regulated during the cell cycle. J Cell Biol 128: 1131-1144.

Sorbi S, Bird ED, Blass JP. 1983. Decreased pyruvate dehydrogenase complex activity in Huntington and Alzheimer brain. Ann Neurol 13: 72-78.

Spanswick D, Smith MA, Mirshamsi S, Routh VH, Ashford MJL. 2000. Insulin activates ATP-sensitive K^+ channels in hypothalamic neurons of lean, but not obese rats. Nat Neurosci 3: 757-758.

Stadtman ER. 1992. Protein oxidation and aging. Science 257: 1220-1224.

Stadtman ER, Berlett BS. 1991. Fenton chemistry. Amino acid oxidation. J Biol Chem 266: 17201-17211.

Stadtman ER, Oliver CN. 1991. Metal-catalized oxidation of proteins. Physiological consequences. J Biol Chem 266: 2005-2008.

Steen E, Terry BM, Rivera EJ, Cannon JL, Neely TR, et al. 2005. Impaired insulin and insulin-like growth factor expression and signaling mechanisms in Alzheimer's disease – is this type 3 diabetes? J Alzheimer Dis 7: 63-80.

Stover JF, Sakowitz OW, Schöning B, Rupprecht S, Kroppenstedt SN, et al. 2003. Norepinephrine infusion increases interleukin-6 in plasma and cerebrospinal fluid of brain-injured rats. Med Sci Monit 9: BR382-BR388.

Sun FF, Fleming WE, Taylor BM. 1993. Degradation of membrane phospholipids in the cultured human astroglial cell line UC-11MG during ATP depletion. Biochem Pharmacol 45: 1149-1155.

Sutherland C, Cohen P. 1994. The α-isoform of glycogen synthase kinase-3 from rabbit skeletal muscle is inactivated by p 70S6 kinase or MAP kinase-activated protein kinase-1 in vitro. FEBS Lett 338: 37-42.

Svennerholm L, Gottfries CG. 1994. Membrane lipids, selectively diminished in Alzheimer brains, suggest synapse loss as a primary event in early-onset form (type I) and demyelination in late-onset form (type II). J Neurochem 62: 1039-1047.

Swaab DF, Raadsheer FC, Endert EF, Hofman MA, Kamphorst WC, et al. 1994. Increases in cortisol levels in aging and Alzheimer's disease in postmortem cerebrospinal fluid. Neuroendocrinology 6: 681-687.

Sweeney G, Klip A. 1998. Regulation of the Na$^+$/K$^+$-ATPase by insulin: Why and how? Mol Cell Biochem 182: 121-133.

Takami K, Terai K, Matsuo A, Walker DG, McGeer PL. 1997. Expression of presenilin-1 and-2 mRNAs in rat and Alzheimer's disease brains. Brain Res 748: 122-130.

Takashima A, Murayama M, Murayama O, Kohno T, Honda T, et al. 1998. Presenilin 1 associates with glycogen kinase-3 β and its substrate tau. Proc Natl Acad Sci USA 95: 9637-9641.

Tarkowski E, Blennow K, Wallin A, Tarkowski A. 1999. Intracerebral production of tumor necrosis factor-α, a local neuroprotective agent, in Alzheimer disease and vascular dementia. J Clin Immun 19: 223-230.

Tham A, Nordberg A, Grissom FE, Carlsson-Skiwirut C, Viitanen M, et al. 1993. Insulin-like growth factors and insulin-like growth factor binding proteins in cerebrospinal fluid and serum of patients with dementia of the Alzheimer type. J Neural Transm (P-D Sect) 5: 165-176.

Tomizawa K, Ohta J, Matsushita M, Moriwaki A, Li ST, et al. 2002. Cdk5/p35 regulates neurotransmitter release through phosphorylation and downregulation of P/Q-type voltage-dependent calcium channel activity. J Neurosci 22: 2590-2597.

Treherne JM, Ashford MLJ. 1991. The regional distribution of sulfonylurea binding sites in rat brain. Neuroscience 40: 523-531.

Turnbow MA, Keller RS, Rice KM, Garnar CW. 1994. Dexamethasone down-regulation of insulin receptor substrate-1 in 3T3 L1 adipocytes. J Biol Chem 269: 2516-2520.

Turnbull S, Tabner BJ, El-Agnaf OM, Moore S, Davies Y, et al. 2001. Alpha-synuclein implicated in Parktinson's disease catalyses the formation of hydrogen peroxide in vitro. Free Radic Biol Med 30: 1163-1170.

Turner M, Moran NF, Kopelman MD. 2002. Subcortical dementia. Br J Psychiatry 180: 148-151.

Uchida N, Honjo Y, Johnson KR, Wheelock MJ, Takeichi M. 1996. The catenin/cadherin adhesion system is localized in synaptic junctions bordering transmitter release zone. J Cell Biol 135: 767-779.

Unger J, McNeill TH, Moxley RT, White M, Moss A, et al. 1989. Distribution of insulin receptor-like immunoreactivity in the rat forebrain. Neuroscience 31: 143-157.

Valverde AM, Teruel T, Navarro P, Benito M, Lorenzo M. 1998. Tumor necrosis factor-α causes insulin receptor substrate-2-mediated insulin resistance and inhibits insulin-induced adipogenesis in fetal brown adipocytes. Endocrinology 139: 1229-1238.

Valverde AN, Mur C, Pon S, Alvarez AM, White MF, et al. 2001. Association of insulin receptor substrate-1 (IRS-1) Y895 with Grb-2 mediates the insulin signalling involved in IRS-1-deficient brown adipocyte mitogenesis. Mol Cell Biol 21: 2269-2280.

van Houten M, Posner BI, Kopriwa BM, Brawer JR. 1979. Insulin binding sites in the rat brain: In vivo localization to the circumventricular organs by quantitative autoradiography. Endocrinology 105: 666-673.

van Houten M, Posner BI, Kopriwa BM, Brawer JR. 1980. Insulin binding sites localized to nerve terminals in rat median eminence and arcuate nucleus. Science 207: 1081-1083.

Vanhaesebroeck B, Alessi DR. 2000. The PIK 3 K-PDK1 connection: More than just a road to PKB. Biochem J 346: 561-576.

Vekrellis K, Ye Z, Qiu WQ, Walsh D, Hartle D, et al. 2000. Neurons regulate extracellular levels of amyloid β-protein via proteolysis by insulin-degrading enzyme. J Neurosci 20: 1657-1665.

Verde C, Pascale MC, Martive G, Lotti LV, Torrisi MR, et al. 1995. Effect of ATP depletion and DTT on the transport of membrane proteins from the endoplasmic reticulum and the intermediate compartment to the Golgi complex. Eur J Cell Biol 67: 267-274.

Villa RF, Gorini A, Hoyer S. 2002. ATPases of synaptic plama membranes from hippocampus after ischemia and recover during aging. Neurochem Res 27: 861-870.

Virkamäki A, Ueki K, Kahn CR. 1999. Protein-protein interaction in insulin signalling and the molecular mechanisms of insulin resistance. J Clin Invest 103: 931-943.

Virsolvy-Vergine A, Leray H, Kuroki S, Lupo B, Dufour M, et al. 1992. Endosulfine, an endogenous peptidic ligand for the sulfonylurea receptor: Purification and partial characterization from ovine brain. Proc Natl Acad Sci USA 89: 6629-6633.

Vlassara H, Brownlee M, Cerami A. 1985. High-affinity H receptor-mediated uptake and degradation of glucose-modified proteins: A potential mechanism for the removal of senescent macromolecules. Proc Natl Acad Sci USA 82: 5588-5592.

Vlassara H, Bucala R, Striker L. 1994. Biology of disease. Pathogenetic effects of advanced glycosylation: Biochemical, biologic, and clinical implications for diabetes and aging. Lab Invest 70: 138-151.

Vogelsberg-Ragaglia V, Schuck T, Trojanowski JQ, Lee VM. 2001. PP2A mRNA expression is quantitatively decreased in Alzheimer's disease hippocampus. Exp Neurol 168: 402-412.

Vosseller K, Sakabe K, Wells L, Hart GW. 2002. Diverse regulation of protein function by 0-GlcNAc: A nuclear and cytoplasmic carbohydrate post-translational modification. Curr Opin Chem Biol 6: 851-857.

Waksman G, Hamel E, Delay-Goyet P, Roques BP. 1986. Neuronal localization of the neutral endopeptidase "enkephalinase" in rat brain revealed by lesions and autoradiography. EMBO J 5: 3165-3166.

Walaas U, Fonnum F. 1980. Biochemical evidence for gluta-mate as a transmitter in hippocampal efferents to the basal forebrain and hypothalamus in the rat brain. Neuroscience 5: 1691-1698.

Walker D, Kim S, McGeer P. 1998. Expression of complement C4 and C9 genes by human astrocytes. Brain Res 809: 31-38.

Walker D, Lue L, Beach T. 2001. Gene expression profiling of amyloid beta peptide-stimulated human post-mortem brain microglia. Neurobiol Aging 22: 957-966.

Wallace W, Ahlers ST, Gotlib J, Bragin V, Sugar J, et al. 1993. Amyloid precursor protein in the cerebral cortex is rapidly and persistently induced by loss of subcortical innervation. Proc Natl Acad Sci USA 90: 8712-8716.

Wallace WC, Akar CA, Lyons EW. 1997a. Amyloid precursor protein potentiates the neurotrophic activity of NGF. Mol Brain Res 52: 201-212.

Wallace WC, Akar CA, Lyons WE, Kole HK, Egan JE, et al. 1997b. Amyloid precursor protein requires the insulin sig-naling pathway for neurotrophic activity. Mol Brain Res 52: 213-227.

Wang JZ, Gong CX, Zaidi T, Grundke-Iqbal I, Iqbal K. 1995. Dephosphorylation of Alzheimer paired helical filaments by protein phosphatase-2A and −2B. J Biol Chem 270: 4854-4860.

Watanabe A, Hasegawa M, Suzuki M, Takio K, Morishima-Kawashima M, et al. 1993. In vivo phosphorylation sites in fetal and adult rat tau. J Biol Chem 268: 25712-25717.

Welch WJ, Suhan JP. 1985. Morphological study of the mam-malian stress response: Characterization of changes in the cytoplasmic organelles, cytoskeleton, and nucleoli, and ap-pearance of intranuclear actin filaments in rat fibroblasts after heat-shock treatment. J Cell Biol 101: 1198-1211.

Welsh GI, Proud CG. 1993. Glycogen synthase kinase-3 is rapidly inactivated in response to insulin and phosphory-lates eukaryotic initiation factor eIF-2B. Biochem J 294: 625-629.

Werther GA, Hogg A, Oldfield BJ, McKinley MJ, Figdor R, et al. 1987. Localization and characterization of insulin receptors in rat brain and pituitary gland using in vitro autoradiography and computerized densitometry. Endocri-nology 121: 1562-1570.

Westerberg E, Deshpande DK, Wieloch T. 1987. Regional differences in arachidonic acid release in rat hippocampal CA1 and CA3 regions during cerebral ischemia. J Cereb Blood Flow Metab 7: 189-192.

White MF, Kahn CR. 1994. The insulin signalling system. J Biol Chem 269: 1-4.

Whittemore SR, Ebendal T, Larkfors L, Olson L, Seiger A, et al. 1986. Development and regional expression of beta nerve growth factor messenger RNA and protein in the rat central nervous system. Proc Natl Acad Sci USA 83: 817-821.

Wild-Bode C, Yamazaki T, Capell A, Leimer U, Steiner H, et al. 1997. Intracellular generation and accumulation of amy-loid β-peptide terminating at amino acid-42. J Biol Chem 272: 16085-16088.

Wilson CA, Doms RW, Lee VMY. 1999. Intracellular APP processing and Aβ production in Alzheimer disease. J Neu-ropathol Exp Neurol 58: 787-794.

Wong KL, Tyce GM. 1983. Glucose and amino acid metabo-licm in rat brain during sustained hypoglycemia. Neuro-chem Res 8: 401-415.

Wu HC, Lee EHY. 1997. Identification of a rat brain gene associated with aging by PCR differential display method. J Mol Neurosci 8: 13-18.

Wurtman RJ. 1992. Choline metabolism as a basis for the selective vulnerability of cholinergic neurons. Trends Neu-rosci 15: 117-122.

Wyllie MG, Gilbert JL. 1979. An investigation into the role of synaptic vesicular Mg^{2+}-ATPase in neurotransmitter re-lease using benzhydryl piperazine. Arch Int Pharmacodyn Ther 241: 4-15.

Xia W, Zhang J, Perez R, Koo EH, Selkoe DJ. 1997. Interaction between amyloid precursor protein and presenilins in mammalian cells: Implications for the pathogenesis of Alz-heimer disease. Proc Natl Acad Sci USA 94: 8208-8213.

Xie L, Helmerhorst E, Taddel K, Plewright B, van Bronswijk W, et al. 2002. Alzheimer's β-amyloid peptides compete for insulin binding to the insulin receptor. J Neurosci 22 (RC221): 1-5.

Xu H, Sweeny D, Wang R, Thinakaran G, Lo AC, et al. 1997. Generation of Alzheimer beta-amyloid protein in the trans-Golgi network in the apparent abscence of vesicle forma-tion. Proc Natl Acad Sci USA 94: 3748-3752.

Yan LJ, Sohal RS. 1998. Mitochondrial adenine nucleotide translocase is modified oxidatively during aging. Proc Natl Acad Sci USA 95: 12896-12901.

Yan SD, Chen X, Fu J, Chen M, Zhu H, et al. 1996. RAGE and amyloid-β peptide neurotoxicity in Alzheimer's disease. Nature 382: 685-691.

Yanagisawa M, Planel E, Ishiguro K, Fugita SC. 1999. Starva-tion induces tau hyperphosphorylation in mouse brain: Implications for Alzheimer's disease. FEBS Lett 461: 329-333.

Yang AJ, Chandswangbhuvana D, Shu T, Henschen A, Glabe CG. 1999. Intracellular accumulation of insoluble, newly snythesized Aβn-42 in amyloid precursor protein-trans-fected cells that have been treated with Aβ 1–42. J Biol Chem 274: 20650-20656.

Yang Y, Turner RS, Gaut JR. 1998. The chaperone BiP/GRP78 binds to amyloid precursor protein and decreases Abeta 40 and Abeta 42 secretion. J Biol Chem 273: 25552-25555.

Yasojima K, Akiyama H, McGeer EG, McGeer PL. 2001a. Reduced neprilysin in high plaque areas of Alzheimer brain: A possible relationship to deficient degradation of β-amlyoid peptide. Neurosci Lett 297: 97-100.

Yasojima K, McGeer EG, McGeer PL. 2001b. Relationship between β-amlyoid peptide generating molecules and neprilysin in Alzheimer disease and normal brain. Brain Res 919: 115-121.

Yki-Jarvinen H, Virkamaki A, Daniels MC, McClain D, Gottschalk WK. 1998. Insulin and glucosamine infusions increase O-linked N-acetyl-glucosamine in skeletal muscle protein in vivo. Metabolism 47: 449-455.

Yoo BC, Kim SH, Cairns N, Fountoulakis M, Lubec G. 2001. Deranged expression of molecular chaperones in brains of patients with Alzheimer's disease. Biochem Biophys Res Commun 280: 249-258.

Youdim MB. 2003. What have we learnt from CDNA microarray gene expression studies about the role of iron in MPTP induced neurodegeneration and Parkinson's disease? J Neural Transm 65 (Suppl.): 73-88.

Youdim MB, Ben-Shachar D, Riederer P. 1993. The possible role of iron in the etiopathology of Parkinson's disease. Mov Disord 8: 1-12.

Yu G, Chen F, Levesque G, Nishimura M, Zhang DM, et al. 1998. The presenilin 1 protein is a component of a high molecular weight intracellular complex that contains β-catenin. J Biol Chem 273: 16470-16475.

Zahniser NR, Goens MB, Hanaway PJ, Vinych JV. 1984. Characterization and regulation of insulin receptors in rat brain. J Neurochem 42: 1354-1362.

Zaia A, Piantanelli L. 2000. Insulin receptors in the brain cortex of aging mice. Mech Ageing Dev 113: 227-232.

Zhang H, Sternberger NH, Rubinstein LJ, Herman MM, Binder LI, et al. 1989. Abnormal processing of multiple proteins in Alzheimer's disease. Proc Natl Acad Sci USA 86: 8045-8049.

Zhao W, Chen H, Xu H, Moore E, Meiri N, et al. 1999. Brain insulin receptors and spatial memory. J Biol Chem 274: 34893-34902.

Zhao WQ, Chen H, Quon MH, Alkon DL. 2004. Insulin and the insulin receptor in experimental models of learning and memory. Eur J Pharmacol 490: 71-81.

Zhao WQ, Feng G, Alkon DL. 2003. Impairment of phosphatase 2A contributes to the prolonged MAP kinase phosphorylation in Alzheimer's disease fibroblasts. Neurobiol Dis 14: 458-469.

Zhou J, Liyanage U, Medina M, Ho C, Simmons AD, et al. 1997. Presenilin 1 interaction in the brain with a novel member of the Armadillo family. Neuroreport 8: 1489-1494.

Zini S, Tremblay E, Pollard H, Moreau J, Ben-Ari Y. 1993. Regional distribution of sulfonylurea receptors in the brain of rodent and primate. Neuroscience 55: 1085-1091.

13 Proteomics Analysis in Alzheimer's Disease: New Insights into Mechanisms of Neurodegeneration

D. A. Butterfield · R. Sultana

Abstract: Redox proteomics is that branch of proteomics in which oxidatively modified proteins are identified using protein separation combined with mass spectrometry and bioinformatics. Using redox proteomics, brain and plasma proteins in Alzheimer's disease (AD) and models thereof that are oxidatively modified have been identified. Most oxidatively modified proteins are dysfunctional, suggesting that in AD the function of oxidatively modified proteins is compromised. Several categories of proteins ranging from those involved in energy metabolism, excitotoxicity, proteasome function, lipid asymmetry and cholinergic function, neuritic function, prevention of neuronal entrance to the cell cycle, hyperphosphorylation of tau, and amyloid beta-peptide production, synaptic function, and pH buffering and carbon dioxide transport are oxidatively dysfunctional in AD brain. Redox proteomics studies of various models of AD recapitulate many of these oxidatively modified proteins. This review outlines these studies and posits that oxidatively dysfunctional proteins are intimately involved in the pathogenesis of AD.

List of Abbreviations: AD, Alzheimer's disease; 2D-PAGE, Two-dimensional gel electrophoresis; 3-NT, 3-Nitrotyrosine; Aβ(1-42), Amyloid beta-peptide 1-42; apoE, apolipoprotein E; APP, Amyloid precursor protein; CAII, Carbonic anhydrase II; CK, Creatine kinase; CNS, Central nervous system; Cpn60, Chaperonin 60; DRP-2, Dihydropyrimidinase-related protein 2; ESI, Electrospray ionization; GAPDH, Glyceraldehyde-3-phosphate dehydrogenase; GS, Glutamine synthetase; GSK3β, Glycogen synthase kinase 3β; HCNP, Hippocampal cholinergic neurostimulating peptide; HNE, 4-Hydroxy-2-nonenal; ICAT, Isotopically coded affinity tags; IEF, Isoelectric focusing; MALDI, Matrix-assisted laser desorption ionization; MCI, Mild cognitive impairment; MS, Mass spectrometry; NFT, Neurofibrillary tangles; PEBP, Phosphatidylethanolamine binding protein; Pin1, Peptidy prolyl csi-trans isomerase; SELDI, Surface-enhanced laser desorption ionization; SNAP, Gamma-soluble NSF attachment protein; TPI, Triosephosphate isomerase; UCH L-1, Ubiquitin carboxy-terminal hydrolase L-1

1 Introduction

Alzheimer's disease (AD), a neurodegenerative disease characterized by progressive loss of memory associated with loss of neuronal cells, is known to affect approximately 5 million individuals in the US and is expected to increase from 11.3 to 16 million people by 2050. The etiology of AD is not known, however based on studies of inherited AD with mutations in the genes for presenilin-1 (PS-1), presenilin-2 (PS-2) (Cruts et al., 1998), and amyloid precursor protein (APP) (Goate et al., 1991; Sherrington et al., 1995), a key role of amyloid β-peptode (Aβ) has been proposed (Selkoe, 2001). In addition, there is also a reported association between AD and allele 4 of the apolipoprotein E (APOE) gene (Levy-Lahad et al., 1995; Slooter et al., 1998), endothelial nitric oxide synthase 3 gene (Dahiyat et al., 1999), and alpha-2-macroglobulin (Blacker et al., 1998).

AD is characterized by progressive loss of memory and the accumulation of extracellular amyloid plaques, intracellular neurofibrillary tangles (NFT), and loss of synaptic connections within selective brain regions. NFT consists of aggregates of hyperphosphorylated microtubule associated protein tau that forms paired helical filaments and related straight filaments (Grundke-Iqbal et al., 1986). Aβ, a 40–42 amino acid peptide derived form proteolytic cleavage of APP by the action of beta- and gamma-secretases, is the main component of SP.

Among the various neurotransmitter systems, cholinergic neurons in the nucleus basalis of Meynert (NBM) are lost early in the course of AD, and the dysfunction of cholinergic neurons is believed to be involved in cognitive deficits in this disease (Coyle et al., 1983).

Several mechanisms have been proposed to underlie AD pathogenesis that includes: amyloid cascade, excitoxicity, oxidative stress, and inflammation. Several lines of evidence suggest an imbalance in the antioxidant and oxidant systems in AD, that suggest oxidative stress is imported in the pathogenesis and/or progression of AD (Markesbery, 1997; Butterfield et al., 2001, 2002b; Lauderback et al., 2001; Butterfield and Lauderback, 2002). Either the oxidants or the products of oxidative stress could modify proteins or activate other pathways that may lead to additional impairment of cellular functions and neuronal loss

(Smith et al., 1994; Markesbery, 1997; Mark et al., 1997; Butterfield et al., 2001, 2002b; Lovell et al., 2001). Oxidative stress is manifested by increased protein oxidation, lipid peroxidation, DNA oxidation, advanced glycation end products, reactive oxygen species (ROS), and reactive nitrogen species (RNS) formation. Protein carbonyls, 4-hydroxy-2-nonenal (HNE), and 3-nitrotyrosine (3-NT) are among the early oxidative markers observed after oxidative insult in a cell (Smith et al., 1996, 1997; Lovell et al., 2001; Butterfield, 2002; Butterfield and Lauderback, 2002; Castegna et al., 2003). Protein carbonyl, HNE, and 3-NT levels were found to be elevated in AD brain (Smith et al., 1994; Hensley et al., 1995; Markesbery and Lovell, 1998; Aksenov et al., 2001; Sultana et al., 2006a,b,c). In addition in AD brain, protein oxidation occurs in Aβ-rich regions like cortex and hippocampus, but is not observed in cerebellum where the Aβ was found to be negligible (Hensley et al., 1995; Sultana et al., 2005b, c). Recently, we found evidence of increased protein oxidation and lipid peroxidation in mild cognitive impairment (MCI) (Keller et al., 2005; Butterfield et al., 2006a,b), approximately 70% of whom convert to AD. Thus, oxidative stress may be one of the earliest biochemical alterations of AD.

Protein carbonyl groups can be introduced into proteins by direct oxidation of certain amino acid side chains, peptide backbone scission, or by Michael addition reactions with products of lipid peroxidation or glycooxidation (Berlett and Stadtman, 1997; Butterfield and Stadtman, 1997). Oxidative stress also could induce additional damage via the overexpression of inducible nitric oxide synthase (iNOS). In cell cultures, the use of vitamin E diminishes Aβ (1–42)-induced toxicity, further supporting a role of oxidative stress in AD pathology (Yatin et al., 2000; Butterfield et al., 2002a; Boyd-Kimball et al., 2004a, b).

Redox proteomics couples two-dimensional gel electrophoresis separation of proteins with mass spectrometric techniques to provide a valuable modality to determine oxidatively modified brain proteins (Butterfield, 2004). Proteomics has enabled us to identify a large number of oxidatively modified proteins in cells, tissues, and other biological samples that were previously undetected by other methods like immunoprecipitation. There are several serious limitations to the use of immunoprecipitation to identify proteins; for example, the necessary availability of the antibody for the protein of interest, knowledge about the proteins of interest, and the time-consuming and laborious nature of the process. Moreover, sometimes a posttranslational modification can change the structure of proteins, thereby preventing the formation of the appropriate antigen–antibody complex.

In this chapter, we discuss the proteomics identification of oxidatively modified brain proteins and their implications in the pathology and biochemistry of AD using proteomics techniques.

2   Methodology

2.1   AD Brain Tissue

Human brain tissue samples were taken at autopsy from AD and control subjects, immediately frozen in liquid nitrogen, and stored at -80°C. The Rapid Autopsy Program of the University of Kentucky Alzheimer's Disease Clinical Center (UK ADCC) resulted in extremely short postmortem intervals (PMIs) (always less than 4 h). All AD subjects displayed progressive intellectual decline and met NINCDS-ADRDA Workgroup criteria for the clinical diagnosis of probable AD (McKhann et al., 1984). Control subjects underwent annual testing of mental status, as a part of the UK ADRC normal volunteer longitudinal aging study did not have and a history of dementia or other neurologic disorders. All control subjects had test scores in the normal range. Neuropathologic evaluation of control brains revealed only age-associated gross and histopathologic alterations. Brain tissues were prepared for proteomics as described (Castegna et al., 2002a; Sultana et al., 2005b, c).

2.2   Two-Dimensional Polyacrylamide Gel Electrophoresis

Two-dimensional gel electrophoresis (2D-PAGE) is a sensitive technique that is widely used to analyze the complex protein mixtures based on two physicochemical properties: the first step in 2D-PAGE involves isoelectric focusing (IEF), which separates proteins based on their isoelectric points (pI); followed by the second dimension that separates proteins based on their relative mobility (Mr) on SDS-PAGE (Rabilloud, 2002). Usually a single spot on the 2D gel corresponds to a single protein (Tilleman et al., 2002). Thus, hundreds of different protein spots of different proteins can be separated, and the pI, molecular weight and protein expression, post- and cotranslational modification, etc., can be obtained in most cases. In addition, 2D-PAGE is used to catalogue proteins and create databases (Kaji et al., 2000). 2D-PAGE is a sensitive, reliable method with high reproducibility, but many challenges still exist. A serious limitation of 2D electrophoresis is the solubilization process for membrane proteins (Santoni et al., 2000), as the ionic detergents necessary are not compatible with IEF, due to the charge introduced to the protein, thereby interfering with its separation in the first dimension. Another limitation of 2D electrophoresis is the inability to detect low-abundance proteins and a third limitation of 2D-PAGE is the low detectability of proteins of high trypsin content, because of the small resulting peptides.

Chaotropic agents, such a urea and thiourea, coupled with nonionic or zwitterionic detergents can be used to solubilize proteins and avoid protein precipitation during the IEF and the SDS gel (Herbert, 1999). The use of immobilized pH IEF strips eliminates the typical cathodic drift associated with previously used tube gels and improves the protein map reproducibility between samples (Molloy, 2000). In addition, the use of narrow range IEF strips enables one to separate proteins over a wide range of pH with a unit pH difference of one. However, the normally employed IEF strip pH range, i.e., three to ten, limits the identification of highly basic proteins. The identification of the low-abundance proteins in a given sample is a limitation, as noted earlier, one that is important when a protein of this group is involved in the pathogenesis of a disease.

Other separation methods, including 2D-HPLC and isotopically coded affinity tags (ICAT), are also used for protein separation. HPLC is often used to separate peptides produced by trypsin digestion of a protein mixture. This technique couples a strong cation-exchange column in tandem with a reverse-phase (RP) column (Stevens et al., 2002; Wagner et al., 2003) that is connected to a mass spectrometer. Analysis of protein expression in two different sets of samples together can be achieved by labeling the samples with different isotopes that bind to cysteine residues using the ICAT method. The isotopic labels assist in the evaluation of mass spectra and the differences in expression of a protein in the two samples (Smolka et al., 2001). Of course, a limitation of the ICAT method is that cysteine-deficient proteins are refractive to this method.

The redox proteomics techniques in our laboratory are used to identify oxidatively modified proteins in AD brain and related models. In this method, we use a parallel analysis in that we couple 2D-PAGE with 2D-immunochemical detection of protein carbonyls derivatized by 2,4-dinitrophenylhydrazine (DNPH), nitrated proteins indexed by 3-NT, or protein adducts of HNE, followed by MS analysis, as shown in ❷ *Figure 13-1*. Proteins containing reactive carbonyl groups/3-NT/HNE in AD and control brain samples are detected by 2D Western blot analysis using specific antibodies. The 2D Western blots and 2D gel images are matched by computer-assisted image analysis, and the antiDNP/nitrotyrosine/HNE immunoreactivity of individual proteins are normalized to their content, obtained by measuring the intensity of colloidal Coomassie Blue staining or SYPRO ruby stained spots. Such analysis allows comparison of levels of oxidatively modified brain proteins in AD versus control subjects.

2.3 Mass Spectrometry and Database Searching

The proteins of interest are excised, exposed to trypsin, eluted from the gel, and subjected to mass spectrometry analysis. The method most commonly used is in-gel digestion of protein with a protease that would not only cleave the protein into small peptides, but also produces sequence-specific proteolysis. These mass fingerprints are characteristic of a particular protein, which facilitates the identification of a particular protein using a suitable database (❷ *Table 13-1*) that compares the experimental masses with theoretical masses of trypsin-generated protein sequences.

Mass spectrometry is essential for the identification of the peptide masses and amino acid sequence for the proteins of interest. Early MS studies could not provide a precise peptide mass due to molecular

◘ Figure 13-1

Protocol for the identification of the oxidized protein by redox proteomics

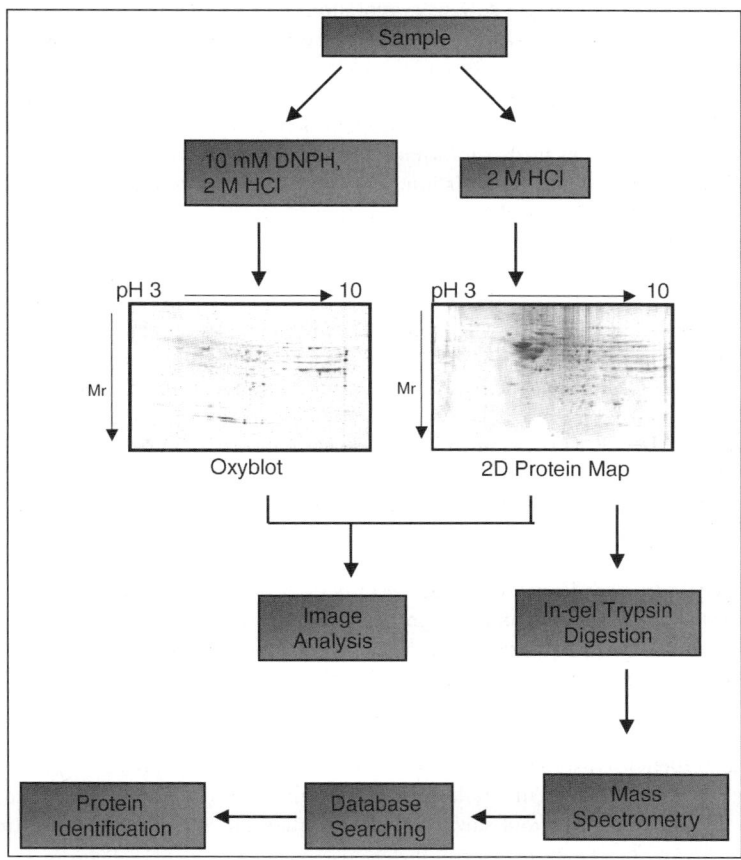

◘ Table 13-1

Mass spectrometry search engines for peptide mass fingerprinting

Mascot http://www.matrixscience.com
MOWSE http://www.hgmp.mrc.ac.uk/Bioinformatics/Webapp/mowse
Profound http://www.prowl.rocketfeller.edu/sgi-bin/profound
MS-fit http://www.prospector.ucsf.edu/ucsfhtlm3.4/msfit.htm
Peptident http://www.expasy.ch.ch/tools/peptident.html

fragmentation of the sensitive ion. However, the introduction of softer ionization techniques helped to overcome the fragmentation, thereby helping to a better identification of proteins. The two most used MS techniques are matrix assisted laser desorption/ionization (MALDI) and electrospray ionization (ESI). In MALDI analysis the peptide sample is mixed with a matrix (usually α-cyano-4-hydroxycinnamic acid or 2,5-dihydroxybenzoic acid) and deposited to a plate that is subjected to laser radiation. The matrix absorbs the energy, which is then transferred to the peptides. The peptides then evaporate as detectable MH^+ ions in an unclear process that is related to the acidic nature of the matrix. In ESI, HPLC is coupled to MS and the liquid sample from the HPLC column enters directly into MS for characterization. Due to the high potential difference between the capillary and the MS instrument, the inlet sample is dispersed as small droplets. These droplets undergo solvent evaporation until droplet fission occurs, due to the high charge-to-surface tension ratio, finally leading to the formation of a single detectable ion per droplet. For better online preseparation of peptides with HPLC and MS, a low salt concentration is important, as salt may interfere with ionization. In addition, reducing the flow time to nanoliters per minute can increase the time for analysis. Tandem MS/MS provides better isolation and fragmentation of a specific ion. Consequently, tandem MS/MS provides further information about the sequence of the protein (Aebersold and Goodlett, 2001). In MS/MS analysis, the isolation of a single ion is achieved by scanning all the ions that were generated from a sample, followed by the application of a wide range of frequencies, except for the resonating frequency of the ion of interest. The final step in MS/MS is fragmentation of the isolated ion, which provides additional information for protein identification or for evaluation of possible protein modification.

An additional proteomics tool in which mass spectrometry is utilized is surface-enhanced laser desorption/ionization time of flight (SELDI-TOF) that couples the classical methods of chromatographic sample preparation with mass spectrometry analysis. This method has potential applications in biomarker analysis, but is limited to proteins below a relatively low molecular weight. We did not use this method in the studies outlined in this review, but SELDI-TOF is described in detail in our recent review (Butterfield et al., 2003).

After mass spectrometric analysis, the correct identity of the protein is determined by online protein databases. The protein sequence database SwissProt is the most commonly used database for protein identification that is based on computer algorithms (Hoogland et al., 2000), and is available through the internet. These databases are described in ❷ *Table 13-1*. These search engines provide a theoretical protease digestion of the proteins contained in the database. The experimental masses obtained by MS are compared with this theoretical digestion and several factors, such as the protein size and the probability of a single peptide to occur in the whole database, are taken into account. The search engine produces a probability score for each entry, which is calculated by a mathematical algorithm that is specific for each search engine. Any hit with a score higher than the one specific for significance of the particular search engine is considered statistically significant and has an excellent legitimate chance to be the protein cut from a given spot. In addition, the molecular weight and PI of the protein are calculated based on the position in the 2D map to avoid any false identification. In many cases, validation of protein identification is achieved by immunochemical means (Castegna et al., 2002a; Poon et al., 2004; Sultana and Butterfield, 2004).

3   Proteomics Studies of Oxidatively Modified Proteins in AD Brain

AD brain showed increased protein oxidation levels as indexed by increased protein carbonyls and 3-NT levels (Hensley et al., 1995; Smith et al., 1997; Aksenov et al., 2001; Butterfield et al., 2001; Butterfield and Lauderback, 2002; Castegna et al., 2002b, 2003). Oxidative modification of proteins generally renders them less active, leading to loss of function (Hensley et al., 1995; Aksenova et al., 1999; Lauderback et al., 2001; Butterfield, 2003). Identification of specifically oxidized proteins in AD brain provides an insight into the role of oxidative stress in AD and also helps to unravel the mechanism(s) associated with AD pathology. Using proteomics, we identified several oxidatively modified proteins in the inferior parietal and hippocampal regions of AD brains: creatine kinase BB (CK), glutamine synthase (GS), ubiquitin carboxy-terminal hydrolase L-1 (UCH L-1), dihydropyrimidinase-related protein 2 (DRP-2), α-enolase, phosphoglycerate mutase 1, γ-soluble NSF attachment protein (SNAP), carbonic anhydrase II (CA-II), and peptidyl prolyl *cis–trans* isomerase (Pin1). Similarly, proteomics has been used to identify 3-NT posttranslationally modified proteins in AD brain: neuropolypeptide h3, triosephosphate isomerase and α-enolase, CA-II, glyceraldehyde 3-phosphate dehydrogenase, voltage-dependent anionic channel protein-1, and ATP synthase alpha chain (Castegna et al., 2002b, 2003; Sultana et al., 2006a, b, c). All these proteins were assigned to different groups based on their functions and linked to the observed AD pathology are delineated (❷ *Table 13-2*).

◘ Table 13-2
Proteomic identification of specifically oxidatively modified proteins in AD brain

Energy-Related Enzymes: Creatine kinase; α-enolase, γ-enolase, triosephosphate isomerase; phosphoglycerate mutase 1, glyceraldehydes 3-phosphate dehydrogenase

Neurotransmitter-Related Proteins: Gamma-soluble *N*-ethylmaleimide-sensitive factor attachment protein, glutamine synthetase; glutamate transportor, EAAT2

Proteasome-Related Proteins: Ubiquitin carboxyl terminal hydrolase I-1; heat shock protein

Cholinergic System: Neuropolypeptide h3

Structural Proteins: Dihydropyrimidinase-related protein 2; β-actin

Cell Cycle/Tau Phosphorylation production: Peptidyl prolyl *cis–trans*-isomerase

pH Regulation Protein: Carbonic anhydrase II

3.1 Proteins Involved in Energy Metabolism

Creatine kinase (BB isoform), α-enolase, and triosephosphate isomerase, glyceraldehyde 3-phosphate dehydrogenase (GADPH), phosphoglycerate mutase 1, and alpha ATPase are enzymes involved in ATP production and energy metabolism, either directly or indirectly. In AD brain, CK, enolase and GADPH activity are decreased, and the oxidation of these proteins suggest the impairment of ATP synthesis (Mazzola and Sirover, 2001; Aksenova et al., 2002; Castegna et al., 2002a; Sultana et al., 2005b). In addition, oxidation of other proteins involved in glucose metabolism, a main source of ATP production in brain, may not only impair glucose uptake and utilization, but may also lead to other deleterious effects in the cell. For example, decreased ATP production would lead to (a) impaired ion-motive ATPase needed to maintain potential gradients, operate pumps, and maintain membrane lipids asymmetry; (b) altered protein synthesis, sorting, and transport; (c) maintenance of synaptic transmission—all of which are hallmarks of AD (Castegna et al., 2002a, b, 2003; Hoyer, 2004; Sultana et al., 2006b, c). Also ATP diminution induces hypothermia leading to abnormal tau hyperphosphorylation through differential inhibition of kinase and phosphatase activities, ion pumps, electrochemical gradients, cell potential, voltage-gated ion channels (Planel et al., 2004).

3.2 Proteins Involved in Glutamate Reuptake or Conversion

HNE, a lipid peroxidation product, has been shown to oxidatively modify the glutamate transporter EAAT2 in AD brain (Lauderback et al., 2001). Further, Aβ(1–42) treated synaptosomes showed oxidative modification of EAAT2 by HNE (Lauderback et al., 2001). HNE has been shown to induce conformational changes in protein that could impair the function of protein (Subramaniam et al., 1997). Therefore, in AD brain oxidative modification of EAAT2 implies the loss of activity of this protein, with consequent accumulation of glutamate in the synaptic cleft. This, in turn, would lead to influx of calcium into the cell via activation of the NMDA and AMPA receptors, leading to cell death (Masliah et al., 1995). Similarly, oxidative inactivation of GS would lead to elevated glutamate levels, increasing chances for excitotoxic neuronal death.

3.3 Altered Synaptic Function

Cognitive deficits in AD individuals can be related to loss of synaptic circuitry in AD brain (Masliah et al., 1994; Scheff and Price, 2003). γ-SNAP is a member of the N-ethylmaleimide-sensitive factor (NSF) attachment proteins (SNAPs) that play an important role in vesicular transport in the constitutive secretory pathway and in neurotransmitter release, hormone secretion, and mitochondrial organization (Beckers et al., 1989; Stenbeck, 1998). Proteomics analysis demonstrates that γ-SNAP is oxidized in AD brain (Sultana et al., 2006b). The oxidation of SNAP conceivably can impair learning and memory processes and altered neurotransmitter systems in AD brain.

3.4 Proteasomal Dysfunction

The proteasomal system is important for proteolytic degradation of damaged or aggregated proteins from a cell. The proteins to be degraded are ubiquitinylated, forming a poly (ubiquitin) chain that serves as a signal for export to the 26S proteosome. Ubiquitin is removed from poly (ubiquitin) chains by UCH L-1 prior to insertion of the damaged protein into the core of the proteasome (Wilkinson et al., 1995). Ubiquitin is in a fixed pool in brain. In AD brain UCH L-1 was found to be oxidized with decreased activity (Castegna et al., 2002a). The decreased UCH L-1 activity could lead to depletion of the free pool of ubiquitin or cause saturation of the proteasome with polyubiquitin chains. Inactivated UCH L-1 itself leads to oxidative stress in brain (Castegna et al., 2004b). In addition, such a scenario also leads to accumulation of the damaged protein and synaptic deterioration and degeneration. A recent in vitro study showed that HNE decreases hydrolase activity of recombinant UCH-L1 (Masliah et al., 1995; Hyun et al., 2002). Both the cross-linked protein and HNE-bound protein can clog the pore of the proteasome (Hyun et al., 2002), rendering it dysfunctional. As shown by others, a dysfunctional proteasome can led to neuronal death (Keller et al., 2000; Shringarpure et al., 2000; Halliwell, 2002; Lang-Rollin et al., 2003). Recently, UCH L-1 was linked to proper synaptic function, implying that oxidative dysfunction of UCH L-1 could alter processes related to learning and memory, a hall mark of AD (Gang et al., 2006).

3.5 Lipid Abnormalities and Cholinergic Failure

Neuropolypeptide h3, a phosphatidylethanolamine binding protein (PEBP) or hippocampal cholinergic neurostimulating peptide (HNCP), likely plays an important role in maintaining phospholipid asymmetry that is crucial to the structure and function of membranes (Daleke and Lyles, 2000). PEBP has been identified as a specifically oxidized protein with reduced activity in AD brain (Castegna et al., 2003). Loss of PEBP activity conceivably could lead to apoptosis by the exposure of phosphatidylserine to the outer bilayer leaflet of the membrane, leading to the cell death. The addition of Aβ(1–42) to synaptosomes leads to loss of phospholipid asymmetry (Mohmmad-Abdul and Butterfield, 2005), as does HNE (Castegna et al., 2004a), which in turn is formed by Aβ(1–42)-induced lipid peroxidation (Lauderback et al., 2001). This is consistent with the proposed role of Aβ(1–42) in oxidative stress and neurodegeneration in AD brain (Butterfield et al., 2001, 2002b; Butterfield and Lauderback, 2002). This enzyme also regulates the level of choline acetyltransferase, an enzyme whose activity was reported to be decreased in AD brain (Davies, 1999). In synaptosomes treated with Aβ(1–42), HNE is bound to cholineacetyltransferase (Butterfield and Lauderback, 2002), suggesting the possible role of oxidation of this protein in cognitive decline in AD.

3.6 Neuritic Abnormalities

DRP-2 is one of the four members of the dihydropyrimidinase-related protein family (DRP-1, -2, -3, and -4) that is involved in axonal outgrowth and pathfinding through transmission and modulation of extracellular signals (Hamajima et al., 1996a, b; Wang and Strittmatter, 1996; Kato et al., 1998). One of the identified extracellular signals is mediated by the proteins of the collapsin-semaphorin family in collaboration with their receptor, neuropilion (He and Tessier-Lavigne, 1997; Kolodkin et al., 1997). Collapsin contributes to axonal pathfinding by inducing growth cone collapse, which repels the outgrowing axon (Luo et al., 1993). It was reported that DRP-2 can induce growth cone collapse (Goshima et al., 1995; Wang and Strittmatter, 1996) by Rho-kinase phosphorylation (Arimura et al., 2000), and binding to tubulin heterodimers and bundled microtubule as carriers to promote microtubules assembly and dynamics (Gu and Ihara, 2000; Fukata et al., 2002). Many neurodegenerative diseases are associated with DRP-2. It was suggested that incorporation of DRP-2 in the NFT decreases cytosolic DRP-2 and leads to abnormal neuritic and axonal growth, thus accelerating the neurotic degeneration in AD (Yoshida et al., 1998). Decreased expression of DRP-2 protein was observed in AD, adult Down's syndrome (DS) (Lubec et al., 1999), fetal DS (Weitzdoerfer et al., 2001), schizophrenia, and affective disorders (Johnston-Wilson et al., 2000). Oxidation of this protein has been observed in AD brain, which conceivably could be related to the observed cognitive impairment in AD (Lubec et al., 1999; Castegna et al., 2002b, 2003). There is always a constant process of developing new synaptic connections to incorporate newly experienced information in brain. Thus, the oxidation of this protein might lead to impaired short-term memory due to inability to access new memory via neuritic processing. Short-term memory loss is one of the earliest clinical signs of AD. Moreover, oxidation of DRP-2 would be predicted to lead to shortening of dendrites and synapse loss, both known in AD brain (Coleman and Flood, 1987).

3.7 Cell Cycle, Tau Hyperphosphorylation, and AB Production

Peptidyl-prolyl isomerases (PPIases) or Pin1 is a chaperone enzyme that catalyzes the conversion of the *cis* to *trans* conformation and vice versa of proteins between given amino acids and a proline (Schutkowski et al., 1998). Also, PPIases have been shown to be necessary for entry into mitosis, and we determined by proteomics that PPIase is oxidized in AD brain (Sultana et al., 2006a, b). This modification can cause dramatic structural modifications, which can affect the properties of targeted proteins. Pin1 possesses a WW domain, which specifically recognizes pSer-Pro and pThr-Pro motifs in which the first amino acid is phosphorylated. Pin1 binds to many proteins implicated in cell-cycle regulation (e.g., p53, Myt1, Wee1, and Cdc25C), and Pin1 also targets tau, a protein forming part of the neuronal cytoskeleton, which is hyperphosphorylated in patients suffering from AD (Zhou et al., 2000). Recent studies show that Pin1 is colocalized with phosphorylated tau and also show an inverse relationship between expression of tau and Pin1 in AD (Holzer et al., 2002; Kurt et al., 2003; Ramakrishnan et al., 2003). In AD brain, cell-cycle machinery is reported to be altered (Arendt, 2003). Pin1 oxidation could, therefore, be involved in the pathogenesis of AD. The oxidized Pin1 showed a decrease in activity both under in vivo and in vitro conditions (Sultana et al., 2006a). Moreover, Pin1 oxidation conceivably could lead to the observed increased accumulation of phosphorylated tau protein in AD brain. For example, Pin 1 couples tau to specific kinases that can phosphorylate tau. Presumably, Pin1 modulates the kinase activities. Moreover, Pin1 couples hyperphosphorylated can to a specific protein phosphotase (PP2A). Consequently, loss of activity of Pin1 due to oxidation conceivably could lead to hyperphosphorylated tau, which as noted alone, could lead to loss of microtubule assembly in AD. This, in turn, could markedly affect antegrade and retrograde transport to and from the synapse to the cell body, providing another mechanism for neuronal death. Further, Lu et al. (1999) showed that the Pin1 could restore the function of tau protein in AD (Lu et al., 1999). Oxidative modification of Pin1 could be one of the initial events that trigger tangle formation, cell cycle related abnormalities, and oxidative damage. A recently described downstream target of pin1 is APP, and this binding appears to be involved in Ag production (Pastorino et al., 2006). Thus, oxidative dysfunction of pin1 conceivable, could be important in two hallmark pathological lesions of AD: NFT and SP.

3.8 pH Maintenance

The activity of enzymes depends on the pH of the cells. Carbonic anhydrase is one enzyme that regulates cellular pH, CO_2 and HCO_3^- transport, and also plays a role in maintaining H_2O and electrolyte balance (Sly and Hu, 1995). The physiological functions of CA-II are regulation of cellular pH regulation, production of CSF, and the synthesis of glucose and lipids (Fernley, 1988). CA-II also catalyzes the reversible hydration of CO_2, a reaction fundamental to many cellular and systemic processes including glycolysis and acid and fluid secretion. The catalytically active form of CA-II is confined to oligodendrocytes and subtypes of protoplasmic astrocytes in the CNS. CA-II deficiency leads to cognitive defects varying from disabilities to severe mental retardation in addition to osteoporosis, renal tubular acidosis, and cerebral calcification. The activity of this protein was observed to be decreased in AD brain compared with the age matched control (Meier-Ruge et al., 1984), and the identification of this protein as one of the oxidized protein by redox proteomics (Sultana etal., 2006b, c) likely explains the observed diminished enzyme activity in AD brain. Further, oxidized CA-II may lead to an imbalance of both the extracellular and intracellular pH in the cell, mitochondrial alterations in oxidative phosphorylation, and impaired synthesis of glucose and lipids. The identification of CA-II as an oxidized protein may add to the progression of AD, as pH plays a very crucial role for enzymes to function. Moreover, pH affects protein aggregation. Thus, dysfunctional CA-II due to its oxidative modification could contribute to the known protein aggregation in AD brain.

3.9 Mitochondrial Dysfunction and Apoptosis

Dysfunction of mitochondria has been reported in AD brain (Bubber et al., 2005). Several studies indicate that Aβ decreases the activity of mitochondrial respiratory chain complexes. ATP synthase alpha and voltage-dependent anion channel proteins were found to be nitrated proteins in AD hippocampus using our proteomics approach. ATP synthase alpha plays an important role in synthesis and release of ATP (Matsuyama et al., 1998). The nitration of this protein could compromise brain ATP synthesis and induce damaging ROS production, which if severe could lead to neuronal death. Voltage-dependent anion channel protein (VDAC-1), an outer pore component of the mitochondrial permeability transition pore (MPTP), plays an essential role in movement of metabolites like ATP in and out of mitochondria by passive diffusion, synaptic communication, and early stages of apoptosis (Shimizu et al., 1999; Yoo et al., 2001a; Cesar Mde and Wilson, 2004). Nitration of this protein suggests an altered MPTP leading to mitochondrial depolarization and altered signal transduction pathways, which is crucial in synaptic transmission and plasticity. In addition, it could also induce apoptotic events leading to cell death.

4   Oxidized Proteins in AD Plasma

Isoforms of fibrinogen gamma-chain precursor protein and of alpha-1-antitrypsin precursor were found to be oxidized in AD plasma by proteomics (Choi et al., 2002), and these proteins had been implicated in the AD pathology. Indeed, identification of oxidized proteins in plasma using proteomics is being touted as an important biomarker for AD (Korolainen et al., 2002).

5   Proteomic Studies on Oxidatively Modified Proteins in Models of AD

Models of AD have been of principal importance to provide insight into the neurochemical and cellular changes associated with this disease. Some of these AD models are discussed later.

5.1 Synaptic Degeneration Model

Synaptic degeneration in AD brains shows a high correlation with decreased cognitive function, which has led to the hypothesis that disappearance of synapses is a key event in early cognitive decline (Crystal et al., 1988; Hamos et al., 1989; Manczak et al., 2004). In particular, it is believed that early synaptic loss in the hippocampal dentate gyrus disrupts the circuitry between the hippocampus and the entorhinal cortex leading to the memory deficits associated with AD (Manczak et al., 2004). Aβ accumulation in synaptic termini may play an important role in oxidative stress and synaptic degeneration (Hensley et al., 1995; Mattson et al., 1998; Gylys et al., 2004). In a recent study, incubation of synaptosomes with Aβ(1–42) showed specific oxidation of β-actin, glial fibrillary acidic protein, and DRP-2 (Boyd-Kimball et al., 2005a). Further, H+-transporting two-sector ATPase, syntaxin-binding protein 1, glutamate dehydrogenase, β-actin, and elongation factor Tu showed a trend toward oxidation (Boyd-Kimball et al., 2005a). The oxidation of these proteins implies loss of a wide variety of cellular functions including cytoskeletal integrity, neuronal communication, energy metabolism, excitotoxicity, and protein synthesis that are pathological hallmark of AD. These findings further suggest a role of oxidative stress in the degeneration of neurons. The accumulation of Aβ at the synapse may increase oxidative stress leading to impaired normal synaptic functions (Mattson et al., 1998; Boyd-Kimball et al., 2005a).

5.2 Aβ(1–42) Induces Specific Protein Oxidation In Vivo

AD brain is characterized by degeneration of the basal forebrain cholinergic neurons, a degeneration that is associated with the cognitive deficits (Whitehouse et al., 1981; Frolich, 2002). Intracerebral injection of Aβ(1–42) into the nucleus basalis magnocellarius of rat brain was performed to mimic a cholinergic model of AD (Giovannini et al., 2002). In this animal model, an extensive protein oxidation was observed in the hippocampus compared with the cortex and nucleus basalis, though all three regions demonstrated oxidative damage (Boyd-Kimball et al., 2005c). Following intracerebral injection of Aβ(1–42), a large number of oxidized brain proteins were identified using proteomics. These oxidatively modified proteins included proteins involved in cellular structure (β-synuclein), signal transduction (14-3-3 zeta), energy metabolism [phosphoglycerate mutase 1 (PGM1), pyruvate dehydrogenase, and glyceraldehyde 3-phosphate dehydrogenase], and stress responses (HSP60) (Boyd-Kimball et al., 2005c). It should be noted that some of these proteins replicate what is seen in AD brain.

The oxidized proteins in common with AD are not discussed in detail in this section of this review, as they were discussed in detail earlier. However, the other oxidized proteins that were not found to be oxidized in human AD sample by proteomics but found in this model are discussed here. The 14-3-3 protein, one of the oxidized proteins in this model, is involved in a number of cellular functions including signal transduction, protein trafficking, and metabolism (Takahashi, 2003; Dougherty and Morrison, 2004). The expression of this protein was reported to be increased in AD brain (Fountoulakis et al., 1999) and CSF (Burkhard et al., 2001). These proteins are also found to be associated with NFT in AD brain (Layfield et al., 1996) and has been shown to act as an effector of tau protein phosphorylation (Hashiguchi et al., 2000). Recently, it has been shown that 14-3-3 zeta acts as a scaffolding protein simultaneously binding to tau and glycogen synthase kinase 3β (GSK3β) in a multiprotein tau phosphorylation complex (Agarwal-Mawal et al., 2003). The oxidation of 14-3-3 zeta in this rat model following Aβ(1–42) injection into NBM could unite both the importance of Aβ(1–42) and the hyperphosphorylation of tau, as observed in AD.

In rat brain, β-synuclein is associated with cholinergic components particularly in the basal forebrain (Li et al., 2002) and plays a role in synaptic vesicle homeostasis. And the oxidation of this protein in cholinergic regions of the brain could lead to loss of synapses and cholinergic deficits documented in AD (Masliah et al., 1994; Frolich, 2002; Giovannini et al., 2002). Recently, β-synuclein has been shown to increase Akt activity by directly interacting with Akt in neuroblastoma cells transfected with β-synuclein. The increase in Akt activity was shown to protect neurons against rotenone, suggesting that β-synuclein may play a protective role in the CNS (Hashimoto et al., 2004). If this is the case, oxidation of β-synuclein could lead to a conformation change in the protein preventing its direct interaction with Akt and abolishing this protective effect.

Chaperonin 60 (Cpn60), [also called heat shock protein 60 (HSP60)], another oxidized protein found in this model is involved in mediating the proper folding and assembly of mitochondrial proteins, especially in response to oxidative stress (Bozner et al., 2002). Fibroblasts from AD patients showed an increase oxidation and decreased expression compared with age matched control (Yoo et al., 2001b; Choi et al., 2003). The oxidation of HSP60 could lead to increased protein misfolding and aggregation, as well as an increased vulnerability to oxidative stress. This is particularly important due to the lack of mechanisms to protect mitochondria from oxidative stress and the vicinity of mitochondrial proteins to reactive oxygen species generated during normal oxidative phosphorylation and more so in concert with mitochondrial dysfunction.

5.3 *Caenorhabditis elegans*

Transgenic *Caenorhabditis elegans* expressing human Aβ(1–42) is an in vivo model of Aβ-associated toxicity and deposition. Transgenic *C. elegans* expressed constitutive human Aβ (1–42) via a body-wall muscle myosin promoter and an Aβ minigene (Yatin et al., 1999). Increased oxidative stress in *C. elegans* expressing Aβ(1–42) is found and methionine 35 of Aβ(1–42) is involved in the mechanism of oxidative damage (Yatin et al., 1999). A temperature-inducible Aβ expression system in the *C. elegans* is an excellent model to study the relationship between Aβ toxicity, fibril formation, and oxidative stress (Drake et al., 2003). We used this model to investigate the proteins oxidized by Aβ(1–42) in vivo using proteomics (Boyd-Kimball et al., 2005b). In this animal model, 16 proteins were identified to be oxidatively modified that are involved in a variety of cellular functions including energy metabolism, protein degradation, cytoskeletal integrity, antioxidant system, signal transduction, and lipid metabolism. Protein oxidation has been shown to alter protein conformation leading to loss of function (Hensley et al., 1995; Subramaniam et al., 1997; Lauderback et al., 2001; Castegna et al., 2002b; Sultana et al., 2006b). Thus, it is likely that oxidation of the proteins identified in this study also leads to loss of function.

5.4 SAMP8 Mice

The SAMP8 mouse strain is a model for accelerated senescence that shows an increase in Aβ deposition at ages 4–12 months at 7 to 20-fold excess compared with Aβ deposition observed in the normal aging mouse brain (Kumar et al., 2001). This large increase in Aβ in the SAMP8 mouse brain is much closer to the estimated 50% increase in Aβ observed in AD brain (Rosenberg, 2000) than is observed in APP transgenic mouse models. SAMP8 mice also show an increased oxidative stress and develop learning and memory deficits by 12 months of age (Butterfield et al., 1997b). Hence, SAMP8 mouse may be an excellent model for studying Aβ toxicity. Indeed, rodent Aβ(1–42) is toxic to neurons after sufficient exposure (Boyd-Kimball et al., 2004a).

In aged mice, oxidized proteins were identified using a combination of MudPIT techniques with a hydrazine biotin-streptavidin (Soreghan et al., 2003). Recently using redox proteomics, alpha enolase, gamma enolase, lactate dehydrogenase 2, alpha spectrin, creatine kinase, and DRP-2 were identified as oxidized protein in SAMP8 mice brain (Poon et al., 2004). Among these proteins alpha enolase, gamma enolase, DRP-2, LDH, and CK are the common oxidized proteins observed in human AD brain and SAMP8. The expression of alpha spectrin and DRP-2 was reported to be altered in AD brain.

Spectrin is a membrane-associated cytoskeletal protein that forms a supporting and organized scaffold for intracellular cohesion with the association of actins (Leto et al., 1988). The breakdown products of α-spectrin are used as markers of apoptosis (Vanderklish and Bahr, 2000). In rat cortical neuronal cultures, Aβ treatment induced α-spectrin breakdown products by activating capases (Harada and Sugimoto, 1999). In SAMP8 mouse brain, α-spectrin showed an increase in oxidation associated with a decrease in expression (Poon et al., 2004). These results suggest that the proteolytic mechanism in apoptosis involves oxidative modification and degradation of α-spectrin (Poon et al., 2004) that could disrupt the cytoskeleton and the structure of cells in brain. This, in turn, would affect intercellular and intracellular communications, and consequently likely contribute to the learning and memory deficits observed in SAMP8 mice. Consistent with the role of oxidative damage in SAMP8 mice (Butterfield et al., 1997a; Farr et al., 2003), α-lipoic acid given to SAMP8 mice, which improves learning and memory, leads to decreased oxidative stress and modulation of the oxidatively modified proteins, as demonstrated by proteomics (Poon et al., 2005a). Antisense oligonucleotide directed against the Aβ region of the APP gene not only improved learning and behavior of aged mice, but also decreased oxidative stress, Aβ levels, and the modulation of specifically oxidized brain proteins in aged SAMP8 mice (Poon et al., 2004, 2005b).

6 Future of Proteomics in AD

Using the redox proteomics approach, a number of common oxidatively modified proteins were found in different regions of AD brain (❯ *Figure 13-2*) and also between animal models and AD brain (❯ *Table 13-3*). The appearance of common oxidized proteins in various animal models and AD brain suggests common mechanisms of disease. The results also support the role of Aβ(1–42) in the induction of oxidative stress and the related AD biochemistry and pathology. The use of animal models together with the proteomics approach can provide insight into the mechanisms of neurodegeneration in AD and could also help in the development of therapeutic approaches to prevent or delay this devastating disease. With the improvement in proteomics technology, other kinds of protein posttranslational modifications can be studied. Functional proteomics and pharmacoproteomics in AD and models thereof, coupled with the proteomics of individuals at risk for AD, will provide further insight into the molecular mechanisms for and treatment of this devastating dementing disorder. Such studies are going on in our laboratory. Indeed, an initial assessment of oxidatively modified brain proteins in PS-1-mediated familial AD revealed UCH L-1 as an oxidatively modified protein, among others (Butterfield et al., 2006a).

■ Figure 13-2

Comparison of oxidatively modified proteins in two different regions of AD brain

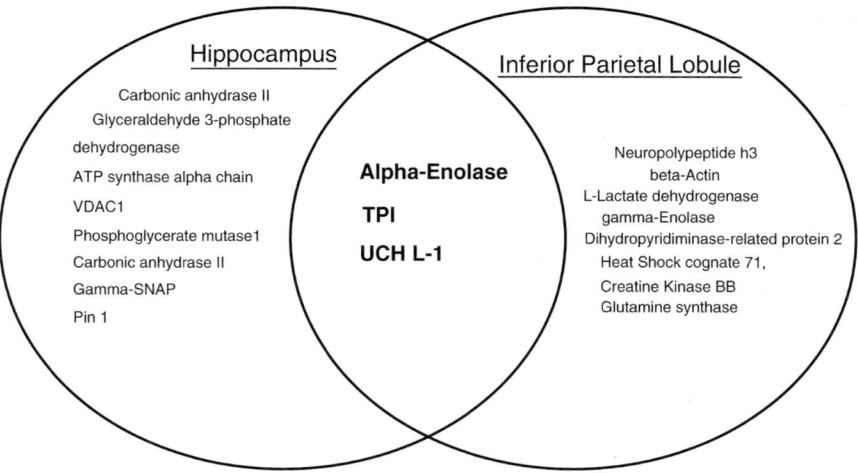

Acknowledgment

This work was supported, in part, by NIH grants [AG-05119; AG-10836].

◼ Table 13-3

Oxidatively modified brian proteins in Alzheimer's Disease and models thereof

S. No	AD Brain	SAMP8 Mice	Cholinergic Deficit Animal Model	C. elegans	Synapsomes
1	Alpha enolase[a]	Alpha enolase[a]	Glyceraldehyde 3-phosphate dehydrogenase[b]	Gluthathione S-transferase (GST)[c]	Beta-actin
2	Dihydropyrimidinase related protein-1 (DRP-2)[d]	Dihydropyrimidinase related protein-1 (DRP-2)[d]	Phosphoglycerate mutase 1 (PGM1)[b]	Creatine kinase[e]	Glial fibrillary acidic protein
3	Gamma enolase[a]	Gamma enolase[a]	Gluthathione S-transferase (GST)[c]	Acyl Co A dehydrogenase	Dihydropyrimidinase related protein-1 (DRP-2)[d]
4	Creatine kinase[e]	Creatine kinase[e]	Beta synuclein	Transketolase	H+-transporting two-sector ATPase
5	Gluthathione S-transferase (GST)[c]	Lactate dehydrogenase	14-3-3 Zeta	Guanine nucleotide binding protein	Syntaxin binding protein 1
6	Phosphoglycerate mutase 1[b]	Alpha spectrin	Pyruvate dehydrogenase	Adenine kinase	Glutamate dehydrogenase
7	Glyceraldehyde 3-phosphate[b]		HSP 60	Proteosome alpha subunit 4	Beta-actin
8	Gamma-SNAP			Myosin light chain 1	Elongation factor Tu
9	Triosephosphate isomerase			Lipid binding protein-6	
10	Carbonic anhydrase II				
11	Peptidyl prolyl cis-trans isomerase				
12	Neuropolypeptide h3				
13	UCHL-1				
14	Glutamine synthetase				

[a]Green: Common proteins between AD brain and SAMP8 mice
[b]Purple: Common proteins between AD brain and in vivo Aβ(1–42)/cholinergic deficit model
[c]Blue: Common proteins between AD brain and in vivo Aβ(1–42)/cholinergic deficit model and C. elegans that express human Aβ(1–42)
[d]Light blue: Common proteins between AD brain, SAMP8 and synaptosomes
[e]Red: Common proteins between AD brain, SAMP8 and C. elegans that express human Aβ(1–42)

References

Aebersold R, Goodlett DR. 2001. Mass spectrometry in proteomics. Chem Rev 101: 269-295.

Agarwal-Mawal A, Qureshi HY, Cafferty PW, Yuan Z, Han D, et al. 2003. 14-3-3 connects glycogen synthase kinase-3 beta to tau within a brain microtubule-associated tau phosphorylation complex. J Biol Chem 278: 12722-12728.

Aksenov MY, Aksenova MV, Butterfield DA, Geddes JW, Markesbery WR. 2001. Protein oxidation in the brain in Alzheimer's disease. Neuroscience 103: 373-383.

Aksenova M, Butterfield DA, Zhang SX, Underwood M, Geddes JW. 2002. Increased protein oxidation and decreased creatine kinase BB expression and activity after spinal cord contusion injury. J Neurotrauma 19: 491-502.

Aksenova MV, Aksenov MY, Payne RM, Trojanowski JQ, Schmidt ML, et al. 1999. Oxidation of cytosolic proteins and expression of creatine kinase BB in frontal lobe in different neurodegenerative disorders. Dement Geriatr Cogn Disord 10: 158-165.

Arendt T. 2003. Synaptic plasticity and cell cycle activation in neurons are alternative effector pathways: The 'Dr. Jekyll and Mr. Hyde concept' of Alzheimer's disease or the yin and yang of neuroplasticity. Prog Neurobiol 71: 83-248.

Arimura N, Inagaki N, Chihara K, Menager C, Nakamura N, et al. 2000. Phosphorylation of collapsin response mediator protein-2 by Rho-kinase. Evidence for two separate signaling pathways for growth cone collapse. J Biol Chem 275: 23973-23980.

Beckers CJ, Block MR, Glick BS, Rothman JE, Balch WE. 1989. Vesicular transport between the endoplasmic reticulum and the Golgi stack requires the NEM-sensitive fusion protein. Nature 339: 397-398.

Berlett BS, Stadtman ER. 1997. Protein oxidation in aging, disease, and oxidative stress. J Biol Chem 272: 20313-20316.

Blacker D, Wilcox MA, Laird NM, Rodes L, Horvath SM, et al. 1998. Alpha-2 macroglobulin is genetically associated with Alzheimer disease. Nat Genet 19: 357-360.

Boyd-Kimball D, Sultana R, Mohmmad-Abdul H, Butterfield DA. 2004a. Rodent Abeta(1–42) exhibits oxidative stress properties similar to those of human Abeta(1–42): Implications for proposed mechanisms of toxicity. J Alzheimers Dis 6: 515-525.

Boyd-Kimball D, Mohmmad Abdul H, Reed T, Sultana S, Butterfield DA. 2004b. Role of phenylalanine 20 in Alzheimer's amyloid β-peptide (1–42)-induced oxidative stress and neurotoxicity. Chem Res Toxicol 17: 1743-1749.

Boyd-Kimball D, Castegna A, Sultana R, Poon HF, Petroze R, et al. 2005a. Proteomic identification of proteins oxidized by Aβ(1–42) in synaptosomes: Implications for Alzheimer's disease. Brain Res 1044: 206-215.

Boyd-Kimball D, Poon HF, Lynn BC, Cai J, Pierce WM, et al. 2006. Proteomic identification of proteins specifically oxidized in Caenorhabditis elegans expressing human Aβ (1–42): Implications for Alzheimer's disease. Neurobiol Aging 27: 1239-1249.

Boyd-Kimball D, Sultana R, Poon HF, Lynn BC, Casamenti F, et al. 2005c. Proteomic identification of proteins specifically oxidized by intracerebral injection of Aβ(1–42) into rat brain: Implications for Alzheimer's disease. Neuroscience 132: 313-324.

Bozner P, Wilson GL, Druzhyna NM, Bryant-Thomas TK, Le Doux SP, et al. 2002. Deficiency of chaperonin 60 in Down's syndrome. J Alzheimers Dis 4: 479-486.

Bubber P, Haroutunian V, Fisch G, Blass JP, Gibson GE. 2005. Mitochondrial abnormalities in Alzheimer brain: Mechanistic implications. Ann Neurol 57: 695-703.

Burkhard PR, Sanchex JC, Landis T, Hochstrasser DF. 2001. CSF detection of the 14-3-3 protein in unselected patients with dementia. Neurology 56: 1528-1533.

Butterfield DA. 2002. Amyloid beta-peptide (1–42)-induced oxidative stress and neurotoxicity: Implications for neurodegeneration in Alzheimer's disease brain. A review. Free Radic Res 36: 1307-1313.

Butterfield DA. 2003. Amyloid beta-peptide [1–42]-associated free radical-induced oxidative stress and neurodegeneration in Alzheimer's disease brain: Mechanisms and consequences. Curr Med Chem 10: 2651-2659.

Butterfield DA. 2004. Proteomics: A new approach to investigate oxidative stress in Alzheimer's disease brain. Brain Res 1000: 1-7.

Butterfield DA, Lauderback CM. 2002. Lipid peroxidation and protein oxidation in Alzheimer's disease brain: Potential causes and consequences involving amyloid beta-peptide-associated free radical oxidative stress. Free Radic Biol Med 32: 1050-1060.

Butterfield DA, Stadtman ER. 1997. Protein oxidation processes in aging brain. Adv Cell Aging Gerontol 2: 161-191.

Butterfield DA, Boyd-Kimball D, Castegna A. 2003. Proteomics in Alzheimer's disease: Insights into potential mechanisms of neurodegeneration. J Neurochem 86: 1313-1327.

Butterfield DA, Castegna A, Drake J, Scapagnini G, Calabrese V. 2002a. Vitamin E and neurodegenerative disorders associated with oxidative stress. Nutr Neurosci 5: 229-239.

Butterfield DA, Castegna A, Lauderback CM, Drake J. 2002b. Evidence that amyloid beta-peptide-induced lipid peroxidation and its sequelae in Alzheimer's disease brain contribute to neuronal death. Neurobiol Aging 23: 655-664.

Butterfield DA, Hensley K, Cole P, Subramaniam R, Aksenov M, et al. 1997a. Oxidatively induced structural alteration of glutamine synthetase assessed by analysis of spin label incorporation kinetics: Relevance to Alzheimer's disease. J Neurochem 68: 2451-2457.

Butterfield DA, Howard BJ, Yatin S, Allen KL, Carney JM. 1997b. Free radical oxidation of brain proteins in accelerated senescence and its modulation by N-tert-butyl-alpha-phenylnitrone. Proc Natl Acad Sci USA 94: 674-678.

Butterfield DA, Poon HF, St. Clair D, Kellor JN, Pieree WM, Klein JB, Markesbary WR. 2006a. Redox proteomics identification of oxidatively modified hippocampal proteins in mild cognitive impairment: Insights into the development of Alzheimer's disease. Neurobiol Dis 22: 223-232.

Butterfield DA, Reed T, Perluigi M, De Marco C, Coccia R, Cini C, Sultana R. 2006b. Elevated protien-bound levels of the lipid peroxidatin product, 4-hydroxy-2-nonenal, in brain form persons with mild cognitive impairment. Neurosci Lett 397: 170-173.

Butterfield DA, Gnjec A, Poon HF, Castegna A, Pierce WM, Klein JB, Martins RN. 2006c. Redox proteomics identification of oxidatively modified brain proteins in inherited Alzheimer's disease: An initial assessment. J Alz Dis 10: 391-397.

Castegna A, Aksenov M, Aksenova M, Thongboonkerd V, Klein JB, et al. 2002a. Proteomic identification of oxidatively modified proteins in Alzheimer's disease brain. Part I: Creatine kinase BB, glutamine synthase, and ubiquitin carboxy-terminal hydrolase L-1. Free Radic Biol Med 33: 562-571.

Castegna A, Aksenov M, Thongboonkerd V, Klein JB., Pierce WM, et al. 2002b. Proteomic identification of oxidatively modified proteins in Alzheimer's disease brain. Part II: dihydropyrimidinase-related protein 2, alpha-enolase and heat shock cognate 71. J Neurochem 82: 1524-1532.

Castegna A, Lauderback CM, Mohmmad-Abdul H, Butterfield DA. 2004a. Modulation of phospholipid asymmetry in synaptosomal membranes by the lipid peroxidation products, 4-hydroxynonenal and acrolein: Implications for Alzheimer's disease. Brain Res 1004: 193-197.

Castegna A, Thongboonkerd V, Klein JB, Lynn B, Markesbery WR, et al. 2003. Proteomic identification of nitrated proteins in Alzheimer's disease brain. J Neurochem 85: 1394-1401.

Castegna A, Thongboonkerd V, Klein J, Lynn BC, Wang YL, et al. 2004b. Proteomic analysis of brain proteins in the gracile axonal dystrophy (gad) mouse, a syndrome that emanates from dysfunctional ubiquitin carboxyl-terminal hydrolase L-1, reveals oxidation of key proteins. J Neurochem 88: 1540-1546.

Cesar Mde C, Wilson JE. 2004. All three isoforms of the voltage-dependent anion channel (VDAC1, VDAC2, and VDAC3) are present in mitochondria from bovine, rabbit, and rat brain. Arch Biochem Biophys 422: 191-196.

Choi J, Malakowsky CA, Talent JM, Conrad CC, Carrll CA, et al. 2003. Anti-apoptotic proteins are oxidized by Abeta25-35 in Alzheimer's fibroblasts. Biochim Biophys Acta 1637: 135-141.

Choi J, Malakowsky CA, Talent JM, Conrad CC, Gracy RW. 2002. Identification of oxidized plasma proteins in Alzheimer's disease. Biochem Biophys Res Commun 293: 1566-1570.

Coleman PD, Flood DG. 1987. Neuron numbers and dendritic extent in normal aging and Alzheimer's disease. Neurobiol Aging 8: 521-545.

Coyle JT, Price DL, DeLong MR. 1983. Alzheimer's disease: A disorder of cortical cholinergic innervation. Science 219: 1184-1190.

Cruts M, van Duijn CM, Backhovens H, Van den Broeck M, Wehnert A, et al. 1998. Estimation of the genetic contribution of presenilin-1 and -2 mutations in a population-based study of presenile Alzheimer disease. Hum Mol Genet 7: 43-51.

Crystal H, Dickson D, Fuld P, Masur D, Scott R, et al. 1988. Clinico-pathologic studies in dementia: Nondemented subjects with pathologically confirmed Alzheimer's disease. Neurology 38: 1682-1687.

Dahiyat M, Cumming A, Harrington C, Wischik C, Xuereb J, et al. 1999. Association between Alzheimer's disease and the NOS3 gene. Ann Neurol 46: 664-667.

Daleke DL, Lyles JV. 2000. Identification and purification of aminophospholipid flippases. Biochim Biophys Acta 1486: 108-127.

Davies P. 1999. Challenging the cholinergic hypothesis in Alzheimer disease. JAMA 281: 1433-1434.

Dougherty MK, Morrison, DK. 2004. Unlocking the code of 14-3-3. J Cell Sci 117: 1875-1884.

Drake J, Link CD, Butterfield DA. 2003. Oxidative stress precedes fibrillar deposition of Alzheimer's disease amyloid beta-peptide (1–42) in a transgenic *Caenorhabditis elegans* model. Neurobiol Aging 24: 415-420.

Farr SA, Poon HF, Dogrukol-Ak D, Drake J, Banks WA, et al. 2003. The antioxidants alpha-lipoic acid and N-acetylcysteine reverse memory impairment and brain oxidative stress in aged SAMP8 mice. J Neurochem 84: 1173-1183.

Fernley RT. 1988. Non-cytoplasmic carbonic anhydrases. Trends Biochem Sci 13: 356-359.

Fountoulakis M, Cairns N, Lubec G. 1999. Increased levels of 14-3-3 gamma and epsilon proteins in brain f patients with Alzheimer's disease and Down syndrome. J Neural Transm Suppl 57: 323-335.

Frolich L. 2002. The cholinergic pathology in Alzheimer's disease – discrepancies between clinical experience and pathophysiological findings. J Neural Transm 109: 1003-1013.

Fukata Y, Itoh TJ, Kimura T, Menager C, Nishimura T, et al. 2002. CRMP-2 binds to tubulin heterodimers to promote microtubule assembly. Nat Cell Biol 4: 583-591.

Giovannini MG, Scali C, Prosperi C, Bellucci A, Vannucchi MG, et al. 2002. Beta-amyloid-induced inflammation and cholinergic hypofunction in the rat brain in vivo: Involvement of the p38MAPK pathway. Neurobiol Dis 11: 257-274.

Goate A, Chartier-Harlin MC, Mullan M, Brown J, Crawford F, et al. 1991. Segregation of a missense mutation in the amyloid precursor protein gene with familial Alzheimer's disease. Nature 349: 704-706.

Gong B, Cao Z, Zheng P, Vitolo OV, Liu S, Staniszewski A, Moolman D, Zhang H, Shelanski M, Arancio O. 2006. Ubiquitin hydrolase UCH-L1 rescues β-amyloid-induced decreases in synaptic function and contextual memory. Cell 126: 775-788.

Goshima Y, Nakamura F, Strittmatter P, Strittmatter SM. 1995. Collapsin-induced growth cone collapse mediated by an intracellular protein related to UNC-33. Nature 376: 509-514.

Grundke-Iqbal I, Iqbal K, Quinlan M, Tung YC, Zaidi MS, et al. 1986. Microtubule-associated protein tau. A component of Alzheimer paired helical filaments. J Biol Chem 261: 6084-6089.

Gu Y, Ihara Y. 2000. Evidence that collapsin response mediator protein-2 is involved in the dynamics of microtubules. J Biol Chem 275: 17917-17920.

Gylys KH, Fein JA, Yang F, Wiley DJ, Miller CA, et al. 2004. Synaptic changes in Alzheimer's disease: Increased amyloid-β and gliosis in surviving terminals is accompanied by decreased PSD-95 fluorescence. Am J Pathol 165: 1809-1817.

Halliwell B. 2002. Hypothesis: Proteasome dysfunction: A primary event in neurodegeneration that leads to nitrative and oxidative stress and subsequent cell death. Ann N Y Acad Sci 962: 182-194.

Hamajima N, Matsuda K, Sakata S, Tamaki N, Sasaki M, et al. 1996. A novel gene family defined by human dihydropyrimidinase and three related proteins with differential tissue distribution. Gene 180: 157-163.

Hamos JE, DeGennaro LJ, Drachman DA. 1989. Synaptic loss in Alzheimer's disease and other dementias. Neurology 39: 355-361.

Harada J, Sugimoto M. 1999. Activation of caspase-3 in beta-amyloid-induced apoptosis of cultured rat cortical neurons. Brain Res 842: 311-323.

Hashiguchi M, Sobue K, Paudel HK. 2000. 14-3-3 zeta is an effector of tau protein phosphorylation. JBC 275: 25247-25254.

Hashimoto M, Bar-on P, Ho G, Takenouchi T, Rockenstein E, et al. 2004. Beta-synuclein regulates Akt activity in neuronal cells: A possible mechanism for neuroprotection in Parkinson's disease. J Biol Chem, Published online.

He Z, Tessier-Lavigne M. 1997. Neuropilin is a receptor for the axonal chemorepellent Semaphorin III. Cell 90: 739-751.

Hensley K, Hall N, Subramaniam R, Cole P, Harris M, et al. 1995. Brain regional correspondence between Alzheimer's disease histopathology and biomarkers of protein oxidation. J Neurochem 65: 2146-2156.

Herbert B. 1999. Advances in protein solubilization for two-dimensional gel electrophoresis. Electrophoresis 20: 660-663.

Holzer M, Gartner U, Stobe A, Hartig W, Gruschka H, et al. 2002. Inverse association of Pin1 and tau accumulation in Alzheimer's disease hippocampus. Acta Neuropathol 104: 471-481.

Hoogland C, Sanchez JC, Tonella L, Binz PA, Bairoch A, et al. 2000. The 1999 SWISS-2DPAGE database update. Nucl Acids Res 28: 286-288.

Hoyer S. 2004. Causes and consequences of disturbances of cerebral glucose metabolism in sporadic Alzheimer disease: Therapeutic implications. Adv Exp Med Biol 541: 135-152.

Hyun DH, Lee MH, Halliwell B, Jenner P. 2002. Proteasomal dysfunction induced by 4-hydroxy-2,3-*trans*-nonenal, an end-product of lipid peroxidation: A mechanism contributing to neurodegeneration? J Neurochem 83: 360-370.

Johnston-Wilson NL, Sims CD, Hofmann JP, Anderson L, Shore AD, et al. 2000. Disease-specific alterations in frontal cortex brain proteins in schizophrenia, bipolar disorder, and major depressive disorder. The Stanley Neuropathology Consortium. Mol Psychiatry 5: 142-149.

Kaji H, Tsuji T, Mawuenyega KG, Wakamiya A, Taoka M, et al. 2000. Profiling of *Caenorhabditis elegans* proteins using two-dimensional gel electrophoresis and matrix assisted laser desorption/ionization-time of flight-mass spectrometry. Electrophoresis 21: 1755-1765.

Kato Y, Hamajima N, Inagaki H, Okamura N, Koji T, et al. 1998. Post-meiotic expression of the mouse dihydropyrimidinase-related protein 3 (DRP-3) gene during spermiogenesis. Mol Reprod Dev 51: 105-111.

Keller JN, Hanni KB, Markesbery WR. 2000. Impaired proteasome function in Alzheimer's disease. J Neurochem 75: 436-439.

Keller JN, Schmitt FA, Scheff SW, Ding Q, Chen Q, et al. 2005. Evidence of increased oxidative damage in subjects with mild cognitive impairment. Neurology 64: 1152-1156.

Kolodkin AL, Levengood DV, Rowe EG, Tai YT, Giger RJ, et al. 1997. Neuropilin is a semaphorin III receptor. Cell 90: 753-762.

Korolainen MA, Goldsteins G, Alafuzoff I, Koistinaho J, Pirttila T. 2002. Proteomic analysis of protein oxidation in Alzheimer's disease brain. Electrophoresis 23: 3428-3433.

Kumar VB, Vyas K, Franko M, Choudhary V, Buddhiraju C, et al. 2001. Molecular cloning, expression, and regulation of hippocampal amyloid precursor protein of senescence accelerated mouse (SAMP8). Biochem Cell Biol 79: 57-67.

Kurt MA, Davies DC, Kidd M, Duff K, Howlett DR. 2003. Hyperphosphorylated tau and paired helical filament-like structures in the brains of mice carrying mutant amyloid precursor protein and mutant presenilin-1 transgenes. Neurobiol Dis 14: 89-97.

Lang-Rollin I, Rideout H, Stefanis L. 2003. Ubiquitinated inclusions and neuronal cell death. Histol Histopathol 18: 509-517.

Lauderback CM, Hackett JM, Huang FF, Keller JN, Szweda LI, et al. 2001. The glial glutamate transporter, GLT-1, is oxidatively modified by 4-hydroxy-2-nonenal in the Alzheimer's disease brain: The role of Abeta1-42. J Neurochem 78: 413-416.

Layfield R, Fergusson J, Aitken A, Lowe J, Landon M, et al. 1996. Neurofibrillary tangles of Alzheimer's disease brains contain 14-3-3 proteins. Neurosci Lett 209: 57-60.

Leto TL, Fortugno-Erikson D, Barton D, Yang-Feng TL, Francke U, et al. 1988. Comparison of nonerythroid alpha-spectrin genes reveals strict homology among diverse species. Mol Cell Biol 8: 1-9.

Levy-Lahad E, Lahad A, Wijsman EM, Bird TD, Schellenberg GD. 1995. Apolipoprotein E genotypes and age of onset in early-onset familial Alzheimer's disease. Ann Neurol 38: 678-680.

Li JY, Henning Jensen P, Dahlstrom A. 2002. Differential localization of a-, b-, and g-synucleins in the rat CNS. Neuroscience 113: 463-478.

Lovell MA, Xie C, Markesbery WR. 2001. Acrolein is increased in Alzheimer's disease brain and is toxic to primary hippocampal cultures. Neurobiol Aging 22: 187-194.

Lu, PJ, Wulf G, Zhou XZ, Davies P, Lu KP. 1999. The prolyl isomerase Pin1 restores the function of Alzheimer-associated phosphorylated tau protein. Nature 399: 784-788.

Lubec G, Nonaka M, Krapfenbauer K, Gratzer M, Cairns N, et al. 1999. Expression of the dihydropyrimidinase related protein 2 (DRP-2) in Down syndrome and Alzheimer's disease brain is downregulated at the mRNA and dysregulated at the protein level. J Neural Transm Suppl 57: 161-177.

Luo Y, Raible D, Raper JA. 1993. Collapsin: A protein in brain that induces the collapse and paralysis of neuronal growth cones. Cell 75: 217-227.

Manczak M, Park BS, Jung Y, Reddy PH. 2004. Differential expression of oxidative phosphorylation genes in patients with Alzheimer's disease: Implications for early mitochondrial dysfunction and oxidative damage. Neuromol Med 5: 147-162.

Mark RJ, Lovell MA, Markesbery WR, Uchida K, Mattson MP. 1997. A role for 4-hydroxynonenal, an aldehydic product of lipid peroxidation, in disruption of ion homeostasis and neuronal death induced by amyloid beta-peptide. J Neurochem 68: 255-264.

Markesbery WR. 1997. Oxidative stress hypothesis in Alzheimer's disease. Free Radic Biol Med 23: 134-147.

Markesbery WR, Lovell MA. 1998. Four-hydroxynonenal, a product of lipid peroxidation, is increased in the brain in Alzheimer's disease. Neurobiol Aging 19: 33-36.

Masliah E, Alford M, De Teresa R, Mallory M, Hansen L. 1995. Deficient glutamate transport is associated with neurodegeneration in Alzheimer's disease. Ann Neurol 40: 759-766.

Masliah E, Mallory M, Hansen L, DeTeresa R, Alford M, et al. 1994. Synaptic and neuritic alterations during the progression of Alzheimer's disease. Neurosci Lett 174: 67-72.

Matsuyama S, Xu Q, Velours J, Reed JC. 1998. The mitochondrial F0F1-ATPase proton pump is required for function of the proapoptotic protein Bax in yeast and mammalian cells. Mol Cell 1: 327-336.

Mattson MP, Partin J, Begley JG. 1998. Amyloid beta-peptide induces apoptosis-related events in synapses and dendrites. Brain Res 807: 167-176.

Mazzola JL, Sirover MA. 2001. Reduction of glyceraldehyde-3-phosphate dehydrogenase activity in Alzheimer's disease and in Huntington's disease fibroblasts. J Neurochem 76: 442-449.

McKhann G, Drachman D, Folstein M, Katzman R, Price D, et al. 1984. Clinical diagnosis of Alzheimer's disease: Report of the NINCDS-ADRDA Work Group under the auspices of Department of Health and Human Services Task Force on Alzheimer's Disease. Neurology 34: 939-944.

Meier-Ruge W, Iwangoff P, Reichlmeier K. 1984. Neurochemical enzyme changes in Alzheimer's and Pick's disease. Arch Gerontol Geriatr 3: 161-165.

Mohmmad-Abdul H, Butterfield DA. 2005. Protection against amyloid beta-peptide (1–42)-induced loss of phospholipid asymmetry in synaptosomal membranes by tricyclodecan-9-xanthogenate (D609) and ferulic acid ethyl ester: Implications for Alzheimer's disease. Biochim Biophys Acta 1741: 140-148.

Molloy MP. 2000. Two-dimensional electrophoresis of membrane proteins using immobilized pH gradients. Anal Biochem 280: 1-10.

Pastorino L, Sun A, Lu PJ, Zhou XZ, Balastik M, Finn G, Wulf G, Lim J, Li SH, Li X, Xia W, Nicholson LK, Lu KP. 2006. The prolyl isomerase Pin1 regulates amyloid precursor protein processing and amyloid beta production. Nature 440: 528-534.

Planel E, Miyasaka T, Launey T, Chui DH, Tanemura K, et al. 2004. Alterations in glucose metabolism induce hypothermia leading to tau hyperphosphorylation through differential inhibition of kinase and phosphatase activities: Implications for Alzheimer's disease. J Neurosci 24: 2401-2411.

Poon HF, Castegna A, Farr SA, Thongboonkerd V, Lynn BC, et al. 2004. Quantitative proteomics analysis of specific protein expression and oxidative modification in aged senescence-accelerated-prone 8 mice brain. Neuroscience 126: 915-926.

Poon HF, Farr S, Thongboonkerd V, Lynn BC, Banks WA, et al. 2005a. Proteomic analysis of specific brain proteins in aged SAMP8 mice treated with alpha-lipoic acid: Implications for aging and age-related neurodegenerative disorders. Neurochem Int 46: 159-168.

Poon HF, Farr SA, Banks WA, Pierce WM, Klein JB, et al. 2005b. Proteomic identification of less oxidized brain proteins in aged senescence-accelerated mice following administration of antisense oligonucleotide directed at the Aβ region of amyloid precursor protein. Mol Brain Res 138: 8-16.

Rabilloud T. 2002. Two-dimensional gel electrophoresis in proteomics: Old, old fashioned, but it still climbs up the mountains. Proteomics 2: 3-10.

Ramakrishnan P, Dickson DW, Davies P. 2003. Pin1 colocalization with phosphorylated tau in Alzheimer's disease and other tauopathies. Neurobiol Dis 14: 251-264.

Rosenberg RN. 2000. The molecular and genetic basis of AD: The end of the beginning: The 2000 Wartenberg lecture. Neurology 54: 2045-2054.

Santoni V, Molloy M, Rabilloud T. 2000. Membrane proteins and proteomics: Un amour impossible? Electrophoresis 21: 1054-1070.

Scheff SW, Price DA. 2003. Synaptic pathology in Alzheimer's disease: A review of ultrastructural studies. Neurobiol Aging 24: 1029-1046.

Schutkowski M, Bernhardt A, Zhou XZ, Shen M, Reimer U, et al. 1998. Role of phosphorylation in determining the backbone dynamics of the serine/threonine-proline motif and Pin1 substrate recognition. Biochemistry 37: 5566-5575.

Selkoe DJ. 2001. Presenilin, Notch, and the genesis and treatment of Alzheimer's disease. Proc Natl Acad Sci USA 98: 11039-11041.

Sherrington R, Rogaev EI, Liang Y, Rogaeva EA, Levesque G, et al. 1995. Cloning of a gene bearing missense mutations in early-onset familial Alzheimer's disease. Nature 375: 754-760.

Shimizu S, Narita M, Tsujimoto Y. 1999. Bcl-2 family proteins regulate the release of apoptogenic cytochrome c by the mitochondrial channel VDAC. Nature 399: 483-487.

Shringarpure R, Grune T, Sitte N, Davies KJ. 2000. 4-Hydroxynonenal-modified amyloid-beta peptide inhibits the proteasome: Possible importance in Alzheimer's disease. Cell Mol Life Sci 57: 1802-1809.

Slooter AJ, Cruts M, Kalmijn S, Hofman A, Breteler MM, et al. 1998. Risk estimates of dementia by apolipoprotein E genotypes from a population-based incidence study: The Rotterdam study. Arch Neurol 55: 964-968.

Sly WS, Hu, PY. 1995. Human carbonic anhydrases and carbonic anhydrase deficiencies. Annu Rev Biochem 64: 375-401.

Smith MA, Richey Harris PL, Sayre LM, Beckman JS, Perry G. 1997. Widespread peroxynitrite-mediated damage in Alzheimer's disease. J Neurosci 17: 2653-2657.

Smith MA, Richey PL, Taneda S, Kutty RK, Sayre LM, et al. 1994. Advanced Maillard reaction end products, free radicals, and protein oxidation in Alzheimer's disease. Ann N Y Acad Sci 738: 447-454.

Smith MA, Sayre LM, Monnier VM, Perry, G. 1996. Oxidative posttranslational modifications in Alzheimer disease. A possible pathogenic role in the formation of senile plaques and neurofibrillary tangles. Mol Chem Neuropathol 28: 41-48.

Smolka MB, Zhou H, Purkayastha S, Aebersold R. 2001. Optimization of the isotope-coded affinity tag-labeling procedure for quantitative proteome analysis. Anal Biochem 297: 25-31.

Soreghan B, Thomas SN, Yang AJ. 2003. Aberrant sphingomyelin/ceramide metabolic-induced neuronal endosomal/lysosomal dysfunction: Potential pathological consequences in age-related neurodegeneration. Adv Drug Deliv Rev 55: 1515-1524.

Stenbeck G. 1998. Soluble NSF-attachment proteins. Int J Biochem Cell Biol 30: 573-577.

Stevens SM Jr, Kem WR, Prokai, L. 2002. Investigation of cytolysin variants by peptide mapping: Enhanced protein characterization using complementary ionization and mass spectrometric techniques. Rapid Commun Mass Spectrom 16: 2094-2101.

Subramaniam R, Roediger F, Jordan B, Mattson MP, Keller JN, et al. 1997. The lipid peroxidation product, 4-hydroxy-2-trans-nonenal, alters the conformation of cortical synaptosomal membrane proteins. J Neurochem 69: 1161-1169.

Sultana R, Butterfield DA. 2004. Oxidatively modified GST and MRP1 in Alzheimer's disease brain: Implications for accumulation of reactive lipid peroxidation products. Neurochem Res 29: 2215-2220.

Sultana R, Boyd-Kimball D, Poon HF, Cai J, Pierce WM, et al. 2006a. Oxidative modification and down-regulation of Pin1 Alzheimer's disease hippocampus: A redox proteomics analysis. Neurobiol Aging 27: 918-925.

Sultana R, Boyd-Kimball D, Poon HF, Cai J, Pierce WM, et al. 2006b. Regional redox proteomics to identify oxidized proteins in Alzheimer's disease brain: A mechanistic approach to understand pathological and biochemical alterations in AD. Neurobiol Aging 27: 1564-1576.

Sultana R, Poon HF, Cai J, Pierce WM, Merchant M, et al. 2006c. Identification of nitrated proteins in Alzheimer's disease brain using a redox proteomics approach. Neurobiol Dis 22: 76-87.

Takahashi Y. 2003. The 14-3-3 proteins: Gene, gene expression, and function. Neurochem Res 28: 1265-1273.

Tilleman K, Stevens I, Spittaels K, Haute CV, Clerens S, et al. 2002. Differential expression of brain proteins in glycogen synthase kinase-3 transgenic mice: A proteomics point of view. Proteomics 2: 94-104.

Vanderklish PW, Bahr BA. 2000. The pathogenic activation of calpain: A marker and mediator of cellular toxicity and disease states. Int J Exp Pathol 81: 323-339.

Wagner Y, Sickmann A, Meyer HE, Daum G. 2003. Multidimensional nano-HPLC for analysis of protein complexes. J Am Soc Mass Spectrom 14: 1003-1011.

Wang LH, Strittmatter SM. 1996. A family of rat CRMP genes is differentially expressed in the nervous system. J Neurosci 16: 6197-6207.

Weitzdoerfer R, Fountoulakis M, Lubec G. 2001. Aberrant expression of dihydropyrimidinase related proteins-2,-3 and -4 in fetal Down syndrome brain. J Neural Transm Suppl 61: 95-107.

Whitehouse PJ, Price DL, Clark AW, Coyle JT, Delong MR. 1981. Alzheimer's disease: Evidence for selective loss of cholinergic neurons in the nucleus basalis. Ann Neurol 10: 122-126.

Wilkinson KD, Tashayev VL, O'Connor LB, Larsen CN, Kasperek E, et al. 1995. Metabolism of the polyubiquitin degradation signal: Structure, mechanism, and role of isopeptidase T. Biochemistry 34: 14535-14546.

Yatin SM, Varadarajan S, Link CD, Butterfield DA. 1999. In vitro and in vivo oxidative stress associated with Alzheimer's amyloid beta-peptide (1–42). Neurobiol Aging 20: 325-330; discussion 339–342.

Yatin SM, Varadarajan S, Butterfield DA. 2000. Vitamin E prevents Alzheimer's amyloid beta-peptide (1–42)-induced neuronal protein oxidation and reactive oxygen species production. J Alzheimers Dis 2: 123-131.

Yoo BC, Fountoulakis M, Cairns N, Lubec G. 2001a. Changes of voltage-dependent anion-selective channel proteins VDAC1 and VDAC2 brain levels in patients with Alzheimer's disease and Down syndrome. Electrophoresis 22: 172-179.

Yoo BC, Kim SH, Cairns N, Fountoulakis M, Lubec G. 2001b. Deranged expression of molecular chaperones in brains of patients with Alzheimer's disease. Biochem Biophys Res Commun 280: 249-258.

Yoshida H, Watanabe A, Ihara Y. 1998. Collapsin response mediator protein-2 is associated with neurofibrillary tangles in Alzheimer's disease. J Biol Chem 273: 9761-9768.

Zhou XZ, Kops O, Werner A, Lu PJ, Shen M, et al. 2000. Pin1-dependent prolyl isomerization regulates dephosphorylation of Cdc25C and tau proteins. Mol Cell 6: 873-883.

14 Vascular Dementia

A. Cagnin · L. Battistin

Abstract: Although vascular dementia (VaD) represents the second most common cause of dementia after Alzheimer's disease (AD) in the elderly, and is referred to as the "silent epidemic of the twenty-first century," there is still a controversy on terminology, classification, and diagnostic criteria of VaD. The diagnosis of VaD resides on diagnostic clinical criteria determining (1) a cognitive impairment, (2) the presence of cerebrovascular disease, and (3) specifically in poststroke or multi-infarct dementia, a temporal relationship between these. The search for a reliable biochemical test helping in the diagnosis of VaD is so far not available. Several vascular risk factors have a role in the development of VaD and their identification and treatment are among the major aspects of VaD management. A new line of research in this field is the study of genetic factors underlying vascular cognitive impairment which are: (1) genes predisposing to cerebrovascular disease, and (2) genes that influence brain tissue responses to cerebrovascular lesions. Evidence in favor of a coexistence of vascular and degenerative components in the pathogenesis of dementia in an elderly population comes from neuropathological and epidemiological studies. There is now a great debate whether VaD and AD are more than common coexisting unrelated pathologies or, instead, represent different results of synergistic pathological mechanisms. Current available medications for the treatment of VaD include acetylcholinesterase inhibitors for mild to moderate cases, and memantine, an NMDA-receptor antagonist. However, therapeutic preventive approaches aiming at reducing incident VaD by targeting patients at risk of cerebrovascular disease (primary prevention), or acting on patients after a stroke (secondary prevention) to prevent stroke recurrence and the progression of brain changes associated with cognitive impairment are the mandatory strategies.

List of Abbreviations: ADDTC, Alzheimer Disease Diagnostic and Treatment Centers; AD, Alzheimer's disease; MID, multi-infarct dementia; sVaD, subcortical vascular dementia; VaD, vascular dementia; VCI, vascular cognitive impairment

1   The Concept of Vascular Cognitive Disorder

In Western countries vascular dementia (VaD) is the second most common cause of dementia after Alzheimer's disease (AD) among the elderly. A meta-analysis of the European studies on incidence of dementia showed that VaD constituted 17.6% of all dementia (Gorelick et al., 1994; Fratiglioni et al., 2000). In Europe and North America, AD is more common than VaD in a 2:1 ratio; in contrast, in Japan and China VaD accounts for almost 50% of all dementia. In a Japanese population the prevalence of VaD may reach 47% as demonstrated by using combined clinical and neuroimaging diagnostic tools (Ikeda et al., 2001). Considering the epidemiology, VaD is referred as the "silent epidemic of the twenty-first century."

Although the increasing interest in the fields of public health as well as in research, there is still a great confusion on terminology, classification, and diagnostic criteria of VaD. From a historical perspective, the term *arteriosclerotic dementia* had been introduced by Binswanger at the end of the nineteenth century to define a condition of cognitive decline caused by a chronic diffuse low cerebral perfusion secondary to atherosclerosis of cerebral vessels (Binswanger, 1894). Until the 1950s and 1960s, dementia in elderly subjects was usually labeled *arteriosclerotic dementia*. In 1974 Hachinski criticized this term as both misleading and inaccurate, and proposed the term *multi-infarct dementia* (MID) to introduce the concept that development of cognitive decline needs an accumulation of cerebral infarct volumes, with preferential cortical involvement, overcoming a critical threshold (Hachinski et al., 1974).

The recognition of cases with cognitive deficits caused by pathology of cerebral small vessels, with vascular lesions in the basal ganglia, midbrain, and diffuse cerebral white matter, led to the identification of a subtype of dementia called subcortical vascular dementia (sVaD). The general term of VaD has then been used as an umbrella to describe MID, sVaD, and dementia due to hypoxia or hemorrhagic encephalopathy. This classification based on the pathophysiology of vascular dementia is summarized in ❷ *Table 14-1*.

More recently, the term "vascular cognitive disorder" has been proposed to comprehend the broad spectrum of syndromes and diseases characterized by dementia and mild cognitive impairment due to cerebrovascular lesions (Sachdev, 1999; Roman et al., 2004). This category encompasses both VaD caused by single or multiple cortical infarcts and VaD caused by subcortical vascular disease and also the new entity of

◘ Table 14-1

Pathophysiological classification of VaD

Large vessel dementia	• Multiple infarct dementia (MID) • Single-strategic infarcts (*mesial-temporal, caudate and thalamus, fronto-cingulate, angular gyrus*)
Small vessel dementia (sVaD)	• Binswanger subcortical encephalopathy • Lacunar state • CADASIL[a] and other hereditary angiopathies • Collagen or inflammatory vascular diseases
Hypoperfusive	• Diffuse hypoxic-ischemic encephalopathy
Hypoxic-ischemic dementia	• Incomplete white matter infarcts
Hemorrhagic dementia	• Subdural hemorrhage • Subarachnoid hemorrhage • Intracerebral hemorrhage

[a]Cerebral autosomal dominant arteriopathy with subcortical infarcts and leukoencephalopathy

vascular cognitive impairment (VCI). The term VCI refers to cases of vascular mild cognitive impairment (MCI) without dementia, by analogy with the concept of amnestic MCI, currently considered the earliest clinically diagnosable stage of AD. In contrast to MCI, VCI is characterized by isolated executive deficits, and not memory loss, without dementia (O'Brien et al., 2003).

2 Diagnostic Criteria

In the 1990s, numerous clinical criteria were formulated for VaD, including DSM–IV criteria for dementia of the vascular type, Alzheimer Disease Diagnostic and Treatment Centers (ADDTC) criteria for ischemic vascular dementia (Chui et al., 1992), the National Institute of Neurological Disorders and Stroke-Association Internationale pour la Recherche et l'Enseignement en Neurosciences (NINDS-AIREN) criteria (Roman et al., 1993), and the ICD-10 code (World Health Organization, 1993) for vascular dementia. All these clinical criteria share three fundamental components: (1) determination of dementia or cognitive impairment, (2) presence of vascular brain injury by clinical history or neuroimaging techniques, and (3) evidence of a relationship between cognitive impairment and vascular injury (e.g., temporal relationship). Unfortunately, these criteria are not interchangeable and may result in up to threefold differences in the number of cases classified as VaD.

The NINDS-AIREN criteria are the most conservative, and are widely used in epidemiologic studies and pharmacological drug trials. NINDS-AIREN criteria define probable VaD as cognitive decline from a previously higher level of performance in memory and two or more cognitive domains (Roman et al., 1993). Evidence of cerebrovascular disease on both clinical examination and neuroimaging is required, as is the evidence of a relationship between the stroke and cognitive decline. The link is supported by at least two of the following: (1) onset of dementia within three months after a recognized stroke, (2) abrupt deterioration in cognition, (3) stepwise deterioration. The neuroimaging criteria support the diagnosis if they fulfil standard requirements involving site and size. For example, leucoencephalopathy must involve at least 25% of the total white matter. According to the NINDS-AIREN criteria, VaD can be grouped into three major categories: (1) MID, (2) dementia due to single strategic ischemic lesion, and (3) sVaD due to small vessel pathology. Other minor subgroups include hemorrhagic and hypoxic dementia. Each group defines a form of dementia with different etiopathology, neuropathology, clinical manifestations, and neuroimaging findings.

When moving away from the well defined concept of poststroke dementia to reach the heterogeneous field of subcortical vascular disease and VCI, we are challenged in our diagnostic work-up by the actual limitations of the available clinical diagnostic criteria and the coexistence, in the elderly, of vascular and degenerative brain pathologies. Two major limitations arise from the application of the available criteria: (1) the focus on memory loss on the neuropsychological assessment, and (2) the radiological criteria based on vascular lesion load necessary to determine VaD.

Memory loss, necessary to fulfil the NINDS-AIREN criteria, may not be present at the beginning of the disease, the executive deficits being predominant (Royall and Roman, 2000; Price et al., 2005). Moreover, in VCI there is no single, classical neurobehavioral and cognitive phenotype since vascular brain injury may affect any region of the brain.

From revision of diagnostic criteria it has been clear that it is difficult to determine the degree of vascular lesions, shown by imaging tools, necessary to contribute to the development of dementia. While it is possible to attribute MRI hyperintensities larger than 1–2 mm in diameter to vascular brain injury, it is often unclear whether these lesions are responsible for the cognitive impairment. The neuroimaging requirements of the NINDS-AIREN criteria are not able to distinguish stroke patients with and without dementia (Ballard et al., 2004). Identification of "critical volumes" for infarcts and white matter hyperintensities and determination of their brain distribution are not correlated with cognitive impairment and may result in a useless effort.

3   Mixed Dementia

The possibility of concomitant AD often confounds the relationship between cerebrovascular disease and VaD. Evidence in favor of a coexistence of vascular and degenerative components in the pathogenesis of dementia in an elderly population with slowly progressive cognitive decline comes from neuropathological and epidemiological studies. Since Alzheimer's first seminal work, vascular pathology, namely arteriosclerosis, endothelial proliferation, and neovascularization, has been often found to be associated with typical neurodegenerative changes, senile plaques, and neurofibrillary tangles. A series of autopsy studies confirmed that among cases of dementia, Alzheimer-related pathology was associated with vascular lesions in nearly one third of cases (Jellinger, 2007). In addition, many epidemiological reports confirmed that the presence of vascular factors increase the risk of developing AD (❷ Table 14-2).

❑ Table 14-2

Common risk factors for Vascular Dementia and Alzheimer's disease

- Age
- Hyperlipidemia
- Diabetes mellitus
- Arterial hypertension
- Hypotension
- Recurrent stroke
- Hyperhomocysteinemia
- Hyperfibrinogenemia
- Cardiac diseases[a]

[a]Atrial fibrillation, congestive heart failure, and other cardiac arrhythmias

It is still controversial whether AD-like pathology and cerebrovascular lesions are coexisting but unrelated pathologies or whether they represent different results of synergistic patho-mechanisms (❷ Figure 14-1). In the near future it will be necessary to clarify the synergism between vascular lesions and degenerative brain changes on one hand, and develop tools to implement in vivo the relative contribution of vascular and degenerative changes to the development of dementia.

☐ **Figure 14-1**

Potential interplay between vascular factors (oligaemia and hypoxia) and amyloid β (Aβ) deposition. Aβ induces vascular dysregulation and promotes atherosclerosis through the release of inflammatory mediators thus contributing to oligaemia. The ischemic process, in turn, induces Aβ formation by enhancing Aβ cleavage and reducing Aβ clearance

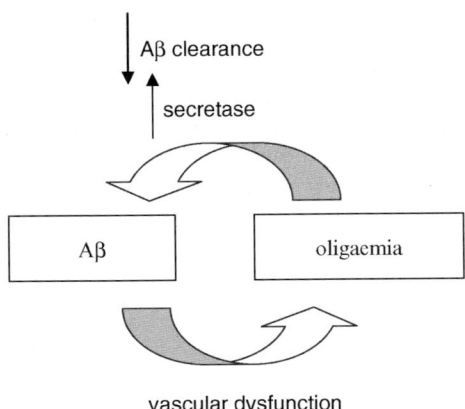

4   Diagnosis: Clinical Evaluation

The presence of cognitive impairment can be established by history and examination, paying careful attention to the detection of neuropsychological deficits other than memory loss. From a general perspective, in VaD there is no a single, classical cognitive and neurobehavioral phenotype since vascular brain injury may affect any region of the brain. In practice, it is widely accepted that at least in sVaD, due to cerebral small vessels disease and hypoperfusion, patients present with a peculiar cognitive syndrome distinct from that seen in AD, and characterized by predominant executive deficits due to the disconnection of frontal lobes from subcortical grey structures (Looi and Sachdev, 1999). Deficits in executive functions involved impairments in volition, planning, purposive action, and effective performance. In the abilities of daily living these deficits translate into mental speed impairment with poor information retrieval and difficulties in tasks requiring shifting of attention and cognitive flexibility. In these patients the performance in tests exploring the long term memory is higher than in AD. It is therefore recommended in the neuropsychological assessment of screening to include tests tailored on the examination of executive and frontal lobe functions such as the CLOX test (Royall et al., 1998) and the Trail Making Test part B (Luria, 1966). However, a recent study in autopsy proven AD and VaD patients aiming at assessing whether cognitive impairment with a predominant involvement of executive functions may represent an in vivo marker of VaD showed that cerebrovascular disease may affect cognition in a variable and less specific pattern than Alzheimer-related pathology (Reed et al., 2007). In fact, the profile of cognitive impairment with predominant executive dysfunction was detected only in a subgroup of patients with cerebrovascular disease.

VaD patients often experience mood changes, such as depression and emotional instability, and other behavioral aspects influenced by subcortical-frontal pathways more frequently than AD patients. The neuropsychiatry burden on VaD patients and caregivers, in combination with a preserved insight of their clinical condition, are a major cause of patients' distress even in the early stage of disease. Characterization and quantification of the behavioral disorders are important steps in the diagnostic process of VaD patients.

5   Diagnostic Tools: Present and Future

The main core of the diagnostic assessment is focused on determination of vascular risk factors and cerebrovascular lesions.

5.1   Genetic Markers

The study of genetic factors underlying vascular cognitive impairment is a new line of research (Leblanc et al., 2006). Leblanc and colleagues suggest that the genes underlying VaD belong to two classes:
(1)   genes predisposing to cerebrovascular disease;
(2)   genes that influence brain tissue responses to cerebrovascular lesions (e.g., the ability to recover from a vascular brain injury).

5.1.1 Class 1

Regarding the first class of genes, only few monogenic forms of cerebrovascular disease have been identified: (1) cerebral autosomal dominant arteriopathy with subcortical infarcts and leucoencephalopathy (CADASIL), (2) hereditary cerebral haemorrhage with amyloidosis (HCHWA). CADASIL, the most common hereditary form of recurrent stroke, is due to a mutation in the *Notch 3* gene, located on chromosome 19. *Notch 3* gene is one of a few genes encoding transmembrane cell surface receptors, named Notch proteins. These receptors control the signals related to cell proliferation, maturation, differentiation, and programmed cell death through cell–cell interaction. Vascular abnormalities are consistent with a narrowing of the microvessels lumen due to an increased thickness of the intima and expansion of extracellular matrix. Arterial vessels show extracellular deposits of granular osmiophilic material (GOM) of 1–2 μm in size. Both intracranial and extracranial arterial vessels may present extracellular GOM, which still remain to be characterized. They may be lipophilic but do not appear to contain *Notch 3* fragments. In addition to GOM, using an antibody against the amino-terminal region of the *Notch 3* receptor, increased immunoreactivity within deposits in the vasculature was apparent in CADASIL patients compared to normal controls.

Familial cerebral amyloid angiopathies are due to mutations in a variety of genes related to the amyloid precursor proteins (APP) and their products. These genes include those encoding for cystatin C, amyloid beta precursor protein, gelsolin, transthyretin, and prion protein. HCHWA is caused by a mutation in the gene for APP. Vascular pathology includes deposition of fibrillar amyloid peptide in the wall vessels of arterioles. Cerebral vascular autoregulation and blood–brain barrier function are deeply impaired. Clinically, intracerebral microbleeds and lobar hemorrhages coexist with ischemic infarcts to the development of dementia.

Other monogenic diseases, such as Fabry's disease, lead to stroke as recognized, although not principal, manifestation. Several other genes responsible for inducing vascular risk factors (i.e., hypertension, stroke, hyperlipidemia, etc.) are under investigation.

5.1.2 Class 2

The second class of genes influencing the brain response to vascular injury is less known and it represents a very recent field of research. It has been suggested that this class of genes may belong to different subgroups. One group may include genes shared by VaD and AD: genes involved in the pathways of amyloid production or elimination (presenilins, APP, and apolipoprotein E-APOE) may confer susceptibility to develop dementia after vascular injury thus explaining, at least in part, the well known interaction and synergism between VaD and AD. Other groups of genes involved in tissue response to brain ischemia are related to proteins such as grow factors, glutamate, and gamma-aminobutyric acid receptors etc, and to a combination of determinants of premorbid level of cognitive performance.

Our group studied a panel of genes supposed to confer susceptibility to develop vascular dementia. The polymorphisms (SNPs) of genes encoding for the following proteins were analyzed: paraoxonase 1 (PON1), an esterase with antioxidant activity on low-density lipoproteins, glutathione S-transferase omega-1 (GSTO1), involved in antioxidant pathways, methylenetetrahydrofolate reductase (MTHFR), related to plasma levels of homocysteine, and APOE. We failed to detect differences in genotype distributions among patients with subcortical VaD and control subjects for the following SNPs: -107 T-C, -161A-G, and Gln192Arg in the PON1 gene; Ala140Asp in the GSTO gene and C677T in the MTHFR gene (data not shown). Interestingly, we found an increased frequency of the APOE4 allele in the VaD group compared with age-matched healthy elderly subjects (VaD with APOE 4+: 18%; controls with APOE 4+: 7%, $p < 0.05$). The association between APOE4 and AD is robust, while its association with VaD has been controversial. Some investigators reported an increased frequency of the APOE 4 allele in VaD or AD with cerebrovascular disease (Slooter et al., 1997), while others did not (Traykov et al., 1999).

5.2   Serum Biological Markers

The diagnosis of VaD relies on diagnostic clinical criteria determining a cognitive impairment, the presence of cerebrovascular disease and, only in the case of poststroke dementia or multi-infarct dementia, a temporal relationship between these. The search for a reliable biochemical marker helping in the diagnosis of VaD is so far not available. In fact, whereas the levels of tau and beta-42 peptide in the cerebral spinal fluid are sensitive markers of AD, they yield the lowest specificity in VaD. Recently our group studied the level of antibodies (Abs) against heparan sulfate (HS), a constituent of the perivascular basement membrane and endothelial cells, as potential biomarkers in VaD and in cerebrovascular disease (Briani et al., 2005). We assessed serum titres of anti-HS Abs in patients with VaD and in patients with combined AD and cerebrovascular disease (mixed dementia). Moreover, plasma levels of homocysteine, an independent risk factor for the development of dementia as well as for cerebrovascular disease, were also determined. High anti-HS Abs titers were present in only one patient with VaD and in two patients with mixed dementia, as well as in two control patients with other neurological diseases (stroke and epilepsy). We did not find differences in the homocysteine level between VaD, AD, and control group. The findings suggested that neither elevated anti-HS Abs titres nor increased homocysteinemia may be useful biological markers for identifying patients with VaD or cerebrovascular disease.

An active research field is focused on the search for biological markers of VaD in CSF. Promising results have been obtained measuring the CSF level of matrix-degrading metalloproteinases (MMPs), including gelatinase A (MMP-2), gelatinase B (MMP-9). MMPs are secreted by astrocytes, endothelial cells, microglia, and neurons. Gelatinases attack the basal lamina macromolecules, causing the proteolytic disruption of the blood–brain barrier and have been implicated in the breakdown of myelin. Cerebral ischemia causes an elevation of gelatinase A and B at different time points after the stroke.

Gelatinase molecules in CSF are secreted by endogenous brain cells and infiltrating inflammatory cells or may derive from the blood compartment after extravasation. The hypothesis is that small vessel disease and brain ischemia are associated with adjacent demyelination and inflammatory processes with increased MMPs. There has been recent evidence that the CSF level of MMP-9 is significantly increased in VaD patients compared to those with AD or to healthy elderly subjects (Adair et al., 2004). The increase of MMP-9 is not specific for VaD and it has been reported in a variety of neuroinflammatory diseases. However, it may contribute to distinguish VaD from AD.

5.3   Neuroimaging: Surrogate Markers of Brain/Vascular Pathology

Computed tomography (CT) and magnetic resonance imaging (MRI) brain scans acquired an important role in the diagnostic work-up of dementia, increasing in vivo diagnostic accuracy of both VaD and AD. Neuroimaging investigations are crucial because they define the extent and the type of vascular lesions, and identify the presence of nonvascular pathology such as global or regional atrophy. Brain MRI is superior to CT for detecting white matter lesions and lacunes. However, MRI is not widely available in clinical setting and future research needs to validate scoring scales to quantify vascular pathology with both imaging tools and to address whether CT can substitute MRI.

The use of semiquantitative scales to measure brain vascular lesion load have been developed to meet the NINDS-AIREN diagnostic radiological requirements. It has been demonstrated that the proposed NINDS-AIREN radiological criteria are not accurate (Ballard et al., 2004). In fact, it is well established that patients with similar lesion load of vascular pathology may range from normal cognitive performance to dementia. The MRI screening of elderly subjects showed that 15% to 25% of subjects have MRI signs of small vessel disease (Vermeer et al., 2002) and only a part of these develop cognitive decline. It appears that there are variables other than number, location, and type of vascular lesions determining cognitive impairment. For example, it has been demonstrated that hippocampal and cortical atrophy can predict dementia better than quantification of lesion load in subcortical VaD (Fein et al., 2000).

Functional brain imaging assessing cerebral blood flow (CBF), using Single Photon Emission Computed Tomography (SPECT), and glucose metabolism, using Positron Emission Tomography (PET), may provide additional information particularly for small vessel disease in which the relationship between cognition and lesion burden appears to be complex. Subcortical small infarcts can deteriorate the cognitive performance by deafferenting cortical regions. Therefore, a functional assessment of cortical and subcortical CBF and glucose metabolism can predict cognitive decline in small vessel disease better than structural neuroimaging. In a combined study using MRI, FDG-PET, and 99mTc- HMPAO (hexamethylpropyleneamine oxime) SPECT scans in patients with cerebral microangiopathy with and without neuropsychological deficits, cognitive impairment correlated with the degree of hypometabolism/hypoperfusion and with global brain atrophy, but not with vascular lesions in white matter and subcortical grey regions (Sabri et al., 1999).

6   Therapy

6.1   Prevention Strategies

These therapeutic approaches aim at reducing incident VaD by targeting patients at risk of cerebrovascular disease but still free from cognitive impairment (primary prevention), or acting on patients after a stroke (secondary prevention) to prevent stroke recurrence and the progression of brain changes associated with cognitive impairment. In both strategies the main objective is the prevention or management of vascular risk factors including arterial hypertension, hyperlipidemia, coronary heart disease and arrhythmia, carotid atherosclerosis, and hyperhomocysteinemia (O'Brien et al., 2003).

Treatment of systolic hypertension has given the most robust results. In the Systolic Hypertension in Europe (Syst-Eur) Study it has been demonstrated that the use of antihypertensive agents blocking the calcium channel could decrease the incidence of dementia in cognitive intact elderly subjects (Forette et al., 2002). In a recent meta-analysis of three trials there was no convincing evidence that blood pressure lowering prevents the development of cognitive impairment in hypertensive patients without prior cerebrovascular disease (McGuinness et al., 2006). Nonetheless, the authors pointed out that the results may be confounded by the high degree of drop-out in the active group and the number of placebo patients given active treatment during the course of the trial. In secondary prevention, the use of perindopril, an angiotensin-converting enzyme inhibitor, has been shown to reduce the risk of dementia in patients with recurrent stroke with a mechanism not fully dependent on the hypertension lowering effect (Tzourio et al., 2003). In contrast, in the acute phase of stroke, the importance of avoiding or treating hypoxic ischemic disorders for prevention of incident vascular cognitive impairment is well established (Moroney et al., 1996).

Statins are lipid-lowering agents acting as inhibitors of the 3-hydroxy-3-methyl-glutaryl-CoA reductase, representing the rate-limiting enzyme in cholesterol biosynthesis. Statins are effective in the prevention of vascular events as secondary prevention following myocardial infarction and ischemic stroke (Collins et al., 2002) and in primary prevention of myocardial infarction and stroke in high risk patients with elevated blood pressure (Sever et al., 2003). Statins seem of potential benefit in the prevention and treatment of dementia through mechanisms not completely dependent on the lipid-lowering effect. While there is evidence that the use of statins in patients without dementia is not effective in reducing incident dementia (Zandi et al., 2005), in patients with brain at risk or in secondary prevention of vascular cognitive impairment seems promising but requires further investigations (Suribhatla et al., 2005).

Elevated plasma homocysteine level is associated with an increased risk to develop cerebrovascular disease, VaD, and also AD (McCaddon et al., 2001; Homocysteine Studies Collaboration. 2002). To date, intervention trials with folate and vitamin B12 (alone or in combination) gave inconclusive results due to methodological issues (Malouf et al., 2003). A few studies assessing the effects of these vitamins in slowing the progression of AD, MCI and evaluating the effect on cognitive deficit in patients with hyperhomocysteinemia are in progress.

6.2   The Cholinergic Neurotransmission

Ascending cholinergic projections from the basal forebrain provide one of the most massive and widespread neural inputs of the cerebral cortex. Cholinergic axons innervate all cortical areas and therefore influence all aspects of cognition and behavior. However, experimental and clinical studies demonstrated that the cholinergic pathways may influence mostly processes related to attention, memory, and synaptic plasticity. The administration of anticholinergic agents and the induction of selective lesions of the cholinergic neurons are able to induce amnesia and impairment on attention.

A cholinomimetic strategy by increasing cholinergic activity should improve cognitive abilities as long as a cholinergic deficit is present. The presence of cholinergic deficit in VaD, similar to that observed in AD, is well documented as demonstrated by reductions in several cholinergic markers in neocortex, hippocampus, and striatum and in acetylcholine levels in CSF. A decrease of the cholinergic neurotransmission is present in VaD also without concomitant AD pathology. In fact, white matter lesions are also accompanied by damage to cholinergic fibres, likewise in CADASIL, in which AD pathology does not coexist. The cholinergic deficits in VaD may result from ischemia of the basal forebrain cholinergic nuclei supplied by penetrating arteries particularly vulnerable to the effects of arterial hypertension and diabetes. In addition, localized strokes in the white matter may interfere with the cholinergic bundles projecting to the neocortex from the nucleus basalis and traveling within the white matter of the cerebral hemispheres (Selden et al., 1998). Furthermore, hippocampal CA1 neurones are selectively vulnerable to ischemia leading to hippocampal atrophy, a neuropathological finding typical in brains of VAD patients without concomitant Alzheimer-like pathology (Vinters et al., 2000).

6.3   Cholinesterase Inhibitors

The most efficacious intervention to increase the cholinergic tone has been the use of cholinesterase inhibitors (ChEI), rather than nicotinic or muscarinic agonists. Experimental data suggest that recommended doses of ChEI allow for a twofold increase of acetylcholine levels in the cortex of mild to moderate AD patients thus nearly restoring normal levels of the neurotransmitter. Clinically significant results related to the pharmacologic effect of these drugs have been discovered on cognition, particularly on attention and concentration in AD. Three of the central ChEI agents approved for treatment of mild to moderate AD, donepezil, rivastigmine, and galantamine, have been used also in VaD in order to achieve some benefit in improving or stabilizing cognitive deficits.

One of the first systematic attempts to target VaD with ChEI using rivastigmine produced encouraging results. More recently, positive findings have been reported in the treatment of vascular and mixed dementia with galantamine (Erkinjuntti et al., 2002), donepezil (Black et al., 2003; Wilkinson et al., 2003), and rivastigmine up to 12 months (Moretti et al., 2004). The largest clinical trial evaluating pure VaD to date, with a total of 1,219 patients treated for 24 weeks, was performed using donepezil and demonstrated a statistically significant improvement in cognitive function while effects on global and functional outcomes were mixed, with no evidence of clear dose response (Malouf and Birks, 2005). It is of interest to note that cognitive decline in untreated VaD patients was less severe than what is known from previous trials in placebo-treated AD patients in a time frame of 24 weeks (Black et al., 2004). A 6-month trial in a more heterogeneous population represented by pure VaD and mixed dementia patients using galantamine gave similar results (Erkinjuntti et al., 2002) and in an open-label study the cognitive effects were maintained for at least 12 months with a score on the scale for cognitive performance still close to baseline at the end of the trial (Erkinjuntti et al., 2003). Of note, in the pure VaD subgroup the authors did not find the same encouraging results than in the subgroup with AD and concurrent cerebrovascular disease. In a recent, although not complete, systematic review on galantamine the available data suggest some benefit over placebo in global cognition and executive functions in VaD only in one of the two trials considered. Safety profile was characterized by a high rate of gastrointestinal side effects (Craig and Birks, 2006).

In small open-label studies of patients with subcortical VaD, rivastigmine improved cognition as well as behavior (Moretti et al., 2004). These data wait for further confirmation from larger ongoing double-blind controlled trials (Craig and Birks, 2005).

The results of these controlled trials are in favor of a modest efficacy of ChEI in VaD (pure or mixed) in improving cognitive and global functioning (Erkinjuntti et al., 2004). Despite these positive findings, at present, treatment with ChEI has not been so far approved by regulatory agencies.

6.4 Memantine

Memantine is a voltage-dependent, moderate affinity, noncompetitive N-methyl-D-aspartate (NMDA) receptor antagonist. It blocks the effects of pathologically elevated tonic levels of glutamate that may lead to neuronal dysfunction. In the Alzheimer brain, memantine reduces the background Ca++ influx through the NMDA receptors. When the levels of glutamate in the synaptic cleft are transiently increased as required for learning and memory processes, the resultant depolarization of the postsynaptic membrane is sufficient to displace memantine and allow Ca++ influx as a result of stronger stimulation. Memantine selectively reduces the elevated background Ca++ influx through the NMDA receptors in dementia, but due to its voltage-dependency it does not affect the Ca++ flow under physiological stimuli. (Danysz et al., 2000). By doing so, memantine is able to optimize the noise to signal ratio in neurophysiological processes of learning.

Memantine significantly slowed the rate of functional and cognitive deterioration in patients with moderate to severe AD in a 28-week US randomized, double blind, placebo controlled study (Resberg et al., 2003). On the basis of the hypothesis of glutamate-induced neurotoxicity in cerebral ischemia, the efficacy of memantine (20 mg/die) has been evaluated in patients with mild to moderate VaD in two randomized, placebo-controlled 6-month trials (Orgogozo et al., 2002; Wilcock et al., 2002). In both studies the group on memantine performed significantly better than the placebo group on scales assessing global cognition. Interestingly, the patients who obtained the largest benefit were those more cognitively compromised at baseline and those affected by cerebral small vessel pathology without cortical infarctions. In conclusion, the modest beneficial effect of memantine on cognition in patients with mild to moderate VaD is not clinically discernable at 6 months (McShane et al., 2006).

6.5 Other Treatment Options

Nimodipine is a dihydropyridine calcium antagonist with multiple mechanisms of action. Nimodipine has an effect on age-related microangiopathy as vasoactive agent not interfering with the autoregulation of cerebral blood flow. In addition, it may contribute to neuroprotection by blocking L-type calcium receptors. Up to 2001, there were insufficient data in favor of nimodipine as symptomatic treatment for VaD, although small open labelled trials pointed towards a putative modest benefit in selected patients with subcortical VaD (Lopez-Arieta, 2001). The first exploratory randomized, double-blind, controlled trial of nimodipine focusing on subcortical VaD has recently been published. The results of this study favor the nimodipine-treated group that showed a better performance on verbal fluency and global cognitive function assessed with MMSE compared with the placebo group. Furthermore, the high drop-out rates in the placebo group suggest that nimodipine might protect against cardiovascular comorbidities (Pantoni et al., 2005).

Few studies explored the efficacy of pharmacotherapy on isolated cognitive deficit. A recent study on neglect showed that an improvement on leftward space exploration in right-hemisphere damage patients could be obtained by using guanfacine, a noradrenergic agonist that modulates dorsolateral prefrontal cortex (Malhotra et al., 2006).

7   Conclusions

The diagnosis of VaD remains problematic and there are a few limitations of current clinical diagnostic criteria. There is lack of consensus regarding both the clinical and pathological definitions. The possibility of concomitant AD usually arises but cannot be confidently assessed until autopsy. The future development of a method or a combination of biomarkers to define the independent severity of cerebrovascular disease and AD pathology contributing to brain injury in dementia remains an important priority.

ChEI and NMDA receptor antagonist represent promising treatments for patients with VaD or VCI. Although a significant effect has been reached on cognition or global function by using ChEI in VaD patients, particularly in the subcortical vascular dementia subgroup, to date it is still unclear whether the clinical effect size is sufficient to justify clinical routine use of these treatments in vascular dementia.

Recognition of the condition of "at risk brains" or vascular mild cognitive impairment not fully converted into dementia is crucial to implement preventive strategies. Correction of vascular risk factors may lead to reduced lesion load of vascular pathology thus decreasing the incidence of VaD.

Acknowledgments

This work was supported in part by a grant from the Italian Ministry of Health within the Finalized Research Projects 2006–2007 (type II, no. 107).

References

Adair J, Charlie J, Dencoff JE, Kaye JA, Quinn JF, et al. 2004. Measurement of gelatinase B (MMP-9) in cerebrospinal fluid of patients with vascular dementia and Alzheimer disease. Brain 35: 159-162.

Ballard CG, Burton EJ, Barber R, Stephens S, Kenny RA, et al. 2004. Ninds airen neuroimaging criteria do not distinguish stroke patients with and without dementia. Neurology 63: 983-988.

Binswanger O. 1894. Die Abgrenzung der allgemeinen progressiven paralyse. Berl Klein Wochenschr 31: 1102-1205, 1137-1139, 1180-1186.

Black S, Roman GC, Geldmacher DS, Salloway S, Hecker J, et al. 2004. Efficacy and tolerability of donepezil in vascular dementia: Positive results of a 24 week, multicenter, international, randomized, placebo-controlled clinical trial. Stroke 34: 2323-2332.

Briani C, Cagnin A, Gallo L, Toffanin E, Varagnolo M, et al. 2005. Anti-heparan sulphate antibodies and homocysteine: Markers of vascular pathology? J Neurol Sci 229-230: 215-218.

Chui HC, Victoroff JI, Margolin D, Jagust W, Shankle R, et al. 1992. Criteria for the diagnosis of ischemic vascular dementia proposed by the state of California Alzheimer disease diagnostic and treatment centers (ADDTC). Neurology 42: 473-480.

Craig D, Birks J. 2005. Rivastigmine for vascular cognitive impairment. Cochrane Database Syst Rev 2: CD004744.

Craig D, Birks J. 2006. Galantamine for vascular cognitive impairment. Cochrane Database Syst Rev 1: CD004746.

Danysz W, Parsons CG, Mobius HJ, Stoffler A, Quack G. 2000. Neuroprotective and symptomatological action of memantine relevant for Alzheimer's disease: A unified glutamatergic hypothesis on the mechanism of action. Neurotoxicity Res 2: 85-97.

Erkinjuntti T, Roman G, Gauthier S, Feldman H, Rockwood K. 2004. Emerging therapies for vascular dementia and vascular cognitive impairment. Stroke 35: 1010-1017.

Erkinjuntti T, Kurz A, Gauthier S, Bullock R, Lilienfeld S, et al. 2002. Efficacy of galantamine in probable vascular dementia and Alzheimer's disease combined with cerebrovascular disease: A randomized trial. Lancet 359: 1283-1290.

Erkinjuntti T, Kurz A, Small GW, Bullock R, Lilienfeld S, et al. 2003. An open-label extension trial of galantamine in patients with probable vascular dementia and mixed dementia. Clin Ther 25: 1765-1782.

Fein G, Di Sclafani V, Tanabe J, Cardenas V, Weiner MW, et al. 2000. Hippocampal and cortical atrophy predict dementia in subcortical ischemic vascular disease. Neurology 55: 1626-1635.

Forette F, Seux ML, Staessen JA, Thijs L, Babarskiene MR, et al. 2002. The prevention of dementia with antihypertensive treatment: New evidence from the systolic hypertension in Europe (syst-eur) study. Arch Intern Med 162: 2046-2052.

Fratiglioni L, Launer LJ, Andersen K, Breteler MM, Copeland JR, et al. 2000. Incidence of dementia and major subtypes in Europe: A collaborative study of population-based cohorts. Neurologic diseases in the elderly research group. Neurology 54: S4-S9.

Gorelick PB, Roman G, Mangone CA. 1994. Vascular dementia. Handbook of Neuroepidemiology, Gorelick PB, Alter MA, editors. New York: Mecel Dekker; pp. 197-214.

Hachinski VC, Lassen MA, Marshall J. 1974. Multi-infarct dementia. A cause of mental deterioration in elderly. Lancet 2: 207-210.

Homocysteine Studies Collaboration. 2002. Homocysteine and risk of ischemic heart disease and stroke: A meta-analysis. JAMA 288: 2015-2022.

Ikeda M, Hokoishi K, Maki N, Nebu A, Tachibana N, et al. 2001. Increased prevalence of vascular dementia in Japan. Neurology 57: 839-844.

Jellinger KA. 2007. The enigma of vascular cognitive disorder and vascular dementia. Acta Neuropathol 113: 349-388.

Leblanc GG, Meschia JF, Stuss DT, Hachinski V. 2006. Genetics of vascular cognitive impairment. The opportunity and the challenges. Stroke 37: 248-255.

Looi JCL, Sachdev PS. 1999. Differentiation of vascular dementia from AD on neuropsychological tests. Neurology 53: 670-678.

Lopez-Arieta BJ. 2001. Nimodipine for primary degenerative, mixed and vascular dementia. Cochrane Database Syst Rev 1: CD000147.

Luria AR. 1966. Higher cortical function in man. New York: Basic Books.

Malhotra PA, Parton AD, Greenwood R, Husain M. 2006. Noradrenergic modulation of space exploration in visual neglect. Ann Neurol 59: 186-190.

Malouf R, Birks J. 2005. Donepezil for vascular cognitive impairment. Cochrane Database Syst Rev 1: CD004395.

Malouf M, Grimley EJ, Arcosa SA. 2003. Folic acid with and without vitamin B12 for cognition and dementia. Cochrane Database Syst Rev 4: CD004514.

McCaddon A, Hudson P, Davies G, Hughes A, Williams JH, et al. 2001. Homocysteine and cognitive decline in healthy elderly. Dement Geriatr Cogn Disord 12: 309-313.

McGuinness B, Todd S, Passmore P, Bullock R. 2006. The effects of blood pressure lowering on development of cognitive impairment and dementia in patients without apparent prior cerebrovascular disease. Cochrane Database Syst Rev 19: CD004034.

McShane R, Areosa Sastre A, Minakaran N. 2006. Memantine for dementia. Cochrane Database Syst Rev 3: CD003154.

Moretti R, Torre P, Antonello RM, Cazzato G. 2004. Rivastigmne in subcortical vascular dementia: A comparison trial on efficacy and tolerability for 12 months follow-up. Eur J Neurol 8: 361-362.

Moroney JT, Bagiella E, Desmond DW, Paik MC, Stern Y, et al. 1996. Risk factors for incident dementia after stroke: Role of hypoxic ischemic disorders. Stroke 27: 1283-1289.

O'Brien JT, Erkinjuntti T, Reisberg B, Roman G, Sawada T, et al. 2003. Vascular cognitive impairment. Lancet Neurol 2: 89-98.

Orgogozo JM, Rigaud AS, Stoffler A, Mobius HJ, Forette F. 2002. Efficacy and safety of memantine in patients with mild to moderate vascular dementia: A randomized, placebo-controlled trial (MMM300). Stroke 33: 1834-1839.

Pantoni L, del Ser T, Soglian AG, Amigoni S, Spadari G, et al. 2005. Efficacy and safety of nimodipine in subcortical vascular dementia: A randomized placebo-controlled trial. Stroke 36: 619-624.

Price CC, Jefferson AL, Merino JG, Heilman KM, Libon DJ. 2005. Subcortical vascular dementia: Integrating neuropsychological and neuroradiologic data. Neurology 65: 376-382.

Reed BR, Mungas DM, Kramer JH, Ellis W, Vinters HV, et al. 2007. Profiles of neuropsychological impairment in autopsy-defined Alzheimer's disease and cerebrovascular disease. Brain 130: 731-739.

Resberg B, Doody R, Stoffler A, Schmitt F, Ferris S, et al. 2003. Memantine in moderate to severe Alzheimer's disease. N Engl J Med 348: 1333-1341.

Roman GC, Sachdev P, Royall DR, Bullock RA, Orgogozo JM, et al. 2004. Vascular cognitive disorder: A new diagnostic category updating vascular cognitive impairment and vascular dementia. Neurol Sci 226: 81-87.

Roman GC, Tatemichi TK, Erkinjutti T, Cummings JL, Masdeu JC, et al. 1993. Vascular dementia: Diagnostic criteria for research studies. Report of the ninds-airen international workshop. Neurology 43: 250-260.

Royall DR, Roman GC. 2000. Differentiation of vascular dementia from AD on neuropsychological tests. Neurology 55: 604-606.

Royall DR, Coder JA, Polk M. 1998. Clox: An executive clock drawing task. J Neurol Neurosurg Psychiatry 64: 588-594.

Sabri O, Ringerstein EB, Hellwig D, Schneider R, Schreckenberger M, et al. 1999. Neuropsychological impairment correlates with hypoperfusion and hypometabolism but not with severity of white matter lesions on MRI in patients with cerebral microangiopathy. Stroke 30: 556-566.

Sachdev P. 1999. Vascular cognitive disorder. Int J Geriatr Psychiatry 14: 402-403.

Selden NR, Gitelman DR, Salamon-Murayama N, Parrsh TB, Mesulam MM. 1998. Trajectories of cholinergic pathways within the cerebral hemispheres of the human brain. Brain 121: 2249-2257.

Sever PS, Dahlof D, Poulter N, Wedel H, Beevers G, et al. 2003. Prevention of coronary and stroke events with atorvastatin in hypertensive patients who have average or lower-than-average cholesterol concentrations, in the Anglo-Scandinavian cardiac outcomes trial lipid lowering arm (ASCOT-LLA): A multicentre controlled trial. Lancet 361: 1149-1158.

Slooter AJC, Tang MX, van Duijn CM, Stern Y, Ott A, et al. 1997. Apolipoprotein E 4 and the risk of dementia with stroke: A population-based investigation. JAMA 277: 818-821.

Suribhatla S, Dennis MS, Potter JF. 2005. A study of statin use in the prevention of cognitive impairment of vascular origin in the UK. J Neurol Sci 229-230: 147-150.

Traykov L, Rigaud AS, Caputo L, Couderc R, Coste J, et al. 1999. Apolipoprotein E phenotypes in demented and cognitively impaired patients with and without cerebrovascular disease. Eur J Neurology 6: 415-421.

Tzourio C, Anderson C, Chapman N, Woodward M, Neal B, et al. 2003. Effects of blood pressure lowering with perindopril and indapamide therapy on dementia and cognitive decline in patients with cerebrovascular disease. Arch Intern Med 163: 1069-1075.

Vermeer SE, Koudstaal PJ, Oudkerk M, Hofman A, Breteler MM. 2002. Prevalence and risk factors of silent brain infarcts in the population based Rotterdam scan study. Stroke 33: 21-25.

Vinters HV, Ellis WG, Zarow C, Zaias BW, Jagst WJ, et al. 2000. Neuropathologic substrates of ischemic vascular dementia. J Neuropathol Exp Neurol 60: 658-659.

Wilcock G, Mobius HJ, Stoffler A, MMM 500 group. 2002. A double-blind, placebo-controlled multicentre study of memantine in mild to moderate vascular dementia (MMM500). Int Clin Psychopharmacol 17: 297-305.

Wilkinson D, Dody R, Helme R, Taubman K, Minzer J, et al. 2003. Donepezil 308 study group. Donepezil in vascular dementia: A randomized, placebo-controlled study. Neurology 61: 479-486.

World Health Organization 1993. International classification of diseases, Tenth Revision. Geneva: World Health Organization.

Zandi PP, Sparks DL, Khachaturian AS, Tschanz J, Norton M, et al. 2005. Cache county study investigators. Do statins reduce risk of incident dementia and Alzheimer's disease? The Cache County Study. Arch Gen Psych 62: 217-224.

15   Lewy Body Disorders

K. A. Jellinger

Abstract: Lewy body disorders are a heterogenous group of neurodegenerative diseases, the morphological hallmark of which are Lewy bodies, neuronal intracytoplasmic aggregations containing misfolded fibrillar α-synuclein (AS) and a variety of other chemical substances. Lewy bodies, regardless of their location (cortical or subcortical) and the phenotype of proteinopathies (amyloidopathy, synucleinopathy, tauopathy) in which they occur, have a rather uniform biochemical and immunohistochemical profile, but may show some molecular biological and developmental diversities. Current knowledge about the molecular basis of their principal constituent, AS, a small presynaptic, natively unfolded protein that undergoes multiple conformational and translational modifications during Lewy body formation as well as the major biochemical components of these inclusions are reviewed. However, the ultimate causes and molecular pathways of their formation and the relevant role of AS with regard to neurotoxicity and neuroprotection and its final role in dysfunction and death of involved neurons are to be clarified. Although Lewy bodies and AS-positive dystrophic (Lewy) neurites are present in a large number of human diseases as well as in the aged brain, the major disorders hallmarked by these AS-positive inclusions are sporadic Parkinson disease (brainstem type of Lewy body disease) with and without dementia and different subtypes (phenotypes) of dementia with Lewy bodies. The essential clinical, neuropathological, and biochemical data of these disorders and their nosological relations are critically reviewed. Their pathological diagnosis is established by validated consensus criteria based on semiquantitative assessment of cortical and subcortical Lewy bodies as their common hallmarks, and they are accompanied by multisystem degeneration with neuronal loss, gliosis, and complex biochemical deficiencies with and without Alzheimer pathologies, the clinical impact of which is still under discussion. Considerable overlap between both these two most frequent forms of α-synucleinopathies and co-occurrence of AS, amyloid β-peptide, and tau deposits in various neurodegenerative and aging disorders suggest collision of several proteinopathies with cross-seeding and synergistic interactions, confirmed by experimental and in vitro studies. Finally, the relations between the different proteinopathies with respect to their pathogenesis are discussed, but their pathobiological relations are still to be elucidated. Despite a wealth of experimental data, there is still no consensus on whether oligomers, protofibrils, or mature fibrils (or some combination of these) of various proteins are the important toxic species that mediate neurodegeneration, nor do we understand the mechanisms by which they compromise the function and viability of selective vulnerable cells.

List of Abbreviations: 3-NT, 3-nitrotyrosin; Ach, acetylcholine; AChE, Acetylcholinesterase; AD, Alzheimer's disease; AGE, advanced glycation end products; ALS, amyotrophic lateral sclerosis; APP, amyloid precursor protein; AS, α-synuclein; Aβ, β-amyloid; BA, Brodmann area; BDNF, brain-derived neurotrophic factor; C, carboxy-; CAA, cerebral amyloid angiopathy; CAB, calbindin; CERAD, Consortium for Establishing the Diagnosis of Alzheimer's disease; ChAT, choline acetyl transferase; CN, caudate nucleus; COX, cytochrome c oxidase; CSF, cerebrospinal fluid; DA, dopamine; DAT, dopamine transporter; DLB, dementia with Lewy bodies; dmX, dorsal motor nucleus of the vagus nerve; ERK, extracellular signal-related kinase; FTD, frontotemporal dementia; GABA, γ-amino butyric acid; GAD, glutamate decarboxylase; GCI, glial cytoplasmic inclusions; GP, globus pallidus; GPi, globus pallidus internus; 8-HOG, 8-hydroxy-2-deoxy-guanidine; LB, Lewy body; LBD, Lewy body disease; LBV/AD, Lewy body variant of Alzheimer's disease; Lc, locus ceruleus; LN, Lewy neurite; MAPK, mitogen-activated protein kinase; MHC, major histocompatibility complex; MPTP, 1-methyl-4-phenyl-1,2,3,6-tetrahydropyridin; mRNA, messenger RNA; mtRNA, mitochondrial ribonucleic acid; MSA, multiple system atrophy; NAC, nonamyloid component; nAChr, nicotinic acetylcholine receptor; NBM, nucleus basalis of Meynert; NFT, neurofibrillary tangles; NM, neuromelanin; NMDA, N-methyl-D-aspartate; NO, nitric oxide; NPV, negative predictable values; Pael-R, parkin-associated endothelin receptor; PB, pale body; PDC, Parkinson dementia complex; PDD, Parkinson's disease with dementia; PH-8, phenylalanine hydroxylase; PiD, Pick's disease; PPE, proenkephalin; PPNc, nucleus tegmenti pedunculopontinus pars compacta; PPV, positive predictable values; RAC, Raclopride; RNA, ribonucleic acid; ROS, reactive oxygen species; SDS, sodium dodecyl sulfate; SMA, supplementary motor area; SN, substantia nigra; SNc, substantia nigra pars compacta; SNr, substantia nigra pars reticulata; SPECT, single-photon-emission computer tomography; STN, subthalamic nucleus; tg, transgenic; TH, tyrosine hydroxylase; Ub, ubiquitin; UPP, ubiquitin-proteasomal pathway; VIM, ventral intermediate thalamus; VMAT, vesicular monoamine transporter; VTA, ventral tegmental area

1 Introduction

Lewy bodies (LBs), intraneuronal spherical cytoplasmic inclusions containing misfolded α-synuclein (AS) and other proteins, have been observed in a variety of neurodegenerative diseases summarized as LB disorders. AS is not only the major component of LBs but also occurs in intraneuritic LBs, dystrophic Lewy neurites (LN), in neuronal and glial inclusions in Parkinson's disease (PD), dementia with Lewy bodies (DLB), LB dysphagia, and multiple system atrophy (MSA), in axonal swellings that typify neuroaxonal dystrophies, e.g., Hallervorden-Spatz disease or panthothenic kinase-related neurodegeneration, and others, like Lewy-like inclusions in motor neuron disease, in sporadic and familial Alzheimer's disease (AD), Down's syndrome, Guam parkinsonian-dementia complex (PDC), subacute sclerosing panencephalitis, Meige syndrome, ataxia-telangiectasia, human aging, and others, as essential or coincidental morphological features. All these disorders sharing common pathological lesions composed of aggregates of conformationally and posttranslationally modified AS in selective vulnerable populations of neurons and glia are summarized as synucleinopathies (for review Galvin et al., 2001; Goedert, 2001; Dickson, 2003; Ma et al., 2003; Jellinger, 2003b; Norris et al., 2004; Goedert and Spillantini, 2005; McNaught and Olanow, 2006).

It has been shown that recombinant AS can form filaments in vitro that are structurally and anti-genetically similar to those extracted from the diseased human brain (Serpell et al., 2000) and in transgenic (tg) mice expressing mutant human AS, models of α-synucleinopathies, and that AS staining is more widespread than ubiquitin (Ub) staining making it the best marker for LBs (Munoz, 1999; Gomez-Tortosa et al., 2000; Schneider et al., 2002). A plethora of evidence points toward the culpability of AS in the pathogenesis of PD including: (1) linkage of AS mutations to familial forms of PD, (2) triplication of the AS locus causing PD, and (3) overexpression of AS in transgenic mice and *Drosophila* leading to PD-like phenotypes. Studies of purified AS have revealed its ability to interact with diverse molecules including monoamines. Mono-amine metabolism is associated with oxidative stress conditions that may contribute to DA–AS interactions, promoting aggregation and neuronal damage. In addition, multiple candidate analysis identified AS as a definite susceptibility gene for sporadic PD (Mizuta et al., 2006), multiple regions of AS being associated with PD (Mueller et al., 2005). The present chapter reviews only synucleinopathies histologically featured by the presence of LBs (❷ *Table 15-1*), including the pathology of PD, while other synucleinopathies, e.g., MSA, are described in other chapters.

2 α-Synuclein

2.1 Molecular Basis

AS belongs to a family of abundant brain proteins whose physiological functions are only incompletely determined. The synuclein family consists of three members: α-, β-, and γ-synuclein (Goedert, 2001; Spillantini, 2003; Tofaris and Spillantini, 2005), which show differential localization in the CNS (Li et al., 2002). Only AS is associated with the filamentous inclusions (Goedert, 2001). The AS gene is located on chromosome 4, β- and γ-synuclein genes on chromosomes 5 and 10, respectively. The synuclein proteins range from 127 to 140 amino acids in length and their sequences are 55%–62% identical. The AS gene is approximately 112 kb in length and consists of two exons, of which exon 5 makes up the coding region of the protein (❷ *Figure 15-1*). Alternatively spliced forms lacking exon 3 or 5 exist, but their mRNA is not abundant in brain. Synucleins are characterized by the presence of imperfect 11 amino acid terminal repeats that contain the KTKEGV consensus sequence and are separated by 5–8 amino acids. These seven unique 11-mer repeat sequences, of which six amino acids, Lys-Thr-Lys-Glu-Gly-Val, are conserved, are important for α-helix formation and reversible lipid-binding function (Bussell and Eliezer, 2003). The repeat region is followed by a hydrophobic stretch and a negatively charged carboxy (c)-terminal region; the hydrophobic stretch includes the nonamyloid component (NAC) peptide, which is critical for fibril formation. As AS is mainly present in presynaptic nerve terminals, it has to be transported to nerve terminal regions from neuronal perikarya by axonal transport. AS is a small presynaptic heat-stable, natively unfolded protein with potential for protein–protein interactions that have the tendency for self-oligomerization and

◘ Table 15-1
Lewy body disorders

1. Lewy body diseases
   Parkinson disease (brainstem type of LB disease)
     Sporadic
     Familial with AS mutations
     Familial with other mutations
     Pure autonomic failure
     Lewy body dysphagia
   Dementia with Lewy bodies (DLB)
     "Pure" form—transitional/limbic
     "Pure" form—neocortical
   Lewy body variant of Alzheimer's disease (LBV-AD)
   Sporadic AD
   Familial Alzheimer's disease (AD) with APP mutation
     with PS-1 mutation
     with other mutations
     Familial British dementia
   Normal aging
   Down's syndrome

2. Other disorders with Lewy bodies
   Neuroaxonal dystrophies
     Neurodegeneration with brain iron accumulation type I (Hallervorden-Spatz disease)
   Motor neuron disease
     Amyotrophic lateral sclerosis – sporadic
                                    – familial
   ALS-dementia complex of Guam
   Tauopathies
     Frontotemporal degeneration/dementia
     Pick's disease
     Progressive supranuclear palsy
     Corticobasal degeneration
   Prion diseases
   Ataxia telangiectatica
   Meige's syndrome

self-association to form pathological, insoluble fibrils, and known mutations in the molecule favor fibril formation (Trojanowski, 2003; Beyer, 2006). Unmodified AS protein that becomes structured upon binding to lipid membranes, the synaptic localization mediated by lipid rafts (Fortin et al., 2004), is degraded by the 205 proteasome in an Ub-independent manner (Tofaris et al., 2001), while in pathological lesions it is mono- or diubiquitinated (Hasegawa et al., 2002). There are interactions between fatty acids and AS (Lucke et al., 2006) and AS redistribution to neuromelanin lipid in the substantia nigra (SN) early in PD (Halliday et al., 2005). Proteasome binding to insoluble AS filaments and soluble AS oligomers results in marked inhibition of its chymotrypsin-like hydrolytic activity through a noncompetitive mechanism that is mimicked by model β-amyloid (Aβ) peptide aggregates. Endogenous ligands of aggregated AS-like heat-shock protein (HSP)70 and glyceraldehyde-6-phosphate dehydrogenase bind filaments and inhibit their antiproteasomal activity and, thus, suppressing their toxicity, may have a protective role (Auluck et al., 2002; Klucken et al., 2004). The inhibitory effect of amyloid aggregates may thus be amenable to modulation by endogenous chaperones (Lindersson et al., 2004). AS is in the sodium dodecyl sulfate (SDS)-soluble fractions of diseased brains of humans and tg mice expressing A53T mutant human AS (Kotzbauer et al., 2005).

◘ Figure 15-1

Diagrammatic scheme of the SNCA gene, normal and alternative RNA transcripts, and functional domains of α-synuclein (modified from Dickson, 2001). PD, Parkinson's disease; DLB, dementia with Lewy bodies; LBV/AD, Lewy body variant of Alzheimer's disease; MSA, multiple system atrophy

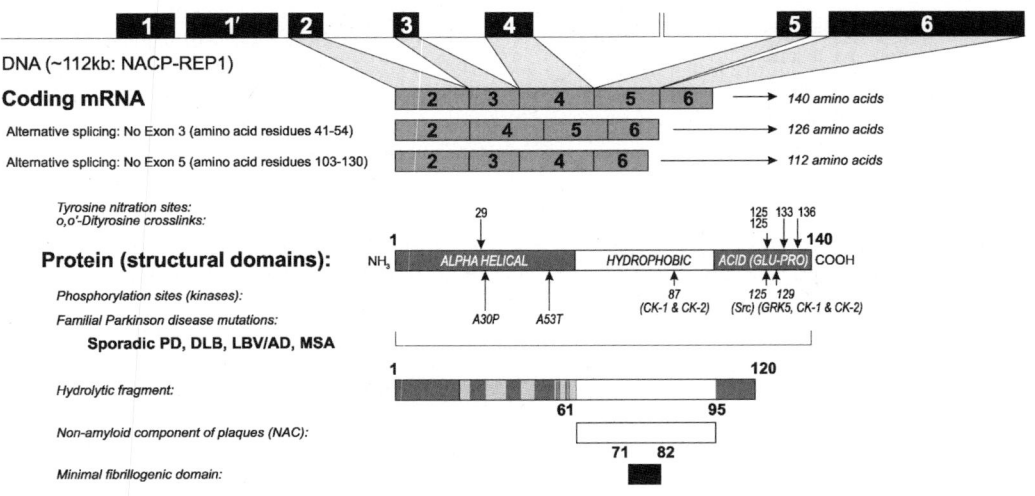

2.1.1 Aggregation of Synuclein

AS exists physiologically in both soluble and membrane-bound states in unstructured and α-helical conformations, respectively. The physiological function of AS appears to require its translocation between these subcellular compartments and interconversion between the two conformations. Abnormal processing of AS is predicted to lead to pathological changes in its binding properties and function.

The frequency of AS aggregation in human diseases has given rise to extensive studies of the process in vitro. These studies have revealed that this aggregation represents a nucleation-dependent process (Wood et al., 1999). The transition from monomeric AS to filamentous aggregates is characterized by a lag phase during which a buildup of soluble nucleation-competent oligomeric AS species takes places. The rapid filament growth does not occur until these structures have reached a critical centration. In addition, in vitro ubiquitination of fibrillar AS recapitulated the pattern observed in human disease and the mouse model, supporting the hypothesis that ubiquitination of AS is not required for inclusion formation and that AS fibrillation precedes its ubiquitination (Sampathu et al., 2003). Ub is a small heat-shock protein that is involved in a number of physiologic processes, most notably energy-dependent degradation of denatured or abnormal proteins through the proteasome (Petrucelli and Dawson, 2004). Other heat-shock proteins, e.g., Hsp90, converging with Ub in filamentous AS inclusions indicate its predominant role in AS inclusions in PD and related synucleinopathies (Uryu et al., 2006). Abnormal proteins destined for degradation by the proteasome have polyubiquitin chains conjugated to their peptide backbone. Ub is also conjugated to proteins through other acceptor sites without generating a polyubiquitin chain, and these ubiquitinated proteins are not targeted. The kinetics of AS fibrillation is consistent with a nucleation-dependent mechanism in which the critical early stage of the structural transformation involves a partially folded intermediate. Although the basis for the toxic effects of aggregated AS are unknown, it has been proposed that transient oligomers are responsible, possibly by forming pores in membranes. The molecular basis for AS aggregation/fibrillation, including factors that either accelerate or inhibit fibrillation, effects of molecular crowding, oxidation, point mutations, and lipid membranes, as well as the variety of conformational and oligomeric states that AS can adopt, have been reviewed recently. It is apparent that neuronal cells must have a very fine balance of factors that control the levels and potential aggregation of AS (Fink, 2006).

Elevated amounts of highly soluble oligomers of AS are formed in both aged tg mice and human PD and DLB brains, and their promotion by unsaturated fatty acids in vivo is suggested to precede the insoluble aggregates associated with neurodegeneration (Welch and Yuan, 2003; Sharon et al., 2003a). Helical, membrane-bound AS is unlikely to contribute to aggregation and fibrillation and is inhibited by lipid-binding on phospholipid vesicles (Zhu and Fink, 2003). Accumulation of a highly modified 22–24 kDa AS species found in LB diseases is ubiquitinated and represents a partly phosphorylated protein. Lysosomal proteolysis plays a pivotal role in regulating AS turnover; accumulations were elevated in conditions where macroautophagy was impaired (Webb et al., 2003). Proteolytic enzymes may also catalyze the degradation of AS. Degradation of mutant AS is impaired by chaperon-mediated autophagy (Cuervo et al., 2004). Iwata et al. (2003) showed that neurosin prevented AS polymerization by reducing the amount of monomer in vitro and also by generating fragmented ASs that themselves inhibited the polymerization. Neurosin is a mitochondrial protein that is released to the cytosol on cellular stress. Because kDa unmodified AS is degraded by the proteasome in an Ub-independent manner, accumulation of the 22–24 kDa protein is considered disease specific, and C-terminal-truncated AS, presumably a result of aberrant proteolysis, is found only in association with AS aggregates. Hence, ubiquitination of AS in LBs is not associated with impairment of proteasome function, which, however, may play a role in the initiation or progression of the neurodegenerative process in sporadic PD (McNaught et al., 2006). There is a difference among monomeric AS, AS protofibrils (oligomers), and AS fibrils. Protofibrillar AS, in contrast to the monomeric and fibrillar forms, binds synthetic vesicles very tightly via a β-sheet-rich structure and transiently permeabilizes these vesicles (Giasson et al., 1999). Volles et al. (2001) directly observed the destruction of vesicular membranes by protofibrillar AS using atomic force microscopy. Thus, oligomers of AS may be more toxic to nigral neurons than fibrils. In diseases, where it aggregates, AS changes conformation and aggregates with fibrils of β-sheet structure similar to other amyloid proteins; the interaction between Aβ peptide and AS suggest molecular overlapping pathologies of AD and LB disease (Mandal et al., 2006). AS deposits are selectively phosphorylated at Ser129 and Tyr125 residues (Okochi et al., 2000; Ellis et al., 2001; Fujiwara et al., 2002) in neurodegenerative disorders, underscoring the importance of phosphorylation of filamentous proteins in their pathogenesis (Iwatsubo, 2003). Despite differences in the molecular composition of the structural elements of these filamentous lesions as well as the brain regions and cell types they affect, growing evidence supports the notion that these different misfolded proteins engage common downstream pathogenic targets, although the different localizations of hallmark protein aggregations (extracellular Aβ deposits in AD, cytoplasmic LBs in PD, or intranuclear inclusions in Huntington's disease/HD) may suggest that each protein strikes a single cellular domain, thereby precluding the possibility of common pathogenesis. In addition to showing common morphological and chemical features, the inclusion bodies colocalize with several of the same proteins, including various hallmark chaperones and components of the degrading UPS (Sherman and Goldberg, 2001). This might reflect an irreversible sequestration and subsequent loss of functions and/or a failed attempt to refold and degrade aggregated proteins (Muchowski and Wacker, 2005). The cellular response to these aggregates includes (1) the recruitment of chaperones or proteins involved in the folding of nascent translational products and in the resolubilization of aggregated polypeptides, and (2) the ubiquitination of aggregates, suggesting cellular attempts to degrade deposits of these mutant proteins via the ubiquitin-proteasomal pathway (UPP) (Myung et al., 2001). However, ubiquitination of proteins in inclusions may be part of a signal transduction mechanism in afflicted neurons and not reflect abortive proteolysis (Layfield et al., 2003). It has been accepted by many that misfolding and deposition impairs a gain of function that ultimately leads to neuronal death (Hardy and Selkoe, 2002), although there are opponents who challenge this dogma (Lee et al., 2004).

Among AS proaggregatory proteins, whose dysregulation may contribute to disease progression, the brain-specific p25α was identified as a candidate that preferentially binds AS in its aggregated state (Lindersson et al., 2004). p25α is normally only expressed in oligodendrocytes in contrast to AS which is normally only expressed in neurons. This expression is changed in synucleinopathies: In PD and DLB, p25α was detectable in neuronal LBs along with AS, while in MSA degenerating oligodendrocytes displayed accumulation of p25α and dystopically expressed AS in the glial cytoplasmic inclusions (GCIs)

(Lindersson et al., 2004; Jellinger, 2006b). These data suggest that p25α plays a proaggregatory role in neurodegenerative disorders hallmarked by AS aggregates (Lindersson et al., 2005). P25α immunoreactivity has been demonstrated in cortical and—inconsistently—in subcortical LBs as well as in oligodendroglial GCIs but not in neuronal AS-positive inclusions in MSA (Baker et al., 2006; Jellinger, 2006b). Sequence-specific truncation of AS, favored in human AS compared with murine AS, is enhanced in neuronal cells and promotes its pathological aggregation (Lee et al., 2003b). Selective nitration of AS in its amino-terminal part and the presence of nitrated AS in LBs, LNs, and glial inclusions are linked with oxidative damage in neurodegeneration (Duda et al., 2000; Giasson et al., 2000), while cleavage of AS by calpain suggests its potential role in degradation of fibrillized and nitrated species of AS (Mishizen-Eberz et al., 2005). However, the presence of AS-positive deposits in DLB lacking 3-nitrotyrosin (3-NT)-immunoreactivity suggests that nitration is not a prerequisite for AS deposition (Gomez-Tortosa et al., 2000; Gomez-Tortosa et al., 2001).

Oxidative stress is also important for the initiation of AS oligomer formation and may also facilitate the fibrillization of AS by either promoting fibril formation or stabilizing formed fibrils or their precursors, or both (Ischiropoulos, 2003). Dimeric dityrosine cross-linking of AS is facilitated by its oxidation (Krishnan et al., 2003). This dimer forms a core of AS fibril and aggregate formation. AS facilitates the toxicity of oxidized catechol metabolites (Hasegawa et al., 2006), and oxidative damage is known to occur in the nigral neurons in PD. Nitric oxide (NO) increases the number of AS clusters, and inhibitors of NO synthase block this increase, supporting the hypothesis that NO is involved in the enhancement of the number of AS clusters. Thus, AS is involved in synaptic plasticity by augmenting transmitter release from the presynaptic terminal (Liu et al., 2004). Increased AS content in neurons is associated with relatively low levels of oxidative stress (Quilty et al., 2006). Numerous in vitro studies have shown that AS oligomer formation and fibril formation are enhanced by oxidative stress (Hashimoto et al., 1999; Kang and Kim, 2003; Paik et al., 2003), pesticides and herbicides (Sherer et al., 2003), metals and metal proteins (Nielsen et al., 2001; Martinez et al., 2003; Yamin et al., 2003; Lee et al., 2003a), polyunsaturated fatty acids (Sharon et al., 2003a), and polyamines (Antony et al., 2003). In the MPTP model of parkinsonism, adenosine diphosphate (ADP) depletion and ROS overproduction occur soon after MPT application, subjecting the intoxicated cells to an early energy crisis and OS. Among the essential molecular pathways that are pivotal in triggering cell-death cascades, alterations in ATP synthesis and ROS production lead to the demise of the affected neurons (Przedborski et al., 2004).

Chronic systemic treatment of rats with rotenone, a pesticide known to inhibit mitochondria, causes selective nigrostriatal dopaminergic degeneration with associated inclusions containing fibrillar AS (Betarbet et al., 2000). Rotenone treatment may induce an increase in oxidative stress in the dopaminergic neurons, which in turn may facilitate fibrillization of AS, providing a link between oxidative stress and pathogenesis of synucleinopathies (Giasson et al., 2000; Giasson and Lee, 2000). Moreover, Rotenone exposure, may promote the formation of LBs because rats treated with the insecticide rotenone develop LB-like AS inclusions (Betarbet et al., 2000). Both antioxidant components and small molecule inhibitors have potent antifibrillogenic and fibril-destabilizing effects on AS fibrils in vitro (Masuda et al., 2006; Ono and Yamada, 2006), and inhibitors of AS oligomerization and toxicity may have importance for future therapeutic strategies for PD and related disorders (Hashimoto et al., 2004; Amer et al., 2006).

AS is natively unfolded in solution, but can be induced to form either an α-helical or a β-sheet structure depending on its concentration and the solution conditions; the β-sheet structure is cytotoxic (Kessler et al., 2003). As an aggregated fibril form of AS is the major component of LBs, fibril formation and aggregation of AS have emerged as important mechanisms of AS neurotoxicity. This process is stimulated by several factors like the pathogenic missense mutations A30P and A53T (Giasson et al., 1999; Narhi et al., 1999; Conway et al., 2000), increased concentration of AS, decreased pH, elevated temperature (Hashimoto et al., 1998a, 1999), proteolytic truncations of the acidic C terminus (Crowther et al., 1998; Serpell et al., 2000), molecular crowding (Uversky and Fink, 2002), phosphorylation of Ser129 (Fujiwara et al., 2002), oxidative modifications like nitrations (Souza et al., 2000), and DA conjugation (Conway et al., 2001) or altered degradation of the protein through dysfunction of the UBI proteasome pathway (see Sidhu et al., 2004a, b). The reverse inhibitory effects on AS aggregation have also been demonstrated because the aggregation-incompetent synucleins, β- and γ-synuclein, may block the process (Hashimoto et al., 2001; Uversky et al., 2002). In vivo and in vitro studies showed that overexpression of AS, its aggregation, and interaction with other proteins are the most critical factors affecting the survival of neurons.

2.1.2 Transgenic Models

The identification of AS as a genetic component in PD and DLB (Polymeropoulos et al., 1997; Kruger et al., 1998; Singleton et al., 2003; Farrer et al., 2004; Zarranz et al., 2004) has prompted the generation of AS tg animals that involve overexpression of wild-type and mutant AS in both neurons and oligodendrocytes (see Barbieri et al., 2001; Dauer and Przedborski, 2003; Maries et al., 2003; Orth and Tabrizi, 2003; Jellinger, 2003f). Mice overexpressing the normal or mutated forms of AS display a highly variable phenotype (Hashimoto et al., 2003a). A major determinant of severity seems to be the promoter used for the transgene (Fernagut and Chesselet, 2004). Most mice overexpressing AS under control of the prion promoter, and some of those overexpressing it under control of the Thy 1 promoter, display severe anomalies including massive pathology in motoneurons that are not primarily affected in PD (van der Putten et al., 2000; Giasson et al., 2002; Lee et al., 2002b). By contrast, mice overexpressing AS under control of the tyrosine hydroxylase (TH) or platelet-derived growth factor 3 promoters do not display motoneuron pathology (Masliah et al., 2000; Matsuoka et al., 2001; Rathke-Hartlieb et al., 2001; Richfield et al., 2002). Two such mouse lines present a neurochemical deficit in the nigrostriatal pathway and behavioral anomalies (Masliah et al., 2000; Richfield et al., 2002). In view of the indisputable role of AS in both familial and sporadic PD, as ascertained by its accumulation in LBs, these mice provide a compelling model to investigate the role of these cytoplasmic inclusions. Furthermore, age-related loss of nigrostriatal dopaminergic neurons is accelerated in mice overexpressing AS—an effect that is worsened by exposure to pesticides (Thiruchelvam et al., 2004). AS expression in the SN of MPTP-lesioned nonhuman primates is enhanced, suggesting that protein elevation within cell bodies may be a late feature of neurons that have endured toxic stress (Purisai et al., 2005), Neuronal overexpression of wild-type or mutant human A53T AS in mice induces movement disorders and dopamine (DA) loss similar to human PD and leads to cytoplasmic and nuclear inclusions containing AS and Ub which, however, differ from typical LBs (Kahle et al., 2000; Masliah et al., 2000; van der Putten et al., 2000). Overexpression of human wild-type or mutant AS induces dopamine-related and caspase-independent apoptosis in dopaminergic neurons (Zhou et al., 2006), and AS tg mice develop mitochondrial degeneration and cell death (Martin et al., 2006). It disrupts vesicular pH and leads to a marked increase in the levels of cytosolic catechol species, an effect that may in turn trigger cellular oxyradical damage. Although multiple molecular mechanisms may be responsible for the perturbation of cytosolic catecholamine homeostasis, this study provides critical evidence about how AS might exert its cytotoxicity and selectively damage catecholaminergic cells (Mosharov et al., 2006). Experimental models of PD have shown a mechanistic chain of events between altered AS, proteasome impairment and formation of neuronal inclusions, and catecholamine cell death (Giorgi et al., 2006). Recent studies further indicate that overexpression of AS via the recombinant adenoassociated viral (rAAV) vector system apparently recapitulates several important features of PD and DLB brains, and this AS overexpression model is likely to be ideal for the study of the pathogenesis of PD and DLB. This model is also useful for gene therapy research (Mochizuki et al., 2006). Progressive degeneration of the nigral dopaminergic neurons, a hallmark pathological feature for any PD model, was not observed in AS A30P tg mice (Dawson, 2000; Gomez-Isla et al., 2003), and two others overexpressing AS under control of the murine prion promoter revealed severe neurodegeneration outside of the substantia nigra (SN) (Giasson et al., 2002; Lee and Lee, 2002). Aβ enhances AS accumulation and neuronal deficits in a tg mouse model linking AD and PD (Masliah et al., 2001). Lack of nigral pathology was seen in tg mice expressing human wild-type AS (A30P, A531) driven by the TH promoter (Matsuoka et al., 2001), whereas degeneration and progressive loss of nigrostriatal neurons with reduction of striatal DA and TH levels were observed in adult rats overexpressing A53T-mutated human AS (Kirik et al., 2002). In A53T-overexpressing double-mt mice, inclusion formation and neurodegeneration are parkin independent (von Coelln et al., 2006). Other PD models in rat using recombinant adenoviral and lentiviral vectors targeted overexpression of AS in SN and induced LB-like pathology and degeneration of the nigrostriatal system (Kirik et al., 2002; Klein et al., 2002; Lo Bianco et al., 2002). Overexpression of wild-type and mutant AS in the fruit fly (*Drosophila*) reproduced many features of human PD with a progressive, age-dependent loss of a subset of DA neurons with ʟ-dopa-responsive progressive motor dysfunction (Feany and Bender, 2000; Bonini and Fortini, 2003). AS phosphorylation controls neurotoxicity and inclusion formation in this *Drosophila* model of PD (Chen and Feany, 2005).

The overexpression or induction of the molecular chaperone hsp70 in AS tg flies prevents this neuron loss, implicating the misfolding or aggregation of AS in disease pathogenesis (Auluck et al., 2002). Some neurons accumulated intracellular aggregates containing AS fibers that closely resemble LBs, resulting from interference with endogenous chaperone activity that accelerates AS toxicity (Auluck et al., 2002). Stereotactic injection of lentiviral vectors encoding wild-type and A30P mutant human AS into the brain caused overexpression of AS-induced abnormal accumulation of AS in cell bodies and neurites, AS-positive neuritic varicosities, and cytoplasmic inclusions that stained with Ub and AS antibodies, later associated with degenerative morphology and a significant loss of AS-positive cells. This indicates that overexpression of AS can induce Lewy-like pathology and neurodegeneration which is not restricted to dopaminergic cells (Lauwers et al., 2003). Recent studies of human brains confirmed AS mRNA expression in SN neurons with little labeling in other basal ganglia (Solano et al., 2000). AS-knockout mice reported by Abeliovich et al. (2000) were viable and fertile, exhibited intact brain architecture, and possessed a normal complement of dopaminergic cell bodies, fibers, and synapses. Nigrostriatal terminals of AS(−/−) mice displayed a standard pattern of DA discharge and reuptake in response to simple electrical stimulation. They exhibited an increased release with paired stimuli. AS(−/−) mice displayed a reduction in striatal DA and an attenuation of DA-dependent locomotor response to amphetamine. These authors concluded that AS was an essential presynaptic, activity-dependent negative regulator. AS-null mice produced by Dauer et al. (2002) were resistant to 1-methyl-4-phenyl-1,2,3,6-tetrahydropyridine (MPTP) neurotoxicity.

2.2 The Function of α-Synuclein

The function of AS is not clear; it may be involved in synaptic vesicle transport, may play an important role in neuronal plasticity (Kaplan et al., 2003), and as an activity-dependent negative regulator of DA release and biosynthesis (Clayton and George, 1999; Abeliovich et al., 2000; Perez et al., 2002). Reviews of the molecular composition of AS and speculations about its function have appeared recently (Uversky and Fink, 2002; Alves da Costa, 2003; Dev et al., 2003; Kaplan et al., 2003; Maries et al., 2003; Spillantini, 2003; Mochizuki and Hattori, 2005; Sidhu et al., 2004a; Bennett, 2005). Axonal transport of human AS slows with aging but is not affected by familial PD-linked mutations (Li et al., 2004). Although AS is the major component of LBs and related inclusions, the mechanism by which its dysfunction leads to cell degeneration is poorly understood. In neurodegenerative disorders, AS is suggested to play a dual role by being involved not only in synaptic function but also in Aβ fibrillogenesis (Kaplan et al., 2003). Downregulation of serine protease neurosin causes accumulation of AS in cultured cells (Iwata et al., 2003).

AS, both wild type and mutant, is neurotoxic when overexpressed in nigral neurons. Kirik et al. (2002) transduced the AS gene into nigral neurons by local injection using an adeno-associated virus vector. They found that AS-overexpressed nigral neurons showed AS-positive cytoplasmic inclusions and swollen, dystrophic neurites. Striatal DA was reduced by 30%–80% and TH activity 40%–50% by 8 weeks after the transduction.

Overexpression of AS in vitro impairs mitochondrial function and leads to increased OS (Hsu et al., 2000). Conversely, administration of MPTP to AS-overexpressing mice reduced grossly deformed mitochondria (Song et al., 2004), and oxidative damage produced by both rotenone and chronic MPTP results in oxidative damage and AS aggregates closely resembling Lewy bodies (Betarbet et al., 2000; Fornai et al., 2005).

Mutations in parkin, which encodes a ubiquitin E3 ligand and causes recessive early onset PD, are associated with marked mitochondrial abnormalities and less resistance to OS induced by paraquat (Greene et al., 2003; Pesah et al., 2004). Parkin, associated with the outer mitochondrial membrane and protecting against mitochondrial swelling and release of cytochrome c induced by a variety of ROS, is important in the degradation of oxidatively damaged proteins. Several studies have implicated mtDNA abnormalities and mitochondrial dysfunction in PD pathogenesis (Valente et al., 2004), and a link between mitochondrial dysfunction, OS, and protein degradation is beginning increasingly prominent in PD pathogenesis.

Aggregations of AS may become cytotoxic and can produce neuronal death due to production of oxidative stress and mitochondrial energy deficit by interaction with cytochrome c oxidase (COX) (Hsu et al., 2000; Kanda et al., 2000; Goedert, 2001; Elkon et al., 2002; Mouradian, 2002; Jenner, 2003). On the other hand, monomeric AS efficiently prevents lipid oxidation, whereas fibrillar AS had no such effect, suggesting that the prevention of unsaturated lipid oxidation by AS requires that it bind to the lipid membrane. The antioxidant function of AS is attributed to its facile oxidation via the formation of methionine sulfoxide, as shown by mass spectrometry. These findings suggest that the inhibition of lipid oxidation by AS may be a physiological function of the protein (Zhu et al., 2006). The phosphorylation of AS, aggresomes formed by AS and synphilin-1, and the involvement of the Ub-proteolytic system in its degradation may be the structural manifestation of a cytoprotective process designed to eliminate damaged cellular elements (Lee et al., 2002a; Tanaka et al., 2004). It may render the mutated or damaged protein less toxic than the soluble form in order to delay the onset of neuronal degeneration. AS has been shown to protect neurons against oxidative stress and excitotoxicity, it may inhibit apoptosis induced by a wide variety of stimuli or modulate the expression levels or activity of proteins implicated in the induction and control of apoptosis (Shults, 2006), and its protective effect is increased by Bcl-2 overexpression. There is ample evidence that, overall, AS in itself is not a toxic compound, but that it can be converted to a toxic compound in the presence of specific components of the dopaminergic systems, thereby conferring the selectivity for degeneration for the SNpc dopaminergic neurons, characteristic of PD (Lotharius and Brundin, 2002; Perez and Hastings, 2004; Sidhu et al., 2004a, b). Through its peculiar biochemical properties, AS is able to interact differently with numerous molecular systems, contributing in more than one way to the pathophysiology of PD. Considering AS as a central regulator of the dopamine life cycle, and as an effector of dopamine-dependent oxidative stress, paves the way for a more accurate understanding of both the physiology and the pathology of dopamine neurotransmission. However, at both low and over-expressed levels in the cell, AS resulted in cytotoxicity, probably caused by decreased Bcl-2 and increased Bax expression, which is followed by cytochrome c release, caspase activation, microglia-mediated inflammatory response, and regulation of the mitogen- activated protein kinase (MAPK) pathway (Ferrer et al., 2001; Seo et al., 2002) by reducing the amount of available active MAPK (Iwata et al., 2001). AS and its neurotoxic fragment inhibit dopamine uptake into rat striatal synaptosomes, indicating its important role in alteration of DA synaptic homeostasis, probably by NO-mediated DAT alteration (Adamczyk et al., 2006). Parkin accelerates the degradation of AS via the activation of the nonproteasomal protease, calpain, but not caspase, and the cytoprotective effect of parkin on AS cytotoxicity was significantly inhibited by the presence of calpain-specific inhibitors (Kim et al., 2003b). β-Synuclein displays an antiapoptotic p-53-dependent phenotype and protects neurons from 6-hydroxydopamine-induced caspase-3 activation, restoring the anti-apoptotic function of AS (Da Costa et al., 2003). AS may further interfere with extracellular signal-regulated kinase (ERK) by dysregulating caveolin I expression and reestablishing the normal pattern of neuronal outgrowth. Disruption of scaffold proteins, such as caveolin and JNK-interacting protein (JIP), which are necessary to integrate signaling pathways, may lead to cycles of synapse loss and aberrant sprouting. This is significant because both caveolar dysfunction and altered axonal plasticity might be important in the pathogenesis of various neurodegenerative disorders (Hashimoto and Masliah, 2003; Hashimoto et al., 2003b). AS accumulates within SN neurons and is entrapped in pigment granules during neuromelanin (NM) biosynthesis, i.e., before melanin depletion which is characteristic of PD (Fasano et al., 2003). Overexpression of AS is now known to precipitate PD (Kirik et al., 2003; Lauwers et al., 2003; Singleton et al., 2003; Farrer et al., 2004), although there is no increase in AS transcription in this region in PD (Kingsbury et al., 2004). Further, acidic lipids enhance metal-catalyzed AS aggregation (Lee et al., 2003a; Sharon et al., 2003a, b), and thus changes in the NM lipid pool may contribute to intracellular conditions favoring neurodegeneration in PD. Altered protein handling appears to be a central factor in the pathogenic process occurring in the various hereditary and sporadic forms of PD. This suggests that manipulation of proteolytic systems is a rational approach in the development of neuroprotective therapies that could modify the pathological course of PD (McNaught and Olanow, 2006).

3   Lewy Bodies

3.1   Types of Lewy Bodies and Related Aggregates

The characteristic findings in LB diseases (LBD) are LBs occurring in two types, the classical brainstem and the cortical types. Classical LBs are spherical intraneuronal cytoplasmic inclusions 8–30 μm in diameter with hyaline eosinophilic core, developing concentric lamellar bands and a narrow pale stained halo (Lowe et al., 2002). Most LBs are single, some neurons may contain multiple or polymorphic inclusions (❯ *Figure 15-2a*). In some regions of the brain, such as in the dorsal motor nucleus of vagus (dmX), similar occlusions within neuronal processes are referred to as intraneuritic LBs. They can be detected in conventional histopathological preparations and should be distinguished from LNs which are not visible in routine histology but are seen in AS-immunostains. Ultrastructurally, classical LBs are nonmembrane-bound, granulofilamentous structures, composed of radially arranged 7–20 nm intermediate filaments associated with granular electron-dense material and vesicular structures, the core showing densely packed filaments and dense granular material, whereas the periphery has radially arranged 10-nm filaments (Forno, 1996) (❯ *Figure 15-2b*).

◻ Figure 15-2

AS immunohistochemistry. (a) Lewy body in substantia nigra neuron with peripheral ring-like AS-positive immunoreaction. (b) Electron microscopy of nigral LB showing a central electron-dense filamentous core with a loosely fibrillary rim (original magnification x2500). (c) Multiple round homogenous Lewy bodies in cortex of cingulate gyrus. (d) Dystrophic AS-immunoreactive neurites in CA2/3 sectors of Amon's horn

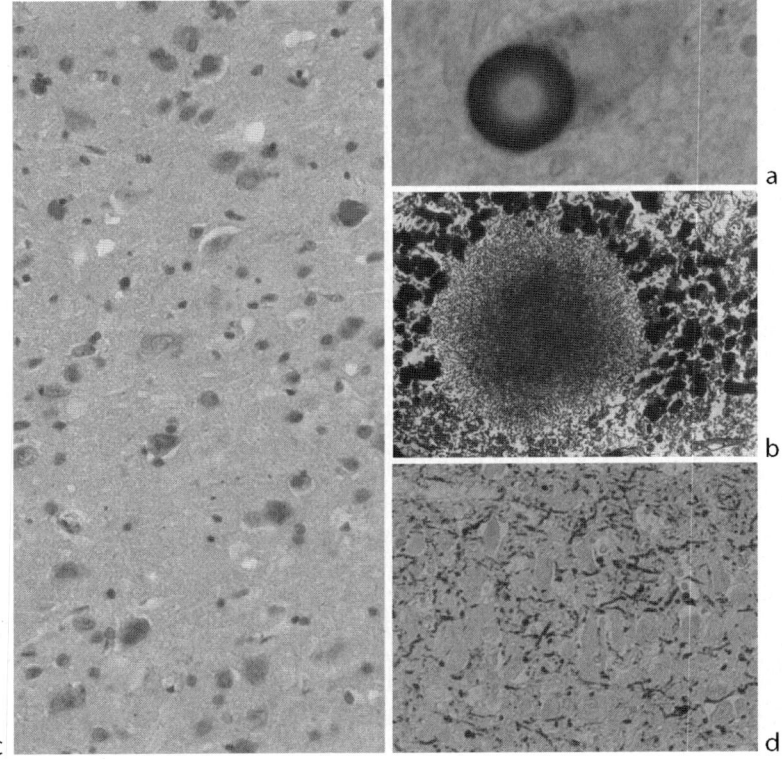

Neurons most vulnerable to LBs include the monoaminergic neurons in the substantia nigra pars compacta (SNc), locus ceruleus (LC), dmn X, cholinergic neurons in the basal forebrain, the oculomotor complex, dorsal raphe, and subpeduncular nuclei of the pons. LBs are rarely detected in basal ganglia, thalamus, and pontine basis but are common in the posterior and lateral hypothalamus, in the brainstem reticular formation, and olfactory bulb, while the cerebellar cortex is preserved. In the spinal cord, the neurons of the intermediolateral cell column are involved. LBs are found in autonomic (sympathetic and parasympathetic) ganglia, including submucosal ganglia of the lower esophagus and colon, in the adrenal, and in enteric, cardiac, and pelvic neuronal plexuses (see ❯ *Table 15-2*).

◘ Table 15-2

Distribution of Lewy bodies in Parkinson's disease

Affected region[a]	Major (putative) neuromediators[b]	Frequent	Rare
Cerebral cortex	Multiple		+
Anterior cingulate gyrus		++	
Temporal, insular		+	
Allocortex (parahippocampal)			+
Amygdala, central accessory cort. nucleus		++	
Basal nuclei			+
Nucleus basalis of Meynert	Acetylcholine	++	
Thalamus, midline nuclei		++	
Nucleus amygdalae		++	
Hypothalamus, lateral nucleus		++	
Lateral posterior nucleus	Norepinephrine		+
Paraventricular nucleus	Multiple		+
Tuberomamillar nucleus		++	
Subthalamic nucleus	Dopamine		+
Periaqueductal gray	Multiple		+
Substantia nigra zona compacta	Dopamine	++	
Nucleus parabrachialis pigmentosus	Dopamine	++	
Nucleus paranigralis	Dopamine	++	
Westphal-Edinger nucleus	Acetylcholine	+	
Darkschewitsch nucleus	Acetylcholine		+
Supratrochlear nucleus	Serotonin	+	
Nucleus tegmenti pedunculopontinus	Acetylcholine	+	
Central pontine gray	Multiple		+
Locus ceruleus	Norepinephrine	++	
Nucleus subceruleus	Norepinephrine	+	
Nucleus pontis centralis oralis	Serotonin	+	
Central superior nucleus of raphe	Serotonin		+
Processus griseum pontis supralemniscalis	?	+	
Dorsal motor nucleus of vagus	Norepinephrine	++	
Nucleus of Roller	?		+
Nucleus gigantocellularis	Serotonin		+
Nucleus paragigantocellularis lateralis	?		+
Nucleus medullae oblongatae centralis	Serotonin		+
C1 and C2 groups in medulla oblongata	Epinephrine	++	
Olfactory bulb			
Spinal cord, intermediolateral column	Multiple		+
Spinal cord, intermediomedial column	Acetylcholine		+
Spinal cord, anterior horn			+
Autonomic (symp. and parasymp.) ganglia	Catecholamines	++	
Enteric nerve plexuses	Catecholamines		+
Adrenal medulla	Catecholamines		+

Modified from Jellinger and Mizuno (2003)

[a]Nomenclature of the brain stem nuclei according to Olszewski and Baxter (1982)

[b]Nieuwenhuys et al. (1988)

+ mild; ++ severe

Cortical LBs, eosinophilic, rounded, angular or reniform structures without obvious halo, are poorly visible using conventional histological staining methods, while immunohistochemistry using antibodies against AS is the preferred method for detecting LBs (Munoz, 1999) and has greater sensitivity in detecting LBs in cerebral cortex (❷ *Figure 15-2d*). Ultrastructurally, they are poorly organized, composed of felt-like arranged granulofibrillary structures (Forno, 1996). They are found in small nonpyramidal neurons in lower layers in the frontal, insular, temporal multimodal association and limbic cortices; regions with densest accumulation are the insular cortex, amygdala, the parahippocampal, and cingulate gyri. LBs are rarely detected in the primary cortices. AS immunohistochemistry of cortical LBs is demonstrable in up to 100% of PD and DLB brains (McKeith et al., 1996), particularly involving the cingulate gyrus, hippocampus, amygdala, and middle temporal gyrus. Similar lesions, rounded areas of granular pale-staining eosinophilic material in brainstem neurons, referred to as "pale bodies" (PBs) consist of disorganized fibrils interspersed with vacuoles and granular matter. Although they have been considered as precursors of LBs (Dale et al., 1992), their pathogenic relationship remains unclear. Whereas ultrastructural and staining differences were considered as evidence of separate aggregation processes (Pappolla et al., 1988), their relationship is supported by the fact that PBs are always accompanied by LBs, and they correlate with them in abundance and outnumber LBs in nonsymptomatic cases of PD and those with short duration of disease (Dale et al., 1992), and also by the fact that both contain not only three common protein complexes (AS, Ub, and p62) but also in similar abundance (Kuusisto et al., 2003). Other forms of intraneuronal aggregates have been reported to accompany LBs, including "LB-like matter" and "diffuse," "cloud-like" or "irregularly shaped" AS-immunoreactive accumulations (Wakabayashi et al., 1998a, b; Gomez-Tortosa et al., 2000). Although suggested to be precursors of PBs and/or LBs, these relationships await to be established. On autopsy, the majority of patients with PARK2 mutations lack LB pathology, and autopsy information is not available on patients with mutations in the genes encoding UCH-L1, DJ-1, and PTEN-induced kinase 1 (Giasson and Lee, 2003). Thus, it is unclear whether these patients represent LB disease or simply an overlapping clinical phenotype of genetically distinct disorders.

3.2　Constitution of Lewy Bodies and Aggresomes

While abundant evidence suggests that abnormal fibrillation of AS may be pivotal for the pathological protein aggregates in LBs (Lotharius and Brundin, 2002), it is not clear how the AS fibrils accumulate to form these complex inclusions (Jellinger, 2003d). Significant reduction of AS mtRNA expression in melanized nigral neurons in PD brain negatively correlated with disease duration. These data and negative AS mRNA expression in glia suggest that LB formation is unlikely to be the result of overexpression of AS (Kingsbury et al., 2004). The presence of Ub, a common feature in neuronal inclusions (Alves-Rodrigues et al., 1998), suggests a dysfunction in Ub-dependent proteolysis that may contribute to inclusion formation via the accumulation of poly-Ub-conjugated proteins (Chung et al., 2001a). The aggregation of proteins and formation of inclusion bodies in several neurodegenerative disorders, such as amyotrophic lateral sclerosis (ALS), Huntington's disease (HD), and PD, have been associated with the aggresome-related process and consequently with the centrosome (McNaught et al., 2002b; Corcoran et al., 2004; Olanow et al., 2004). The aggresomes are microtubule-dependent pericentriolar inclusion bodies, which can be a common cellular response to large amounts of misfolded or aberrant proteins (Kopito, 2000). Formation of the aggresomes is considered to occur when the degradation capacity of the Ub-proteasome system is exceeded either by increased production or by inadequate proteolysis of aberrant proteins (McNaught et al., 2002b). All these nondegraded proteins dispersed in the cytoplasm are then retrogradely transported along microtubules to the centrosome, where they form an enlarged inclusion of aggregated-ubiquitinated proteins surrounded by cytoskeleton elements, known as the aggresome (Kopito, 2000; McNaught et al., 2002b; Olanow et al., 2004). In neural soma and processes in PD and DLB, numerous discrete ubiquitinated aggregates appear to be transported from peripheral sites to the centrosome, where they are sequestered to form LBs. LBs in PD are immunoreactive for γ-tubulin (McNaught et al., 2004) suggesting that aberrant proteins of LBs are related to the centrosome, also termed the microtubule-organizing center, which is a small perinuclear organelle composed of a pair of centrioles surrounded by the centrosome matrix or pericentriolar material. Its matrix contains a variety of proteins, such as proteins responsible for the proteasome-proteolytic pathway, cell-cycle regulatory system, and for the nucleation of microtubules. Thus, the centrosome seems to play an important role in the formation of these complex inclusion bodies during the neurodegenerative processes. NEDD-8 (neuronal precursor cell expressed developmentally downregulated 8), a Ub-like protein that controls vital biological events through its conjugation to members of the cullin family, components of certain Ub e3 lipases, have been shown to be incorporated into LBs and LNs in PD, in Ub-inclusions in MSA, and other neurodegenerative disorders. These findings suggest that NEDD-8 is involved in the formation of various ubiquitinated inclusions via the UPS (Mori et al., 2005).

However, it is poorly defined how the pathological accumulation/aggregation of Ub-containing proteins is related to that of AS during LB formation. The aberrant accumulation or ectopic expression of CDK5 and mitogen-activated protein kinase (MAPK), normally not found in neurons and glia, may lead to the formation of pathological cytoskeletal inclusions (Nakamura et al., 1998). The presence of transglutaminase-catalyzed epsilon N^e(γ-glutaminyl-) lysine cross-linked in the halo of LBs suggests that its activity leads to AS-aggregation (Junn and Mouradian, 2003; Junn et al., 2003). Recently, a novel constituent of several types of neuropathological inclusions, Ub-binding protein p62/sequestosome I, was identified (Kuusisto et al., 2001), which is also a major component of hepatocytic inclusions such as Mallory bodies (Stumptner et al., 2002). In normal physiology, p62 appears to function as a modulator of atypical protein kinase C (Wooten et al., 2001; Chang et al., 2002). Although its function in pathological aggregates is unknown, studies on neurofibrillary tangles (NFTs) and Mallory bodies indicate that it is incorporated relatively early during their formation (Kuusisto et al., 2002; Stumptner et al., 2002). There is evidence that aggregation-prone proteins are likely to recruit UPS components in an attempt to clear proteins from failing proteasomes. Furthermore, UCH-L1 accumulation is likely to play a pathological role in inclusion formation in PD (Ardley et al., 2004). There is a close relation between aggresomes—cytoplasmic inclusions formed as a cytoprotective response to sequester and degrade potentially toxic abnormal proteins—and LBs (McNaught et al., 2002b), and aggresomes formed by AS and synphylin-1 have been found to be cytoprotective (Tanaka et al., 2004). A decrease in proteosomal activity has been found in PD brains, whereas in rats, inhibition of proteasomes causes nigral degeneration with inclusion formation (McNaught and Jenner, 2001; McNaught et al., 2002a). Inhibition of proteasomal function or generation of misfolded proteins causes the formation of aggresome/LB-like inclusions and cytotoxicity in dopaminergic neurons in culture. Proteolytic stress and mitochondrial dysfunction, especially decreased activity of complex I, may reduce proteasome activity through oxidative modification and the aggregation with other oxidative proteins which may play a role in degeneration of dopaminergic neurons (Giasson and Lee, 2003; Maruyama et al., 2003; McNaught and Olanow, 2003; Tretter et al., 2004). Thus, LB formation may be an aggresome-related event in response to increasing levels of abnormal proteins in neurons (McNaught et al., 2002b). These processes are considered to form a complex cascade of interrelated events that lead to neuron death, e.g., by way of apoptosis. However, current views on pathogenic mechanisms in PD may not be as exact as commonly proposed, and further essential steps toward understanding pathogenesis are needed (Jenner and Olanow, 2006).

3.3 Components of Lewy Bodies

Both classical and cortical LBs share immunocytochemical characteristics (❯ *Table 15-3*). There are a number of commercial monoclonal antibodies to AS, including Syn-1 and LB509, which are useful for diagnostic immunohistochemistry. In addition to the particular antibody used, antigen retrieval methods have been shown to be particularly important in the detection of pathologic forms of AS. For example, formic acid and protease pretreatment of tissue sections enhances the signal-to-noise ratio (Takeda et al., 1998; Hamilton, 2000). This is related to the fact that AS is normally abundant in synaptic termini, and immunostaining of normal tissue for AS reveals extensive fine punctate immunoreactivity of gray matter. The aforementioned pretreatments abolish much or all of the normal AS and permit selective visualization of pathologic AS. Biochemical studies have demonstrated that this property is similar to the protease resistance that characterizes prion disorders (Kahle et al., 2001). They are labeled by antibodies to

◘ Table 15-3

Major biochemical components of LBs

α-Synuclein (major component)
α-B-crystallin
Ubiquitin
Phosphorylated neurofilament proteins
Synaptophysin
Chromogranin A
Synphylin
Synphylin-1
γ-Tubulin
P25α (tubulin binding protein)
Parkin
Pael-R (parkin-associated endothelin receptor-like receptors)
Calbindin
Torsin A
Gelsolin-related amyloid
Amyloid β-peptide (Aβ)
Amyloid precursor protein (APP)
Actin-like protein
Ubiquitin-pathway associated enzymes
α-B-crystallin
α-Microglobulin
Cu/Zn superoxide dismutase
Tau proteins
MAP-1B
MAP-2
MAP-9
Lipids
Calmodulin
Tubulin
Tyrosine hydroxylase
14-3-3 protein
Redox-active iron
Cytochrome c
Advanced glycation end products (AGE)
Dorfin, an E3 ubiquitin kinase p62 protein
Cyclin B
Redox-active iron
Vesicular monoamine transporter 2 (VMAT2)

neurofilaments, but most neurofilament antibodies decorate only a subset of LBs. Phosphorylated neurofilament proteins are present in both core and periphery. In addition, antibodies to neurofilament epitopes colocalize to Ub (77%–85% of anti-Ub positive LBs show double labeling with antiphosphorylated neurofilament antibodies) (Galvin et al., 1997).

LBs and related cytoplasmic inclusions express CDK5, a proline-directed protein kinase involved in cell-cycle regulation that probably catalyzes in vitro phosphorylation of neurofilament proteins.

LBs are structurally unstable. Their precise biochemical components are still unknown, but in addition to AS, Ub, and parkin, a large number of proteins have been identified (Lowe et al., 2002; Katsuse et al., 2003a). They include synaptophysin, chromogranin A, gelsolin-related amyloid protein, Aβ peptide (Aβ), actin-like protein, Ub-pathway associated enzymes (UCHI-1, proteases, ligases, kinases), α-microglobulin, immunoglobulin S, α-B-crystallin (in about 10% of cortical LBs), probably mediating the aggregation of microfilaments, Cu/Zn superoxide dismutase, cytosolic and microtubule-associated proteins including tau, MAP-1β, MAP-2, MAP-5 (Jensen et al., 2000), calbindin (CAB), γ-tubulin, TH, G-7 and G-9 proteins, lipids, and redox-active iron. LBs contain 14-3-3 protein, which is involved in numerous signal transduction pathways and interacts with AS (Kawamoto et al., 2002; Ubl et al., 2002; Berg et al., 2003) and torsin A, a protein with homologies to yeast heat-shock protein 104, which may serve as a chaperone for misfolded proteins that require refolding or degradation (McLean et al., 2002). In addition, proteasome subunits and hsp70 chaperone colocalize with AS in the fibrillar inclusions (❷ *Table 15-3*). However, they are absent in the protofibrils, suggesting that the cellular fibrillation also involves nonfibrillar intermediates and the microtubule-dependent inclusion process is required for the protofibril-to-fibril conversion in cells (Lee and Lee, 2002). Furthermore, immunoreactivity to cytochrome *c*, located in the inner membrane of mitochondria and released upon apoptotic stimuli into the cytoplasm, is detected in LBs. They further contain advanced glycation end products (AGEs), markers of transitional metal-induced oxidative stress that induce protein cross-linking and free radical formation (Munch et al., 2000), and cyclin B, the ectopic expression of which indicates changes in cell-cycle regulation through ERK activation leading to apoptosis of postmitotic neurons (Lee et al., 2003c). LBs and Lewy neurites in SN of PD and DLB are immunoreactive for vesicular monoamine transporter 2 (VMAT2), which is responsible for packing DA into vesicles and reducing the effects of neurotoxins by sequestering them into vesicles. Together with the presence of a few vesicle-linked proteins such as synaptophysin and chromogranin A, its involvement suggests the association of this protein with the degeneration of nigral neurons in these disorders (Yamamoto et al., 2006b).

The major structural component of LBs is AS, and in vitro studies have demonstrated that AS, in particular proteinase K-resistant AS (Neumann et al., 2004), aggregates form filaments similar to those seen in LBs. Extracted filaments from SNc in PD are labeled by antibodies against the C-terminal region of AS (Crowther et al., 2000). Multiple AS species including phosphorylated and C-terminal truncated species are present in both LBs and the soluble fraction from brains of patients with DLB. The enrichment of the C-terminal truncated and high-molecular mass forms raise the possibility that they might play a role in their pathogenesis (Chilcote et al., 2003). AS is more compact and is in closer association in LBs than in the neuropil, and its N terminus shows a close intermolecular interaction to Ub suggesting that AS adopts an altered three-dimensional structure and undergoes N-terminal ubiquitination (Sharma et al., 2001). The mechanism of its aggregation that may serve as a nidus for LB formation in vivo has not yet been clarified, while in vitro it is modulated by various factors, e.g., metals, with iron promoting and magnesium inhibiting aggregation (Golts et al., 2002). It has been shown that even small amounts of di- and trivalent metals, including iron, increase the rate of AS fibrillation (Uversky et al., 2001). LBs contain redox-active Fe (Castellani et al., 2000) suggesting that this mechanism may also occur in vivo. Recently, it was shown that in the presence of Fe, AS stimulated the production of hydrogen peroxide in vitro via Fenton chemistry (Turnbull et al., 2001), possibly via the binding of Fe(II) to the protein (Golts et al., 2002). While the relationship between the aggregation state of the protein and free radical production is unclear, this may represent an Fe-mediated mechanism in which hydroxyl radicals may be produced inside the nigral neurons. Oxidative stress may induce aggregation of AS (Hashimoto and Masliah, 1999), leading to fibril formation (Jellinger, 2002). Aggregated AS in diseased brains displays evidence of oxidative damage (Giasson et al., 2000), suggesting a mechanistic link between Fe, oxidative stress, protein aggregation, and cell death in PD and other synucleinopathies. Colocalization of AS and parkin carrying a Ub ligase protein (F3) activity of glycosylated AS and its E2-containing partner UbcH7 suggests that Parkin plays a role in the posttranslational procession and ubiquitination of AS and may be required during LB formation (Shimura et al., 2001; Schlossmacher et al., 2002). Several candidates have been identified as substrates of parkin and shown to be present in LBs: AS, p38 (Corti et al., 2003), and synaptotagmin XI (Huynh et al., 2003). Pael-R, a newly identified substrate for parkin, is localized in the core of LBs, while parkin and AS are accumulated in the halo, cell bodies, and processes. This suggests the involvement of Pael-R (parkin-associated endothelial receptor) in LB formation, and a protective role of parkin in Pael-R induced neurotoxicity in PD (Murakami et al., 2004). Engelender et al. (1999) found a novel protein, which they termed synphilin-1, interacting with AS in vivo using a yeast two-hybrid screen. Cotransfection of both proteins (but not control proteins) in HEK293 cells yielded cytoplasmic eosinophilic inclusions. Synphilin-1 is a synaptic vesicle protein and is also present in LBs (Wakabayashi et al., 2000a). Interestingly, synphilin-1 is a substrate of parkin and is polyubiquitinated by it (Chung et al., 2001b).

Both in vivo and in vitro experimental models are available to study neurodegenerative diseases which feature cellular inclusions, allowing to make inferences on their pathogenesis (Ciechanover and Brundin, 2003; Fornai et al., 2005). These studies and the presence of Ub and other specific proteins of the UPS in these inclusions reinforced the notion that the UPP plays a pivotal role in their formation in PD and other synucleinopathies (McNaught et al., 2003; Wenning and Jellinger, 2005). LBs, the morphological hallmark of PD and DLB, have a distinct central parkin and Ub-positive domain, with AS in the periphery (Gai et al., 2000), but it is incorporated into LBs and dystrophic neurites before ubiquitination. However, age-related defects in the 26/20S proteasome (McNaught et al., 2001) or the relentless production of abnormal proteins could exceed the degradation capacity of the UPP aggresome and cause poorly degraded proteins, promoting the formation of insoluble fibrillary inclusions in selected neurons, e.g., dopamine neurons in PD brain. While the activities of proteasome peptides in striatum are preserved in PD, but reduced in MSA (Furukawa et al., 2002), proteasome activator PA 28 is low in substantia nigra (SN) of PD brain, suggesting that failure in the UPS to degrade unwanted proteins may underlie its vulnerability and degeneration (McNaught et al., 2003). While the role of parkin, the inherited mutations of its encoding gene cause early onset parkinsonism, is unclear, recent studies suggest a progressive loss of parkin function in dopaminergic neurons during aging and sporadic PD, associated with E2 Ub ligase function (LaVoie et al., 2005).

Microtubule-associated protein 1B is also a component of cortical LBs and binds AS filaments (Jensen et al., 2000). Murine AS interacts with Tat-binding protein 1, a subunit of the 700-kDa proteasome activator (PA700) (Ghee et al., 2000).

Natively unfolded tubulin polymerization-promoting protein TPPP/p25 is enriched in filamentous AS-bearing LBs in PD and DLB, and in GCIs in MSA, but not associated with p-tau in inclusions in tauopathies and, thus, is considered a common marker of synucleinopathies (Kovacs et al., 2004).

AS binds to the C-terminal tail of the human dopamine transporter (DAT), and this AS–DAT complex formation facilitates membrane clustering of the transporter, thereby accelerating cellular DA uptake and DA-induced cellular apoptosis (Lee et al., 2001). This may be a mechanism of AS-mediated neurotoxicity, as the selective vulnerability of dopaminergic neurons in PD has been ascribed to oxidative stress as a result of cellular overaccumulation of DA or DA-like molecules. Since only the DA neurons contain significant amounts of DA, this has been hypothesized to account for the selective vulnerability of SN neurons. However, DA itself may not be toxic at physiologic relevant doses, so it is probable that other DA metabolites, e.g., the DA metabolite 3,4-dihydroxyphenylacetaldehyde, may play a major role in AS aggregation and provide a plausible link between DA production and metabolism, AS aggregation, and PD pathogenesis (Galvin, 2006).

Dorfin, a perinuclearly located E3 Ub ligase, with a distribution pattern parallel to that of Ub, has been shown to play a crucial role in the formation of inclusions in synucleinopathies and ALS by ubiquitination of synphilin-1 (Ito et al., 2003). However, it is not specifically implicated in the formation of different neuronal inclusions (Iseki et al., 2003), and direct binding of AS with dorfin has not yet been shown (Hishikawa et al., 2003). The absence of heparan sulfate polyglycans in LBs and LNs suggests that AS fibrillation and stabilization occur independently of glycosaminoglycans (van Horssen et al., 2004). Accumulation of axonal transport substances, like amyloid precursor protein (APP), chromogranin A, synphilin-1, and AS, in cortical LB stages 1–4 suggests that chronic axonal transport blockages is implicated in the development of LBs (Katsuse et al., 2003b). Synphilin-1, detected in normal human brain and in synucleinopathies (Murray et al., 2003), is an AS-interacting protein, the coexpression of which with Parkin in the central and peripheral parts of LBs (Chung et al., 2001b; Lee et al., 2002a) probably is an early requisite step prior to the subsequent but abundant trapping of AS (Junn and Mouradian, 2003). A ubiquitous posttranslational modification, such as casein kinase II-mediated phosphorylation, can regulate inclusion body formation in the context of AS and synphylin-1 interaction (Lee et al., 2004a). Inclusion bodies generated by tg mitochondrial cybrids, created by introducing the mitochondrial genes from PD patients into neuroblastoma cells that lack mitochondrial DNA, showed similar staining as LBs. They stained with resin, thioflavin S, and antibodies to AS, Ub, parkin, synphylin-1, neurofilament, β-tubulin, the proteasome, nitrotyrosine, and cytochrome *c* (Trimmer et al., 2004). Aggresomes formed by AS and synphilin-1 are cytoprotective (Tanaka et al., 2004). While the density of the major components of LBs (full-length AS with similar epitope mapping) suggests that a pathway leading from normal soluble to abnormal misfolded filamentous proteins is central for their pathogenesis, regardless of the primary disorder, there are conformational differences in AS between neuronal and glial aggregates showing nonuniform mapping for its epitopes. In DLB and PD brains, substantial amounts of detergent-soluble and insoluble AS were detected which differed from both controls and patients with MSA, the latter having significantly higher levels of soluble AS. These differences may result from different processing of AS in neurons and oligodendrocytes bearing the typical GCIs in MSA (Campbell et al., 2000, 2001). Detergent treatment of pathological MSA AS aggregates, but not recombinant AS, yields discrete AS-positive species with annular morphologies. The ability of the pathological AS to form annular aggregates may be an important factor contributing to the toxicity of the protein in disease that may have implications in designing therapeutic strategies aimed at detoxing AS aggregates (Pountney et al., 2004).

3.4 Formation of Lewy Bodies

Cortical LBs show diffuse AS-and Ub labeling; subcortical LBs have a distinct central Ub-positive domain, while AS primarily occurs in the periphery, suggesting that these two proteins do not have the same compartment, with ubiquitination being the later event (Gai et al., 1992). Recent studies divided AS-positive cortical LBs into six developmental stages (Katsuse et al., 2003b). Accumulation of AS was first observed as granular components in the neuronal cytoplasm (stage 1), then filamentous components became intermixed (stage 2), followed by accumulation of AS-forming round LBs composed of dense granulo filamentous components and spreading to their dendrites (stages 3 and 4). Then, the shape of LBs becomes deformed with a loose periphery (stage 5), and finally, they degrade to extracellular LBs composed of loose filamentous components with reduced immunoreactivity and involved astroglial processes (stage 6). These findings suggest that filamentous compartments are formed with the accumulation of AS during the development of cortical LBs. Some studies suggest homogenous deposition of AS and Ub in the central core of the LB (Galvin et al., 2001), whereas triple immunolabeling indicates heterogeneity, due to conformational changes revealing different epitopes in relation to LB evolution (Sakamoto et al., 2002).

Intraneuronal LB pathology begins in the axonal terminal before involving the cell body and finally the dendrite (Katsuse et al., 2003b). AS first accumulates in the neuronal cytoplasm without filamentous components. LBs and LNs then form, composed of granulo-filamentous components, before the inclusions are degraded to extracellular LBs after the death and disappearance of the involved neuron (Forno, 1996). The processes of tumor necrosis factor (TNF) α- and oxidase synthase-positive astroglia are in close association (Katsuse et al., 2003a).

The course of LB-related α-synucleinopathy in the brainstem, with initial formation of LNs but without LBs or neuronal cell loss, and later development of intraneuronal inclusions and neuronal degeneration (Braak et al., 2002, 2003; Jellinger, 2003a), is in line with increasing evidence supporting the central role for LNs in LB neurodegeneration and engenders a "neuritic dystrophic hypothesis" (Duda, 2004).

Recent studies on the morphogenesis of classical LBs suggest that punctate perikaryal AS, PBs, and LBs originate as successive stages of a complex aggregation process. Punctate AS-positivity appears in perikaryal areas as the earliest histologically observable sign of protein aggregation. An initial intraneuronal appearance of dustlike particles related to neuromelanin or lipofuscin is suggested to be cross-linked to AS (Fasano et al., 2003), with homogeneous deposition of AS and Ub in the central core (Galvin et al., 2001). The next step is the appearance of a larger body within the AS-containing granules, representing their coalescence into a clump-like mass whereby p62 is incorporated. These masses enlarge by assimilating the less aggregated AS, giving rise to what appears as PB. Concomitantly, immunoreactivity for Ub increases, likely due to the sequestration of Ub-conjugated proteins into the growing PB which may adopt varied morphologies, e.g., globular, arch-like, or irregular, possibly reflecting differences in the intraneuronal environment or the maturation stage. Starting peripherally, the PB undergoes progressive condensation, leading to the emergence of one or more spherical foci. By further condensation and incorporation of AS-positive material, these foci develop into classical LBs, while the remains of the PB eventually disappear (Kuusisto et al., 2003). Three-dimensional phase contrast radiography of LBs in midbrain of PD gave striking evidence that several LBs are agglomerated by dim edges in a neuron (Koh et al., 2006). In contrast to AD plaques, no evidence was found that inflammatory mechanisms (activated microglial cells or complement factors) are involved in cortical LB formation (Rozemuller et al., 2000).

3.5 Lewy Bodies and Tau

Tau is a soluble low-molecular weight cytoskeletal-associated protein that binds and stabilizes microtubules, promoting microtubule assembly and thereby influencing axonal transport (Bonini and Fortini, 2003). The function of tau is negatively regulated by phosphorylation and the tau protein in inclusions is abnormally phosphorylated, resulting in loss of function (the disruption of axonal transport) and gain-of-toxic function (formation of tau aggregates) at the same time. In the adult human brain, there are six tau isoforms that differ by the presence of either three (3-R) or four (4-R) C-terminal tandem repeated sequences of 31 or 32 amino acids, which are encoded by exons 9–12 (Goedert, 2003). Hyperphosphory-lated tau proteins aggregate within neurons as intracellular bundles referred to as paired helical filaments that may be involved in neuron dysfunction and death (Tolnay and Probst, 1999). Filamentous neuronal and glial tau inclusions associated with the degeneration of affected brain areas are the morphological hallmarks of "tauopathies," a group of heterogeneous dementias and movement disorders for which the intracellular accumulation of abnormal tau filaments appear to represent common mechanisms of disease (Goedert, 2003). The recent discovery of multiple mutations in MAPT, the gene that encodes tau, which are pathogenetic for a group of autosomal dominant multisystem neurodegenerative disorders, collectively referred to as frontotemporal dementia (FTD) and parkinsonism (FTDP-17) linked to chromosome 17, provided unequivocal evidence for the role of tau abnormalities in the onset and progression of neurode-generative disease (Spillantini et al., 1998). Several mutations of MAPT alter the biophysical properties of tau, leading to enhanced fibrillation in vitro, further supporting the gain-of-toxic-function hypothesis. The ability of many cofactors to induce tau fibrillation may partly explain the occurrence of tau pathology in many different neurodegenerative diseases including α-synucleinopathies.

Double immunostaining of LBs using antibodies to AS and a panel of monoclonal antibodies to phosphorylated and nonphosphorylated tau epitopes reveals tau immunoreactivity in LBs in the medulla of 80% of PD and DLB patients (Arima et al., 1999), in tg mice (Maries et al., 2003), and almost 100% in AD patients (Marui et al., 2004). Tau tends to coaggregate with AS in neurons most vulnerable to NFT formation, e.g., those of the LC and the basal nucleus of Meynert (NBM) (Ishizawa et al., 2003). The common co-occurrence of AS- and AD-type pathology suggests a combination of lesions related to protein dysmetabolism possibly with a synergistic protein–protein interaction.

Numerous genetic models of α-synucleinopathies and tauopathies were generated in a variety of species, ranging from nematodes and the fruit fly *Drosophila* to tg mice, by overexpressing wild-type or mutant tau protein and AS, or expressing proteins thought to be involved in the regulation of tau phosphorylation and LB formation (Jellinger, 2003f; Lauwers et al., 2003; Maries et al., 2003). Although they are incomplete replicas of either PD or LB disease, tg mice expressing mutant AS, recapitulate many features of LB disease, including AS fibrillation and aggregation in association with neurodegeneration and progressive motor dysfunction. In contrast, the use of combinatorial genetic models with APP and AS accelerates the formation of tau pathology, suggesting that tau may be part of a final common pathway for neurodegeneration (Lewis et al., 2001; Duda et al., 2002a, b; Giasson et al., 2003). The co-occurrence of A53T AS and tau neuropathology in mouse models suggests that mutations in the AS gene induce the pathological accumulations of tau observed in familial PD caused by the A53T AS mutation (Polymeropoulos et al., 1997; Duda et al., 2002b; Kotzbauer et al., 2005). Also mutations of the LRRK2 gene show phenotypic heterogeneity, including variable combinations of AS and tau pathology (Paisan-Ruiz et al., 2004; Wszolek et al., 2004; Zimprich et al., 2004).

The combination of tau and AS has been shown to systematically promote fibrillation of both proteins (Giasson et al., 2003). However, AS is likely to be only one of several cofactors as "pathological chaperones" that can induce the fibrillation of tau. Recent studies showed that AS induces polymerization of purified tubulin into microtubules. Mutant forms of AS lose this potential. The binding site of AS to tubulin was identified, and colocalization of AS with microtubules was shown in cultured cells. The demonstration of microtubule-polymerizing activity of AS suggests a striking resemblance between AS and tau; both have the same physiological function and pathological features, making abnormal structures in diseased brains known as synucleinopathies and tauopathies (Alim et al., 2004). Abnormal accumulation of AS is also associated with plaque formation in AD and with LB formation in PD, DLB, and AD, and recent studies in DLB showed that the presence of Aβ deposits in the cerebral cortex was associated with extensive AS lesions and higher levels of insoluble Aβ, suggesting that Aβ enhances the development of AS lesions in LB disease (Pletnikova et al., 2005). This may indicate some synergistic reactions between AS and Aβ, as has been suggested for AS and tau (Iseki et al., 2003; Wenning and Jellinger, 2005; Galpern and Lang, 2006), and points to a potential role for this molecule in synaptic damage and neurotoxicity through amyloid-like fibril formation and mitochondrial dysfunction, linking AS with the pathogenesis of these disorders (Hashimoto and Masliah, 1999). However, LBs in DLB and AD immunohistochemically showed different morphology in the coexistence with NFTs. Immunoelectron microscopy showed that LBs and LB-related neurites in DLB were composed of AS-positive Lewy filaments, while in AD they were formed by the aggregation of AS on the paired helical filament (PH). These findings suggest that Lewy pathology in AD may develop only in its most favorable region in PD and DLB, i.e., in the amygdala (Harding et al., 2002c; Popescu et al., 2004), and that it develops mainly in the neuronal cell body and not in the axonal terminal. It is further suggested that it may develop secondary to the development of Alzheimer's pathology with pathomechanisms different from that in DLB (Marui et al., 2004).

3.6 Dystrophic Neurites

LNs, coarse dystrophic neurites, are inclusions in the axonal processes of neurons showing similar immunohistochemistry (❷ *Figure 15-2d*). They are most frequent in the accessory cortical nuclei of the amygdala, CA 2/3 region of the hippocampus, in many brainstem nuclei, the olfactory bulb, intermediolateral columns of the spinal cord, sympathetic and parasympathetic ganglia, enteric, cardiac, and pelvic neurons, and the adrenal medulla, indicating involvement of multiple systems (Gai et al., 1995; Braak et al., 1999; Braak and Braak, 2000; Braak et al., 2003; Jellinger, 2003a). Absence of TH immunoreactivity suggests that many of these neuritic processes are not derived from dopaminergic neurons (Jellinger and Mizuno, 2003).

4 Lewy Bodies and Neuronal Cell Death

Deposits of insoluble proteinaceous fibrils may contribute to dysfunction or death of the involved cells (Trojanowski, 2003). However, the biological significance of these cytoplasmic inclusions, especially the role they may play in neurodegeneration, is still enigmatic. LBs, which are the sequelae of frustraneous proteolytic degradation of abnormal cytoskeletal elements, may represent—similar to other cellular inclusions, such as NFTs in AD and Pick bodies—end products or reactions to unknown neuronal degenerative processes associated with disturbances of axonal protein transport and which may lead to cell death. Current evidence suggests that complex I inhibitions may be a central cause of sporadic PD and that its derangement causes AS aggregation that contributes to the demise of neurons through impairment in protein handling and detoxification (Dauer and Przedborski, 2003; Dawson and Dawson, 2003; Tretter et al., 2004).

An unresolved question is whether the formation of LBs within neurons exerts a toxic or protective influence on the cell (Goldberg and Lansbury, 2000). It has long been discussed whether LB formation is a cause of nigral neuronal death or represents a survival effort of the neurons. The concentration and aggregation of proteins have been demonstrated in model systems to be detrimental to the cell (Feany and Bender, 2000; Ostrerova-Golts et al., 2000). On the other hand, there appears to be no correlation between the density of LB formation and cell loss (Gomez-Isla et al., 1999; Gomez-Tortosa et al., 1999; Henderson et al., 2000a), and the low number of neurons containing LBs in any brain region (less than 5% of the total neuronal number) would not be expected to result in a significantly altered synaptic function. Indeed, nuclear size, an indicator of RNA synthesis, does not vary in SN cells containing LBs compared to those without (Gertz et al., 1994), suggesting that LBs do not disturb cell metabolism. LBs have also been proposed to represent a protective reflex within the cell, since fibrillar inclusions may sequester toxic species, diverting AS from toxic assembly pathways (Goldberg and Lansbury, 2000), thereby protecting the cell (Saha et al., 2000). Fragmentation of Golgi apparatus seen in 5% of PD nigral neurons with LB, in 3% of these without LBs, and in 19% of neurons containing pale bodies suggests that the cytotoxicity of AS-positive aggregates is reduced in the process of LB formation (Fujita et al., 2006). Nevertheless, significant intracellular protein aggregation and LB formation are pathological processes, reflecting changes in the cellular environment.

LB formation may well be a survival effort of neurons by entrapping toxic proteins in an insoluble form, but there is no proof for this hypothesis. Rideout et al. (2001) addressed this question experimentally. They applied pharmacological inhibitors of the proteasome lactacystin, or ZIE(O-tBu)-A-leucinal (PSI), to dopaminergic PC12 cells. Proteasomal inhibition caused a dose-dependent increase in death of both naive and neuronally differentiated PC12 cells. A percentage of the surviving cells contained discrete cytoplasmic ubiquitinated inclusions. The ubiquitinated inclusions were present only within surviving cells, and their number was increased if cell death was prevented. Their results suggest that the formation of inclusion bodies is not the cause of cell death but rather the survival effort of the cells. Tanaka et al. (2004) also addressed this issue, forming aggresomes in 293T cells by coexpressing AS and synphilin-1. Inhibition of proteasome activity in these cells resulted in the formation of juxtanuclear aggregates with characteristics of aggresomes. Apoptotic cell death was correlated with the presence or absence of aggresomes in these cells. Aggresomes were found in 60% of nonapoptotic cells but in only 10% of apoptotic cells. They concluded that aggresome formation played a neuroprotective role. Thus, LB formation may also be a survival effort of neurons rather than a cause of nigral neuronal death in PD.

Neurodegeneration in LBD has been related to a cascade of multiple noxious factors, including formation of free radicals, lipid peroxidation, oxidative and proteolytic stress, protein–iron interaction, Fe-dysregulation, mitochondrial dysfunction and nuclear RNA deficits, disorders of calcium homeostasis, impaired bioenergetics, inhibition or loss of neuroprotective mechanisms, perturbations in protein degradation systems such as proteasome and autophagy, excitotoxicity from increased glutamatergic input to SN, or an interaction between these factors (❷ *Figure 15-3*) (see Betarbet et al., 2002; Jellinger and Mizuno, 2003; Jenner, 2003; Andersen, 2004; Emerit et al., 2004; Simon et al., 2004; Przedborski, 2005).

In PD, many biochemical changes indicate compromised antioxidant systems, suggested to underlie cellular vulnerability to progressive OS, which generates excessive ROS or free radicals selectively in SN with subsequent cell damage (Double et al., 2002; Dauer and Przedborski, 2003; Faucheux et al., 2003; Linert et al., 2006).

● Significant increase of iron in the SNp with a shift of $Fe(II):Fe(II)$ of 2:1 as compared to 1:2 in controls. Sequestration of redox-active iron and aberrant accumulation of ferric iron causing formation of OH radicals via the Fenton reaction suggest that the iron-catalyzed oxidative reaction plays a significant role in AS aggregation in vivo. Neuromelanin, a product of dopamine auto-oxidation, is capable of forming a complex with iron, thereby potentiating the generation of free radicals and the aggregation of AS. Loss of soluble AS, by its aggregation, can increase dopamine synthesis with accompanying increased generation of reactive metabolites, finally leading to degeneration (Double et al., 2002; Perez et al., 2002; Gotz et al., 2004).

● Sequestration of redox-active iron in LBs of PD SN substantiates the OS hypothesis, while the absence of redox-active iron in neocortical LBs highlights a fundamental difference between cortical and brain stem LBs (Castellani et al., 2000; Faucheux et al., 2003).

◻ Figure 15-3

Hypothetic scheme of pathogenesis of Parkinson's disease. Chr, chromosome; Uch-L1, ubiquitin pathway-associated enzyme; NO, nitric oxide; MPTP, 1-methyl-4-phenyl-1,2,3,6-tetrahydropyridine; 6-OHDA, 6-hydroxy-dopamine; TaClo, 1-trichloromethyl-1,2,3,4 tetra-hydro-β-carboline; Hyperphosph, hyperphosphorylated; GSH, glutathione; AGE, advanced glycation end products; ATP, adenosine triphosphate

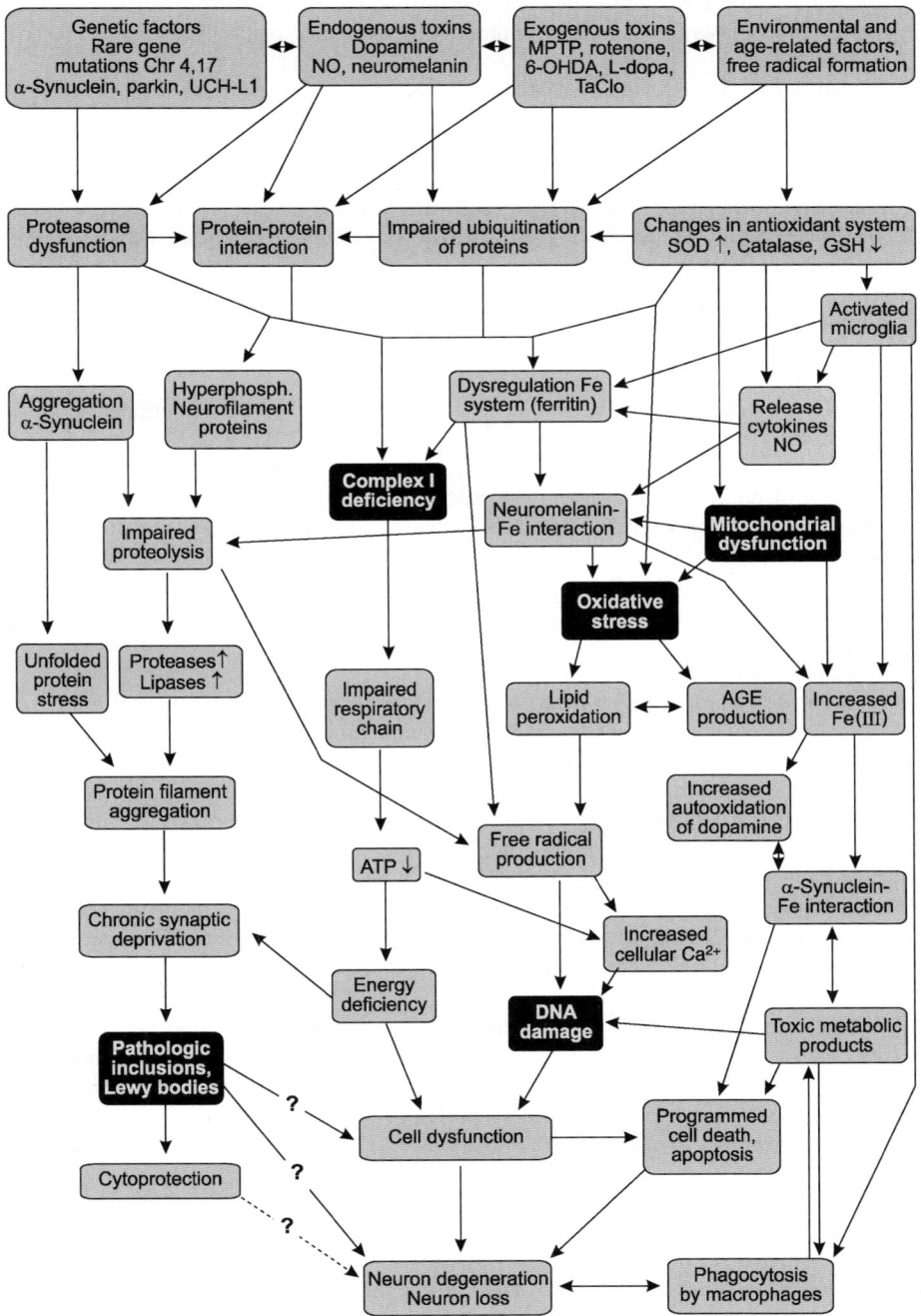

- Glutathione GSH, an important compound of antioxidative defense and protein repair, and glutathione peroxidase activity (destroys H_2O_2) are both decreased in SN of PD patients and in incidental LBD (preclinical PD), probably preceding both complex I and dopamine loss (Jha et al., 2000).

- Superoxide dismutase (SOD), an enzyme indicative of superoxide generation, shows increase of both isoenzymes (Cu-Zn-SOD and Mn-SOD) in PD SN suggesting increased superoxide generation. The recent finding of oxidative modification and aggregation of Cu-Zn-SOD in spontaneous AD and PD suggests that these disorders may share common or overlapping pathogenic mechanisms with ALS (Choi et al., 2005).

- Postmortem studies reported increased basal levels of thiobarbituric acid-reactive substances, which are secondary products of lipid peroxidation, in SN of PD coupled with a decrease in polyunsaturated fatty acids, the substrates for lipid peroxidation.

- Increase of intracellular 8-hydroxydeoxyguanosine (8-HOG) produced by free-radical damage in DNA in SN neurons corresponds to its degeneration pattern (Zhang et al., 1999). Increased peripheral 8-HOG levels in MSA, AD, and ALS suggest that systemic DNA/RNA oxidation is commonly observed in these diseases (Kikuchi et al., 2002).

- Nitric oxide (NO) as a free radical may induce increased lipid peroxidation, release of Fe2+, damage to DNA, inhibition of cytochrome c oxidase (COX) and SOD, and damage mitochondrial function by inhibiting complexes II, III, and IV and is often increased in several neurodegenerative diseases (Stewart and Heales, 2003).

- Peroxynitrite, formed by reduced SOD, induces aggregation of AS in situ and nitrated AS is found in the core of LBs (Fujiwara et al., 2002; Dauer and Przedborski, 2003) indicating its involvement in damaging structural proteins.

- Cross-linking of AS by AGEs has been observed in PD and in incidental LB disease suggesting that AGE-promoted LB formation may reflect early disease-specific changes rather than late epiphenomena (Munch et al., 2000). Widespread accumulation of nitrated AS in LBs provides evidence to directly linked oxidative and nitrative damage. Demonstration of hydroxynonenal (HNE) products in LBs indicates that peroxidation may play a critical role in their formation (Castellani et al., 2000). AS has been shown to produce neuronal death due to OS and promotion of mitochondrial defects (Hsu et al., 2000; Saha et al., 2000).

- Colocalization of AS and 3-nitrotyrosine (3-NT), a marker of protein nitration, through oxidative mechanisms has been observed in LBs and dystrophic neurites in DLB and GCIs in MSA, whereas most "pale bodies" and Lewy neurites in hippocampus lack 3-NT immunoreactivity, suggesting that nitration is not a prerequisite for AS deposition (Gomez-Tortosa et al., 2000).

- Genetic studies have implicated OS in PD pathogenesis indicating OS response (Bonifati et al., 2003; Canet-Aviles et al., 2004; Martinat et al., 2004; Taira et al., 2004; Kim et al., 2005).

- In the MPTP model of parkinsonism, ADP depletion and ROS overproduction occur soon after MPT (1-methyl-4-phenyl-1,2,3,6-tetrahydropyridine) application, subjecting the intoxicated cells to an early energy crisis and OS. Among the essential molecular pathways that are pivotal in triggering cell-death cascades, alterations in ATP synthesis and ROS production lead to the demise of the affected neurons (Przedborski et al., 2004).

- Chronic systemic treatment of rats with rotenone, a pesticide known to inhibit mitochondria, causes selective nigrostriatal dopaminergic degeneration with associated inclusions containing fibrillar AS (Betarbet et al., 2000). Rotenone treatment may induce an increase in oxidative stress in the dopaminergic neurons, which in turn may facilitate fibrillization of AS, providing a link between oxidative stress and pathogenesis of synucleinopathies (Giasson and Lee, 2000; Giasson et al., 2000).

Many of the above factors may participate in the pathogenesis of several experimental models of PD, indicating a multicomponent process involving OS, mitochondrial dysfunction, complex I inhibition, etc. (Dauer and Przedborski, 2003; Jellinger, 2003).

Although the causes of neuronal death in DLB are enigmatic, several mechanisms are being considered, including programmed versus passive cell death (i.e., apoptosis versus necrosis) and autophagy (see Jellinger, 2003c, d). All these forms appear to form part of a continuum between apoptosis and necrosis, depending on the severity of the insult and other circumstances. In human postmortem brains of patients with neurodegenerative disorders dying neurons are present, some of which display the features of apoptosis and increased expression of both pro- and antiapoptotic proteins indicating a disturbed balance between these death-related factors, which may be associated with incomplete cell-cycle activation in neurons (Jellinger, 2003c, d; Tatton et al., 2003). DNA fragmentation and definite histological features of apoptosis in neurons bearing LBs are extremely rare, whereas they are more frequent in glial cells (see Wullner et al., 1999; Jellinger, 2000, 2001, 2003c). There were no significant differences in the expression of most apoptosis-related proteins in PD, DLB, and controls between cortical and subcortical neurons with and without LBs. No activation of caspase-3, the effector enzyme in the apotosis cascade, or stress proteins was seen in involved neurons, whereas reactive astroglia and microglia showed upregulation of several apoptosis-related proteins and caspase-3. These and other data suggest that DNA fragmentation in neuro-degenerative disorders indicates programmed cell death, only exceptionally following the classic pathways of apoptosis but rather reflecting the combined action of DNA repair and accelerated DNA damage within susceptible neuron populations (Jellinger, 2001, 2003c, d; Graeber and Moran, 2002). Neurons in a proapoptotic environment may show increased vulnerability to metabolic disturbances related to pathogenetic factors suspected in these disorders, e.g., oxidative stress and mitochondrial impairment preceding DNA fragmentation, which may be early events in programmed cell death and are promoted by various triggers. Moreover, in most neurodegenerative disorders only a minority of cells with such inclusions display signs of programmed cell death, while there may be other forms of cell demise that are neither classical apoptosis nor necrosis with occasional autophagy. A block of the typical apoptotical demise may have profound implications in vivo, as persistence within the nervous system of damaged, but "undead" cells, followed by delayed lysis.

Whether programmed cell death or similar forms actually occur in human neurodegenerative disorders remains controversial, and the possibility has been neither confirmed by numerous studies nor definitely excluded (see Jellinger, 2000, 2001, 2003c, d; Przedborski, 2005). The role of various molecular processes in neurodegeneration, including abnormal protein degradation and aggregation, within the course of cellular dysfunction and death in these disorders needs to be further elucidated.

5 Lewy Body Disorders

LB-related disorders form a diverse group of neurodegenerative diseases representing diffuse proteinopathies caused by misfolded AS protein that forms amyloid-like filamentous aggregations in many brain regions. LBs, however, occur in many other neurodegeneraive disorders of various etiology (see ❷ *Table 15-1*). The distribution of LBs alone does not permit accurate prediction of the clinical presentation. Thus, an accurate clinicopathologic diagnosis of the LBDs at the present time requires clinical information (McKeith et al., 2004). In the absence of clinical information, a descriptive pathological diagnosis is all that is possible with certainty.

5.1 Parkinson's Disease

5.1.1 Definition

PD or brainstem type of LB disease is the most common neurodegenerative movement disorder in the elderly, with progressive degeneration of the dopaminergic nigrostriatal system and other neuronal networks caused by loss of pigmented neurons in the SNc and the association of many subcortical nuclei with widespread occurrence of LBs and LNs (see Jellinger and Mizuno, 2003; Jellinger, 2005).

Accepted clinical criteria for the diagnosis of possible and probable PD (Gelb et al., 1999) have a high sensitivity for detecting parkinsonism but have a specificity of only 75% for identifying idiopathic PD and to differentiate it from other LB disorders, particularly DLB (Litvan et al., 1998). For the diagnosis of definite PD, histopathological confirmation is required. Several clinicopathological studies have shown that LBD, including PD, represents 60%–83% of cases, whereas other degenerative disorders masquerading as PD, such as DLB, MSA, and progressive supranuclear palsy (PSP), account for 15%–30%. Awareness of the high rate of misdiagnosis and refinements in the clinical diagnostic criteria of PD seem to have improved accuracy of diagnosis to 85% (Hughes et al., 1992, 2001; Schrag et al., 2002). In this chapter, only the neuropathology of sporadic LB-related PD or brainstem type of LBD is reviewed (otherwise see chapter Riederer (pp. .....).

5.1.2 Neuropathology

Macroscopically, the brain is unremarkable or may show mild cortical atrophy and enlargement of the ventricular system. On cut surface, pallor of the SN and LC is evident. PD is histopathologically characterized by the presence of LBs and LNs in association with variable neuron loss in the midbrain and other subcortical nuclei. In addition to degeneration of the DA-containing neurons in the SNc, other cell groups including the LC and the NBM are affected.

There is severe depletion of melanized neurons (45%–66%) and of dopaminergic neurons immuno-reactive for TH (60%–85%) in the A9 group of SNc, particularly in the ventrolateral tier (area A, 91%–97%) projecting to the striatum, followed by the medioventral, dorsal, and lateral areas. Susceptibility of dopaminergic neurons depends on their distribution within compartments of SN defined by CAB immuno-reactivity. The CAB-rich matrix is separated from five CAB-poor zones or nigrosomes that show greater cell loss in the caudal and mediolateral region. From there it spreads to other nigrosomes and finally to the matrix along a caudo-rostral, latero-medial, and ventro-dorsal progression (Damier et al., 1999). This temporospatial disorder corresponds to a somatotopic pattern of dopaminergic terminal loss in striatum, which is more severe in the dorsal and caudal putamen than in the caudate nucleus (CN) (Kish et al., 1988). The degree of SNc cell loss and of the resulting reduction of TH and DAT immunoreactivity in the putamen followed by the CN and nucleus accumbens show close correlation to the duration and severity of motor dysfunction (Jellinger and Paulus, 1992; Ma et al., 1997; Zarow et al., 2003), and recent analysis of the degree of cell loss in brainstem catecholamine cell groups revealed that only the A9 SN had consistent significant cell loss early in the disease course with greater A9 cell loss correlating with increasing disease duration (Halliday et al., 2005).

The A10 group of dopaminergic neurons—ventral tegmental area (VTA), nucleus parabrachialis, and parabrachialis pigmentosus—projecting to cortical and limbic areas (mesocortic–limbic system) shows less severe involvement (40%–50% cell loss) (McRitchie et al., 1997), while the peri-retrorubral A8 region, which contains only few dopaminergic but CAB-rich neurons, and the central periventricular gray show little (20%–32%, cell loss in A8) or no degeneration. Cell depletion in these nuclei does not correlate with the duration of illness (Damier et al., 1999). With an inverse pattern to the contents of CAB and other calcium-binding proteins, SNc lesions in human PD are similar to those produced by the neurotoxin MPTP, one of the most widely used models of PD (Varastet et al., 1994), whereas others reported greater cell loss in LC than in SN in PD and AD (Zarow et al., 2003).

There is a similar distribution of reduced intensity of DAT mRNA in the remaining SNc neurons and decreased AS mRNA expression in SN and cortex of PD brains (Neystat et al., 1999), with loss of the vesicular monoamine transporter VMAT2 (a dopaminergic neuronal marker) in striatum, orbitofrontal cortex, and amygdala but not in the SN in early stages of PD (Ciliax et al., 1999).

Although LBs are not specific to PD and may occur in a variety of conditions, a positive diagnosis of PD can usually be made by inspecting two unilateral sections from the midpart of the SN and finding LBs. If no LBs are found, two further sections should be examined. If LBs are not seen in either SN or LC, then the diagnosis of PD of the LB type can be excluded. In case of cell loss from the SN and LC in the absence of LBs, an alternative cause of parkinsonism should be perused (Jellinger and Mizuno, 2003). Recent studies further demonstrated the early and widespread occurrence of argyrophilic, AS-positive, tau-negative glial inclusions in both oligodendroglia and astrocytes in PD brain (Arai et al., 1999; Hishikawa et al., 2001; Mochizuki et al., 2002).

Neuronal loss is accompanied by extracellular release of NM with uptake into macrophages, very rare neuronophagia or phagocytosis of neurons by macrophages, and increase of major histocompatibility complex (MHC) class II positive microglia which may release proinflammatory cytokines and other substances that mediate immune reactions. Large amounts of activated microglia and perivascular iron-laden macrophages, free melanin, and extraneuronal LBs are associated with reactive astrocytosis. These changes are either related to cell demise or suggest a role of neuroinflammation in the pathogenesis of PD (Orr et al., 2002; Hunot and Hirsch, 2003; Imamura et al., 2003; Herrera et al., 2005). They are actively associated with damaged neurons and neurites but may also represent a source of trophic factors with upregulation of neurotrophins in response to signals received from failing nigral neurons and may protect against reactive oxygen species and glutamate (Knott et al., 2002; Teismann et al., 2003b). In vivo demonstration of parallel changes in microglial activation and corresponding dopaminergic terminal loss in the affected nigrostriatal pathology in early PD by PET scan supports that neuroinflammatory response to intrinsic microglia contribute significantly to the progressive degenerative process of the disease (Ouchi et al., 2005).

5.1.3 Neuronal Vulnerability

Evidence suggests that selective vulnerability of SN neurons rich in NM and caspase-3, which have high expression of DAT mRNA (Uhl, 1998), is unrelated to their intrinsic capacity of DA synthesis (Kingsbury et al., 1999). They have low levels of calcium-binding proteins, which have a neuroprotective role by buffering effects of Ca^{2+} influx into cells, and weaker trophic support by neurotrophins, brain-derived neurotrophic factor (BDNF), and glycolytic enzymes which may trigger neurodegeneration (Howells et al., 2000). On the contrary, neurons in the subthalamic nucleus (STN) and GABAergic neurons in the substantia nigra pars recticulata (SNr) that are rich in calcium-binding proteins (calcineurin and paralbumin) are either not affected or involved only in terminal stages of PD. Decrease of caspase-3-positive pigmented SNc neurons in PD (reduction of 76% compared to controls) suggests that the distribution of this apoptosis-related enzyme may also contribute to their regional vulnerability (Hartmann et al., 2000). COX-2, the rate-limiting enzyme in prostaglandin synthesis, the inhibition of which prevents the formation of the oxidant species dopamine-quinone, is upregulated in many DA neurons of both PD and MPTP mice. This supports the critical role for COX-2 in both the pathogenesis and selectivity of the neurodegenerative process (Teismann et al., 2003a). The regional increase of 8-hydroxy-2-deoxy-guanidine (HOG) corresponding to the pattern of degeneration in PD suggests oxidative damage to neuronal cytoplasm, supporting the hypothesis that oxidative stress is a major pathogenic factor in PD (Betarbet et al., 2002; Jenner, 2003). In summary, the majority of midbrain neurons severely affected in PD are: (1) melanized cells located in the densely populated ventral tier of SNc; (2) rich in DAT and apoptosis-related enzymes but poor in glycolytic enzymes and calcium-binding proteins (e.g., CAB) and (3) arborize densely in the striatum and sparsely in the extrastriatal structures. Conversely, most dopaminergic neurons that resist degeneration in PD are: (1) located in the scantily populated dorsal tier of SNc; (2) contain CAB and glycolytic enzymes but are poor in DAT; and (3) arborize profusely in the extrastriatal components of the basal ganglia and sparsely in the striatum. Dopaminergic neurogenesis, intracellular and extracellular substances, and interactions among these factors have been discussed as underlying causes of selective death of dopaminergic neurons in PD (Barzilai and Melamed, 2003). Furthermore, local dysfunction of the BBB in the midbrain may be a causal factor in PD (Kortekaas et al., 2005).

Recent studies showed that early brainstem cell loss in PD is mainly confined to the A9 neurons in the SN and associated with more widespread LB formation, consistent with current diagnostic criteria for PD (Gelb et al., 1999). In this region of greatest vulnerability to PD, intracellular NM lipid changes occur prior to evidence of LB formation, indicating an early pathogenic event. Changes in intracellular mechanisms are considered critical to the neurodegeneration observed in PD. Recent data suggest an integral change in NM as a prelude to PD neurodegeneration in vulnerable A9 neurons. The increased concentration of neuronal AS and NM in A9 neurons may predispose these neurons to precipitate AS around NM lipid under oxidative conditions. These changes may trigger a cascade of events leading to larger intracellular aggregates of AS and the dispersement of NM within the cell. A potential key component to these changes may be the observed increase in iron in NM in PD, which is thought to increase oxidative load within the microenvironment of the pigment granule (Double et al., 2002; Faucheux et al., 2003; Zecca et al., 2003). In the remaining healthy A9 neurons (no change in their size or NM amount), unstained intracellular NM has an increased optical density which is associated with a loss of cholesterol and an aggregation of AS to NM lipid, while a proportion of A9 neurons are more affected with large AS accumulations in LBs; at early disease stages LBs are not confined to the A9 cell group with substantial cell loss, and also occur in a number of reasonably preserved brain regions, including the amygdala which is affected early in PD (Harding et al., 2002b; Braak et al., 2003). No changes in cell size and pigment density and intracellular changes in AS location and lipid components were observed in healthy A10 neurons in PD compared with controls. Pigmented A9 neurons in later stages of degeneration with obvious LB formation had a significant reduction in intracellular pigment. A9 neurons of normal morphological appearance and no characteristic pathology in PD exhibited significantly increased pigment density associated with a concentration of AS to the lipid component of the pigment and a loss of associated cholesterol. These changes in vulnerable but apparently healthy A9 neurons occurred without any change in cell size or in the amount of intracellular pigment compared with controls (Halliday et al., 2005). The increase in NM density is consistent with previously reported increases in oxidation and iron-loading reactions known to precipitate AS (Faucheux et al., 2003; Zecca et al., 2003). Whereas other NM-containing brainstem neurons (A6 and A2) are relatively preserved in early PD and do not exhibit the same intracellular changes seen in the A9 cells (Halliday et al., 2005). The selectivity of the early NM lipid changes in A9 neurons are likely to be related to their early vulnerability in PD.

5.1.4 Development of LB-Related Pathology

In incidental LBD, suggested to represent a sub/preclinical form of PD, AS-immunoreactive LNs and LBs have been demonstrated in many reticular and raphe brainstem nuclei, the ceruleus–subceruleus and glossopharyngeus–vagus complex, and the olfactory bulb in the absence of both nigral changes and sensory motor PD symptoms, suggesting early involvement of these systems with ascending progression (Del Tredici et al., 2002). A recently proposed staging of brain pathology related to sporadic/idiopathic PD (Braak et al., 2002, 2003, 2004, 2006) distinguishes presymptomatic stages 1 and 2 with LB pathology confined to the "gain setting system" of the lower brainstem (Holstege, 1996) and olfactory bulb, with the NBM, SNc, and other regions being preserved. In stage 3, the LC, the central nucleus of the amygdala, the nuclei of the basal forebrain, and the posterolateral and posteromedial SNc are the focus of initally subtle and, then, severe cytoskeletal changes and neuronal depletion, while the cortex is preserved. In stage 4, the anteromedial temporal mesocortex and neocortex are additionally affected. Stages 3 and 4 can be correlated with clinically symptomatic stages, while in terminal stages 5 and 6, the pathological process reaches the telencephalic cortex, first affecting the sensory association cortex and prefrontal areas, later progressing to the primary sensory and motor areas or involving almost the whole neocortex (❷ *Figure 15-2*). The initial lesions in the lower brainstem and olfactory system preceding classical PD-related pathology may explain early autonomic and olfactory impairment that may precede the somatomotor dysfunctions (Przuntek et al., 2004). Recent studies largely confirmed this staging, showing that all brains of subjects with clinical PD showed AS-positive inclusions and neuronal losses in medullary and pontine nuclei and SN, with additional lesions in NBM (90%), limbic cortex (60%), cingulate area (46%), CA 2/3 hippocampal area (36%), isocortex (29%), and amygdala (25%), corresponding to pathology stages 4–6 (Jellinger, 2003a, 2004b). Recent studies of 21 PD patients by six observers from five different institutions revealed highly significant inter- and intra-rater reliability, and supported the suitability of the staging procedure of AS pathology for application in routine neuropathology and brain banking (Müller et al., 2005). However, some of the clinical PD cases corresponded to Braak stage 3, often considered as "preclinical," while others without demonstrable involvement of medullary nuclei showed extensive LB-pathology in other brain areas, suggesting deviations from the proposed stereotypic expansion pattern, while incidental LB pathology may affect solely the LC and SN (Jellinger, 2003a; Parkkinen et al., 2003). Preclinical cases might have developed extremely late-onset PD compared with other cases (Halliday et al., 2006). The late stages 5 and 6 of LB pathology (involvement of sensory association and prefrontal areas, later of primary sensory and motor areas) suggest overlap or transition between PD and DLB (Jellinger, 2003a, 2004a; Saito et al., 2004) (❷ *Figure 15-4*).

■ Figure 15-4

Progress and distribution pattern of PD-related neuronal pathology. ab, accessory basal nucleus of amygdala; ac, accessory cortical nucleus of amygdala; ad, anterodorsal nucleus of amygdala; am, anteromedial nucleus of thalamus; ba, basal nucleus of amygdala; bn, basal nucleus of Meynert; ca, caudate nucleus; ca1, first Ammon's horn sector; ca2, second Ammon's horn sector; cc, corpus callosum; ce, central nuclei of amygdala; cg, central gray of mesencephalon; cl, claustrum; cm, centrum medianum; co, cortical nuclei of amygdala; cr, central nucleus of raphe; db, nucleus of diagonal band; dm, dorsomedial hypothalamic nucleus; dr, dorsal nucleus of raphe; ds, decussation of superior cerebellar peduncles; dv, dorsal nuclear complex of vagal nerve; en, entorhinal region; fn, facial motor nucleus; fo, fornix; gi, gigantocellular reticular nucleus; gr, granular nucleus of amygdala; hn, hypoglossal motor nucleus; in, infundibular nucleus; ir, intermediate reticular zone; la, lateral nucleus of amygdala; lc, locus ceruleus; ld, laterodorsal nucleus of thalamus; lg, lateral geniculate body; li, nucleus limitans thalami; lt, lateral nuclei of thalamus; mb, medial lemniscus; md, mediodorsal nuclei of thalamus; me, medial nuclei of amygdala; mf, medial longitudinal fascicle; mg, medial geniculate body; ml, medial lemniscus; mm, medial mamillary nucleus; ms, medial septal nucleus; mt, mamillothalamic tract; mv, dorsal motor nucleus of vagal nerve; os, oliva superior; ot, optic tract; pa, anterior pallidum, external pallidum; pf, parafascicular nucleus; ph, posterior hypothalamic nucleus; pi, internal pallidum; po, periventricular nuclei of thalamus; pr, prepositus nucleus; pu, putamen; pv, paraventricular nucleus; re, reticular nucleus of thalamus; rh, reticular nuclei of thalamus; ru, nucleus ruber; sb, subiculum; sc, superior cerebellar peduncle; sn, substantia nigra; sn, subthalamic nucleus; so, supraoptic nucleus; sp, subpeduncular nucleus; st, nucleus of stria terminalis; su, subthalamic nucleus; te, transentorhinal region; tl, lateral tuberal nucleus; tm, tuberomamillary nucleus; tp, tegmental pedunculopontine nucleus; vm, ventromedial hypothalamic nucleus; vt, dopaminergic nuclei of ventral tegmentum (paranigral nucleus and pigmented parabrachial nucleus); zi, zona incerta

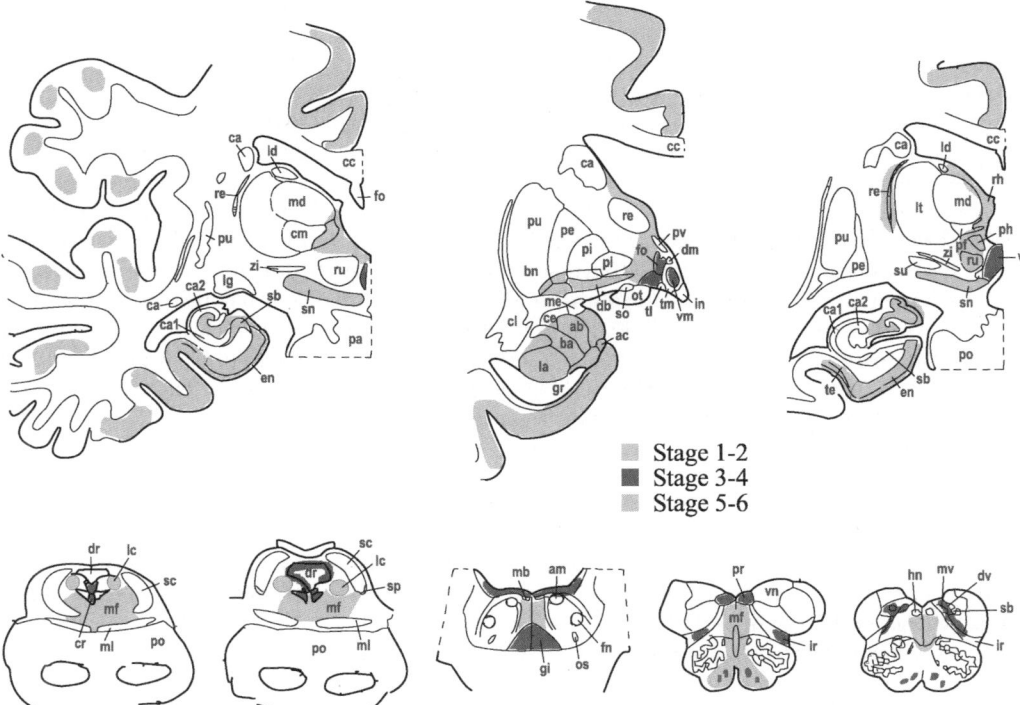

5.1.5 Progression of Dopaminergic Lesions

Dopaminergic fibers, originating from different cell groups in the ventral mesencephalon (i.e., SNc [A9 cell group], VTA, and the retrorubral area [A8]), strongly innervate the striatum and the frontal cortex, including the prefrontal cortex and the motor and premotor areas (Berger et al., 1991). The degeneration of the nigrostriatal system causes DA denervation in the striatum progressing from the ventrorostral to the posterior putamen and CN. It has been estimated that the onset of clinical motor symptoms in PD occurs with loss of about 50% of SNc neurons, reduction of striatal DA uptake by 57%–80%, and DAT loss of 56% (Bernheimer et al., 1973; Morrish et al., 1998; Rinne et al., 2001). These changes are preceeded by a preclinical phase, the duration and progression of which are still under discussion. From single-photon-emission computer tomography (SPECT) and PET studies, the estimated preclinical periods range from 4.6 to 6.6 years, with an annual decline of striatal DA uptake of 8%–10% and of DAT between 5.7 and 6.4 years or 10%–13%, while other β-CIT SPECT data were not suggestive of substantial changes of striatal DA in the first 5–7 years after symptom onset (Pirker et al., 2003). Reduction of striatal DA by 57%–80% and DAT loss of 56% cause motor symptoms (Rinne et al., 2001; Nurmi et al., 2003). Thus, about 50% of dopaminergic striatal innervation appears sufficient for normal motor function. PET studies indicate a slower progression of striatal and SN DA storage (^{18}F-dopa uptake) with ropinirol versus levodopa administration (Rakshi et al., 2002; Brooks et al., 2003; Whone et al., 2003). However, medullary structures as the induction site of AS-SN pathology have been questioned, as large parts of the nervous system including the spinal cord and the peripheral autonomic system have not been examined in incidental LB disease. Dopaminergic denervation of the striatum causes severe loss of dendrites on type I medium spiny neurons, the principal source of dopaminergic input from SN (McNeill et al., 1988). Dysfunction of neostriatal medium spiny neurons (MSNs) may underlie late-stage motor complications of PD. Shortened dendrite length of MSNs was similar in four regions of neostriatum in late-stage PD, with the greatest loss in caudal putamen. These structural changes in MSN may contribute to late-stage motor complications of PD (Zaja-Milatovic et al., 2005). Together with abundant AS pathology in striatum (Duda et al., 2002a, b), ultrastructural findings in PD (Lach et al., 1992), and progressive loss of TH- and DAT-immunoreactive nigrostriatal fibers, this suggests transsynaptic degeneration as a possible substrate for the severity of motor deficits and decreased efficacy of dopaminomimetic treatment in late stages of PD (Ito et al., 1996). Highly selective loss of pyramidal neurons in the presupplementary motor area (SMA), along with dopaminergic basal ganglia dysfunction (MacDonald and Halliday, 2002) is associated with progressive loss of D_2 receptors in the prefrontal cortex (Kaasinen et al., 2003). Longitudinal evaluation of cerebral morphological changes in PD using voxel-based morphometry analysis by MRI showed progressive volume decrease in limbic, paralimbic, and neocortical associative temporooccipital regions in PD patients with no dementia, and particularly involvement of neocortical regions in PDD (Ramirez-Ruiz et al., 2005). Occipital hypoperfusion in PD without dementia correlates with impaired cortical visual processing (Abe et al., 2003) and posterior cingulate metabolic changes occur in PD patients without dementia (Camicioli et al., 2004).

5.1.6 Symptom-Related Lesion Pattern

PD shows a marked heterogeneity in the clinical phenotype (Foltynie et al., 2002). The major clinical subtypes of PD show specific morphologic patterns of pathophysiological importance (Jellinger, 2002, 2004a; Jellinger and Mizuno, 2003).

In the "akinetic-rigid type", accounting for about 50% of all patients, the ventrolateral SNc which projects to the dorsal putamen degenerates more severely than the medial part projecting to the CN and anterior putamen. There is a ventromedial gradient loss of TH- and DAT-immunoreactive fibers and endings (both frequently colocalized in synaptic vesicles and plasma membranes) progressing from dorsal to ventral putamen (Morrish et al., 1996), with predominant involvement of the metenkephalin- and substance P-rich acetylcholinesterase (AChE)-poor striosomes of putamen projecting to the severely involved ventrolateral SNc. Preservation of the CAB-positive somatostatin-rich matrix, showing increased somatostatin mRNA expression (Eve et al., 1997) and projecting to γ-amino butyric acid (GABA) neurons of SNr and motor thalamus, and of periventricular islands of the CN and nucleus accumbens suggests that the endings richest in DAT are most sensitive to degeneration (Miller et al., 1997). Studies of CN biopsies in PD patients have found reduced TH immunostaining and revealed differences in substance P and metenkephalin levels, with both being normal or variably reduced; low metenkephalin immunostaining correlated with the severity of clinical motor symptoms (De Ceballos and Lopez-Lozano, 1999). Reduced dopaminergic input to the putamen causes increased activity of the GABAergic "indirect" striatal efferent loop via SNr and internal globus pallidus (Gpi) to the ventrolateral thalamus projecting to the cortex (❷ *Figure 15-5*). The excessive excitatory glutamatergic drive from the STN and the Gpi/SNr leads to an akinetic-rigid syndrome via reduced cortical activation. Bradykinesia is functionally related to impairment of cortical and subcortical systems that regulate kinematic parameters of movement, and overactivation of prefrontal cortical areas as compensatory mechanisms and impaired

❏ Figure 15-5

Schematic diagram of basal ganglia-thalamocortical circuitry under normal conditions and in Parkinson disease. Cortex: SM, supplementary motor field; PM, premotor field; MR, motor cortex; GP, postcentral gyrus; GC, gyrus cinguli; HI, hippocampus; CE, entorhinal cortex. Basal ganglia: SNpc, SN zona compacta; SNpr, SN reticulata; STN, subthalamic nucleus; A8, retrorubral field; GPi, internal globus pallidus; GPe, external globus pallidus; PAN, pedunculopontine nucleus; VL, VLM ventrolateral/medial thalamus; Vm, medioventral thalamus; CS, superior colliculus; MPT, mesopontine tegmentum; ACh, acetylcholine; SP, substance P; ENK, enkephalin; DA, dopamine; 1, nigrostriatal dopaminergic pathway; 2, striato-nigral pathway; 3, "indirect" loop; 4, "direct" loop; 5, motor or complex loop; 6, thalamocortical pathway; 7, pallido-subthalamic pathway; +, excitatory, −, inhibitory

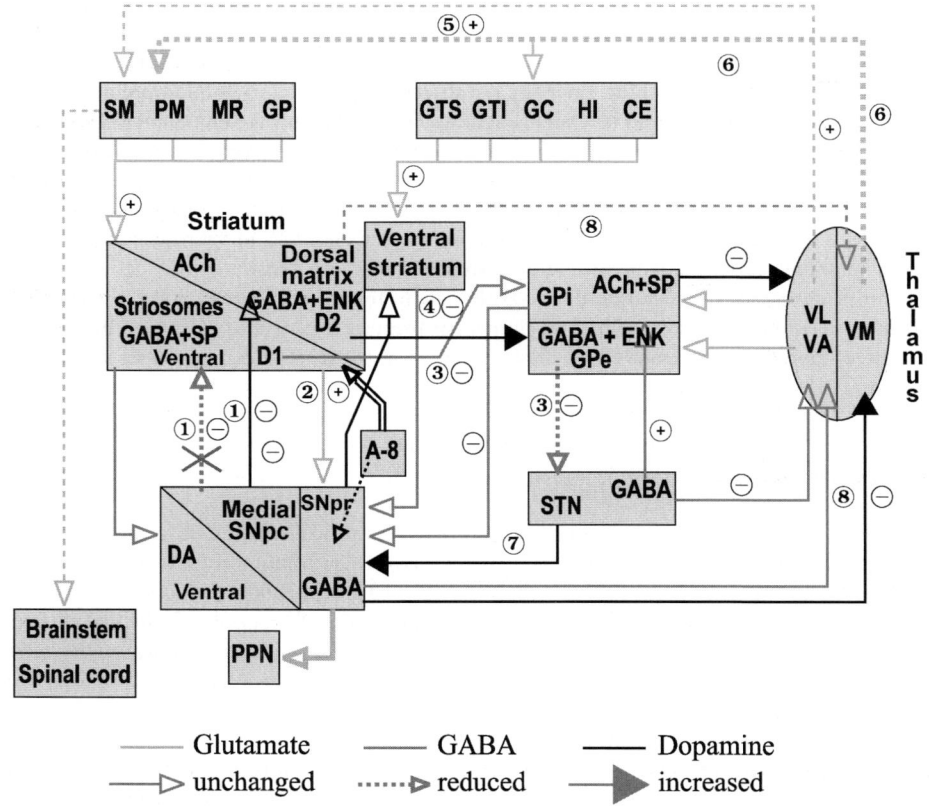

suppression (Turner et al., 2003). Increased GABAergic activity reduced by levodopa treatment disappears in the course of disease and may induce changes in N-methyl-D-aspartate (NMDA) receptors and glutamatergic synapses, favoring motor complications (Chase and Oh, 2000). The uncoupling of receptor systems is considered a major cause of drug resistance and adverse levodopa effects (Bezard et al., 2001). Free endogenous DA may induce a relative hyperstimulation of dopaminergic receptors. Subsequent internalization of DA receptors in response to excess stimulation may account for the development of motor fluctuations and dyskinesias after levodopa treatment, which also have been related to increased proproenkephalin (PPE)-B mRNA levels in the striatum (Calon et al., 2002; Henry et al., 2003).

The "tremor-dominant" type of PD, occurring in about 25% of patients, shows less severe total cell loss and less severe depletion of the lateral SNc, but damage to the retrorubral A8 field that is usually preserved in akinetic-rigid PD (McRitchie et al., 1997; Damier et al., 1999). It projects to the matrix of the dorsolateral striatum and ventromedial thalamus, influencing the striatal efflux via SNc and thalamus to the prefrontal cortex (❷ *Figure 15-3*). Functional neuroimaging in resting tremor suggests increased activity of the ventral intermediate thalamus (VIM) and dysfunction of cerebellar connections (Deuschl et al., 2000; Lozza et al., 2002; Timmermann et al., 2003), which has considerable implications for stereotactic treatment of tremor (deep stimulation of VIM) (see Deuschl and Volkmann, 2002; Rehncrona et al., 2003).

Insular LB burden has been associated with orthostatic hypotension in PD (Papapetropoulos and Mash, 2006).

5.1.7 Involvement of Extranigral Systems

Parkinson's disease is a multisystem disorder with involvement of many extranigral systems. Lesions in PD are not random, but region specific, and not all neurons of a given transmitter type are vulnerable to LBs. The following neuronal systems are involved (see Jellinger, 2002).

Mesocortical dopaminergic system: Degeneration of dopaminergic fibers originating from the medial SNc, VTA, and retrorubral area, which innervate the striatum, prefrontal motor and premotor areas, and thalamus, causes dysfunction of both the basal ganglia and cerebral cortex (reduction of TH-immunoreactivity in the prefrontal cortex). It has been related to cognitive and behavioral impairment in PD.

Noradrenergic system: LC, the main source of noradrenergic innervation of brainstem, hippocampus, and neocortex, showing cell loss of 40%–50% in PD, is more severely affected in patients with depression and dementia or both and can approach the loss seen in AD. Cell depletion in the rostral part of LC, which projects to temporal cortex and hippocampus, correlates with the density of AD pathology, possibly consistent with retrograde degeneration due to pathology in their cortical target areas, while LC cell loss in PD without relationship to cortical pathology suggests primary degeneration (Hoogendijk et al., 1995). Deprivation of noradrenergic innervation has been related to depression, dementia, and autonomic dysfunction (Gerlach et al., 2003). The noradrenergic dorsal vagal nucleus shows little changes in PD (5%–17% cell loss), but early and frequent involvement by AS pathology with severe loss of substance P-immunoreactive neurons and little loss of noradrenergic cells (Halliday et al., 1990). The adrenergic nuclei A1 and A2 in the medulla oblongta suffer no cell loss but depletion of noradrenalin synthesizing cells in the C-11 area (Gai et al., 1994).

Serotonergic system: The dorsal raphe nucleus or nucleus supratrochlearis and the central raphe nucleus, giving rise to ascending serotonergic pathways, suffer loss of TH-immunoreactive neurons. While its phenylalanine hydroxylase (PH-8)-positive cells are unaffected, there is 50% reduction of PH-8 serotonin-synthesizing cells in caudal midbrain and pons (Halliday et al., 1990), causing reduction of serotonin and its receptors in the striatum and frontal cortex. PET and SPECT studies confirmed the reduction of serotonin transporters in both striatum and midbrain, which are not affected in early stages of PD (Kerenyi et al., 2003; Kim et al., 2003a).

Cholinergic system: The magnocellular part of the NBM, projecting to neocortex and other cholinergic brainstem nuclei, in PD shows an average cell loss of 30%–40%, not correlating with age or duration of illness. Neuronal loss is less severe in PD without dementia and aged controls (0%–40%) than in AD (50%–70%) and most severe in Parkinsons's disease with dementia (PDD) (56%–90%) with decrease of cholinergic innervation of the cortex and hippocampus and often accompanied by severe cortical AD lesions. This suggests a critical threshold of NBM cell loss with equivalent cortical cholinergic denervation before dementia becomes apparent. Considerable variability of both NBM cell loss and decrease of cholinergic cortical markers, irrespective of cortical lesions, suggest primary degeneration of the cholinergic forebrain system in PD (Jellinger, 2002), whereas in vivo investigations of acetylcholine activity in the cortex and amygdala with preserved activity and glucose consumption in the NBM suggest secondary (retrograde) degeneration (Herholz et al., 2004), or as shown in AD, due to defective retrograde transport of NGF (Mufson et al., 1995).

The 'nucleus tegmenti pedunculopontinus' pars compacta (PPNc), a cholinergic loop nucleus in the dorsolateral part of the caudal mesencephalic tegmentum and center for balance between cholinergic and dopaminergic basal ganglia functions in PD suffers 36%–57% cell loss, strongly correlated with SN cell depletion but not with the patient's age, duration of illness, and LB counts (Jellinger and Mizuno, 2003). Unaltered parameters in the thalamus and STN suggest that the PPNc lesion in PD is a retrograde degeneration rather than a part of systemic degeneration. Overactivity of the PPNc in MPTP parkinsonism indicates dysfunction in the tegmento-nigro-subthalamic-cortical loops, contributing to disorders of locomotion, gait, and posture, the sleeping–waking cycle, and cognitive impairment (Parent and Cossette, 2001; Parent et al., 2001). Significant thalamic presynaptic cholinergic deficits have been observed in cases of combined cortical and subcortical neurodegeneration in which dementia developed after prolonged parkinsonism (Ziabreva et al., 2006).

The "Westphal-Edinger nucleus", a visceral subdividision of the oculomotor complex giving rise to cholinergic fibers to the ciliary ganglion regulating pupilloconstriction, suffers 55% neuronal loss in PD which, together with damage to nuclei in the periaqueductal gray and nucleus interstitialis of Cajal, explains neuro-ophthalmic and REM sleep dysfunctions in PD.

Pathological lesions in the 'amygdala', which is interconnected with the prefrontal cortex, hippocampus, basal forebrain, and brainstem, regulating behavioral and autonomic functions, in PD mainly involve the accessory cortical and lateral nuclei compared with the basal and lateral nuclei which are involved more in AD (Braak et al., 1996; Harding et al., 2002c). They may be involved in endocrine and autonomic dysfunctions, with little impact on mental impairment in PD.

Other systems involved in PD are the reticular brainstem nuclei controlling somatomotor and autonomic systems (Benarroch et al., 2000; Braak et al., 2003), posterolateral hypothalamus, the center median-parafascicular thalamus (Henderson et al., 2000a, b), the intralaminar thalamic nuclei (Bacci et al., 2004), and the intermediolateral nuclei and Clark's columns in the spinal cord.

The *histaminergic* fibers in the SN are increased but are thinned due to enlarged vesicles that compensate the deficit of DA or a putative growth inhibitory factor (Anichtchik et al., 2000), with increased histamine but not its metabolites found in PD striatum and SN, which is not seen in MSA (Rinne et al., 2002).

The *GABAergic system* shows reduced activity of glutamate decarboxylase (GAD) in basal ganglia and cerebral cortex, a 50% reduction of GAD mRNA in Gpe which exerts inhibitory influence on the SN (Nisbet et al., 1996), and reduction of GABA receptors due to degeneration of dopaminergic neurons which increases GABAergic activity in early stages of PD but disappears in the course of illness and introduction of levodopa therapy.

Involvement of the *peptidergic system* is variable, with reduction of enkephalin, somatostation, and neuropeptide receptors in basal ganglia, but increase of somatostatin and mRNA expression, reduced substance P in the striatum, mild reduction of GABA in the thalamus, and decrease of NMDA receptors in CN (Gerlach et al., 2003), while increased synthesis of PPE in striatal neurons is related to levodopa-induced complications (Chase and Oh, 2000).

The decrease of dopamine content in the SNc in PD disrupts the delicate balance between neurotransmitters, e.g., noradrenaline, serotonin (5-HT), acetylcholine, GABA, and glutamate, and neuropeptides, e.g., enkephalins and substance P, in the brain, especially within the basal ganglia circuit. This causes a variety of neurotransmission changes that eventually influence the motor cortex, and as a result affect movement control and autonomic functions. In addition, levodopa and dopamine agonists might also alter mRNA transcripts of neurotransmitters and neuropeptides in the basal ganglia, thus contributing to the imbalance in neurotransmission (Gilgun-Sherki et al., 2004).

5.1.8 Distinction Between PD and Aging Brain

The distribution of SNc differs from age-related lesions preferentially involving the dorsal tier of SNc that is involved only in later stages of PD (Fearnley and Lees, 1994). Recent studies showed 35%–41% reduction in the total numbers of pigmented SN cells with severe loss of DAT-immunoreactive neurons in the aged (Ma et al., 1999a), which may be compensated by neuronal hypertrophy (Cabello et al., 2002). The age-related neuronal decrease was estimated at 5.8% per decade in total cell numbers, 7.4% in cell density, and 4.4% in neuronal volume (Ma et al., 1999b). However, another study did not detect any significant loss of TH-immunoreactive neurons in the SN or other midbrain nuclei in very old subjects (Kubis et al., 2000). While NM shows a continuous acceleration in SNc with aging, in PD it is severely depleted compared to age-matched controls, which correlates with dropout of SNc neurons (Zecca et al., 2001).

5.1.9 Pathology of Dementia in PD

Cognitive impairment in PD with an average prevalence and incidence of 40% (see Aarsland, 2006) may be caused by four types of changes (❷ *Table 15-4*): pathology in subcortical structures, degeneration of the medial SN and nuclei of other ascending pathways, cortical and hippocampal LB pathology, and coincident

❏ Table 15-4
Brain lesions associated with mental impairment in parkinsonism (from Jellinger, 2000)

Dysfunction of subcortico-cortical neuronal systems
Degeneration of nigrostriatal dopaminergic pathway → deafferentation of strio-(pre)frontal loops
Cell loss in medial substantia nigra and VTA → mesocorticolimbic dopamine deficiency
Degeneration of noradrenergic systems (locus ceruleus)
Degeneration of serotonergic systems (dorsal raphe nuclei)
Degeneration of cholinergic systems (nucleus basalis of Meynert)
Degeneration of amygdala + limbic system (thalamus, hippocampus)
Combined degeneration of subcortical ascending systems
Cortical pathologies
Cortical and limbic Lbs/LNs (93%–100% in PD, 100% in DLB)
Limbic and/or neocortical neuritic Alzheimer-type
Lesions + synapse loss
Combination of Alzheimer and Lewy body pathologies
(Limbic system)
Combination of cortical, subcortical, and other pathologies

AD pathology in cortical and limbic structures or a combination of these lesions (Jellinger, 2006a). Evidence that subcortical PD pathology may account for cognitive decline is given by decreased cortical choline acetyltransferase (ChAT) activity in the prefrontal cortex related to severe loss of large cholinergic neurons in the basal forebrain, dopaminergic hypofunction, and decreased numbers of dopaminergic D_1 receptors in the CN even in the absence of AD pathology (Hurtig et al., 2000). Cognitive impairments in early PD are accompanied by reduction in activity in the frontostriatal circuitry (Lewis et al., 2003), while executive and attentional dysfunctions are unlikely to be a consequence of CN atrophy as assessed by MRI (Almeida et al., 2003). Volumetric MRI suggests progressive hippocampal atrophy that increases from PD to PDD and AD (Camicioli et al., 2003), and progressive volume decrease of neocortical regions in PDD (Ramirez-Ruiz et al., 2005), while cholinergic function is more severely affected in PDD than in AD (Bohnen et al., 2003). Recent experimental data suggest that the nigrostriatal pathway and SNc are essential components of a memory system that processes learning and that works independently of the hippocampal system (Da Cunha et al., 2003).

PDD was highly correlated with a loss of response to dopaminergic drugs and also to reduction of D_3 receptors that was most severe in the nucleus accumbens, central striatum, and globus pallidus (GP) (Joyce et al., 2002). Nondopaminergic motor features are also frequent in PDD (Burn et al., 2003). Cognitive impairment is often correlated with the density of LNs and neuritic degeneration in the hippocampus and periamygdaloid cortex (Churchyard and Lees, 1997; Mattila et al., 1999), suggesting that disruption of the limbic loop contributes to dementia. Damage in the amygdala causes its "disconnection" from key regions in a similar manner as described for the hippocampus in AD (Braak et al., 1996). Although a few cortical LBs are found in virtually all PD cases, they are usually not widespread and not numerous in the temporal lobe of PD without dementia (Hurtig et al., 2000). While some authors have reported correlations between cognitive impairment and the numbers of cortical LBs (Mattila et al., 2000; Kovari et al., 2003), others have not found such an association (Harding and Halliday, 2001; Colosimo et al., 2003). PD patients without dementia have AD pathology largely restricted to the limbic system, corresponding to Braak stage 4 or less, whereas PDD cases often have severe neuritic AD lesions (Delacourte et al., 2002). Some studies suggest diffuse or transitional DLB with mild-to-moderate AD pathology as the major pathological substrate (Apaydin et al., 2002); others revealed significant correlation between cognitive impairment and widespread neuritic AD pathology that also led to significantly shorter survival (Jellinger et al., 2002). The density of both limbic LBs and neuritic plaques correlated well with dementia severity, suggesting that both pathologies independently or synergistically contribute to dementia (Harding and Halliday, 2001) or may have common origins with mutual triggering. The prevalence of Aβ deposition in the cerebral cortex in PD and its contribution to cognitive impairment are under discussion (Mastaglia et al., 2003). In conclusion, the main changes underlying dementia in PD seem to be LB pathology in the cerebral cortex and limbic structures, frequent coexistence of neuritic AD pathology, and associated neuronal and synaptic dysfunction/loss (Emre, 2003). In contrast to some authors reporting neocortical atrophy in PD, recent quantitative stereological studies found no global loss of neocortical neurons in PD but could not exclude that local neuron loss in specific subpopulations of neocortical neurons or cell loss in small but essential neocortical subregions may be part of the structural defects of PD (Pedersen et al., 2005). Voxel-based morphometric MRI studies showed neocortical volume reduction as the most relevant finding in PDD (Ramirez-Ruiz et al., 2005).

5.1.10 Dementia with Lewy Bodies

DLB is a relatively new term (McKeith et al., 1996) for a progressive dementia syndrome in the elderly associated clinically with the core neuropsychiatric features of fluctuating cognition, visual hallucinations, and cognitive impairment, in association with parkinsonism. The pathological features include a variable burden of α-synucleinopathy and AD-type pathology (McKeith et al., 1996; Ince and McKeith, 2003; McKeith et al., 2005; O'Brien et al., 2006). The syndrome emerged initially as an atypical variant of AD with particular pathological and clinical features (Okazaki et al., 1961; Kosaka, 1990; Ince et al., 1998). It has had several diagnostic labels, including diffuse LBD (Dickson et al., 1987; Lennox et al., 1989; Kosaka, 1990; Hely et al., 1996; Klatka et al., 1996; Kosaka and Iseki, 1996b), LB dementia (Gibb et al., 1987; Perry et al., 1990a, 1994, 1995; Brown, 1999; Louis et al., 1995; McKeith et al., 1995a, b), LB disease with dementia (Cummings, 1995), senile dementia of LB type (Perry et al., 1990c; McKeith et al., 1994), dementia associated with cortical LBs (Byrne et al., 1991), cortical LB dementia (Gibb et al., 1989), AD with LBs (Gearing et al., 1999), LB variant of AD (Hansen et al., 1989; Hansen et al., 1990; Forstl et al., 1993; Samuel et al., 1996; Weiner et al., 1996; Brown et al., 1998b; Connor et al., 1998; Perneczky et al., 2005), and cortical LBD (Burn, 2004). The workshop on DLB proposed the present name, DLBs (Kosaka and Iseki, 1996a; Ince et al., 1998; Ballard et al., 1999a, b; Gomez-Isla et al., 1999; Gomez-Tortosa et al., 1999; Del Ser et al., 2001; Rosenberg et al., 2001), and proposed clinical and pathological diagnostic criteria (see McKeith et al., 1996). For further diagnostic criteria see McKeith et al. (2005).

5.1.11 Diagnostic Concepts (DLB Versus PDD)

Close clinical and pathological similarities are being recognized between DLB and PDD (Holmes et al., 1999). Both are clinically determined syndromes, although consensus clinical criteria have been proposed and validated for PDD (Emre, 2003). An arbitrary "one year rule" has until now been used to separate DLB from PDD, onset of dementia within 12 months of parkinsonism qualifies as DLB and more than 12 months of parkinsonism before dementia as PDD (McKeith et al., 1996), but this approach has considerable limitations (Burn, 2004; McKeith et al., 2004). Pathologically, both disorders have LBs, the characteristic feature of LBD, but there are as yet no defined morphological criteria that separate the disorders either from each other or from PD without dementia. Moreover, autopsy statistics of patients with clinically diagnosed PDD and DLB show heterogeneity in terms of distribution of LBs, as well as AD and vascular pathology (Gibb et al., 1987; Hansen et al., 1990; Baba et al., 1998; McKeith et al., 2000; Jellinger, 2002, 2003e). Reclassification and pathological subtypes of DLB and its relations to PD and AD have been discussed by Iseki (2004).

5.1.12 Epidemiology

The rate of incidence and prevalence of DLB is unknown. In population-based clinical studies of people aged over 65 years, the prevalence of DLB and PDD was reported to be 0.3% and 0.7%, respectively, which suggests that each could account for up to 10% of all dementia cases, a proportion consistent with DLB rates of 10%–15% from hospital-based autopsy series. A community-based study of people aged over 65 years found that 5% met consensus criteria for DLB (3.3% probable, 1.7% possible), representing 22% of all dementia cases (Rahkonen et al., 2003), similar to other clinical estimates (Shergill et al., 1994; Stevens et al., 2002), and consistent with estimates of LB prevalence in a dementia case register followed up to autopsy (Litvan et al., 1998). In a recent autopsy series, DLB represents the second most frequent cause of dementia in the elderly after AD, accounting for 7%–30%, with a mean of 15% (McKeith et al., 2000; Ince and McKeith, 2003; Mosimann and McKeith, 2003; Brayne et al., 2006). One population-based autopsy study found that LBs were evenly distributed between demented and nondemented individuals, which could be interpreted as evidence for a substantial proof of preclinical cases (Neuropathology-Group, 2001). DLB has been described as a syndrome in the elderly but onset prior to 65 years is not unknown. No classical epidemiological studies to investigate age, sex variation, and potential risk factors for DLB have been reported. The disorders may have a slight male preponderance compared with AD.

5.1.13 Genetics

No single gene determinant of DLB has been described. A few families with autosomal dominant inheritance were described (Wakabayashi et al., 1995; Ishikawa et al., 1997; Brett et al., 2002; Galvin et al., 2002; Tsuang et al., 2002), but genetic investigation has been limited and so far negative, and in most patients with DLB there are no genetic mutations in the AS gene or other PD genes (McKeith et al., 2004). However, there is recent evidence that DLB and familial PD are related to a new mutation, E46K, of AS (Zarranz et al., 2004), and mutations in the β-synuclein gene may predispose to familial DLB (Ohtake et al., 2004). Furthermore, mutations in glucocerebrosidase have been shown to be an important risk factor for LBD (Goker-Alpan et al., 2006). In addition, positive associations between DLB have been described with apolipoprotein E ε4 (Galasko et al., 1994; Harrington et al., 1994; Benjamin et al., 1995; Lippa et al., 1995; Kawanishi et al., 1996; Olichney et al., 1996; Lamb et al., 1998; Singleton et al., 2002), iNOS 2A (Xu et al., 2000), CYP2D6B (Saitoh et al., 1995; Tanaka et al., 1998; Furuno et al., 2001), which was confirmed for PD but not for DLB (Atkinson et al., 1999), and butyrlcholinesterase K (Singleton et al., 1998), while a negative association has been described for a series of genetic markers and risk factors (see Ince and McKeith, 2003). A common NURR1 polymorphism is associated with PD and DLB (Zheng et al., 2003). The frequent occurrence of LBs in familial and sporadic AD (Hamilton, 2000; Dickson, 2001; Lippa, 2003; Parkkinen et al., 2003; Saito et al., 2003; Trembath et al., 2003) may indicate a link with the genetic basis of AD.

5.1.14 Clinical Phenomena

The core clinical features of DLB include cognitive decline with dysexecutive and visuospatial–visuo-constructional components, psychomotor speed, and, in contrast to AD, often with relatively preserved mnemonic functions early in the disease course (Walker et al., 1997a; Ballard et al., 1999a, b; Mori et al., 2000; Doubleday et al., 2002; Collerton et al., 2003). Cognitive decline is more prevalent at disease onset, and parkinsonism less prevalent. Visual hallucinations are a core diagnostic feature but are also present in PDD. These hallucinations are well formed, recurrent, and frequently comprise people or animals. They have been related to retinal involvement with presence of AS-positive pale inclusions in the outer plexiform layers of the retina, ultrastructural changes of the cones, and changes in the distribution of AS (Maurage et al., 2003). Similar retinal changes have been observed in an AS-tg mouse model (Masliah et al., 2000). Fluctuations in cognitive functions which may vary over minutes, hours, or days occur in 50%–75% of patients and are associated with shifting degrees of attention and alertness (Walker et al., 2000a, b). Perception, attention, working memory, and fluctuating attention are disproportionally impaired in DLB compared with AD (Ballard et al., 2001a; Calderon et al., 2001; Ferman et al., 2002; Bradshaw et al., 2004; Salmon and Hamilton, 2006), but also occur in PDD (Ballard et al., 2002), and no definite differences in cognitive tasks between DLB and PDD subjects were found (Horimoto et al., 2003; Noe et al., 2004; O'Brien et al., 2006), while the cognitive decline is faster in LBD + AD than in pure AD (Olichney et al., 1998). Neurological signs include extrapyramidal movement disorders strongly reminiscent of PD in up to 70% of DLB patients (Ballard et al., 1997). In up to 25% of autopsy-confirmed cases, however, there is no record of parkinsonism, which is not necessary for the clinical diagnosis of DLB. The severity of parkinsonism is usually equal to that in without dementia PD patients (Aarsland et al., 2001) and shows similar progression rates in motor scores (Ballard et al., 2000). The pattern of extrapyramidal signs in DLB shows greater postural instability and facial impassivity, gait difficulty, and less tremors (Jankovic et al., 1990; Burn et al., 2002). They may be found at presentation in 25% of cases, especially in those with mild AD pathology (Jellinger et al., 2002) and develop in a further 50% during the course of illness. Supporting clinical features are REM sleep behavior disorders with parasomnia and vivid or frightening dreams and complex motor behavior during REM sleep as well as sleep–wakefulness disorders, hypersomnolence, and narcolepsy, which may contribute to the fluctuation typical of DLB (Boeve et al., 1998, 2001; Turner et al., 2000; Ferman et al., 2002, 2004), but these disorders are also seen in PD (Arnulf et al., 2000).

Autonomic abnormalities include orthostatic hypotension, "dizziness," and (pre)syncope that are high-risk factors for falls (McKeith et al., 1992; Kuzuhara and Yoshimura, 1993; Imamura et al., 2000; Kaufmann et al., 2001; Kenny and Allen, 2006). Urinary incontinence has been reported early in the course of DLB (Del Ser et al., 1996).

Neuroleptic sensitivity to both conventional and atypical neuroleptics, used to control behavioral and psychiatric symptoms, may become severe and life-threatening (McKeith et al., 1992, 1995a, b; Aarsland et al., 2003). A 50% prevalence of depressive symptom in DLB is similar to that in PD and higher than in AD.

Age at onset is 50–83 years and age at death 68–92 years (Holmes et al., 1999). There is no considerable difference in disease progression between DLB and AD (Ballard et al., 2001b), although some patients with DLB have a very rapid disease course (Armstrong et al., 1991; Lopez et al., 2000). There is conflicting evidence whether there is any difference between DLB and AD in survival from onset until death or that survival is worse in DLB (Cercy and Bylsma, 1997; Walker et al., 2000c). In a cohort of autopsy-proven DLB patients (66% DDLB and 33% Lewy body variant of Alzheimer's disease (LBV/AD)), dementia, fluctuating cognition, hallucinations at symptom onset, and shorter latencies to dementia onset strongly predicted shorter survival ($p = 0.027$). Similar negative effects on disease duration had increasing coexistent AD pathology: survival in patients with definite AD mean 5.2 years versus no AD mean 7.2 years ($p < 0.007$) (Jellinger et al., 2003). Anosmia is more frequent in DLB than in AD, but addition of impaired odor detection to the consensus criteria did not improve the overall performance (McShane et al., 2001).

5.1.15 Clinical Diagnostic Criteria for DLB

The Consensus Consortium on DLB characterized "probable" DLB as comprising a progressive dementia with the presence of two or more of the three core clinical features and the variable presence of several supportive features (❯ *Table 15-5*). Most of these clinical symptoms can be present in AD and other

❏ Table 15-5
Consensus guidelines for the cinical diagnosis of dementia with Lewy bodies

The central feature is progressive cognitive decline of sufficient magnitude to interfere with normal social or occupational function. Prominent or persistent memory impairment may not occur in the early stages but is usually evident with progression. Deficits of attention, of frontal-subcortical skills, and visuospatial ability may be prominent
Two of the following core features are essential for probable DLB, one is essential for possible DLB
Fluctuating cognition with pronounced variations in attention and alertness
Recurrent visual hallucinations typically well formed and detailed
Spontaneous motor features of parkinsonism
Features supportive of the diagnosis are
Repeated falls
Syncope
Transient loss of consciousness
Neuroleptic sensitivity
Systematized delusions
Hallucinations in other modalities
A diagnosis of DLB is less likely in the presence of
Stroke disease with focal neurological signs or on brain imaging
Evidence of any physical illness, or other brain disorder, sufficient to account for the clinical picture

McKeith et al. (1996). These criteria are proposed as being able to predict with high likelihood that dementia is associated with cortical Lewy bodies. They are potentially applicable to patients with idiopathic Parkinson's disease who subsequently develop dementia. These criteria do not exclude the presence of concomitant Alzheimer pathology, and many patients may simultaneously meet guidelines for the clinical diagnosis of Alzheimer's disease

dementia syndromes. The validity and reliability of the current clinical criteria for diagnosing DLB are highly variable because of different methods and case mixes (see Litvan et al., 2003). In general, specificity is high, ranging between 84% and 100%, but sensitivity of case detection is limited and ranges from 18% to 100% with PPV (positive predictable values) between 0% and 100% and NPV (negative predictable values) between 43% and 93% (McKeith et al., 2004; Burn et al., 2005). The presence of visual hallucinations in early-stage dementia has been shown to be most specific to DLB (99%), and visuospatial impairment was the most sensitive (74%), differentiating DLB from AD, while lack of visuospatial impairment had a negative predictive value of 90% (Tiraboschi et al., 2006). Other studies showed lower sensitivity, especially when there was concurrent AD (Lopez et al., 2002). The clinical diagnostic accuracy was higher for DLB patients with low (75%) compared to high (39%) Braak stages (Merdes et al., 2003). Case validation is compromised by the lack of defined neuropathological criteria for DLB and the presence of LBs in a high number of cases at autopsy with non-DLB clinical presentations such as AS pathology in AD (Hamilton, 2000; Dickson, 2001; DeLucia et al., 2002; Jellinger, 2003a; Trembath et al., 2003).

5.1.16 Laboratory and Neuroimaging Findings

No clinically useful tests have been reported based on plasma and cerebrospinal fluid (CSF) analysis for the diagnosis of DLB; however, some studies have been reported on CSF tau and Aβ in differentiating AD from

DLB. CSF tau, especially phosphorylated (p) tau may be lower in DLB than in AD, although both show decreased levels of $A\beta_{42}$ (Kanemaru et al., 2000; Andreasen et al., 2001; Parnetti et al., 2001; Tschampa et al., 2001; Buerger et al., 2002; Bibl et al., 2006). CSF interleukin-6 levels did not distinguish between both diseases (Gomez-Tortosa et al., 2003), while the native form of AS is not found in CSF (Jakowec et al., 1998). The standard EEG may show early slowing and transient temporal slow-wave activity (Yamamoto and Imai, 1988; Briel et al., 1999; Barber et al., 2000a). Neuroimaging investigations can be helpful in supporting the clinical diagnosis (Small, 2004). Changes associated with DLB include preservation of the hippocampal and medial temporal lobe volume on MRI (Barber et al., 1999, 2000b, c, 2001), except for DLB cases with severe medial temporal atrophy due to concomitant AD pathology (Lippa et al., 1998; Harvey et al., 1999). Several studies suggested that there are no differences in the degree of brain atrophy in DLB compared with other dementia syndromes (O'Brien et al., 2001). MRI morphometry showed preservation of the medial temporal lobe, hippocampus, and amygdala in DLB relative to AD (Hashimoto et al., 1998b), and significant gray matter loss was also observed in the thalamus of AD patients compared to DLB (Burton et al., 2002). This was considered a potential *antemortem* marker in distinguishing DLB from AD. Other features of generalized atrophy (Barber et al., 1999), white matter changes (Barber et al., 2000c), and rates of progression of brain atrophy (O'Brien et al., 2001) are not helpful in differential diagnosis. Diffusion tensor MRI (DT-MRI) studies in DLB showed abnormalities in the corpus callosum, pericallosal areas, and the frontal, parietal, occipital and, less prominently, temporal white matter, and in the caudate nucleus and the putamen of patients compared with controls. The average grey matter volume was lower in patients than in controls. These findings of concomitant grey matter atrophy and white matter abnormalities in regions with a high prevalence of long connecting fiber tracts might suggest the presence of neurodegeneration involving associative cortices. The modest involvement of the temporal lobe fits with the relative preservation of global neuropsychological measures and memory tasks in the early stage of DLB. The selective involvement of parietal, frontal, and occipital lobes might explain some of the clinical and neuropsychological features of DLB, providing a possible distinctive marker for this disease. The abnormalities found in the subcortical grey matter may indicate that DLB and PD share a similar nigrostriatal involvement caused by common pathophysiological mechanisms (Bozzali et al., 2005). Cognitive impairment in PD and DLB is unlikely to be a direct consequence of CN atrophy as assessed on MRI (Almeida et al., 2003).

SPECT and PET studies either showed a diffuse glucose hypometabolism in the entire cortex with relative sparing of the primary sensory-motor cortex, which seems to be a typical pattern for DLB distinction from AD (Mirzaei et al., 2003), or selective occipital hypoperfusion in DLB (Ishii et al., 1998; Imamura et al., 2001; Lobotesis et al., 2001; Okamura et al., 2001), but they are not associated with occipital lobe atrophy in MRI (Middelkoop et al., 2001).

DAT, a marker of striatonigral degeneration, the loss of which in the caudate and putamen, detected by dopaminergic SPECT (Donnemiller et al., 1997; Ransmayr et al., 2001), and use of presynaptic and postsynaptic ligands may be useful in distinguishing DLB from AD (Walker et al., 1997b; Colloby et al., 2002; Gilman et al., 2005). A sensitivity of 83% and specificity of 100% have been reported for the association of an abnormal scan with an autopsy diagnosis of DLB (McKeith et al., 1995b).

Regional correlation of pre- and postsynaptic dopaminergic functions in the striatum, using voxel-based PET analysis in PD, showed decline of presynaptic functions in the posterior and dorsolateral part of the putamen with relative inverse increase of raclopride (RAC) uptake as a marker of postsynaptic dopaminergic function in those regions. In DLB, presynaptic function was declined in the dorsolateral striatum, predominantly in the posterior part as in PD, but the change extended to the CN, and the degree was more severe and less asymmetric, while RAC uptake was relatively increased. This analysis may be useful to understand the alterations of dopaminergic functions in the progression of LBD (Suzuki et al., 2003).

5.1.17 Neuropathology of DLB

There are no characteristic macroscopic abnormalities except for some degree of diffuse cerebral atrophy and variable pallor of the SN (Dickson, 2002a, b; Ince and McKeith, 2003). There is no accepted "gold standard" for the pathological diagnosis of DLB as has been proposed for AD. The hallmark is α-synucleinopathy manifested as LBs of the classical and cortical types and neuritic degeneration. Accord-ing to the Consensus Pathological Guidelines (McKeith et al., 1996; McKeith et al., 2005; see also Dickson, 2006), LBs are scored semiquantitatively according to the severity and anatomical distribution, separating brainstem predominant (PD), limbic (or transitional), and neocortical types (❷ *Table 15-6*). They did not

❏ Table 15-6

Consensus pathological guidelines for scoring cortical LB deposition

Cortical region	Brodmann area	Anatomy	Score		
Entorhinal cortex	29	Medial flank of collateral sulcus	0	1	2
Cingulate gyrus	24	Whole gyral cortex	0	1	2
Mid-frontal cortex	8/9	Lateral flank of superior frontal sulcus	0	1	2
Mid-temporal cortex	21	Inferior surface of superior temporal sulcus	0	1	2
Inferior parietal lobule	40	Lateral flank of parietal sulcus	0	1	2

For each region, Lewy bodies are counted from the depth of the sulcus to the lip. Counts are not made over the crest of the gyri except for the cingulate gyrus. Lewy bodies are predominantly located in deeper cortical layers (layers 5 and 6). In each region, a count of up to five Lewy bodies in the cortical ribbon gives a score of 1 in the table. Counts greater than 5 score as 2. The sum of the five areas is used to derive the category of cortical spread (maximum score 10)

provide diagnostic criteria as it is sometimes mistakenly assumed. It was hoped that this method could be adapted to be included in the CERAD protocol (Mirra et al., 1991), although this has not yet been achieved. The system of scoring cortical LB pathology is based on five cortical areas: transentorhinal cortex (Brodmann area (BA) 29), anterior cingulate cortex (BA 24), mid-frontal gyrus (BA 8/9), superior temporal gyrus (BA 21), and inferior parietal lobule (BA 40). This protocol has been simplified by excluding the frontal region because of the common finding of occasional LBs in this region in PD in the absence of dementia (Harding and Halliday, 2001). The Consensus Pathological Guidelines were formulated prior to the discovery of the role of AS in LB formation and are, therefore, based on Ub immunohistochemistry or conventional staining, e.g., hematoxylin and eosin (H&E), but recent studies demonstrated that AS immunohistochemistry is much more sensitive for both LBs and neuritic lesions (Gomez-Tortosa et al., 2000; Schneider et al., 2002). Recent studies of differential expression of AS isoforms in DLB revealed just marginally elevated AS levels in comparison to AD and aged control brains; AS-isoform 112 was markedly increased, whereas AS-140 levels were significantly diminished in both DLB and AD in comparison with controls. These data indicate differential overexpression of AS-112 in DLB—a finding that could be of importance in its pathogenesis (Beyer et al., 2004). Recent studies found a reduction of AS mRNA in DLB (44.9% of control). The abundance of the Triton-soluble fraction (bioavailable protein) was not altered, but there was an increase of the Triton-insoluble component (likely representing aggregates). Evaluation of several chaperones showed that HSP70 mRNA was increased in DLB, whereas the mRNAs for HSP90 and HDJ1 were unchanged. HSP70 accumulated in the Triton-soluble fraction, whereas HSP90 and HDJ1 proteins accumulated in the Triton-insoluble fraction. These observations suggest that sporadic DLB is not associated with overexpression of AS. Rather, the persistence of normal soluble AS protein levels, despite the reduction of its mRNA, suggests a primary defect in clearance of the protein. However, this reduced clearance cannot be attributed to a failure of chaperone expression, because their mRNA is unchanged or increased in the DLB brain (Cantuti-Castelvetri et al., 2005). Triton X-100-insoluble AS was enriched nearly twofold in the temporal cortex of patients with DLB compared to age-matched controls. By contrast, the total amount of AS protein was unchanged. Surprisingly, the degree of Triton X-100-insoluble AS did not correlate with either the duration of illness or the number of LBs counted using stereological methods from an adjacent block of tissue. However, the Triton X-100-soluble fraction of AS did correlate strongly with the expression of several heat-shock proteins (HSPs) in DLB but not control cases, suggesting a coordinated HSP response in DLB neocortex (Klucken et al., 2006).

The anatomical distribution of LBs in DLB does not follow the hierarchical spread of NFTs (Braak et al., 1996; Marui et al., 2002), although they exhibit similar spatial patterns to SPs and NFTs, while in some instances, clusters of LBs appeared to be more closely related spatially to clusters of SPs than NFTs (Armstrong et al., 1998). LBs are commonly encountered in pigmented midbrain and brainstem nuclei, the dmnX, the basal forebrain, and in limbic cortical regions, especially the amygdala, entorhinal and anterior cingulate cortices (❷ *Figure 15-2c*), and in the deep layers of other neocortical areas, while rare in the occipital cortex; they affect various subgroups of neurons including pyramidal cells and GABAergic interneurons (Smith et al., 1995; Wakabayashi et al., 1995; Rezaie et al., 1996; Gomez-Tortosa et al., 2000) (see Ince and McKeith, 2003), whereas cortical neurons expressing calcium-binding proteins are spared (Gomez-Tortosa et al., 2001). Reduced interneuronal lipofuscin content not confined to neurons developing LBs was found in DLB (Drach et al., 1998).

Patients with widespread cortical LBs have a condition called "diffuse" DLB (DDLB), while brains with cortical LBs confined to the limbic lobe are classified as "transitional DLB" (Ince and McKeith, 2003). The upper cerebral cortex frequently shows spongiform changes with loss of neurons and apical dendrites, but its cause remains undetermined. It has been demonstrated that up to 70% of pyramidal cells may be lost in DLB (Wakabayashi et al., 1995), and the subsequent loss of apical dendritic permeation into the upper cortical layers may be a substrate for spongiosis. Hypometabolism in the primary and association visual cortices in DLB, which is not present in AD, may be related to white matter vacuolation in the occipital lobe (Higuchi et al., 2000), but this awaits further confirmation.

The SN and other subcortical nuclei show lesions with neuronal loss, LBs, and LNs, the distribution pattern and immunohistochemistry of which is often indistinguishable from that of sporadic PD (Braak et al., 2003; Jellinger, 2003a), except for an occasionally more severe neuronal loss in the dorsolateral compared with the ventral parts of SNc, and more frequent involvement by LBs of the limbic system, in particular of the CA2/3 subareas of Ammon's horn and the neocortex in DLB (Jellinger, 2004, 2006a). These data suggest close morphologic and pathogenic relationships between both disorders. They cause degeneration of the nigrostriatal projection demonstrable by dopaminergic presynaptic ligand SPECT that distinguishes DLB from AD (Walker et al., 2002).

DLB is frequently associated with AD pathology of variable intensity and extent (Ince and McKeith, 2003; McKeith et al., 2004). A subgroup with minimal diffuse Aβ deposition and neuritic AD lesions absent or restricted to the hippocampus with Braak stages 3 or 4 corresponding to plaque-predominant AD (Hansen et al., 1993) is referred to as "pure" DLB (Ince and McKeith, 2003). They may show a preponderance of diffuse plaques with different proportions of SPs between DLB less $Aβ_{1-40}$ and AD [more frequent Aβ (1-40) than (1-42)] deposits (Mann et al., 1998; Lippa et al., 1999b). While the Aβ load and SP density were consistently higher in AD and DLB than in controls (McKenzie et al., 1996), the Aβ pathology of DLB cases with frequent single large clusters differs from that in "pure" AD (Armstrong et al., 2000). Neuritic plaques are frequently present at a burden which is equivalent to that in AD (Perry et al., 1995) and would fit to the diagnosis of definite AD using the CERAD protocol (Hansen and Samuel, 1997; Hansen et al., 1998), while many cases have predominantly diffuse amyloid plaques with few neuritic elements (Apaydin et al., 2002), or only have minimal cerebral amyloid deposition (Hely et al., 1996). LBD and LBV/AD were found to differ from AD in that NPs generally did not contain PHF units unless they are accompanied by neocortical NFTs (Samuel et al., 1997b), whereas large amounts of PHF-tau was present in AD cases with NFTs (Strong et al., 1995). Kosaka and Iseki (1996b) proposed to divide DLDB into a "common form," characterized by abundant neocortical SPs with significant NFTs in medial temporal cortex, and a "pure DLDB" in which AD pathology is minimal. However, this classification depends on the criteria used to define AD, which would lead to different proportions within each category (Hansen and Samuel, 1997). Patients with significant neuritic AD pathology sufficient to warrant a diagnosis of definite AD as well as LBs may be considered as two groups: (1) those with the clinical features of DLB should be diagnosed as "Lewy body variant of AD" (LBV/AD) (Hansen and Samuel, 1997; Brown et al., 1998a, b; Perneczky et al., 2005) and (2) those with more prominent Alzheimer pathology and minimal LBs (e.g., limited to the amygdala) and a clinical picture of typical AD should be diagnosed as "AD with incidental LBs" (Del Ser et al., 2001).

Most DLB brains have an excess of AD-typical hyperphosphorylated tau protein in the hippocampus when compared with age-matched controls and without dementia PD patients. They occupy higher Braak stages of AD pathology than age-matched controls but lower stages than brains with pure AD (Hansen and Samuel, 1997; Harding and Halliday, 1998). Among 96 cases of autopsy-proven DLB, 63% was classified as "pure" DLB without considerable neuritic AD lesions and 37% was associated with severe AD pathology (Braak stages 5 and 6). LBV/AD was present in 33% of the cases with limbic and neocortical type of LBD (Jellinger, 2003e, g). Reclassification of 51 Japanese autopsy cases gave similar results: 66.6% fulfilled the definition of DLB and met the criteria for LB stages I–IV (Marui et al., 2002), while 33.3% met AD criteria (Braak stages 4–6, CERAD C), which did not meet these LB stages. All DLB brains showed severe SN cell loss, while the AD brains revealed little or, rarely, moderate SN damage (Marui et al., 2004). Lewy pathology in the brainstem-predominant type was defined below stage 1 (Marui et al., 2002).

Recent biochemical evaluation of tau, Aβ, and AS overlap in a group of sporadic DLB cases (mean age 73.1 ± 8.1 years) showed that all brains were associated with important deposits of AS, $A\beta_{42}$, and tau that were similar in quality to those in AD, confirming morphological data indicating less severe NFT pathology in DLB versus AD (mean Braak stages 4.1 versus 5.1). Tau pathology was less severe in DLB (+AD) than in "pure" AD, indicating a close relationship between both diseases (❯ *Figure 15-6*).

☐ **Figure 15-6**
Biochemical evaluation of tau, Aβ, and AS overlap. All brain samples in this study were fully characterized for tau, Aβ, and AS pathologies. Average scores and standard deviations are shown. Compared with age-matched AD patients (*dark bars*), DLB + AD patients (*light bars*) show lesser tau pathology evaluated either by tau staging (*left*) or Braak staging (*middle*). DLB + AD patients had as much Apx-42 deposits as AD patients (*right*). Modified from Deramecourt et al. (2006)

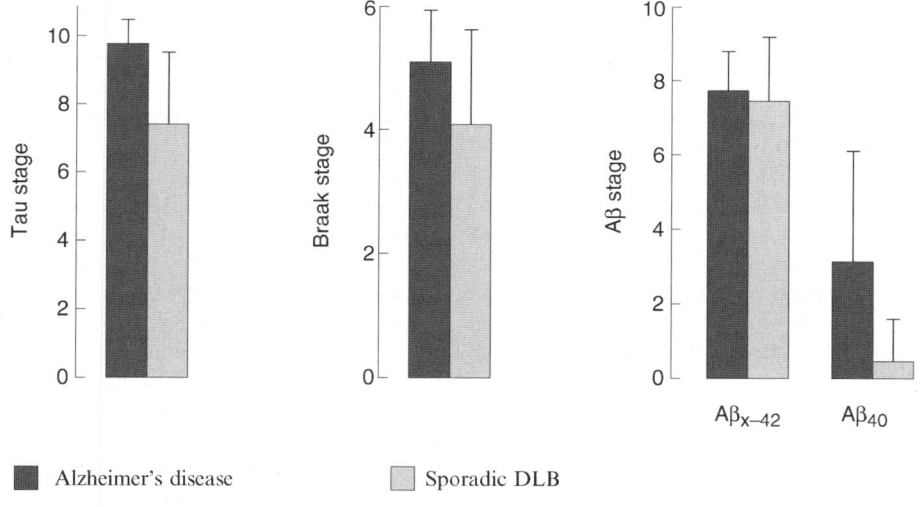

"Pure" DLB cases usually show no significant differences in neocortical synapse density and synapse protein (synaptophysin) immunoreactivity versus controls, while severe synapse protein loss comparable to AD is seen in LBV/AD (Samuel et al., 1997a; Hansen et al., 1998). It may correlate with dementia but has no value in discriminating with dementia disorders. Despite comparable LB counts and ChAT losses, the DLBD patients had significantly less dementia than the LBV/AD patients, which suggests that neocortical

LBs and ChAT depletion contribute to cognitive impairment in LBD, and that concomitant AD pathology promotes increased severity of dementia compared to DLBD (Samuel et al., 1997a). Immunohistochemistry showed that the number of choline acetyltransferase (ChAT)-immunoreactive neurons in DLB brains was significantly lower than the numbers in AD and nondemented (control) brains. No significant difference in the number of ChAT-positive neurons was found between AD and control brains, but their size was significantly smaller in AD and DLB. These results show that cholinergic neurons of the Ch1–2 regions are more severely affected in DLB than in AD, even in DLB cases that did not fulfill the neuropathologic criteria for definite AD. Furthermore, some LBs were observed in the Ch1–2 regions. Thus, cholinergic neuronal loss in the Ch1–2 regions might be specific to the pathology of DLB. Taking the distribution of cholinergic fibers in the hippocampus into consideration, these data suggest a possibility that hippocampal cholinergic projection is involved in Lewy-related neurites in the CA2–3 regions, the origin of which remains unclear (Fujishiro et al., 2006). Other lesions include cerebral amyloid angiopathy (CAA) present especially in cases with significant plaque formation (Wu et al., 1992).

Hippocampal pathology in DLB is usually less prominent than in AD, and neuronal loss in the perforant pathway is often milder and more variable than that in AD (Lippa et al., 1997).

The presence of diffuse neuritic lesions in DLB and PD was first demonstrated in the CA 2/3 region of the hippocampus (Dickson et al., 1991, 1994; Kim et al., 1995); they were not found in AD and were initially regarded as a means to discriminate DLBD from more restricted forms of LBD, but this was not proven. Recent studies showed that LB and LN involvement of the CA 2/3 region of the hippocampus differs between DLB and PD (Parkkinen et al., 2003; Jellinger, 2003a). Hippocampal CA 2/3 neuronal loss is a consistent finding in families with SNCA missense mutations (Polymeropoulos et al., 1997; Farrer et al., 2004). The decrease of synaptic markers in the hippocampal–entorhinal formation in AD was more severe than in DLB; it was detected in a subset of neurites in plaques and was involved in the formation of cortical LNs (Wakabayashi et al., 1994). The occurrence of spongiform changes in the transentorhinal cortex and amygdala are closely related with Ub-positive granular structures in the hippocampus due to degeneration of terminal or distal axons of the perforant pathway caused by degeneration of the large pyramidal neurons (Iseki et al., 1997). They are most severe in the neocortical form but absent in DLB + AD (Marui et al., 2004). Disturbances of the nigro-amygdaloid connections related to degeneration of terminal or distal axons of LB-bearing dopaminergic SN neurons could be responsible for psychotic symptoms in DLB patients (Iseki et al., 2000). One study suggested a selective loss of frontal projecting neurons in DLB versus those projecting to temporal regions in AD (Harding et al., 2002b).

While previous studies in PD and DLB suggested comparative preservation of the striatum (Forno, 1996), the use of novel antibodies to AS showed abundant striatal pathology in LBDs with highest density of lesions in patients with a combination of AD and DLB or pure DLBs, suggesting that these AS aggregates may contribute to the parkinsonism seen in these disorders (Duda et al., 2002a). MRT revealed atrophy of the putamen in DLB but not in AD, while no correlation was found between degree of volume loss and severity of parkinsonism (Cousins et al., 2003) or the reduction in CN volume with cognitive deficits (Almeida et al., 2003). According to recent studies, DLB brains showed a significantly higher burden of (diffuse) amyloid plaques in the putamen and caudate nucleus and slightly more severe tau pathology than PDD brains despite similar neuritic Braak stages. Phases of Aβ development in DLB brains often, but inconsistently, correlated with both neuritic Braak stages and severity of striatal Aβ load, while these correlations were almost never seen in PDD cases with Alzheimer lesions. They also revealed a higher burden of AS lesions (both Lewy neurites and LBs) than PDD cases that commonly had a paucity of all three types of lesions. The globus pallidus was virtually spared in both phenotypes (Jellinger and Attems, 2006). The cerebellum in DLB and PD shows Aβ-positive round inclusions in the white matter, most of which are located in the Purkinje cell axons and consist of granulofilamentous structures; they were not observed in MSA (Mori et al., 2003). In DLB cases, there were LBs and neurites, as well as dystrophic neurons in the ventrolateral medulla, but the number of catecholaminergic and serotonergic neurons was not significantly reduced. All these groups were depleted in MSA. There were LB pathology and dystrophic neurons in the raphe in all DLB cases. Cell numbers were reduced in both the raphe obscurus and raphe pallidus. These data suggest that although DLB affects medullary autonomic areas it does so less severely than MSA, particularly in the case of the ventrolateral medulla (VLM), which controls sympathetic outputs maintaining arterial pressure. In DLB, orthostatic hypotension may be due primarily to involvement of sympathetic ganglion neurons rather than VLM neurons (Benarroch et al., 2005).

Glial lesions in DLB include thorn-shaped astrocytes or coiled bodies immunoreactive for AS (Terada et al., 2000). The role of microglia and inflammatory pathology in the evolution of DLB is unresolved (Iseki et al., 2000; Rozemuller et al., 2000; Togo et al., 2001), but it has been shown that some LB-bearing neurons were involved due to the presence of interleukin-1α and TNFα-positive microglia, and most extracellular LBs were associated with the presence of TNFα- and iNOS-positive astroglia in DLB brains showing severe LB pathology. These findings support the concept that cytokines and nitric oxide (NO) are significantly implicated in neuronal damage and death including LB formation in DLB brains (Katsuse et al., 2003a).

5.1.18 PD, DLB, and AD: One or More Entities or Phenotypes?

There are no differences in LB density in any brain area among DLB patients with cognitive changes or parkinsonism, and neither in the correlations of LB density with Braak stage, frequency of neuritic plaques, nor between LBs in cortex and SN, suggesting that DLB should not be considered a severe form of PD (Gomez-Isla et al., 1999).

Dementias associated with (cortical) LBs are traditionally classified as DLB, characterized by parkinsonism, visual hallucinations, visuospatial impairment, and cognitive fluctuations, or PDD (see McKeith et al., 1996, 2005). There are both clinical and neurobiological similarities between both disorders (McKeith and Mosimann, 2004). The consensus guidelines for DLB (McKeith et al., 1996, 1999) suggest that PD patients who develop dementia more than 12 months after the initial motor symptoms should be diagnosed as PDD rather than DLB, although the proposed recent changes to the consensus criteria (McKeith et al., 2005) may alter this to make the diagnosis according to whether parkinsonism or cognitive impairment arises first. This distinction has been criticized by those who regard the different clinical presentations as simply representing different points on a common spectrum of LB disease (see Aarsland et al., 2004). According to recent diagnostic criteria, other than age at onset, temporal course of dementia, and possibly levodopa responsibility, no major differences between DLB and PPD have been found in any variable examined, including clinical and cognitive profile.

Whereas LB densities, in general, cannot separate DLB from PDD and temporal lobe LBs are usually rare in PD without dementia, the severity and duration of dementia appears to be related to both increasing parahippocampal LB densities and neuritic plaque grade. A screening algorithm suggesting that semiquantitative LB density thresholds in the parahippocampus may distinguish DLB patients with or without dementia from other pathologies (Harding and Halliday, 2001) awaits further confirmation.

LB pathology in PD stages 5 and 6 (Braak et al., 2002, 2003) overlap with that of DLB stages (Marui et al., 2002), suggesting that PD shows a pathological continuity with DLB. However, there are deviations between DLB and sporadic PD in the severity and distribution pattern of lesions of SNc in DLB (occasionally less severe and involving dorsolateral more than ventromedial nuclei) and a more frequent involvement of the CA 2/3 subarea of hippocampus in DLB. Despite the morphological similarity of subcortical neuronal lesions and in the quality and distribution pattern of subcortical and cortical LBs in the different types of LBD, suggesting morphological and pathogenic relations between LBD and both PD and AD, their nosological relationship remains to be clarified.

The relative contributions of LBs, AS, Alzheimer-type pathology, neuronal loss, and neurochemical deficits as determinants of dementia in both PD and DLB remain unresolved, although some recent studies suggest that Lewy-related pathology is more strongly associated than AD-type changes (Hurtig et al., 2000; Apaydin et al., 2002; Kovari et al., 2003; Braak et al., 2005). The spreading pattern of AS pathology with onset in the lower brainstem and progression to the midbrain, dorsal forebrain, amygdala, limbic cortex, and final extension to the neocortex in DLB is very similar to that seen in sporadic PD (Braak et al., 2003; Jellinger, 2004). The late stages 5 and 6 of LB pathology (involvement of sensory association and prefrontal, primary sensory, and motor areas) suggest transition between PD and DLB (Saito et al., 2003; Jellinger, 2004). DLB has been classified as the limbic and neocortical type according to the degree of AS pathology, and, on the other hand, as the pure form, common form, and AD form according to the degree of Alzheimer pathology (Brown et al., 1998a; Hansen et al., 1990; Perneczky et al., 2005).

The question whether PD, PDD, and DLB are different disorders or represent a single entity with distinct clinical subtypes has recently been discussed controversially (Duyckaerts, 2002; Lees, 2002; Aarsland et al., 2004; Burn, 2004; Mori, 2005; Takahashi and Wakabayashi, 2005; Tsuboi and Dickson, 2005). Differences in the proportion of $A\beta_{40}$ deposits, the lack of relationship of LB formation to the amount of AD lesions (Gomez-Isla et al., 1999), differences in the Lewy pathology between AD and DLB (Marui et al., 2004), the distribution of neuritic AD pathology, tau proteins, cholinergic biochemistry, genetic differences in ApoE ε4 and ε2 frequency (Martinoli et al., 1995; Lippa et al., 1999a), and absence in AD of the CP206 gene observed in PD and DLB (Tanaka et al., 1998) argue for a separation of DLB and AD. The presence of AS-positive lesions or LBs in up to 14% of normal subjects over age 40 more frequently in individuals with dementia, and in 7%–71% of sporadic AD (Parkkinen et al., 2003) with considerable numbers of LBs and LNs in the amygdala, even in the absence of subcortical LBs (Lippa et al., 1998; Arai et al., 2001; Dickson, 2001; Jellinger, 2003a), and more severe LB pathology in DLB cases with severe AD lesion than in DDLB and various clinical, biochemical, and morphological overlaps between PD, DLB, and AD suggest that the process of LB formation is triggered, at least in part, by AD pathology (Brown et al., 1998a, b; Marui et al., 2002; Iseki et al., 2003). This is in accordance with recent findings of significantly more frequent diffuse Aβ plaques and more frequent tau and AS pathologies in caudate nucleus and putamen in DLB compared to PDD, irrespective of the cortical neuritic Alzheimer lesions (Jellinger and Attems, 2006). The identification of a 35-amino acid fragment of AS in the amyloid plaques in DLB brain have raised the possibility that Aβ and AS interact with each other. Investigation of the molecular interaction of AS with Aβ 40 and/or Aβ 42 using multidimensional NMR spectroscopy showed that NMR data indicate specific sites of interaction between membrane-bound AS with Aβ peptide and vice versa. These Aβ–AS interactions were demonstrated by reduced amide peak intensity or change in chemical shift of amide proton of the interacting proteins. On the basis of these results, a plausible molecular mechanism of overlapping pathocascade of AD and PD in DLB due to interactions between AS and Aβ is suggested. These molecular interactions between Aβ and AS may lead to the onset of DLB (Mandal et al., 2006). These lesions represent a collision of two processes, pathological tau protein and AS, which may not only occur within the same brain but also in the same region or within a single cell (Arai et al., 2001; Dickson, 2001). These lesions and the recent biochemical data on tau and Aβ (Deramecourt et al., 2006) (❯ *Figure 15-6*) challenge the view of LBD as a distinct entity.

5.1.19 Associated Vascular Pathology

The frequency of associated cerebrovascular pathology in DLB has been suggested to be less common than in AD. In a cohort of 96 DLB cases, cerebrovascular lesions of various intensity (lacunes, amyloid angiopathy, old ischemic infarcts, and hemorrhages) were less frequent than in both PD and AD (31.6% versus 44.0% and 48%) but only slightly more frequent than in age-matched controls (32.8%). Acute ischemic strokes and hemorrhages present in PD, controls, and AD (4.1% versus 3.3% versus 7%), were not observed in DLB (Jellinger, 2003e). These data suggest less susceptibility to stroke and death from stroke in the DLB population, probably due to autonomic disorders and orthostatic hypotension. On the other hand, like in PD and AD, cognitive impairment in DLB appears to be independent from coexisting vascular pathology and is mainly related to AD or LB pathologies or a combination of both (Dickson, 2002b; Jellinger, 2003f; Mosimann and McKeith, 2003).

5.1.20 Biochemistry

Cholinergic System   Neocortical ChAT levels in DLB are similar to that in PDD and lower than in AD (Dickson et al., 1987; Perry et al., 1990b, 1994; Francis et al., 2006). Although neuronal loss in the cholinergic basal forebrain is consistently found in PDD, DLB, and AD (Dickson, 2001), early and more widespread cholinergic losses differentiate DLB and LBV/AD from AD (Tiraboschi et al., 2002; Ince and McKeith, 2003). However, there is only a modest correlation between synaptophysin and ChAT activity, indicating that, unlike AD, neurochemical markers do not appear to be correlated with cognitive decline in LBV/AD (Sabbagh et al., 1999). On the other hand, cortical ChAT activity has been shown to parallel NBM pathology (Lippa et al., 1999b). Cortical cholinergic activity is lowest in hallucinating compared to nonhallucinating DLB cases, while serotonin activity is relatively preserved (Perry et al., 1995). Loss of striatal ChAT in DLB (Langlais et al., 1993), probably due to pathology of intrinsic local circuits, may correlate with milder extrapyramidal clinical symptoms even in DLB patients who have equivalent loss of dopaminergic SN neurons compared with those with PD (Perry et al., 1990b). In DLB and PD, the muscarinic M_1 subtype is elevated in the neocortex (Perry et al., 2003) due to upregulation of postsynaptic receptors. DLB contrasts with AD in that muscarinic M_1 and M_2 receptors are differentially affected (Shiozaki et al., 1999), which may reflect differences in the underlying extent of pathology in the cholinergic projection from the basal forebrain. There are significant deficits in α6/α3 neuronal nicotinic acetylcholine receptors (nAChRs) in thalamic nuclei, caudate, and putamen in DLB that may be related to the loss of dopaminergic neurons and may contribute to neuropsychiatric features, e.g., impaired consciousness in DLB (Ray et al., 2004). In contrast to AD, where cholinergic deficits are more prominent in the basal lateral group of the amygdala, in DLB the nuclei of the basal, lateral, and cortical medial groups are severely involved, which may be due to the involvement of both basal forebrain and brainstem cholinergic nuclei (Sahin et al., 2006). Both α4 and α7 nAChR subunits are decreased in several brain areas in both AD and DLB, whereas the number of astrocytes immunoreactive for α7 nAChRs, probably associated with inflammatory mechanisms related to degenerative processes, was increased in the hippocampus and entorhinal cortex in AD but not in controls and DLB (Teaktong et al., 2004). Normal receptor coupling via G protein suggests that the cholinergic projective neurons are intact in DLB and provide a rationale for cholinergic therapy (Duda, 2004).

Previous studies suggested close association of alterations of the nicotinic markers with primary histopathological changes, i.e., dopaminergic cell loss in PD and DLB, amyloid SPs and NFTs in subiculum and entorhinal areas, and that downregulation of the receptor may precede neurodegeneration (Perry et al., 1995). Reduction of nicotinic acetylcholine receptor (nAChr) binding in the putamen did not correlate with AS expression in both PD and DLB (Martin-Ruiz et al., 2002) and also does not appear to be related to SPs and NFTs in AD or DLB, thus not being a reliable marker of cognitive loss in these disorders (Sabbagh et al., 2001), while other researchers showed a different pattern of nAChr loss and of (^{3}H) nicotine binding in AD and DLB (Court et al., 2000; Rei et al., 2000).

Dopaminergic System   DLB is associated with disruption of the dopaminergic input to the striatum, e.g., reduced DA and homovanillic acid/DA ratio as seen in autopsy tissue of DLB (Langlais et al., 1993; Perry et al., 1990a). DAT is low in DLB in comparison to AD and is demonstrable using FP-CIT-SPECT imaging. Autopsy studies showed no alterations in D_1, D_2, or D_3 subtypes in DLB patients who have been treated with levodopa (Piggott et al., 1999; Perry et al., 2003). AD with LB pathology and antemortem psychosis is associated with selective alteration in striatal DA receptor density, and particularly the D_3 receptor may be an important target of neurologeptic treatment of psychosis (Sweet et al., 2001). The absence of D_1 upregulation in DLB, in contrast to PD, may reflect differences in the nature and extent of basal ganglia pathology.

Clinico-Pathological Relations   The number of cortical LBs is not robustly correlated with either the severity or the duration of dementia (Gomez-Tortosa et al., 1999; Harding and Halliday, 2001), although associations have been reported with LB and plaque densities in midfrontal cortex (Samuel et al., 1996) and in the entorhinal and cingulate cortex (Kovari et al., 2003). Some authors suggest that LB formation may affect the feedback corticocortical pathway and the afferent cortical projections, whereas NFT formation may affect the feed forward corticocortical pathway (Armstrong et al., 1997), whereas others suggested LNs and neurotransmitter deficits to be more likely linked with clinical symptoms (Gomez-Isla et al., 1999; Perry et al., 2003). Neurodegeneration within the cholinergic system may mediate many motor problems, e.g., postural instability and balance problems, as well as determining the neuropsychiatric symptomatology of both PDD and DLB (Burn et al., 2003, 2005). There seems to be no significant cortical pathology associated with fluctuating cognition, which has been related to involvement of intralaminar thalamic nuclei (Henderson et al., 2000a). However, increased numbers of LBs in the anterior and inferior temporal lobe have been associated with the presence of well-formed visual hallucinations (Harding et al., 2002a, b). These areas are particularly implicated in the generation of complex visual impact and their pathological involvement perhaps contributes to the vivid and complex character of hallucinations in DLB, contrasting with the very simple visual symptoms associated with occipital-lobe lesions. LB pathology in the primary visual pathways showed no significant differences between DLB patients with or without visual hallucinations (Yamamoto et al., 2006a). Parkinsonism is related to the degree of cell loss and pathology in the nigrostriatal pathway. Depression is related to extensive loss of the serotonin transport and reuptake sites in both the cortex and striatum. Although paradoxically this is lower in patients without depression, which is consistent with the concept that depression is associated with compensatory regenerative activity in serotonergic neurons (Perry et al., 2003). Increased $\alpha2$-adrenergic receptor binding in LC projection areas in DLB in response to loss of LC noradrenergic innervation could lower the threshold for increased agitation in response to noradrenergic outflow in both DLB and AD (Leverenz et al., 2001). Disorders of REM sleep and related symptoms have been related to neuronal loss in the catecholaminergic brainstem nuclei, in particular the SN and LC, that inhibit cholinergic neurons in the pedunculopontine nucleus mediating atonia during REM sleep (Turner et al., 2000). Autonomic abnormalities are presumed to arise from AS pathology affecting peripheral nerves, autonomic ganglia, and the sympathetic innervation of the myocardium (Arai et al., 2000). Failure in the upregulation of D_2 receptors in response to neuroleptic blockade or reduced dopaminergic innervation may be the crucial factors for neuroleptic sensitivity in DLB (Piggott et al., 1998). The additional neuritic AD pathology affects the clinical presentation and course of DLB. Patients with few NFTs show more core clinical features of DLB, whereas those with many tangles show a pattern more like AD (Walker et al., 2000c; Merdes et al., 2003). The substantial cortical dysfunction found clinically in patients with limited cortical cell loss, negligible NFT counts, but numerous cortical LBs and LNs, suggests that they themselves are associated with much functional neuronal impairment, but elucidation of the pathophysiology of DLB is an important challenge for future research.

5.2.21 Pathogenesis

The pathogenesis of DLB concerns three major issues. The first is the molecular pathogenesis of AS aggregation and inclusion body formation. The second concerns the molecular pathogenesis of Alzheimer-type pathology, albeit in a more limited distribution and reduced severity compared with AD. The third relates to the different patterns of anatomical susceptibility in the nervous system within the spectrum of the α-synucleinopathies.

5.2.22 Differential Diagnosis

The concept of DLB is essentially syndromatic. The main differential diagnosis, on the basis of clinical overlap with other disorders, have been AD, PD, PDD, vascular dementia, and Creutzfeldt-Jakob disease (see Ince and McKeith, 2003; McKeith et al., 2004, 2005; O'Brien et al., 2006).

Management of DLB   The therapeutic implications of the cholinergic pathology so far identified in DLB can be summarized as follows: Cortical cholinergic deficits with functionally intact muscarinic M_1 receptors suggest a potential value for cholinergic replacement therapy. Cortical cholinergic deficits in DLB relate more to psychiatric than cognitive symptoms and cholinergic replacement therapy may be effective in alleviating these.

There is consistent evidence that cholinesterase inhibitors are more effective in DLB than in AD (Samuel et al., 2000; Mosimann and McKeith, 2003; McKeith et al., 2004; O'Brien et al., 2006). Fluctuating cognition, visual hallucinations, apathy, anxiety, and sleep disturbances are significantly improved with cholinesterase inhibitors, which, therefore, are deemed to be the first-line treatment for DLB (Grace et al., 2001; Aarsland et al., 2002). No information is yet available about their use in combination with anti-parkinsonian or atypical antipsychotic agents, but sensitivity reactions have been documented with most atypical neuroleptics and they should be used with great caution (McKeith et al., 1995b; Walker et al., 1999; Aarsland et al., 2003; Burn et al., 2005).

6   Lewy Bodies in Aging and Alzheimer's Disease

LBs occur in the SN and brainstem in up to about 10% of neurologically normal aged individuals; some of them may represent "incidental" LBD or incipient synucleinopathy of no or little clinical relevance or suggesting that (subclinical) PD is present concurrently in some subjects (Del Tredici et al., 2002). AS pathology (both cortical and subcortical) has been reported in up to 14% of subjects over the age of 40, with an average of 11% and 23% in individuals without dementia and clinical dementia, respectively, suggesting that LB pathology may be important for the pathogenesis of dementia (Parkkinen et al., 2003; Saito et al., 2003), while cortical LBs are common in both familial and sporadic forms of AD estimates ranging from 7% to 71% of sporadic AD (Hamilton, 2000; Lippa, 2003; Trembath et al., 2003). Recent studies reported localized LB-related synucleinopathy in various brain areas in up to 30% of aged subjects without neuropsychiatric disorders or extrapyramidal symptoms (Jellinger, 2003a, 2004b; Parkkinen et al., 2003; Saito et al., 2003; Mikolaenko et al., 2005; Parkkinen et al., 2005). Those with involvement of the medullary, pontine, and midbrain nuclei also can be considered "incidental LB disease." It appears unclear whether in these cases extrapyramidal symptoms were not mentioned in the case histories, or whether they represent pre/subclinical forms of LB disease. The distribution or load of AS pathology did not permit a dependable postmortem diagnosis of extrapyramidal symptoms or cognitive impairment. Some neurologically unimpaired cases had a reasonable burden of AS pathology in both brainstem and cortical areas, suggesting that AS-positive structures are not definite markers of neuronal dysfunction (Parkkinen et al., 2005). Among centenarians, AS-positive structures throughout the brain, with relatively well-preserved SN, were found in 34.8%, without relation to neuritic Braak stages (Ding et al., 2006). A recent study of the incidence of LB-related α-synucleinopathy in 1,241 consecutive elderly patients distinguished negative stage 0 (87.3%), stage I (incidental LBs, 12%), stage II (LBs without attributable clinical symptoms, 3.8%), stage III (PD without dementia, 8.1%), stage IV (DLB transitional form, 2.1%), and stage V (DLB neocortical form, 1.9%) (Saito et al., 2004). Numerous LBs and LNs in the amygdala have been observed in over 60% of familial AD including those with mutations in the APP, PS-1, and PS-2 genes, in over 50% of aged Down's syndrome cases, and in 15%–60% of AD, even in the absence of subcortical LBs (Dickson, 2001; DeLucia et al., 2002; Trembath et al., 2003; for review, see Lippa, 2003).

Recent studies observed AS-immunoreactive LBs in the amygdala in 15.8% of AD, as compared to 38.3% in LBV/AD. The Braak stages and ApoE ε allele frequency did not differ between AD cases with and without LB in the amygdala (DeLucia et al., 2002), while AD patients with LB lesions in the amygdala were significantly older than those without such lesions (Jellinger, 2003e, 2004b). These data suggest that LBs occur in the amygdala and in other brain areas only in some AD cases, which are distinct from DLB, and PD showing a different involvement pattern (Jellinger, 2003a, b, 2004b; Parkkinen et al., 2003; Saito et al., 2003). In contrast to a recent series of AD cases (Mikolaenko et al., 2005), others observed no correlations between the presence and absence of LB-related α-synucleinopathy and the severity of neuritic AD pathology (Jellinger, 2004b), but the relationship between both types of lesions with aging needs further clarification. Whether LBs are considered a pathological feature related to the primary disease or as indicator of a second disorder is undetermined. Since the amygdala is heavily involved with neuritic pathology in AD, these findings may suggest that the process of LB formation is triggered, at least in part, by AD pathology (Iseki et al., 2003). A similar phenomenon has been described in Guam PDC (Yamazaki et al., 2000). However, the identification of cases with a reasonably high burden of AS pathology in both brainstem and cortical areas without clinical symptoms suggests that there must be certain unrecognized pathologies of decisive importance that mediate the biological abnormality (Jellinger, 2004a, b; Parkkinen et al., 2005). Therefore, as the role of AS pathology in the pathogenesis of extrapyramidal and cognitive impairment remains unclear, and a reasonable burden of AS pathology can be found in various clinical conditions, it does not appear feasible or even possible to achieve optimal clinicopathological correlations in the diagnosis of PD, PDD, and DLB (McKeith et al., 2004). It is reasonable to propose that the aggregation of AS protein as an inclusion is not a definite marker of neuronal dysfunction and may even represent a protective response triggered in affected neurons.

In autopsy cases without PD-associated symptoms, AS pathology was encountered in autonomic nuclei of the thoracic spinal cord, brain stem, and olfactory nerves in all cases, in sacral parasympathetic nuclei in almost all, in the myenteric plexus of oesophagus and sympathetic ganglia in 80%, and in the vagus nerve in 75% (Bloch et al., 2006). In PDD, AS-positive neurites were found in the nerve fibers of the dermis, skin, adrenal glands, and sympathetic ganglia. PD cases were frequently associated with LBs in the adrenal gland (100%) and in the skin (75%), while in DLB, association with LB-related α-synucleinopathy in the skin was far less frequent (36%). Thus, skin biopsy may have diagnostic significance in PD with advanced autonomic failure (Ikemura et al., 2006).

The evidence suggests that particular neurons have a limited repertoire of response to injury and that in the setting of neurodegeneration certain neurons display fibrillary aggregates that are typical of two different disease processes. The frequent occurrence of various tau-positive but AS-negative structures in neurons, astroglia, and grains in DLB may also indicate that different cytoskeletal abnormalities form a link between some neurodegenerative disorders (Iseki et al., 2002). These lesions represent a collision of two processes, pathological tau protein and AS, which may not only occur within the same brain but also in the same region or within the same cell in the human brain (Iseki et al., 1999, 2000; Arima et al., 2000; Arai et al., 2001) and tg mice (Maries et al., 2003). The coincubations of tau and AS have been shown to synergistically promote fibrillations of both proteins (Giasson et al., 2003). Recent evidence suggests that tau and AS might constitute a unique class of unstructured proteins that assemble predominantly into homopolymeric (rather than heteropolymerinc) fibrils, which deposit mainly in separate amyloid inclusions, but occasionally deposit together. Thus, the ability of tau and AS to affect each other directly or indirectly might contribute to the overlap in the clinical and pathological features of tauopathies and synucleinopathies (Lee et al., 2004b). The frequent co-occurrence of Aβ-positive (plaques) and AS-positive lesions in DLB, AD, and aged brains also suggests synergistic interactions; and data from experimental animals demonstrated that Aβ deposits promote AS lesions (Masliah et al., 2001). However, it is unclear whether there is a common underlying pathophysiology in these diseases, or if the LB merely represents a common final pathology leading to or preventing neuronal degeneration.

7   Lewy Bodies in Other Disorders

Although traditionally considered the hallmark pathological feature of PD and unusual in other diseases, the increasing use of AS immunohistochemistry has facilitated the realization that LBs are common in a variety of CNS disorders (❱ *Tables 15-7* and ❱ *15-8*). Some patients with pure autonomic failure (PAF) show LBs in the autonomic ganglia (Boeve et al., 1998; Arai et al., 2000; Arnulf et al., 2000; Kaufmann et al., 2001). LBs are also seen as the pathological substrate in rapid eye movement sleep behavioral disorder (RBD) that is often associated with PD, DLB, MSA, or other disorders (❱ *Table 15-7*). Neuropathology of the majority of cases revealed LBD, particularly neocortical and rarely limbic type, or MSA, occasionally associated with argyrophilic grains or AD pathology (Boeve et al., 2003). Recent studies in familial PD have revealed pathological overlaps of both sporadic LBD with predominantly neuronal inclusions (PD) and glial inclusions (MSA) (Gwinn-Hardy et al., 2000; Duda et al., 2002a, b; Lantos and Quinn, 2003). However, in addition to AS-immunoreactive glia in many cases of late-stage LBD–PD, DLB, and PAF (Arai et al., 1999, 2000; Wakabayashi et al., 2000b; Hishikawa et al., 2001), it may also occur in nonfamilial synucleinopathies. Glial involvement that appears to be a late and secondary manifestation (Wakabayashi et al., 2000b) is an early and pathogenetically important presentation in MSA and its distribution similar to that of LBs suggests that they have a common pathology and pathogenetic background (Lantos and Quinn, 2003). Neurons containing LBs in Guam PDC are rare (seen in only 10% of PDC cases), with no immunoreactive neuropil threads in the CA 2/3 region of the hippocampus (M. Takahashi, personal communication).

◻ Table 15-7
Hypothetical spectrum of Lewy body disorders

Syndrome	Neocortex	Limbic cortex and nbM	Substantia nigra	Dorsal motor neuron of X	Lateral gray horn	Myenteric and sympathetic ganglia
DLB	++ to +++	+++	+ to ++	+ to +++	?	?
PD	+ to ++	++ to +++	+++	+++	+ to ++	+ to ++
PAP	0	++	+	+++	++ to +++	+++
LB dysphagia	0	0	0 to +	0	?	+++

Adapted from INce and McKeith (2003)

n, nucleus;

LB, Lewy body;

nbM, nucleus basalis of Meynert;

PD, Parkinson's disease;

PAF, pure autonomic failure;

?, no data available

◻ Table 15-8
Conditions associated with Lewy bodies

Disorder	Relative frequency of LBs
Ageing and Alzheimer's disease	
Sporadic Alzheimer's disease	Often
Familial Alzheimer's disease	Often
Down's syndrome	Often
Tauopathies	
Pick's disease	Rarely
Progressive supranuclear palsy	Rarely
Corticobasal degeneration	Rarely
Parkinson–dementia complex of Guam	Rarely
Synucleinopathies	
Parkinson's disease	Always
Dementia with Lewy bodies	Always
Multiple system atrophy	Sometimes
Neurodegeneration with brain iron (Hallervorden-Spatz disease)	Sometimes
Pure autonomic failure	Sometimes
Meige's syndrome	Sometimes
Frontotemporal lobar degeneration	Rarely
Motor neuron disease (*amyotrophic lateral sclerosis*)	Rarely
Miscellaneous disorders	
Neuroaxonal dystrophy	Rarely
Subacute sclerosing panencephalitis	Rarely
Ataxia telangiectasia	Rarely

Adapted from Lippa (2003)

There is increasing evidence that LBs have clinical or biological significance in neurodegenerative diseases that are not primary synucleinopathies, e.g., tauopathies including FTD and classical Pick's disease PiD (Takauchi et al., 1995; Henderson et al., 2001; Popescu and Lippa, 2001; Mori et al., 2002; Popescu et al., 2004), where the amygdala is often heavily involved, but cytoplasmic inclusions are not present. AD and PiD often reveal double-labeling of amygdaloidal LBs with NFTs and Pick bodies (Lippa, 2003). LBs have also been found in the brains of patients with PSP (Jellinger et al., 1980; Mori et al., 1986; Gearing et al., 1994), which showed no increased numbers of LBs compared to normal controls (Tsuboi et al., 2001). Guamanian PDC is another example of a hybrid of tauopathy and synucleinopathy in which AS deposition has recently been detected in both amygdala and the cerebellum (Yamazaki et al., 2000; Forman et al., 2002; Sebeo et al., 2004), where it involves both Purkinje cells and astrocytes.

8   Collision of Several Proteinopathies

Evidence suggests that particular neurons have a limited repertoire or response to injury and that in the setting of neurodegeneration, certain neurons display fibrillary aggregates that are typical of two or more different disease processes. Multiple initiating events, either common to many neurodegenerative diseases or specific to individual disease conditions, may lead to cleavage of cytoskeletal proteins. The same genetic process may give rise to tau or synuclein accumulation susceptibility/pathology, but an individual's response to other genetic or environmental factors may influence the pathological process. The frequent occurrence of various tau-positive structures in neurons, astroglia, and grains in DLB and other LB diseases may also indicate that different cytoskeletal abnormalities form a link between some neurodegenerative disorders (Iseki et al., 2002). AS and/ or tau may be indirectly, inconsistently, or unpredictably related to the pathological process (bystander pathology). These lesions represent a collision of two processes, pathological tau protein and AS, which may not only occur within the same brain but even in the same region (usually amygdala) or within the same cell in the human brain (Iseki et al., 1999, 2002; Arai et al., 2000; Arima et al., 2000) and in tg mice (Maries et al., 2003). Recent genetic data indicating interactions of AS and tau genotypes in PD (Mamah et al., 2005) are consistent with in vitro experiments that revealed tau-mediated fibrillation of AS proteins of low concentrations (dose threshold effect) (Giasson et al., 2003). Moreover, tau and AS synergistically promote and propagate the polymerization of each other into amyloid fibrils, but each protein preferentially formed homopolymeric rather than heteropolymeric filaments. These in vitro findings are further supported by studies of tg mice engineered to express mutant human A53T AS (see Jellinger, 2003f; Maries et al., 2003). These mice develop a robust synucleinopathy phenotype with tau inclusions. Although tau and AS are located in different compartments in normal neurons, both proteins accumulate in the same compartment in disease. Recent evidence suggests that tau and AS might constitute a unique class of unstructured proteins that assemble predominantly into homopolymeric (rather than heteropolymeric) fibrils, which deposit mainly in separate amyloid inclusions, but occasionally deposit together. Thus, the ability of tau and AS to affect each other directly or indirectly might contribute to the overlap in the clinical and pathological features of tauopathies and synucleinopathies (Lee et al., 2004b). These and other data implicate fibrillization of AS and tau in the pathogenesis of PD and other synucleinopathies and suggest that distinct amyloidogenic proteins may cross-seed each other in neurodegenerative diseases. Because both proteins have long half-lives, they may reach a critical threshold for fibrillation through a two-step mechanism, involving initiation followed by propagation. The initiation may most likely involve the formation of amyloidogenic AS "seeds" that prime tau to acquire a confirmation that favors polymerization. However, it is unclear, whether there is a common underlying pathophysiology in these disorders or if the LB and tau deposits merely represent a common final pathology leading to or even preventing neuronal degeneration.

9   Conclusions

In summary, LBs occur commonly in a variety of neurodegenerative diseases. When present in high numbers, LBs are likely to be of biological or clinical significance. Evidence from biochemical and immunohistochemical studies indicates that the morphology of AS filaments in neuronal LBs show basic similarities, regardless of the underlying disease. Although investigators argue whether LBs represent a concurrent disease state, all LBs studied thus far contain the full-length AS protein and have uniform exposure of AS epitopes, regardless of whether the primary disease is an amyloidopathy, synucleinopathy, or tauopathy. Therefore, one might envision LB formation in nonsynucleinopathies as a common reaction or process generated in response to differing biological triggers.

Further studies are needed to elucidate a number of open questions that are currently discussed: (1) Are cortical and subcortical LBs identical inclusions or are there differences in their molecular biological/ neurochemical qualities and their development? (2) Are LBs neurotoxic causing dysfunction of involved cells, e.g., by oxidative stress, or are they neuroprotective by aggregating toxic (soluble) proteins in fibrillary insoluble form, thus allowing survival of neurons associated with LBs? (3) What are the relations between AS and tau that may coexist in the same cells, with frequent occurrence of LBs in tauopathies and their pathologies in synucleinopathies/LB disorders? (4) What are the biological relations between the different forms of synucleinopathies, e.g., LB disorders and those marked by GCIs (MSA), which also occasionally may co-occur?

References

Aarsland D. 2006. Dementia in Parkinson's disease. Dementia with Lewy Bodies and Parkinson's Disease Dementia. Brien JO, Ames D, McKeith I, Chiu E, editors. London: Taylor & Francis; pp. 221-239.

Aarsland D, Ballard C, Larsen JP, McKeith I, O'Brien J, et al. 2003. Marked neuroleptic sensitivity in dementia with Lewy bodies and Parkinson's disease. Nordic J Psychiatry 57: 94.

Aarsland D, Ballard C, McKeith I, Perry RH, Larsen JP. 2001. Comparison of extrapyramidal signs in dementia with Lewy bodies and Parkinson's disease. J Neuropsychiatry Clin Neurosci 13: 374-379.

Aarsland D, Ballard CG, Halliday G. 2004. Are Parkinson's disease with dementia and dementia with Lewy bodies the same entity? J Geriatr Psychiatry Neurol 17: 137-145.

Aarsland D, Laake K, Larsen JP, Janvin C. 2002. Donepezil for cognitive impairment in Parkinson's disease: A randomised controlled study. J Neurol Neurosurg Psychiatry 72: 708-712.

Abe Y, Kachi T, Kato T, Arahata Y, Yamada T, et al. 2003. Occipital hypoperfusion in Parkinson's disease without dementia: Correlation to impaired cortical visual processing. J Neurol Neurosurg Psychiatry 74: 419-422.

Abeliovich A, Schmitz Y, Farinas I, Choi-Lundberg D, Ho WH, et al. 2000. Mice lacking alpha-synuclein display functional deficits in the nigrostriatal dopamine system. Neuron 25: 239-252.

Adamczyk A, Kazmierczak A, Strosznajder JB. 2006. α-Synuclein and its neurotoxic fragment inhibit dopamine uptake into rat striatal synaptosomes. Relationship to nitric oxide. Neurochem Int 49: 407-412.

Alim MA, Ma QL, Takeda K, Aizawa T, Matsubara M, et al. 2004. Demonstration of a role for alpha-synuclein as a functional microtubule-associated protein. J Alzheimers Dis 6: 35-42; discussion 443–449.

Almeida OP, Burton EJ, McKeith I, Gholkar A, Burn D, et al. 2003. MRI study of caudate nucleus volume in Parkinson's disease with and without dementia with Lewy bodies and Alzheimer's disease. Dement Geriatr Cogn Disord 16: 57-63.

Alves da Costa C. 2003. Recent advances on alpha-synuclein cell biology: Functions and dysfunctions. Curr Mol Med 3: 7-24.

Alves-Rodrigues A, Gregori L, Figueiredo-Pereira ME. 1998. Ubiquitin, cellular inclusions and their role in neurodegeneration. Trends Neurosci 21: 516-520.

Amer DA, Irvine GB, El-Agnaf OM. 2006. Inhibitors of α-synuclein oligomerization and toxicity: A future therapeutic strategy for Parkinson's disease and related disorders. Exp Brain Res 173: 223-233.

Andersen JK. 2004. Iron dysregulation and Parkinson's disease. J Alzheimers Dis 6: S47-S52.

Andreasen N, Minthon L, Davidsson P, Vanmechelen E, Vanderstichele H, et al. 2001. Evaluation of CSF-tau and CSF-Abeta42 as diagnostic markers for Alzheimer disease in clinical practice. Arch Neurol 58: 373-379.

Anichtchik OV, Rinne JO, Kalimo H, Panula P. 2000. An altered histaminergic innervation of the substantia nigra in Parkinson's disease. Exp Neurol 163: 20-30.

Antony T, Hoyer W, Cherny D, Heim G, Jovin TM, et al. 2003. Cellular polyamines promote the aggregation of alpha-synuclein. J Biol Chem 278: 3235-3240.

Apaydin H, Ahlskog JE, Parisi JE, Boeve BF, Dickson DW. 2002. Parkinson disease neuropathology: Later-developing dementia and loss of the levodopa response. Arch Neurol 59: 102-112.

Arai K, Kato N, Kashiwado K, Hattori T. 2000. Pure autonomic failure in association with human alpha-synucleinopathy. Neurosci Lett 296: 171-173.

Arai T, Ueda K, Ikeda K, Akiyama H, Haga C, et al. 1999. Argyrophilic glial inclusions in the midbrain of patients with Parkinson's disease and diffuse Lewy body disease are immunopositive for NACP/alpha-synuclein. Neurosci Lett 259: 83-86.

Arai Y, Yamazaki M, Mori O, Muramatsu H, Asano G, et al. 2001. Alpha-synuclein-positive structures in cases with sporadic Alzheimer's disease: Morphology and its relationship to tau aggregation. Brain Res 888: 287-296.

Ardley HC, Scott GB, Rose SA, Tan NG, Robinson PA. 2004. UCH-L1 aggresome formation in response to proteasome impairment indicates a role in inclusion formation in Parkinson's disease. J Neurochem 90: 379-391.

Arima K, Hirai S, Sunohara N, Aoto K, Izumiyama Y, et al. 1999. Cellular co-localization of phosphorylated tau- and NACP/alpha-synuclein-epitopes in Lewy bodies in sporadic Parkinson's disease and in dementia with Lewy bodies. Brain Res 843: 53-61.

Arima K, Mizutani T, Alim MA, Tonozuka-Uehara H, Izumiyama Y, et al. 2000. NACP/alpha-synuclein and tau constitute two distinctive subsets of filaments in the same neuronal inclusions in brains from a family of parkinsonism and dementia with Lewy bodies: Double-immunolabeling fluorescence and electron microscopic studies. Acta Neuropathol (Berl) 100: 115-121.

Armstrong RA, Cairns NJ, Lantos PL. 1997. Laminar distribution of cortical Lewy bodies and neurofibrillary tangles in dementia with Lewy bodies. Neurosci Res Comm 21: 145-152.

Armstrong RA, Cairns NJ, Lantos PL. 1998. The spatial patterns of Lewy bodies, senile plaques, and neurofibrillary tangles in dementia with Lewy bodies. Exp Neurol 150: 122-127.

Armstrong RA, Cairns NJ, Lantos PL. 2000. Beta-amyloid deposition in the temporal lobe of patients with dementia with Lewy bodies: Comparison with non-demented cases and Alzheimer's disease. Dement Geriatr Cogn Disord 11: 187-192.

Armstrong TP, Hansen LA, Salmon DP, Masliah E, Pay M, et al. 1991. Rapidly progressive dementia in a patient with the Lewy body variant of Alzheimer's disease. Neurology 41: 1178-1180.

Arnulf I, Bonnet AM, Damier P, Bejjani BP, Seilhean D, et al. 2000. Hallucinations, REM sleep, and Parkinson's disease: A medical hypothesis. Neurology 55: 281-288.

Atkinson A, Singleton AB, Steward A, Ince PG, Perry RH, et al. 1999. CYP2D6 is associated with Parkinson's disease but not with dementia with Lewy bodies or Alzheimer's disease. Pharmacogenetics 9: 31-35.

Auluck PK, Chan HY, Trojanowski JQ, Lee VM, Bonini NM. 2002. Chaperone suppression of alpha-synuclein toxicity in a Drosophila model for Parkinson's disease. Science 295: 865-868.

Baba M, Nakajo S, Tu PH, Tomita T, Nakaya K, et al. 1998. Aggregation of alpha-synuclein in Lewy bodies of sporadic Parkinson's disease and dementia with Lewy bodies. Am J Pathol 152: 879-884.

Bacci JJ, Kachidian P, Kerkerian-Le Goff L, Salin P. 2004. Intralaminar thalamic nuclei lesions: Widespread impact on dopamine denervation-mediated cellular defects in the rat basal ganglia. J Neuropathol Exp Neurol 63: 20-31.

Baker KG, Huang Y, McCann H, Gai WP, Jensen PH, et al. 2006. P25α immunoreactive but α-synuclein immunonegative neuronal inclusions in multiple system atrophy. Acta Neuropathol (Berl) 111: 193-195.

Ballard CG, Aarsland D, McKeith I, O'Brien J, Gray A, et al. 2002. Fluctuations in attention: PD dementia vs DLB with parkinsonism. Neurology 59: 1714-1720.

Ballard CG, Ayre G, O'Brien J, Sahgal A, McKeith IG, et al. 1999a. Simple standardised neuropsychological assessments aid in the differential diagnosis of dementia with Lewy bodies from Alzheimer's disease and vascular dementia. Dement Geriatr Cogn Disord 10: 104-108.

Ballard CG, Holmes C, McKeith I, Neill D, O'Brien J, et al. 1999b. Psychiatric morbidity in dementia with Lewy bodies: A prospective clinical and neuropathological comparative study with Alzheimer's disease. Am J Psychiatry 156: 1039-1045.

Ballard CG, McKeith I, Burn D, Harrison R, O'Brien J, et al. 1997. The UPDRS scale as a means of identifying extrapyramidal signs in patients suffering from dementia with Lewy bodies. Acta Neurol Scand 96: 366-371.

Ballard CG, O'Brien J, Gray A, Cormack F, Ayre G, et al. 2001a. Attention and fluctuating attention in patients with dementia with Lewy bodies and Alzheimer disease. Arch Neurol 58: 977-982.

Ballard CG, O'Brien J, Morris CM, Barber R, Swann A, et al. 2001b. The progression of cognitive impairment in dementia with Lewy bodies, vascular dementia and Alzheimer's disease. Int J Geriatr Psychiatry 16: 499-503.

Ballard CG, O'Brien J, Swann A, Neill D, Lantos P, et al. 2000. One year follow-up of parkinsonism in dementia with Lewy bodies. Dement Geriatr Cogn Disord 11: 219-222.

Barber PA, Varma AR, Lloyd JJ, Haworth B, Snowden JS, et al. 2000a. The electroencephalogram in dementia with Lewy bodies. Acta Neurol Scand 101: 53-56.

Barber R, Ballard C, McKeith IG, Gholkar A, O'Brien JT. 2000b. MRI volumetric study of dementia with Lewy bodies: A comparison with AD and vascular dementia. Neurology 54: 1304-1309.

Barber R, Gholkar A, Scheltens P, Ballard C, McKeith IG, et al. 2000c. MRI volumetric correlates of white matter lesions in dementia with Lewy bodies and Alzheimer's disease. Int J Geriatr Psychiatry 15: 911-916.

Barber R, Gholkar A, Scheltens P, Ballard C, McKeith IG, et al. 1999. Medial temporal lobe atrophy on MRI in dementia with Lewy bodies. Neurology 52: 1153-1158.

Barber R, McKeith IG, Ballard C, Gholkar A, O'Brien JT. 2001. A comparison of medial and lateral temporal lobe atrophy in dementia with Lewy bodies and Alzheimer's disease: Magnetic resonance imaging volumetric study. Dement Geriatr Cogn Disord 12: 198-205.

Barbieri S, Hofele K, Wiederhold KH, Probst A, Mistl C, et al. 2001. Mouse models of alpha-synucleinopathy and Lewy pathology. Alpha-synuclein expression in transgenic mice. Neuropathology and Genetics of Dementia. Tolnay M, Probst A, editors. New York: Kluwer Academic/Plenum; pp. 147-166.

Barzilai A, Melamed E. 2003. Molecular mechanisms of selective dopaminergic neuronal death in Parkinson's disease. Trends Mol Med 9: 126-132.

Benarroch EE, Schmeichel AM, Low PA, Boeve BF, Sandroni P, et al. 2005. Involvement of medullary regions controlling sympathetic output in Lewy body disease. Brain 128: 338-344.

Benarroch EE, Schmeichel AM, Parisi JE. 2000. Involvement of the ventrolateral medulla in parkinsonism with autonomic failure. Neurology 54: 963-968.

Benjamin R, Leake A, Ince PG, Perry RH, McKeith IG, et al. 1995. Effects of apolipoprotein E genotype on cortical neuropathology in senile dementia of the Lewy body and Alzheimer's disease. Neurodegeneration 4: 443-448.

Bennett MC. 2005. The role of α-synuclein in neurodegenerative diseases. Pharmacol Ther 105: 311-331.

Berg D, Riess O, Bornemann A. 2003. Specification of 14-3-3 proteins in Lewy bodies. Ann Neurol 54: 135.

Berger B, Gaspar P, Verney C. 1991. Dopaminergic innervation of the cerebral cortex: Unexpected differences between rodents and primates. Trends Neurosci 14: 21-27.

Bernheimer H, Birkmayer W, Hornykiewicz O, Jellinger K, Seitelberger F. 1973. Brain dopamine and the syndromes of Parkinson and Huntington. Clinical, morphological and neurochemical correlations. J Neurol Sci 20: 415-455.

Betarbet R, Sherer TB, Di Monte DA, Greenamyre JT. 2002. Mechanistic approaches to Parkinson's disease pathogenesis. Brain Pathol 12: 499-510.

Betarbet R, Sherer TB, Mac Kenzie G, Garcia-Osuna M, Panov AV, et al. 2000. Chronic systemic pesticide exposure reproduces features of Parkinson's disease. Nat Neurosci 3: 1301-1306.

Beyer K. 2006. α-Synuclein structure, posttranslational modification and alternative splicing as aggregation enhancers. Acta Neuropathol (Berl) 112: 237-251.

Beyer K, Lao JI, Carrato C, Mate JL, Lopez D, et al. 2004. Differential expression of alpha-synuclein isoforms in dementia with Lewy bodies. Neuropathol Appl Neurobiol 30: 601-607.

Bezard E, Brotchie JM, Gross CE. 2001. Pathophysiology of levodopa-induced dyskinesia: Potential for new therapies. Nat Rev Neurosci 2: 577-588.

Bibl M, Mollenhauer B, Esselmann H, Lewczuk P, Klafki HW, et al. 2006. CSF amyloid-β-peptides in Alzheimer's disease, dementia with Lewy bodies and Parkinson's disease dementia. Brain 129: 1177-1187.

Bloch A, Probst A, Bissig H, Adams H, Tolnay M. 2006. α-Synuclein pathology of the spinal and peripheral autonomic nervous system in neurologically unimpaired elderly subjects. Neuropathol Appl Neurobiol 32: 284-295.

Boeve BF, Silber MH, Ferman TJ, Kokmen E, Smith GE, et al. 1998. REM sleep behavior disorder and degenerative dementia: An association likely reflecting Lewy body disease. Neurology 51: 363-370.

Boeve BF, Silber MH, Ferman TJ, Lucas JA, Parisi JE. 2001. Association of REM sleep behavior disorder and neurodegenerative disease may reflect an underlying synucleinopathy. Mov Disord 16: 622-630.

Boeve BF, Silber MH, Parisi JE, Dickson DW, Ferman TJ, et al. 2003. Synucleinopathy pathology and REM sleep behavior disorder plus dementia or parkinsonism. Neurology 61: 40-45.

Bohnen NI, Kaufer DI, Ivanco LS, Lopresti B, Koeppe RA, et al. 2003. Cortical cholinergic function is more severely affected in parkinsonian dementia than in Alzheimer disease: An in vivo positron emission tomographic study. Arch Neurol 60: 1745-1748.

Bonifati V, Rizzu P, van Baren MJ, Schaap O, Breedveld GJ, et al. 2003. Mutations in the DJ-1 gene associated with autosomal recessive early-onset parkinsonism. Science 299: 256-259.

Bonini NM, Fortini ME. 2003. Human neurodegenerative disease modeling using Drosophila. Annu Rev Neurosci 26: 627-656.

Bozzali M, Falini A, Cercignani M, Baglio F, Farina E, et al. 2005. Brain tissue damage in dementia with Lewy bodies: An in vivo diffusion tensor MRI study. Brain 128: 1595-1604.

Braak H, Braak E. 2000. Pathoanatomy of Parkinson's disease. J Neurol 247 (Suppl. 2): II3-II10.

Braak H, Braak E, Yilmazer D, de Vos RA, Jansen EN, et al. 1996. Pattern of brain destruction in Parkinson's and Alzheimer's diseases. J Neural Transm 103: 455-490.

Braak H, Del Tredici K, Bratzke H, Hamm-Clement J, Sandmann-Keil D, et al. 2002. Staging of the intracerebral inclusion body pathology associated with idiopathic

Parkinson's disease (preclinical and clinical stages). J Neurol 249: III1-III5.

Braak H, Del Tredici K, Rub U, de Vos RA, Jansen Steur EN, et al. 2003. Staging of brain pathology related to sporadic Parkinson's disease. Neurobiol Aging 24: 197-211.

Braak H, Ghebremedhin E, Rub U, Bratzke H, Del Tredici K. 2004. Stages in the development of Parkinson's disease-related pathology. Cell Tissue Res 318: 121-134.

Braak H, Müller CM, Rüb U, Ackermann H, Bratzke H, et al. 2006. Pathology associated with sporadic Parkinson's disease—where does it end? J Neural Transm Suppl 70: 89-97.

Braak H, Rub U, Jansen Steur EN, Del Tredici K, de Vos RA. 2005. Cognitive status correlates with neuropathologic stage in Parkinson disease. Neurology 64: 1404-1410.

Braak H, Sandmann-Keil D, Gai W, Braak E. 1999. Extensive axonal Lewy neurites in Parkinson's disease: a novel pathological feature revealed by alpha-synuclein immunocytochemistry. Neurosci Lett 265: 67-69.

Bradshaw J, Saling M, Hopwood M, Anderson V, Brodtmann A. 2004. Fluctuating cognition in dementia with Lewy bodies and Alzheimer's disease is qualitatively distinct. J Neurol Neurosurg Psychiatry 75: 382-387.

Brayne C, Zaccai J, McCracken C. 2006. Epidemiology of dementia with Lewy bodies. Dementia with Lewy Bodies and Parkinson's Disease Dementia. Brien JO, Ames D, McKeith I, Chiu E, editors. London: Taylor & Francis; pp. 33-47.

Brett FM, Henson C, Staunton H. 2002. Familial diffuse Lewy body disease, eye movement abnormalities, and distribution of pathology. Arch Neurol 59: 464-467.

Briel RC, McKeith IG, Barker WA, Hewitt Y, Perry RH, et al. 1999. EEG findings in dementia with Lewy bodies and Alzheimer's disease. J Neurol Neurosurg Psychiatry 66: 401-403.

Brooks DJ, Frey KA, Marek KL, Oakes D, Paty D, et al. 2003. Assessment of neuroimaging techniques as biomarkers of the progression of Parkinson's disease. Exp Neurol 184 (Suppl. 1): S68-S79.

Brown DF. 1999. Lewy body dementia. Ann Med 31: 188-196.

Brown DF, Dababo MA, Bigio EH, Risser RC, Eagan KP, et al. 1998a. Neuropathologic evidence that the Lewy body variant of Alzheimer disease represents coexistence of Alzheimer disease and idiopathic Parkinson disease. J Neuropathol Exp Neurol 57: 39-46.

Brown DF, Risser RC, Bigio EH, Tripp P, Stiegler A, et al. 1998b. Neocortical synapse density and Braak stage in the Lewy body variant of Alzheimer disease: A comparison with classic Alzheimer disease and normal aging. J Neuropathol Exp Neurol 57: 955-960.

Buerger K, Zinkowski R, Teipel SJ, Tapiola T, Arai H, et al. 2002. Differential diagnosis of Alzheimer disease with cerebrospinal fluid levels of tau protein phosphorylated at threonine 231. Arch Neurol 59: 1267-1272.

Burn DJ. 2004. Cortical Lewy body disease. J Neurol Neurosurg Psychiatry 75: 175-178.

Burn DJ, Mosimann UP, McKeith IG. 2005. Clinical diagnosis of dementia with Lewy bodies. Current Clinical Neurology: Atypical Parkinsonian Disorders. Litvan I, editor. Totowa, NJ: Humana Press Inc.; pp. 361-371.

Burn DJ, Rowan EN, McKeith IG, O'Brien JT, Minett T, 2002. Is parkinsonism phenotype predictive of dementia? A cross-sectional study. 7th International Congress of Parkinson's Disease and Movement Disorders. Miamai, FL; p. S 22.

Burn DJ, Rowan EN, Minett T, Sanders J, Myint P, et al. 2003. Extrapyramidal features in Parkinson's disease with and without dementia and dementia with Lewy bodies: A cross-sectional comparative study. Mov Disord 18: 884-889.

Burton EJ, Karas G, Paling SM, Barber R, Williams ED, et al. 2002. Patterns of cerebral atrophy in dementia with Lewy bodies using voxel-based morphometry. Neuroimage 17: 618-630.

Bussell R Jr, Eliezer D. 2003. A structural and functional role for 11-mer repeats in alpha-synuclein and other exchangeable lipid binding proteins. J Mol Biol 329: 763-778.

Byrne EJ, Lennox G, Godwin-Austen LB. 1991. Dementia associated with cortical Lewy bodies: proposed diagnostic criteria. Dementia 2: 283-284.

Cabello CR, Thune JJ, Pakkenberg H, Pakkenberg B. 2002. Ageing of substantia nigra in umans: cell loss may be compensated by hypertrophy. Neuropathol Appl Neurobiol 28: 283-291.

Calderon J, Perry RJ, Erzinclioglu SW, Berrios GE, Dening TR, et al. 2001. Perception, attention, and working memory are disproportionately impaired in dementia with Lewy bodies compared with Alzheimer's disease. J Neurol Neurosurg Psychiatry 70: 157-164.

Calon F, Birdi S, Rajput AH, Hornykiewicz O, Bedard PJ, et al. 2002. Increase of preproenkephalin mRNA levels in the putamen of Parkinson disease patients with levodopa-induced dyskinesias. J Neuropathol Exp Neurol 61: 186-196.

Camicioli R, Moore MM, Kinney A, Corbridge E, Glassberg K, et al. 2003. Parkinson's disease is associated with hippocampal atrophy. Mov Disord 18: 784-790.

Camicioli RM, Korzan JR, Foster SL, Fisher NJ, Emery DJ, et al. 2004. Posterior cingulate metabolic changes occur in Parkinson's disease patients without dementia. Neurosci Lett 354: 177-180.

Campbell BC, Li QX, Culvenor JG, Jakala P, Cappai R, et al. 2000. Accumulation of insoluble alpha-synuclein in dementia with Lewy bodies. Neurobiol Dis 7: 192-200.

Campbell BC, McLean CA, Culvenor JG, Gai WP, Blumbergs PC, et al. 2001. The solubility of alpha-synuclein in multiple system atrophy differs from that of dementia with Lewy bodies and Parkinson's disease. J Neurochem 76: 87-96.

Canet-Aviles RM, Wilson MA, Miller DW, Ahmad R, McLendon C, et al. 2004. The Parkinson's disease protein DJ-1 is neuroprotective due to cysteine-sulfinic acid-driven mitochondrial localization. Proc Natl Acad Sci USA 101: 9103-9108.

Cantuti-Castelvetri I, Klucken J, Ingelsson M, Ramasamy K, McLean PJ, et al. 2005. α-synuclein and chaperones in dementia with Lewy bodies. J Neuropathol Exp Neurol 64: 1058-1066.

Castellani RJ, Siedlak SL, Perry G, Smith MA. 2000. Sequestration of iron by Lewy bodies in Parkinson's disease. Acta Neuropathol (Berl) 100: 111-114.

Cercy SP, Bylsma FW. 1997. Lewy bodies and progressive dementia: a critical review and meta-analysis. J Int Neuropsychol Soc 3: 179-194.

Chang S, Kim JH, Shin J. 2002. p62 Forms a ternary complex with PKCzeta and PAR-4 and antagonizes PAR-4-induced PKCzeta inhibition. FEBS Lett 510: 57-61.

Chase TN, Oh JD. 2000. Striatal mechanisms and pathogenesis of parkinsonian signs and motor complications. Ann Neurol 47: S122-S129.

Chen L, Feany MB. 2005. α-Synuclein phosphorylation controls neurotoxicity and inclusion formation in a Drosophila model of Parkinson disease. Nat Neurosci 8: 657-663.

Chilcote TJ, de Laat R, Keim P, Huang J, Barbour R, 2003. Comparison of alpha-synuclein species in Lewy bodies and the soluble fraction of diffuse Lewy body disease brain (abstract). Soceity for Neuroscience. 33rd Annual Meeting, Society for Neuroscience, New Orleans; pp. Program No. 297.13.

Choi J, Rees HD, Weintraub ST, Levey AI, Chin LS, Li L. 2005. Oxidative modifications and aggregation of Cu,Zn-superoxide dismutase associated with Alzheimer and Parkinson diseases. J Biol Chem 280: 11648-11655.

Chung KK, Dawson VL, Dawson TM. 2001a. The role of the ubiquitin-proteasomal pathway in Parkinson's disease and other neurodegenerative disorders. Trends Neurosci 24: S7-S14.

Chung KK, Zhang Y, Lim KL, Tanaka Y, Huang H, et al. 2001b. Parkin ubiquitinates the alpha-synuclein-interacting protein, synphilin-1: Implications for Lewy-body formation in Parkinson disease. Nat Med 7: 1144-1150.

Churchyard A, Lees AJ. 1997. The relationship between dementia and direct involvement of the hippocampus and amygdala in Parkinson's disease. Neurology 49: 1570-1576.

Ciechanover A, Brundin P. 2003. The ubiquitin proteasome system in neurodegenerative diseases: Sometimes the chicken, sometimes the egg. Neuron 40: 427-446.

Ciliax BJ, Drash GW, Staley JK, Haber S, Mobley CJ, et al. 1999. Immunocytochemical localization of the dopamine transporter in human brain. J Comp Neurol 409: 38-56.

Clayton DF, George JM. 1999. Synucleins in synaptic plasticity and neurodegenerative disorders. J Neurosci Res 58: 120-129.

Collerton D, Burn D, McKeith I, O'Brien J. 2003. Systematic review and meta-analysis show that dementia with Lewy bodies is a visual-perceptual and attentional-executive dementia. Dement Geriatr Cogn Disord 16: 229-237.

Colloby SJ, Fenwick JD, Williams ED, Paling SM, Lobotesis K, et al. 2002. A comparison of (99m)Tc-HMPAO SPET changes in dementia with Lewy bodies and Alzheimer's disease using statistical parametric mapping. Eur J Nucl Med Mol Imaging 29: 615-622.

Colosimo C, Hughes AJ, Kilford L, Lees AJ. 2003. Lewy body cortical involvement may not always predict dementia in Parkinson's disease. J Neurol Neurosurg Psychiatry 74: 852-856.

Connor DJ, Salmon DP, Sandy TJ, Galasko D, Hansen LA, et al. 1998. Cognitive profiles of autopsy-confirmed Lewy body variant vs pure Alzheimer disease. Arch Neurol 55: 994-1000.

Conway KA, Lee SJ, Rochet JC, Ding TT, Williamson RE, et al. 2000. Acceleration of oligomerization, not fibrillization, is a shared property of both alpha-synuclein mutations linked to early-onset Parkinson's disease: Implications for pathogenesis and therapy. Proc Natl Acad Sci USA 97: 571-576.

Conway KA, Rochet JC, Bieganski RM, Lansbury PT Jr. 2001. Kinetic stabilization of the alpha-synuclein protofibril by a dopamine-alpha-synuclein adduct. Science 294: 1346-1349.

Corcoran LJ, Mitchison TJ, Liu Q. 2004. A novel action of histone deacetylase inhibitors in a protein aggresome disease model. Curr Biol 14: 488-492.

Corti O, Hampe C, Koutnikova H, Darios F, Jacquier S, et al. 2003. The p38 subunit of the aminoacyl-tRNA synthetase complex is a Parkin substrate: Linking protein biosynthesis and neurodegeneration. Hum Mol Genet 12: 1427-1437.

Court JA, Piggott MA, Lloyd S, Cookson N, Ballard CG, et al. 2000. Nicotine binding in human striatum: Elevation in schizophrenia and reductions in dementia with Lewy bodies. Parkinson's disease and Alzheimer's disease and in relation to neuroleptic medication. Neuroscience 98: 79-87.

Cousins DA, Burton EJ, Burn D, Gholkar A, McKeith IG, et al. 2003. Atrophy of the putamen in dementia with Lewy bodies but not Alzheimer's disease: An MRI study. Neurology 61: 1191-1195.

Crowther RA, Jakes R, Spillantini MG, Goedert M. 1998. Synthetic filaments assembled from C-terminally truncated alpha-synuclein. FEBS Lett 436: 309-312.

Crowther RA, Daniel SE, Goedert M. 2000. Characterisation of isolated alpha-synuclein filaments from substantia nigra of Parkinson's disease brain. Neurosci Lett 292: 128-130.

Cuervo AM, Stefanis L, Fredenburg R, Lansbury PT, Sulzer D. 2004. Impaired degradation of mutant α-synuclein by chaperone-mediated autophagy. Science 305: 1292-1295.

Cummings JL. 1995. Lewy body diseases with dementia: Pathophysiology and treatment. Brain Cogn 28: 266-280.

Da Costa CA, Masliah E, Checler F. 2003. Beta-synuclein displays an antiapoptotic p53-dependent phenotype and protects neurons from 6-hydroxydopamine-induced caspase 3 activation: Cross-talk with alpha-synuclein and implication for Parkinson's disease. J Biol Chem 278: 37330-37335.

Da Cunha C, Wietzikoski S, Wietzikoski EC, Miyoshi E, Ferro MM, et al. 2003. Evidence for the substantia nigra pars compacta as an essential omponent of a memory system independent of the hippocampal memory system. Neurobiol Learn Mem 79: 236-242.

Dale GE, Probst A, Luthert P, Martin J, Anderton BH, et al. 1992. Relationships between Lewy bodies and pale bodies in Parkinson's disease. Acta Neuropathol (Berl) 83: 525-529.

Damier P, Hirsch EC, Agid Y, Graybiel AM. 1999. The substantia nigra of the human brain. II. Patterns of loss of dopamine-containing neurons in Parkinson's disease. Brain 122: 1437-1448.

Dauer W, Przedborski S. 2003. Parkinson's disease: Mechanisms and models. Neuron 39: 889-909.

Dauer W, Kholodilov N, Vila M, Trillat AC, Goodchild R, et al. 2002. Resistance of alpha -synuclein null mice to the parkinsonian neurotoxin MPTP. Proc Natl Acad Sci USA 99: 14524-14529.

Dawson TM, Dawson VL. 2003. Molecular pathways of neurodegeneration in Parkinson's disease., Science 302: 819-822.

Dawson VL. 2000. Neurobiology: Of flies and mice. Science 288: 631-632.

De Ceballos ML, Lopez-Lozano JJ. 1999. Subgroups of parkinsonian patients differentiated by peptidergic immunostaining of caudate nucleus biopsies. Peptides 20: 249-257.

De Lucia MW, Cookson N, Dickson DW. 2002. Synuclein-immunoreactive Lewy bodies are detected in the amygdala in less than 20% of Alzheimer's disease (AD) cases (abstr.). J Neuropathol Exp Neurol 61: 454.

Del Ser T, Hachinski V, Merskey H, Munoz DG. 2001. Clinical and pathologic features of two groups of patients with dementia with Lewy bodies: Effect of coexisting Alzheimer-type lesion load. Alzheimer Dis Assoc Disord 15: 31-44.

Del Ser T, Munoz DG, Hachinski V. 1996. Temporal pattern of cognitive decline and incontinence is different in Alzheimer's disease and diffuse Lewy body disease. Neurology 46: 682-686.

Del Tredici K, Rub U, De Vos RA, Bohl JR, Braak H. 2002. Where does parkinson disease pathology begin in the brain? J Neuropathol Exp Neurol 61: 413-426.

Delacourte A, Sergeant N, Champain D, Wattez A, Maurage CA, et al. 2002. Nonoverlapping but synergetic tau and APP pathologies in sporadic Alzheimer's disease. Neurology 59: 398-407.

Deramecourt V, Bombois S, Maurage CA, Ghestem A, Drobecq H, Vanmechelen E, Lebert F, Pasquier F, Delacourte A. 2006. Biochemical staging of synucleinopathy and amyloidopathy in dementia with Lewy bodies. J Neuropathol Exp Neurol 65: 278-288.

Deuschl G, Volkmann J. 2002. Tremors: Differential diagnosis, pathophysiology, and therapy. Parkinson's Disease and Movement Disorders, 4th edn. Jankovic JJ, Tolosa E, editors. Philadelphia: Lippincott Williams & Wilkins; pp. 270-290.

Deuschl G, Wenzelburger R, Loffler K, Raethjen J, Stolze H. 2000. Essential tremor and cerebellar dysfunction clinical and kinematic analysis of intention tremor. Brain 123 (Pt. 8): 1568-1580.

Dev KK, Hofele K, Barbieri S, Buchman VL, Putten van der H. 2003. Part II: Alpha-synuclein and its molecular pathophysiological role in neurodegenerative disease. Neuropharmacology 45: 14-44.

Dickson D. 2003. Synucleinopathies. Neurodegeneration: The Molecular Pathology of Dementia and Movement Disorders. Dickson DW, editor. Basel: ISN Neuropath Press; pp. 155-223.

Dickson DW. 2001. Alpha-synuclein and the Lewy body disorders. Curr Opin Neurol 14: 423-432.

Dickson DW. 2002a. Dementia with Lewy bodies: Neuropathology. J Geriatr Psychiatry Neurol 15: 210-216.

Dickson DW. 2002b. Neuropathology of parkinsonian disorders. Parkinson's Disease and Movement Disorders. Jankovic JJ, Tolosa E, editors. Philadelphia: Lippincott Williams & Wilkins; pp. 256-269.

Dickson DW. 2006. Neuropathology and pathogenesis of dementia with Lewy bodies. Dementia with Lewy Bodies and Parkinson's Disease Dementia. Brien JO, Ames D, McKeith I, Chiu E, editors. London: Taylor & Francis; pp. 149-166.

Dickson DW, Davies P, Mayeux R, Crystal H, Horoupian DS, et al. 1987. Diffuse Lewy body disease. Neuropathological and biochemical studies of six patients. Acta Neuropathol (Berl) 75: 8-15.

Dickson DW, Ruan D, Crystal H, Mark MH, Davies P, et al. 1991. Hippocampal degeneration differentiates diffuse Lewy body disease (DLBD) from Alzheimer's disease: Light and electron microscopic immunocytochemistry of CA2-3 neurites specific to DLBD. Neurology 41: 1402-1409.

Dickson DW, Schmidt ML, Lee VM, Zhao ML, et al. 1994. Immunoreactivity profile of hippocampal CA2/3 neurites in diffuse Lewy body disease. Acta Neuropathol (Berl) 87: 269-276.

Ding ZT, Wang Y, Jiang YP, Hashizume Y, Yoshida M, et al. 2006. Characteristics of α-synucleinopathy in centenarians. Acta Neuropathol (Berl) 111: 450-458.

Donnemiller E, Heilmann J, Wenning GK, Berger W, Decristoforo C, et al. 1997. Brain perfusion scintigraphy with 99mTc-HMPAO or 99mTc-ECD and 123I-beta-CIT single-photon emission tomography in dementia of the Alzheimer-type and diffuse Lewy body disease. Eur J Nucl Med 24: 320-325.

Double KL, Ben-Shachar D, Youdim MB, Zecca L, Riederer P, et al. 2002. Influence of neuromelanin on oxidative pathways within the human substantia nigra. Neurotoxicol Teratol 24: 621-628.

Doubleday EK, Snowden JS, Varma AR, Neary D. 2002. Qualitative performance characteristics differentiate dementia with Lewy bodies and Alzheimer's disease. J Neurol Neurosurg Psychiatry 72: 602-607.

Drach LM, Bohl J, Wach S, Schlote W, Goebel HH. 1998. Reduced intraneuronal lipofuscin content in dementia with Lewy bodies compared with Alzheimer's disease and controls. Dement Geriatr Cogn Disord 9: 1-5.

Duda JE. 2004. Pathology and neurotransmitter abnormalities of dementia with Lewy bodies. Dement Geriatr Cogn Disord 17 (Suppl. 1): 3-14.

Duda JE, Lee VM, Trojanowski JQ. 2000. Neuropathology of synuclein aggregates. J Neurosci Res 61: 121-127.

Duda JE, Giasson BI, Mabon ME, Lee VM, Trojanowski JQ. 2002a. Novel antibodies to synuclein show abundant striatal pathology in Lewy body diseases. Ann Neurol 52: 205-210.

Duda JE, Giasson BI, Mabon ME, Miller DC, Golbe LI, et al. 2002b. Concurrence of alpha-synuclein and tau brain pathology in the Contursi kindred. Acta Neuropathol (Berl) 104: 7-11.

Duyckaerts C. 2002. Is Parkinson's disease with dementia and dementia with Lewy body the same disease? Mov Disord 17 (Suppl. 5): S9.

Elkon H, Don J, Melamed E, Ziv I, Shirvan A, et al. 2002. Mutant and wild-type alpha-synuclein interact with mitochondrial cytochrome C oxidase. J Mol Neurosci 18: 229-238.

Ellis CE, Schwartzberg PL, Grider TL, Fink DW, Nussbaum RL. 2001. Alpha-synuclein is phosphorylated by members of the Src family of protein-tyrosine kinases. J Biol Chem 276: 3879-3884.

Emerit J, Edeas M, Bricaire F. 2004. Neurodegenerative diseases and oxidative stress. Biomed Pharmacother 58: 39-46.

Emre M. 2003. Dementia associated with Parkinson's disease. Lancet Neurol 2: 229-237.

Engelender S, Kaminsky Z, Guo X, Sharp AH, Amaravi RK, et al. 1999. Synphilin-1 associates with alpha-synuclein and promotes the formation of cytosolic inclusions. Nat Genet 22: 110-1104.

Eve DJ, Nisbet AP, Kingsbury AE, Temlett J, Marsden CD, et al. 1997. Selective increase in somatostatin mRNA expression in human basal ganglia in Parkinson's disease. Brain Res Mol Brain Res 50: 59-70.

Farrer M, Kachergus J, Forno L, Lincoln S, Wang DS, et al. 2004. Comparison of kindreds with parkinsonism and alpha-synuclein genomic multiplications. Ann Neurol 55: 174-179.

Fasano M, Giraudo S, Coha S, Bergamasco B, Lopiano L. 2003. Residual substantia nigra neuromelanin in Parkinson's disease is cross-linked to alpha-synuclein. Neurochem Int 42: 603-606.

Faucheux BA, Martin ME, Beaumont C, Hauw JJ, Agid Y, et al. 2003. Neuromelanin associated redox-active iron is increased in the substantia nigra of patients with Parkinson's disease. J Neurochem 86: 1142-1148.

Feany MB, Bender WW. 2000. A Drosophila model of Parkinson's disease. Nature 404: 394-398.

Fearnley JM, Lees AJ. 1994. Pathology of Parkinson's disease. Neurodegenerative Diseases. Calne DB, editor. Philadelphia: Saunders; pp. 545-554.

Ferman TJ, Boeve BF, Smith GE, Silber MH, Lucas JA, et al. 2002. Dementia with Lewy bodies may present as dementia and REM sleep behavior disorder without parkinsonism or hallucinations. J Int Neuropsychol Soc 8: 907-914.

Ferman TJ, Smith GE, Boeve BF, Ivnik RJ, Petersen RC, et al. 2004. DLB fluctuations: Specific features that reliably differentiate DLB from AD and normal aging. Neurology 62: 181-187.

Fernagut PO, Chesselet MF. 2004. Alpha-synuclein and transgenic mouse models. Neurobiol Dis 17: 123-130.

Ferrer I, Blanco R, Carmona M, Puig B, Barrachina M, et al. 2001. Active, phosphorylation-dependent mitogen-activated protein kinase (MAPK/ERK), stress-activated protein kinase/c-Jun N-terminal kinase (SAPK/JNK), and p38 kinase expression in Parkinson's disease and Dementia with Lewy bodies. J Neural Transm 108: 1383-1396.

Fink AL. 2006. The aggregation and fibrillation of α-synuclein. Acc Chem Res 39: 628-634.

Foltynie T, Brayne C, Barker RA. 2002. The heterogeneity of idiopathic Parkinson's disease. J Neurol 249: 138-145.

Forman MS, Schmidt ML, Kasturi S, Perl DP, Lee VM, et al. 2002. Tau and alpha-synuclein pathology in amygdala of Parkinsonism-dementia complex patients of Guam. Am J Pathol 160: 1725-1731.

Fornai F, Soldani P, Lazzeri G, di Poggio AB, Biagioni F, et al. 2005. Neuronal inclusions in degenerative disorders Do they represent static features or a key to understand the dynamics of the disease? Brain Res Bull 65: 275-290.

Forno LS. 1996. Neuropathology of Parkinson's disease. J Neuropathol Exp Neurol 55: 259-272.

Forstl H, Burns A, Luthert P, Cairns N, Levy R. 1993. The Lewy-body variant of Alzheimer's disease. Clinical and pathological findings. Br J Psychiatry 162: 385-392.

Fortin DL, Troyer MD, Nakamura K, Kubo S, Anthony MD, et al. 2004. Lipid rafts mediate the synaptic localization of alpha-synuclein. J Neurosci 24: 6715-6723.

Francis PT, Perry EK, Piggott MA, Duda JE. 2006. Neurochemical pathology of dementia with Lewy bodies. Dementia with Lewy Bodies and Parkinson's Disease Dementia. Brien JO, Ames D, McKeith I, Chiu E, editors. London: Taylor & Francis; pp. 167-175.

Fujishiro H, Umegaki H, Isojima D, Akatsu H, Iguchi A, et al. 2006. Depletion of cholinergic neurons in the nucleus of the medial septum and the vertical limb of the diagonal band in dementia with Lewy bodies. Acta Neuropathol (Berl) 111: 109-114.

Fujita Y, Ohama E, Takatama M, Al-Sarraj S, Okamoto K. 2006. Fragmentation of Golgi apparatus of nigral neurons with α-synuclein-positive inclusions in patients with Parkinson's disease. Acta Neuropathol (Berl) 112: 261-265.

Fujiwara H, Hasegawa M, Dohmae N, Kawashima A, Masliah E, et al. 2002. Alpha-Synuclein is phosphorylated in synucleinopathy lesions. Nat Cell Biol 4: 160-164.

Furukawa Y, Vigouroux S, Wong H, Guttman M, Rajput AH, et al. 2002. Brain proteasomal function in sporadic Parkinson's disease and related disorders. Ann Neurol 51: 779-782.

Furuno T, Kawanishi C, Iseki E, Onishi H, Sugiyama N, et al. 2001. No evidence of an association between CYP2D6 polymorphisms among Japanese and dementia with Lewy bodies. Psychiatry Clin Neurosci 55: 89-92.

Gai WP, Blessing WW, Blumbergs PC. 1995. Ubiquitin-positive degenerating neurites in the brainstem in Parkinson's disease. Brain 118 (Pt. 6): 1447-1459.

Gai WP, Blumbergs PC, Geffen LB, Blessing WW. 1992. Age-related loss of dorsal vagal neurons in Parkinson's disease. Neurology 42: 2106-2111.

Gai WP, Vickers JC, Blumbergs PC, Blessing WW. 1994. Loss of non-phosphorylated neurofilament immunoreactivity, with preservation of tyrosine hydroxylase, in surviving substantia nigra neurons in Parkinson's disease. J Neurol Neurosurg Psychiatry 57: 1039-1046.

Gai WP, Yuan HX, Li XQ, Power JT, Blumbergs PC, et al. 2000. In situ and in vitro study of colocalization and segregation of α-synuclein, ubiquitin, and lipids in Lewy bodies. Exp Neurol 166: 324-333.

Galasko D, Saitoh T, Xia Y, Thal LJ, Katzman R, et al. 1994. The apolipoprotein E allele ε 4 is overrepresented in patients with the Lewy body variant of Alzheimer's disease. Neurology 44: 1950-1951.

Galpern WR, Lang AE. 2006. Interface between tauopathies and synucleinopathies: A tale of two proteins. Ann Neurol 59: 449-458.

Galvin JE. 2006. Interaction of α-synuclein and dopamine metabolites in the pathogenesis of Parkinson's disease: A case for the selective vulnerability of the substantia nigra. Acta Neuropathol (Berl) 112: 115-126.

Galvin JE, Lee VM, Baba M, Mann DM, Dickson DW, et al. 1997. Monoclonal antibodies to purified cortical Lewy bodies recognize the mid-size neurofilament subunit. Ann Neurol 42: 595-603.

Galvin JE, Lee SL, Perry A, Havlioglu N, McKeel DW, et al. 2002. Familial dementia with Lewy bodies: Clinicopathologic analysis of two kindreds. Neurology 59: 1079-1082.

Galvin JE, Lee VM, Trojanowski JQ. 2001. Synucleinopathies: Clinical and pathological implications. Arch Neurol 58: 186-1890.

Gearing M, Lynn M, Mirra SS. 1999. Neurofibrillary pathology in Alzheimer disease with Lewy bodies: Two subgroups. Arch Neurol 56: 203-208.

Gearing M, Olson DA, Watts RL, Mirra SS. 1994. Progressive supranuclear palsy: Neuropathologic and clinical heterogeneity. Neurology 44: 1015-1024.

Gelb DJ, Oliver E, Gilman S. 1999. Diagnostic criteria for Parkinson disease. Arch Neurol 56: 33-39.

Gerlach M, Reichmann H, Riederer P. 2003. Die Parkinson-Krankheit. 3rd edn. Grundlagen, Klinik, Therapie, Wien-New York: Springer-Verlag.

Gertz HJ, Siegers A, Kuchinke J. 1994. Stability of cell size and nucleolar size in Lewy body containing neurons of substantia nigra in Parkinson's disease. Brain Res 637: 339-341.

Ghee M, Fournier A, Mallet J. 2000. Rat alpha-synuclein interacts with Tat binding protein 1, a component of the 26S proteasomal complex. J Neurochem 75: 2221-2224.

Giasson BI, Lee VM. 2000. A new link between pesticides and Parkinson's disease. Nat Neurosci 3: 1227-1228.

Giasson BI, Lee VM. 2003. Are ubiquitination pathways central to Parkinson's disease? Cell 114: 1-8.

Giasson BI, Duda JE, Quinn SM, Zhang B, Trojanowski JQ, et al. 2002. Neuronal alpha-synucleinopathy with severe movement disorder in mice expressing A53T human alpha-synuclein. Neuron 34: 521-533.

Giasson BI, Duda JE, Murray IV, Chen Q, Souza JM, et al. 2000. Oxidative damage linked to neurodegeneration by selective alpha-synuclein nitration in synucleinopathy lesions. Science 290: 985-989.

Giasson BI, Forman MS, Higuchi M, Golbe LI, Graves CL, et al. 2003. Initiation and synergistic fibrillization of tau and alpha-synuclein. Science 300: 636-640.

Giasson BI, Uryu K, Trojanowski JQ, Lee VM. 1999. Mutant and wild type human alpha-synucleins assemble into elongated filaments with distinct morphologies in vitro. J Biol Chem 274: 7619-7622.

Gibb WR, Esiri MM, Lees AJ. 1987. Clinical and pathological features of diffuse cortical Lewy body disease (Lewy body dementia). Brain 110 (Pt. 5): 1131-1153.

Gibb WR, Luthert PJ, Janota I, Lantos PL. 1989. Cortical Lewy body dementia: Clinical features and classification. J Neurol Neurosurg Psychiatry 52: 185-192.

Gilgun-Sherki Y, Hellmann M, Melamed E, Offen D. 2004. The role of neurotransmitters and neuropeptides in Parkinson's disease: Implications for therapy. Drug Future 29: 1261-1272.

Gilman S, Koeppe RA, Little R, An H, Junck L, et al. 2005. Differentiation of Alzheimer's disease from dementia with Lewy bodies utilizing positron emission tomography with [(18)F]fluorodeoxyglucose and neuropsychological testing. Exp Neurol 191 (Suppl. 1): S95-S103.

Giorgi FS, di Poggio AB, Battaglia G, Pellegrini G, Murri L, et al. 2006. A short overview on the role of α-synuclein and proteasome in experimental models of Parkinson's disease. J Neural Transm Suppl. 70: 105-109.

Goedert M. 2001. Alpha-synuclein and neurodegenerative diseases. Nat Rev Neurosci 2: 492-501.

Goedert M. 2003. Introduction to the tauopathies. Neurodegeneration: The Molecular Pathology of Dementia and Movement Disorders. Dickson DW, editor. Basel: ISN Press; pp. 82-85.

Goedert M, Spillantini MG. 2005. Neurodegenerative alpha-synucleinopathies. Current Clinical Neurology: Atypical Parkinsonian Disorders. Litvan I, editor. Totowa, NJ: Humana Press Inc.; pp. 77-94.

Goker-Alpan O, Giasson BI, Eblan MJ, Nguyen J, Hurtig HI, et al. 2006. Glucocerebrosidase mutations are an important risk factor for Lewy body disorders. Neurology 67: 908-910.

Goldberg MS, Lansbury PT Jr. 2000. Is there a cause-and-effect relationship between alpha-synuclein fibrillization and Parkinson's disease? Nat Cell Biol 2: E115-E119.

Golts N, Snyder H, Frasier M, Theisler C, Choi P, et al. 2002. Magnesium inhibits spontaneous and iron-induced aggregation of alpha-synuclein. J Biol Chem 277: 16116-16123.

Gomez-Isla T, Growdon WB, McNamara M, Newell K, Gomez-Tortosa E, et al. 1999. Clinicopathologic correlates in temporal cortex in dementia with Lewy bodies. Neurology 53: 2003-2009.

Gomez-Isla T, Irizarry MC, Mariash A, Cheung B, Soto O, et al. 2003. Motor dysfunction and gliosis with preserved dopaminergic markers in human alpha-synuclein A30P transgenic mice. Neurobiol Aging 24: 245-258.

Gomez-Tortosa E, Gonzalo I, Fanjul S, Sainz MJ, Cantarero S, et al. 2003. Cerebrospinal fluid markers in dementia with Lewy bodies compared with Alzheimer disease. Arch Neurol 60: 1218-1222.

Gomez-Tortosa E, Newell K, Irizarry MC, Albert M, Growdon JH, et al. 1999. Clinical and quantitative pathologic correlates of dementia with Lewy bodies. Neurology 53: 1284-1291.

Gomez-Tortosa E, Newell K, Irizarry MC, Sanders JL, Hyman BT. 2000. Alpha-synuclein immunoreactivity in dementia with Lewy bodies: Morphological staging and comparison with ubiquitin immunostaining. Acta Neuropathol (Berl) 99: 352-357.

Gomez-Tortosa E, Sanders JL, Newell K, Hyman BT. 2001. Cortical neurons expressing calcium binding proteins are spared in dementia with Lewy bodies. Acta Neuropathol (Berl) 101: 36-42.

Gotz ME, Double K, Gerlach M, Youdim MB, Riederer P, 2004. The relevance of iron in the pathogenesis of Parkinson's disease. Ann N Y Acad Sci 1012: 193-208.

Grace J, Daniel S, Stevens T, Shankar KK, Walker Z, et al. 2001. Long-Term use of rivastigmine in patients with dementia with Lewy bodies: An open-label trial. Int Psychogeriatr 13: 199-205.

Graeber MB, Moran LB. 2002. Mechanisms of cell death in neurodegenerative diseases: Fashion, fiction, and facts. Brain Pathol 12: 385-390.

Greene JC, Whitworth AJ, Kuo I, Andrews LA, Feany MB, et al. 2003. Mitochondrial pathology and apoptotic muscle degeneration in Drosophila parkin mutants. Proc Natl Acad Sci USA 100: 4078-4083.

Gwinn-Hardy K, Mehta ND, Farrer M, Maraganore D, et al. 2000. Distinctive neuropathology revealed by alpha-synuclein antibodies in hereditary parkinsonism and dementia linked to chromosome 4p. Acta Neuropathol (Berl) 99: 663-672.

Halliday GM, Del Tredici K, Braak H. 2006. Critical appraisal of brain pathology staging related to presymptomatic and symptomatic cases of sporadic Parkinson's disease. J Neural Transm (Suppl. 70): 99-103.

Halliday G, Ophof A, Broe M, Jensen PH, Kettle E, et al. 2005. α-Synuclein redistributes to neuromelanin lipid in the substantia nigra early in Parkinson's disease. Brain 128: 2654-2664.

Halliday GM, Li YW, Blumbergs PC, Joh TH, Cotton RG, et al. 1990. Neuropathology of immunohistochemically identified brainstem neurons in Parkinson's disease. Ann Neurol 27: 373-385.

Hamilton RL. 2000. Lewy bodies in Alzheimer's disease: A neuropathological review of 145 cases using alpha-synuclein immunohistochemistry. Brain Pathol 10: 378-384.

Hansen L, Salmon D, Galasko D, Masliah E, Katzman R, et al. 1990. The Lewy body variant of Alzheimer's disease: A clinical and pathologic entity. Neurology 40: 1-8.

Hansen LA, Samuel W. 1997. Criteria for Alzheimer's disease and the nosology of dementia with Lewy bodies. Neurology 48: 126-132.

Hansen LA, Daniel SE, Wilcock GK, Love S. 1998. Frontal cortical synaptophysin in Lewy body diseases: Relation to Alzheimer's disease and dementia. J Neurol Neurosurg Psychiatry 64: 653-656.

Hansen LA, Masliah E, Galasko D, Terry RD. 1993. Plaque-only Alzheimer disease is usually the Lewy body variant, and vice versa. J Neuropathol Exp Neurol 52: 648-654.

Hansen LA, Masliah E, Terry RD, Mirra SS. 1989. A neuropathological subset of Alzheimer's disease with concomitant Lewy body disease and spongiform change. Acta Neuropathol (Berl) 78: 194-201.

Harding AJ, Halliday GM. 1998. Simplified neuropathological diagnosis of dementia with Lewy bodies. Neuropathol Appl Neurobiol 24: 195-201.

Harding AJ, Halliday GM. 2001. Cortical Lewy body pathology in the diagnosis of dementia. Acta Neuropathol (Berl) 102: 355-363.

Harding AJ, Broe GA, Halliday GM. 2002a. Visual hallucinations in Lewy body disease relate to Lewy bodies in the temporal lobe. Brain 125: 391-403.

Harding AJ, Lakay B, Halliday GM. 2002b. Selective hippocampal neuron loss in dementia with Lewy bodies. Ann Neurol 51: 125-128.

Harding AJ, Stimson E, Henderson JM, Halliday GM. 2002c. Clinical correlates of selective pathology in the amygdala of patients with Parkinson's disease. Brain 125: 2431-2445.

Hardy J, Selkoe DJ. 2002. The amyloid hypothesis of Alzheimer's disease: Progress and problems on the road to therapeutics. Science 297: 353-356.

Harrington CR, Louwagie J, Rossau R, Vanmechelen E, Perry RH, et al. 1994. Influence of apolipoprotein E genotype on senile dementia of the Alzheimer and Lewy body types. Significance for etiological theories of Alzheimer's disease. Am J Pathol 145: 1472-1484.

Hartmann A, Hunot S, Michel PP, Muriel MP, Vyas S, et al. 2000. Caspase-3: A vulnerability factor and final effector in apoptotic death of dopaminergic neurons in Parkinson's disease. Proc Natl Acad Sci USA 97: 2875-2880.

Harvey GT, Hughes J, McKeith IG, Briel R, Ballard C, et al. 1999. Magnetic resonance imaging differences between dementia with Lewy bodies and Alzheimer's disease: A pilot study. Psychol Med 29: 181-187.

Hasegawa M, Fujiwara H, Nonaka T, Wakabayashi K, Takahashi H, et al. 2002. Phosphorylated alpha-synuclein is ubiquitinated in alpha-synucleinopathy lesions. J Biol Chem 277: 49071-49076.

Hasegawa T, Matsuzaki-Kobayashi M, Takeda A, Sugeno N, Kikuchi A, et al. 2006. α-synuclein facilitates the toxicity of oxidized catechol metabolites: Implications for selective neurodegeneration in Parkinson's disease. FEBS Lett 580: 2147-2152.

Hashimoto M, Hsu LJ, Sisk A, Xia Y, Takeda A, et al. 1998a. Human recombinant NACP/alpha-synuclein is aggregated and fibrillated in vitro: Relevance for Lewy body disease. Brain Res 799: 301-306.

Hashimoto M, Kitagaki H, Imamura T, Hirono N, Shimomura T, et al. 1998b. Medial temporal and whole-brain atrophy in dementia with Lewy bodies: A volumetric MRI study. Neurology 51: 357-362.

Hashimoto M, Hsu LJ, Xia Y, Takeda A, Sisk A, et al. 1999. Oxidative stress induces amyloid-like aggregate formation of NACP/alpha-synuclein in vitro. Neuroreport 10: 717-721.

Hashimoto M, Kawahara K, Bar-On P, Rockenstein E, Crews L, et al. 2004. The role of α-synuclein assembly and metabolism in the pathogenesis of Lewy body disease. J Mol Neurosci 24: 343-352.

Hashimoto M, Rockenstein E, Mante M, Mallory M, Masliah E. 2001. Beta-synuclein inhibits alpha-synuclein aggregation: A possible role as an anti-parkinsonian factor. Neuron 32: 213-223.

Hashimoto M, Rockenstein E, Masliah E. 2003a. Transgenic models of alpha-synuclein pathology: Past, present, and future. Ann N Y Acad Sci 991: 171-188.

Hashimoto M, Takenouchi T, Rockenstein E, Masliah E. 2003b. Alpha-synuclein up-regulates expression of caveolin-1 and down-regulates extracellular signal-regulated kinase activity in B103 neuroblastoma cells: Role in the pathogenesis of Parkinson's disease. J Neurochem 85: 1468-1479.

Hashimoto M, Masliah E. 1999. Alpha-synuclein in Lewy body disease and Alzheimer's disease. Brain Pathol 9: 707-720.

Hashimoto M, Masliah E. 2003. Cycles of aberrant synaptic sprouting and neurodegeneration in Alzheimer's and dementia with Lewy bodies. Neurochem Res 28: 1743-1756.

Hely MA, Reid WG, Halliday GM, McRitchie DA, Leicester J, et al. 1996. Diffuse Lewy body disease: Cinical features in nine cases without coexistent Alzheimer's disease. J Neurol Neurosurg Psychiatry 60: 531-538.

Henderson JM, Carpenter K, Cartwright H, Halliday GM. 2000a. Loss of thalamic intralaminar nuclei in progressive supranuclear palsy and Parkinson's disease: Clinical and therapeutic implications. Brain 123 (Pt. 7): 1410-1421.

Henderson JM, Carpenter K, Cartwright H, Halliday GM. 2000b. Degeneration of the centre median-parafascicular complex in Parkinson's disease. Ann Neurol 47: 345-352.

Henderson JM, Gai WP, Hely MA, Reid WG, Walker GL, et al. 2001. Parkinson's disease with late Pick's dementia. Mov Disord 16: 311-319.

Henry B, Duty S, Fox SH, Crossman AR, Brotchie JM. 2003. Increased striatal pre-proenkephalin B expression is associated with dyskinesia in Parkinson's disease. Exp Neurol 183: 458-468.

Herholz K, Weisenbach S, Zundorf G, Lenz O, Schroder H, et al. 2004. In vivo study of acetylcholine esterase in basal forebrain, amygdala, and cortex in mild to moderate Alzheimer disease. Neuroimage 21: 136-143.

Herrera AJ, Tomas-Camardiel M, Venero JL, Cano J, Machado A. 2005. Inflammatory process as a determinant factor for the degeneration of substantia nigra dopaminergic neurons. J Neural Transm 112: 111-119.

Higuchi M, Tashiro M, Arai H, Okamura N, Hara S, et al. 2000. Glucose hypometabolism and neuropathological correlates in brains of dementia with Lewy bodies. Exp Neurol 162: 247-256.

Hishikawa N, Hashizume Y, Yoshida M, Sobue G. 2001. Widespread occurrence of argyrophilic glial inclusions in Parkinson's disease. Neuropathol Appl Neurobiol 27: 362-372.

Hishikawa N, Niwa J, Doyu M, Ito T, Ishigaki S, et al. 2003. Dorfin localizes to the ubiquitylated inclusions in Parkinson's disease, dementia with Lewy bodies, multiple system atrophy, and amyotrophic lateral sclerosis. Am J Pathol 163: 609-619.

Holmes C, Cairns N, Lantos P, Mann A. 1999. Validity of current clinical criteria for Alzheimer's disease, vascular dementia and dementia with Lewy bodies. Br J Psychiatry 174: 45-50.

Holstege G. 1996. The somatic motor system. Prog Brain Res 107: 9-26.

Hoogendijk WJ, Pool CW, Troost D, van Zwieten E, Swaab DF. 1995. Image analyser-assisted morphometry of the locus coeruleus in Alzheimer's disease, Parkinson's disease and amyotrophic lateral sclerosis. Brain 118: 131-143.

Horimoto Y, Matsumoto M, Nakazawa H, Yuasa H, Morishita M, et al. 2003. Cognitive conditions of pathologically confirmed dementia with Lewy bodies and Parkinson's disease with dementia. J Neurol Sci 216: 105-108.

Howells DW, Porritt MJ, Wong JY, Batchelor PE, Kalnins R, et al. 2000. Reduced BDNF mRNA expression in the Parkinson's disease substantia nigra. Exp Neurol 166: 127-135.

Hsu LJ, Sagara Y, Arroyo A, Rockenstein E, Sisk A, et al. 2000. Alpha-synuclein promotes mitochondrial deficit and oxidative stress. Am J Pathol 157: 401-410.

Hughes AJ, Daniel SE, Kilford L, Lees AJ. 1992. Accuracy of clinical diagnosis of idiopathic Parkinson's disease: A clinico-pathological study of 100 cases. J Neurol Neurosurg Psychiatry 55: 181-184.

Hughes AJ, Daniel SE, Lees AJ. 2001. Improved accuracy of clinical diagnosis of Lewy body Parkinson's disease. Neurology 57: 1497-1499.

Hunot S, Hirsch EC. 2003. Neuroinflammatory processes in Parkinson's disease. Ann Neurol 53: S49-S58; discussion S58–S60.

Hurtig HI, Trojanowski JQ, J, Galvin Ewbank D, Schmidt ML, et al. 2000. Alpha-synuclein cortical Lewy bodies correlate with dementia in Parkinson's disease. Neurology 54: 1916-1921.

Huynh DP, Scoles DR, Nguyen D, Pulst SM. 2003. The autosomal recessive juvenile Parkinson disease gene product, parkin, interacts with and ubiquitinates synaptotagmin XI. Hum Mol Genet 12: 2587-2597.

Ikemura M, Saito Y, Sengoku R, Fumimura Y, Arai T, et al. 2006. Lewy body-related α-synucleinopathy involves cutaneous nerves in human aging (Abstract). Brain Pathol 16 (Suppl. 1): S57.

Imamura K, Hishikawa N, Sawada M, Nagatsu T, Yoshida M, et al. 2003. Distribution of major histocompatibility complex class II-positive microglia and cytokine profile of Parkinson's disease brains. Acta Neuropathol (Berl) 106: 518-526.

Imamura T, Hirono N, Hashimoto M, Kazui H, Tanimukai S, et al. 2000. Fall-related injuries in dementia with Lewy bodies (DLB) and Alzheimer's disease. Eur J Neurol 7: 77-79.

Imamura T, Ishii K, Hirono N, Hashimoto M, Tanimukai S, et al. 2001. Occipital glucose metabolism in dementia with Lewy bodies with and without Parkinsonism: A study using positron emission tomography. Dement Geriatr Cogn Disord 12: 194-197.

Ince PG, McKeith IG. 2003. Dementia with Lewy bodies. Neurodegeneration: The Molecular Pathology of Dementia and Movement Disorders. Dickson DW, editor. Basel: ISN Neuropath Press.

Ince PG, Perry EK, Morris CM. 1998. Dementia with Lewy bodies. A distinct non-Alzheimer dementia syndrome? Brain Pathol 8: 299-324.

Ischiropoulos H. 2003. Oxidative modifications of alpha-synuclein. Ann N Y Acad Sci 991: 93-100.

Iseki E. 2004. Dementia with Lewy bodies: Reclassification of pathological subtypes and boundary with Parkinson's disease or Alzheimer's disease. Neuropathology 24: 72-78.

Iseki E, Li F, Odawara T, Kosaka K. 1997. Hippocampal pathology in diffuse Lewy body disease using ubiquitin immunohistochemistry. J Neurol Sci 149: 165-169.

Iseki E, Marui W, Akiyama H, Ueda K, Kosaka K. 2000. Degeneration process of Lewy bodies in the brains of patients with dementia with Lewy bodies using alpha-synuclein-immunohistochemistry. Neurosci Lett 286: 69-73.

Iseki E, Marui W, Kosaka K, Ueda K. 1999. Frequent coexistence of Lewy bodies and neurofibrillary tangles in the same neurons of patients with diffuse Lewy body disease. Neurosci Lett 265: 9-12.

Iseki E, Takayama N, Furukawa Y, Marui W, Nakai T, et al. 2002. Immunohistochemical study of synphilin-1 in brains of patients with dementia with Lewy bodies - synphilin-1 is non-specifically implicated in the formation of different neuronal cytoskeletal inclusions. Neurosci Lett 326: 211-215.

Iseki E, Togo T, Suzuki K, Katsuse O, Marui W, et al. 2003. Dementia with Lewy bodies from the perspective of tauopathy. Acta Neuropathol (Berl) 105: 265-270.

Ishii K, Imamura T, Sasaki M, Yamaji S, Sakamoto S, et al. 1998. Regional cerebral glucose metabolism in dementia with Lewy bodies and Alzheimer's disease. Neurology 51: 125-130.

Ishikawa A, Takahashi H, Tanaka H, Hayashi T, Tsuji S. 1997. Clinical features of familial diffuse Lewy body disease. Eur Neurol 38 (Suppl. 1): 34-38.

Ishizawa T, Mattila P, Davies P, Wang D, Dickson DW. 2003. Colocalization of tau and alpha-synuclein epitopes in Lewy bodies. J Neuropathol Exp Neurol 62: 389-397.

Ito H, Kusaka H, Matsumoto S, Imai T. 1996. Striatal efferent involvement and its correlation to levodopa efficacy in patients with multiple system atrophy. Neurology 47: 1291-1299.

Ito T, Niwa J, Hishikawa N, Ishigaki S, Doyu M, et al. 2003. Dorfin localizes to Lewy bodies and ubiquitylates synphilin-1. J Biol Chem 278: 29106-29114.

Iwata A, Maruyama M, Akagi T, Hashikawa T, Kanazawa I, et al. 2003. Alpha-synuclein degradation by serine protease neurosin: Implication for pathogenesis of synucleinopathies. Hum Mol Genet 12: 2625-2635.

Iwata A, Maruyama M, Kanazawa I, Nukina N. 2001. Alpha-Synuclein affects the MAPK pathway and accelerates cell death. J Biol Chem 276: 45320-45329.

Iwatsubo T. 2003. Aggregation of alpha-synuclein in the pathogenesis of Parkinson's disease, J Neurol 250 (Suppl. 3): III11-III14.

Jakowec MW, Petzinger GM, Sastry S, Donaldson DM, McCormack A, et al. 1998. The native form of alpha-synuclein is not found in the cerebrospinal fluid of patients with Parkinson's disease or normal controls. Neurosci Lett 253: 13-16.

Jankovic J, McDermott M, Carter J, Gauthier S, Goetz C, et al. 1990. Variable expression of Parkinson's disease: A base-line analysis of the DATATOP cohort. The Parkinson Study Group. Neurology 40: 1529-1534.

Jellinger K, Riederer P, Tomonaga M. 1980. Progressive supranuclear palsy: clinico-pathological and biochemical studies. J Neural Transm Suppl 76: 111–128.

Jellinger KA. 2000. Cell death mechanisms in Parkinson's disease. J Neural Transm 107: 1-29.

Jellinger KA. 2000. Morphological substrates of mental dysfunction in Lewy body disease: An update. J Neural Transm Suppl 59: 185-212.

Jellinger KA. 2001. Cell death mechanisms in neurodegeneration. J Cell Mol Med 5: 1-17.

Jellinger KA. 2002. Recent developments in the pathology of Parkinson's disease. J Neural Transm Suppl 62: 347–376.

Jellinger KA. 2003a. Alpha-synuclein pathology in Parkinson and Alzheimer disease brain: Incidence and topographic distribution: a pilot study. Acta Neuropathol 106: 191-201.

Jellinger KA. 2003b. Neuropathological spectrum of synucleinopathies. Mov Disord 18 (Suppl. 6): S2-S12.

Jellinger KA. 2003c. Apoptosis versus non-apoptotic mechanisms in neurodegeneration. Neuroinflammation: Mechanisms and Management, 2nd edn. Wood PL, editor. Totowa, NJ: Humana Press; pp. 29-88.

Jellinger KA. 2003d. General aspects of neurodegeneration. J Neural Transm 65: 101–144.

Jellinger KA. 2003e. Prevalence of cerebrovascular lesions in Parkinson's disease. A postmortem study. Acta Neuropathol 105: 415-419.

Jellinger KA. 2003f. Experimental models of synucleinopathies. Neurodegeneration. Dickson DW, editor. Basel: ISN Neuropath Press; pp. 215-223.

Jellinger KA. 2003g. Prevalence of vascular lesions in dementia with Lewy bodies. A postmortem study. J Neural Transm 110: 771-778.

Jellinger KA. 2004a. Neuropathology of movement disorders. Youman's Neurological Surgery, 5th edn., Vol. 3. Winn HR, editor. Philadelphia: Saunders; pp. 2699-2782.

Jellinger KA. 2004b. Lewy body-related α-synucleinopathy in the aged human brain. J Neural Transm 111: 1219-1235.

Jellinger KA. 2005. The pathology of Parkinson's disease: recent advances. Scientific Basis of Treatment of Parkinson's Disease, 2nd edn. Galvez-Jimenez N, editor. London: Taylor & Francis Medical Books; pp. 53-85.

Jellinger KA. 2006. Neuropathology of dementia in Parkinson's disease. Ann Neurol 59: 727.

Jellinger KA. 2006b. P25α immunoreactivity in multiple system atrophy and Parkinson disease. Acta Neuropathol (Berl) 112: 112.

Jellinger KA, Attems J. 2006. Does striatal pathology distinguish Parkinson disease with dementia and dementia with Lewy bodies? Acta Neuropathol (Berl) 112: 253-260.

Jellinger KA, Mizuno Y. 2003. Parkinson disease. Neurodegeneration: The Molecular Pathology of Dementia and

Movement Disorders. Dickson DW, editor. Basel: ISN Neuropath Press; pp. 159-185.

Jellinger KA, Paulus W. 1992. Clinico-pathological correlations in Parkinson's disease. Clin Neurol Neurosurg 94 (Suppl.): S86-S88.

Jellinger KA, Seppi K, Wenning GK. 2003. Clinical and neuropathological correlates of Lewy body disease. Acta Neuropathol (Berl) 106: 188-189.

Jenner P. 2003. Oxidative stress in Parkinson's disease. Ann Neurol 53: S26-S36; discussion S36-S38.

Jenner P, Olanow CW. 2006. The pathogenesis of cell death in Parkinson's disease. Neurology 66: S24-S36.

Jensen PH, Islam K, Kenney J, Nielsen MS, Power J, et al. 2000. Microtubule-associated protein 1B is a component of cortical Lewy bodies and binds alpha-synuclein filaments. J Biol Chem 275: 21500-21507.

Jha N, Jurma O, Lalli G, Liu Y, Pettus EH, et al. 2000. Glutathione depletion in PC12 results in selective inhibition of mitochondrial complex I activity. Implications for Parkinson's disease. J Biol Chem 275: 26096-26101.

Joyce JN, Ryoo HL, Beach TB, Caviness JN, Stacy M, et al. 2002. Loss of response to levodopa in Parkinson's disease and co-occurrence with dementia: Role of D3 and not D2 receptors. Brain Res 955: 138-152.

Junn E, Mouradian MM. 2003. A proposed model for the genesis of Lewy bodies (abstract). Soceity for Neuroscience. 33rd Annual Meeting, Society for Neuroscience, New Orleans, pp. Program No. 297.14.

Junn E, Ronchetti RD, Quezado MM, Kim SY, Mouradian MM. 2003. Tissue transglutaminase-induced aggregation of alpha-synuclein: Implications for Lewy body formation in Parkinson's disease and dementia with Lewy bodies. Proc Natl Acad Sci USA 100: 2047-2052.

Kaasinen V, Aalto S, Nagren K, Hietala J, Sonninen P, et al. 2003. Extrastriatal dopamine D(2) receptors in Parkinson's disease: A longitudinal study. J Neural Transm 110: 591-601.

Kahle PJ, Neumann M, Ozmen L, Muller V, Jacobsen H, et al. 2000. Subcellular localization of wild-type and Parkinson's disease-associated mutant alpha-synuclein in human and transgenic mouse brain. J Neurosci 20: 6365-6373.

Kahle PJ, Neumann M, Ozmen L, Muller V, Odoy S, et al. 2001. Selective insolubility of alpha-synuclein in human Lewy body diseases is recapitulated in a transgenic mouse model. Am J Pathol 159: 2215-2225.

Kanda S, Bishop JF, Eglitis MA, Yang Y, Mouradian MM. 2000. Enhanced vulnerability to oxidative stress by alpha-synuclein mutations and C-terminal truncation. Neuroscience 97: 279-284.

Kanemaru K, Kameda N, Yamanouchi H. 2000. Decreased CSF amyloid beta42 and normal tau levels in dementia with Lewy bodies. Neurology 54: 1875-1876.

Kang JH, Kim KS. 2003. Enhanced oligomerization of the alpha-synuclein mutant by the Cu,Zn-superoxide dismutase and hydrogen peroxide system. Mol Cells 15: 87-93.

Kaplan B, Ratner V, Haas E. 2003. Alpha-synuclein: Its biological function and role in neurodegenerative diseases. J Mol Neurosci 20: 83-92.

Katsuse O, Iseki E, Kosaka K. 2003a. Immunohistochemical study of the expression of cytokines and nitric oxide synthases in brains of patients with dementia with Lewy bodies. Neuropathology 23: 9-15.

Katsuse O, Iseki E, Marui W, Kosaka K. 2003b. Developmental stages of cortical Lewy bodies and their relation to axonal transport blockage in brains of patients with dementia with Lewy bodies. J Neurol Sci 211: 29-35.

Kaufmann H, Hague K, Perl D. 2001. Accumulation of alpha-synuclein in autonomic nerves in pure autonomic failure. Neurology 56: 980-981.

Kawamoto Y, Akiguchi I, Nakamura S, Honjyo Y, Shibasaki H, et al. 2002. 14-3-3 proteins in Lewy bodies in Parkinson disease and diffuse Lewy body disease brains. J Neuropathol Exp Neurol 61: 245-253.

Kawanishi C, Suzuki K, Odawara T, Iseki E, Onishi H, et al. 1996. Neuropathological evaluation and apolipoprotein E gene polymorphism analysis in diffuse Lewy body disease. J Neurol Sci 136: 140-142.

Kenny RA, Allen LM. 2006. Autonomic dysfunction in dementia with Lewy bodies. Dementia with Lewy Bodies and Parkinson's Disease Dementia. Brien JO, Ames D, McKeith I, Chiu E, editors. London: Taylor & Francis; pp. 107-127.

Kerenyi L, Ricaurte GA, Schretlen DJ, McCann U, Varga J, et al. 2003. Positron emission tomography of striatal serotonin transporters in Parkinson disease. Arch Neurol 60: 1223-1229.

Kessler JC, Rochet JC, Lansbury PT Jr. 2003. The N-terminal repeat domain of alpha-synuclein inhibits beta-sheet and amyloid fibril formation. Biochemistry 42: 672-678.

Kikuchi A, Takeda A, Onodera H, Kimpara T, Hisanaga K, et al. 2002. Systemic increase of oxidative nucleic acid damage in Parkinson's disease and multiple system atrophy. Neurobiol Dis 9: 244-248.

Kim H, Gearing M, Mirra SS. 1995. Ubiquitin-positive CA2/3 neurites in hippocampus coexist with cortical Lewy bodies. Neurology 45: 1768-1770.

Kim RH, Smith PD, Aleyasin H, Hayley S, Mount MP, et al. 2005. Hypersensitivity of DJ-1-deficient mice to 1-methyl-4-phenyl-1,2,3,6-tetrahydropyrindine (MPTP) and oxidative stress. Proc Natl Acad Sci USA 102: 5215-5220.

Kim SE, Choi JY, Choe YS, Choi Y, Lee WY. 2003a. Serotonin transporters in the midbrain of Parkinson's disease patients: a study with 123I-beta-CIT SPECT. J Nucl Med 44: 870-876.

Kim SJ, Sung JY, Um JW, Hattori N, Mizuno Y, et al. 2003b. Parkin cleaves intracellular alpha-synuclein inclusions via the activation of calpain. J Biol Chem 278: 41890-41899.

Kingsbury AE, Daniel SE, Sangha H, Eisen S, Lees AJ, et al. 2004. Alteration in alpha-synuclein mRNA expression in Parkinson's disease. Mov Disord 19: 162-170.

Kingsbury AE, Marsden CD, Foster OJ. 1999. The vulnerability of nigral neurons to Parkinson's disease is unrelated to their intrinsic capacity for dopamine synthesis: An in situ hybridization study. Mov Disord 14: 206-218.

Kirik D, Annett LE, Burger C, Muzyczka N, Mandel RJ, et al. 2003. Nigrostriatal alpha-synucleinopathy induced by viral vector-mediated overexpression of human alpha-synuclein: A new primate model of Parkinson's disease. Proc Natl Acad Sci USA 100: 2884-2889.

Kirik D, Rosenblad C, Burger C, Lundberg C, Johansen TE, et al. 2002. Parkinson-like neurodegeneration induced by targeted overexpression of alpha-synuclein in the nigrostriatal system. J Neurosci 22: 2780-2791.

Kish SJ, Shannak K, Hornykiewicz O. 1988. Uneven pattern of dopamine loss in the riatum of patients with idiopathic Parkinson's disease. Pathophysiologic and clinical implications. N Engl J Med 318: 876-880.

Klatka LA, Louis ED, Schiffer RB. 1996. Psychiatric features in diffuse Lewy body disease: A clinicopathologic study using Alzheimer's disease and Parkinson's disease comparison groups. Neurology 47: 1148-1152.

Klein RL, King MA, Hamby ME, Meyer EM. 2002. Dopaminergic cell loss induced by human A30P alpha-synuclein gene transfer to the rat substantia nigra. Hum Gene Ther 13: 605-612.

Klucken J, Ingelsson M, Shin Y, Irizarry MC, Hedley-Whyte ET, et al. 2006. Clinical and biochemical correlates of insoluble α-synuclein in dementia with Lewy bodies. Acta Neuropathol (Berl) 111: 101-108.

Klucken J, Shin Y, Masliah E, Hyman BT, McLean PJ. 2004. Hsp70 reduces α-synuclein aggregation and toxicity. J Biol Chem 279: 25497-25502.

Knott C, Stern G, Kingsbury A, Welcher AA, Wilkin GP. 2002. Elevated glial brain-derived neurotrophic factor in Parkinson's disease nigra. Parkinsonism Relat Disord 8: 329-341.

Koh SB, Suh SI, Lee DH, Kim AR, Oh CH, et al. 2006. Phase contrast radiography of Lewy bodies in Parkinson disease. Neuroimage 32: 566-569.

Kopito RR. 2000. Aggresomes, inclusion bodies and protein aggregation. Trends Cell Biol 10: 524-530.

Kortekaas R, Leenders KL, van Oostrom JC, Vaalburg W, Bart J, et al. 2005. Blood-brain barrier dysfunction in parkinsonian midbrain in vivo. Ann Neurol 57: 176-179.

Kosaka K. 1990. Diffuse Lewy body disease in Japan. J Neurol 237: 197-204.

Kosaka K, Iseki E. 1996a. Dementia with Lewy bodies. Curr Opin Neurol 9: 271-275.

Kosaka K, Iseki E. 1996b. Diffuse Lewy body disease within the spectrum of Lewy body disease. Dementia with Lewy Bodies. Perry RH, McKeith IG, Perry EK, editor. Cambridge: Cambridge Universtity Press; pp. 238-247.

Kotzbauer PT, Giasson BI, Kravitz AV, Golbe L, Mark MH, Trojanowski JQ, Lee VM-Y. 2005. Fibrillization of alpha-synuclein and tau in familial Parkinson's disease caused by the A53T alpha-synuclein mutation. Exp Neurol 187: 279-288.

Kovacs GG, Laszlo L, Kovacs J, Jensen PH, Lindersson E, et al. 2004. Natively unfolded tubulin polymerization promoting protein TPPP/p25 is a common marker of alpha-synucleinopathies. Neurobiol Dis 17: 155-162.

Kovari E, Gold G, Herrmann FR, Canuto A, Hof PR, et al. 2003. Lewy body densities in the entorhinal and anterior cingulate cortex predict cognitive deficits in Parkinson's disease. Acta Neuropathol (Berl) 106: 83-88.

Krishnan S, Chi EY, Wood SJ, Kendrick BS, Li C, et al. 2003. Oxidative dimer formation is the critical rate-limiting step for Parkinson's disease alpha-synuclein fibrillogenesis. Biochemistry 42: 829-837.

Kruger R, Kuhn W, Muller T, Woitalla D, Graeber M, et al. 1998. Ala30Pro mutation in the gene encoding alpha-synuclein in Parkinson's disease. Nat Genet 18: 106-108.

Kubis N, Faucheux BA, Ransmayr G, Damier P, Duyckaerts C, et al. 2000. Preservation of midbrain catecholaminergic neurons in very old human subjects. Brain 123: 366-373.

Kuusisto E, Parkkinen L, Alafuzoff I. 2003. Morphogenesis of Lewy bodies: Dissimilar incorporation of alpha-synuclein, ubiquitin, and p62. J Neuropathol Exp Neurol 62: 1241-1253.

Kuusisto E, Salminen A, Alafuzoff I. 2001. Ubiquitin-binding protein p62 is present in neuronal and glial inclusions in human tauopathies and synucleinopathies. Neuroreport 12: 2085-2090.

Kuusisto E, Salminen A, Alafuzoff I. 2002. Early accumulation of p62 in neurofibrillary tangles in Alzheimer's disease: Possible role in tangle formation. Neuropathol Appl Neurobiol 28: 228-237.

Kuzuhara S, Yoshimura M. 1993. Clinical and neuropathological aspects of diffuse Lewy body disease in the elderly. Adv Neurol 60: 464-469.

La Voie MJ, Ostaszewski BL, Weihofen A, Schlossmacher MG, Selkoe DJ. 2005. Dopamine covalently modifies and functionally inactivates parkin. Nat Med 11: 1214-1221.

Lach B, Grimes D, Benoit B, Minkiewicz-Janda A. 1992. Caudate nucleus pathology in Parkinson's disease:

Ultrastructural and biochemical findings in biopsy material. Acta Neuropathol (Berl) 83: 352-360.

Lamb H, Christie J, Singleton AB, Leake A, Perry RH, et al. 1998. Apolipoprotein E and alpha-1 antichymotrypsin polymorphism genotyping in Alzheimer's disease and in dementia with Lewy bodies. Distinctions between diseases. Neurology 50: 388-391.

Langlais PJ, Thal L, Hansen L, Galasko D, Alford M, et al. 1993. Neurotransmitters in basal ganglia and cortex of Alzheimer's disease with and without Lewy bodies. Neurology 43: 1927-1934.

Lantos PL, Quinn N. 2003. Multiple system atrophy. Neurodegeneration: The Molecular Pathology of Dementia and Movement Disorders. Dickson DW, editor. Los Angeles: ISN Press; pp. 203-214.

Lauwers E, Debyser Z, Van Dorpe J, De Strooper B, Nuttin B, et al. 2003. Neuropathology and neurodegeneration in rodent brain induced by lentiviral vector-mediated over-expression of alpha-synuclein. Brain Pathol 13: 364-372.

Layfield R, Cavey JR, Lowe J. 2003. Role of ubiquitin-mediated proteolysis in the pathogenesis of neurodegenerative disorders. Ageing Res Rev 2: 343-356.

Lee HG, Casadesus G, Zhu X, Takeda A, Perry G, et al. 2004. Challenging the amyloid cascade hypothesis: Senile plaques and amyloid-β as protective adaptations to Alzheimer disease. Ann N Y Acad Sci 1019: 1-4.

Lee EN, Lee SY, Lee D, Kim J, Paik SR. 2003a. Lipid interaction of alpha-synuclein during the metal-catalyzed oxidation in the presence of Cu2+ and H2O2. J Neurochem 84: 1128-1142.

Lee FJ, Liu F, Pristupa ZB, Niznik HB. 2001. Direct binding and functional coupling of alpha-synuclein to the dopamine transporters accelerate dopamine-induced apoptosis. FASEB J 15: 916-926.

Lee G, Junn E, Tanaka M, Kim YM, Mouradian MM. 2002a. Synphilin-1 degradation by the ubiquitin-proteasome pathway and effects on cell survival. J Neurochem 83: 346-352.

Lee MK, Stirling W, Xu Y, Xu X, Qui D, et al. 2002b. Human alpha-synuclein-harboring familial Parkinson's disease-linked Ala-53 Thr mutation causes neurodegenerative disease with alpha-synuclein aggregation in transgenic mice. Proc Natl Acad Sci USA 99: 8968-8973.

Lee G, Tanaka M, Park K, Lee SS, Kim YM, et al. 2004a. Casein kinase II-mediated phosphorylation regulates alpha-synuclein/synphilin-1 interaction and inclusion body formation. J Biol Chem 279: 6834-6839.

Lee VM, Giasson BI, Trojanowski JQ. 2004b. More than just two peas in a pod: Common amyloidogenic properties of tau and alpha-synuclein in neurodegenerative diseases. Trends Neurosci 27: 129-134.

Lee HJ, Lee SJ. 2002. Characterization of cytoplasmic alpha-synuclein aggregates. Fibril formation is tightly linked to the inclusion-forming process in cells. J Biol Chem 277: 48976-48983.

Lee MK, Li W, West N, Pletnikova O, Troncoso JC, et al. 2003b. Truncated alpha-synuclein is generated in vivo and potentiates alpha-synuclein aggregation (abstract). Soceity for Neuroscience. 33rd Annual Meeting, Society for Neuroscience, Soceity for Neuroscience, New Orleans, pp. Program No. 132.1.

Lee SS, Kim YM, Junn E, Lee G, Park KH, et al. 2003c. Cell cycle aberrations by alpha-synuclein over-expression and cyclin B immunoreactivity in Lewy bodies. Neurobiol Aging 24: 687-696.

Lees AJ. 2002. Parkinson's disease with dementia (PDD) is the same clinicopathological entity as dementia with Lewy bodies (DLB) (abstract). Mov Disord 17 (Suppl. 5): S9.

Lennox G, Lowe JS, Godwin-Austen RB, Landon M, Mayer RJ. 1989. Diffuse Lewy body disease: An important differential diagnosis in dementia with extrapyramidal features. Prog Clin Biol Res 317: 121-130.

Leverenz JB, Miller MA, Dobie DJ, Peskind ER, Raskind MA. 2001. Increased alpha 2-adrenergic receptor binding in locus coeruleus projection areas in dementia with Lewy bodies. Neurobiol Aging 22: 555-561.

Lewis J, Dickson DW, Lin WL, Chisholm L, Corral A, et al. 2001. Enhanced neurofibrillary degeneration in transgenic mice expressing mutant tau and APP. Science 293: 1487-1491.

Lewis SJ, Dove A, Robbins TW, Barker RA, Owen AM. 2003. Cognitive impairments in early Parkinson's disease are accompanied by reductions in activity in frontostriatal neural circuitry. J Neurosci 23: 6351-6366.

Li JY, Henning Jensen P, Dahlstrom A. 2002. Differential localization of alpha-, beta- and gamma-synucleins in the rat CNS. Neuroscience 113: 463-478.

Li W, Hoffman PN, Stirling W, Price DL, Lee MK. 2004. Axonal transport of human alpha-synuclein slows with aging but is not affected by familial Parkinson's disease-linked mutations. J Neurochem 88: 401-410.

Lindersson E, Beedholm R, Hojrup P, Moos T, Gai W, et al. 2004. Proteasomal inhibition by alpha-synuclein filaments and oligomers. J Biol Chem 279: 12924-12934.

Lindersson E, Lundvig D, Petersen C, Madsen P, Nyengaard JR, et al. 2005. p25α stimulates α-synuclein aggregation and is co-localized with aggregated α-synuclein in α-synucleinopathies. J Biol Chem 280: 5703-5715.

Linert W, Jameson GNL, Jameson RF, Jellinger KA. 2006. The chemical interplay between catecholamines and metal ions in neurological diseases. Metal Ions in Life Sciences, Vol.1.

Sigel A, Sigel H, Sigel RKO, editors. Hoboken, NJ: John Wiley & Sons; pp. 281-320.

Lippa CF. 2003. Lewy bodies in conditions other than disorders of alpha-synuclein. Neurodegeneration: The Molecular Pathology of Dementia and Movement Disorders. Dickson DW, editor. Basel: ISN Neuropath Press; pp. 200-202.

Lippa CF, Johnson R, Smith TW. 1998. The medial temporal lobe in dementia with Lewy bodies: A comparative study with Alzheimer's disease. Ann Neurol 43: 102-106.

Lippa CF, Pulaski-Salo D, Dickson DW, Smith TW. 1997. Alzheimer's disease, Lewy body disease and aging: A comparative study of the perforant pathway. J Neurol Sci 147:161-166.

Lippa CF, Smith TW, Saunders AM, Crook R, Pulaski-Salo D, et al. 1995. Apolipoprotein E genotype and Lewy body disease. Neurology 45: 97-103.

Lippa CF, Ozawa K, Mann DM, Ishii K, Smith TW, et al. 1999a. Deposition of beta-amyloid subtypes 40 and 42 differentiates dementia with Lewy bodies from Alzheimer disease. Arch Neurol 56: 1111-1118.

Lippa CF, Smith TW, Perry E. 1999b. Dementia with Lewy bodies: Choline acetyltransferase parallels nucleus basalis pathology. J Neural Transm 106: 525-535.

Litvan I, Bhatia KP, Burn DJ, Goetz CG, Lang AE, et al. 2003. SIC Task Force appraisal of clinical diagnostic criteria for parkinsonian disorders. Mov Disord 18: 467-486.

Litvan I, Mac Intyre A, Goetz CG, Wenning GK, Jellinger K, et al. 1998. Accuracy of the clinical diagnoses of Lewy body disease, Parkinson disease, and dementia with Lewy bodies: A clinicopathologic study. Arch Neurol 55: 969-978.

Liu S, Ninan I, Antonova I, Battaglia F, Trinchese F, et al. 2004. α-Synuclein produces a long-lasting increase in neurotransmitter release. EMBO J 23: 4506-4516.

Lo Bianco C, Ridet JL, Schneider BL, Deglon N, Aebischer P. 2002. Alpha -synucleinopathy and selective dopaminergic neuron loss in a rat lentiviral-based model of Parkinson's disease. Proc Natl Acad Sci USA 99: 10813-10818.

Lobotesis K, Fenwick JD, Phipps A, Ryman A, Swann A, et al. 2001. Occipital hypoperfusion on SPECT in dementia with Lewy bodies but not AD. Neurology 56: 643-649.

Lopez OL, Becker JT, Kaufer DI, Hamilton RL, Sweet RA, et al. 2002. Research evaluation and prospective diagnosis of dementia with Lewy bodies. Arch Neurol 59: 43-46.

Lopez OL, Wisniewski S, Hamilton RL, Becker JT, Kaufer DI, et al. 2000. Predictors of progression in patients with AD and Lewy bodies. Neurology 54: 1774-1779.

Lotharius J, Brundin P. 2002. Pathogenesis of Parkinson's disease: Dopamine, vesicles and alpha-synuclein. Nat Rev Neurosci 3: 932-942.

Louis ED, Goldman JE, Powers JM, Fahn S. 1995. Parkinsonian features of eight pathologically diagnosed cases of diffuse Lewy body disease. Mov Disord 10: 188-194.

Lowe J, Lennox GG, Leigh PN. 2002. Disorders of movement and system degenerations. Greenfield's Neuropathology, 7th edn. Graham DI, Lantos PL, editor. London: E. Arnold; pp. 325-430.

Lozza C, Marie RM, Baron JC. 2002. The metabolic substrates of bradykinesia and tremor in uncomplicated Parkinson's disease. Neuroimage 17: 688-699.

Lucke C, Gantz DL, Klimtchuk E, Hamilton JA. 2006. Interactions between fatty acids and α-synuclein. J Lipid Res 47: 1714-1724.

Ma QL, Chan P, Yoshii M, Ueda K. 2003. Alpha-synuclein aggregation and neurodegenerative diseases. J Alzheimers Dis 5: 139-148.

Ma SY, Ciliax BJ, Stebbins G, Jaffar S, Joyce JN, et al. 1999a. Dopamine transporter-immunoreactive neurons decrease with age in the human substantia nigra. J Comp Neurol 409: 25-37.

Ma SY, Roytt M, Collan Y, Rinne JO. 1999b. Unbiased morphometrical measurements show loss of pigmented nigral neurones with ageing. Neuropathol Appl Neurobiol 25: 394-399.

Ma SY, Roytta M, Rinne JO, Collan Y, Rinne UK. 1997. Correlation between neuromorphometry in the substantia nigra and clinical features in Parkinson's disease using disector counts. J Neurol Sci 151: 83-87.

Mac Donald V, Halliday GM. 2002. Selective loss of pyramidal neurons in the pre-supplementary motor cortex in Parkinson's disease. Mov Disord 17: 1166-1173.

Mamah CE, Lesnick TG, Lincoln SJ, Strain KJ, de Andrade M, et al. 2005. Interaction of alpha-synuclein and tau genotypes in Parkinson's disease. Ann Neurol 57: 439-443.

Mandal PK, Pettegrew JW, Masliah E, Hamilton RL, Mandal R. 2006. Interaction between Aβ-peptide and α-synuclein: Molecular mechanisms in overlapping pathology of Alzheimer's and Parkinson's in dementia with Lewy body disease. Neurochem Res 31: 1153-1162.

Mann DM, Brown SM, Owen F, Baba M, Iwatsubo T. 1998. Amyloid beta protein (A beta) deposition in dementia with Lewy bodies: Predominance of A beta 42(43) and paucity of A beta 40 compared with sporadic Alzheimer's disease. Neuropathol Appl Neurobiol 24: 187-194.

Maries E, Dass B, Collier TJ, Kordower JH, Steece-Collier K. 2003. The role of alpha-synuclein in Parkinson's disease: insights from animal models. Nat Rev Neurosci 4: 727-738.

Martin LJ, Pan Y, Price AC, Sterling W, Copeland NG, et al. 2006. Parkinson's disease α-synuclein transgenic mice develop neuronal mitochondrial degeneration and cell death. J Neurosci 26: 41-50.

Martinat C, Shendelman S, Jonason A, Leete T, Beal MF, et al. 2004. Sensitivity to oxidative stress in DJ-1-deficient dopamine neurons: An ES-derived cell model of primary parkinsonism. PLoS Biol 2: e327.

Martinez J, Moeller I, Erdjument-Bromage H, Tempst P, Lauring B. 2003. Parkinson's disease-associated alpha-synuclein is a calmodulin substrate. J Biol Chem 278: 17379-17387.

Martinoli MG, Trojanowski JQ, Schmidt ML, Arnold SE, Fujiwara TM, et al. 1995. Association of apolipoprotein epsilon 4 allele and neuropathologic findings in patients with dementia. Acta Neuropathol (Berl) 90: 239-243.

Martin-Ruiz C, Lawrence S, Piggott M, Kuryatov A, Lindstrom J, et al. 2002. Nicotinic receptors in the putamen of patients with dementia with Lewy bodies and Parkinson's disease: Relation to changes in alpha-synuclein expression. Neurosci Lett 335: 134-138.

Marui W, Iseki E, Kato M, Akatsu H, Kosaka K. 2004. Pathological entity of dementia with Lewy bodies and its differentiation from Alzheimer's disease. Acta Neuropathol 108: 121-128.

Marui W, Iseki E, Nakai T, Miura S, Kato M, et al. 2002. Progression and staging of Lewy pathology in brains from patients with dementia with Lewy bodies. J Neurol Sci 195: 153-159.

Maruyama W, Nagai M, Naoi M. 2003. Involvement of mitochondrial dysfunction and proteasome inactivation in the pathogenesis of Parkinson's disease (abstract). Society for Neuroscience. 33rd Annual Meeting, Society for Neuroscience, New Orleans, pp. Program No. 204.16.

Masliah E, Rockenstein E, Veinbergs I, Mallory M, Hashimoto M, et al. 2000. Dopaminergic loss and inclusion body formation in alpha-synuclein mice: Implications for neurodegenerative disorders. Science 287: 1265-1269.

Masliah E, Rockenstein E, Veinbergs I, Sagara Y, Mallory M, et al. 2001. Beta-amyloid peptides enhance alpha-synuclein accumulation and neuronal deficits in a transgenic mouse model linking Alzheimer's disease and Parkinson's disease. Proc Natl Acad Sci USA 98: 12245-12250.

Masuda M, Suzuki N, Taniguchi S, Oikawa T, Nonaka T, et al. 2006. Small molecule inhibitors of α-synuclein filament assembly. Biochemistry 45: 6085-6094.

Mastaglia FL, Johnsen RD, Byrnes ML, Kakulas BA. 2003. Prevalence of amyloid-beta deposition in the cerebral cortex in Parkinson's disease. Mov Disord 18: 81-86.

Matsuoka Y, Vila M, Lincoln S, McCormack A, Picciano M, et al. 2001. Lack of nigral pathology in transgenic mice expressing human alpha-synuclein driven by the tyrosine hydroxylase promoter. Neurobiol Dis 8: 535-539.

Mattila PM, Rinne JO, Helenius H, Dickson DW, Roytta M. 2000. Alpha-synuclein-immunoreactive cortical Lewy bodies are associated with cognitive impairment in Parkinson's disease. Acta Neuropathol (Berl) 100: 285-290.

Mattila PM, Rinne JO, Helenius H, Roytta M. 1999. Neuritic degeneration in the hippocampus and amygdala in Parkinson's disease in relation to Alzheimer pathology. Acta Neuropathol (Berl) 98: 157-164.

Maurage CA, Ruchoux MM, de Vos R, Surguchov A, Destee A. 2003. Retinal involvement in dementia with Lewy bodies: A clue to hallucinations? Ann Neurol 54: 542-547.

McKeith IG, Dickson DW, Lowe J, Emre M, O'Brien JT, et al. 2005. Diagnosis and management of dementia with Lewy bodies: Third report of the DLB Consortium. Neurology 65: 1863-1872.

McKeith I, Mintze J, Aarsland D, Burn D, Chiu H, et al. 2004. Dementia with Lewy bodies. Lancet Neurol 3: 19-28.

McKeith IG, Perry RH, Fairbairn AF, Jabeen S, Perry EK. 1992. Operational criteria for senile dementia of Lewy body type (SDLT). Psychol Med 22: 911-922.

McKeith IG, Ballard CG, Harrison RW. 1995a. Neuroleptic sensitivity to risperidone in Lewy body dementia. Lancet 346: 699.

McKeith IG, Galasko D, Wilcock GK, Byrne EJ. 1995b. Lewy body dementia—diagnosis and treatment. Br J Psychiatry 167: 709-717.

McKeith IG, Ballard CG, Perry RH, Ince PG, O'Brien JT, et al. 2000. Prospective validation of consensus criteria for the diagnosis of dementia with Lewy bodies. Neurology 54: 1050-1058.

McKeith IG, Fairbairn AF, Perry RH, Thompson P. 1994. The clinical diagnosis and misdiagnosis of senile dementia of Lewy body type (SDLT). Br J Psychiatry 165: 324-332.

McKeith IG, Galasko D, Kosaka K, Perry EK, Dickson DW, et al. 1996. Consensus guidelines for the clinical and pathologic diagnosis of dementia with Lewy bodies (DLB): Report of the consortium on DLB international workshop. Neurology 47: 1113-1124.

McKeith IG, Mosimann UP. 2004. Dementia with Lewy bodies and Parkinson's disease. Parkinsonism Relat Disord 10 (Suppl. 1): S15-18.

McKeith IG, Perry EK, Perry RH. 1999. Report of the second dementia with Lewy body international workshop: Diagnosis and treatment. Consortium on Dementia with Lewy Bodies. Neurology 53: 902-905.

McKenzie JE, Edwards RJ, Gentleman SM, Ince PG, Perry RH, et al. 1996. A quantitative comparison of plaque types in Alzheimer's disease and senile dementia of the Lewy body type. Acta Neuropathol (Berl) 91: 526-529.

McLean PJ, Kawamata H, Shariff S, Hewett J, Sharma N, et al. 2002. TorsinA and heat shock proteins act as molecular chaperones: Suppression of alpha-synuclein aggregation. J Neurochem 83: 846-854.

McNaught KS, Belizaire R, Isacson O, Jenner P, Olanow CW. 2003. Altered proteasomal function in sporadic Parkinson's disease. Exp Neurol 179: 38-46.

McNaught KS, Jackson T, JnoBaptiste R, Kapustin A, Olanow CW. 2006. Proteasomal dysfunction in sporadic Parkinson's disease. Neurology 66: S37-49.

McNaught KS, Jenner P. 2001. Proteasomal function is impaired in substantia nigra in Parkinson's disease. Neurosci Lett 297: 191-194.

McNaught KS, Olanow CW. 2003. Proteolytic stress: A unifying concept for the etiopathogenesis of Parkinson's disease. Ann Neurol 53 (Suppl. 3): S73-S84.

McNaught KS, Olanow CW, Halliwell B, Isacson O, Jenner P. 2001. Failure of the ubiquitin-proteasome system in Parkinson's disease. Nat Rev Neurosci 2: 589-594.

McNaught KS, Olanow CW. 2006. Protein aggregation in the pathogenesis of familial and sporadic Parkinson's disease. Neurobiol Aging 27: 530-545.

McNaught KS, Bjorklund LM, Belizaire R, Isacson O, Jenner P, et al. 2002a. Proteasome inhibition causes nigral degeneration with inclusion bodies in rats. Neuroreport 13: 1437-1441.

McNaught KS, Perl DP, Brownell AL, Olanow CW. 2004. Systemic exposure to proteasome inhibitors causes a progressive model of Parkinson's disease. Ann Neurol 56: 149-162.

McNaught KS, Shashidharan P, Perl DP, Jenner P, Olanow CW. 2002b. Aggresome-related biogenesis of Lewy bodies. Eur J Neurosci 16: 2136-2148.

McNeill TH, Brown SA, Rafols JA, Shoulson I. 1988. Atrophy of medium spiny I riatal dendrites in advanced Parkinson's disease. Brain Res 455: 148-152.

McRitchie DA, Cartwright HR, Halliday GM. 1997. Specific A10 dopaminergic nuclei in the midbrain degenerate in Parkinson's disease. Exp Neurol 144: 202-213.

McShane RH, Nagy Z, Esiri MM, King E, Joachim C, et al. 2001. Anosmia in dementia is associated with Lewy bodies rather than Alzheimer's pathology. J Neurol Neurosurg Psychiatry 70: 739-743.

Merdes AR, Hansen LA, Jeste DV, Galasko D, Hofstetter CR, et al. 2003. Influence of Alzheimer pathology on clinical diagnostic accuracy in dementia with Lewy bodies. Neurology 60: 1586-1590.

Middelkoop HA, Flier van der WM, Burton EJ, Lloyd AJ, Paling S, et al. 2001. Dementia with Lewy bodies and AD are not associated with occipital lobe atrophy on MRI. Neurology 57: 2117-2120.

Mikolaenko I, Pletnikova O, Kawas CH, O'Brien R, Resnick SM, Crain B, Troncoso JC. 2005. α-Synuclein lesions in normal aging, Parkinson disease, and Alzheimer disease: Evidence from the Baltimore Longitudinal Study of Aging (BLSA). J Neuropathol Exp Neurol 64: 156-162

Miller GW, Staley JK, Heilman CJ, Perez JT, Mash DC, et al. 1997. Immunochemical analysis of dopamine transporter protein in Parkinson's disease. Ann Neurol 41: 530-539.

Mirra SS, Heyman A, McKeel D, Sumi SM, Crain BJ, et al. 1991. The consortium to establish a registry for Alzheimer's disease (CERAD). Part II. Standardization of the neuropathologic assessment of Alzheimer's disease. Neurology 41: 4794-4786.

Mirzaei S, Rodrigues M, Koehn H, Knoll P, Bruecke T. 2003. Metabolic impairment of brain metabolism in patients with Lewy body dementia. Eur J Neurol 10: 573-575.

Mishizen-Eberz AJ, Norris EH, Giasson BI, Hodara R, Ischiropoulos H, et al. 2005. Cleavage of α-Synuclein by calpain: Potential role in degradation of fibrillized and nitrated species of α-synuclein. Biochemistry 44: 7818-7829.

Mizuta I, Satake W, Nakabayashi Y, Ito C, Suzuki S, et al. 2006. Multiple candidate gene analysis identifies α-synuclein as a susceptibility gene for sporadic Parkinson's disease. Hum Mol Genet 15: 1151-1158.

Mochizuki A, Komatsuzaki Y, Shoji S. 2002. Association of Lewy bodies and glial cytoplasmic inclusions in the brain of Parkinson's disease. Acta Neuropathol (Berl) 104: 534-537.

Mochizuki I, Hattori N. 2005. Alpha-synuclein, nigral degeneration and parkinsonism. Scientific Basis for the Treatment of Parkinson's Disease. Galvez-Jimenez N, editor. London & New York: Taylor & Francis; pp. 87-120.

Mochizuki H, Yamada M, Mizuno Y. 2006. α-Synuclein overexpression model. J Neural Transm Suppl. 70: 281-284.

Mori H. 2005. Pathological substrate of dementia in Parkinson's disease—its relation to DLB and DLBD. Parkinsonism Relat Disord 11 (Suppl 1:) S41-45.

Mori E, Shimomura T, Fujimori M, Hirono N, Imamura T, et al. 2000. Visuoperceptual impairment in dementia with Lewy bodies. Arch Neurol 57: 489-493.

Mori F, Hayashi S, Yamagishi S, Yoshimoto M, Yagihashi S, et al. 2002. Pick's disease: Alpha- and beta-synuclein-immunoreactive Pick bodies in the dentate gyrus. Acta Neuropathol (Berl) 104: 455-461.

Mori F, Nishie M, Piao YS, Kito K, Kamitani T, et al. 2005. Accumulation of NEDD8 in neuronal and glial inclusions of neurodegenerative disorders. Neuropathol Appl Neurobiol 31: 53-61.

Mori F, Piao YS, Hayashi S, Fujiwara H, Hasegawa M, et al. 2003. Alpha-synuclein accumulates in Purkinje cells in Lewy body disease but not in multiple system atrophy. J Neuropathol Exp Neurol 62: 812-819.

Mori H, Yoshimura M, Tomonaga M, Yamanouchi H. 1986. Progressive supranuclear palsy with Lewy bodies. Acta Neuropathol (Berl) 71: 344-346.

Morrish PK, Rakshi JS, Bailey DL, Sawle GV, Brooks DJ. 1998. Measuring the rate of progression and estimating the preclinical period of Parkinson's disease with [18F]dopa PET. J Neurol Neurosurg Psychiatry 64: 314-319.

Morrish PK, Sawle GV, Brooks DJ. 1996. Regional changes in [18F]dopa metabolism in the striatum in Parkinson's disease. Brain 119: 2097-2103.

Mosharov EV, Staal RG, Bove J, Prou D, Hananiya A, et al. 2006. α-Synuclein overexpression increases cytosolic catecholamine concentration. J Neurosci 26: 9304-9311.

Mosimann UP, McKeith IG. 2003. Dementia with Lewy bodies and Parkinson's disease dementia: Two synucleinopathies. ACNR 3: 8-16.

Mouradian MM. 2002. Recent advances in the genetics and pathogenesis of Parkinson disease. Neurology 58: 179-185.

Muchowski PJ, Wacker JL. 2005. Modulation of neurodegeneration by molecular chaperones. Nat Rev Neurosci 6: 11-22.

Mueller JC, Fuchs J, Hofer A, Zimprich A, Lichtner P, et al. 2005. Multiple regions of α-synuclein are associated with Parkinson's disease. Ann Neurol 57: 535-541.

Mufson EJ, Conner JM, Kordower JH. 1995. Nerve growth factor in Alzheimer's disease: Defective retrograde transport to nucleus basalis. Neuroreport 6: 1063-1066.

Munch G, Luth HJ, Wong A, Arendt T, Hirsch E, et al. 2000. Crosslinking of alpha-synuclein by advanced glycation endproducts: An early pathophysiological step in Lewy body formation? J Chem Neuroanat 20: 253-257.

Müller CM, de Vos RA, Maurage CA, Thal DR, Tolnay M, et al. 2005. Staging of sporadic Parkinson disease-related α-synuclein pathology: Inter- and intra-rater reliability. J Neuropathol Exp Neurol 64: 623-628.

Munoz DG. 1999. Stains for the differential diagnosis of degenerative dementias. Biotech Histochem 74: 311-320.

Murakami T, Shoji M, Imai Y, Inoue H, Kawarabayashi T, et al. 2004. Pael-R is accumulated in Lewy bodies of Parkinson's disease. Ann Neurol 55: 439-442.

Murray IJ, Medford MA, Guan HP, Rueter SM, Trojanowski JQ, et al. 2003. Synphilin in normal human brains and in synucleinopathies: Studies with new antibodies. Acta Neuropathol (Berl) 105: 177-184.

Myung J, Kim KB, Crews CM. 2001. The ubiquitin-proteasome pathway and proteasome inhibitors. Med Res Rev 21: 245-273.

Nakamura S, Kawamoto Y, Nakano S, Akiguchi I, Kimura J. 1998. Cyclin-dependent kinase 5 and mitogen-activated protein kinase in glial cytoplasmic inclusions in multiple system atrophy. J Neuropathol Exp Neurol 57: 690-698.

Narhi L, Wood SJ, Steavenson S, Jiang Y, Wu GM, et al. 1999. Both familial Parkinson's disease mutations accelerate alpha-synuclein aggregation. J Biol Chem 274: 9843-9846.

Neumann M, Muller V, Kretzschmar HA, Haass C, Kahle PJ. 2004. Regional distribution of proteinase K-resistant α-synuclein correlates with Lewy body disease stage. J Neuropathol Exp Neurol 63: 1225-1235.

Neuropathology-Group. 2001. Pathological correlates of late-onset dementia in a multicentre, community-based population in England and Wales. Neuropathology Group of the Medical Research Council Cognitive Function and Ageing Study (MRC CFAS). Lancet 357: 169–175.

Neystat M, Lynch T, Przedborski S, Kholodilov N, Rzhetskaya M, et al. 1999. Alpha-synuclein expression in substantia nigra and cortex in Parkinson's disease. Mov Disord 14: 417-422.

Nielsen MS, Vorum H, Lindersson E, Jensen PH. 2001. Ca2+ binding to alpha-synuclein regulates ligand binding and oligomerization. J Biol Chem 276: 22680-22684.

Nisbet AP, Eve DJ, Kingsbury AE, Daniel SE, Marsden CD, et al. 1996. Glutamate decarboxylase-67 messenger RNA expression in normal human basal ganglia and in Parkinson's disease. Neuroscience 75: 389-406.

Noe E, Marder K, Bell KL, Jacobs DM, Manly JJ, et al. 2004. Comparison of dementia with Lewy bodies to Alzheimer's disease and Parkinson's disease with dementia. Mov Disord 19: 60-67.

Norris EH, Giasson BI, Lee VM. 2004. Alpha-synuclein: Normal function and role in neurodegenerative diseases. Curr Top Dev Biol 60: 17-54.

Nurmi E, Bergman J, Eskola O, Solin O, Vahlberg T, et al. 2003. Progression of dopaminergic hypofunction in striatal subregions in Parkinson's disease using [18F]CFT PET. Synapse 48: 109-115.

O'Brien J, Ames D, McKeith I, Chiu E. 2006. Dementia with Lewy Bodies and Parkinson's Disease Dementia. London: Taylor & Francis.

O'Brien JT, Paling S, Barber R, Williams ED, Ballard C, et al. 2001. Progressive brain atrophy on serial MRI in dementia with Lewy bodies, AD, and vascular dementia. Neurology 56: 1386-1388.

Ohtake H, Limprasert P, Fan Y, Onodera O, Kakita A, et al. 2004. Beta-synuclein gene alterations in dementia with Lewy bodies. Neurology 63: 805-811.

Okamura N, Arai H, Higuchi M, Tashiro M, Matsui T, et al. 2001. [F-18]FDG-PET study in dementia with Lewy bodies and Aizheimer's disease. Progr Neuro Psychopharmacol Biol Psychiatry 25: 447-456.

Okazaki H, Lipkin LE, Aronson SM. 1961. Diffuse intracytoplasmic ganglionic inclusions (Lewy type) associated with progressive dementia and quadriparesis in flexion. J Neuropathol Exp Neurol 20: 237-244.

Okochi M, Walter J, Koyama A, Nakajo S, Baba M, et al. 2000. Constitutive phosphorylation of the Parkinson's disease associated alpha-synuclein. J Biol Chem 275: 390-397.

Olanow CW, Perl DP, DeMartino GN, McNaught KS. 2004. Lewy-body formation is an aggresome-related process: A hypothesis. Lancet Neurol 3: 496-503.

Olichney JM, Galasko D, Salmon DP, Hofstetter CR, Hansen LA, et al. 1998. Cognitive decline is faster in Lewy body variant than in Alzheimer's disease. Neurology 51: 351-357.

Olichney JM, Hansen LA, Galasko D, Saitoh T, Hofstetter CR, et al. 1996. The apolipoprotein E epsilon 4 allele is associated with increased neuritic plaques and cerebral amyloid angiopathy in Alzheimer's disease and Lewy body variant. Neurology 47: 190-196.

Ono K, Yamada M. 2006. Antioxidant compounds have potent anti-fibrillogenic and fibril-destabilizing effects for α-synuclein fibrils in vitro. J Neurochem 97: 105-115.

Orr CF, Rowe DB, Halliday GM. 2002. An inflammatory review of Parkinson's disease. Prog Neurobiol 68: 325-340.

Orth M, Tabrizi SJ. 2003. Models of Parkinson's disease. Mov Disord 18: 729-737.

Ostrerova-Golts N, Petrucelli L, Hardy J, Lee JM, Farer M, et al. 2000. The A53T alpha-synuclein mutation increases iron-dependent aggregation and toxicity. J Neurosci 20: 6048-6054.

Ouchi Y, Yoshikawa E, Sekine Y, Futatsubashi M, Kanno T, et al. 2005. Microglial activation and dopamine terminal loss in early Parkinson's disease. Ann Neurol 57: 168-175.

Paik SR, Lee D, Cho HJ, Lee EN, Chang CS. 2003. Oxidized glutathione stimulated the amyloid formation of alpha-synuclein. FEBS Lett 537: 63-67.

Paisan-Ruiz C, Jain S, Evans EW, Gilks WP, Simon J, et al. 2004. Cloning of the gene containing mutations that cause PARK8-linked Parkinson's disease. Neuron 44: 595-600.

Papapetropoulos S, Mash DC. 2006. Insular pathology in Parkinson's disease patients with orthostatic hypotension. Parkinsonism Relat Disord.

Pappolla MA, Shank DL, Alzofon J, Dudley AW. 1988. Colloid (hyaline) inclusion bodies in the central nervous system: Their presence in the substantia nigra is diagnostic of Parkinson's disease. Hum Pathol 19: 27-31.

Parent A, Cossette M. 2001. Extrastriatal dopamine and Parkinson's disease. Adv Neurol 86: 45-54.

Parent A, Levesque M, Parent M. 2001. A re-evaluation of the current model of the basal ganglia. Parkinsonism Relat Disord 7: 193–198.

Parkkinen L, Kauppinen T, Pirttila T, Autere JM, Alafuzoff I. 2005. Alpha-synuclein pathology does not predict extrapyramidal symptoms or dementia. Ann Neurol 57: 82-91.

Parkkinen L, Pirttila T, Tervahauta M, Alafuzoff I. 2005. Widespread and abundant α-synuclein pathology in a neurologically unimpaired subject. Neuropathology 25: 304-314.

Parkkinen L, Soininen H, Alafuzoff I. 2003. Regional distribution of alpha-synuclein pathology in unimpaired aging and Alzheimer disease. J Neuropathol Exp Neurol 62: 363-367.

Parnetti L, Lanari A, Amici S, Gallai V, Vanmechelen E, et al. 2001. SF phosphorylated tau is a possible marker for discriminating Alzheimer's disease from dementia with Lewy bodies. Phospho-Tau International Study Group. Neurol Sci 22: 77-78.

Pedersen KM, Marner L, Pakkenberg H, Pakkenberg B. 2005. No global loss of neocortical neurons in Parkinson's disease: A quantitative stereological study. Mov Disord 20: 164-171.

Perneczky R, Mosch D, Neumann M, Kretzschmar H, Muller U, et al. 2005. The Alzheimer variant of Lewy body disease: A pathologically confirmed case-control study. Dement Geriatr Cogn Disord 20: 89-94.

Perez RG, Hastings TG. 2004. Could a loss of α-synuclein function put dopaminergic neurons at risk? J Neurochem 89: 1318-1324.

Perez RG, Waymire JC, Lin E, Liu JJ, Guo F, et al. 2002. A role for alpha-synuclein in the regulation of dopamine biosynthesis. J Neurosci 22: 3090-3099.

Perry EK, Haroutunian V, Davis KL, Levy R, Lantos P, et al. 1994. Neocortical cholinergic activities differentiate Lewy body dementia from classical Alzheimer's disease. Neuroreport 5: 747-749.

Perry EK, Marshall E, Perry RH, Irving D, Smith CJ, et al. 1990a. Cholinergic and dopaminergic activities in senile dementia of Lewy body type. Alzheimer Dis Assoc Disord 4: 87-95.

Perry EK, Smith CJ, Court JA, Perry RH. 1990b. Cholinergic nicotinic and muscarinic receptors in dementia of Alzheimer, Parkinson and Lewy body types. J Neural Transm Park Dis Dement Sect 2: 149-158.

Perry RH, Irving D, Blessed G, Fairbairn A, Perry EK. 1990c. Senile dementia of Lewy body type. A clinically and neuropathologically distinct form of Lewy body dementia in the elderly. J Neurol Sci 95: 119-139.

Perry EK, Morris CM, Court JA, Cheng A, Fairbairn AF, et al. 1995. Alteration in nicotine binding sites in Parkinson's disease, Lewy body dementia and Alzheimer's disease: Possible index of early neuropathology. Neuroscience 64: 385-395.

Perry EK, Piggott MA, Johnson M. 2003. Neurotransmitter correlates of neuropsychiatric symptoms in dementia with Lewy bodies. Mental and Behavioral Dysfunction in Movement Disorder. Bedard M-A, Agid Y, Chouinard S, Fahn S, Korczyn AD, et al. editors. Totowa, NJ: Humana Press, Inc.; pp. 285-294.

Pesah Y, Pham T, Burgess H, Middlebrooks B, Verstreken P, et al. 2004. Drosophila parkin mutants have decreased mass and cell size and increased sensitivity to oxygen radical stress. Development 131: 2183-2194.

Petrucelli L, Dawson TM. 2004. Mechanism of neurodegenerative disease: Role of the ubiquitin proteasome system. Ann Med 36: 315-320.

Piggott MA, Marshall EF, Thomas N, Lloyd S, Court JA, et al. 1999. Striatal dopaminergic markers in dementia with Lewy bodies, Alzheimer's and Parkinson's diseases: Rostrocaudal distribution. Brain 122 (Pt. 8): 1449-1468.

Piggott MA, Perry EK, Marshall EF, McKeith IG, Johnson M, et al. 1998. Nigrostriatal dopaminergic activities in dementia with Lewy bodies in relation to neuroleptic sensitivity: Comparisons with Parkinson's disease. Biol Psychiatry 44: 765-774.

Pirker W, Holler I, Gerschlager W, Asenbaum S, Zettinig G, et al. 2003. Measuring the rate of progression of Parkinson's disease over a 5-year period with beta-CIT SPECT. Mov Disord 18: 1266-1272.

Pletnikova O, West N, Lee MK, Rudow GL, Skolasky RL, et al. 2005. Aβ deposition is associated with enhanced cortical α-synuclein lesions in Lewy body diseases. Neurobiol Aging 26: 1183-1192.

Polymeropoulos MH, Lavedan C, Leroy E, Ide SE, Dehejia A, et al. 1997. Mutation in the alpha-synuclein gene identified in families with Parkinson's disease. Science 276: 2045-2047.

Popescu A, Lippa CF. 2001. Lewy bodies in the amygdala: alpha-synuclein expression is increased in specific neurodegenerative diseases (abstract). Neurology 56 (Suppl. 3): A177-A178.

Popescu A, Lippa CF, Lee VM, Trojanowski JQ. 2004. Lewy bodies in the amygdala: Increase of alpha-synuclein aggregates in neurodegenerative diseases with tau-based inclusions. Arch Neurol 61: 1915-1919.

Pountney DL, Lowe R, Quilty M, Vickers JC, Voelcker NH, et al. 2004. Annular alpha-synuclein species from purified multiple system atrophy inclusions. J Neurochem 90: 502-512.

Przedborski S. 2005. Pathogenesis of nigral cell death in Parkinson's disease. Parkinsonism Relat Disord 11 (Suppl 1:) S3-S7.

Przedborski S, Tieu K, Perier C, Vila M. 2004. MPTP as a mitochondrial neurotoxic model of Parkinson's disease. J Bioenerg Biomembr 36: 375-379.

Przuntek H, Muller T, Riederer P. 2004. Diagnostic staging of Parkinson's disease: Conceptual aspects. J Neural Transm 111: 201-216.

Purisai MG, McCormack AL, Langston WJ, Johnston LC, Di Monte DA. 2005. α-Synuclein expression in the substantia nigra of MPTP-lesioned non-human primates. Neurobiol Dis 20: 898-906.

Putten van der H, Wiederhold KH, Probst A, Barbieri S, Mistl C, et al. 2000. Neuropathology in Science expressing human alpha-synuclein. J Neurosci 20: 6021-6029.

Quilty MC, King AE, Gai WP, Pountney DL, West AK, et al. 2006. α-Synuclein is upregulated in neurones in response to chronic oxidative stress and is associated with neuroprotection. Exp Neurol 199: 249-256.

Rahkonen T, Eloniemi-Sulkava U, Rissanen S, Vatanen A, Viramo P, et al. 2003. Dementia with Lewy bodies according to the consensus criteria in a general population aged 75 years or older. J Neurol Neurosurg Psychiatry 74: 720-724.

Ramirez-Ruiz B, Marti MJ, Tolosa E, Bartres-Faz D, Summerfield C, et al. 2005. Longitudinal evaluation of cerebral morphological changes in Parkinson's disease with and without dementia. J Neurol 252: 1345-1352.

Rakshi JS, Pavese N, Uema T, Ito K, Morrish PK, et al. 2002. A omparison of the progression of early Parkinson's disease in patients started on ropinirole or L-dopa: An 18F-dopa PET study. J Neural Transm 109: 1433-1443.

Ransmayr G, Seppi K, Donnemiller E, Luginger E, Marksteiner J, et al. 2001. Striatal dopamine transporter function in dementia with Lewy bodies and Parkinson's disease. Eur J Nucl Med 28: 1523-1528.

Rathke-Hartlieb S, Kahle PJ, Neumann M, Ozmen L, Haid S, et al. 2001. Sensitivity to MPTP is not increased in Parkinson's disease-associated mutant alpha-synuclein transgenic mice. J Neurochem 77: 1181-1184.

Ray M, Bohr I, McIntosh JM, Ballard C, McKeith I, et al. 2004. Involvement of α6/α3 neuronal nicotinic acetylcholine receptors in neuropsychiatric features of dementia with Lewy bodies: [^{125}I]- α-conotoxin MII binding in the thalamus and striatum. Neurosci Lett 372: 220-225.

Rehncrona S, Johnels B, Widner H, Tornqvist AL, Hariz M, et al. 2003. Long-term efficacy of thalamic deep brain stimulation for tremor: Double-blind assessments. Mov Disord 18: 163-170.

Rei RT, Sabbagh MN, Corey-Bloom J, Tiraboschi P, Thal LJ. 2000. Nicotinic receptor losses in dementia with Lewy bodies: Comparisons with Alzheimer's disease. Neurobiol Aging 21: 741-746.

Rezaie P, Cairns NJ, Chadwick A, Lantos PL. 1996. Lewy bodies are located preferentially in limbic areas in diffuse Lewy body disease. Neurosci Lett 212: 111-114.

Richfield EK, Thiruchelvam MJ, Cory-Slechta DA, Wuertzer C, Gainetdinov RR, et al. 2002. Behavioral and neurochemical effects of wild-type and mutated human alpha-synuclein in transgenic mice. Exp Neurol 175: 35-48.

Rideout HJ, Larsen KE, Sulzer D, Stefanis L. 2001. Proteasomal inhibition leads to formation of ubiquitin/alpha-synuclein-immunoreactive inclusions in PC12 cells. J Neurochem 78: 899-908.

Rinne JO, Anichtchik OV, Eriksson KS, Kaslin J, Tuomisto L, et al. 2002. Increased brain histamine levels in Parkinson's disease but not in multiple system atrophy. J Neurochem 81: 954-960.

Rinne JO, Nurmi E, Ruottinen HM, Bergman J, Eskola O, et al. 2001. [F-18]FDOPA and [F-18]CFT are both sensitive PET markers to detect presynaptic dopaminergic hypofunction in early Parkinson disease. Synapse 40: 193-200.

Rosenberg CK, Cummings TJ, Saunders AM, Widico C, McIntyre LM, et al. 2001. Dementia with Lewy bodies and Alzheimer's disease. Acta Neuropathol (Berl) 102: 621-626.

Rozemuller AJ, Eikelenboom P, Theeuwes JW, Jansen Steur EN, de Vos RA. 2000. Activated microglial cells and complement factors are unrelated to cortical Lewy bodies. Acta Neuropathol (Berl) 100: 701-708.

Sabbagh MN, Corey-Bloom J, Tiraboschi P, Thomas R, Masliah E, et al. 1999. Neurochemical markers do not correlate with cognitive decline in the Lewy body variant of Alzheimer disease. Arch Neurol 56: 1458-1461.

Sabbagh MN, Reid RT, Hansen LA, Alford M, Thal LJ. 2001. Correlation of nicotinic receptor binding with clinical and neuropathological changes in Alzheimer's disease and dementia with Lewy bodies. J Neural Transm 108: 1149-1157.

Saha AR, Ninkina NN, Hanger DP, Anderton BH, Davies AM, et al. 2000. Induction of neuronal death by alpha-synuclein. Eur J Neurosci 12: 3073-3077.

Sahin HA, Emre M, Ziabreva I, Perry E, Celasun B, et al. 2006. The distribution pattern of pathology and cholinergic deficits in amygdaloid complex in Alzheimer's disease and dementia with Lewy bodies. Acta Neuropathol (Berl) 111: 115-125.

Saito Y, Kawashima A, Ruberu NN, Fujiwara H, Koyama S, et al. 2003. Accumulation of phosphorylated alpha-synuclein in aging human brain. J Neuropathol Exp Neurol 62: 644-654.

Saito Y, Ruberu NN, Sawabe M, Arai T, Kazama H, et al. 2004. Lewy body-related alpha-synucleinopathy in aging. J Neuropathol Exp Neurol 63: 742-749.

Saitoh T, Xia Y, Chen X, Masliah E, Galasko D, et al. 1995. The CYP2D6B mutant allele is overrepresented in the Lewy body variant of Alzheimer's disease. Ann Neurol 37: 110-112.

Sakamoto M, Uchihara T, Hayashi M, Nakamura A, Kikuchi E, et al. 2002. Heterogeneity of nigral and cortical Lewy bodies differentiated by amplified triple-labeling for alpha-synuclein, ubiquitin, and thiazin red. Exp Neurol 177: 88-94.

Salmon DP, Hamilton JM. 2006. Neuropsychological features of dementia with Lewy bodies. Dementia with Lewy Bodies and Parkinson's Disease Dementia. Brien JO, Ames D, McKeith I, Chiu E, editors. London: Taylor & Francis; pp. 48-72.

Sampathu DM, Giasson BI, Pawlyk AC, Trojanowski JQ, Lee VM. 2003. Ubiquitination of alpha-synuclein is not required for formation of pathological inclusions in alpha-synucleinopathies. Am J Pathol 163: 91-100.

Samuel W, Alford M, Hofstetter CR, Hansen L. 1997a. Dementia with Lewy bodies versus pure Alzheimer disease: Differences in cognition, neuropathology, cholinergic dysfunction, and synapse density. J Neuropathol Exp Neurol 56: 499-508.

Samuel W, Crowder R, Hofstetter CR, Hansen L. 1997b. Neuritic plaques in the Lewy body variant of Alzheimer disease lack paired helical filaments. Neurosci Lett 223: 73-76.

Samuel W, Caligiuri M, Galasko D, Lacro J, Marini M, et al. 2000. Better cognitive and psychopathologic response to donepezil in patients prospectively diagnosed as dementia with Lewy bodies: A preliminary study. Int J Geriatr Psychiatry 15: 794-802.

Samuel W, Galasko D, Masliah E, Hansen LA. 1996. Neocortical Lewy body counts correlate with dementia in the Lewy body variant of Alzheimer's disease. J Neuropathol Exp Neurol 55: 44-52.

Schlossmacher MG, Frosch MP, Gai WP, Medina M, Sharma N, et al. 2002. Parkin localizes to the Lewy bodies of Parkinson disease and dementia with Lewy bodies. Am J Pathol 160: 1655-1667.

Schneider JA, Bienias JL, Gilley DW, Kvarnberg DE, Mufson EJ, et al. 2002. Improved detection of substantia nigra pathology in Alzheimer's disease. J Histochem Cytochem 50: 99-106.

Schrag A, Ben-Shlomo Y, Quinn N. 2002. How valid is the clinical diagnosis of Parkinson's disease in the community? J Neurol Neurosurg Psychiatry 73: 529-534.

Sebeo J, Hof PR, Perl DP. 2004. Occurrence of alpha-synuclein pathology in the cerebellum of Guamanian patients with parkinsonism-dementia complex. Acta Neuropathol 107: 497-503.

Seo JH, Rah JC, Choi SH, Shin JK, Min K, et al. 2002. Alpha-synuclein regulates neuronal survival via Bcl-2 family expression and PI3/Akt kinase pathway. FASEB J 16: 1826-1828.

Serpell LC, Berriman J, Jakes R, Godert M, Crowther RA. 2000. Fiber diffraction of synthetic alpha-synuclein filaments shows amyloid-like cross-beta conformation. Proc Natl Acad Sci USA 97: 4897-4902.

Sharma N, McLean PJ, Kawamata H, Irizarry MC, Hyman BT. 2001. Alpha-synuclein has an altered conformation and shows a tight intermolecular interaction with ubiquitin in Lewy bodies. Acta Neuropathol (Berl) 102: 329-334.

Sharon R, Bar-Joseph I, Frosch MP, Walsh DM, Hamilton JA, et al. 2003a. The formation of highly soluble oligomers of alpha-synuclein is regulated by fatty acids and enhanced in Parkinson's disease. Neuron 37: 583-595.

Sharon R, Bar-Joseph I, Mirick GE, Serhan CN, Selkoe DJ. 2003b. Altered fatty acid composition of dopaminergic neurons expressing alpha-synuclein and human brains with alpha-synucleinopathies. J Biol Chem 278: 49874-49881.

Sherer TB, Kim JH, Betarbet R, Greenamyre JT. 2003. Subcutaneous rotenone exposure causes highly selective dopaminergic degeneration and alpha-synuclein aggregation. Exp Neurol 179: 9-16.

Shergill S, Mullen E, D'Ath P, Katona C. 1994. What is the clinical prevalence of Lewy body dementia. Int J Geriatr Psychiatry 9: 907-912.

Sherman MY, Goldberg AL. 2001. Cellular defenses against unfolded proteins: A cell biologist thinks about neurodegenerative diseases. Neuron 29: 15-32.

Shimura H, Schlossmacher MG, Hattori N, Frosch MP, Trockenbacher A, et al. 2001. Ubiquitination of a new form of alpha-synuclein by parkin from human brain: Implications for Parkinson's disease. Science 293: 263-269.

Shiozaki K, Iseki E, Uchiyama H, Watanabe Y, Haga T, et al. 1999. Alterations of muscarinic acetylcholine receptor subtypes in diffuse Lewy body disease: Relation to Alzheimer's disease. J Neurol Neurosurg Psychiatry 67: 209-213.

Shults CW. 2006. Lewy bodies. Proc Natl Acad Sci USA 103: 1661-1668.

Sidhu A, Wersinger C, Moussa CE, Vernier P. 2004a. The role of α-synuclein in both neuroprotection and neurodegeneration. Ann N Y Acad Sci 1035: 250-270.

Sidhu A, Wersinger C, Vernier P. 2004b. Does α-synuclein modulate dopaminergic synaptic content and tone at the synapse? FASEB J 18: 637-647.

Simon DK, Lin MT, Zheng L, Liu GJ, Ahn CH, et al. 2004. Somatic mitochondrial DNA mutations in cortex and substantia nigra in aging and Parkinson's disease. Neurobiol Aging 25: 71-81.

Singleton AB, Farrer M, Johnson J, Singleton A, Hague S, et al. 2003. Alpha-synuclein locus triplication causes Parkinson's disease. Science 302: 841.

Singleton AB, Gibson AM, Edwardson JA, McKeith IG, Morris CM. 1998. Butyrylcholinesterase K: An association with dementia with Lewy bodies. Lancet 351: 1818.

Singleton AB, Wharton A, O'Brien KK, Walker MP, McKeith IG, et al. 2002. Clinical and neuropathological correlates of apolipoprotein E genotype in dementia with Lewy bodies. Dement Geriatr Cogn Disord 14: 167-175.

Small GW. 2004. Neuroimaging as a diagnostic tool in dementia with Lewy bodies. Dement Geriatr Cogn Disord 17 (Suppl. 1): 25-31.

Smith MC, Mallory M, Hansen LA, Ge N, Masliah E. 1995. Fragmentation of the neuronal cytoskeleton in the Lewy body variant of Alzheimer's disease. Neuroreport 6: 673-679.

Solano SM, Miller DW, Augood SJ, Young AB, Penney JB Jr. 2000. Expression of alpha-synuclein, parkin, and ubiquitin carboxy-terminal hydrolase L1 mRNA in human brain: Genes associated with familial Parkinson's disease. Ann Neurol 47: 201-210.

Song DD, Shults CW, Sisk A, Rockenstein E, Masliah E. 2004. Enhanced substantia nigra mitochondrial pathology in human α-synuclein transgenic mice after treatment with MPTP. Exp Neurol 186: 158-172.

Souza JM, Giasson BI, Chen Q, Lee VM, Ischiropoulos H. 2000. Dityrosine cross-linking promotes formation of stable alpha–synuclein polymers. Implication of nitrative and oxidative stress in the pathogenesis of neurodegenerative synucleinopathies. J Biol Chem 275: 18344-18349.

Spillantini MG. 2003. Introduction to the synucleinopathies. Neurodegeneration: The Molecular Pathology of Dementia and Movement Disorders. Dickson DW, editor. Basel: ISN Press; pp. 156-158.

Spillantini MG, Murrell JR, Goedert M, Farlow MR, Klug A, et al. 1998. Mutation in the tau gene in familial multiple system tauopathy with presenile dementia. Proc Natl Acad Sci USA 95: 7737-7741.

Stevens T, Livingston G, Kitchen G, Manela M, Walker Z, et al. 2002. Islington study of dementia subtypes in the community. Br J Psychiatry 180: 270-276.

Stewart VC, Heales SJ. 2003. Nitric oxide-induced mitochondrial dysfunction: Implications for neurodegeneration. Free Radic Biol Med 34: 287-303.

Strong C, Anderton BH, Perry RH, Perry EK, Ince PG, et al. 1995. Abnormally phosphorylated tau protein in senile dementia of Lewy body type and Alzheimer disease: Evidence that the disorders are distinct. Alzheimer Dis Assoc Disord 9: 218-222.

Stumptner C, Fuchsbichler A, Heid H, Zatloukal K, Denk H. 2002. Mallory body: A disease-associated type of sequestosome. Hepatology 35: 1053-1062.

Suzuki M, Inoue K, Mishina M, Mitani K, Ishiwata K, 2003. Regional correlation of pre- and post-synaptic dopaminergic function in striatum of dementia with Lewy bodies: The comparison with Parkinson's disease—a voxel based analysis (abstract). Society for Neuroscience. 33rd Annual Meeting, Society for Neuroscience, New Orleans, pp. Program No. 917.9.

Sweet RA, Hamilton RL, Healy MT, Wisniewski SR, Henteleff R, et al. 2001. Alterations of striatal dopamine receptor binding in Alzheimer disease are associated with Lewy body pathology and antemortem psychosis. Arch Neurol 58: 466-472.

Taira T, Saito Y, Niki T, Iguchi-Ariga SM, Takahashi K, et al. 2004. DJ-1 has a role in antioxidative stress to prevent cell death. EMBO Rep 5: 213-218.

Takahashi H, Wakabayashi K. 2005. Controversy: Is Parkinson's disease a single disease entity? Yes. Parkinsonism Relat Disord 11 (Suppl. 1): S31-S37.

Takauchi S, Yamauchi S, Morimura Y, Ohara K, Morita Y, et al. 1995. Coexistence of Pick bodies and atypical Lewy bodies in the locus ceruleus neurons of Pick's disease. Acta Neuropathol (Berl) 90: 93-100.

Takeda A, Hashimoto M, Mallory M, Sundsumo M, Hansen L, et al. 1998. Abnormal distribution of the non-Abeta component of Alzheimer's disease amyloid precursor/ alpha-synuclein in Lewy body disease as revealed by proteinase K and formic acid pretreatment. Lab Invest 78: 1169-1177.

Tanaka M, Kim YM, Lee G, Junn E, Iwatsubo T, et al. 2004. Aggresomes formed by alpha-synuclein and synphilin-1 are cytoprotective. J Biol Chem 279: 4625-4631.

Tanaka S, Chen X, Xia Y, Kang DE, Matoh N, et al. 1998. Association of CYP2D microsatellite polymorphism with Lewy body variant of Alzheimer's disease. Neurology 50: 1556-1562.

Tatton WG, Chalmers-Redman R, Brown D, Tatton N. 2003. Apoptosis in Parkinson's disease: Signals for neuronal degradation. Ann Neurol 53 (Suppl. 3): S61-S70.

Teaktong T, Graham AJ, Court JA, Perry RH, Jaros E, et al. 2004. Nicotinic acetylcholine receptor immunohistochemistry in Alzheimer's disease and dementia with Lewy bodies: Differential neuronal and astroglial pathology. J Neurol Sci 225: 39-49.

Teismann P, Tieu K, Choi DK, Wu DC, Naini A, et al. 2003a. Cyclooxygenase-2 is instrumental in Parkinson's disease neurodegeneration. Proc Natl Acad Sci USA 100: 5473-5478.

Teismann P, Tieu K, Cohen O, Choi DK, Wu du C, et al. 2003b. Pathogenic role of glial cells in Parkinson's disease. Mov Disord 18: 121-129.

Terada S, Ishizu H, Haraguchi T, Takehisa Y, Tanabe Y, et al. 2000. Tau-negative astrocytic star-like inclusions and coiled bodies in dementia with Lewy bodies. Acta Neuropathol (Berl) 100: 464-468.

Thiruchelvam MJ, Powers JM, Cory-Slechta DA, Richfield EK. 2004. Risk factors for dopaminergic neuron loss in human alpha-synuclein transgenic mice. Eur J Neurosci 19: 845-854.

Timmermann L, Gross J, Dirks M, Volkmann J, Freund HJ, et al. 2003. The cerebral oscillatory network of parkinsonian resting tremor. Brain 126: 199-212.

Tiraboschi P, Hansen LA, Alford M, Merdes A, Masliah E, et al. 2002. Early and widespread cholinergic losses differentiate dementia with Lewy bodies from Alzheimer disease. Arch Gen Psychiatry 59: 946-951.

Tiraboschi P, Salmon DP, Hansen LA, Hofstetter RC, Thal LJ, et al. 2006. What best differentiates Lewy body from Alzheimer's disease in early-stage dementia? Brain 129: 729-735.

Tofaris GK, Layfield R, Spillantini MG. 2001. Alpha-synuclein metabolism and aggregation is linked to ubiquitin-independent degradation by the proteasome. FEBS Lett 509: 22-26.

Tofaris GK, Spillantini MG. 2005. α-Synuclein dysfunction in Lewy body diseases. Mov Disord 20 (Suppl. 12:) S37-S44.

Togo T, Iseki E, Marui W, Akiyama H, Ueda K, et al. 2001. Glial involvement in the degeneration process of Lewy body-bearing neurons and the degradation process of Lewy bodies in brains of dementia with Lewy bodies. J Neurol Sci 184: 71-75.

Tolnay M, Probst A. 1999. REVIEW: Tau protein pathology in Alzheimer's disease and related disorders. Neuropathol Appl Neurobiol 25: 171-187.

Trembath Y, Rosenberg C, Ervin JF, Schmechel DE, Gaskell P, et al. 2003. Lewy body pathology is a frequent co-pathology in familial Alzheimer's disease. Acta Neuropathol 105: 484-488.

Tretter L, Sipos I, Adam-Vizi V. 2004. Initiation of neuronal damage by complex I deficiency and oxidative stress in Parkinson's disease. Neurochem Res 29: 569-577.

Trimmer PA, Borland MK, Keeney PM, Bennett JP, et al. 2004. Parkinson's disease transgenic mitochondrial cybrids generate Lewy inclusion bodies. J Neurochem 88: 800-812.

Trojanowski JQ. 2003. Protein aggregation. Neurodegeneration: The Molecular Pathology of Dementia and Movement Disorders. Dickson DW, editor. Los Angeles: ISN Press; pp. 11-13.

Tschampa HJ, Schulz-Schaeffer W, Wiltfang J, Poser S, Otto M, et al. 2001. Decreased CSF amyloid beta42 and normal tau levels in dementia with Lewy bodies. Neurology 56: 576.

Tsuang DW, Dalan AM, Eugenio CJ, Poorkaj P, Limprasert P, et al. 2002. Familial dementia with Lewy bodies: a clinical and neuropathological study of 2 families. Arch Neurol 59: 1622-1630.

Tsuboi Y, Ahlskog JE, Apaydin H, Parisi JE, Dickson DW. 2001. Lewy bodies are not increased in progressive supranuclear palsy compared with normal controls. Neurology 57: 1675-1678.

Tsuboi Y, Dickson DW. 2005. Dementia with Lewy bodies and Parkinson's disease with dementia: Are they different? Parkinsonism Relat Disord 11 (Suppl. 1): S47-S51.

Turnbull S, Tabner BJ, El-Agnaf OM, Moore S, Davies Y, et al. 2001. Alpha-synuclein implicated in Parkinson's disease catalyses the formation of hydrogen peroxide in vitro. Free Radic Biol Med 30: 1163-1170.

Turner RS, D'Amato CJ, Chervin RD, Blaivas M. 2000. The pathology of REM sleep behavior disorder with comorbid Lewy body dementia. Neurology 55: 1730-1732.

Turner RS, Grafton ST, McIntosh AR, De Long MR, Hoffman JM. 2003. The functional anatomy of parkinsonian bradykinesia. Neuroimage 19: 163-179.

Ubl A, Berg D, Holzmann C, Kruger R, Berger K, et al. 2002. 14-3-3 protein is a component of Lewy bodies in Parkinson's disease-mutation analysis and association studies of 14-3-3 eta. Brain Res Mol Brain Res 108: 33-39.

Uhl GR. 1998. Hypothesis: The role of dopaminergic transporters in selective vulnerability of cells in Parkinson's disease. Ann Neurol 43: 555-560.

Uryu K, Richter-Landsberg C, Welch W, Sun E, Goldbaum O, et al. 2006. Convergence of heat shock protein 90 with ubiquitin in filamentous α-synuclein inclusions of α-synucleinopathies. Am J Pathol 168: 947-961.

Uversky VN, Fink AL. 2002. Amino acid determinants of alpha-synuclein aggregation: Putting together pieces of the puzzle. FEBS Lett 522: 9-13.

Uversky VN, Li J, Fink AL. 2001. Metal-triggered structural transformations, aggregation, and fibrillation of human alpha-synuclein. A possible molecular NK between Parkinson's disease and heavy metal exposure. J Biol Chem 276: 44284-44296.

Uversky VN, Li J, Souillac P, Millett IS, Doniach S, et al. 2002. Biophysical properties of the synucleins and their propensities to fibrillate: Inhibition of alpha-synuclein assembly by beta- and gamma-synucleins. J Biol Chem 277: 11970-11978.

Valente EM, Abou-Sleiman PM, Caputo V, Muqit MM, Harvey K, et al. 2004. Hereditary early-onset Parkinson's disease caused by mutations in PINK1. Science 304: 1158-1160.

van Horssen J, de Vos RA, Steur EN, David G, Wesseling P, et al. 2004. Absence of heparan sulfate proteoglycans in Lewy bodies and Lewy neurites in Parkinson's disease brains. J Alzheimers Dis 6: 469-474.

Varastet M, Riche D, Maziere M, Hantraye P. 1994. Chronic MPTP treatment reproduces in baboons the differential vulnerability of mesencephalic dopaminergic neurons observed in Parkinson's disease. Neuroscience 63: 47-56.

Volles MJ, Lee SJ, Rochet JC, Shtilerman MD, Ding TT, et al. 2001. Vesicle permeabilization by protofibrillar alpha-synuclein: Implications for the pathogenesis and treatment of Parkinson's disease. Biochemistry 40: 7812-7819.

von Coelln R, Thomas B, Andrabi SA, Lim KL, Savitt JM, et al. 2006. Inclusion body formation and neurodegeneration are parkin independent in a mouse model of α-synucleinopathy. J Neurosci 26: 3685-3696.

Wakabayashi K, Engelender S, Yoshimoto M, Tsuji S, Ross CA, et al. 2000a. Synphilin-1 is present in Lewy bodies in Parkinson's disease. Ann Neurol 47: 521-523.

Wakabayashi K, Hayashi S, Yoshimoto M, Kudo H, Takahashi H. 2000b. NACP/alpha-synuclein-positive filamentous inclusions in astrocytes and oligodendrocytes of Parkinson's disease brains. Acta Neuropathol (Berl) 99: 14-20.

Wakabayashi K, Hansen LA, Masliah E. 1995. Cortical Lewy body-containing neurons are pyramidal cells: Laser confocal imaging of double-immunolabeled sections with anti-ubiquitin and SMI32. Acta Neuropathol (Berl) 89: 404-408.

Wakabayashi K, Hayashi S, Ishikawa A, Hayashi T, Okuizumi K, et al. 1998a. Autosomal dominant diffuse Lewy body disease. Acta Neuropathol (Berl) 96: 207-210.

Wakabayashi K, Hayashi S, Kakita A, Yamada M, Toyoshima Y, et al. 1998b. Accumulation of alpha-synuclein/NACP is a cytopathological feature common to Lewy body disease and multiple system atrophy. Acta Neuropathol (Berl) 96: 445-452.

Wakabayashi K, Honer WG, Masliah E. 1994. Synapse alterations in the hippocampal-entorhinal formation in Alzheimer's disease with and without Lewy body disease. Brain Res 667: 24-32.

Walker MP, Ayre GA, Cummings JL, Wesnes K, McKeith IG, et al. 2000a. The clinician assessment of fluctuation and the one day fluctuation assessment scale. Two methods to assess fluctuating confusion in dementia. Br J Psychiatry 177: 252-256.

Walker MP, Ayre GA, Perry EK, Wesnes K, McKeith IG, et al. 2000b. Quantification and characterization of fluctuating cognition in dementia with Lewy bodies and Alzheimer's disease. Dement Geriatr Cogn Disord 11: 327-335.

Walker Z, Allen RL, Shergill S, Mullan E, Katona CL. 2000c. Three years survival in patients with a clinical diagnosis of dementia with Lewy bodies. Int J Geriatr Psychiatry 15: 267-273.

Walker Z, Allen RL, Shergill S, Katona CL. 1997a. Neuropsychological performance in Lewy body dementia and Alzheimer's disease. Br J Psychiatry 170: 156-158.

Walker Z, Costa DC, Janssen AG, Walker RW, Livingstone G, et al. 1997b. Dementia with Lewy bodies: A study of post-synaptic dopaminergic receptors with iodine-123 iodobenzamide single-photon emission tomography. Eur J Nucl Med 24: 609-614.

Walker Z, Costa DC, Walker RW, Shaw K, Gacinovic S, et al. 2002. Differentiation of dementia with Lewy bodies from Alzheimer's disease using a dopaminergic presynaptic ligand. J Neurol Neurosurg Psychiatry 73: 134-140.

Walker Z, Grace J, Overshot R, Satarasinghe S, Swan A, et al. 1999. Olanzapine in dementia with Lewy bodies: A clinical study. Int J Geriatr Psychiatry 14: 459-466.

Webb JL, Ravikumar B, Atkins J, Skepper JN, Rubinsztein DC. 2003. Alpha-synuclein is degraded by both autophagy and the proteasome. J Biol Chem 278: 25009-25013.

Weiner MF, Risser RC, Cullum CM, Honig L, White C, et al. 1996. Alzheimer's disease and its Lewy body variant: A clinical analysis of postmortem verified cases. Am J Psychiatry 153: 1269-1273.

Welch K, Yuan J. 2003. Alpha-synuclein oligomerization: a role for lipids? Trends Neurosci 26: 517-519.

Wenning GK, Jellinger KA. 2005. The role of α-synuclein and tau in neurodegenerative movement disorders. Curr Opin Neurol 18: 357-362.

Whone AL, Watts RL, Stoessl AJ, Davis M, Reske S, et al. 2003. Slower progression of Parkinson's disease with ropinirole versus levodopa: The REAL-PET study. Ann Neurol 54: 93-101.

Wood SJ, Wypych J, Steavenson S, Louis JC, Citron M, et al. 1999. Alpha-synuclein fibrillogenesis is nucleation-dependent. Implications for the pathogenesis of Parkinson's disease. J Biol Chem 274: 19509-19512.

Wooten MW, Seibenhener ML, Mamidipudi V, Diaz-Meco MT, Barker PA, et al. 2001. The atypical protein kinase C-interacting protein p62 is a scaffold for NF-kappaB activation by nerve growth factor. J Biol Chem 276: 7709-7712.

Wszolek ZK, Pfeiffer RF, Tsuboi Y, Uitti RJ, McComb RD, et al. 2004. Autosomal dominant parkinsonism associated with variable synuclein and tau pathology. Neurology 62: 1619-1622.

Wu E, Lipton RB, Dickson DW. 1992. Amyloid angiopathy in diffuse Lewy body disease. Neurology 42: 2131-2135.

Wullner U, Kornhuber J, Weller M, Schulz JB, Loschmann PA, et al. 1999. Cell death and apoptosis regulating proteins in Parkinson's disease: a cautionary note. Acta Neuropathol (Berl) 97: 408-412.

Xu W, Liu L, Emson P, Harrington CR, McKeith IG, et al. 2000. The CCTTT polymorphism in the NOS2A gene is associated with dementia with Lewy bodies. Neuroreport 11: 297-299.

Yamamoto T, Imai T. 1988. A case of diffuse Lewy body and Alzheimer's diseases with periodic synchronous discharges. J Neuropathol Exp Neurol 47: 536-548.

Yamamoto R, Iseki E, Murayama N, Minegishi M, Marui W, et al. 2006a. Investigation of Lewy pathology in the visual pathway of brains of dementia with Lewy bodies. J Neurol Sci 246: 95-101.

Yamamoto S, Fukae J, Mori H, Mizuno Y, Hattori N. 2006b. Positive immunoreactivity for vesicular monoamine transporter 2 in Lewy bodies and Lewy neurites in substantia nigra. Neurosci Lett 396: 187-191.

Yamazaki M, Arai Y, Baba M, Iwatsubo T, Mori O, et al. 2000. Alpha-synuclein inclusions in amygdala in the brains of patients with the parkinsonism-dementia complex of Guam. J Neuropathol Exp Neurol 59: 585-591.

Yamin G, Glaser CB, Uversky VN, Fink AL. 2003. Certain metals trigger fibrillation of methionine-oxidized alpha-synuclein. J Biol Chem 278: 27630-27635.

Zaja-Milatovic S, Milatovic D, Schantz AM, Zhang J, Montine KS, et al. 2005. Dendritic degeneration in neostriatal medium spiny neurons in Parkinson disease. Neurology 64: 545-547.

Zarow C, Lyness SA, Mortimer JA, Chui HC. 2003. Neuronal loss is greater in the locus coeruleus than nucleus basalis and substantia nigra in Alzheimer and Parkinson diseases. Arch Neurol 60: 337-341.

Zarranz JJ, Alegre J, Gomez-Esteban JC, Lezcano E, Ros R, et al. 2004. The new mutation, E46K, of alpha-synuclein causes Parkinson and Lewy body dementia. Ann Neurol 55: 164-173.

Zecca L, Gallorini M, Schunemann V, Trautwein AX, Gerlach M, et al. 2001. Iron, neuromelanin and ferritin content in the substantia nigra of normal subjects at different ages: Consequences for iron storage and neurodegenerative processes. J Neurochem 76: 1766-1773.

Zecca L, Zucca FA, Wilms H, Sulzer D. 2003. Neuromelanin of the substantia nigra: A neuronal black hole with protective and toxic characteristics. Trends Neurosci 26: 578-580.

Zhang J, Perry G, Smith MA, Robertson D, Olson SJ, et al. 1999. Parkinson's disease is associated with oxidative damage to cytoplasmic DNA and RNA in substantia nigra neurons. Am J Pathol 154: 1423-1429.

Zheng K, Heydari B, Simon DK. 2003. A common NURR1 polymorphism associated with Parkinson disease and diffuse Lewy body disease. Arch Neurol 60: 722-725.

Zhou ZD, Yap BP, Gung AY, Leong SM, Ang ST, et al. 2006. Dopamine-related and caspase-independent apoptosis in dopaminergic neurons induced by overexpression of human wild type or mutant α-synuclein. Exp Cell Res 312: 156-170.

Zhu M, Fink AL. 2003. Lipid binding inhibits alpha-synuclein fibril formation. J Biol Chem 278: 16873-16877.

Zhu M, Qin ZJ, Hu D, Munishkina LA, Fink AL. 2006. α-Synuclein can function as an antioxidant preventing oxidation of unsaturated lipid in vesicles. Biochemistry 45: 8135-8142.

Ziabreva I, Ballard CG, Aarsland D, Larsen JP, McKeith IG, et al. 2006. Lewy body disease: Thalamic cholinergic activity related to dementia and parkinsonism. Neurobiol Aging 27: 433-438.

Zimprich A, Biskup S, Leitner P, Lichtner P, Farrer M, et al. 2004. Mutations in LRRK2 cause autosomal-dominant parkinsonism with pleomorphic pathology. Neuron 44: 601-607.

16 Gene Hunting by Substractive Hybridization in Down Syndrome Correlation with Proteomics Analysis

O. Golubnitschaja · M. Fountoulakis

Abstract: Down syndrome (DS) or trisomy 21 is a complex genetic and metabolic disorder and the most common genetic cause of mental retardation with an incidence of 1 in about 700 live births. DS is associated with facial dysmorphology, congenital heart defects, high accidence of acute megakaryoblastic leukemias in childhood, immunologic disorders, thyroid dysfunction, frequent diabetes mellitus, etc. Around the third decade of life patients with DS usually develop neuropathological features of Alzheimer's disease (AD).

For a long time, "gene dosage effect" was believed to underlie the abnormalities characteristic for DS; recent studies, however, have shown this theory to be too simplistic. In spite of plenty of projects devoted to DS, molecular pathomechanisms leading to singular DS complications remain quite unclear. Research in this field has accelerated in recent years, applying highly sensitive molecular biology instruments and technologies. Thereby, subtractive hybridization (SH) and brain proteomics (BP) belong to the group of most effective tools developed for the identification of disease-specific gene expression—so-called "disease proteome/transcriptome."

Here we report an overview about SH and BP results achieved through *ex vivo* comparative analysis of human fetal DS brains versus controls. The functional alterations found are stretched on the whole complex of cellular events in DS regarding brain development, cell reproduction, DNA replication, quality control and repair, synthesis of gene products, stress response, antitumor controlling mechanisms, central metabolic processes like metabolism of glucose and amino acids, function of single subcellular structures such as mitochondria, energy metabolism, neutralization of reactive oxygen species, general brain detoxification, and complex neuronal function. Affected molecular pathways demonstrate a clear predisposition of DS for neurodegeneration, cancer, and AD. Gene expression alterations overlapping between DS and AD pathomechanisms might be further considered as potential molecular markers for early diagnosis of AD predisposition also in non-DS individuals.

List of Abbreviations: AD, Alzheimer's Disease; CNS, Central Nervous System; DS, Down Syndrome; GABA, Gamma-Amino-Butyric Acid; SH, Subtractive Hybridization

1   Introduction

Down syndrome (DS) or trisomy 21 is a complex genetic and metabolic disorder and the most common genetic cause of mental retardation with an incidence of 1 in about 700 live births. DS is associated with facial dysmorphology, congenital heart defects, high incidence of acute megakaryoblastic leukemias in childhood, immunologic disorders, thyroid dysfunction, frequent *Diabetes mellitus*, etc. Around the third decade of live patients with DS usually develop neuropathological features of Alzheimer's disease. For a long time the so-called Gene dosage effect, i.e., the excessive expression of genes encoded on chromosome 21 (altogether 127 known, 98 predicted, and 59 pseudo-genes as reported by Hattori et al., 2000), has been believed to underlie the abnormalities characteristic for DS. Recent studies collectively demonstrated, however, a high variability and great complexity of gene regulation in the DS critical region, indicating the theory, that 50% over-expression of these genes should explain the DS phenomenon, to be too simplistic. In spite of plenty of projects devoted to DS, molecular pathomechanisms leading to the single DS complications remain quite unclear. Research in this field in recent years has accelerated applying highly sensitive molecular biology instruments and technologies. Thereby, subtractive hybridization (SH) belongs to the most effective tools developed for the identification of alterations in disease specific gene expression.

SH is a smart gene hunting technique, which has been applied in the investigation of DS for the first time by Labudova (1998a, b). Only 11 articles can be found in "Pubmed," which deal with ex vivo investigation of DS fetal brain using and originating from SH, with the only exception being the one big project devoted to DS fetal brain abnormalities on transcription level. In the present article, the application of SH in the investigation of the DS fetal brain is described by the investigator, who has established the technology and performed comparative studies using SH (please note, O. Golubnitschaja has published earlier as O. Labudova).

2   The Gene Hunting Method of Subtractive Hybridization

Subtraction is a concept taken from mathematics: one pool is compared with another one. Both pools consist of a complementary part and a noncomplementary part of gene transcripts. This noncomplementary part—the difference between the gene pools—is the aim of the investigation and considered as the result of the subtraction (❷ *Figure 16-1*).

◘ Figure 16-1

Principle of subtractive hybridization: Each mRNA-pool consists of individual gene transcripts, which are compared by SH. The result of the subtraction is the difference between the two pools compared considering both qualitative (present or absent) and quantitative (amount of copies per unit weight of mRNA in the corresponding pool) differences in expression of individual transcripts

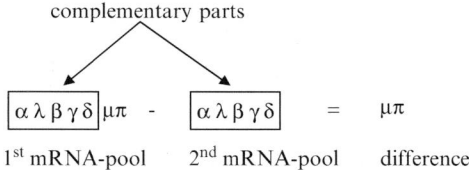

2.1   The Construction of a Subtractive Library

The construction of a subtractive library is principally based on the cloning of transcripts (in the form of cDNA) of those genes, which are either activated (upregulated) or suppressed (downregulated) under pathologic conditions. In order to identify these genes, two pools of transcripts are used: a complete pool of mRNA expressed in patients and a complete pool of mRNA expressed in controls. These two transcript pools, so called induced and noninduced, undergo a molecular biological comparison with subsequent subtraction of identical gene transcripts resulting in the isolation of differentially expressed transcripts, which are either activated or suppressed due to the disease. Although the cause of gene activation is different in individual cases, an "induced pool" includes some additional transcripts of activated ("induced") genes in comparison with the "noninduced pool" of mRNA. This is the reason for the term "induced pool of mRNA." In our study, we have considered both following hypotheses: (1) individuals with DS have some additionally activated genes, and (2) individuals with DS have some suppressed genes in fetal brain compared to controls. Consequently, we have considered the mRNA-pool isolated from fetal brain of DS individuals once as "induced" and in parallel experiments as "noninduced" pool of transcripts. The construction of the subtractive libraries was performed as follows: 10 μg of mRNA isolated from DS and control brain tissue were biotinylated by UV-irradiation at 360 nm according to the instructions supplied by the provider of the subtractor kit (Invitrogen, Cat.No K4320-01). In order to avoid false positive results of the subtractive hybridization, mRNA of each "noninduced pool" was taken for the hybridization in excess. One microgram of each "induced" mRNA pool was subjected to a reverse transcriptase reaction with subsequent denaturation of the mRNA-template. Each newly synthesized cDNA-pool ("induced pool") was hybridized with the corresponding "non-induced" biotinylated mRNA-pool at 68°C for 48 h. The hybridization mixture was incubated with streptavidin, whereby all biotinylated molecules (both "noninduced" mRNAs and mRNA/cDNA hybrids) became covalently bound with streptavidin. The streptavidin/nucleic acid complexes were removed by repeated phenol-chloroform extractions and subtracted differentially expressed cDNAs were precipitated with alcohol (Sive and John, 1988). For each compared pair of DS/control brains, we have subtracted both induced "DS-genes" (induced gene transcripts in DS compared to the corresponding control tissue) and induced "normal genes" (suppressed gene transcripts in DS compared to the corresponding control tissue). The gene transcription abnormalities were compared in two different pairs of sex-matched DS/control brains. Altogether, four subtractive libraries were constructed with two libraries (first with induced and second with suppressed "DS-genes") from each pair of comparison.

2.2 Cloning of Subtracted cDNAs

The second strand cDNA synthesis was performed using a cDNA synthesis kit (Boehringer Mannheim, Cat. No 1117831). In order to increase the efficiency for the cloning of the subtracted molecules, they were ligated with *Not*I-linkers followed by *Not*I-digestion. These *Not*I-linked cDNAs were further ligated to the *Not*I-site of the pSPORT1 cloning vector (GIBCO), and the recombinant molecules were used for the transformation of highly competent INVαF *Escherichia coli* cells (Invitrogen, Cat. No C2020-03), which were plated onto selection media.

2.3 Identification of Differentially Expressed Transcripts

Recombinant plasmid DNAs were purified using QiaFilter Plasmid Midi System (Qiagen, Cat. No 12245), analyzed for the length of inserted fragments using the method of restriction analysis and sequenced by MWG-Biotech, Germany. Homologies were determined by computer assisted analysis comparing our data of subtracted sequences with International Gene Banks (EMBL and SWISS-PROT). Alignments were prepared using the specialized computer program "DNASIS"- (MWG-Biotech). According to our experience, only partial cDNA fragments of genes could be usually subtracted most probably due to aggressive isolation conditions partially destroying the native cDNA structure. In order to avoid false positive results of the subtractive hybridization, for further considerations in this project, only those genes were taken into account as differentially expressed, the cDNAs of which were independently subtracted from both individual DS/control pairs of comparison. In order to assure the accuracy of the SH results, a further quantification of either specific transcripts or corresponding proteins was carried out. Of special value are the results that were obtained by the application of two independent methods, SH and proteomics. Proteomics studies changes in the protein levels and modifications that result from various diseases like DS. Proteomics usually involves protein separation by two-dimensional (2-D) electrophoresis and protein identification by mass spectrometry (Fountoulakis, 2001). ❷ *Figure 16-2* shows the 2-D electrophoresis analysis of fetal brain. Spots representing proteins for which deranged levels were found by proteomics technologies are indicated together with the protein identities. The correlation of SH results with the proteomics data is described and discussed later. A dysregulation of those genes, whose functions might be potentially relevant for DS pathology, was therefore, verified step-by-step, both on transcription and/or translation levels.

3   Functional Groups of Subtracted Genes

Both up- and downregulation of a single gene expression rate in a cell or tissue are generally accompanied by regulatory chain reactions in functionally dependent pathways with concomitant alterations in the regulation of other genes. This was observed when we used the SH approach to investigate different pathologic conditions, like DS (the publications are cited later), perinatal asphyxia (Labudova et al., 1999a), normal-tension glaucoma (Golubnitschaja-Labudova et al., 2000), etc. A well-performed subtraction usually results in the identification of functionally dependent gene groups. Based on identified subtracted cDNAs, the following functional gene groups were selected for DS: (1) nuclear factors in respect to DNA-replication/repair and transcription factors, (2) enzymes of intermediary metabolism, (3) neuronal abnormality and Alzheimer's disease (AD) relevant genes, (4) hypothetical protein products. All these gene groups are obviously relevant to the abnormalities in DS brain development and are discussed later.

□ Figure 16-2

Two-dimensional electrophoresis analysis of fetal brain. The spots are indicated, for which changed levels were found in DS compared to the control brain

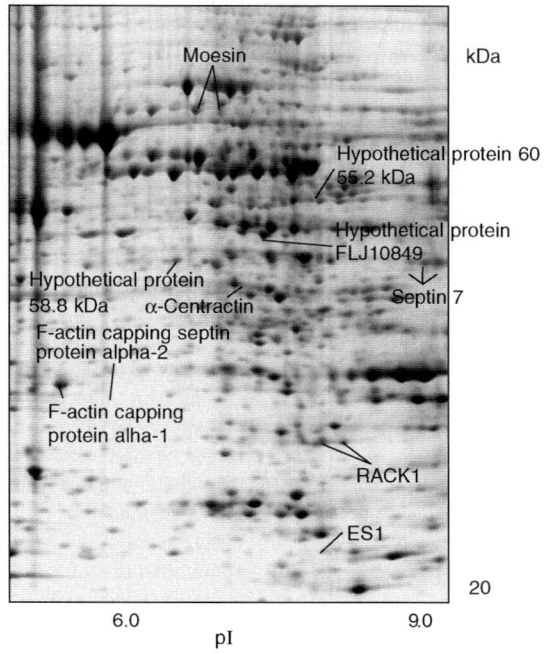

3.1 DNA-Replication/Repair and Transcription Factors

The subtracted cDNAs demonstrating 100% homology to human nuclear factors, known to play a role in DNA replication, repair, and transcription, form the most abundant group of genes with an altered regulation in DS found in our laboratory and are listed in ❍ *Table 16-1*. Expression alterations in this group of genes are crucial for the fetal brain development and should be considered particularly in relation to the DS pathology. Our results show a dramatic difference in the regulation of a group of factors essential for the performance of cell cycle control, DNA repair, replication, gene transcription, cell proliferation, differentiation, and apoptosis. Thereby, the key role is played by pleiotropic factors involved in central cellular processes. Thus, a dysregulation of both thioredoxin and nuclear factor kappa B leads to a limited signal transduction via dependent redox systems and consequently to impairments in de novo DNA synthesis, general gene expression, cell proliferation, inflammation, immune response, apoptosis, etc. (Hirota et al., 1997; Labudova et al., 1998a, b). The downregulation of both factors may substantially contribute to a defect in the metabolism of reactive oxygen species and an increased neural apoptosis well documented for fetal DS brain (Busciglio and Yankner, 1995). A remarkable increase in the expression of DNA repair genes has been demonstrated in the brain of DS individuals, and is compatible with the proposed oxidative DNA damage under trisomy 21 (Hermon et al., 1998; Fang-Kircher et al., 1999).

    The chromodomain helicase DNA binding protein 9—a member of the chromatin remodeling protein family regulating gene expression, DNA repair, and replication—also belongs to the pleiotropic factors. This is already the second aberrantly expressed chromatin remodeling protein, which becomes reported in relation to DS. The first one—WDR9—has been mapped to chromosome 21 within a DS critical region and also demonstrates a transcription regulatory activity (Huang et al., 2003). A functional specificity of chromodomain helicase DNA binding protein 9 is not understood yet.

◘ Table 16-1

DNA-replication/repair and transcription factors

Gene name	Accession number	Alterations in fetal DS brain	Gene function	References
Chromodomain helicase DNA binding protein 9	NP_079410.3	Downregulated	Chromatin remodeling proteins family regulating gene expression, DNA repair and replication	Unpublished results
JunD	X56681	downregulated	Transcription factor involved in cell cycle control	Labudova et al. (1998a, 1999b)
HOXA 13	U82827	Downregulated	Transcription factor of developmental genes family	Labudova et al. (1999b)
NF-kappa B	Z47747	Downregulated	Nuclear, transcription, and signaling factor in the anti-apoptotic pathway	Labudova et al. (1999b)
NF-kappa B interacting protein	**Q5I1X4**	Downregulated	Essential factor in NF-kappa B dependent signaling pathway	Unpublished results
RNA binding motif protein 15	CAI19077	Upregulated	Oncoprotein candidate involved in RNA processing, Hox and Ras/Map kinase signaling, proliferation/differentiation of megakaryoblasts, and development of megakaryoblastic leukemia	Unpublished results
Scleraxis	U58681	Downregulated	Transcription factor of the basic helix-loop-helix family involved in regulation of cell growth and differentiation	Yeghiazaryan et al. (1999)
Similar to pre-pro-megakaryocyte potentiating factor precursor	Q96KJ4	Upregulated	Oncoprotein candidate involved in megakaryoblastic leukemia	Unpublished results
SRm300	Q9UHA8	Downregulated	Splicing coactivator subunit	unpublished results
Thioredoxin	X54539	Downregulated	Cellular red-ox status, DNA synthesis de novo, replication and repair	Kitzmueller et al. (1999)
U1-70K SnRNP	Y00886	Downregulated	Essential splicing factor involved in RNA–protein complex of sliceosomes	Kitzmueller et al. (1999); Seidl et al. (2001)
XRCC1	M36089	Upregulated	X-ray repair cross-complementing gene 1 is essential in DNA repair machinery preventing micronuclei building and sister chromatid exchange	Kitzmueller et al. (1999)

The DS type of neurodegeneration has been proposed to be the consequence of an aberrant gene expression regulation. Here we provide arguments for the functional importance of abnormal gene expression in relation to the DS brain pathology. Thus, an aberrant regulation of HOXA 13—a member of the developmental gene family—indicates a potential contribution of this gene to the DS developmental abnormalities. The downregulation of both junD and scleraxis leads to an aberrant regulation of cell growth and might be functionally connected to the deteriorated mental status of DS individuals. Interestingly, junD protein expression has been found to be remarkably reduced also in distinct regions of adult DS brain, whereas no alterations in the target expression could be seen in AD brain when compared to controls (Labudova et al., 1998a). This finding indicates a DS specific junD dysregulation with no functional linkage to the AD type of neurodegeneration. In contrast, measurements of the scleraxis expression levels in adult DS and AD brain revealed a strong similarity in gene regulation: correlating well between transcription and translation levels, the significantly reduced target expression has been shown in frontal, temporal, parietal, and occipital lobes as well as cerebellum of patients with DS and AD compared to controls (Yeghiazaryan et al., 1999). Therefore, the downregulated scleraxis expression argues for and might be implicated in the similarity of DS and AD types of neuropathology.

Collectively, the downregulation of both splicing regulators SRm300 and U1-70K SnRNP may strongly contribute to an impairment of the splicing machinery during DS fetal brain development. Additional comparative measurements of U1-70K SnRNP levels in adult DS brain tissue demonstrated a coordinated reduction of gene expression on transcription and translation levels in frontal, temporal, parietal, and occipital lobes as well as cerebellum compared to controls (Seidl et al., 2001). The suppression of U1-70K SnRNP, therefore, seems to be specific for DS. Both splicing regulators SRm300 and U1-70K SnRNP have a broad gene specificity, influencing, therefore, an expression of a wide spectrum of genes, the list of which currently remains speculative.

As can be seen in ❷ Table 16-1, there are two candidates with a potential oncoprotein activity involved in the development of megakaryoblastic leukemia (MKL), also termed megakaryocytic acute leukemia. MKL 1 is the chromosome 22-encoded gene initially identified due to its t(1;22) translocation and fusion with the chromosome 1-encoded RNA binding motif protein 15 (RBM 15) (Ma et al., 2001). The RBM 15-MKL 1 fusion protein is believed to possess oncogenic properties and triggers the acute megakaryoblastic leukemias in infants and young children. An occurrence of the acute megakaryoblastic leukemia is known to be about 20-fold higher in individuals with DS compared to the general population (reviewed in Zipursky et al., 1992). The existence of molecular pathomechanisms promoting the acute megakaryoblastic leukemia in DS has not been clearly documented yet. The simultaneous upregulation of both RBM 15 and the pre-pro-megakaryocyte potentiating factor precursor in fetal DS brain tissue described here for the first time contributes to our further understanding of these mechanisms.

3.2   Enzymes of Intermediary Metabolism

We classified the genes involved in intermediary metabolism into four subgroups, which functionally deal with (1) glucose metabolism, (2) mitochondrial function, (3) methionine/ homocysteine pathway, and (4) detoxification of reactive aldehydes; they are listed in ❷ Table 16-2. The genes labeled with an asterisk (*) were identified in DS fetal brain by independently applied methods, SH and proteomics.

3.2.1 Glucose Metabolism

Impaired glucose metabolism is well documented in DS (reviewed in Schapiro et al., 1992). Moreover, DS belongs to those neurodegenerative disorders that are known to be associated with *Diabetes mellitus*, namely, increased insulin resistance and obesity, disturbed insulin sensitivity, and excessive or impaired insulin secretion (reviewed in Ristow, 2004). In literature, one can see attempts to explain these facts arguing with obesity usually observed in DS individuals. Our results demonstrate, however, the dysregulation of key enzymes of glucose metabolism already during fetal development, suggesting the DS conditioned alterations in gene expression regulation as the primary cause, rather than problems with high body weight in adulthood. Further, we have measured the activity of both the glucose-6-phosphate isomerase (PGI) and the glucose metabolism rate limiting enzyme phosphofructokinase (PFK) in brain tissue of adult patients with DS and AD versus controls. No significant alterations were found in the activity of PFK in both DS and AD brain tissue. Therefore, no visible "gene dosage effect" in relation to PFK activity in DS could be seen, although the PFK gene is encoded on trisomic chromosome 21. In contrast, a decreased activity of PGI

◻ Table 16-2
Enzymes of intermediary metabolism

Gene name	Accession number	Alterations in fetal DS brain	Gene function	References
Glucose-6-phosphate isomerase	P06744	Downregulated	Glucose metabolism	Labudova et al. (1999d)
Phosphoglycerate kinase	V00572	Upregulated	Glucose metabolism	Labudova et al. (1999b, 1999c)
*Glyceraldehyde-3-phosphate dehydrogenase	J02642	Upregulated	Glucose metabolism	Labudova et al. (1999b), Lubec et al. (1999a)
*Aldehyde dehydrogenase 1	K03000	Downregulated	Detoxification of reactive aldehydes	Lubec et al. (1999b)
S-adenosyl-L-homocysteine hydrolase	P23526	Downregulated	Homocysteine/methionine metabolism	Labudova et al. (1999b)
*Ubiquinol-cytochrome-c-reductase	S00680	Downregulated	Mitochondrial electron transport	Kim et al. (2000)

in frontal and temporal lobes as well as cerebellum, and an increased one in occipital lobe of DS brain was detected; no difference in PFK activity in AD brain, compared to controls, could be seen (Labudova et al., 1999d). These findings indicate a disease specific dysregulation of PGI in DS with no relevance for AD type of neuropathology.

The SH identification of excessive phosphoglycerate kinase (PGK) transcripts in fetal DS brain has been further complemented with the data of the specific kinase activity in adult DS brain, which was significantly enhanced in frontal, temporal, and occipital lobes as well as cerebellum compared to controls (Labudova et al., 1999c). It is noteworthy that the activity of PGK in the same regions of AD brain was nonsignificantly increased, indicating the specificity of PGK upregulation in DS.

Glyceraldehyde 3-phosphate dehydrogenase (G3PD)—a key enzyme in carbohydrate metabolism—was identified by SH as highly upregulated in the transcription level in fetal DS brain. Proteomics data confirmed the significant upregulation of the protein in adult DS brain, which was particularly remarkable in the frontal cortex (Lubec et al., 1999a). Although slightly enhanced in the brain of patients with AD, the enzymatic activity of G3PD was significantly elevated only in the DS brain compared to controls, indicating further differences in the neurochemical pathomechanisms between DS and AD.

3.2.2 Mitochondrial Metabolism

Closely related to each other, impaired glucose metabolism and defects in mitochondrial respiratory enzymes could synergistically contribute to the impairment of energy metabolism observed in DS (Stocchi et al., 1985). Recently, an increased mitochondrial superoxide production in young DS compared with normally developed children has been demonstrated (Capone et al., 2002). Although of great importance, this evidencecannot be explained due to trisomy 21 only, as no significant difference between mitochondrial superoxide production levels measured in DS and in children with cognitive impairment of unknown etiology has been found. The decrease of ubiquinol-cytochrome-c-reductase expression in the transcription level in fetal DS brain reported here correlates well with proteomics data published earlier for distinct regions of the adult DS brain (Kim et al., 2000), indicating the disease-related suppression of this gene in DS. The ubiquinol-cytochrome-c-reductase has an essential function as mitochondrial electron transporter, transferring electrons from the CoQH2-cytochrome-c-reductase complex. The reduced ubiquinol-cytochrome-c-reductase activity can substantially contribute to both the generation of reactive oxygen species and the impairment of energy metabolism in DS.

3.2.3 Methionine/Homocysteine Pathway

The deterioration on this type of metabolism has been shown to play a dual role in DS pathology. On the one hand, there is a growing body of evidence for the functional link between an increased risk for DS children at birth and a decreased activity of key enzymes of the methionine/homocysteine pathway—methylenetetrahydrofolate reductase (MTHFR) and methionine synthase reductase (MTRR)—conditioned by gene polymorphism. This is consequently associated with methyl deficiency, DNA hypomethylation, and spontaneous chromosome abnormalities observed in DS (James et al., 1999; Hobbs et al., 2000; Pogribna et al., 2001). On the other hand, impairments in the methionine/homocysteine pathway—folate deficiency as well as reduced levels of homocysteine—are also well documented for DS individuals (Chadefaux et al., 1988). The key enzyme converting homocysteine into cysteine, the cystathione β-synthase (CBS), is located on chromosome 21 and overexpressed in DS individuals (Chadefaux et al., 1985). Consequently, the intracellular levels of homocysteine, methionine, glutathione, S-adenosyl-methionine, and S-adenosyl-homocysteine are decreased, whereas cystathionine, cysteine, and adenosine levels are significantly increased compared to controls (Pogribna et al., 2001).

S-adenosylhomocysteine hydrolase catabolizes the S-adenosylhomocysteine molecule, converting it into adenosine and homocysteine. It is a key enzyme of the intracellular methylation metabolism and a powerful inhibitor of most transmethylation reactions linked to adenosine and L-homocysteine. We found this gene to be downregulated in fetal DS brain, which conforms to data about intermediates in methionine/homocysteine metabolism published earlier. The role of this gene in the pathogenesis of DS has not been elucidated yet. We consider this finding as an important inducement that should promote further study of its role in DS, because the suppression of S-adenosylhomocysteineAdenosylhomocysteine hydrolase substantially contributes to a general imbalance in the methionine/homocysteine pathway, and might have a broad spectrum of pathophysiological consequences due to a shifted cellular methylation status.

3.2.4 Detoxification of Reactive Aldehydes

Aldehydes are generated during retinoic acid biosynthesis and the metabolism of amino acids, lipids, carbohydrates, and drugs. Aldehydes are highly reactive molecules that are intermediates or products involved in a broad spectrum of physiological, biological, and pharmacological processes. Aldehyde dehydrogenases are $NAD(P)^{(+)}$-dependent enzymes, which oxidize a wide range of aldehydes to their corresponding carboxylic acids. Therefore, aldehyde dehydrogenase is one of the central molecules of intermediary metabolism involved in plenty of intracellular processes including drug metabolism and metabolism of neurotransmitters such as dopamine, norepinephrine, serotonine, and GABA (Murphy et al., 2003; Slominski et al., 2003; Panoutsopoulos et al., 2004). The detoxification of reactive aldehydes in human brain is essential for a normal function of CNS. An aldehyde dehydrogenase deficiency causes mental retardation (Willemsen et al., 2001) and has been implicated in AD neuropathology (Kamino et al., 2000).

SH revealed a significant suppression of the aldehyde dehydrogenase gene transcription in fetal DS brain. The aldehyde dehydrogenase deficiency in DS has been confirmed also by the proteomic analysis of the adult DS brain (Lubec et al., 1999b). The discovered aldehyde dehydrogenase deficiency in both DS and AD represents an important overlap in their molecular pathomechanisms and can explain the accumulation of abnormally modified proteins and other harmful bioactive products found in aging DS brain tissue, similar to AD.

3.3 Neuronal Abnormality and AD Relevant Genes

This group involves genes, whose protein product activity is particularly relevant for the neuronal function impaired in DS (❷ *Table 16-3*). The neuronal function is one of the most intensively studied subjects in DS: plenty of relevant molecules such as neurotrophic signaling factors, receptors, channels, transporters, etc. underwent qualitative and quantitative investigations in human neurones as well as in the murine model of

◻ Table 16-3

Neuronal abnormality and AD relevant genes

Gene name	Accession number	Alterations in fetal DS brain	Gene function	References
Alpha-1A-adrenergic receptor	JH0447	Upregulated	Neuronal function	Unpublished results
Alzheimer's beta-amyloid precursor interaction protein	BAC2257 8.1	Upregulated	Neuronal function	Unpublished results
apolipoprotein E	E08423	Upregulated	Neuronal function	Kitzmueller et al. (1999)
2'3'-cyclic nucleotide-3'-phosphodiesterase	M19650,	Downregulated	Brain development and signal transduction	Labudova et al. (1999b)
(cGMP-inhibited 3'5'cyclic nucleotide phosphodiesterase A, CGI-PDE A)	(Q14432)			
Neural cell adhesion molecule L1	P32004	Downregulated	L1CAM mediates adhesion properties of cells	Unpublished results
Neuronal acetylcholine receptor protein alpha 7	AAB4011 4.1	Downregulated	Neuronal function	Unpublished results
Neuronal alpha 1A Ca^{2+}channel	Af004883	Upregulated	Neuronal function	Labudova et al. (1999b)
Serotonin receptor 4	Q13639	Downregulated	Neuronal function	Unpublished results
Thyroid stimulating hormone receptor	BC063613	Upregulated	Endocrine function	Labudova et al. (1999e)
Vasopressin		Upregulated	Neuronal function	Labudova et al. (1998b)

DS. Collectively, these investigations demonstrated how multifaceted the neural function in human beings is and how closely related to each other the functions of molecules relevant for neural activity are. Here we describe the findings obtained by SH, which provide an image of the complexity of the neuropathology in DS.

3.3.1 Brain Development

Developmental changes in maturing brain do not linearly proceed from immature during the prenatal development over childhood to mature in adults and, therefore, should be considered separately. Mono-amine transmitters such as serotonin, dopamine, and norepinephrine act as maturation promoting factors in brain development prior to their role as neurotransmitters in the mature brain. Thereby serotonin acts as a growth factor during embryogenesis and serotonin receptor activity regulates the correct sequence of events in brain structuring (reviewed in Sodhi and Sanders-Bush, 2004). Furthermore, the activation of the serotonin receptor has been shown to positively affect the growth of stunted neurones through stabilization and elaboration of the cytoskeleton of neurones as well as inhibition of apoptosis (Azmitia, 2001). Downregulation of the serotonin receptor demonstrated here for fetal DS brain can result in increased neuronal death and reduced neuronal cell density as well as maturation capacity, which was well documented for DS brain.

Furthermore, being dysregulated, the gene encoding neural cell adhesion molecule L1 (L1CAM) is known to contribute to abnormal prenatal brain development and has been identified among the downregulated cDNAs in fetal DS brain compared to controls. The impaired function of L1CAM during development of the nervous system results in neurites outgrowth, fasciculation, neural cell migration, and causes mental retardation (Takahashi et al., 1997). It is noteworthy that another member of the neural cell adhesion molecule family named NCAM2 has been mapped to 21q21 between markers D21S18 and D21S282. Being slightly overexpressed, this gene has been proposed to be a good candidate for involvement in DS pathology (Paoloni-Giacobino et al., 1997). This imbalance in regulation of the neural cell adhesion molecule family may play an important role in neural abnormalities observed during DS fetal brain development.

3.3.2 Abnormalities Implicated in AD Type of Neurodegeneration

The following molecules encoded by cDNAs subtracted as abnormally expressed from fetal DS brain have been implicated in the development of the AD type of neurodegeneration: alpha 1A adrenergic receptor (Hong et al., 2001), Alzheimer's beta-amyloid precursor interaction protein (reviewed in Snow and Wight, 1989), apolipoprotein E (Diedrich et al., 1991; Namba et al., 1991), 2'3'-cyclic nucleotide-3'-phosphodiesterase (Reinikainen et al., 1989; Vlkolinsky et al., 2001), L1CAM (Jin et al., 2004), neuronal acetylcholine receptor protein alpha 7 (Nagele et al., 2002), neuronal Ca^{2+} channel (Branconnier et al., 1992), serotonine receptor 4 (Crews et al., 1994), thyroid stimulating hormone receptor (Labudova et al., 1999e), vasopressin (Labudova et al., 1998b).

The dysregulation of AD relevant genes already during fetal brain development in DS indicates the predisposition of DS individuals for the onset of both aging and AD diseases extremely early in their life compared to the general population.

3.4 Hypothetical Protein Products

This is one of the most intriguing groups of genes, the cDNAs of which were subtracted from fetal DS brain (❿ *Table 16-4*). The cellular functions of their protein products are not clear yet, and currently we can only speculate about their role in fetal brain as well as their importance for DS pathology. An investigation of a physiological role of these protein products, however, might open new perspectives in our understanding of the phenomenon of DS.

❑ Table 16-4
Hypothetical protein products

Gene name	Accession number	Alterations in fetal DS brain	Gene function	References
Protein product with homology to solute carrier family 27	BAC11578 with homology to CAH71420.1	Downregulated	Fatty acid transporter	Unpublished results
Novel protein (HT036)	CAI14893.1	Upregulated	The function is not known yet.	Unpublished results
Protein product homologous to extraneuronal monoamine transporter	XP_498158	Downregulated	Organic cation transporter, similar to solute carrier family 22 member 3	Unpublished results

4 Concluding Remarks

Summarizing our data presented in this chapter, we would like to stress the multitudinousness of functional groups of genes found to be differentially expressed in fetal DS brain compared to controls.

These functional alterations are stretched on the whole complex of cellular events in DS regarding brain development, cell reproduction, DNA replication, quality control and repair, synthesis of gene products, stress response, anti-tumor controlling mechanisms, central metabolic processes like metabolism of glucose and amino acids, function of single subcellular structures such as mitochondria, energy metabolism, neutralization of reactive oxygen species, general brain detoxification, complex neuronal function, etc. Thereby, multiple pathways might be considered as a clear predisposition for some typical features of DS pathology such as neurodegeneration. Although seen as early as during fetal development, the DS pathomechanisms potentially leading to neurodegeneration demonstrate only a partial overlap with those postulated for AD. On the one hand, the difference could be explained as being a result of the differential gene expression in fetal compared to adult human brain. On the other hand, this could represent some unique features of DS-specific neuropathology. Gene expression alterations overlapping between DS and AD pathomechanisms might be further considered as potential molecular markers for early diagnosis of AD predisposition in non-DS individuals also.

References

Azmitia EC. 2001. Neuronal instability: implications for Rett's syndrome. Brain Dev 23: S1-S10.

Branconnier RJ, Branconnier ME, Walshe TM, McCarthy C, Morse PA. 1992. Blocking the Ca(2+)-activated cytotoxic mechanisms of cholinergic neuronal death: A novel treatment strategy for Alzheimer's disease. Psychopharmacol Bull 28: 175-181.

Busciglio J, Yankner BA. 1995. Apoptosis and increased generation of reactive oxygen species in Down's syndrome neurons in vitro. Nature 378: 776-779.

Capone G, Kim P, Jovanovich S, Payne L, Freund L, et al. 2002. Evidence for increased mitochondrial superoxide production in Down syndrome. Life Sci 70: 2885-2895.

Chadefaux B, Rethore MO, Raoul O, Ceballos I, Poissonnier M, et al. 1985. Cystathionine beta synthase: Gene dosage effect in trisomy 21. Biochem Biophys Res Commun 128: 40-44.

Chadefaux B, Ceballos I, Hamet M, Coude M, Poissonnier M, et al. 1988. Is absence of atheroma in Down syndrome due to decreased homocysteine levels? Lancet 2: 741.

Crews FT, Kurian P, Freund G. 1994. Cholinergic and serotonergic stimulation of phosphoinositide hydrolysis is decreased in Alzheimer's disease. Life Sci 55: 1993-2002.

Diedrich JF, Minnigan H, Carp RI, Whitaker JN, Race R, et al. 1991. Neuropathological changes in scrapie and Alzheimer's disease are associated with increased expression of apolipoprotein E and cathepsin D in astrocytes. J Virol 65: 4759-4768.

Fang-Kircher SG, Labudova O, Kitzmueller E, Rink H, Cairns N, et al. 1999. Increased steady state mRNA levels of DNA-repair genes XRCC1, ERCC2 and ERCC3 in brain of patients with Down syndrome. Life Sci 64: 1689-1699.

Fountoulakis M. 2001. Proteomics: Current technologies and applications in neurological disorders and toxicology. Amino Acids 21: 363-381.

Golubnitschaja-Labudova O, Liu R, Decker C, Zhu P, Haefliger IO, et al. 2000. Altered gene expression in lymphocytes of patients with normal-tension glaucoma. Curr Eye Res 21: 867-876.

Hattori M, Fujiyama A, Taylor TD, Watanabe H, Yada T, Park HS, Toyoda A, Ishii K, Totoki Y, Choi DK, Groner Y, Soeda E, Ohki M, Takagi T, Sakaki Y, Taudien S, Blechschmidt K, Polley A, Menzel U, Delabar J, Kumpf K, Lehmann R, Patterson D, Reichwald K, Rump A, Schillhabel M, Schudy A, Zimmermann W, Rosenthal A, Kudoh J, Schibuya K, Kawasaki K, Asakawa S, Shintani A, Sasaki T, Nagamine K, Mitsuyama S, Antonarakis SE, Minoshima S, Shimizu N, Nordsiek G, Hornischer K, Brant P, Scharfe M, Schon O, Desario A, Reichelt J, Kauer G, Blocker H, Ramser J, Beck A, Klages S, Hennig S, Riesselmann L, Dagand E, Haaf T, Wehrmeyer S, Borzym K, Gardiner K, Nizetic D, Francis F, Lehrach H, Reinhardt R, Yaspo ML. 2000. The DNA sequence of human chromosome 21. Nature 405: 311-319.

Hermon M, Cairns N, Egly JM, Fery A, Labudova O, et al. 1998. Expression of DNA excision-repair-cross-complementing proteins p80 and p89 in brain of patients with Down Syndrome and Alzheimer's disease. Neurosci Lett 251: 45-48.

Hirota K, Matsui M, Iwata S, Nishiyama A, Mori K, et al. 1997. AP-1 transcriptional activity is regulated by a direct association between thioredoxin and Ref-1. Proc Natl Acad Sci USA 94: 3633-3638.

Hobbs CA, Sherman SL, Yi P, Hopkins SE, Torfs CP, et al. 2000. Polymorphisms in genes involved in folate metabolism as maternal risk factors for Down syndrome. Am J Hum Genet 67: 623-630.

Hong CJ, Wang YC, Liu TY, Liu HC, Tsai SJ. 2001. A study of alpha-adrenoceptor gene polymorphisms and Alzheimer disease. J Neural Transm 108: 445-450.

Huang H, Rambaldi I, Daniels E, Featherstone M. 2003. Expression of the Wdr9 gene and protein products during mouse development. Dev Dyn 227: 608-614.

James SJ, Pogribna M, Pogribny IP, Melnyk S, Hine RJ, et al. 1999. Abnormal folate metabolism and mutation in the methylenetetrahydrofolate reductase gene may be maternal risk factors for Down syndrome. Am J Clin Nutr 70: 495-501.

Jin K, Peel AL, Mao XO, Xie L, Cottrell BA, et al. 2004. Increased hippocampal neurogenesis in Alzheimer's disease. Proc Natl Acad Sci USA 101: 343-347.

Kamino K, Nagasaka K, Imagawa M, Yamamoto H, Yoneda H, et al. 2000. Deficiency in mitochondrial aldehyde dehydrogenase increases the risk for late-onset Alzheimer's disease in the Japanese population. Biochem Biophys Res Commun 273: 192-196.

Kim SH, Vlkolinsky R, Cairns N, Lubec G. 2000. Decreased levels of complex III core protein 1 and complex V beta chain in brains from patients with Alzheimer's disease and Down syndrome. Cell Mol Life Sci 57: 1810-1816.

Labudova O, Krapfenbauer K, Moenkemann H, Rink H, Kitzmuller E, et al. 1998a. Decreased transcription factor junD in brains of patients with Down syndrome. Neurosci Lett 252: 159-162.

Labudova O, Fang-Kircher S, Cairns N, Moenkemann H, Yeghiazaryan K, et al. 1998b. Brain vasopressin levels in Down syndrome and Alzheimer's disease. Brain Res 806: 55-59.

Labudova O, Schuller E, Yeghiazarjan K, Kitzmueller E, Hoeger H, et al. 1999a. Genes involved in the pathophysiology of perinatal asphyxia. Life Sci 64: 1831-1838.

Labudova O, Kitzmueller E, Rink H, Cairns N, Lubec G. 1999b. Gene expression in fetal Down syndrome brain as revealed by subtractive hybridization. J Neural Transm Suppl 57: 125-136.

Labudova O, Kitzmueller E, Rink H, Cairns N, Lubec G. 1999c. Increased phosphoglycerate kinase in the brains of patients with Down's syndrome but not with Alzheimer's disease. Clin Sci (Lond) 96: 279-285.

Labudova O, Cairns N, Kitzmuller E, Lubec G. 1999d. Impaired brain glucose metabolism in patients with Down syndrome. J Neural Transm Suppl 57: 247-256.

Labudova O, Cairns N, Koeck T, Kitzmueller E, Rink H, et al. 1999e. Thyroid stimulating hormone-receptor overexpression in brain of patients with Down syndrome and Alzheimer's disease. Life Sci 64: 1037-1044.

Lubec G, Labudova O, Cairns N, Fountoulakis M. 1999a. Increased glyceraldehyde 3-phosphate dehydrogenase levels in the brain of patients with Down's syndrome. Neurosci Lett 260: 141-145.

Lubec G, Labudova O, Cairns N, Berndt P, Langen H, et al. 1999b. Reduced aldehyde dehydrogenase levels in the brain of patients with Down syndrome. J Neural Transm Suppl57: 21-40.

Ma Z, Morris SW, Valentine V, Li M, Herbrick JA, et al. 2001. Fusion of two novel genes, RBM15 and MKL1, in the t (1;22)(p13;q13) of acute megakaryoblastic leukemia. Nat Genet 28: 220-221.

Murphy TC, Amarnath V, Gibson KM, Picklo MJ Sr. 2003. Oxidation of 4-hydroxy-2-nonenal by succinic semialdehyde dehydrogenase (ALDH5A). J Neurochem 86: 298-305.

Nagele RG, D'Andrea MR, Anderson WJ, Wang HY. 2002. Intracellular accumulation of beta-amyloid(1–42) in neurons is facilitated by the alpha 7 nicotinic acetylcholine receptor in Alzheimer's disease. Neuroscience 110: 199-211.

Namba Y, Tomonaga M, Kawasaki H, Otomo E, Ikeda K. 1991. Apolipoprotein E immunoreactivity in cerebral amyloid deposits and neurofibrillary tangles in Alzheimer's disease and kuru plaque amyloid in Creutzfeldt-Jakob disease. Brain Res 541: 163-166.

Panoutsopoulos GI, Kouretas D, Beedham C. 2004. Contribution of aldehyde oxidase, xanthine oxidase, and aldehyde dehydrogenase on the oxidation of aromatic aldehydes. Chem Res Toxicol 17: 1368-1376.

Paoloni-Giacobino A, Chen H, Antonarakis SE. 1997. Cloning of a novel human neural cell adhesion molecule gene (NCAM2) that maps to chromosome region 21q21 and is potentially involved in Down syndrome. Genomics 43: 43-51.

Pogribna M, Melnyk S, Pogribny I, Chango A, Yi P, et al. 2001. Homocysteine metabolism in children with Down syndrome: In vitro modulation. Am J Hum Genet 69: 88-95.

Reinikainen KJ, Pitkanen A, Riekkinen PJ. 1989. 2',3'-cyclic nucleotide-3'-phosphodiesterase activity as an index of myelin in the post-mortem brains of patients with Alzheimer's disease. Neurosci Lett 106: 229-232.

Ristow M. 2004. Neurodegenerative disorders associated with diabetes mellitus. J Mol Med 82: 510-529.

Schapiro MB, Haxby JV, Grady CL. 1992. Nature of mental retardation and dementia in Down syndrome: Study with PET, CT, and neuropsychology. Neurobiol Aging 13: 723-734.

Seidl R, Labudova O, Krapfenbauer K, Henriksson EW, Craft J, et al. 2001. Deficient brain snRNP70K in patients with Down syndrome. Electrophoresis 22: 43-48.

Sive HL, St John T. 1988. A simple subtractive hybridization technique employing photoactivatable biotin and phenol extraction. Nucleic Acids Res 16: 10937.

Slominski A, Pisarchik A, Semak I, Sweatman T, Wortsman J. 2003. Characterization of the serotoninergic system in the C57BL/6 mouse skin. Eur J Biochem 270: 3335-3344.

Snow AD, Wight TN. 1989. Proteoglycans in the pathogenesis of Alzheimer's disease and other amyloidoses. Neurobiol Aging 10: 481-497.

Sodhi MS, Sanders-Bush E. 2004. Serotonin and brain development. Int Rev Neurobiol 59: 111-174.

Stocchi V, Magnani M, Cucchiarini L, Novelli G, Dallapiccola B. 1985. Red blood cell adenine nucleotides abnormalities in Down syndrome. Am J Med Genet 20: 131-135.

Takahashi S, Makita Y, Okamoto N, Miyamoto A, Oki J. 1997. L1CAM mutation in a Japanese family with X-linked hydrocephalus: A study for genetic counseling. Brain Dev 19: 559-562.

Vlkolinsky R, Cairns N, Fountoulakis M, Lubec G. 2001. Decreased brain levels of 2',3'-cyclic nucleotide-3'-phosphodiesterase in Down syndrome and Alzheimer's disease. Neurobiol Aging 22: 547-553.

Willemsen MA, Ijlst L, Steijlen PM, Rotteveel JJ, de Jong JG, et al. 2001. Clinical, biochemical and molecular genetic characteristics of 19 patients with the Sjogren-Larsson syndrome. Brain 124: 1426-1437.

Yeghiazaryan K, Turhani-Schatzmann D, Labudova O, Schuller E, Olson EN, Cairns N, Lubec G. 1999. Downregulation of the transcription factor scleraxis in brain of patients with Down syndrome. J Neural Transm Suppl 57: 305-314.

Zipursky A, Poon A, Doyle J. 1992. Leukemia in Down syndrome: A review. Pediatr Hematol Oncol 9: 139-149.

17   HIV Dementia: A Neurodegenerative Disorder with Viral Etiology

E. Koutsilieri · G. Arendt · E. Neuen-Jacob · C. Scheller · E. Grünblatt · P. Riederer

Abstract: Infection of the central nervous system (CNS) by human immunodeficiency virus (HIV) commonly results in neuropsychiatric dysfunction known, in the most severe form, as HIV dementia. Despite decreased mortality in patients who are treated with highly active antiretroviral therapy and a decrease in the incidence of HIV dementia, HIV encephalitis, the pathological correlate of HIV dementia persists, suggesting the need of continued research focused on the pathophysiology of HIV infection in CNS.

The purpose of this article is to provide a selective overview on the neurobiology of HIV, to emphasize common pathogenetic processes with "classical" neurodegenerative disorders, and discuss new treatment strategies for HIV/CNS disease.

List of Abbreviations: AIDS, Acquired immunodeficiency syndrome; CNS, Central nervous system; cAMP, Cyclic adenosine monophosphate; HAART, Highly active antiretroviral therapy; HIV, Human immunodeficiency virus; HIVE, HIV encephalitis; MIP, Macrophage inflammatory protein; MP/MG, Macrophages/microglia; MRAM, Most rapid alternating movements; NMDA, N-methyl-D-aspartate; PD, Parkinson's disease; PET, Positron emission tomography; SIV, Simian immunodeficiency virus; WHO, World Health Organization

1 Introduction

Human immunodeficiency virus type-1 (HIV-1 or HIV) is a retrovirus that infects CD4+ lymphocytes and mononuclear phagocytes (monocytes, macrophages, and microglia) causing profound immunosuppression that eventually may develop into acquired immunodeficiency syndrome (AIDS). HIV enters the nervous system early after infection primarily through infected monocytes (Kolson et al., 1998). However, productive infection is rarely detectable before immunosuppression. Over time, central nervous system (CNS) infection results in neurotransmission dysfunction and HIV encephalitis (HIVE) expressed as neuropsychiatric disorders that define HIV dementia.

According to the World Health Organization (WHO), nearly 40 million people worldwide are infected with HIV (WHO, Report on the Global AIDS Epidemic, Update, December 2003), and it is estimated that one third of the adults and more than one half of the children will develop a dementing illness. In general, HIV dementia does not develop until advanced HIV infection. Typically, patients will have had other AIDS-defining illnesses before the onset of the neuropsychiatric syndromes. Occasionally, however, HIV dementia can be present before profound immunosuppression. For example, the prevalence of HIV dementia was 0.4% during the asymptomatic phase of the infection (Miller et al., 1990) but raised to 16% among patients with symptomatic HIV infection (McArthur, 1987). Before the introduction of highly active antiretroviral therapy (HAART) in 1996, the cumulative risk of developing dementia was estimated to be 15%–20% (McArthur et al., 1993). Subtle forms exist in at least 30% of symptomatic HIV-seropositive adults (Janssen et al., 1989; Sacktor et al., 2002). HIVE was observed with a frequency of 20%–25% (Neuen-Jacob et al., 1993b; Masliah et al., 2000). It is noteworthy that HIV has been the leading cause of dementia in people less than 60 years of age (McArthur et al., 1993). The clinical prognosis of dementia used to be particularly poor before the introduction of HAART. People died on average 6 months after onset (Harrison and McArthur, 1995).

After HAART, a 60% reduction in death rates was reported in USA between 1996 and 1998 (Palella et al., 1998). HAART has also been associated with a decreased incidence of HIV dementia (Brodt et al., 1997; Sacktor et al., 2002) as well as with an increased survival following dementia diagnosis (Dore et al., 2003). However, despite these advances, therapeutic failures with HAART still occur in about half of the patients (Fatkenheuer et al., 1997). Cognitive impairment persists among individuals with advanced infection, and the prevalence of HIV dementia appears increased (Dore et al., 2003). As penetration of several antiretroviral agents is prevented by the blood–brain barrier, complete eradication of the virus is not possible. The brain serves as a reservoir for viral replication, representing a sanctuary where resistances may develop (Ellis et al., 2000; Eggers et al., 2001).

It is interesting that in the era of HAART, the morphological spectrum of HIV-induced diseases changed (Gray et al., 2003). New and more severe forms of HIVE emerged including chronic "burnt out" forms of HIVE with a loss of pathognomonic features and a severe leukoencephalopathy with intense perivascular macrophage/lymphocytic infiltration and extensive white matter destruction. The latter might be due to an exaggerated response from a newly reconstituted immune system and/or breakdown of the blood–brain barrier, possibly mediated by increasingly toxic combinations of drugs (Gray et al., 2003; Langford et al., 2003).

2 Pathogenetic Processes: From CNS Viral Infection to Neurodegeneration

To date, the pathogenesis of HIV dementia is multifactorial and so far not completely understood. As mentioned previously, HIV enters the CNS early in the course of infection. To enter a cell, HIV must bind to CD4 receptor, found on T lymphocytes, monocytes, macrophages, microglia, and some dendritic cells, and subsequently to one of α- or β-chemokine coreceptors. CCR5, the chemokine receptor for macrophage inflammatory protein (MIP)-1α and MIP-1β, is the primary coreceptor used by most HIV isolates recovered from the CNS (for review see Albright et al., 2003). The most accepted model for entry of HIV into CNS, also known as the "Trojan horse" hypothesis, suggests that HIV enters the brain through the infiltration of infected monocytes that differentiate into macrophages (Liu et al., 2000). After crossing the blood–brain barrier into the CNS, macrophages produce viral particles and infect neighboring microglia. As neurons lack CD4 receptors and thus are not infected by HIV, neurodegeneration is most likely to be mediated through indirect mechanisms. Two predominant models have been proposed to explain CNS destruction during HIV infection, both focusing on the role of brain macrophages/microglia (MP/MG), as the only productively infected cells in the brain. The first model describes neurotoxicity mediated by the virus itself or viral proteins. The second model depicts toxicity of endogenous compounds released by MP/MG as part of the host response to virus infection (for review see Koutsilieri et al., 2002b). The latter gains with time more acceptance as the presence of activated MP/MG correlate better with degree of dementia than viral load (Glass et al., 1995).

Viral proteins that have been reported to be neurotoxic include the HIV-envelope glycoprotein 160 (gp 160) that is cleaved into gp 120 and gp 41, the transactivator protein Tat, as well as the accessory proteins Nef and Vpr. It has been shown that gp 120 is neurotoxic and may induce neuronal and dendritic alterations (Brenneman et al., 1988). Moreover, gp 120 expression in transgenic mice leads to dendritic and neuronal degeneration, as well as neuronal loss, astrocytosis, and microglial activation (Toggas et al., 1994). On the other hand, there is interference with signal transduction pathways in neurons and/or glia cells that may lead to disturbances of neuronal function and neural degeneration (Levi et al., 1993; Wyss-Coray et al., 1996). Gp 41 may impair excitatory amino acid transport in astrocytes (Kort, 1998) and induce nitric oxide (Adamson et al., 1999). It was reported that the metabolism of uninfected cells can be influenced by uptake of the secreted viral Tat (Kolson et al., 1993) and that the nerve growth factor and Tat can act synergistically in activating HIV gene expression (Ensoli et al., 1994). Further, Tat toxicity was shown to be dependent on a polyamine site on the N-methyl-D-aspartate (NMDA) receptor (Prendergast et al., 2002). Both Tat and Nef increase production of quinolinic acid (Smith et al., 2001). Vpr has been shown to form ion channels (Piller et al., 1998) and cause apoptosis in human neurons (Patel et al., 2002).

Activated MP/MG, whether HIV infected, are known to produce a long repertoire of proinflammatory cytokines and other soluble factors that are candidate neurotoxins. Such toxic products have been commonly found in "classical" neurodegenerative disorders such as Alzheimer's disease (AD) and Parkinson's disease (PD). This sustained overproduction and secretion of toxic products by activated immunocompetent cells, leading to injury or death to nearby neurons, is the unifying issue in all neurodegenerative disorders, including HIV dementia (for review see Rausch and Davis, 2001). For example, elevated levels of IL-1, IL-6, TNF-α, and certain isoforms of TGF-β are common findings during neurodegeneration. TNF-α has apoptotic effects on neurons, possibly by reducing gene expression of Bcl-2 (Pulliam et al., 1998). IL-1 alters neuronal signaling by inhibiting synaptic transmission (Xiong et al., 2000). IL-6 enhances excitotoxic processes associated with neuronal death (Qiu and Gruol, 2003), and TGF-β induces structural and functional impairments in the brain (Buckwalter et al., 2002; Wyss-Coray et al., 2002)—just to mention few of the functions that these factors exert for cellular destruction. Further, activated MP/MG are major sources of reactive oxygen intermediates as well as mediators of excitotoxicity through, e.g., the secretion of quinolinic acid and glutamate (Heyes et al., 1991; Koutsilieri et al., 1999) associated with neurodegeneration (Piani et al., 1991; Lipton, 1998). Consequently, MP/MG have been discussed as a main contributor in neurodegenerative processes. We have to remember that under physiological conditions, MP/MG serve major homeostatic and reparative functions in CNS. However, when the immune reaction is inadequate or too prolonged as in the case of the neurodegenerative disorders, cell destruction follows (for review see Koutsilieri et al., 2002a). The very interesting issue is that HIV dementia has a totally distinct etiology from other neurodegenerative disorders, namely a viral infection. Moreover, CNS lesions and their distribution are different among various neurodegenerative disorders. However, pathogenetic processes are similar and suggest that pathways to neuronal death may be common in CNS diseases. It is our task to recognize the parallels in the pathogenetic processes of disparate CNS diseases and to utilize these insights for further research in order to obtain a holistic view on the pathways leading to neuronal death.

3 Neuropathology

On the basis of neuropathological criteria, the morphological substrate of HIV dementia is represented by two major pathological patterns: HIVE and HIV leukoencephalopathy (Budka et al., 1991). Macroscopically, the brains often are atrophic and show widened sulci and mild to moderate dilatation of the lateral ventricles corresponding with neuroradiological findings. Histologically, HIVE is characterized by multiple foci composed of microglia, mononucleated, and multinucleated macrophages (❷ *Figures 17-1* and ❷ *17-2*),

❑ Figure 17-1

Typical microglial nodule with some multinucleated giant cells in the white matter. H&E

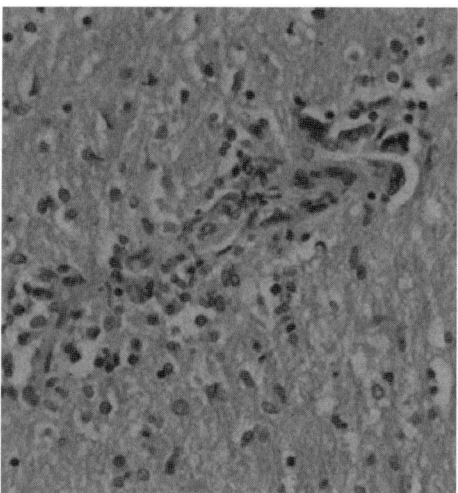

often in a perivascular distribution. Lymphocytic infiltrates are usually lacking. HIV leukoencephalopathy is defined by a more diffuse white matter damage including bilaterally symmetrical myelin pallor, reactive astrocytosis, and the presence of macrophages. However, overlap forms may occur. The morphological hallmark in both manifestations is the presence of multinucleated giant cells (❷ *Figures 17-3* and ❷ *17-4*) that have been known to harbor HIV and are regarded as major target cells for HIV infection in the brain (Budka, 1986; Michaels et al., 1988; Orenstein et al., 1988). Diffuse astrocytosis also is an important component of HIVE as well as HIV leukoencephalopathy and has been reported to precede other morphological changes in the brain (McArthur et al., 1989). More rarely, vacuolar leukoencephalopathy and diffuse polio-dystrophy have been described in patients with HIV dementia. However, these conditions

■ Figure 17-2

Strong immunoreactivity against the monoclonal anti macrophage antibody CD68 in multinucleated giant cells, microglial cells, and macrophages, preferentially located in perivascular distribution

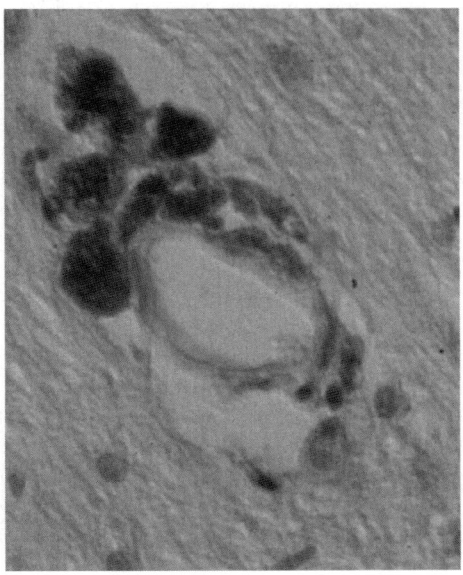

■ Figure 17-3

Typical multinucleated giant cell as hallmark for HIV-1 infection of the brain. Haematoxylin–Eosin (HE)

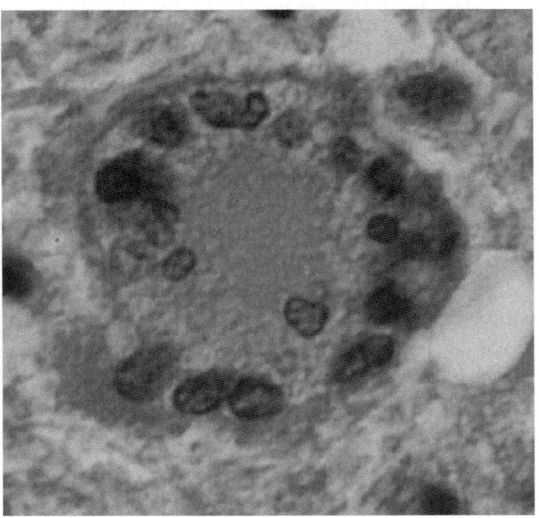

are not HIV specific, and the direct relationship to HIV infection of the brain has not been fully elucidated (Budka, 1991). The presence of HIV in the brain of patients with HIVE has been visualized by means of Southern blot analysis and in situ hybridization, immunohistochemistry, electron microscopy (Gabuzda et al., 1986; Koenig et al., 1986; Stoler et al., 1986; Budka et al., 1987; Pumarola-Sune et al., 1987; Price et al., 1988; Budka, 1990; Kure et al., 1990; Neuen-Jacob et al., 1993a, b), and finally by polymerase chain reaction (An et al., 1996).

☐ Figure 17-4
HIV-1 core proteins are visualized by immunostaining with the mouse Mab against HIV-1 p24 at the surface of a multinucleated giant cell

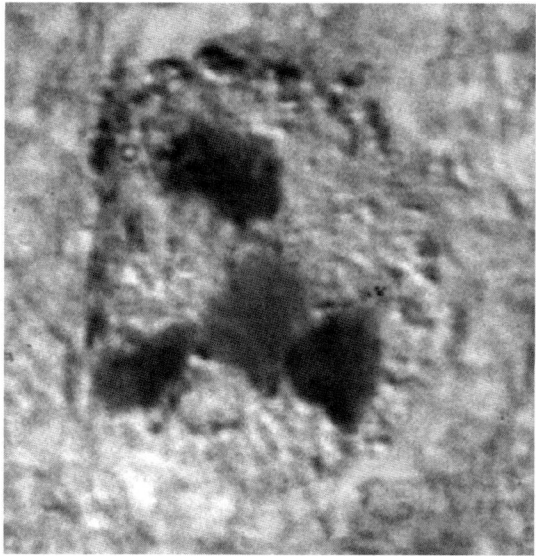

4 Dopaminergic Dysfunction in HIV Dementia

Many of the clinical features in HIV patients as well as electrophysiological, pathological, virological, radiological, and metabolic findings have been attributed to abnormalities in the dopaminergic system.

The symptoms of HIV dementia resemble those of PD and comprise cognitive deficits like brady-phrenia, memory disturbances and apathy, as well as psychomotor slowing and depressive mood. Psycho-motor slowing often is the first sign of HIV dementia and may occur even in the absence and many years before cognitive decline (Arendt et al., 1990, 1994a, 2001; von Giesen et al., 2002). Motor performance of HIV carriers strongly resembles the test results of patients with defined basal ganglia disease, such as PD, Wilson's, and Huntington's disease, i.e., slow reaction time (Dunlop et al., 1992), impairment of rapid alternating index finger movements and extensions, also indicating altered brain development in children and postural imbalance (Arendt et al., 1994b). These pathological electrophysiological motor results may or may not correlate with cerebrospinal fluid viral load pointing out different stages of HIV brain infection.

High sensitivity to dopamine receptor-blocking substances was observed in HIV positive patients provoking severe rigidity and hypokinetic symptoms after exposure to low-dose neuroleptic therapy (Hollander et al., 1985; Hriso et al., 1991).

Specific electrophysiological tests can detect patients at risk for developing HIV dementia very early in the course of the disease, even years before deficits can be detected clinically or are described by the patients themselves. Motor test comprise simple reaction times (Dunlop et al., 1992) as well as most rapid alternating movements (MRAM) and contraction times (Arendt et al., 2001). Patients with pathological motor performance reveal basal ganglia deficits even years after their first pathological motor test and later on clinically overt dementia as described above.

Radiological correlates of basal ganglia dysfunction are smaller volumes (Aylward et al., 1993) or signal hyperintensity in magnetic resonance scans (Filippi et al., 1998). In children, bilateral symmetrical calcification of the basal ganglia were seen in computed tomography (CT) (Belman et al., 1986) as correlate of HIV-associated brain disease turning out to be a calcific arteriopathy in neuropathology studies.

Positron emission tomography (PET) studies were contradictory; all authors reported basal ganglia involvement, but some of them found hypermetabolism, others hypometabolism. Studies of Rottenberg et al. (1996) as well as von Giesen et al. (2000) described an initial hypermetabolism followed by hypometabolism in advancing disease. von Giesen et al. (2000) speculated that there could be an initial dopaminergic receptor function upregulation by the virus and an overshooting cellular response leading to a final breakdown of its metabolism. This theory remains to be proven by quantitative receptor function analysis or by molecular imaging procedures.

Magnetic resonance imaging data, so far available, are also conflicting. Some authors (Meyerhoff et al., 1996) showed a decreased N-acetyl-aspartate/choline (NAA/Cho) ratio in the basal ganglia indicative for neuronal loss, others described an increased Cho/creatine ratio indicating inflammation and gliosis as described also in neuropathology studies of early HIV-related CNS disease (von Giesen et al., 2001), whereas Chang et al. (1999) found such abnormalities in the basal ganglia only in patients with dementive symptoms.

Other authors reported alterations of biogenic amines in the cerebrospinal fluid and found significantly reduced dopamine levels in patients with overt clinical symptoms (Larsson et al., 1991; Berger et al., 1994).

There are also neuropathology studies supporting the involvement of dopaminergic pathways in HIV dementia (Brew et al., 1995). The basal ganglia show a topographic predilection for HIVE. Kure et al. (1990) found the highest degree of gp 41-positive microglial cells in the globus pallidus, followed by striatum and thalamus. Neuen-Jacob et al. (1993a) reported that the deep gray matter, in particular putamen and thalamus, was involved in every case with HIVE, and indeed in this location the highest levels of HIV p24-positive multinucleated giant cells and microglial cells were demonstrable. Interestingly, the morphological changes are most severe in patients with i.v. drug abuse (unpublished own data), indicating that an altered dopamine metabolism might contribute to the pathogenesis of HIVE. Furthermore, the basal ganglia may represent the only manifestation of HIVE (❯ *Figure 17-5*) in cases with early HIV dementia

❑ Figure 17-5

Microglial nodule with single multinucleated giant cells in the putamen in a case with early HIVE

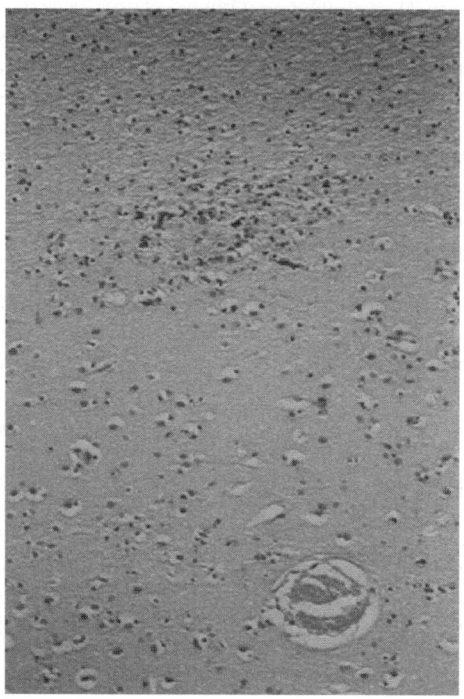

(Neuen-Jacob et al., 1993a). Reyes et al. (1991) described an average decline of 25% in the number of neuronal cell bodies in the substantia nigra in AIDS patients, but the authors did not detect Lewy bodies, a common feature of PD.

Although it is clear that HIV affects dopaminergic systems, the influence of dopamine on viral infection has not been adequately assessed. Accumulating evidence indicates that dopamine may adversely affect the viral infection and may be a pathogenetic factor for HIV dementia. For example, dopamine was shown to induce toxicity on HIV-exposed neuronal cells and to be involved in the regulation of HIV gene expression in various cells of CNS and immune system (Sawaya et al., 1996; Koutsilieri et al., 1997; Rohr et al., 1999). It was reported by our group that dopamine markedly activates HIV in chronically infected T lymphoblasts (Scheller et al., 2000) and that this effect is linked to oxidative stress processes, suggesting that dopamine exposure may modulate cellular vulnerability to HIV. Moreover, we also found that in simian immunodeficiency virus (SIV)-infected rhesus monkeys, dopaminergic substances caused marked neurodegenerative changes and increased brain viral load and TNF-α expression, accelerating the SIV-induced pathology in the CNS of macaques (Czub et al., 2001, 2004).

5 Therapeutic Approaches

Therapeutic interventions for HIV dementia have focused primarily on the use of HAART. However, several factors speak for specific adjunctive medications: (1) the increasing number of people living with HIV dementia following HAART, (2) the variability in CNS penetration of antiretroviral agents, (3) the persistence of HIVE in the era of HAART treatment, (4) the early immune activation and initiation of neurodegenerative processes, leading, in absence of specific treatment, to brain destruction.

In clinical trials, adjuvant therapies have included the NMDA antagonist memantine, the calcium channel blocker nifidepine, the TNF-α antagonist pentoxifylline, and the antioxidants thioctic acid and selegiline (for review see McArthur et al., 2003). The results have been rather disappointing. Selegiline, a dopaminergic substance with various other effects (for review see Gerlach et al., 1992) has been reported as the only substance that improved cognition (Dana Consortium, 1998; Sacktor et al., 2000). In accordance, in rhesus monkeys infected with SIV, selegiline could restore levels of choline acetyltransferase, a biochemical marker for cognition (Koutsilieri et al., 2000, 2001). However, this substance in the same monkeys aggravated SIV-induced neuropathology, increased CNS viral load, and induced vacuolization (Czub et al., 2001) as well as increased TNF-α brain expression (Czub et al., 2004). Further, in a retroviral rat model for neurodegeneration, selegiline increased the number of brain cells expressing major histocompatibility complex class II molecules and increased CNS viral load (Czub et al., 1999), suggesting that this substance potentiates CNS pathology in retroviral infections.

Several potential therapeutic agents have been tested in cell cultures and animals for their protective/rescuing capacity following neurotoxicity induced by HIV or viral proteins. The anticonvulsant, valproate, has recently been shown to protect neurotoxicity induced by HIV gp 120 on cortical neurons as well as in SCID mice with HIVE (Dou et al., 2003). Lithium treatment of both cells and animals exposed to gp 120 had a positive outcome based on the activation of phosphatidylinositol kinase/AKT pathway (Everall et al., 2002). Further, pharmacological classes of substances discussed for their therapeutic potency in neurodegeneration include antioxidants, anti-inflammatory, and NMDA blockers. Nitric oxide synthase inhibitors showed already some beneficial results in PD and Huntington's disease models (Maragos and Silverstein, 1995; Matthews et al., 1997). IL-10 suppresses microglia (O'Keefe et al., 1999; Tan et al., 1999), which can block TNF-α release and nitric oxide production (Tan et al., 2000a, b). NMDA blockers have been shown to attenuate neuronal injury related with excitotoxic processes and were of the first agents used in cultures against immunodeficiency-induced toxicity (Lipton, 1998). Finally, as HIV protein Tat was reported to reduce cyclic adenosine monophosphate (cAMP) synthesis in rat microglia (Patrizio et al., 2001), and cAMP has been shown to exert a neuroprotective role in several models of brain pathologies, a reinforced cAMP signaling induced by several agents may be an interesting therapeutic approach to limit glia-associated damage of neuronal cells.

6 Epilogue

HAART has dramatically altered the HIV pandemic in the developed world. Most patients treated with HAART maintain clinically undetectable plasma virus loads with concomitant decreases in mortality and morbidity. Nonetheless, HIVE persists causing devastating complications in CNS during HIV infection. Research on the interaction between immune and neuronal cells in CNS and elucidation of the intracellular pathways associated with neurodegeneration during HIV infection is essential not only to gain insight into the pathophysiological basis for neuronal death but also to devise specific pharmacological strategies to ameliorate neuronal degeneration in HIV dementia.

Acknowledgments

E.K, G.A., E.N., C.S., and P.R. were supported by a grant from the Bundesministerium für Bildung, Wi*ff*ssenschaft, Forschung und Technologie, Germany (01 KI 0211 Competence Network HIV/AIDS).

References

Adamson DC, McArthur JC, Dawson TM, Dawson VL. 1999. Rate and severity of HIV-associated dementia (HAD): Correlations with Gp41 and iNOS. Mol Med 5: 98-109.

Albright AV, Soldan SS, Gonzalez-Scarano F. 2003. Pathogenesis of human immunodeficiency virus-induced neurological disease. J Neurovirol 9: 222-227.

An S, Ciardi A, Giometto B, Scaravilli T, Gray F, et al. 1996. Investigation on the expression of major histocompatibility class II and cytokines and detection of HIV-1 DNA within brains of asymptomatic and symptomatic HIV-1-positive patients. Acta Neuropathol 91: 494-503.

Arendt G, Hefter H, Elsing C, Strohmeyer G, Freund HJ. 1990. Motor dysfunction in HIV-infected patients without clinically detectable central-nervous deficit. J Neurol 237: 362-368.

Arendt G, Hefter H, Hilperath F, von Giesen HJ, Strohmeyer G, et al. 1994a. Motor analysis predicts progression in HIV-associated brain disease. J Neurol Sci 123: 180-185.

Arendt G, Maecker HP, Purrmann J, Homberg V. 1994b. Control of posture in patients with neurologically asymptomatic HIV infection and patients with beginning HIV-1-related encephalopathy. Arch Neurol 51: 1232-1235.

Arendt G, von Giesen HJ, Hefter H, Theisen A. 2001. Therapeutic effects of nucleoside analogues on psychomotor slowing in HIV infection. AIDS 15: 493-500.

Aylward EH, Henderer JD, McArthur JC, Brettschneider PD, Harris GJ, et al. 1993. Reduced basal ganglia volume in HIV-1-associated dementia: Results from quantitative neuroimaging. Neurology 43: 2099-2104.

Belman AL, Lantos G, Horoupian D, Novick BE, Ultmann MH, et al. 1986. AIDS: Calcification of the basal ganglia in infants and children. Neurology 36: 1192-1199.

Berger J, Kumar M, Kumar A, Fernandez J, Levin B. 1994. Cerebrospinal fluid dopamine in HIV-1 infection. AIDS 8: 67-71.

Brenneman D, Westbrook G, Fitzgerald S, Ennist D, Elkins K, et al. 1988. Neuronal cell killing by the envelope protein of HIV and its prevention by vasoactive intestinal peptide. Nature 335: 639-642.

Brew B, Rosenblum M, Cronin K, Price R. 1995. AIDS dementia complex and HIV-1 brain infection: Clinical-virological correlations. Ann Neurol 38: 563-570.

Brodt HR, Kamps BS, Gute P, Knupp B, Staszewski S, et al. 1997. Changing incidence of AIDS-defining illnesses in the era of antiretroviral combination therapy. AIDS 11: 1731-1738.

Buckwalter M, Pepper JP, Gaertner RF, Von Euw D, Lacombe P, et al. 2002. Molecular and functional dissection of TGF-beta1-induced cerebrovascular abnormalities in transgenic mice. Ann N Y Acad Sci 977: 87-95.

Budka H. 1986. Multinucleated giant cells in brain: A hallmark of the acquired immune deficiency syndrome (AIDS). Acta Neuropathol (Berl) 69: 253-258.

Budka H. 1990. Human immunodeficiency virus (HIV) envelope and core proteins in CNS tissues of patients with the acquired immune deficiency syndrome (AIDS). Acta Neuropathol (Berl) 79: 611-619.

Budka H. 1991. Neuropathology of human immunodeficiency virus infection. Brain Pathol 1: 163-175.

Budka H, Costanzi G, Cristina S, Lechi A, Parravicini C, et al. 1987. Brain pathology induced by infection with the human immunodeficiency virus (HIV). A histological, immunocytochemical, and electron microscopical study of 100 autopsy cases. Acta Neuropathol (Berl) 75: 185-198.

Budka H, Wiley C, Kleihues P, Artigas J, Asbury A, et al. 1991. HIV-associated disease of the nervous system: Review of nomenclature and proposal for neuropathology-based terminology—Consensus report. Brain Pathol 1: 143-152.

Chang L, Ernst T, Leonido-Yee M, Walot I, Singer E. 1999. Cerebral metabolite abnormalities correlate with clinical severity of HIV-1 cognitive motor complex. Neurology 52: 100-108.

Czub M, Czub S, Gosztonyi G, Koutsilieri E, Sopper S, et al. 1999. Effects of Selegiline in a retroviral rat model for neurodegenerative disease. J Neurovirol 5: 458-464.

Czub S, Czub M, Koutsilieri E, Sopper S, Villinger F, et al. 2004. Modulation of simian immunodeficiency virus neuropathology by dopaminergic drugs. Acta Neuropathol (Berl) 107: 216-226.

Czub S, Koutsilieri E, Sopper S, Czub M, Stahl-Hennig C, et al. 2001. Enhancement of central nervous system pathology in early simian immunodeficiency virus infection by dopaminergic drugs. Acta Neuropathol (Berl) 101: 85-91.

Dana Consortium. 1998. A randomized, double-blind, placebo-controlled trial of deprenyl and thioctic acid in human immunodeficiency virus-associated cognitive impairment. Dana Consortium on the Therapy of HIV Dementia and Related Cognitive Disorders. Neurology 50: 645-651.

Dore GJ, McDonald A, Li Y, Kaldor JM, Brew BJ. 2003. Marked improvement in survival following AIDS dementia complex in the era of highly active antiretroviral therapy. AIDS 17: 1539-1545.

Dou H, Birusingh K, Faraci J, Gorantla S, Poluektova LY, et al. 2003. Neuroprotective activities of sodium valproate in a murine model of human immunodeficiency virus-1 encephalitis. J Neurosci 23: 9162-9170.

Dunlop O, Bjorklund RA, Abdelnoor M, Myrvang B. 1992. Five different tests of reaction time evaluated in HIV seropositive men. Acta Neurol Scand 86: 260-266.

Eggers C, Hertogs K, Stürenburg H, Stellbrink H, Lunzen Jv. 2001. Viral persistence in the CSF under HAART despite adequate plasma response associated with symptomatic HIV-1 infection of the CNS, 8th Conference on Retroviruses and Opportunistic Infections. Chicago USA.

Ellis RJ, Gamst AC, Capparelli E, Spector SA, Hsia K, et al. 2000. Cerebrospinal fluid HIV RNA originates from both local CNS and systemic sources. Neurology 54: 927-936.

Ensoli F, Ensoli B, Thiele CJ. 1994. HIV-1 gene expression and replication in neuronal and glial cell lines with immature phenotype: Effects of nerve growth factor. Virology 200: 668-676.

Everall IP, Bell C, Mallory M, Langford D, Adame A, et al. 2002. Lithium ameliorates HIV-gp120-mediated neurotoxicity. Mol Cell Neurosci 21: 493-501.

Fatkenheuer G, Theisen A, Rockstroh J, Grabow T, Wicke C, et al. 1997. Virological treatment failure of protease inhibitor therapy in an unselected cohort of HIV-infected patients. AIDS 11: F113-F116.

Filippi CG, Sze G, Farber SJ, Shahmanesh M, Selwyn PA. 1998. Regression of HIV encephalopathy and basal ganglia signal intensity abnormality at MR imaging in patients with AIDS after the initiation of protease inhibitor therapy. Radiology 206: 491-498.

Gabuzda D, Ho D, Monte Sdl, Hirsch M, Rote T, et al. 1986. Immunohistochemical identification of HTLV-III antigen in brains of patients with AIDS. Ann Neurol 20: 289-295.

Gerlach M, Riederer P, Youdim MB. 1992. The molecular pharmacology of L-deprenyl. Eur J Pharmacol 226: 97-108.

Glass J, Fedor H, Wesselingh S, McArthur J. 1995. Immunocytochemical quantitation of human immunodeficiency virus in the brain: Correlations with dementia. Ann Neurol 38: 755-762.

Gray F, Chretien F, Vallat-Decouvelaere AV, Scaravilli F. 2003. The changing pattern of HIV neuropathology in the HAART era. J Neuropathol Exp Neurol 62: 429-440.

Harrison MJG, McArthur J. 1995. HIV associated dementia complex. AIDS and Neurology. New York: Churchill Livingstone; pp. 31-64.

Heyes MP, Brew BJ, Martin A, Price RW, Salazar AM, et al. 1991. Quinolinic acid in cerebrospinal fluid and serum in HIV-1 infection: Relationship to clinical and neurological status. Ann Neurol 29: 202-209.

Hollander H, Golden J, Mendelson T, Cortland D. 1985. Extrapyramidal symptoms in AIDS patients given low-dose metoclopramide or chlorpromazine [letter]. Lancet 2: 1186

Hriso E, Kuhn T, Masdeu JC, Grundman M. 1991. Extrapyramidal symptoms due to dopamine-blocking agents in patients with AIDS encephalopathy. Am J Psychiatry 148: 1558-1561.

Janssen RS, Saykin AJ, Cannon L, Campbell J, Pinsky PF, et al. 1989. Neurological and neuropsychological manifestations of HIV-1 infection: Association with AIDS-related complex but not asymptomatic HIV-1 infection. Ann Neurol 26: 592-600.

Koenig S, Gendelman HE, Orenstein JM, Dal Canto MC, Pezeshkpour GH, et al. 1986. Detection of AIDS virus in macrophages in brain tissue from AIDS patients with encephalopathy. Science 233: 1089-1093.

Kolson D, Buchhalter J, Collman R, Hellmig B, Farrell C, et al. 1993. HIV-1 Tat alters normal organization of neurons and astrocytes in primary rodent brain cell cultures: RGD sequence dependence. AIDS Res Hum Retroviruses 9: 677-685.

Kolson DL, Lavi E, Gonzalez-Scarano F. 1998. The effects of human immunodeficiency virus in the central nervous system. Adv Virus Res 50: 1-47.

Kort JJ. 1998. Impairment of excitatory amino acid transport in astroglial cells infected with the human immunodeficiency virus type 1. AIDS Res Hum Retroviruses 14: 1329-1339.

Koutsilieri E, Czub S, Scheller C, Sopper S, Tatschner T, et al. 2000. Brain choline acetyltransferase reduction in SIV infection. An index of early dementia? Neuroreport 11: 2391-2393.

Koutsilieri E, Gotz ME, Sopper S, Sauer U, Demuth M, et al. 1997. Regulation of glutathione and cell toxicity following exposure to neurotropic substances and human immunodeficiency virus-1 in vitro. J Neurovirol 3: 342-349.

Koutsilieri E, Scheller C, Sopper S, Gotz ME, Gerlach M, et al. 2001. Selegiline completely restores choline acetyltransferase activity deficits in simian immunodeficiency infection. Eur J Pharmacol 411: R1-R2.

Koutsilieri E, Scheller C, Tribl F, Riederer P. 2002a. Degeneration of neuronal cells due to oxidative stress–microglial contribution. Parkinsonism Relat Disord 8: 401-406.

Koutsilieri E, Sopper S, Heinemann T, Scheller C, Lan J, et al. 1999. Involvement of microglia in cerebrospinal fluid glutamate increase in SIV-infected rhesus monkeys (Macaca mulatta). AIDS Res Hum Retroviruses 15: 471-477.

Koutsilieri E, Sopper S, Scheller C, ter Meulen V, Riederer P. 2002b. Involvement of dopamine in the progression of AIDS Dementia Complex. J Neural Transm 109: 399-410.

Kure K, Weidenheim KM, Lyman WD, Dickson DW. 1990. Morphology and distribution of HIV-1 gp41-positive microglia in subacute AIDS encephalitis. Pattern of involvement resembling a multisystem degeneration. Acta Neuropathol 80: 393-400.

Langford TD, Letendre SL, Larrea GJ, Masliah E. 2003. Changing patterns in the neuropathogenesis of HIV during the HAART era. Brain Pathol 13: 195-210.

Larsson M, Hagberg L, Forsman A, Norkrans G. 1991. Cerebrospinal fluid catecholamine metabolites in HIV-infected patients. J Neurosci Res 28: 406-409.

Levi G, Patrizio M, Bernardo A, Petrucci T, Agresti C. 1993. Human immunodeficiency virus coat protein gp120 inhibits the beta-adrenergic regulation of astroglial and microglial functions. Proc Natl Acad Sci USA 90: 1541-1545.

Lipton SA. 1998. Neuronal injury associated with HIV-1: approaches to treatment. Annu Rev Pharmacol Toxicol 38: 159-177.

Liu Y, Tang XP, McArthur JC, Scott J, Gartner S. 2000. Analysis of human immunodeficiency virus type 1 gp160 sequences from a patient with HIV dementia: Evidence

for monocyte trafficking into brain. J Neurovirol 6 (Suppl. 1): S70-S81.

Maragos WF, Silverstein FS. 1995. Inhibition of nitric oxide synthase activity attenuates striatal malonate lesions in rats. J Neurochem 64: 2362-2365.

Masliah E, De Teresa RM, Mallory ME, Hansen LA. 2000. Changes in pathological findings at autopsy in AIDS cases for the last 15 years. AIDS 14: 69-74.

Matthews RT, Yang L, Beal MF. 1997. S-Methylthiocitrulline, a neuronal nitric oxide synthase inhibitor, protects against malonate and MPTP neurotoxicity. Exp Neurol 143: 282-286.

McArthur J. 1987. Neurologic manifestations of AIDS. Medicine 66: 407-437.

McArthur JC, Becker PS, Parisi JE, Trapp B, Selnes OA, et al. 1989. Neuropathological changes in early HIV-1 dementia. Ann Neurol 26: 681-684.

McArthur JC, Haughey N, Gartner S, Conant K, Pardo C, et al. 2003. Human immunodeficiency virus-associated dementia: An evolving disease. J Neurovirol 9: 205-221.

McArthur JC, Hoover DR, Bacellar H, Miller EN, Cohen BA, et al. 1993. Dementia in AIDS patients: Incidence and risk factors. Multicenter AIDS Cohort Study. Neurology 43: 2245-2252.

Meyerhoff DJ, Weiner MW, Fein G. 1996. Deep gray matter structures in HIV infection: A proton MR spectroscopic study. AJNR Am J Neuroradiol 17: 973-978.

Michaels J, Price RW, Rosenblum MK. 1988. Microglia in the giant cell encephalitis of acquired immune deficiency syndrome: Proliferation, infection and fusion. Acta Neuropathol (Berl) 76: 373-379.

Miller EN, Selnes OA, McArthur JC, Satz P, Becker JT, et al. 1990. Neuropsychological performance in HIV-1-infected homosexual men: The Multicenter AIDS Cohort Study (MACS). Neurology 40: 197-203.

Neuen-Jacob E, Arendt G, Wendtland B, Jacob B, Schneeweis M, et al. 1993a. Frequency and topographical distribution of CD68-positive macrophages and HIV-1 core proteins in HIV-associated brain lesions. Clin Neuropathol 12: 315-324.

Neuen-Jacob E, Figge C, Arendt G, Wendtland B, Jacob B, et al. 1993b. Neuropathological studies in the brains of AIDS patients with opportunistic diseases. Int J Legal Med 105: 339-350.

O'Keefe GM, Nguyen VT, Benveniste EN. 1999. Class II transactivator and class II MHC gene expression in microglia: Modulation by the cytokines TGF-beta IL-4, IL-13 and IL-10. Eur J Immunol 29: 1275-1285.

Orenstein JM, Meltzer MS, Phipps T, Gendelman HE. 1988. Cytoplasmic assembly and accumulation of human immunodeficiency virus types 1 and 2 in recombinant human colony-stimulating factor-1-treated human monocytes: An ultrastructural study. J Virol 62: 2578-2586.

Palella FJ Jr, Delaney KM, Moorman AC, Loveless MO, Fuhrer J, et al. 1998. Declining morbidity and mortality among patients with advanced human immunodeficiency virus infection. HIV Outpatient Study Investigators. N Engl J Med 338: 853-860.

Patel CA, Mukhtar M, Harley S, Kulkosky J, Pomerantz RJ. 2002. Lentiviral expression of HIV-1 Vpr induces apoptosis in human neurons. J Neurovirol 8: 86-99.

Patrizio M, Colucci M, Levi G. 2001. Human immunodeficiency virus type 1 Tat protein decreases cyclic AMP synthesis in rat microglia cultures. J Neurochem 77: 399-407.

Piani D, Frei K, Do KQ, Cuenod M, Fontana A. 1991. Murine brain macrophages induced NMDA receptor mediated neurotoxicity in vitro by secreting glutamate. Neurosci Lett 133: 159-162.

Piller SC, Jans P, Gage PW, Jans DA. 1998. Extracellular HIV-1 virus protein R causes a large inward current and cell death in cultured hippocampal neurons: Implications for AIDS pathology. Proc Natl Acad Sci USA 95: 4595-4600.

Prendergast MA, Rogers DT, Mulholland PJ, Littleton JM, Wilkins LH Jr, et al. 2002. Neurotoxic effects of the human immunodeficiency virus type-1 transcription factor Tat require function of a polyamine sensitive-site on the N-methyl-D-aspartate receptor. Brain Res 954: 300-307.

Price RW, Brew B, Sidtis J, Rosenblum M, Scheck AC, et al. 1988. The brain in AIDS: central nervous system HIV-1 infection and AIDS dementia complex. Science 239: 586-592.

Pulliam L, Zhou M, Stubblebine M, Bitler CM. 1998. Differential modulation of cell death proteins in human brain cells by tumor necrosis factor alpha and platelet activating factor. J Neurosci Res 54: 530-538.

Pumarola-Sune T, Navia BA, Cordon-Cardo C, Cho ES, Price RW. 1987. HIV antigen in the brains of patients with the AIDS dementia complex. Ann Neurol 21: 490-496.

Qiu Z, Gruol DL. 2003. Interleukin-6, beta-amyloid peptide and NMDA interactions in rat cortical neurons. J Neuroimmunol 139: 51-57.

Rausch DM, Davis MR. 2001. HIV in the CNS: Pathogenic relationships to systemic HIV disease and other CNS diseases. J Neurovirol 7: 85-96.

Reyes MG, Faraldi F, Senseng CS, Flowers C, Fariello R. 1991. Nigral degeneration in acquired immune deficiency syndrome (AIDS). Acta Neuropathol (Berl) 82: 39-44.

Rohr O, Sawaya BE, Lecestre D, Aunis D, Schaeffer E. 1999. Dopamine stimulates expression of the human immunodeficiency virus type 1 via NF-kappaB in cells of the immune system. Nucleic Acids Res 27: 3291-3299.

Rottenberg D, Sidtis J, Strother S, Schaper K, Anderson J, et al. 1996. Abnormal cerebral glucose metabolism in HIV-1 seropositive subjects with and without dementia. J Nucl Med 37: 1133-1141.

Sacktor N, McDermott MP, Marder K, Schifitto G, Selnes OA, et al. 2002. HIV-associated cognitive impairment before and after the advent of combination therapy. J Neurovirol 8: 136-142.

Sacktor N, Schifitto G, McDermott MP, Marder K, McArthur JC, et al. 2000. Transdermal selegiline in HIV-associated cognitive impairment: Pilot, placebo-controlled study. Neurology 54: 233-235.

Sawaya B, Rohr O, Aunis D, Schaeffer E. 1996. Chicken ovalbumin upstream promoter transcription factor, a transcriptional activator of HIV-1 gene expression in hum brain cells. J Biol Chem 271: 23572-23576.

Scheller C, Sopper S, Jassoy C, ter Meulen V, Riederer P, et al. 2000. Dopamine activates HIV in chronically infected T lymphoblasts. J Neural Transm 107: 1483-1489.

Smith DG, Guillemin GJ, Pemberton L, Kerr S, Nath A, et al. 2001. Quinolinic acid is produced by macrophages stimulated by platelet activating factor Nef and Tat. J Neurovirol 7: 56-60.

Stoler MH, Eskin TA, Benn S, Angerer RC, Angerer LM. 1986. Human T-cell lymphotropic virus type III infection of the central nervous system. A preliminary in situ analysis. JAMA 256: 2360-2364.

Tan J, Town T, Mori T, Wu Y, Saxe M, et al. 2000a. CD45 opposes beta-amyloid peptide-induced microglial activation via inhibition of p44/42 mitogen-activated protein kinase. J Neurosci 20: 7587-7594.

Tan J, Town T, Mullan M. 2000b. CD45 inhibits CD40L-induced microglial activation via negative regulation of the Src/p44/42 MAPK pathway. J Biol Chem 275: 37224-37231.

Tan J, Town T, Saxe M, Paris D, Wu Y, et al. 1999. Ligation of microglial CD40 results in p44/42 mitogen-activated protein kinase-dependent TNF-alpha production that is opposed by TGF-beta 1 and IL-10. J Immunol 163: 6614-6621.

Toggas S, Masliah E, Rockenstein E, Rall G, Abraham C, et al. 1994. Central nervous system damage produced by expression of the HIV-1 coat protein gp 120 in transgenic mice. Nature 367: 188-197.

von Giesen HJ, Antke C, Hefter H, Wenserski F, Seitz RJ, et al. 2000. Potential time course of human immunodeficiency virus type 1-associated minor motor deficits: Electrophysiologic and positron emission tomography findings [In Process Citation]. Arch Neurol 57: 1601-1607.

von Giesen HJ, Koller H, Theisen A, Arendt G. 2002. Therapeutic effects of nonnucleoside reverse transcriptase inhibitors on the central nervous system in HIV-1-infected patients. J Acquir Immune Defic Syndr 29: 363-367.

von Giesen HJ, Wittsack HJ, Wenserski F, Koller H, Hefter H, et al. 2001. Basal ganglia metabolite abnormalities in minor motor disorders associated with human immunodeficiency virus type 1. Arch Neurol 58: 1281-1286.

Wyss-Coray T, Masliah E, Toggas S, Rockenstein E, Brooker M, et al. 1996. Dysregulation of signal transduction pathways as a potential mechanism of nervous system alterations in HIV-1 gp120 transgenic mice and humans with HIV-1 encephalitis. J Clin Invest 97: 789-798.

Wyss-Coray T, Yan F, Lin AH, Lambris JD, Alexander JJ, et al. 2002. Prominent neurodegeneration and increased plaque formation in complement-inhibited Alzheimer's mice. Proc Natl Acad Sci USA 99: 10837-10842.

Xiong H, Zeng YC, Lewis T, Zheng J, Persidsky Y, et al. 2000. HIV-1 infected mononuclear phagocyte secretory products affect neuronal physiology leading to cellular demise: Relevance for HIV-1-associated dementia. J Neurovirol 6 (Suppl. 1): S14-S23.

18 Multiple Sclerosis and Autoimmune Encephalomyelitis

H. Lassmann

Abstract: Multiple sclerosis (MS) is a chronic inflammatory disease of the central nervous system, which leads to the formation of disseminated demyelinated plaques in the central nervous system. The disease manifests in an early relapsing form, which is followed by a stage of progressive neurological deterioration. In the early stages of the disease focal demyelinated white matter lesions are formed by the influx of new waves of inflammatory cells from the circulation. In this phase peripheral immunosuppression or modulation of the immune response reduces the appearance of new lesions and ameliorates disease. In contrast, in the progressive stage, inflammation becomes trapped within the brain and spinal cord and gives rise to diffuse (mainly axonal) damage in the global white matter and profound demyelination in the cerebral cortex. The mechanisms, involved in the formation of MS lesions, are complex and heterogeneous, involving cytotoxic T-lymphocytes, specific autoantibodies and toxic products of activated macrophages and microglia cells. The mechanisms of tissue injury, involved in the formation of MS lesions are in part reflected in different models of autoimmune encephalomyelitis. However, many features of MS pathology and pathogenesis are not covered by this model.

List of Abbreviations: Apo-E, apolipoprotein E; CNS, central nervous system; CNTF, ciliary neurotrophic factor; EAE, experimental autoimmune encephalomyelitis; GFAP, glia fibrillary acidic protein; HLA-D, human leukocyte antigen-D; MAG, myelin associated glycoprotein; MBP, myelin basic protein; MHC, major histocompatibility complex; MOG, myelin oligodendrocyte glycoprotein; MRI, magnetic resonance imaging; MS, multiple sclerosis; NAA, *N*-acetylaspartate; PLP, proteolipid protein; PPMS, primary progressive MS; RRMS, relapsing/remitting MS; SPMS, secondary progressive MS; TNF, tumour necrosis factor

1 Multiple Sclerosis

Multiple sclerosis (MS) is the most common cause of nontraumatic disability in young adults, affecting approximately one million worldwide (Anderson et al., 1992). It is a chronic disease of the central nervous system (CNS), leading to major neurological disability in most patients over a course of several years or decades. For this reason, MS is not only a major burden for the individual patients but also has a profound socioeconomic impact. It was estimated that the annual costs of MS, including health care costs as well as the costs for personal services, alterations to home, purchase of special equipment, and earning loss amount to US $2.5 million over the lifetime of each individual patient (Whetten Goldstein et al., 1998). The benefit of currently approved immunomodulatory and immunosuppressive treatment is clearly established, but its effect is rather limited. Similarly limited is currently our understanding of the pathogenesis of the disease and the biological basis of the treatment response. For these reasons, intense research efforts are performed to elucidate the cause of the disease and the pathogenetic mechanisms, underlying the formation of its lesions.

1.1 Epidemiology and Clinical Manifestation of Multiple Sclerosis

MS is a disease that is not evenly distributed around the world. Its highest prevalence is found in northern Europe, northern America, and the southern parts of Australia, while the numbers of patients are much lower in areas of Mediterranean or tropical climate (Compston, 1998). One explanation for this observation is that a cold climate favors the development of the disease. This, however, cannot be the only reason, since even in areas at the same latitude and with identical climate, the prevalence of MS can be profoundly different. A second factor appears to be the genetic background. Populations with northern European or Scandinavian background appear to be most susceptible for developing MS (Compston, 2004). This is not only the case in Europe but also in other regions in the world. There too the percentage of inhabitants with northern European ancestors to a large extent determines the global risk in the population to develop MS. Supporting a major effect of the genetic background on MS susceptibility is also the fact that the disease is rare in populations with oriental or central African origin.

Yet, there are in addition to the described north/south gradient of MS prevalence other observations, which suggest that besides the genetic background, environmental factors may play a role in disease induction (Compston, 2004). Migration studies, although so far based on relatively small populations, indicate that the increased risk to develop MS is retained, when immigration from a high-risk to a low-risk area occurs after puberty. On the contrary, when the immigration takes place before the age of 6 years, the risk of developing MS changes to that from the low-risk area (Dean, 1967; Elian et al., 1990; Kahana et al., 1994). These data suggest that the exposure to an exogenous agent between the age of 6 years and puberty may play a role in determining the risk to develop MS. What this exogenous factor could be is so far undetermined, but virus infections occurring during childhood and puberty are attractive candidates. In conclusion, epidemiological investigations point toward two major factors, which determine the risk to develop MS: the genetic background of the patients and a possible environmental trigger.

MS is a disease with a broad and heterogenous spectrum of clinical courses and highly variable neurological presentation (Lublin and Reingold, 1996). Most patients start with a relapsing/remitting disease (RRMS), characterized by exacerbations, which are followed by complete or incomplete recovery. After several years, this relapsing course converts into a state of slow but continuous progression, the so-called secondary progressive phase of the disease (SPMS). Ten to fifteen percent of the patients miss the relapsing phase of the disease, but reveal uninterrupted slow progression immediately from the onset. This form of the disease is called primary progressive multiple sclerosis (PPMS). A small percentage of patients may present with a few moderate relapses of the disease and later recover (benign MS) or may show a rapidly progressive fulminate disease, which leads to the patients death within a few months after onset (acute MS).

Relapsing of MS generally starts during young adulthood with highest incidence of disease onset at an age between 20 and 30 years and the patients enter the progressive phase 10–15 years later. Females are twice as often affected compared to males. In contrast, patients with primary progressive disease do not show this gender preference and overall have a later disease onset, which matches the age of onset of the progressive stage in patients with SPMS. Furthermore, the speed of progression and the general development of disease in PPMS patients are in average very similar to that of the progressive phase seen in patients with SPMS (Confavreux et al., 2000).

These clinical data suggest that there are two different pathogenetic processes in MS patients that are responsible for the induction of relapses or progression respectively and which develop at least in part independently from each other (Confavreux et al., 2000). This view is further supported by the response to treatment in MS patients. While anti-inflammatory, immunomodulatory, or immunosuppressive therapies are effective during the acute and relapsing stage of the disease, they exert little benefit, when patients have entered the progressive phase (Leary and Thompson, 2003). From these observations, it has been suggested that the pathogenesis of MS is driven by two potentially independent factors, brain inflammation that induces the formation of new lesions in relapsing patients, and neurodegeneration that leads to diffuse brain damage in patients with progressive MS (Kalman and Leist, 2003; Owens, 2003).

The clinical manifestation and the type of neurological deficit are also highly variable and heterogenous between individual MS patients. Lesions in the CNS in MS may arise at any location within the patient's brain or spinal cord. The quality of neurological deficit, however, mainly depends on the location of the lesion in the CNS. It is thus not surprising that some patients present with motor deficits, while in others, sensory symptoms, visual deficits, or cerebellar signs are dominant. Furthermore, within the same patient, different lesions that occur at different time points in disease development are generally located at different places in the brain and spinal cord and are thus reflected by different neurological deficits. This dissemination of the disease in time (=chronic relapsing disease) and space (=different location of lesions within the neuraxis) is the basis of the clinical diagnosis of MS (Poser et al., 1983; McDonald et al., 2001).

1.2 Genetics of Multiple Sclerosis

There is no doubt that the genetic background of a given individual to a large extent determines the susceptibility to develop MS and modifies disease manifestation and severity (Kalman and Lublin, 1999; Herrera and Ebers, 2003; Compston, 2004). This is best exemplified in twin studies. When one monocygotic twin is affected by MS, the risk for the other twin to get the disease is 30%, which is around 600 times higher compared to the risk in the general population. However, this result also means that two-thirds of monocygotic twins remain unaffected. This suggests that in addition to genes other, possibly environmental factors are likely to be involved in disease induction.

Within MS families, disease transmission does not follow classical Mendelian traits. Instead, with increasing genetic distance from the affected individual, the disease risk drops dramatically (monozygotic twins: 30%; dizygotic twins: 2%–5%; Compston, 2004). Such results most likely reflect a highly polygenic situation, which means that multiple different genes with low individual impact are responsible for the global disease risk. What genes could then be involved in regulating disease incidence and severity in MS?

This question can be addressed by two fundamentally different means: a global genetic screening of MS families and a candidate gene approach in either a global cross-sectional MS population or within MS families. The first approach was used in several large international collaborative studies based in the USA, in Canada, and in the United Kingdom (Sawcer et al., 1996; Ebers et al., 1996; Haines et al., 1996). In these studies, multiple different chromosomal regions were identified, which showed linkage with disease susceptibility or severity. Although on the basis of enormous number of patients and nonaffected family members, the results of these studies were rather disappointing. They found consistent linkage with MS susceptibility in a region of chromosome 6p, which contained the MHC region. This confirms previous data from candidate gene approaches, which clearly show an association between certain HLA-D genotypes and MS (Olerup and Hillert, 1991). The other chromosomal regions, however, which turned up in these studies, were to a large part divergent between the different studies, and even a meta-analysis of the data, pooled from several different studies, did not come up with conclusive answers (GAMES and The Transatlantic Multiple Sclerosis Genetics Cooperative, 2003).

To overcome this situation, a whole genome approach, based on a 0.5 cM map of microsatellite markers, was recently initiated using pooled DNA from MS patients, unrelated controls, and trio families (affected individuals and their parents; Compston, 2004). This study is so far not completed, but first results confirm the association of MS with the HLA-D region on chromosome 6p. Additional genomic regions on 1p, 17q, and 19q, which in part already appeared in the first screens, were also positive in this study. It is hoped that with this approach, when it is completed, these genomic regions will be definitely confirmed and new regions will be identified.

Cross-sectional, population-based studies, comparing unrelated MS patients with controls, already in the 1970s provided evidence for an association between disease susceptibility and the HLA-D genotype (Olerup and Hillert, 1991). Thus, the association with the class II major histocompatibility antigen genes DR 15 and DQ6 are conformed now in many studies. In addition, in selected ethnic groups, such as for instance in Sardinia, an association with DR 4 is well documented (Marrosu et al., 1992). Regarding non-MHC genes, an association has been found with various genes, coding for immunologically relevant proteins (CTLA4, IgVH, Il-1Ra, Il-4R, ICAM-1, MCP-3, CCR5, IFNa, Il-2, Il-10; Ligers et al., 1999; Compston and Coles, 2002; Compston, 2004) or proteins of the nervous tissue, which may be relevant for determining the extent of immune-mediated injury (Apo-E4; Fazekas et al., 2000; CNTF; Giess et al., 2002; mitochondrial DNA; Kellar-Wood et al., 1994; Mojon et al., 1999). Many other obvious candidate genes have been excluded (Compston and Coles, 2002). Unfortunately, all associations detected so far with exception of the HLA-D are very weak and have to be confirmed in independent studies in larger sets of patients.

The interpretation of all these results is that there is no dominant single gene, which confers the susceptibility risk for MS and that there are most likely multiple genes, each of which individually contribute to the global risk only to a minor degree. These genes may either be involved in regulating the immune process, thus determining disease susceptibility or resistance, or they may influence the reaction of the target tissue during immune-mediated attack, thus influencing disease severity.

1.3 Magnetic Resonance Imaging and Spectroscopy: A Tool to Study the Development of Brain Injury in Multiple Sclerosis

Within recent years, major advancements have been achieved in nuclear magnetic resonance (NMR) imaging to detect pathological alterations in the CNS in living patients. MS lesions are particularly suitable for detection and monitoring by magnetic resonance imaging (MRI) techniques, since they are located within the white matter and the destruction of myelin sheaths, which is the pathological hallmark of MS lesions, leads to major changes in extracellular fluid diffusion properties. Thus, white matter plaques in MS patients can easily be detected and the change in structural properties can be longitudinally studied over the course of the disease (Miller et al., 1997).

MRI and spectroscopy (MRS) provide various different tools to study the dynamic development of tissue injury in MS. Conventional T2-weighted images and similar but more sophisticated techniques, such as diffusion-weighted images or magnetization transfer ratios, are very sensitive to detect lesions in the white matter, the specificity regarding the type of tissue injury, however is limited. As they measure in essence the water diffusion in the extracellular space, inflammation-induced edema as well as myelin and/or axonal loss are reflected by increased signals within the lesions. The extent of tissue destruction can more readily be imaged in T-1 weighted sequences, which show profound decrease in signal intensities in areas of massive myelin and axonal loss (the so-called T-1-black holes; Brück et al., 1997; van Walderveen et al., 1998; Bitsch et al., 2001). The status of blood–brain barrier (BBB) dysfunction can be visualized by injecting the paramagnetic tracer gadolinium DTPA, which accumulates at sites of acute vasogenic edema (Rovaris and Filippi, 2000). Finally, the extent of axonal injury and loss can be detected by MRS, determining the levels of N-acetylaspartate (NAA) within the tissue (Arnold et al., 1992).

All of these imaging techniques, however, have their limitations. There is so far no specific procedure available, which allows to estimate the true extent of demyelination, remyelination, or axonal loss separately. Furthermore, gadolinium leakage, even when high dosage of the tracer and long perfusion times are used, only detects massive BBB damage at sites of acute inflammation but does not pick up more subtle changes of BBB disturbance in chronic lesions. Also in contrast to general belief, NAA detected in MRS is not necessarily a marker for neuronal or axonal loss. NAA is synthesized and concentrated within mitochondria (Lu et al., 2004). Thus, a transient mitochondrial dysfunction may lead to profound reduction in NAA tissue levels, which however is reversible with time of lesion development (Narayanan et al., 2001). Besides these problems in interpretation of MRI and MRS data, these techniques provided valuable information regarding the dynamic changes within the brain of MS patients, both related to the development of individual lesions as well as the detection of accumulating brain damage in the global course of the disease.

When MS patients are followed by serial MRI scans during the acute and relapsing stage of the disease, the appearance of new focal white matter lesions is a major hallmark of the pathological process. In many instances, these lesions start by the appearance of focal areas of Gd-DTPA leakage as a marker for acute BBB damage (McDonald and Miller, 1996). This is followed by the appearance of focal areas of T-2 signal abnormalities, resulting from tissue edema, demyelination, and axonal loss. Some, but not all, lesions also show profound hypointensity in T-1-weighted lesions, reflecting massive widening of the extracellular space due to edema demyelination and profound axonal injury (Brück et al., 1997; van Walderveen et al., 1998). The extent of T-1 hypointensity correlates in part with the degree of axonal loss within plaques. In addition, T-1 hypointensity may in part revert during further development in the lesions. Some data suggest that this recovery of T-1 signal may at least in part be explained by remyelination of the lesions (Bitsch et al., 2001).

Besides this common pattern of lesion evolution, in other MS patients alterations in the tissue may occur, which precede massive BBB disturbance, visualized by gadolinium enhancement (Filippi et al., 1998). In this situation, subtle MRI changes can be seen in areas where at later time points classical T-2 lesions are found. Sometimes gadolinium enhancement may follow the appearance of new zones of T-2 signal abnormalities. These data suggest that lesion formation in MS brains may be accomplished by different mechanisms, a view which is also supported by heterogenous mechanisms of plaque formation seen in pathology (see below).

All the above described alterations account for the formation of focal-demyelinated plaques, which is the dominant feature of MS pathology during the acute or relapsing stage of the disease. In addition, however, a diffuse process of tissue injury is present in MS patients, which is most dramatic in patients with progressive disease and affects the brain in a global sense. In MRI images, this process is reflected by global atrophy of the brain and spinal cord, which is associated with profound diffuse abnormalities within the so-called "normal" white matter and the cortex (Losseff et al., 1996; Chard et al., 2002). This is also reflected by a loss of NAA, which not only occurs in established focal white matter plaques but also diffusely affects the global brain tissue (De Stefano et al., 1999). As diffuse alterations of the global nervous tissue together with atrophy are most pronounced in patients with progressive disease, they have for long been considered as a secondary consequence of axonal destruction within the plaques. Nevertheless, they can—with more subtle technologies—also be detected already at very early stages of the disease, in some patients even at the time of first clinical presentation (Filippi et al., 2003).

Taken together these MRI studies suggest that the CNS of MS patients is affected by two distinctly different processes, which may in part develop independently from each other. One is the formation of local plaques within the white matter, which occurs on the background of BBB damage and is most likely a consequence of an inflammatory process. The second is a diffuse brain injury, which also starts early in the disease but is most prominent in patients with progressive MS. This diffuse injury is not associated with BBB injury detectable by gadolinium enhancement and was thus considered to reflect a "neurodegenerative" component of the disease that may develop independently from inflammation.

1.4 Pathology and Pathogenesis of Multiple Sclerosis

The key features of MS pathology have been described more than 100 years ago (Charcot, 1868). MS is a chronic inflammatory disease of the CNS, which is associated with the formation of large plaques of primary demyelination, relative axonal preservation, and astrocytic scar formation (❷ *Table 18-1*). Demyelinated

❑ Table 18-1
The essential pathological features of MS and their functional consequences

Inflammation	Reversible within hours or days	Nitric oxide-mediated conduction block; function is rapidly restored when inflammation subsides and the concentration of nitric oxide drops
Demyelination	Partially reversible within days to weeks	Conduction block in acutely demyelinated axons (no Na$^+$ channels in demyelinated internode)
		Slowing of conduction velocity in chronically demyelinated axons (redistribution of Na$^+$ channels leads to restoration of conduction with decreased conduction velocity)Remyelination restores normal conduction in axons
Axonal loss	Irreversible	Axonal destruction by macrophages in actively demyelinating lesions (identified mediators: proteases, nitric oxide, excitotoxins)
		Slow burning axonal injury and loss in chronic established lesions (lack of trophic support by oligodendrocytes?)
		Axonal destruction leads to irreversible conduction failure, which is clinically manifest, when the threshold of functional compensation is passed

plaques may appear at any location of the CNS, certain areas of the neuraxis, however, are preferentially affected (Lumsden, 1970). These are the periventricular white matter, the subcortical white matter, the optic nerves and tracts, the cerebellar peduncles, and the spinal cord. In the latter, the cervical portions are most commonly affected, leading to triangular fan-shaped lesions, which have their base at the subpial surface and extend with their tip into the depth of the medulla spinalis (Fog, 1950).

The most important reason for this uneven distribution of MS lesions is that the plaques are in general centered by a small vein or venule (Rindfleisch, 1863) and seem to grow either by confluence of small perivenular lesions or by radial expansion. The perivenous nature of the lesions is best seen in the extension of larger periventricular plaques, which frequently show small, fingerlike expansions in the periphery that follow into the deep white matter the course of small veins with perivascular inflammation (the so-called "Dawson fingers"; Dawson, 1916). Thus, perivenous inflammation is apparently the initial factor, which precipitates a new focal MS lesion, and areas of the CNS with a high density of postcapillary veins and venules are therefore more likely to be affected than areas with low venous density.

1.4.1 Inflammation

Inflammation is present in the CNS in all MS patients irrespective of the stage of the disease or the activity of the lesions. The inflammatory infiltrates are concentrated around small veins and venules and, in particular in active lesions, they also disperse into the parenchyma. There they are found in highest density in areas of active demyelination and tissue injury. Inflammatory infiltrates in MS mainly consist of lymphocytes and macrophages. In rare cases of acute and fulminate disease, an additional component of granulocytes and eosinophils can be present (Lassmann et al., 2001; Lucchinetti et al., 2002).

Within the lymphocyte population, T lymphocytes dominate. Class I MHC-restricted CD8[+] T cells generally outnumber CD4[+] cells (Gay et al., 1997). Furthermore, clonal expansion, which may reflect local antigen-driven expansion, is found in the CD8[+] in much higher incidence than in the CD4[+] T cell population (Babbe et al., 2000). This T cell infiltration is associated with up-regulation of MHC molecules on leukocytes as well as on local glia cells and neurons. Class I molecules are found in active lesions of acute and chronic MS on all cell types of the CNS, including oligodendrocytes, astrocytes, neurons, and the axons as well as on microglia and endothelial cells (Hoftberger et al., 2004), while the expression of class II molecules is mainly restricted to microglia, macrophages, and lymphocytes. In addition, CD1 molecules, which are required for the presentation of complex glycolipids to T lymphocytes, can be found on macrophages and astrocytes (Battistini et al., 1996; Hoftberger et al., 2004). MHC expression is further accompanied in active lesions by the expression of costimulatory molecules such as B-7, CTLA4, CD40, and CD40-ligand (Windhagen et al., 1995; Gerritse et al., 1996). Thus, the whole armament of antigen recognition for T lymphocytes is present within the active lesions of MS and in principle all cells of the nervous system can be target for class I MHC-mediated cytotoxicity by CD8[+] T lymphocytes. Is there also evidence that T cells indeed recognize a specific antigen within MS lesions?

This question can only be addressed by indirect means. As mentioned above, clonal expansion of T cell populations is present in the CNS of MS patients (Babbe et al., 2000). T lymphocytes reactive against myelin proteins are found in the peripheral circulation of patients in higher frequencies than in controls (Olsson et al., 1990; Kerlero de Rosbo et al., 1993), and studies on the T cell receptor usage of cells, isolated from MS plaques, showed similar complementary-determining region-3 sequences compared to those of myelin basic protein (MBP) reactive T cells, isolated from the peripheral circulation (Oksenberg et al., 1993). The potential of local antigen recognition by T cells was further supported by the observation that MBP/class II complexes were found on antigen presenting cells within MS plaques (Krogsgaard et al., 2000). Thus, it is likely that the T cells in MS lesions are not innocent bystander cells but indeed recognize their specific (auto?)-antigen within the CNS.

Besides T cells, B lymphocytes and plasma cells are also present within MS plaques. They comprise a rather small fraction of total lymphocytes. As with T cells also B cells are present in higher numbers in active compared to inactive lesions, they too are present in the perivascular infiltrates and—in particular in active lesions—are dispersed within the parenchyma (Esiri, 1977, 1980). In contrast, plasma cells are mainly located within the perivascular connective tissue spaces and in the meninges. Their dispersion into the parenchyma is restricted to very active and destructive lesions. Plasma cells produce different immuno-globulin subclasses (IgM, IgG, and IgA; Mussini et al., 1977) and are responsible for intrathecal immunoglobulin production, which is a diagnostically useful marker in the cerebrospinal fluid (CSF) of MS patients.

The vast majority of inflammatory cells in MS lesions are macrophages. Several different cell types contribute to this macrophage pool in the lesions. Some are recruited from circulating monocytes. These cells express the chemokine receptors CCR1 and CCR5 as well as the proinflammatory calcium-binding members of the S-100 family MRP 8 and MRP 14 (Brück et al., 1995; Trebst et al., 2001; Mahad et al., 2004). Other cells reveal immunological features of (immature?) dendritic cells, being reactive for the chemokine receptor CCR7, microglia markers (IBA-1 and CD68) as well as costimulatory molecules (CD86), and MHC class II antigens (Kivisäkk et al., 2004). The largest subpopulation of macrophages in the lesions is apparently derived from resident microglia. In initial lesions, most of these cells reveal a process bearing phenotype (Gay et al., 1997) and are negative for the markers for migrating monocytes (CCR1 and CCR5) but express MHC class II, microglia markers (CD68, complement receptor 3), and Fc-receptors (Ulvestad et al., 1994; Trebst et al., 2001). At later stages of plaque formation, these cells change to a round and foamy macrophage phenotype, which additionally up-regulate CCR5 and CCR8 (Trebst et al., 2001, 2003; Mahad et al., 2004). As shown already in the earliest descriptions of MS pathology, macrophages are intimately attached to injured myelin in actively demyelinating lesions and are essential in the removal and degradation of myelin debris.

This leukocyte infiltration is orchestrated by a large set of molecules, which are involved in the regulation of the inflammatory reaction and in the induction of tissue injury. Many different pro- and anti-inflammatory cytokines or their respective mRNAs have been detected within active MS plaques by immunocytochemistry and in situ hybridization (Merrill, 1992; Cannella and Raine, 1995), although so far no clear-cut expression pattern that correlates with the stage or the activity of the lesions has been identified. This finding, although surprising at the first glance, is not completely unexpected considering the short half-life and expression window of these cytokines in experimental inflammatory brain lesions. Thus in chronic lesions, like those arising in MS, a timely orchestrated expression pattern of such molecules within the whole lesions is unlikely.

There is, however, some indirect evidence that brain inflammation in MS is driven by proinflammatory cytokines of the Th-1 type, such as for instance γ-interferon, interleukin-2, and tumor necrosis factor-($TNF-\alpha$) and $TNF-\beta$. The composition of the inflammatory infiltrates, consisting mainly of T cells, B cells, and macrophages, as well as the expression patterns of chemokines and their receptors (Sorensen et al., 1999; Huang et al., 2000; Simpson et al., 2000a, b) are consistent with that found in a Th-1 driven delayed hypersensitivity response. From a clinical point of view, the most compelling evidence for MS being a Th-1 cytokine driven disease is the increased relapse frequency following γ-interferon treatment of the patients (Panitch et al., 1987) and the partial therapeutic effects, seen with treatment strategies that induce a shift from Th-1 to Th-2 immune responses (Hohlfeld, 1997).

Having said that, the authors have warranted some words of caution. The involvement of Th-1 cytokines not necessarily implies that the cells, which are driving the disease, have to be $CD4^+$ T-helper 1 cells. Class I MHC-restricted cytotoxic T cells can be polarized in a similar way, producing Th-1 cytokines (Cabarrocas et al., 2003). Remembering the above described dominance and clonal expansion of $CD8^+$ lymphocytes in MS lesions, it is likely that these cells and not the class I MHC-restricted $CD4^+$ cells are the major driving force. The other caveat is that not all MS lesions follow this pattern of inflammation. In a subset of MS patients with fulminate disease and in particular in patients with neuromyelitis optica, the inflammatory reaction is dominated by the infiltration of the tissue by granulocytes, many of them being eosinophilic granulocytes. In these cases, inflammation is associated with profound expression of Th-2 type chemokine receptors (Lucchinetti et al., 2002). Thus, it may be dangerous to think that Th-1 reactions are invariably deleterious and Th-2 reactions are always good for MS patient's brain.

In addition to MHC-antigens, cytokines, and chemokines, there are other molecules that play an important role in the induction and propagation of the inflammatory response. Adhesion molecules are instrumental in guiding the way for leukocytes through the endothelial barrier and within the tissue (Springer, 1994). Several different adhesion molecules have been identified on endothelial cells, leukocytes, and glia cells within MS lesions (Sobel et al., 1990; Dore-Duffy et al., 1993; Washington et al., 1994; Cannella and Raine, 1995). The most important in a clinical setting appear to be $\alpha4$ integrins. Antibodies against this molecule are a very effective anti-inflammatory agent in experimental models of brain inflammation (Yednock et al., 1992) and appear to block inflammation and the formation of new lesions in treatment studies of MS (Tubridy et al., 1999). Other molecules, which are involved in the migration of leukocytes into inflammatory foci, are matrix metalloproteinases (MMPs; Cuzner and Opdenakker, 1999). Not unexpectedly, these molecules too are present within MS lesions (Cuzner et al., 1996; Maeda and Sobel,

1996). Some data suggest that part of the therapeutic action of β-interferon in MS is due to down-modulation of the expression of MMPs.

All these data provide clear evidence that an inflammatory process is present within focal white matter lesions in the MS brain and spinal cord. This inflammatory process is consistent with a T cell-mediated immune response, which leads to macrophage recruitment and microglia activation in the plaques. Clinical observations, that immunomodulatory or immunosuppressive treatment can effectively inhibit the formation of new white matter lesions in MS patients, provide compelling evidence that this inflammatory reaction is a driving force in disease pathogenesis (Hohlfeld, 1997). This view has recently been challenged in a study, which describes the pathology of a newly formed lesion in the brain stem of a patient, which apparently developed within few hours before death (Barnett and Prineas, 2004). In this particular brain, stem plaque degeneration of oligodendrocytes was found in the absence of any inflammatory reaction. The authors conclude that the initial stage of plaque formation develops independent from inflammation, inferring that inflammation is a secondary consequence of tissue damage in MS. This is an unusual finding in MS, which has never been observed before, despite the existence of numerous and detailed previous neuropathology studies. It is also not clear, whether the lesion, described in the study is a genuine MS plaque or is related to other possible pathologies, such as brain stem hypoxia, which may occur in a terminally ill patient. For these reasons, it is still valid to conclude that the formation of focal white matter plaques in MS is driven by inflammation. Whether inflammation is also responsible for the diffuse brain damage, seen mainly in the progressive stage of the disease, will be discussed below.

1.4.2 Sequence of Active Demyelination

The presence of plaques with primary demyelination is an essential diagnostic criterion in the neuropathology of MS. Primary demyelination means that myelin sheaths are removed from axons in a segmental fashion, leaving the axon itself intact (Charcot, 1868; Marburg, 1906). Having said that, however, the authors have to modify this statement. Primary demyelination in the absence of any axonal damage never exists in real life. In other words, whenever there is demyelination, there is also some degree of axonal injury and loss. The extent of this axonal loss can be highly variable between different demyelinating diseases, between different lesions in the same disease or even the same patient. It has, however, to be emphasized that in demyelinating diseases axons are relatively spared in relation to the complete destruction of myelin.

Active plaques in the CNS of MS patients are defined by the presence of ongoing myelin destruction (Brück et al., 1995; ❷ *Table 18-2*). This means that myelin sheaths are structurally damaged and that

□ Table 18-2

Determination of lesional activity in demyelinating multiple sclerosis plaques

Structural dissolution of myelin and oligodendrocytes	Structural changes of myelin (vesicular disruption, invasion of macrophages into myelin sheaths, partial removal of myelin by macrophages) Oligodendrocyte apoptosis or necrosis	Immediate
Early myelin degradation products in macrophages	Macrophages with lysosomal myelin remnants immunoreactive for MOG, MAG, MBP, and PLP	1–2 days
Intermediate myelin degradation products in macrophages	Macrophages with lysosomal myelin remnants immunoreactive for MBP and PLP but not for MOG and MAG	1–2 weeks
Late stages of myelin degradation	Foamy macrophages with neutral lipid inclusions	Weeks to months
Old (burnt out) lesion	No macrophages present within the lesions	More than several months

remnants of myelin are taken up by macrophages and digested within their lysosomes. This implies that active MS lesions can only be identified, when destruction of oligodendrocytes or myelin is directly seen and macrophages are present, which contain myelin debris in the early stage of degradation. The process of myelin degradation is complex and takes a long time. When a fragment of myelin is taken up by a macrophage, minor myelin proteins, such as myelin-associated glycoprotein (MAG) or myelin oligodendrocyte glycoprotein (MOG), are degraded first, followed by the major structural myelin proteins such as proteolipid protein (PLP) or MBP (Brück et al., 1995). Although no respective data exist for the human situation, in experimental demyelinating lesions a loss of MAG and MOG from the lysosomal debris in macrophages takes 1–2 days, while MBP and PLP are degraded within 1 week to 10 days. In parallel with protein removal, the lipids are further degraded into cholesterol esters and triglycerides. This material, which can be detected by using lipid soluble dyes (Oil red O), remains in macrophages until these cells are removed from the brain lesions and can thus persist within the lesions for months. Knowing this, the authors require the presence of macrophages for the identification of active lesions in MS, which contain the earliest stages of myelin degradation (Brück et al., 1995). When such stringent criteria are used, the incidence of active plaques in a collection of MS brains is much lower than generally claimed.

Actively demyelinating plaques appear in different phenotypes (Lassmann et al., 1998). *Acute plaques* are lesions with simultaneous demyelination throughout the whole lesion. Thus, such plaques are entirely filled with macrophages with early myelin degradation products, intermingled demyelinated axons as well as axons still containing fully or partially preserved myelin. Such lesions are rare and mainly found in patients with acute or with severe exacerbations in chronic MS. *Chronic active lesions* contain a demyelinated core, which is surrounded by a rim of macrophages with early myelin degradation products. Such lesions apparently grow by radial expansion around an inactive plaque center and are the dominating type found in exacerbations of chronic relapsing MS. In *inactive plaques*, no evidence for ongoing myelin destruction and no macrophages with early degradation products are found. Such plaques, however, still may contain T cells, B cells, and macrophages with myelin degradation products. So-called *burnt out inactive plaques* just show demyelination and glia scaring with only little residual inflammation.

There are, however, some other plaque types, which do not fit into this simple classification of lesional activity. By studying the pathology of MRI identified brain abnormalities, some lesions were found, which consisted of inflammatory cells, microglia activation, and some reduction of myelin density in the absence of overt demyelination or macrophages with myelin degradation products (De Groot et al., 2001). It was proposed that such lesions may be "preactive plaques." Although it seems logical that initial lesions in MS brains show alterations like those described above, there are several doubts that all such alterations will subsequently develop into classical MS plaques. Such lesions were found in very high frequency (up to 30% of all lesions identified). In addition, inflammation and microglia activation is seen in a rather ubiquitous distribution throughout the whole "normal" white matter in patients with chronic progressive MS. If all these alterations represent "preactive" lesions, numerous new plaques will have to arise in that stage of the disease. This is apparently not the case in most patients.

Another type of lesion, which is difficult to classify, is the slowly expanding demyelinated plaque in patients with progressive MS (Prineas et al., 2001). These are inactive plaques that are surrounded by a small zone of activated microglia and macrophages that express MHC class II molecules and contain neutral lipid degradation products. This alone does not allow to decide whether demyelination is currently active or just occurred several days or weeks before. Such lesions may, however, contain small numbers of macrophages, which still contain earlier myelin degradation products; their unequivocal identification, however, is difficult and requires detailed analysis of very well-preserved tissue.

1.4.3 What is the Primary Target: Myelin or Oligodendrocytes?

The study of oligodendrocytes in MS lesions was for long hampered by the lack of suitable markers, which allowed an unequivocal identification of these cells in archival pathological material. Burnt out inactive plaques in general show a very low density of cells, and these cells mainly consist of fibrillary astrocytes. There are hardly any cells, which structurally resemble oligodendrocytes. It was therefore taken as a dogma for long time that oligodendrocytes are completely destroyed in the lesions and remyelination is absent. When first markers for oligodendrocytes became available in the 1980s, it was thus surprising to find that in some MS lesions numerous oligodendrocytes were present and that this was associated with quite extensive remyelination (Raine et al., 1981; Prineas et al., 1984, 1989; Brück et al., 1994). On the basis of these primary observations, detailed and systematic studies followed. In summarizing the respective results, it is now clear that in most active MS lesions demyelination and oligodendrocyte injury occurs in parallel, although the extent of oligodendrocyte injury and loss is highly variable between different patients and lesions (Prineas et al., 1989; Lucchinetti et al., 1999). In some cases, demyelination may occur with only very minor original loss of oligodendrocytes, while in others all oligodendrocytes are destroyed in parallel or even prior to myelin destruction. Thus in some patients, the myelin sheath appears to be the primary target of the immune response, while in others it is the oligodendrocytes. As will be discussed below, this may be explained by the interindividual heterogeneity in the immunological mechanisms, leading to tissue destruction in different MS patients.

It is not clear so far what is the fate of oligodendrocytes, which originally survive the demyelinating attack and are abundantly present at least in a subset of MS lesions (Lucchinetti et al., 1999; Wolswijk, 2000). Current wisdom from experimental studies tells us that terminally differentiated oligodendrocytes cannot dedifferentiate into new myelin forming cells (Blakemore and Keirstead, 1999; Keirstead and Blakemore, 1999) and should therefore be gradually removed from the tissue by apoptotic cell death. Whether this is really the case in the inflammatory environment of MS lesions, which contains a broad cocktail of inflammatory mediators and growth factors, awaits final proof.

There is, however, good evidence that glial progenitor cells are present within demyelinated MS lesions, which can differentiate into maturity and start the remyelinating process (Wolswijk, 2002). Indeed many active lesions of MS reveal a peripheral zone of demyelinating activity in which besides myelin most of the oligodendrocytes are lost. However, closely adjacent to the active zone, new oligodendrocytes appear and remyelination is rapid and extensive. Yet, in other cases, the formation of new oligodendrocytes is nearly absent and remyelination is sparse (Lucchinetti et al., 1999). The possible mechanisms behind this observation will be discussed in the chapter on remyelination.

1.4.4 The Mechanisms of Demyelination and Oligodendrocyte Injury in Multiple Sclerosis are Manifold and Heterogenous Between Patients

Years of detailed immunological research on autoimmune and virus-induced models of inflammatory demyelination have shown that different effector pathways of the immune system can lead to myelin and/or oligodendrocyte damage (❯ Table 18-3). Many of these pathways have in the meantime also been identified within the MS lesions themselves. What, however, turned out in recent years is that not all pathways are utilized by the same patient at the same time. In contrast, when large numbers of lesions from different patients are studied, a profound heterogeneity in the immunopathology of the lesions becomes apparent (Lucchinetti et al., 2000).

Within an inflammatory lesion, the tissue can be injured by several different means. *Cytotoxic class I MHC-restricted T cells* can directly attack target cells, which present the specific antigen in the context with MHC class I molecules. Cell injury in such a situation can either be mediated through the release of perforin and granzymes from the cytotoxic granules of T cells. Alternatively, T cells can kill their target through the activation of death receptors of the TNF-receptor family (Fas, TNF-R1, TRAIL, and others). Within MS lesions, oligodendrocytes express MHC class I molecules and are thus a potential target for direct T cell-mediated killing (Hoftberger et al., 2004). Furthermore, class I-restricted T cells can be found in the lesions, which are in close apposition with oligodendrocytes and show a polar orientation of the cytotoxic granules toward the surface of the attached oligodendrocytes (Lassmann, unpublished). Such a pattern of T cell-target cell interaction closely resembles the formation of a so-called immunological synapse. Besides the ganzyme/perforin pathway, also a direct destruction though the activation of death receptors appears to be possible in MS. Oligodendrocytes have been shown to express Fas and TNF-receptors, while Fas-ligand has been found on inflammatory cells and microglia (D'Souza et al., 1996; Dowling et al., 1996;

◘ Table 18-3
Possible mechanisms of demyelination and tissue injury in multiple sclerosis

T cell cytotoxicity	Interaction of CD8[+], granzyme B positive T cells with oligodendrocytes and axons; MHC-Class I expression on neurons and glia; expression patterns of death receptors (Fas, TNF-R) and death ligands (Fas) within the lesions
Antibodies	Presence of anti-MOG antibodies with demyelinating potential in a subset of MS patients; deposition of immunoglobulins and complement C9neo antigen on degenerating myelin in a subset of patients
Macrophage mediated	Association between activated macrophages and myelin or axons in active lesions
Proteases	Expression of MMPs within active MS lesions
Reactive oxygen species	Oxidative mitochondrial damage in active MS lesions
Reactive nitrogen species	Presence of i-NOS and nitrotyrosin in active MS lesions
Cytotoxic cytokines	Presence of TNF-α and its receptors in active MS lesions
Excitotoxins	Expression patterns of enzymes, involved in glutamate metabolism suggest a dysregulation of glutamate homeostasis
Gliotoxic factors	Description of an MS-specific gliotoxic factor in cerebrospinal fluid and urine of MS patients

Bonetti and Raine, 1997). Death receptor-mediated killing of oligodendrocytes may, however, be counteracted by the activation of the NFκB pathway in these cells in the active stage of demyelination (Bonetti et al., 1999).

Another way to induce demyelination and tissue injury is through the opsonization of respective targets by specific antibodies, which then initiate the destruction via the activation of complement or through macrophages, binding the immune complex by their Fc-receptors. The prerequisite of antibody-mediated tissue injury is that the respective target antigen is accessible. This means that the BBB has to be compromised, allowing antibodies to enter the CNS compartment. Alternatively, antibodies can be produced intrathecally. Furthermore, the target epitope has to be exposed either on the extracellular surface of the cells or on antigens present in the extracellular matrix (Linington et al., 1988). In MS patients, antibodies against a variety of nervous system targets, including myelin, neuronal, or axonal proteins are induced, but few of these targets so far meet the criteria outlined above. One possible candidate is MOG, which is expressed on the extracellular surface of oligodendrocytes and myelin sheaths. When, however, antibody titers against MOG are determined by ELISA or Western blot, most antibodies are directed against linear epitopes of MOG, which are not accessible for antibody binding in vivo. Thus, MOG antibodies, found in high incidence in MS patients (Berger et al., 2003), in general do not qualify for a potentially demyelinating function. There are, however, conformational epitopes within the naturally folded MOG protein, which are accessible for antibody binding in vivo and are thus potentially pathogenic target epitopes. To detect such antibodies in patients requires testing with MOG-transfected cell lines. When such a test system is used, only a subfraction of MS patients reveal the presence of potentially demyelinating MOG antibodies (Haase et al., 2001). Another potentially pathogenic target for antibody-mediated injury is AN-2 (NG-2); a complex glycoprotein expressed on glial progenitor cells. Antibodies against this protein can also be found in a subset of MS patients (Niehaus et al., 2000).

Several lines of evidence suggest that antibody-mediated demyelination may play an important pathogenic role, at least in a subset of MS patients. Deposition of immunoglobulin at sites of the interaction between myelin sheaths and macrophages has been shown in actively demyelinating MS plaques (Prineas and Graham, 1981), and myelin is taken up in macrophages by coated pits and vesicles, suggesting a receptor-mediated phagocytic process (Prineas et al., 1984). Furthermore, complement components are enriched in active lesions of a subset of MS patients and C9neo antigen, a marker for the activation of the lytic complement component is deposited on the surface of damaged myelin and within lysosomal degradation products within macrophages (Compston et al., 1989; Storch et al., 1998a; Lucchinetti et al., 2000). The deposition of activated complement at sites of active demyelination, however, not necessarily implies an antibody-mediated mechanism of demyelination. Complement may also be activated in an antibody independent way by the interaction of MOG with complement C1q (Johns and Bernard, 1997).

One study directly implied that MOG antibodies may induce demyelination in MS lesions. Performing immunocytochemistry on glutaraldehyde fixed and plastic embedded MS lesions, the authors found binding of biotinlylated linear MOG peptides on degenerating myelin sheaths and concluded that these peptides are bound by MOG antibodies, attached to the damaged myelin sheaths (Genain et al., 1999). There are, however, several doubts regarding the validity of these findings. The linear MOG peptides, which have been used in this study, are not exposed on the surface of the intact molecule and are thus not accessible for antibody binding in vivo. Peptide binding was found throughout all portions of the damaged myelin sheaths, which is not compatible with the selective surface location of the MOG molecule. Furthermore, no immunoreactivity was found in plasma cells within the lesions. Trying to reproduce these findings on unfixed frozen sections of tissue from MS and autoimmune encephalomyelitis, we found some plasma cells, reactive for MOG, but did not find binding of the peptides to damaged myelin.

Thus, all the data suggest that antibody and complement-mediated demyelination may occur in a subset of MS patients. Although MOG is a potential target for demyelinating antibodies, the importance of MOG antibodies in the pathogenesis of MS is so far unresolved, and other potential targets may become apparent in the future.

Demyelination can also be induced by toxic products of activated macrophages and microglia cells. Many of the possible toxic candidates or their synthetic pathways have been identified in MS lesions, including proteases (Cuzner et al., 1996; Maeda and Sobel, 1996), complement components (Compston et al., 1989), cytotoxic cytokines, such as TNF-α (Hofman et al., 1989; Selmaj et al., 1991), as well as reactive oxygen or nitrogen species (De Groot et al., 1997; Smith et al., 1999; Lu et al., 2000; Liu et al., 2001). Furthermore, excitotoxins, which may either be produced by activated microglia (Lipton, 1998) or liberated from partially damaged neurons and axons, can destroy oligodendrocytes and their progenitor cells (Fern and Moller, 2000), possibly in an AMPA-receptor-mediated pathway (Pitt et al., 2000). An imbalanced glutamate homeostasis in MS lesions is also suggested by increased glutaminase expression in macrophages and a reduced expression of glutamate dehydrogenase in oligodendrocytes (Werner et al., 2001).

All these macrophage toxins are nonspecific mediators of tissue injury in an inflammatory condition. Nevertheless, when acting in the white matter of the CNS, a pseudo-selective damage of MS and oligodendrocytes is the rule. The reason for this phenomenon is provided by in vitro studies, which show that MS and oligodendrocytes are in general more susceptible for the damage induced by these factors in comparison to axons, neurons, or astrocytes.

Besides these classical immune-mediated forms of tissue damage, other mechanisms have recently been identified to play a role in the induction of MS lesions. In a subset of MS patients, the pattern of myelin destruction in active lesions closely resembles that found in acute stroke, thus suggesting the involvement of a hypoxia-like tissue injury. This pattern of tissue injury is also associated with profound and selective expression of hypoxia-inducible factor 1α (Aboul-Enein et al., 2003). There are several possible explanations for hypoxia-like lesions in inflammatory conditions. Microcirculation can be disturbed within inflamed lesions either through edema-induced increased tissue pressure or through vasculitic damage of microvessels. Yet, there is so far little hard evidence for microvessel pathology and a disturbance of microcirculation in MS lesions. Another more attractive hypothesis is that an excessive production of reactive oxygen and nitrogen species may induce mitochondrial damage (Bolanos et al., 1997; Beltran et al., 2000), and by this mechanisms result in a state of histotoxic hypoxia (Aboul-Enein and Lassmann, 2004).

Finally, in discussing mechanisms of tissue injury, one has also to consider that the susceptibility of the target tissue for immune-mediated destruction may in part be determined by the genetic background of the patients. As discussed above, disease severity in MS may be linked to genetic factors acting in the CNS tissue such as for instance the APO-E genotype (Fazekas et al., 2000) or a deletion of the ciliary neurotrophic factor (Giess et al., 2002). To what extent such genetic factors influence the pathological phenotype of the lesions has to be determined in the future.

Considering the multitude of potential effector mechanisms, which may be responsible for demyelination and tissue damage in MS, the question arise, whether they all act in parallel in the same patient and lesion or whether individual patients show a preference for distinct destructive pathways. By analyzing a large set of biopsies and autopsies of MS patients, it turned out that the latter is apparently the case. A profound interindividual heterogeneity in the immunopathological patterns of demyelination was found and it was possible to define four major different pathways of tissue injury (Lucchinetti et al., 2000). The first, which is most likely the basic pathogenetic pathway, present in all patients, is a T cell and macrophage-dominated inflammatory reaction in which cytotoxic T lymphocytes and activated macrophages alone are associated with demyelination and tissue injury. In other patient subgroups, this basic pattern of pathology is modified by additional mechanisms. In one subtype, massive deposition of immunoglobulin and complement is present at the sites of active tissue damage, suggesting an antibody-mediated mechanism of tissue injury. In another subgroup of patients, all active lesions show a pathological phenotype of hypoxia-like tissue injury. Finally in a small subgroup of patients, the inflammatory reaction is associated with exceptionally severe oligodendrocyte damage, suggesting that these cells may have a problem coping the immune-mediated attack. It is attractive to speculate that these differences in the immunopathology of the lesions are determined by the genetic background of the patients.

1.4.5 Remyelination

Already in 1906, Otto Marburg described that in active plaques of acute MS abundant axons can be found, which are surrounded by very thin myelin sheaths, which he was only able to visualize by osmic acid impregnation (Marburg, 1906). By comparing the features of these fibers with those present in experimental conditions of demyelination in the peripheral nervous system, he suggested that they may represent attempts of remyelination. Despite these early observations, it has for long time been considered that remyelination does not occur in MS. However, with the introduction of electron microscopy in neuropathology, it became clear that remyelination can be present in particular at the edge of chronic inactive plaques (Suzuki et al., 1969). The presence of remyelination in MS lesions was finally proven in systematic studies by Prineas et al. (1989, 1993b), clearly documenting the presence of thinly myelinated fibers with shortened internodes as well as the cytoplasmic expression of myelin proteins within reactive oligodendrocytes.

Remyelination may not only be present in a limited area at the edge of plaques but may also be present within the total plaque area. Such lesions were first described as "Markschattenherde" (Schlesinger, 1909), which at that time were considered as areas of incomplete demyelination, but the detailed light and electron microscopic analysis of such "shadow plaques" clearly show the presence of uniformly thin myelin sheaths with shortened internodes (Lassmann, 1983; Prineas et al., 1993a, b). It is thus established now that remyelination not only occurs in MS but can also reflect repair of entire lesions.

Our own studies show that the incidence of shadow plaques is highly variable between different MS patients (Lassmann, 1983). In some patients, more than 80% of all plaques within the brain are remyelinated, while in others remyelination is sparse or even absent. Several different reasons may account for this interindividual heterogeneity in the capacity of remyelination (Franklin, 2002).

According to experimental data, remyelination seems to be accomplished by glial progenitor cells, which proliferate and differentiate into mature myelinating oligodendrocytes (Keirstead and Blakemore, 1999). The presence of oligodendrocyte progenitor cells within MS plaques has been convincingly shown by several techniques (Wolswijk, 2002; Lucchinetti et al., 1999). It is, however, important to note that the mere presence of progenitor cells not necessarily has to result in the formation of new myelin sheaths. They can be present in MS lesions in considerable numbers also in the absence of any remyelination (Chang et al., 2002). These observations suggest that progenitor cells may be arrested in their further differentiation by the local plaque environment.

Thus, the question arises, why remyelination is efficient in some patients, while it fails in others. One reason may reside in the mechanisms of tissue injury occurring during the phase of active demyelination. In a subset of MS plaques, recruitment of new oligodendrocytes is rapid and efficient, while in another subset this does not occur (Lucchinetti et al., 1999). Certain immune effector molecules, such as for instance excitotoxins, destroy progenitor cells as effectively or even more efficiently than mature oligodendrocytes (Fern and Moller, 2000). Furthermore, some MS patients develop an antibody response against AN-2, a protein expressed on progenitor cells (Niehaus et al., 2000). Thus in such a situation, progenitor cells may be damaged or lost during the stage of active demyelination and may no more be available for remyelination. The available pool of progenitor cells may also become exhausted, when new waves of demyelination occur in previously remyelinated lesions (Ludwin, 1980). That remyelinated shadow plaques may be sites of a new demyelinating attack in MS patients can frequently be seen in MS pathology (Prineas et al., 1993a). It was further suggested that remyelination may be blocked by cytokines in the plaque environment, involving the notch/jagged pathway (John et al., 2002), although this observation was not confirmed in a subsequent investigation (Storch M., unpublished). Finally, the axons in MS plaques may become impermissive for the induction of remyelination. Polysialylated neuronal cell adhesion molecule, an axonal molecule, which actively suppresses myelination, is reexpressed on demyelinated axons in some MS plaques (Charles et al., 2002).

The process of remyelination is not only controlled by neuronal- or glial-derived growth factors but also by immune mediators. Experimental data show that the blockade of macrophage function within demyelinating lesions effectively blocks the recruitment and differentiation of oligodendrocyte progenitor cells as well as subsequent remyelination (Kotter et al., 2001). In line with this observation, active remyelination is mainly found in MS lesions, which still show pronounced inflammatory cell infiltration (Prineas et al., 1989; Brück et al., 1993; Lucchinetti et al., 1999). Thus, remyelination mainly occurs at early stages of plaque development, while it is sparse in established "burnt out" plaques. This implies that massive and effective immunosuppression in MS patients may not only reduce or block the formation of new lesions but may simultaneously impair reparation.

Considering all the above discussed data, the authors have now well established the fact that spontaneous remyelination occurs in some MS patients and that in these patients remyelination is effective and complete. In other patients, however, remyelination is blocked and there are several different reasons why this blockade can occur. This is currently a very unsatisfactory situation for the design of therapeutic strategies to improve remyelination in MS patients. There are no clinical tools available so far, which allow to identify those patients who need such a therapy. Even if this becomes available in the future, the different mechanisms involved in the blockade of remyelination may render it very difficult to find a therapeutic strategy, which solves the problem in the individual patient.

1.4.6 Axonal Injury and Destruction

Although in focal white matter lesions of MS, primary demyelination with relative preservation of axons is typical, it is known for a long time that axons are not unaffected in this disease (for review see Kornek and Lassmann, 1999). In fact most chronic MS plaques show a reduction of axonal density, which in average amounts to an axonal loss of 50%–60% (Bjartmar et al., 2000). This feature of MS pathology is highly relevant. While the functional impairment induced by inflammation and demyelination is in part reversible, axonal loss, when it exceeds the threshold of compensation, results in irreversible neurological deficit.

Axonal damage in MS plaques occurs in two different patterns, which may reflect different pathogenetic mechanisms. In actively demyelinating lesions profound axonal injury occurs on the background of the inflammatory process. This is reflected by focal axonal accumulation of proteins, such as for instance β-amyloid precursor protein or ion channels, which are shifted along the axon by fast axonal transport (Ferguson et al., 1997; Kornek et al., 2000, 2001). This stage of axonal injury may still be reversible or is followed by axonal transection and the formation of axonal end bulbs (Trapp et al., 1998). Thin caliber axons appear to be more severely affected in the lesions than thick ones (Evangelou et al., 2001). The close topical association between injured axons and microglia suggests that axonal damage is mainly mediated through toxic products of activated macrophages (Trapp et al., 1998; Kornek et al., 2000). In addition, however, cytotoxic T cells may play a role in axonal injury, since its extent correlated with the density of $CD8^+$ T cell infiltration (Bitsch et al., 2000) and a close apposition of cytotoxic T cells to axons can sometimes be found in the lesions (Neumann et al., 2002).

In addition to the extensive axonal injury in actively demyelinating lesions, there is also a slow burning axonal destruction within established chronic demyelinated plaques, which however is absent in remyelinated shadow plaques (Kornek et al., 2000). These data suggest that chronic demyelination may increase the susceptibility of axons to degenerate in response to noxious stimuli. Although it is not clear so far, whether chronic axonal injury in demyelinated plaques is also driven by the inflammatory response, its extent correlates significantly with the number of activated microglia in the lesions (Kornek et al., 2000).

As with the destruction of myelin multiple different immunological mechanisms may induce axonal injury. As mentioned above, activated microglia cells and possibly also cytotoxic T cells seem to be involved. Regarding toxic immune mediators, the authors so far have proteolytic enzymes (Anthony et al., 1998), excitotoxins (Smith et al., 2000), and reactive nitrogen radicals (Smith and Lassmann, 2002) have been identified. Elegant in vitro studies showed that nitric oxide (NO) can induce functional conduction block in axons (Redford et al., 1997; Kapoor et al., 1999), which possibly explains rapid functional recovery in patients, when inflammation in the lesions subsides. Beyond that, however, irreversible degeneration can also be induced by NO in axons that are under metabolic stress for instance induced by repetitive stimulation (Smith et al., 2001). This suggests that the toxic action of NO is only partially due to a direct effect but may be indirectly mediated through energy failure within the tissue. This could be accomplished by NO-mediated mitochondrial injury.

Following the initial trigger, axonal degeneration appears to follow a pathway, which is not specific for MS but is present in a similar way also in white matter stroke. A major step in the further propagation of axonal damage is the uncontrolled influx of sodium (Bechtold et al., 2004), followed by a reverse operation of the sodium/calcium exchanger (Stys, 1998; Petty and Wettstein, 1999). This together with aberrant expression of voltage-gated calcium channels at sites of axonal injury (Kornek et al., 2001) may result in elevated intra-axonal calcium concentrations that activate calcium-dependent proteases. The final result of this scenario is the degradation of axonal cytoskeleton, disturbance of axonal transport, and subsequent axonal disintegration.

In the living MS patient, the extent of axonal damage is reflected by the hypointensity of T-1 signals in conventional MRI scans (van Walderveen et al., 1998) and by the loss of NAA, measured in MRS (Arnold et al., 1992). Interestingly, however, NAA loss may not only reflect axonal loss but also functional impairment of axonal mitochondria. In MS patients, treated with β-interferon, a marked increase in the tissue levels of NAA has been observed in comparison to untreated patients and more surprisingly in comparison to the pretreatment levels in the same patient (Narayanan et al., 2001). As it is unlikely that β-interferon promotes axonal regeneration in MS, it appears that it is the reduction of inflammation-induced functional impairment of axons, which is responsible for this effect. Neurodegeneration and atrophy is also in part reflected by increased levels of neurofilaments or neurofilament antibodies in the CSF (Eikelenboom et al., 2003). However, as repeated CSF punctures during the course of the disease can hardly be obtained for ethical reasons, the clinical value of such determinations is limited.

1.4.7 Pathological Alterations in the Cerebral Cortex of Multiple Sclerosis Patients

MS is generally considered a disease of the white matter of the CNS. The reason for this view is that demyelinated lesions can easily be detected in the white but not in the gray matter. This is also the case for MRI, which is up to now not sensitive enough to visualize the loss of myelin sheaths in the cerebral or cerebellar cortex. Early studies on the pathology of MS, however, noted that demyelinated plaques may also be present in the cortex (Sander, 1898; Brownell and Hughes, 1962; Lumsden, 1970), but even in these studies conventional stains for myelin were not sensitive enough to grasp the real extent of cortical pathology. This situation changed, when sensitive immunocytochemical methods for the detection of myelin antigens in tissue sections were applied in the investigation of cortical pathology in MS (Peterson et al., 2001). Depending on their location and shape, different types of cortical lesions can be distinguished (Kidd et al., 1999; Peterson et al., 2001; Bo et al., 2003a). Corticosubcortical compound lesions are plaques, which extend from the subcortical white matter into adjacent gray matter areas. Perivascular intracortical lesions are small plaques, centered by a small vein or venule in a similar way as classical white matter plaques. The most frequent type of cortical demyelination, however, is a band-like subpial demyelination, which may span over a distance of several gyri and extend in variable depth into the deeper cortical layers. These lesions are clearly related topographically toward the pia mater and the CSF space, and it is likely that meningeal inflammation is an important factor in the development of such lesions (Bo et al., 2003a). Cortical lesions are mainly present in patients with primary or SP MS and are rare or absent in acute or relapsing MS. In progressive MS cortical involvement may be extensive, involving up to 40% of the total cortical area.

On a structural basis, cortical-demyelinated plaques are fundamentally different from those present in the white matter (Bo et al., 2003b). Leukocyte infiltration within the tissue parenchyma is generally minor, but there is profound microglia activation. T cells and B cells are mainly located within the meninges or in the perivascular space of large drainage vessels. In contrast to white matter plaques, microglia retain their process bearing phenotype. Classical foamy macrophages are absent. Activated microglia cells express class II MHC molecules and inducible nitric oxide synthase (i-NOS) but only exceptionally contain early myelin degradation products. This suggests that demyelination in cortical lesions progresses at a very low speed.

Similar as in white matter lesions, primary demyelination with relative axonal sparing is the hallmark of cortical plaques, although some acute axonal injury, axonal transection and neuronal injury and apoptosis can be seen (Peterson et al., 2001; Bo et al., 2003).

The strict topographical orientation of cortical lesions toward the meningeal surface of the brain and the absent tissue infiltration of the lesions by T cells and B cells indicate that a soluble factor, produced within the meninges or the cerebrospinal fluid, may directly or indirectly be responsible for demyelination. It could either be a factor, which activates microglia to produce myelinotoxic mediators, or it may exert myelinotoxic properties by itself. Recently, a gliotoxic factor has been described, which was specifically present in the CSF of MS patients but not of controls (Menard et al., 1998). Whether this factor alone or in combination with other immune mediators is responsible for cortical demyelination has to be determined in the future.

1.4.8 The Pathological Substrate of Diffuse Brain Injury in Progressive Multiple Sclerosis

So far only the focal abnormalities within demyelinated plaques of MS patients have been discussed. In addition there are, however, diffuse abnormalities, which affect the brain in a global sense (Allen and McKeown, 1979; Allen et al., 2001). These diffuse alterations of the "normal" white matter are particularly prominent in patients with progressive MS, although they seem to start already at much earlier stages of the disease (Filippi et al., 2003) and gradually increase in severity over the following years or decades.

All these diffuse abnormalities occur on the background of an inflammatory process, although this inflammatory reaction appears to be different from that found in focal plaques. It is characterized by a low-grade inflammation with small perivascular infiltrates of T cells, B cells, and plasma cells. In addition, some of the leukocytes are present in a diffusely dispersed manner throughout the whole tissue. This inflammatory process is not accompanied by major BBB disturbance, as reflected by the absence of gadolinium leakage in magnetic resonance scans and by only a low-grade extravasation of serum proteins, detected by immunocytochemistry in the tissue.

The most striking alterations within the tissue is a general reduction of myelin density, visualized by conventional myelin stains, which is reflected by closer inspection by some loss of myelinated fibers and reactive astrocytic gliosis (Allen and McKeown, 1979; Allen et al., 2001). We found by immunocytochemistry for axonal proteins the presence of some axonal spheroids and end bulbs in the absence of primary demyelination. This "degenerative" process was associated with a diffuse activation of microglia, which highly express class II MHC molecules and in part show cytoplasmic reactivity for i-NOS.

For a long time, these diffuse abnormalities of the white matter have been considered a secondary consequence of tissue damage within the focal-demyelinated plaques. Axonal destruction in focal plaques in the CNS inevitably results in subsequent Wallerian degeneration of the fiber tracts, which traverse the respective lesions. That Wallerian degeneration is a factor, which in part is responsible for adjacent white matter damage is clearly evident in brain and spinal cord areas, where defined tracts can be outlined. There are a number of well-documented examples for such a mechanism. In the spinal cord, axonal loss is not restricted to demyelinated plaques but is present in nearly identical quantity in rostral or caudal portions of the respective tracts, in areas, which are not affected by focal demyelination (Ganter et al., 1999; Lovas et al.,

2000). Another example is provided in the corpus callosum. There the regional callosal atrophy correlates well with the plaque load in the adjacent brain hemispheres (Evangelou et al., 2000). Furthermore, morphological and biochemical alterations within the "normal" white matter have also been explained in part by the presence of focal-demyelinated lesions, which were too small to be detected by gross inspection of the brain (Allen and McKeown, 1979).

All these explanations have recently been challenged by detailed magnetic resonance studies. They showed that such diffuse white matter abnormalities can occur at early stages of MS or in PPMS, when the plaque load within the brain is much too small to account for the extent of diffuse damage (De Stefano et al., 1999; Filippi et al., 2003). This is also similar in patients with primary progressive disease, in whom diffuse white matter abnormalities are most prominent and focal-demyelinated lesions are generally rare and small. These observations lead to the concept that the diffuse white matter injury may be due to a "neurodegenerative" process, which is not directly related to the events taking place within the focal white matter plaques (Kalman and Leist, 2003; Owens, 2003). It remains so far unresolved, whether diffuse brain injury reflects a degenerative process with secondary inflammation, or whether it too is driven by the chronic inflammatory response.

1.5 A Unifying Concept of Multiple Sclerosis Pathogenesis

It is obviously difficult to assemble all the partly divergent data, coming from clinical, radiology, immunology, neurochemical, and pathology studies on MS, into a general concept of multiple pathogenesis. Despite these problems, a pattern is emerging which may explain many of the findings. Clearly, there are two different components, which contribute to the disease in a partly independent manner: the formation of focal-demyelinated plaques in the white matter, which is particularly prominent in the acute and early relapsing phase of the disease and the diffuse global brain injury, which dominates in patients with progressive disease. Both pathologies occur on the background of an inflammatory reaction, the type of inflammation, however, is different (❷ *Table 18-4*).

❏ **Table 18-4**
Pathological differences between relapsing and progressive multiple sclerosis

Focal new foci of inflammation mainly in the white matter	Diffuse compartmentalized inflammation of the whole brain and spinal cord
New formation of focal-demyelinated plaques with variable axonal injury mainly located in the white matter	Chronic demyelinated plaques in the white matter, some with slow radial expansion
Heterogenous mechanisms of demyelination and tissue injury within the lesions	
Little or no lesions in the cerebral cortex	Massive demyelination in the cerebral and cerebellar cortex
Little or no diffuse damage in the "normal" white matter	Diffuse inflammation, microglia activation and axonal injury within the "normal" white matter

Ample evidence suggests that focal white matter lesions are initiated by new waves of inflammation, which enter the CNS compartment from the circulation. This is associated with profound BBB damage and leads to the activation of local effector cells, such as microglia. Depending on the genetic background of the patient, different pathways of tissue injury may preferentially be used, resulting in an interindividual heterogeneity in the type of immune-mediated injury and in the extent and quality of myelin, oligodendrocyte, or axonal damage. The massive reduction of the formation of new focal lesions by immunomodulatory or immunosuppressive treatments, as seen in MRI studies performed in the course of clinical trials, strongly argue for a central role of the inflammatory process during this phase of the disease. Autoimmune reactions seem to play a role in the pathogenesis of the lesions but may also be involved in neuroprotection, remyelination, and repair (Diemel et al., 1998; Kerschensteiner et al., 1999; Moalem et al., 1999; Kotter et al., 2001). It is, however, not settled whether autoimmunity alone is sufficient or whether infectious agents are necessary to propagate the immune response either in the peripheral immune system or in the CNS itself.

The second component is the diffuse injury, which affects the entire brain. This process too starts very early in the course of the disease but becomes more and more prominent in patients with progressive disease. It is reflected by axonal injury in the global white matter and cortical demyelination, which develops in a slowly progressive manner. Also this process occurs on the background of inflammation, but this inflammatory reaction seems to be trapped and compartmentalized within the CNS. It may be the result of a gradual accumulation of inflammatory cells, which occurs slowly over time in the absence of BBB damage. Being trapped within the brain, it may become independent from the control of the peripheral immune system and may by itself form new lymphatic-like structures within the brain compartment (Serafini et al., 2004). Such a compartmentalized immune reaction, which develops behind a closed BBB, will be reached by peripheral immunomodulatory or immunosuppressive treatment only to a very limited degree. Reactive oxygen and nitrogen species, produced by chronically activated microglia, may play a major role in the induction of tissue damage in this chronic stage, in part possibly through damage of mitochondria. Consistent with this hypothesis are recent data obtained in microarray studies on MS tissue from patients with progressive disease, which provide evidence for mitochondrial damage (McDonough et al., 2003) as well as a consistent up-regulation of genes, which are involved in hypoxic preconditioning (Graumann et al., 2003).

2 Experimental Autoimmune Encephalomyelitis

Neuroparalytic complications, following immunization against neurotropic viruses in the early decades of the twentieth century (Remlinger, 1928), paved the way for the development of experimental autoimmune encephalomyelitis (EAE), which became one of most intensively studied experimental autoimmune disease. These complications occurred following vaccination with virus that was propagated in brain tissue. Pioneer work by Rivers et al. (1933) subsequently established that sensitization of suitable animals with brain tissue itself induces an inflammatory demyelinating disease, which in many respects resembles MS.

In the meantime, most of our knowledge on basic mechanisms of autoimmunity and immune regulation as well as on the mechanisms of brain inflammation and inflammation-induced brain damage have been uncovered through the systematic analysis of this model. To provide a full account of all knowledge, developed in this model, would by far exceed the purpose of this chapter. Thus, in the following, only some aspects that seem to be directly relevant to multiple pathogenesis will be addressed.

2.1 Basic Pathogenetic Principles of EAE

EAE is a nervous system-specific autoimmune disease, which is principally mediated by T lymphocytes. The formal proof of EAE, being a T cell-mediated autoimmune disease, has been achieved by the isolation and propagation of monospecific T lymphocyte lines or clones, which, when transferred into naive syngeneic animals, can induce the disease (Ben Nun et al., 1981). It has for long been believed that EAE is an autoimmune disease against myelin proteins such as MBP, PLP, or MOG, but recently it turned out that brain inflammation can also be induced by T cells directed against proteins expressed in astrocytes (S-100, GFAP, Kojima et al., 1994) or neurons (β-amyloid precursor protein; Furlan et al., 2003). When encephalitogenic T cells enter the brain in the course of EAE, they recruit circulating monocytes and activate local microglia. These effector cells are critically involved in the induction of inflammatory tissue damage in the CNS (Berger et al., 1997).

Originally, encephalitogenic T cells were found to be Th-1 polarized and class II MHC restricted (Ben Nun et al., 1981). This means that the cells, which are able to enter the brain and induce inflammation, are T cells, which recognize their antigenic peptide in the context of class II molecules and produce proinflammatory cytokines such as γ-interferon, TNF-α, or interleukin-2. Later, however, it became clear that class I MHC-restricted T cells, which recognize an antigen present within the brain, also are able to mediate brain inflammation (Oldstone and Southern, 1993; Evans et al., 1996; Huseby et al., 2001; Cabarrocas et al., 2003) and produce a similar spectrum of proinflammatory cytokines.

Antibodies, directed against brain antigens, alone do not induce autoimmune encephalitis, since their potential to pass the intact BBB is very limited. In addition, antibodies need activated effector mechanisms, such as complement or activated macrophages to implement tissue injury. However, in the course of a T cell-mediated encephalitis, tissue damage can be massively augmented by the additional presence of autoantibodies, which are directed against target antigens, expressed on the surface of CNS cells or on components of the extracellular matrix. As an example, immunization with CNS tissue or myelin elicits a demyelinating antibody response, which is mainly directed against MOG (Linington and Lassmann, 1987). When such antibodies are present in T cell-mediated inflammatory brain lesions, extensive demyelination ensues (Linington et al., 1988). This implies that the full spectrum of inflammatory demyelination is most reproducibly induced by active sensitization of genetically susceptible animals with whole brain tissue, myelin, or MOG, and the pathogenesis of the lesions is mediated by a concerted action of T cell-mediated and antibody reactions.

2.2 Immune Surveillance and the Induction of CNS Inflammation

Much of what we know today about immune surveillance of the normal brain and the induction of brain inflammation has been unraveled in models of autoimmune encephalomyelitis and, in particular, by the use of monospecific encephalitogenic T cell lines. When autoreactive T cells, directed against a CNS antigen, are transferred into naive syngeneic recipients, a transient, self-limiting inflammatory disease of the brain and spinal cord is induced, which follows a standardized and predictable course (Ben Nun et al., 1981). The capability of autoreactive T cells to mediate brain inflammation critically depends on their activation state at the time of transfer. Only activated T cells get access to the brain, while resting T cells do not induce disease (Wekerle et al., 1986). Immediately after transfer, a few of the donor T cells enter the CNS apparently through the intact BBB. This early invasion of T cells into the brain is antigen independent but activation dependent (Wekerle et al., 1986; Hickey et al., 1991). Thus, in principle, activated T cells can enter the normal brain irrespective of their reactivity against brain antigens or against targets, which are not present within the CNS. The vast majority of transferred T cells, however, initially home into the peripheral lymphatic tissue. Within the next 2 days, these transferred T cells become reeducated in the lymph nodes and spleen and acquire a new "migratory" phenotype (Flugel et al., 2001). This migratory phenotype is characterized by a down-regulation of activation antigens (e.g., CD134) and a massive up-regulation of certain chemokine receptors. It apparently allows the cells to readily enter nonlymphatic tissues, including the CNS. This occurs about 3–4 days after the original transfer. When the bulk of transferred T cells reach the CNS, they only start the full-blown inflammatory process, when they find their specific antigen and become reactivated (Kawakami et al., 2004). In the early stages of brain inflammation, the vast majority of T cells, present in the inflammatory infiltrates, are derived from the original pool of transferred autoreactive cells (Bauer et al., 1998; Flugel et al., 2001). Within the next days, other leukocytes are recruited into the lesions, consisting of other T cell populations, B lymphocytes, and monocytes.

T cells are effectively cleared from inflammatory lesions of the brain and spinal cord by apoptotic cell death (Pender et al., 1991). At the peak of clinical disease and inflammation, up to 30% of T cells show morphological features of programmed cell death (Schmied et al., 1993). Apoptotic destruction is not only restricted to the pool of autoreactive T cells but also affects the secondarily recruited lymphocytes, which are not reactive for brain antigens (Bauer et al., 1998). A similar mechanism may also account for the removal of B lymphocytes from inflammatory brain lesions (White et al., 2000).

This basic principle of immune surveillance of the brain, which has been developed in the model of EAE, has major general implications. It first shows that all T cells irrespective of their antigen reactivity have access to the brain, provided they are activated in the peripheral immune system. Thus, in the course of a peripheral infection activated T cells may enter the brain and search for their specific antigen. Only, when these cells find their specific antigen in the CNS, they get reactivated and start the inflammatory cascade. In addition, CNS, in contrast to other organs, has developed a very efficient mechanism to terminate inflammation through the apoptotic destruction of T cells (Gold et al., 1997). Although not proven so far, it has been suggested that a defect in the removal of T cells by programmed cell death may in part be responsible for chronic inflammation in MS (Pender, 1998).

It is now well established that the migration of leukocytes through endothelial barriers is mainly controlled by the expression of adhesion molecules, chemokines, and their receptors (Springer, 1994; Luster, 1998). In line with this concept, brain inflammation in EAE can be blocked by systemic application of substances, which block adhesion molecules or chemokine receptors, or in gene-targeted animals with defects of these molecules (Ransohoff, 1999). Pharmacological agents that interfere with leukocyte migration through endothelial barriers are attractive candidates for anti-inflammatory therapy not only in MS but also in other inflammatory diseases.

So far the basic mechanisms of brain inflammation are well characterized in diseases, which are induced and mediated by class II restricted T cells. Whether the same principles account for brain inflammation mediated by CD8$^+$ T cells or for the migration of other cells, like B cells and monocytes, through the BBB has to be determined in the future.

2.3 Multiple and Diverse Immunological Mechanisms are Involved in Tissue Damage

Inflammation in the CNS leads to tissue injury and/or demyelination in a complex interplay between T cells, antibodies, and activated effector cells. Many of the potential mechanisms involved in this process have been elucidated in detailed studies in EAE models.

Class II-restricted T cells are mainly involved in the induction of the inflammatory process and the activation of effector cells. Although CD4$^+$ T cells can be in principle cytotoxic by themselves (Sun and Wekerle, 1986), their cytotoxic function in inflammatory brain lesions is limited. This is due to the fact that MHC class II expression, which is a requirement for specific cytotoxic killing of T cells, is limited in the brain tissue in general to macrophages and microglia. The situation is different in models of brain inflammation, induced by class I MHC-restricted T cells. Autoreactive CD8$^+$ T cells can induce brain inflammation, although so far only few models are available and the mechanisms of tissue injury in these models is not sufficiently clarified (Huseby et al., 2001; Sun et al., 2001). More detailed information is available from one model of brain inflammation, which was induced by hemagglutinin reactive T cells in animals, which express hemagglutinin under the control of the GFAP promoter. These animals too develop transient brain inflammation, which is characterized by selective injury and apoptotic destruction of perivascular astrocytes in the absence of bystander damage of other components of the CNS (Cabarrocas et al., 2003). This suggests that class I-restricted T cells can induce exquisitely specific destruction of cells within the CNS, which express the respective target antigen.

Tissue injury can also be mediated by specific autoantibodies. When such antibodies are present and reach the CNS in an inflammatory environment, highly selective tissue injury is induced. This has been shown convincingly for antibodies against MOG (Linington et al., 1988). Other targets for antibody-mediated demyelination, which have been identified so far in in vitro experiments, are glycolipids, such as for instance galactocerebroside (Dubois-Dalcq et al., 1970). Tissue destruction is mediated either through the activation of complement (Piddlesden et al., 1994) or through the interaction with activated macrophages (Brosnan et al., 1977). Antibody-mediated mechanisms of demyelination play a particularly important role in EAE models, induced in rats (Linington et al., 1988; Storch et al., 1998), guinea pigs (Linington and Lassmann, 1987), and primates (Genain et al., 1999) after sensitization with either whole myelin or MOG.

A third pathway to induce tissue injury in inflammatory lesions is mediated toxic products of activated macrophages and microglia. The extent of macrophage recruitment into EAE lesions largely determines the degree of tissue injury and clinical disease (Berger et al., 1997), and deletion of macrophages or their functional blockade ameliorates or completely blocks the disease (Huitinga et al., 1990). Various different effector molecules of activated macrophages have been identified, which are involved in tissue injury. Thus,

EAE can be blocked by targeting proteases (Brosnan et al., 1980; Gijbels et al., 1994; Clements et al., 1997), complement proteins (Piddlesden et al., 1994), TNF-α (Korner et al., 1997) or its receptors (Eugster et al., 1999; Probert et al., 2000), and, variably with success, by inhibiting the formation of reactive oxygen and nitrogen species (Smith et al., 1999; Willenborg et al., 1999).

Not necessarily related to macrophage function is the observation that tissue injury in EAE can effectively be reduced by AMPA-receptor blockers (Pitt et al., 2000; Smith et al., 2000). These data suggest that excitotoxic mechanisms may play an important role in the induction of immune-mediated brain damage. Whether excitoxicity is induced by uncontrolled glutamate release from injured axons or by production of glutamate or other excitotoxins by activated macrophages and microglia is not yet established.

However, the results obtained in these studies are not always clear-cut and easily interpretable. As an example, blocking the production of NO either by pharmacological means or by deleting i-NOS may both ameliorate or potentiate the disease (Willenborg et al., 1999). Although the mechanisms behind these observations are not fully elucidated, it is clear that reactive nitrogen species have multiple roles in the induction and execution of immune responses and inflammation. Thus, NO blockade in the peripheral immune system may augment disease by inhibiting the mechanisms involved in down-regulation of the immune response, but simultaneously may reduce tissue damage in the lesions, provided an efficient blockade is reached within the CNS compartment.

Thus, in conclusion, multiple different immune-mediated effector mechanisms contribute to demyelination and axonal destruction in EAE lesions. These mechanisms in part act in parallel, and therefore blockade of a single mechanism may be overruled by others. However, as will be discussed below, some data suggest that different effector pathways are preferentially used in different EAE models, depending on the mode of sensitization and the genetic background of the animals.

2.4 EAE is Not EAE: The Diversity of Different Models

Pathology and pathogenesis of EAE is profoundly different in different models of the disease, depending on the method of disease induction, the nature of the sensitizing antigen, or the species or strain of the animal. As mentioned above, disease can either be induced by passive transfer of T cells or by active sensitization. Passive transfer of T cells in general results in an acute, self-limiting inflammatory disease in the CNS with little or no demyelination or tissue injury. By active sensitization, massive demyelination can be induced, provided the sensitizing medium contains antigens, which can induce both encephalitogenic T cells and demyelinating antibodies. This is best achieved presently by using recombinant MOG for sensitization. MBP too elicits after active sensitization a T cell as well as an antibody response. Yet, MBP is located at the cytoplasmic face of the myelin and oligodendrocyte membrane and is thus not accessible for antibodies in vivo in intact myelin sheaths. Thus, animals actively sensitized with MBP develop a dominantly inflammatory disease with little or no demyelination.

Another aspect, which largely determines the outcome of EAE, is the species or strain of the experimental animals. Although no systematic study is available, which directly compares the pathology of EAE in different animal species, it is evident that the pathology of the lesions is very different between mice and other animal species, even when identical sensitization procedures are used. In most mouse strains, a pure T cell-mediated autoimmune response against myelin proteins, such as MBP, PLP, or MOG, results in severe brain inflammation with significant destruction of myelin sheaths and axons. In contrast, in other animal species, such as rats, guinea pigs, and primates, pure T cell-mediated brain inflammation is generally not associated with tissue damage. In addition, auto antibodies are highly effective in promoting demyelination in rats, guinea pigs, and primates, but rather inefficient in most mouse strains. The reasons for these differences are currently unknown, but they may in part be related to mechanisms of effector cell activation and differences in the efficacy of the complement system.

Due to this high degree of heterogeneity between different EAE models and also the heterogeneity of disease pathogenesis in MS patients, the discussion whether EAE is a proper animal model of MS is invalid. EAE can, however, serve as a very useful model to study single aspects of MS pathogenesis in a highly controlled background.

2.5 The Immunological and Pathological Phenotype of EAE is Genetically Determined

Not unlike the situation in MS, certain strains of animals are susceptible for EAE while others are resistant, and EAE susceptibility is genetically determined (Sundvall et al., 1995; Encinas et al., 1996; Olsson et al., 2000). Being a Th-1-mediated autoimmune disease, it is not surprising that disease susceptibility for EAE is to a major part controlled by MHC class II molecules (Weissert et al., 1998). However, even in MHC congenic rat strains profound differences in EAE susceptibility can be seen, which are regulated by genes in the non-MHC background genome. As MS, EAE too is a highly polygenic disease and the risk is conferred by multiple genes, each of them having only a mild to moderate impact (Sundvall et al., 1995; Encinas et al., 1996; Olsson et al., 2000). Several of these genomic regions, which have been shown to confer disease susceptibility for EAE, are synthetic to regions which appeared in genomic screens of MS patients. So far only the gene regions are known, but it is expected that individual genes which are associated with disease susceptibility in EAE will be identified soon.

Besides regulating disease susceptibility, the genetic background may also modify the disease phenotype. As discussed above, most antibodies against MOG are directed against linear peptides of the molecule, which are not accessible for antibodies on the intact protein in vivo. Only a specific subgroup of antibodies, directed against a conformational epitope of the MOG molecule can induce demyelination. What epitope of MOG is recognized by antibodies after sensitization is at least in part controlled by genes in the MHC region (Bourquin et al., 2003). Following sensitization with MOG, mice with an H-2^s background develop demyelinating antibodies against the conformational MOG epitope and show profound antibody-mediated demyelination in the CNS. In contrast, in mice with a H-2^b background only nondemyelinating antibodies against linear peptide epitopes are formed, and the lesions in the CNS only show T cell-mediated and macrophage-mediated tissue injuries. This is the first example in EAE, showing that the dominant immunopathological pathway of tissue destruction is directly regulated by a single gene region. What gene(s) in the MHC are responsible for this effect are not clear yet, but it has to be considered that the genes for MOG itself as well as for several MOG homologs are located within the MHC region. Other genes within the MHC region appear to control macrophage recruitment versus microglia activation in EAE lesions (Storch et al., 2002).

Another observation, which may be directly relevant for MS is that mice with a deletion of the gene for ciliary neurotrophic factor have more severe EAE and more demyelination and oligodendrocyte injury within the lesions compared to wild-type controls (Linker et al., 2002). This observation may have a direct counterpart in MS. About 3% of the Western population have a deletion of the CNTF gene. This alone does not affect life expectancy or induce neurological disease. However, MS patients with a deletion of the CNTF gene suffer from an earlier disease onset and more severe disease, compared to those without this gene defect (Giess et al., 2002). These results show in principle that genes, which regulate the homeostasis of the target tissue, may influence disease outcome in EAE as well as MS.

Besides these examples, which appear to be directly relevant for MS, EAE has been used in many studies as a model to test the effect of the deletion or transgenic overexpression of different genes on clinical disease incidence and phenotype (Owens et al., 2001). Although these studies have provided many interesting data on the regulation of autoimmunity or the mechanisms of immune-mediated tissue damage, it would go beyond the scope of this chapter to discuss them in detail here.

2.6 What Aspects of Multiple Sclerosis Have So Far Not Been Reproduced in EAE Models?

As discussed before, EAE serves as an excellent model to study basic aspects of brain inflammation and immune-mediated tissue injury. Furthermore, in certain EAE models, focal inflammatory demyelinating plaques in the CNS develop, which in most essential aspects of pathology reflect focal white matter lesions in MS patients (Lassmann, 1983; Storch et al., 1998b). There are, however, several important features of MS pathology and pathogenesis, which so far are not covered in EAE models (Lassmann and Ransohoff, 2004).

With the exception of some highly artificial genetic models, a spontaneous occurrence of EAE in the absence of active sensitization has never been observed. In addition, EAE in general tends to be a self-limiting disease, which terminates either after one acute episode or after several months of chronic disease. EAE can thus give no answers regarding the mechanisms, which trigger MS and what is responsible for propagating a chronic disease, which lasts for years or decades.

A second problem relates to the question, what T cell populations are responsible for induction and propagation of the disease. The driving force of the disease in nearly all EAE models is a T cell-mediated autoimmune reaction, accomplished by class II-restricted T cells, which are polarized toward the Th-1 phenotype. It is these cells which dominate the inflammatory lesions at least at early stages of the disease and are clonally expanded. This contrasts the situation in MS, where class I-restricted cells dominate in all stages of the lesions and preferentially show clonal expansion. It has been notoriously difficult so far to induce class I-restricted T cell-mediated autoimmunity against CNS tissue. Whether this is due to technical reasons or whether there is a general conceptional problem is up to now unresolved.

Finally, and most importantly, EAE so far only reproduces the events, which occur in focal white matter lesions in MS. No model up to now has shown the slowly progressive cortical demyelination and the massive diffuse white matter injury, which are the key features of the pathology of progressive MS. This is unfortunate, since it precludes the systematic study of pathogenetic events, which are instrumental during this phase of MS. As far as the decades of previous EAE research are concerned, it may turn out that such diffuse brain injury will never appear on classical Th-1-mediated EAE models.

Thus, in conclusion, EAE is a perfect model to study basic mechanisms of brain inflammation, which are highly relevant for MS pathogenesis. The herterogeneity of both MS as well as EAE, however, precludes that a single EAE model is representative for MS in a global sense. When EAE is used to study certain aspects of MS pathogenesis, it is of key importance to select out of a large spectrum the right model, which is most suitable to address the specific question. Finally, the pathogenesis of MS is far more complex than that of EAE and key aspects of MS are not covered in the available models. This implies that in the context of therapeutic studies, EAE is a good model for proof of principle that a certain disease mechanism is important and that it can be addressed with a specific therapeutic agent. Whether it is beneficial in MS patients can only be decided after having performed proper clinical trials.

References

Aboul-Enein F, Lassmann H. 2004. Mitochondrial damage and histotoxic hypoxia: A pathway of tissue injury in inflammatory brain disease? Acta Neuropathol, submitted.

Aboul-Enein F, Rauschka H, Kornek B, Stadelmann C, Stefferl A, et al. 2003. Preferential loss of myelin associated glycoprotein reflects hypoxia-like white matter damage in stroke and inflammatory brain diseases. J Neuropathol Exp Neurol 62: 25-33.

Allen IV, McKeown SR. 1979. A histological, histochemical and biochemical study of the macroscopically normal white matter in multiple sclerosis. J Neurol Sci 41: 81-91.

Allen IV, McQuid S, Miradkhur M, Nevin G. 2001. Pathological abnormalities in the normal-appearing white matter in multiple sclerosis. Neurol Sci 22: 141-144.

Anderson DW, Ellenberg JH, Leventhal CM, Reingold SC, Rodriguez M, et al. 1992. Revised estimate of the prevalence of multiple sclerosis in the United States. Ann Neurol 31: 333-336.

Anthony DC, Miller KM, Fearn S, Townsend MJ, Opdenakker G, et al. 1998. Matrix metalloproteinase expression in an experimentally-induced DTH model of multiple sclerosis in the rat CNS. J Neuroimmunol 87: 62-72.

Arnold DL, Matthews PM, Francis GS, O'Connor J, Antel JP. 1992. Proton magnetic resonance spectroscopic imaging for metabolic characterization of demyelinating plaques. Ann Neurol 31: 235-241.

Babbe H, Roers A, Waisman A, Lassmann H, Goebels N, et al. 2000. Clonal expansion of CD8+T cells dominate the T cell infiltrate in active multiple sclerosis lesions as shown by micromanipulation and single cell polymerase chain reaction. J Exp Med 192: 393-404.

Barnett MH, Prineas JW. 2004. Relapsing and remitting multiple sclerosis: Pathology of the newly forming lesion. Ann Neurol 55: 458-468.

Battistini L, Fischer FR, Raine CS, Brosnan CF. 1996. CD1b is expressed in multiple sclerosis lesions. J Neuroimmunol 67: 145-151.

Bauer J, Bradl M, Hickey WF, Forss-Petter S, Breitschopf H, et al. 1998. T cell apoptosis in inflammatory brain lesions. Destruction of T cells does not depend on antigen recognition. Am J Pathol 153: 715-724.

Bechtold DA, Kapoor R, Smith K. 2004. Axonal protection using lecainide in experimental autoimmune encephalomyelitis. Ann Neurol 55: 607-616.

Beltran B, Mathur A, Duchen MR, Erusalimsky JD, Moncada S. 2000. The effect of nitric oxide on cell respiration: A key to understanding its role in cell survival or death. Proc Natl Acad Sci USA 97: 14602-14607.

Ben Nun A, Wekele H, Cohen IR. 1981. The rapid isolation of clonable antigen-specific T lymphocyte lines capable of mediating autoimmune encephalomyelitis. Eur J Immunol 11: 195-199.

Berger T, Rubner P, Schautzer F, Egg R, Ulmer H, et al. 2003. Antimyelin antibodies as a predictor of clinically definite multiple sclerosis after the first demyelinating event. New Engl J Med 349: 139-145.

Berger T, Weerth S, Kojima K, Linington C, Wekerle H, et al. 1997. Experimental autoimmune encephalomyelitis: The antigen specificity of T-lymphocytes determines the topography of lesions in the central and peripheral nervous system. Lab Invest 76: 355-364

Bitsch A, Kuhlmann T, Stadelmann C, Lassmann H, Lucchinetti C, et al. 2001. A longitudinal MRI study of histopathologically defined hypointense multiple sclerosis lesions. Ann Neurol 49: 793-796.

Bitsch A, Schuchardt J, Bunkowski S, Kuhlmann T, Bruck W. 2000. Acute axonal injury in multiple sclerosis. Correlation with demyelination and inflammation. Brain 123: 1174-1183.

Bjartmar C, Kidd G, Mork S, Rudick R, Trapp BD. 2000. Neurological disability correlates with spinal cord axonal loss and reduced N-acetyl aspartate in chronic multiple sclerosis patients. Ann Neurol 48: 893-901.

Blakemore WF, Keirstead HS. 1999. The origin of remyelinating cells in the central nervous system. J Neuroimmunol 98: 69-76.

Bo L, Vedeler CA, Nyland H, Trapp BD, Mork SJ. 2003b. Intracortical multiple sclerosis lesions are not associated with increased lymphocyte infiltration. Mult Scler 9: 323-331.

Bo L, Vedeler CA, Nyland HI, Trapp BD, Mork SJ. 2003a. Subpial demyelination in the cerebral cortex of multiple sclerosis. J Neuropathol Exp Neurol 62: 723-732.

Bolanos JP, Almeida A, Stewart V, Peuchen S, Land JM, et al. 1997. Nitric oxide-mediated mitochondrial damage in the brain: Mechanisms and implications for neurodegenerative diseases. J Neurochem 68: 2227-2240.

Bonetti B, Raine CS. 1997. Multiple sclerosis: Oligodendrocytes display cell death-related molecules in situ but do not undergo apoptosis. Ann Neurol 42: 74-84.

Bonetti B, Stegagno C, Cannella B, Rizzuto N, Moretto G, et al. 1999. Activation of NF-kappaB and c-jun transcription factors in multiple sclerosis lesions. Implications for oligodendrocyte pathology. Am J Pathol 155: 1433-1438.

Bourquin C, Schubart A, Tobollik S, Mather I, Ogg S, et al. 2003. Selective unresponsiveness to conformational B cell epitopes of the myelin oligodendrocyte glycoprotein in H-2^b mice. J Immunol 171: 455-461.

Brosnan CF, Cammer W, Norton WT, Bloom BR. 1980. Proteinase inhibitors suppress the development of experimental allergic encephalomyelitis. Nature 285: 235-237.

Brosnan CF, Stoner GL, Bloom BR, Wisniewski HM. 1977. Studies on demyelination by activated lymphocytes in the rabbit eye. II. Antibody dependent cell mediated demyelination. J Immunol 118: 2103-2111.

Brownell B, Hughes JT. 1962. The distribution of plaques in the cerebrum in multiple sclerosis. J Neurol Neurosurg Psychiatry 25: 315-320.

Brück W, Bitsch A, Kolenda H, Brück Y, Stiefel M, et al. 1997. Inflammatory central nervous system demyelination: Correlation of magnetic resonance imaging findings with lesion pathology. Ann Neurol 42: 783-793.

Brück W, Porada P, Poser S, Riechmann P, Hanefeld F, et al. 1995. Monocyte/macrophage differentiation in early multiple sclerosis lesions. Ann Neurol 38: 788-796.

Brück W, Schmied M, Suchanek G, Brück Y, Breitschopf H, et al. 1994. Oligodendrocytes in the early course of multiple sclerosis. Ann Neurol 35: 65-73.

Cabarrocas J, Bauer J, Piaggio E, Liblau R, Lassmann H. 2003. Effective and selective immune surveillance of the brain by MHC class I-restricted cytotoxic T lymphocytes. Eur J Immunol 33: 1174-1182.

Cannella B, Raine CS. 1995. The adhesion molecule and cytokine profile of multiple sclerosis lesions. Ann Neurol 37: 424-435.

Chang A, Tourtellotte WW, Rudick R, Trapp BD. 2002. Premyelinating oligodendrocytes in chronic lesions of multiple sclerosis. New Engl J Med 346: 165-173.

Charcot JM. 1868. Histologie de la sclerose en plaque. Gaz Hopital (Paris) 41: 554-566.

Chard DT, Griffin CM, Parker GJ, Kapoor R, Thompson AJ, et al. 2002. Brain atrophy in clinically early relapsing-remitting multiple sclerosis. Brain 125: 327-337.

Charles P, Reynolds R, Seilhean D, Rougon G, Aigrot MS, et al. 2002. Re-expression of PSA-NCAM by demyelinated axons: An inhibitor or remyelination in multiple sclerosis? Brain 125: 1972-1979.

Clements JM, Cossins JA, Wells GM, Corkill DJ, Helfrich K, et al. 1997. Matrix metalloproteinase expression during experimental autoimmune encephalomyelitis and effects of a combined matrix metalloproteinase and tumour necrosis factor-alpha inhibitor. J Neuroimmunol 74: 85-94.

Compston A, Coles A. 2002. Multiple sclerosis. Lancet 359: 1221-1231.

Compston DA, Morgan BP, Campbell AK, Wilkins P, Cole G, et al. 1989. Immunocytochemical localization of the terminal complement complex in multiple sclerosis. Neuropathol Appl Neurobiol 15: 307-316.

Compston DAS. 1998. Distribution of multiple sclerosis. Mc Alpine's Multiple Sclerosis, 3rd edn. Compston DAS, editor. London: Churchill Livingstone; pp. 63-100.

Compston DAS. 2004. Genetic susceptibility and epidemiology. Myelin Biology and Disorders, Vol. 2. Lazzarini RA, editor. Amsterdam: Elsevier; pp. 701-731.

Confavreux C, Vukusic S, Moreau T, Adeleine P. 2000. Relapses and progression of disability in multiple sclerosis. New Engl J Med 343: 1430-1438.

Cuzner ML, Gveric D, Strand C, Loughlin AJ, Paemen L, et al. 1996. The expression of tissue-type plasminogen-activator, matrix metalloproteinases and endogenous inhibitors in the central nervous system in multiple sclerosis: Comparison of stages of lesion evolution. J Neuropathol Exp Neurol 55: 1194-1204.

Cuzner ML, Opdenakker G. 1999. Plasminogen activators and matrix metalloproteinases, mediators of extracellular proteolysis in inflammatory demyelination of the central nervous system. J Neuroimmunol 94: 1-14.

D'Souza SD, Bonetti B, Balasingam V, Cashman NR, Barker BA, et al. 1996. Multiple sclerosis: Fas signaling in oligodendrocyte death. J Exp Med 184: 2361-2370.

Dawson JW. 1916. The histology of disseminated sclerosis. Trans R Soc 50: 517-540.

De Groot CJ, Bergers E, Kamphorst W, Ravid R, Polman CH, et al. 2001. Post-mortem MRI-guided sampling of multiple sclerosis brain lesions: Increased yield of active demyelinating and (p)reactive lesions. Brain 124: 1635-1645.

De Groot CJ, Ruuls SR, Theeuwes JW, Dijkstra CD, van der Valk P. 1997. Immunocytochemical characterization of the expression of inducible and constitutive isoforms of nitric oxide synthase in demyelinating multiple sclerosis lesions. J Neuropathol Exp Neurol 56: 10-20.

De Stefano N, Narayanan S, Matthews PM, Francis GS, Antel JP, et al. 1999. In vivo evidence for axonal dysfunction remote from focal cerebral demyelination of the type seen in multiple sclerosis. Brain 122: 1933-1999.

Dean G. 1967. Annual incidence, prevalence and mortality of MS in white South African-born and in white immigrants to South Africa. Br Med J 2: 724-730.

Diemel LT, Copelman CA, Cuzner ML. 1998. Macrophages in CNS remyelination: Friend or foe? Neurochem Res 23: 341-347.

Dore-Duffy P, Washington R, Dragovic L. 1993. Expression of endothelial cell activation antigens in microvessels from patients with multiple sclerosis. Adv Exp Med Biol 331: 243-248.

Dowling P, Shang G, Raval S, Menona J, Cook S, et al. 1996. Involvement of the CD95 (APO1/Fas) receptor/ligand system in multiple sclerosis brain. J Exp Med 184: 1513-1518.

Dubois-Dalcq M, Niedieck B, Buyse M. 1970. Action of anti-cerebroside sera on myelinated nervous tissue cultures. Pathol Europea 5: 331-347.

Ebers GC, Kukay K, Bulman DE, Sadovnik AD, Rice G, et al. 1996. A full genome search in multiple sclerosis. Nat Genet 13: 472-476.

Eikelenboom M, Petzold A, Lazeron RH, Silber E, Sharief M, et al. 2003. Multiple sclerosis: Neurofilament light chain antibodies are correlated to cerebral atrophy. Neurology 60: 219-223.

Elian M, Nightingale S, Dean G. 1990. Multiple sclerosis among United Kingdom born children of immigrants from the West Indies. J Neurol Neurosurg Psychiatry 53: 906-911.

Encinas JA, Weiner HL, Kuchroo VK. 1996. Inheritance of susceptibility to experimental autoimmune encephalomyelitis. J Neurosci Res 45: 655-669.

Esiri MM. 1977. Immunoglobulin-containing cells in multiple sclerosis plaques. Lancet 2: 478-480.

Esiri MM. 1980. Multiple sclerosis: A quantitative and qualitative study of immunoglobulin-containing cells in the central nervous system. Neuropathol Appl Neurobiol 6: 9-21.

Eugster HP, Frei K, Bachmann R, Bluethmann H, Lassmann H, et al. 1999. Severity of symptoms and demyelination in MOG-induced EAE depends on TNFR1. Eur J Immunol 29: 626-632.

Evangelou N, Konz D, Esiri MM, Smith S, Palace J, et al. 2000. Regional axonal loss in the corpus callosum correlates with cerebral white matter lesion volume and distribution in multiple sclerosis. Brain 123: 1845-1849.

Evangelou N, Konz D, Esiri MM, Smith S, Palace J, et al. 2001. Size-selective neuronal changes in the anterior optic pathways suggest a differential susceptibility to injury in multiple sclerosis. Brain 124: 1813-1820.

Evans CF, Horwitz MS, Hobbs MV, Oldstone MB. 1996. Viral infection of transgenic mice expressing a viral protein in oligodendrocytes leads to chronic central nervous system autoimmune disease. J Exp Med 184: 2371-2384.

Fazekas F, Strasser Fuchs S, Schmidt H, Enzinger C, Ropele S, et al. 2000. Apolipoprotein E genotype related differences in brain lesions of multiple sclerosis. J Neurol Neurosurg Psychiatry 69: 25-28.

Ferguson B, Matyszak MK, Esiri MM, Perry VH. 1997. Axonal damage in acute multiple sclerosis lesions. Brain 120: 393-399.

Fern R, Moller T. 2000. Rapid ischemic cell death in immature oligodendrocytes: A fatal glutamate release feedback loop. J Neurosci 20: 34-42.

Filippi M, Bozzali M, Rovaris M, Gonen O, Kesavadas C, et al. 2003. Evidence for widespread axonal damage at the earliest clinical stage of multiple sclerosis. Brain 126: 433-437.

Filippi M, Rocca MA, Martinc G, Horsfield MA, Comi G. 1998. Magnetization transfer changes in the normal appearing white matter precede the appearance of enhancing lesions in patients with multiple sclerosis. Ann Neurol 43: 809-814.

Flugel A, Berkowicz T, Ritter T, Labeur M, Jenne DE, et al. 2001. Migratory activity and functional changes of green fluorescent effector cells before and during experimental autoimmune encephalomyelitis. Immunity 14: 547-560.

Fog T. 1950. Topographic distribution of plaques in the spinal cord of multiple sclerosis. Arch Neurol 63: 382-414.

Franklin RJ. 2002. Why does remyelination fail in multiple sclerosis ? Nat Rev Neurosci 3: 705-714.

Furlan R, Brambilla E, Sanvito F, Roccatagliata C, Olivieri S, et al. 2003. Vaccination with amyloid-beta peptide indices autoimmune encephalomyelitis in C57/Bl6 mice. Brain 126: 285-291.

GAMES and The Transatlantic Multiple Sclerosis Genetics Cooperative. 2003. A meta-analysis of genome screens in multiple sclerosis. J Neuro immunol 143: 39-46.

Ganter P, Prince C, Esiri MM. 1999. Spinal cord axonal loss in multiple sclerosis: A post-mortem study. Neuropathol Appl Neurobiol 25: 459-467.

Gay FW, Drye GW, Dick GWA, Esiri MM. 1997. The application of multifactorial cluster analysis in the staging of plaques in early multiple sclerosis: Identification and characterization of the primary demyelinating lesion. Brain 120: 1461-1483.

Genain CP, Cannella B, Hauser SL, Raine CS. 1999. Autoantibodies to MOG mediate myelin damage in MS. Nat Med 5: 170-175.

Gerritse K, Laman JD, Noelle RJ, Aruffo A, Ledbetter JA, et al. 1996. CD40-CD40 ligand interactions in experimental allergic encephalomyelitis and multiple sclerosis. Proc Natl Acad Sci USA 93: 2499-2504.

Giess R, Maurer M, Linker R, Gold R, Warmuth-Metz M, et al. 2002. Association of a null mutation in the CNTF gene with early onset of multiple sclerosis. Arch Neurol 59: 407-409.

Gijbels K, Galardy RE, Steinman L. 1994. Reversal of experimental autoimmune encephalomyelitis with a hydroxamate inhibitor of matrix metalloproteases. J Clin Invest 94: 2177-2182.

Gold R, Hartung HP, Lassmann H. 1997. T-cell apoptosis in autoimmune diseases: Termination of inflammation in the nervous system and other sites with specialized immune-defense mechanisms. Trends Neurosci 20: 399-404.

Graumann U, Reynolds R, Steck AJ, Schaeren-Wiemers N. 2003. Molecular changes in normal appearing white matter in multiple sclerosis are characteristic of neuroprotective mechanisms against hypoxic insult. Brain Pathol 13: 554-573.

Haase CG, Guggenmos J, Brehm U, Andersson M, Olsson T, et al. 2001. The fine specificity of the myelin oligodendrocyte glycoprotein autoantibody response in patients with multiple sclerosis and normal healthy controls. J Neuroimmunol 114: 220-225.

Haines JL and The Multiple Sclerosis Genetics Group. 1996. A complete genomic screen for multiple sclerosis underscores a role for the major histocompatibility complex. Nat Genet 13: 469-471.

Herrera BM, Ebers GC. 2003. Progress in deciphering the genetics of multiple sclerosis. Curr Opin Neurol 16: 253-258.

Hickey WF, Hsu BL, Kimura H. 1991. T lymphocyte entry into the central nervous system. J Neurosci Res 28: 254-260.

Hofman FM, Hinton DR, Johnson K, Merrill JE. 1989. Tumor necrosis factor identified in multiple sclerosis brain. J Exp Med 170: 607-612.

Hoftberger R, Aboul-Enein F, Brueck W, Lucchinetti C, Rodriguez M, et al. 2004. Expression of major histocompatibiltiy complex class I molecules on the different cell types in multiple sclerosis lesions. Brain Pathol 14: 43-50.

Hohlfeld R. 1997. Biotechnological agents for the immunotherapy of multiple sclerosis. Principles, problems and perspectives. Brain 120: 865-916.

Huang D, Han Y, Rani MR, Glabinski A, Trebst C, et al. 2000. Chemokines and chemokine receptors in inflammation of the nervous system: Manifold roles and exquisite regulation. Immunol Rev 177: 52-67.

Huitinga I, Rooijen N, deGroot CJA, Uitdehaag BMJ, Dijkstra CD. 1990. Suppression of experiemntal allergic encephalomyelitis after elimination of macrophages. J Exp Med 172: 1025-1033.

Huseby ES, Liggitt D, Brabb T, Schnabel B, Ohlen C, et al. 2001. A pathogenic role for myelin-specific CD8 (+) T-cells in a model for multiple sclerosis. J Exp Med 194: 669-676.

John GR, Shankar SL, Shafit-Zagardo B, Massimi A, Lee SC, et al. 2002. Multiple sclerosis: Re-expression of a developmental pathway that restricts oligodendrocyte maturation. Nat Med 8: 1115-1121.

Johns TG, Bernard CC. 1997. Binding of complement C1q to myelin oligodendrocyte glycoprotein: A novel mechanism for regulating CNS inflammation. Mol Immunol 34: 33-38.

Kahana E, Zilber N, Abramson JH, Biton Y, Leibowitz Y, et al. 1994. Multiple sclerosis: Genetic versus environmental aetiology: Epidemiology in Israel updated. J Neurol 241: 341-346.

Kalman B, Leist TP. 2003. A mitochondrial component of neurodegeneration in multiple sclerosis. Neuromolecular Med 3: 147-158.

Kalman B, Lublin FD. 1999. The genetics of multiple sclerosis. A review. Biomed Pharmacother 53: 358-370.

Kapoor R, Davies M, Smith KJ. 1999. Temporary axonal conduction block and axonal loss in inflammatory neurological disease. A potential role for nitric oxide? Ann N Y Acad Sci 893: 304-308.

Kawakami N, Lassmann S, Li Z, Odoardi F, Ritter T, et al. 2004. The activation status of neuroantigen specific T cells in the target organ determines the clinical outcome of autoimmune encephalomyelitis. J Exp Med 199: 185-197.

Keirstead HS, Blakemore WF. 1999. The role of oligodendrocytes and oligodendrocyte progenitors in CNS remyelination. Adv Exp Med Biol 468: 183-197.

Kellar-Wood H, Robertson N, Govan GG, Harding AE, Compston DAS. 1994. Leber's hereditary optic neuropathy mitochondrial DNA mutations in multiple sclerosis. Ann Neurol 36: 109-112.

Kerlero de Rosbo N, Milo R, Lees MB, et al., 1993. Reactivity to myelin antigens in multiple sclerosis. J Clin invest 92: 2602-2608.

Kerschensteiner M, Gallmeier E, Behrens L, Leal VV, Misgeld T, et al. 1999. Activated human T cells B cells, and monocytes produce brain-derived neurotrophic factor in vitro and in inflammatory brain lesions: A neuroprotective role of inflammation? J Exp Med 189: 865-870.

Kidd T, Barkhof F, McConnell R, Algra PR, Allen IV, et al. 1999. Cortical lesions in multiple sclerosis. Brain 122: 17-26.

Kivisäkk P, Mahad DJ, Callahan MK, Sikora K, Trebst C, et al. 2004. Intrathecal expression of CCR7 in multiple sclerosis. Implications for central nervous system immunity. Ann Neurol 55: 627-638.

Kojima K, Berger Th, Lassmann H, Hinze-Selch D, Zhang Y, et al. 1994. Experimental autoimmune panencephalitis and uveoretinitis transfered to the Lewis rat by T-lymphocytes specific for the S100β molecule, a calcium binding protein of astroglia. J Exp Med 180: 817-829.

Kornek B, Lassmann H. 1999. Axonal pathology in multiple sclerosis: A historical note. Brain Pathol 9: 651-656.

Kornek B, Storch M, Weissert R, Wallstroem E, Stefferl A, et al. 2000. Multiple sclerosis and chronic autoimmune encephalomyelitis: A comparative quantitative study of axonal injury in active, inactive and remyelinated lesions. Am J Pathol 157: 267-276.

Kornek B, Storch MK, Bauer J, Djamshidian A, Weissert R, et al. 2001. Distribution of calcium channel subunit in dystrophic axons in multiple sclerosis and experimental autoimmune encephalomyelitis. Brain 124: 1114-1124.

Korner H, Lemckert FA, Chaudhri G, Etteldorf S, Sedgwick JD. 1997. Tumor necrosis factor blockade in actively induced experimental autoimmune encephalomyelitis prevents clinical disease despite activated T cell infiltration to the central nervous system. Eur J Immunol 27: 1973-1981.

Kotter MR, Setzu A, Sim FJ, Van-Rooijen N, Franklin RJ. 2001. Macrophage depletion impairs oligodendrocyte remyelination following lysolecithin-induced demyelination. Glia 35: 204-212.

Krogsgaard M, Wucherpfennig KW, Cannella B, Hansen BE, Svejgaard A, et al. 2000. Visualization of myelin basic protein (MBP) T cell epitopes in multiple sclerosis lesions using a monoclonal antibody specific for the human histocompatibility leukocyte antigen (HLA)-DR2-MBP 85-99 complex. J Exp Med 191: 1395-1412.

Lassmann H. 1983. Comparative neuropathology of chronic experimental allergic encephalomyelitis and multiple sclerosis. Springer Schriftenr Neurol 25: 1-135.

Lassmann H, Brück W, Lucchinetti C. 2001. Heterogeneity of multiple sclerosis pathogenesis: Implications for diagnosis and therapy. Trends Mol Med 7: 115-121.

Lassmann H, Raine CS, Antel J, Prineas JW. 1998. Immunopathology of multiple sclerosis: Report on an international meeting held at the Institute of Neurology of the University of Vienna. J Neuroimmunol 86: 213-217.

Lassmann H, Ransohoff RM. 2004. The CD4-Th1 model for multiple sclerosis: A crucial re-appraisal. Trends Immunol 25: 132-137.

Leary SM, Thompson AJ. 2003. Treatment for patients with primary progressive multiple sclerosis. Multiple Sclerosis Therapeutics. Cohen JA, Rudick RA, editors. London: Martin Dunitz; pp. 589-598.

Ligers A, Xu C, Saarinen S, Hillert J, Olerup O. 1999. The CTLA-4 gene is associated with multiple sclerosis. J Neuroimmunol 97: 182-190.

Linington C, Bradl M, Lassmann H, Brunner C, Vass K. 1988. Augmentation of demyelination in rat acute allergic encephalomyelitis by circulating antibodies against a myelin oligodendrocyte glycoprotein. Am J Pathol 130: 443-454.

Linington C, Lassmann H. 1987. Antibody responses in chronic relapsing experimental allergic encephalomyelitis: Correlation of serum demyelinating activity with the antibody titre to the myelin/oligodendrocyte glycoprotein (MOG). J Neuroimmunol 17: 61-69.

Linker RA, Maurer M, Gaupp S, Martini R, Holtmann B, et al. 2002. CNTF is a major protective factor in demyelinating

CNS disease: A neurotrophic cytokine as modulator in neuroinflammation. Nat Med 8: 620-524.

Lipton SA. 1998. Neuronal injury associated with HIV-1: approaches and treatment. Annu Rev Pharmacol Toxicol 38: 159-177.

Liu JS, Zhao ML, Brosnan CF, Lee SC. 2001. Expression of inducible nitric oxide synthase and nitrotyrosine in multiple sclerosis lesions. Am J Pathol 158: 2057-2066.

Losseff NA, Wang L, Lai HM, Yoo DS, Gawne-Cain ML, et al. 1996. Progressive cerebral atrophy in multiple sclerosis. A serial MRI study. Brain 119: 2009-2019.

Lovas G, Szilagyi N, Majtenyi K, Palkovits M, Komoly S. 2000. Axonal changes in chronic demyelinated cervical spinal cord plaques. Brain 123: 308-317.

Lu F, Selak M, O'Connor J, Croul S, Lorenzana C, et al. 2000. Oxidative damage to mitochondrial DNA and activity of mitochondrial enzymes in chronic active lesions of multiple sclerosis. J Neurol Sci 177: 95-103.

Lu ZH, Chakraborty G, Ledden RW, Yahya D, Wu G. 2004. N-acetylaspartate synthase is bimodally expressed in microsomes and mitochondria of brain. Brain Res Mol Brain Res 122: 71-78.

Lublin FD, Reingold SC. 1996. Defining the clinical course of multiple sclerosis: Results of an international survey. Neurology 46: 907-911.

Lucchinetti C, Brück W, Parisi J, Scheithauer B, Rodriguez M, et al. 1999. A quantitative analysis of oligodendrocytes in multiple sclerosis lesions. A study of 117 cases. Brain 122: 2279-2295.

Lucchinetti C, Brück W, Parisi J, Scheithauer B, Rodriguez M, et al. 2000. Heterogeneity of multiple sclerosis lesions: Implications for the pathogenesis of demyelination. Ann Neurol 47: 707-717.

Lucchinetti CF, Mandler R, McGavern D, Brück W, Gleich G, et al. 2002. A role for humoral mechanisms in the pathogenesis of Devic's neuromyelitis optica. Brain 125: 1450-1461.

Ludwin SK. 1980. Chronic demyelination inhibits remyelination in the central nervous system. An analysis of contributing factors. Lab Invest 43: 382-387.

Lumsden CE. 1970. The neuropathology of multiple sclerosis. Handbook of Clinical Neurology, Vol. 9. Vinken PI, Bruyn GW, editors. New York: Elsevier; pp. 217-309.

Luster AD. 1998. Chemokines: Chemotactic cytokines that mediate inflammation. N Engl J Med 338: 436-445.

Maeda A, Sobel RA. 1996. Matrix metalloproteinases in the normal human central nervous system, microglia nodules and multiple sclerosis lesions. J Neuropathol Exp Neurol 55: 300-309.

Mahad DJ, Trebst C, Kivisäkk P, Staugaitis SM, Tucky B, et al. 2004. Expression of chemokine receptors CCR1 and CCR5 reflects differential activation of mononuclear phagocytes

in pattern II and III multiple sclerosis lesions. J Neuropathol Exp Neurol 63: 262-273.

Marburg O. 1906. Die sogenannte "akute Multiple Sklerose." Jahrb Psychiatrie 27: 211-312.

Marrosu MG, Muntoni F, Murru MR, Costa G, Oischedda MP, et al. 1992. HLA-DQB1 genotype in Sardinian multiple sclerosis: Evidence for a key role of DQB1 *0201 and *0302 alleles. Neurology 42: 883-886.

McDonald WI, Compston A, Edan G, Goodkin D, Hartung HP, et al. 2001. Recommended diagnostic criteria for multiple sclerosis. Guidelines from the international panel on the diagnosis of multiple sclerosis. Ann Neurol 50: 121-127.

McDonald WI, Miller DH. 1996. Serial gadolinium-enhanced MRI of the brain and spinal cord in early relapsing-remitting multiple sclerosis. Neurology 46: 373-378.

McDonough J, Dutta R, Gudz T, et al. 2003. Decreases in GABA and mitochondrial genes are implicated in MS cortical pathology through microarray analysis of postmortem MS cortex. Soc Neurosci Abstr 213.212.

Menard A, Pierig R, Pelletier J, Bensa P, Belliveau J, et al. 1998. Detection of a gliotoxic activity in the cerebrospinal fluid from multiple sclerosis patients. Neurosci Lett 245: 49-52.

Merrill JE. 1992. Proinflammatory and antiinflammatory cytokines in multiple sclerosis and central nervous system acquired immunodeficiency syndrome. J Immunother 12: 167-170.

Miller DH, Kesselring J, McDonald WI, Paty DW, Thompson AJ. 1997. Magnetic resonance in multiple sclerosis. Cambridge, United Kingdom: Cambridge University Press.

Moalem G, Leibowitz-Amit R, Yoles E, Mor F, Cohen IR, et al. 1999. Autoimmune T cells protect neurons from secondary degeneration after central nervous system axotomy. Nat Med 5: 49-55.

Mojon D, Fujihara K, Hirano M, Miller C, Lincoff N, et al. 1999. Leber's hereditary optic neuropathy mitochondrial DNA mutations in familial multiple sclerosis. Graefes Arch Clin Exp Ophtalmol 237: 348-350.

Mussini JM, Hauw JJ, Escourolle R. 1977. Immunofluorescence studies of intra cytoplasmatic immunoglobulin binding lymphoid cells in the central nervous system. Report of 32 cases including 19 multiple sclerosis. Acta Neuropathol (Berl) 40: 227-232.

Narayanan D, De Stefano N, Francis GS, Arnoutelis R, Caramanos Z, et al. 2001. Axonal metabolic recovery in multiple sclerosis patients treated with interferon beta-1b. J Neurol 248: 979-986.

Neumann H, Medana I, Bauer J, Lassmann H. 2002. Cytotoxic T lymphocytes in autoimmune and degenerative CNS diseases. Trends Neurosci 25: 313-319.

Niehaus A, Shi J, Grzenkowski M, Diers-Fenger M, Hartung HP, et al. 2000. Patients with active relapsing-remitting multiple sclerosis synthesize antibodies recognizing oligodendrocyte progenitor cell surface protein: Implications for remyelination. Ann Neurol 48: 362-371.

Oksenberg JR, Panzara MA, Begovich AB, Mitchell D, Erlich HA, et al. 1993. Selection for T-cell receptor V beta-D beta-J beta gene rearrangements with specificity for a myelin basic protein peptide in brain lesions of multiple sclerosis. Nature 362: 68-70.

Oldstone MB, Southern PJ. 1993. Trafficking of activated cytotoxic T lymphocytes into the central nervous system: Use of a transgenic model. J Neuroimmunol 46: 25-31.

Olerup O, Hillert J. 1991. HLA class II-associated susceptibility in multiple sclerosis. A critical evaluation. Tissue Antigens 38: 1-15.

Olsson T, Dahlman I, Wallstrom E, Weissert R, Piehl F. 2000. Genetics of rat neuroinflammation. J Neuroimmunol 107: 191-200.

Olsson T, Zhi WW, Höjeberg B, Kostulas V, Yu-Ping J, et al. 1990. Autoreactive T lymphocytes in multiple sclerosis determined by secretion of interferon-γ. J Clin Invest 86: 981-985.

Owens T. 2003. The enigma of multiple sclerosis: Inflammation and neurodegeneration causes heterogenous dysfunction and damage. Curr Opin Neurol 16: 259-265.

Owens T, Wekerle H, Antel J. 2001. Genetic models for CNS inflammation. Nat Med 7: 161-166.

Panitch HS, Hirsch RL, Haley AS, Johnson KP. 1987. Exacerbations of multiple sclerosis in patients treated with gamma interferon. Lancet 1: 893-895.

Pender MP. 1998. Genetically determined failure of activation-induced apoptosis of autoreactive T cells as a cause of multiple sclerosis. Lancet 351: 978-981.

Pender MP, Nguyen KB, McCombe PA, Kerr JF. 1991. Apoptosis in the nervous system in experimental allergic encephalomyelitis. J Neurol Sci 104: 81-87.

Peterson JW, Bo L, Mork S, Chang A, Trapp BD. 2001. Transected neurites, apoptotic neurons and reduced inflammation in cortical multiple sclerosis lesions. Ann Neurol 50: 389-400.

Petty MA, Wettstein JG. 1999. White matter ischemia. Brain Res Rev 31: 58-64.

Piddlesden SJ, Storch M, Hibbs M, Freeman AM, Lassmann H, et al. 1994. Soluble recombinant complement receptor 1 inhibits inflammation and demyelination in antibody-mediated demyelinating experimental allergic encephalomyelitis. J Immunol 152: 5477-5484.

Pitt D, Werner P, Raine CS. 2000. Glutamate excitotoxicity in a model of multiple sclerosis. Nat Med 6: 67-70.

Poser CM, Paty DW, Schinberg L, McDonald WI, Davis FA, et al. 1983. New diagnostic criteria for multiple sclerosis: Guidelines for research protocols. Ann Neurol 13: 227-231.

Prineas JW, Graham JS. 1981. Multiple sclerosis: Capping of surface immunoglobulin G on macrophages engaged in myelin breakdown. Ann Neurol 10: 149-158.

Prineas JW, Barnard RO, Kwon EE, Sharer LR, Cho ES. 1993b. Multiple sclerosis: Remyelination of nascent lesions. Ann Neurol 33: 137-151.

Prineas JW, Barnard RO, Revesz T, Kwon EE, Sharer L, et al. 1993a. Multiple sclerosis. Pathology of recurrent lesions. Brain 116: 681-693.

Prineas JW, Kwon EE, Cho ES, Sharer LR. 1984. Continual breakdown and regeneration of myelin in progressive multiple sclerosis plaques. Ann N Y Acad Sci 436: 11-32.

Prineas JW, Kwon EE, Cho ES, Sharer LR, Barnett MH, et al. 2001. Immunopathology of secondary-progressive multiple sclerosis. Ann Neurol 50: 646-657.

Prineas JW, Kwon EE, Goldenberg PZ, Ilyas AA, Quarles RH, et al. 1989. Multiple sclerosis. Oligodendrocyte proliferation and differentiation in fresh lesions. Lab Invest 61: 489-503.

Probert L, Eugster HP, Akassoglou K, Bauer J, Frei K, et al. 2000. TNFR1 signalling is critical for the development of demyelination and the limitation of T-cell responses during immune-mediated CNS disease. Brain 123: 2005-2019.

Raine CS, Scheinberg L, Waltz JM. 1981. Multiple sclerosis: Oligodendrocyte survival and proliferation in an active established lesion. Lab Invest 45: 534-546.

Ransohoff RM. 1999. Mechanisms of inflammation in MS tissue: Adhesion molecules and chemokines. J Neuroimmunol 98: 57-68.

Redford EJ, Kapoor R, Smith KJ. 1997. Nitric oxide donors reversibly block axonal conduction: Demyelinated axons are especially susceptible. Brain 120: 2149-2157.

Remlinger P. 1928. Les paralysies due traitment antirebique. Ann Insitute Pasteur 55 (Suppl.): 35-68.

Rindfleisch E. 1863. Histologisches Detail zur grauen Degeneration von Gehirn und Rückenmark. Arch Pathol Anat Physiol Klin Med (Virchow) 26: 474-483.

Rivers TM, Sprunt DH, Berry GP. 1933. Observations on attempts to produce acute disseminated encephalomyelitis in monkeys. J Exp Med 58: 39-53.

Rovaris M, Filippi M. 2000. Contrast enhancement and the acute lesion in multiple sclerosis. Neuroimaging Clin N Am 10: 705-716.

Sander M. 1898. Hirnrindenbefunde bei multipler Sklerose. Monatsschr. F Psych u Neurol 4: 429-436.

Sawcer S, Jones HB, Feakes R, Gray J, Smaldon N, et al. 1996. A genome screen in multiple sclerosis reveals susceptibility loci on chromosome 6p21 and 17q22. Nat Genet 13: 464-468.

Schlesinger H. 1909. Zur Frage der akuten multiplen Sklerose und der encephalomyelitis disseminata im Kindesalter. Arb Neurol Inst (Wien) 17: 410-432.

Schmied M, Breitschopf H, Gold R, Zischler H, Rothe G, et al. 1993. Apoptosis of T lymphocytes in experimental autoimmune encephalomyelitis: Evidence for programmed cell death as a mechanism to control inflammation in the brain. Am J Pathol 143: 446-452.

Selmaj K, Raine CS, Cannella B, Brosnan CF. 1991. Identification of lymphotoxin and tumor necrosis factor in multiple sclerosis lesions. J Clin Invest 87: 949-954.

Serafini B, Rosicarelli B, Magliozzi R, Stigliano E, Aloisi F. 2004. Detection of ectopic B-cell follicles with germinal centers in the meninges of patients with secondary progressive multiple sclerosis. Brain Pathol 14: 164-174.

Simpson J, Rezaie P, Newcombe J, Cuzner ML, Male D, et al. 2000a. Expression of the beta-chemokine receptors CCR2, CCR3 and CCR5 in multiple sclerosis central nervous system tissue. J Neuroimmunol 108: 192-200.

Simpson JE, Newcombe J, Cuzner ML, Woodroofe MN. 2000b. Expression of the interferon-gamma-inducible chemokines IP-10 and Mig and their receptor CXCR3, in multiple sclerosis lesions. Neuropathol Appl Neurobiol 26: 133-142.

Smith KJ, Kapoor R, Felts PA. 1999. Demyelination: The role of reactive oxygen and nitrogen species. Brain Pathol 9: 69-92.

Smith KJ, Kapoor R, Hall SM, Davies M. 2001. Electrically active axons degenerate when exposed to nitric oxide. Ann Neurol 49: 470-476.

Smith KJ, Lassmann H. 2002. The role of nitric oxide in multiple sclerosis. Lancet Neurol 1: 232-241.

Smith T, Groom A, Zhu B, Turski L. 2000. Autoimmune encephalomyelitis ameliorated by AMPA antagonists. Nat Med 6: 62-66.

Sobel RA, Mitchell ME, Fondren G. 1990. Intercellular adhesion molecule-1 (ICAM-1) in cellular immune reactions in the human central nervous system. Am J Pathol 136: 1309-1316.

Sorensen TL, Tani M, Jensen J, Pierce V, Lucchinetti C, et al. 1999. Expression of specific chemokines and chemokine receptors in the central nervous system of multiple sclerosis patients. J Clin Invest 103: 807-815.

Springer TA. 1994. Traffic signals for lymphocyte recirculation and leucocyte emigration: The multistep paradigm. Cell 76: 301-314.

Storch MK, Piddlesden S, Haltia M, Iivanainen M, Morgan P, et al. 1998a. Multiple sclerosis: In situ evidence for antibody and complement mediated demyelination. Ann Neurol 43: 465-471.

Storch MK, Stefferl A, Brehm U, Weissert R, Wallström E, et al. 1998b. Autoimmunity to myelin oligodendrocyte glycoprotein in rats mimics the spectrum of multiple sclerosis pathology. Brain Pathol 8: 681-694.

Storch MK, Weissert R, Stefferl A, Birnbacher R, Wallstrom E, et al. 2002. MHC gene related effects on microglia and macrophages in experiemntal autoimmune encephalomyelitis determine the extent of axonal injury. Brain Pathol 12: 287-299.

Stys PK. 1998. Anoxic and ischemic injury of myelinated axons in CNS white matter: From mechanistic concepts to therapies. J Cereb Blood Flow Metab 18: 2-25.

Sun D, Wekerle H. 1986. Ia-restricted encephalitogenic T lymphocytes mediating EAE lyse autoantigen-presenting astrocytes. Nature 320: 70-72.

Sun D, Whitaker JN, Huang Z, Liu D, Coleclough C, et al. 2001. Myelin antigen specific CD8$^+$ T cells are encephalitogenic and produce severe disease in C57BL/6 mice. J Immunol 166: 7579-7587.

Sundvall M, Jirholt J, Yang HT, Jansson L, Engsröm A, et al. 1995. Identification of murine loci associated with susceptibility to chronic experimental autoimmune encephalomyelitis. Nat Genet 10: 313-317.

Suzuki K, Andrews JM, Waltz JM, Terry RD. 1969. Ultrastructura studies of multiple sclerosis. Lab Invest 20: 444-454.

Trapp BD, Peterson J, Ransohoff RM, Rudick R, Mork S. et al. 1998. Axonal transection in the lesions of multiple sclerosis. N Engl J Med 338: 278-285.

Trebst C, Sorensen TL, Kivisakk P, Cathcart MK, Hesselgesser J, et al. 2001. CCR1+/CCR5+mononuclear phagocytes accumulate in the central nervous system of patients with multiple sclerosis. Am J Pathol 159: 1701-1710.

Trebst C, Staugaitis SM, Kivisäkk P, Mahad D, Cathcart MK, et al. 2003. CC chemokine receptor 8 in the central nervous system is associated with phagocytic macrophages. Am J Pathol 162: 427-438.

Tubridy N, Behan PO, Capildeo R, Chaudhuri A, Forbes R, et al. 1999. The effect of anti-alpha4 integrin antibody on brain lesion activity in MS. The UK Antegren Study Group. Neurology 53: 466-472.

Ulvestad E, Williams K, Vedeler C, Antel J, Nyland H, et al. 1994. Reactive microglia in multiple sclerosis lesions have an increased expression of receptors for the Fc part of IgG. J Neurol Sci 121: 125-131.

Van Walderveen MA, Kamphorst W, Scheltens P, van Wasberge JH, Ravid R, et al. 1998. Histopathologic correlate of hypointense lesions on T1-weighted spin-echo MRI in multiple sclerosis. Neurology 50: 1282-1288.

Washington R, Burton J, Todd RF 3rd, Newman W, Dragovic L, et al. 1994. Expression of immunologically relevant endothelial cell activation antigens on isolated central nervous system microvessels from patients with multiple sclerosis. Ann Neurol 35: 89-97.

Weissert R, Wallstrom E, Storch MK, Stefferl A, Lorentzen J, et al. 1998. MHC haplotype-dependent regulation of MOG-induced EAE in rats. J Clin Invest 102: 1265-1273.

Wekerle H, Linington C, Lassmann H, Meyermann R. 1986. Cellular immune reactivity within the CNS. Trends Neurosci 9: 271-277.

Werner P, Pitt P, Raine CS. 2001. Multiple sclerosis: Altered glutamate homeostasis in lesions orrelates with oligodendrocyte and axonal damage. Ann Neurol 50: 169-180.

White CA, Nguyen KB, Pender MP. 2000. B cell apoptosis in the central nervous system in experimental autoimmune encephalomyelitis: Roles of B cell CD95, CD95L and Bcl-2 expression. J Autoimmun 14: 195-204.

Whetten-Goldstein K, Sloan EA, Goldstein LB, Kulas ED (1998). A comprehensive assessment of the cost of multiple sclerosis in the United States. Multiple sclerosis 4: 419-425.

Willenborg DO, Staykova MA, Cowden WB. 1999. Our shifting understanding of the role of nitric oxide in autoimmune encephalomyelitis: A review. J Neuroimmunol 100: 21-35.

Windhagen A, Newcombe J, Dangond F, Strand C, Woodroofe MN, et al. 1995. Expression of costimulatory molecules B7–1 (CD80), B7–2 (CD86), and interleukin 12 cytokine in multiple sclerosis lesions. J Exp Med 182: 1985-1996.

Wolswijk G. 2000. Oligodendrocyte survival, loss and birth in lesions of chronic-stage multiple sclerosis. Brain 123: 105-115.

Wolswijk G. 2002. Oligodendrocyte precursor cells in the demyelinated multiple sclerosis spinal cord. Brain 125: 338-349.

Yednock TA, Cannon C, Fritz LC, Sanchez-Madrid F, Steinman L. 1992. Prevention of experimental autoimmune encephalomyelitis by antibodies against a4β1 integrin. Nature 356: 63-66.

19 Neuropathic Pain

C. Briani · L. Padua · C. Pazzaglia · L. Battistin

Abstract: Neuropathic Pain is a pain deriving from a primary lesion or dysfunction of the nervous system. Different from nociceptive pain, it does not require specific pain receptor stimulation to occur. Symptoms include burning or shooting sensations, abnormal sensitivity to normally painless stimuli (allodynia), or raised sensitivity to painful stimuli (hyperalgesia). Both peripheral and central nervous system disorders can cause neuropathic pain. As to peripheral nervous system, neuropathic pain is a common feature of diabetes, uremia, HIV infection, alcohol abuse, small-fiber painful sensory neuropathy, drugs-induced neuropathy, and trigeminal and postherpetic neuralgia. Central nervous system pain syndromes are mostly secondary to stroke and spinal cord lesions, and occur in up to 8% of patients after a stroke. The pathophysiologic mechanisms of neuropathic pain are not fully understood. Sensitization of nociceptors from regenerating nerve sprouting, spontaneous activity in dorsal root ganglia, ectopic neural firing, and central sensitization seem to play a role. Several neurobiological mechanisms underlie neuropathic pain, among which are increased number of sodium channels and increased glutamate activity. There are different types of neuropathic pain, caused by distinct pathophysiologic mechanisms. Consequently, a drug with one single mechanism of action is not likely to relieve all types of pain or all patients, and often combination therapies are required. Treatment studies for neuropathic pain have been conducted mainly in diabetic, HIV-associated neuropathies, postherpetic and trigeminal neuralgias. Current treatment options include antidepressants and anticonvulsants, which are widely used and represent the first-line agents in the management of neuropathic pain. Analgesic nonnarcotic (tramadol) and narcotics have also proven to be effective.

The authors agreed on using the capital P to define the word Pain, to underlie the deep emotional meaning of the suffering experience which goes beyond the physical symptoms. Pain has the dignity of a true state of life and, as such, we believe it deserves our full respect.

List of Abbreviations: HRQoL, Health Related QoL; IASP, International Association for the Study of Pain; VAS, visual analog scale

1 Neuropathic Pain: Definition

The current definition of neuropathic Pain, as proposed by the International Association for the Study of Pain (IASP) is an "unpleasant sensory and emotional experience associated with actual or potential tissue damage, or described in terms of such damage" (Merksey and Bogduk, 1994). This definition has been criticized by many as vague, particularly the "dysfunction" component. "Dysfunction" blurs the distinction between neuropathic and other possible types of Pain that may result from different underlying mechanisms, allowing several not purely neuropathic disorders to be included. Recently Backonja proposed a new definition of neuropathic Pain, as a Pain which is result of "disease of nervous system, peripheral or central, manifesting with various types and of variable intensities of positive and negative phenomena" (Backonja, 2003a).

Inflammatory and neuropathic Pain were once considered to be distinct entities. However, research over the past decade has brought to light many shared mechanisms, and the distinction between the two is no longer clear. Consideration of mechanisms, symptoms, and the effects of analgesic drugs does not reveal any definitive differentiating factor. Given the present level of understanding, it may not be possible to distinguish between inflammatory and neuropathic Pain in a large number of patients and a satisfying definition of neuropathic Pain is far from being reached.

2 Concept of Pain

Pain represents a physiological response to a noxious stimulus that alerts us to the presence of an actual or potential tissue damage. As such, Pain plays an essential protective biological function. The capacity to experience Pain, in fact, elicits coordinated reflex and behavioral responses to keep tissue damage to a minimum. In contrast, persistent Pain syndromes offer no biological advantage and cause suffering and distress. From being a symptom warning about an actual or potential damage, chronic Pain becomes a disease itself. Such maladaptive Pain typically results from damage to the nervous system – the peripheral nerve, the dorsal root ganglion or dorsal root, or the central nervous system – and is known as neuropathic Pain.

3   Pathophysiologic Classification of Pain

3.1   Nociceptive Pain

Nociceptive Pain is a Pain deriving from chemical, thermal, and mechanical stimulation of Pain receptors of A-delta and C fibers secondary to tissue damage. It is generally acute, typically well localized and self-limiting, resolving along with the healing of the tissue damage. It may be somatic, when it arises from muscle or bone injury, or visceral, when secondary to expansion or other injury of an internal organ. Different from somatic Pain, visceral nociceptive Pain tends to be dull and poorly localized, and usually has a cutaneous referral. Examples of nociceptive Pain are conditions such as osteoarthritis, inflammations, bone fractures or metastasis, bruises, damage of skin, muscle, connective tissue, and viscera.

3.2   Neuropathic Pain

As mentioned above, neuropathic Pain derives from a primary lesion or dysfunction of the peripheral or central nervous system. Different from nociceptive Pain, it does not require specific Pain receptor stimulation to occur. It may persist for months or years after the injury. Neuropathic Pain is therefore chronic, and devoid of any biological function. Conventionally the term chronic Pain is used for Pain lasting more than 3 months, but recently there is consensus in considering chronic any Pain which persists beyond the time of normal healing. Being a chronic process, neuropathic Pain interferes with patients' mood, sleep, physical and cognitive activities, ability to participate in social and recreational life, and thus profoundly affects quality of life. Management of neuropathic Pain therefore involves a multidimensional approach where the medical, cognitive, behavioral, and functional components need to be assessed, quantified, and taken care of. Typical examples of neuropathic Pain syndromes are diabetic peripheral neuropathy, and postherpetic neuralgia.

3.3   Mixed Pain

In some conditions, e.g., cancer and back Pain syndromes with radiculopathy, nociceptive and neuropathic Pain coexist. Moreover, a persistent, untreated nociceptive Pain may lead – through peripheral and central sensitization – to the development of a neuropathic Pain. On the other hand, nervous system dysfunction may be associated with the release of proinflammatory cytokines (Marchand, 2005). Neurogenic inflammation has been considered also to play a role in the pathophysiology of migraine (Moskowitz, 1993) where both neuropathic and nociceptive Pain seem to coexist.

4   Neuropathic Pain: Etiology

Several heterogeneous conditions, deriving from both central (Cohen and Abdi, 2002) and peripheral (Sommer, 2003) neurological diseases, may be responsible for neuropathic Pain.

Among the central nervous system causes of neuropathic Pain, the most common are central poststroke Pain, trauma, spinal cord injuries, phantom limb phenomenon, and multiple sclerosis.

More common are the peripheral nervous system cause of Pain: trigeminal or glossopharyngeal neuralgia, infections (e.g., shingles and postherpetic neuralgia; HIV and hepatitis C-related neuropathy; postviral neuropathy; leprosy; lyme), toxics (alcohol, arsenic, thallium, gold, perhexiline, inorganic mercury, organic solvents, acrylamide, disulfiram), peripheral neuropathy secondary to metabolic (diabetes, glucose intolerance, uremia) or endocrinological (hypo-hyper-thyroidism) dysfunctions, inflammatory or immuno-mediated neuropathies (vasculitic neuropathies, Sjogren's syndrome, chronic inflammatory demyelinating neuropathy, Guillain-Barré syndrome), iatrogenic causes (nucleoside reverse transcriptase inhibitors; chemotherapy-induced neuropathy, among which cisplatin, oxaliplatin, carboplatin, vincristine, taxol, suramin, bortezomib, thalidomide-induced neuropathy), hereditary (e.g., hereditary sensory-autonomic neuropathy-HSAP; Fabry's disease, familial amyloid polyneuropathy, Tangiers's disease), vitamin and nutritional (thiamin, riboflavin, and nicotinic acid) deficiencies or excess (pyridoxine), small fibers as well as idiopathic or cryptogenic sensory polyneuropathy. The Complex Regional Pain Syndromes (CRPS) also represent an expanding cause of neuropathic Pain.

5 Pathophysiological Mechanisms of Neuropathic Pain

Several mechanisms and neurobiological changes, which involve both central and peripheral pathophysiologic processes, play a role in generating and maintaining neuropathic Pain. Peripheral sensitization is a mechanism by which nervous fibers develop spontaneous activity in neurons, lowered activation threshold, ectopic neural firing along the nerve, associated with increased expression of sodium channels, the latter explaining the efficacy of sodium channel blockers drugs (phenytoin, carbamazepine, lidocaine, and tricyclic antidepressants) in relieving neuropathic Pain.

Central sensitization refers to the increased response of central signaling neurons, so that stimuli that are normally not painful become experienced as painful (allodynia), or painful stimuli are experienced as more painful than usual (hyperalgesia). Central mechanisms of neuropathic Pain involve sensitization of sensory neurons in the spinal cord (dorsal root ganglia), spontaneous activity of ectopic foci in dorsal root ganglia, deafferentation or development of new synapses between neurons, failure or reduced activity of the descending inhibitory serotoninergic and noradrenergic pathways. Lesions leading to neuropathic pain directly involve the nociceptive pathway. Increasing evidence from animal studies adds to our understanding of Pain pathophysiology and opens new avenues to study possible tailored therapeutic targets. Moreover, different pathophysiological mechanisms probably underlie different kinds of pain (Truini and Cruccu, 2006). The mechanisms of neuropathic Pain have been recently reviewed (Baron, 2006; Campbell and Meyer, 2006; Truini and Cruccu, 2006).

We will briefly address the main pathophysiological mechanisms underlying neuropathic Pain.

5.1 Neuropathic Pain: Experimental Evidence

Animal models of neuropathic Pain allowed to better understand the neurobiological mechanisms underlying hyperalgesia and allodynia secondary to nerve damage or inflammation. Both the damaged and undamaged nerve fibers express abnormal ectopic discharge and altered gene expression, thus leading to peripheral and central sensitization. The first widely used animal model of neuropathic Pain was chronic constriction injury of the rat sciatic nerve which causes inflammatory response leading to a loss of A-fibers and also C-fibers (Bennett and Xie, 1988). Other models are the partial ligation, where about 33–50% of sciatic nerve fibers are injured with the remaining part of fibers left undamaged (Seltzer et al., 1990), and spinal nerve ligation consisting in section or ligation of one or more of spinal nerves, generally L5 and L6 (Kim and Chung, 1992). The results of the studies from animal models allowed to study Pain-related behavior, allodynia, and hyperalgesia, revealing as every part of the nervous system plays a role in generating neuropathic Pain. Starting from the periphery, after a nerve injury, there is an increase in spontaneous firing not only from nociceptors (Wall et al., 1974), but also in the dorsal root ganglia and along the nerve (Wall and Devor, 1983). Interestingly, this ectopic activity originates both from injured afferent fibers and from uninjured neighboring afferents (Gold, 2000). This increased excitability is the result of an up-regulation of voltage sodium channels, a down regulation of potassium channels, and a reduction in threshold of a family of receptors (transient receptor potential – TRP-channels) involved in detection of heat stimuli. The reduction of threshold and the increase of responsiveness of the high-threshold nociceptor neurons is known as peripheral sensitisation. The hypersensitivity associated with this phenomenon is called "primary hyperalgesia" and is related to the action of nociceptors exposed to injured tissue. An important role in threshold reduction and sensitization is played by the bradykinin, the product of cleavage of kininogen, a substance released by damaged tissues. The bradykinin, reacting with specific receptors, sensitizes nociceptive afferents fibers, playing a role both in peripheral and central sensitization (Wang et al., 2006).

The pro-inflammatory cytokines tumor necrosis factor-alpha (TNFα) and interleukin-1β (IL-1β) have been shown to be upregulated in sciatic nerves and dorsal root ganglia ipsilaterally to experimental nerve injury in a model of chronic constriction injury. Interestingly, the results of inhibition tests using inhibitors to the N-methyl-d-aspartate (NMDA) receptor and to calpain seem to indicate that calpain represents one of the earliest mediators of cytokine upregulation in injured peripheral nerves (Uceyler et al., 2007). Moreover, studies on mice TNF receptor (TNFR) deficient showed that the TNFα mediate different kind of Pain depending on the receptor it binds to (Vogel et al., 2006). Specifically, there was absence of thermal hyperalgesia in mice deficient of TNFR1 and a reduction in mechanical and cold allodynia in mice deficient of TNFR1 or TNFR2 compared to wild-type mice after chronic constriction injury of the sciatic nerve. These findings provide precious information and should be considered when using TNF inhibitors in clinical trials for treatment of Pain.

After peripheral nerve damage a modification in expression of several genes occurs in sensory neurons which may lead to a "phenotypic switch," that is fibers, such as the A fibers, may acquire the capability of enhancing synaptic transmission in the spinal cord and producing central sensitization, tasks generally pertaining to the Pain C-fibers (Woolf, 2004). Modifications that occur in the peripheral nervous system are similar to those present in the central reorganization, expression of the plasticity of the nervous system in response to a noxious stimuli (Woolf and Salter, 2000). The sensory inflow subsequent to a peripheral damage determines a reduction in activation threshold and a hyperexcitability of spinal cord neurons, making it possible that an innocuous stimulus, such as a light touch, elicits Pain. This is the base of the "secondary hyperalgesia" associated with hypersensibility in noninjured tissue.

Central sensitization represents a condition of increased sensitivity of dorsal horn neurons which follows a peripheral noxious stimuli, tissue or nerve injured, leading to a reduced Pain threshold, and a spreading of Pain sensitivity to nondamaged areas. There are two forms of central sensitization. The first is an activity-dependent form, that is induced by afferent activity in nociceptors as result of phosphorylation and modulation of ion-channel receptors, generally as a consequence of A-delta and C-fibers stimuli. It is due to a modification in the phosphorylation of glutamate-activated NMDA receptor, which involves an increased excitability of the cells that can be triggered by inputs that are normally subthreshold. The second is a transcription-dependent form, which refers to long lasting changes in gene regulation of central neurons (Woolf, 2004).

5.2   Nonneuronal Mechanisms in Neuropathic Pain: Immune and Glial Cells

Increasing evidence points to a role of immune mechanisms in the development of neuropathic Pain. Whereas immune cells seem not to play a role in the acute Pain, they do strongly influence neuronal function in chronic Pain situations, where both inflammatory mediators as well as activated immune cells promote a neuroimmune network in response to nervous system damage (Marchand et al., 2005). Several types of immune cells participate in Pain mechanisms, among which are mast cells, neutrophils, macrophages, and T and B lymphocytes. Mast cells are actively involved in the innate immune system. They play a crucial role in allergic diseases and, when activated, release mediators (histamines, proteases, cytokines) capable of sensitizing neurones. Moreover, they initiate an inflammatory cascade that promotes migration of neutrophils, that in turn release mediators and chemokines acting on macrophages and other immune cells. Both T and B lymphocytes are found in animal models of neuropathic Pain, both in the peripheral nerve and in the dorsal root ganglia and spinal cord, suggesting a role of the lymphocytes in generating neuropathic Pain (Hu and McLachlan, 2002). Consistently, athymic rats (which lack T cells) show significantly less allodynia and hyperalgesia after nerve damage (Moalem et al., 2004).

Besides neurons, nonneuronal elements of the nervous system, namely glial cells, also seem to contribute to the initiation and maintenance of chronic Pain. Microglia cells, which represent the resident macrophages of the central nervous system, are capable of releasing, once activated (e.g., after an injury, a microbial attack), proinflammatory cytokines, chemokines, and several other Pain-producing mediators. In animal models, activation of microglia has been shown to induce hyperalgesia and allodynia, which is reversible following pretreatment with minocycline, a potent inhibitor of microglial activation (Raghavendra et al., 2003). The effects of minocycline on neuropathic Pain are associated with reduced production of proinflammatory cytokines. Minocycline is instead not efficacious on neuropathic Pain if the treatment is initiated days after nerve injury. The maintenance of pathological Pain seems instead secondary to activation of astroglial cells, which are not susceptible to minocycline action (Raghavendra et al., 2003). The molecules released by inflammatory and immune cells implicated in Pain mechanisms comprehend a wide variety of mediators, among which are bradykinin (Wang et al., 2006), cytokines (mainly interleukin-1, interleukin-10, TNF), ATP and adenosine, serotonin, eicosanoids, nitric oxide, and neurotrophins. Among the latter, Nerve Growth Factor (NGF) and Brain Derived Neurotrophic Factor (BDNF) seem to be involved in nociception, with NGF and BDNF sensitizing neurons in the peripheral and central nervous system, respectively (Bennett, 2001)

5.3   Central Nervous System Mechanisms of Pain Modulation

The perception of Pain is modulated by endogenous inhibitory pathways descending from the brainstem. The existence of an endogenous nociceptive modulatory system descending from the midbrain periaqueductal gray through the ventromedial medulla to the dorsal horn has been first proposed by Reynolds, following seminal animal studies (Reynolds, 1969). The main neurotransmitters involved in the endogenous inhibitory are noradrenaline and serotonin, as well as endogenous opioids (Pert and Snyder, 1973). Both noradrenaline and serotonin induce membrane hyperpolarization, with a subsequent reduced transmission of primary $A\delta$ and C afferent fibers and increase of inhibitory molecules (GABA and glycine) release from the interneurons. (Yoshimura and Furue, 2006). The efficacy on neuropathic Pain of drugs inhibiting the reuptake of noradrenaline and serotonin, such as the tricyclic antidepressants and the dual inhibitors represent an indirect evidence of the existence of such important modulatory descending pathways.

The role of nociceptin, an endogenous ligand for the opioid receptor-like 1 (ORL1) receptor, in modulating neuropathic has been recently investigated in animal models (Chen and Sommer, 2006). Nociceptin and ORL1 receptor are distributed in several regions of the central nervous system involved in nociception, such as the cerebral cortex, thalamus, periaqueductal gray, and dorsal horns of the spinal cord. Both pro- and antinociceptive effects of nociceptin have been reported in a variety of animal experiments depending on the model, animal species, route of administration, and doses. Recently, Chen and Sommer (Chen and Sommer, 2006) showed experimentally that both nociceptin and ORL1 receptor, normally represented in dorsal root ganglia, increased their expression after peripheral nerve injury and inflammation, indicating that they are active in pain modulation processes already at the level of the first afferent neuron. Moreover, activation of the nociceptin-opioid system in rat sensory neurons has been shown to produce antinociceptive effects in inflammatory pain (Chen and Sommer, 2007).

The activation of descending antinociceptive pathways by opioids has been recently demonstrated using positron emission tomography in healthy subjects after taking the short-acting μ-selective opioid agonist remifentanil (Wagner et al., 2007).

6   Terminology

When dealing with neuropathic Pain, we have to become familiar with terminology.
- Dysesthesia: abnormal sensation on contact with an object
- Paresthesia: spontaneous discomfort (often described as a limb that "has fallen asleep")

- Allodynia: Pain following stimulation by an innocuous stimulus (e.g., light touch), that ordinarily does not elicit a painful response
- Hyperpathia (or hyperalgesia): exaggerated Pain from normally painful stimuli. Primary hyperalgesia, caused by sensitization of C–fibers, occurs within the area of the injury. Secondary hyperalgesia, caused by sensitization of dorsal root ganglia, occurs in the undamaged area surrounding the injury.

7   Neuropathic Pain: Clinical Management

As previously stated, the capacity to experience Pain has a protective role that allows to minimize damage. Once the noxa has been removed the unpleasant sensation disappears; on the contrary a persistent sensation of Pain in the absence of the noxious stimulus, causes severe deterioration of Quality of Life (QoL) and disability.

Neuropathic Pain is perceived by patients as either "spontaneous" or "evoked." The former cannot be attributed to a specific stimulus (Tasker, 1984; Wall, 1991), the latter comes in response to a stimulus. Patients' verbal descriptions of such symptoms, in the presence of neurological examination, can be used to reach a diagnosis of neuropathic Pain.

Over the last two decades, clinical and public health researchers have stressed the need to put the patient at the center of attention during evaluation of outcome. However, when physicians try to see things from the patient's point of view, they do not usually do it in a standardized way and they do not "measure" the patient's perceptions rigorously. This "need to measure" is particularly true when we are dealing with Pain.

Nothing affects life as negatively as Pain. In the "Inferno" (Hell), Dante put Pain in the inscription over the gate of Hell:

Per me si va ne la città dolente,
per me si va ne l'eterno dolore
Through me lies the way to the place of suffering,
Through me lies the way to eternal Pain

Much attention has been given to concepts such as Health Related QoL (HRQoL) (Guyatt et al., 1993) and Patient Satisfaction, which strongly correlate with the patient's expectations for outcome. The concept of Pain measurement is inevitably a part of the evaluation, but is particularly difficult to assess because Pain is experienced subjectively, "inside" the patient. Thus, because of the very subjective nature of Pain, apart from attempts to measure objective physiological effects of Pain on the patient, the easiest way to measure Pain is to ask the patient to describe the Pain and how much it influences his or her life. The need for standardized measures of such items has stimulated an extensive and rigorous process, which has led to the development of validated patient-oriented instruments. These tools, mainly self-administered questionnaires, concentrate on how Pain affects the individual, and let us quantify Pain with measures that are not influenced by the physician's opinion. If our goal is to help our patients, we must understand their needs and their expectations (Amadio, 1993). We must be able to measure Pain with the reliability, sensitivity, validity, and responsiveness possibly associated with objective measures. Several self-administered questionnaires quantitatively measure Pain and related issues, and fulfil the measurement properties for evaluation research. We are therefore able to translate the patients' responses into numbers that can be used for statistical analysis. The questions to be asked must be phrased very carefully, and tested for performance in a target population prior to clinical use, just as a car, a camera, or other devices are tested prior to general release. Clinical epidemiology provides clinicians and researchers with the instruments to evaluate the patient's point of view in a scientific manner (Marshall et al., 1996). In many medical and surgical areas (neurology for neuropathic Pain, oncology for Pain due to cancer, e.g.), outcome assessment for various conditions has already been performed and investigated using patient-oriented tools, and is now widely accepted.

Patient-oriented findings are useful for assessing the evolution of Pain and its impact on patient's life; moreover, they provide data for the improvement of therapeutic strategies in clinical conditions that sometimes may be over-treated with unnecessary side effects.

In order to standardize the diagnostic and therapeutical approaches recently advocated, it is necessary to compare different studies in the same field, to compare studies in different fields, to adopt common measures, to perform multicentric research, to perform multiperspective assessment. In this regard, patient-oriented tools may be very important because one of their features is that the measures have the same value in different situations and in different countries (although in the latter case a rigorous process of translation and validation must be performed).

In order to obtain a suitable multidimensional assessment of Pain the use of validated and standardized patient-oriented measurements should be associated with further standardization of traditional outcome parameters and with new tools for the measurement of both the neurophysiological effects of Pain and the subjective Pain threshold.

Two words of caution are necessary. First of all, the patient-oriented evaluation provides an additional, standardized perspective for the evaluation of the outcome, but it does not reduce the need for the development of new tools. Secondly, certain pathologies that involve cognitive functions cannot be evaluated by these instruments, at least in their present form.

For all of these reasons, Pain descriptors used by patients with neuropathic Pain have been used by clinicians to develop questionnaires that provide a "measure" of Pain.

8   Questionnaires to Measure Neuropathic Pain

One of the most used measures is the visual analog scale (VAS), which is a numeric scale of subjective measurement of Pain intensity, commonly used to measure every kind of Pain and sometimes used in association with other more specific Pain-questionnaire. We included this tool, although it was not developed for neuropathic Pain only, because it is the commonest measure and the "father" of the Pain measurements (Deschamps et al., 1988).

The McGill Pain Questionnaire (MPQ), developed and validated by Melzack (Melzack, 1975), was designed to provide quantitative and multidimensional measures of clinical Pain. It is used to study intensity, characteristics, location, and periodicity of Pain. It has been demonstrated to be of discriminative value, even capable of discriminating between similar Pains such as trigeminal neuralgia and atypical facial Pain.

A "Short Form" of the McGill Pain Questionnaire (MPQ-SF) was designed to obtain information from patients when time is limited. Also in its reduced form, the MPQ-SF is sensitive to clinical changes due to therapy (Melzack, 1987).

The Neuropathic Pain Scale (NPS) was developed in 1997 by Galer and Jensen (Galer and Jensen, 1997) in the attempt to discriminate four diagnostic categories of neuropathic Pain using single descriptors. Only postherpetic neuralgia could be distinguished from the other diagnostic groups (reflex sympathetic dystrophy, diabetic neuropathy, and peripheral nerve injury). The NPS was not used to discriminate between neuropathic Pain and nociceptive Pain symptoms.

In 2001 Michael Bennett (Bennett, 2001), in order to develop a scale that would help differentiating neuropathic Pain from nonneuropathic Pain, created a novel tool for identifying patients in whom neuropathic mechanisms dominate their Pain experience: the Leeds assessment of neuropathic symptoms and signs (LANSS) Pain Scale based on analysis of sensory description and bedside examination of sensory dysfunction, hence a composite measure. Bennett organized his neuropathic Pain scale in form of questions, which are supposed to reflect the essence of neuropathic Pain based on symptoms and physical examination findings of mechanical allodynia and pinprick abnormalities; he did not use detailed sensory testing since the aim of the study was to develop a diagnostic tool that could be easily incorporated into a clinical context.

In 2003 Krause and Backonja (Krause and Backonja, 2003) developed the Neuropathic Pain Questionnaire (NPQ) consisting of 12 items, able to discriminate neuropathic Pain from other kind of Pain and to measure the descriptors. The same authors (Backonja and Krause, 2003) developed a short form of the NPQ. They selected specific items based on the fact that diagnosis of neuropathic Pain is made by the use of positive and negative sensory phenomena and paresthesiae or dysesthesiae. The short form was simpler to use and without statistically significant loss of predictive accuracy.

In 2004 Bouhassira (Bouhassira et al., 2004) published a study describing the development and validation of the Neuropathic Pain Symptom Inventory (NPSI), with the intention to create "a new self-questionnaire specifically designed to evaluate the different symptoms of neuropathic Pain." The NPSI consists of ten descriptors able to discriminate and quantify distinct dimensions of neuropathic Pain.

In the same year Perkins (Perkins et al., 2004) tried to validate a Danish version of the short form of MPQ that could be self-administered. They used the previously developed Danish version of the MPQ as a basis for this questionnaire, trying to demonstrate that the representative words from selected categories of the MPQ were accurate, so that data could be compared across languages and cultures. Unfortunately transcultural translation was not possible because of certain Danish words, so they developed the descriptive Danish Pain questionnaire (BDDPQ) that could be self-administered. In 2005 Zelman (Zelman et al., 2005) validated a modified version of the Brief Pain Inventory (BPI) (Cleeland, 1991) for patients with diabetic polyneuropathy (BPI-DPN). The BPI is a numeric rating scale that measures severity of Pain and its interference with daily function; it has been used in patients with nonneuropathic Pain and, with regard to the neuropathic Pain, it had been used only for herpes zoster. The results of the study showed that scales are able to assess severity and functional interference of neuropathic Pain in those patients.

In 2005, with the purpose of directly comparing the clinical features of neuropathic and nonneuropathic Pain, French Neuropathic Pain Group (Bouhassira et al., 2005) developed a clinician-administered questionnaire named neuropathic Pain diagnostic questionnaire (DN4) consisting of both sensory descriptors and signs related to bedside sensory examination. They found that a small number of items are adequate to discriminate neuropathic Pain from nonneuropathic Pain.

9　Neurophysiological Assessment of Pain

Standard neurophysiological tests are useful to detect peripheral or central sensory damage, but they are not useful to study the nociceptive system. Other neurophysiological techniques, including quantitative sensory tests (QSTs), autonomic tests (ATs), nociceptive reflexes, microneurography (MCNG), and laser-evoked potentials (LEPs) can provide useful data for neuropathic Pain assessment.

9.1　Quantitative Sensory Tests (QSTs)

The thermal stimulation test (TST) obtains an evaluation of the entire peripheral and central portions of the sensory system. Patients with Pain may show altered thresholds for perception of stimulus.

The current perception threshold (CPT) is a neuroselective technique able to evaluate the three types of sensory nerve fibers with stimulus frequency variation; moreover, when suprathreshold stimulus with 250 Hz or 5 Hz frequency is applied it is possible to obtain the nociceptive current perception threshold (NCPT).

The combination of these three tests allows to document and quantify all Pain signs. However, it is better to use these tests in the preliminary assessment, as they are unable to distinguish central from peripheral dysfunction because the abnormalities found are not specifically localized.

9.2　Autonomic Tests

The quantitative sudomotor axon reflex test (QSART) quantitatively assesses function of the postganglionic sudomotor axon and sweat gland by means of an axon reflex. It consists of three phases: Resting Skin Temperatures, Resting Sweat Output, and Stimulated Sweat Output. The QSART is a reliable method to detect pathophysiological change in neuropathic patients. The sympathetic skin response (SSR) investigates sympathetic sudomotor function through somatosympathetic reflexes. This response is due to activation of the postganglionic sudomotor efferent axon, but it is not a quantifiable and reproducible test, so it is not commonly used in the evaluation of neuropathic Pain.

9.3 Microneurography

Microneurography is an invasive technique, which records the activity of single fibers (Torebjork, 1993). Microneurography provides useful information on the physiology of nociceptors; however, as it is difficult and time consuming, its use tends to be limited to research.

9.4 Nociceptive Reflexes

As diagnostic tools, the use of the nociceptive RIII flexion reflex, corneal reflex, and cutaneous silent period in neuropathic Pain is extremely rare.

9.5 Laser Evoked Potentials (LEPs)

At the moment the best neurophysiological technique to study neuropathic pain is represented by laser evoked potentials (LEPs), which selectively excite the free nerve endings (A-delta and C) in the superficial skin layers (Bromm and Treede, 1984). Late LEPs study the A-delta and ultralate LEPs the unmyelinated fibers. Late LEPs have proved to be reliable in assessing damage in a great number of peripheral and central pathologies.

In summary, QSTs give an exact and reproducible assessment of myelinated and unmyelinated small fibers function. The addition of autonomic tests makes it possible to obtain information about the sympathetic system, but the tests that give the best accuracy are, at the moment, the LEPs.

10 Other Pain Measurements

10.1 Bedside Examination

Neurological examination in suspected neuropathic Pain should include quantification and mapping of motor, sensory, and autonomic phenomena in order to identify all signs of neurological dysfunction. Tactile sense is best assessed by a piece of cotton wool, pinprick sense by a pin, thermal sense by warm and cold objects, and vibration sense by a 128-Hz tuning fork. Bedside examination may include quantitative sensory tests, which are easy-to-perform analyses of perception through external stimuli of controlled intensity in an ascending and descending order of magnitude. The most used tools are von Frey hairs and Semmes-Weinstein monofilaments.

In conclusion, we believe that a validated patient-oriented approach to the measurement of Pain, possibly associated with new objective Pain parameters, should be performed whenever Pain plays a crucial role in patient's life, and they should be considered primary outcome measures in the clinical trials where Pain is evaluated.

Nevertheless, further efforts must be done in order to develop and standardize objective measures, probably neurophysiological ones, of Pain to be used in clinical practice and clinical trials and not only in a limited field of research.

11 Pharmacological Therapy

Drugs used to treat neuropathic Pain have different mechanisms of action, including sodium channels or calcium flux modulation, serotonin and noradrenaline reuptake inhibition, μ receptors agonism. Some drugs (e.g., tricyclic antidepressants, TCA) have more than one mechanism. Despite the advent of several new compounds to treat neuropathic Pain, treatment often remains unsatisfactory and combination therapy may be required. Randomized controlled trials are available for the most common neuropathic Pain syndromes, among which are diabetic neuropathy and postherpetic neuralgia. Less is known about the uncommon Pain syndromes or the central Pain syndromes. We will briefly analyze the classes of drugs used to treat neuropathic Pain. For an extensive analysis of all the randomized controlled trials performed in the various neuropathic Pain conditions and the proposed pharmacotherapy recommendations in the different Pain syndromes we refer to a recent review by the European Federation of Neurological Societies (EFNS) (Attal et al., 2006).

11.1   Antidepressants

11.1.1 Tricyclic Antidepressants (TCA)

Several randomized, double-blinded trials have found tricyclic antidepressants (TCA) (amitriptyline, imipramine, desipramine, and clomipramine) superior to placebo in painful diabetic neuropathy (McQuay et al., 1996). The use of antidepressants in neuropathic Pain has recently been addressed in two reviews (Saarto and Wiffen, 2005; Sindrup et al., 2005). TCA have also been shown to be effective in postherpetic neuralgia (Dubinsky et al., 2004). The efficacy of the TCA is likely due to their multimodal mechanisms of action. In fact, they block the reuptake of serotonin and norepinephrine, thus enhancing descending endogenous Pain-modulating networks; moreover, they presumably relieve Pain also by inhibition of the sodium channels, and N-methyl-D-aspartate receptor blockade. The different types of TCA do not seem to differ as to efficacy. It is recommended to start therapy at low dosage (10–25 mg at bedtime), then titrate every 3–7 days by 10 to 25 mg up to 100–150 mg. The use of TCA needs to be monitored for the possible occurrence of anticholinergic side effect (dry mouth, constipation, sedation, orthostatic hypotension, weight gain) and they should not be used in patients with cardiovascular disease, especially recent myocardial infarction, and arrhythmia. They should also be used with caution in patients with glaucoma, urinary retention, elderly, autonomic neuropathy.

11.1.2 Selective Serotonin Reuptake Inhibitors (SSRI)

Selective serotonin reuptake inhibitors (SSRI), which selectively block serotonin uptake, are less effective than TCA (Sindrup et al., 1990; Max et al., 1992) in the treatment of neuropathic Pain. There are differences among drugs of the same class. Paroxetine reduces the Pain of diabetic neuropathy better than placebo, but it is not as effective as the TCA (Sindrup et al., 1990). Citalopram seems equivalent to paroxetine (Sindrup et al., 1992), whereas fluoxetine did not show any benefit in diabetic neuropathy (Max et al., 1992). As a whole, SSRI provide clinically insufficient Pain relief (Saarto and Wiffen, 2005; Sindrup et al., 2005). SSRI adverse effects include asthenia, sweating, constipation, somnolence, dizziness, and sexual dysfunction.

11.1.3 Serotonin and Norepinephrine Reuptake Inhibitors (SNRI)

Dual inhibitors are becoming increasingly interesting drugs in the control of neuropathic Pain, since both serotonin and norepinephrine are involved in Pain modulation in the descending inhibitory pathways of brain and spinal cord. Among Serotonin and Norepinephrine Reuptake Inhibitors (SNRI), duloxetine (60–120 mg/day) has recently been approved in the USA by the Food and Drug Administration (FDA) for treatment of diabetic peripheral neuropathic Pain with a number needed to treat (NNT) of 4.1 compared to placebo (the NNT estimates the number of patients who must receive a given treatment in order to obtain one patient with a specified outcome). The safety and effectiveness of duloxetine were established in two randomized, controlled studies (Goldstein et al., 2005; Raskin et al., 2005). A further randomized controlled trial has recently been published (Wernicke et al., 2006). No studies comparing duloxetine to other drugs are so far available. The optimal dosage of duloxetine is 60 mg/day, with no better response at higher doses

(120 mg/day) and inefficacy at lower doses (20 mg/day) (Goldstein et al., 2005; Raskin et al., 2005). Another SNRI, venlafaxine (225 mg) has been found to relieve neuropathic Pain better than placebo and to be similar to imipramine (150 mg) in a three-period cross-over trial (Sindrup et al., 2003). In a double-blind, randomized, placebo-controlled trial of 244 patients, venlafaxine extended-release (ER) showed a dose-related, clinically significant reduction of Pain in diabetic neuropathy (Rowbotham et al., 2004). For the venlafaxine ER 150–225-mg group, the NNT was 4.5 following 6 weeks of treatment, which is slightly higher than what has been reported for TCA (Sindrup and Jensen, 2000). Since venlafaxine at low doses (75 mg/day) inhibits mostly the reuptake of serotonin, whereas at higher doses the dual serotonin and norepinephrine inhibition is more balanced, the efficacy of venlafaxine at higher doses suggests a role of norepinephrine reuptake in modulating the Pain network. TCA have been shown to relieve peripheral neuropathic Pain in one every two to three patients, whereas SNRI in one every four to five patients (Sindrup et al., 2005). SNRI have, however, much less side effects. Compared to TCA, SNRI do not have the anticholinergic, antihistaminergic, and alpha-1,2 adrenergic-blocking side effects of TCA, and are therefore safer to use particularly in elderly and in patients with cardiovascular problems. The most frequently observed adverse events with duloxetine are nausea, fatigue, vomiting, constipation, somnolence, dry mouth, increased sweating, decreased appetite, and weakness. Venlafaxine adverse events include agitation, diarrhea, gastrointestinal manifestations, and increased liver enzymes. It has also dose-dependent cardio-vascular symptoms, principally hypertension. Side effects are much less common with the ER formulation. In a head-to-head comparison, venlafaxine (225 mg/day) was not better than imipramine (150 mg/day) with respect to tolerability and withdrawal rate for side effects (Sindrup et al., 2003).

11.1.4 Norepinephrine and Dopamine Reuptake Inhibitor

The antidepressant bupropion is a selective norepinephrine and dopamine reuptake inhibitor with no direct actions on the serotonin system. Sustained-release bupropion hydrochloride (150 to 300 mg daily) has been found to be effective in controlling neuropathic Pain in a double-blind randomized controlled trial versus placebo (Semenchuk et al., 2001). Pain relief was significant after 2 weeks of therapy. The lack of significant affinity for muscarinic, histaminergic, or alpha-adrenergic receptors makes bupropion side effect profile substantially different from that of the TCA. The most common side effects include dry mouth, constipation, insomnia, headache, gastrointestinal disturbances, tremor, and dizziness. (Sindrup et al., 2005). Bupropion showed no efficacy in treatment of nonneuropathic chronic low back Pain (Katz et al., 2005).

11.2 Anticonvulsant Drugs

Several anticonvulsant drugs are used in neuropathic Pain (Wiffen et al., 2005a). The precise mechanisms of action remain uncertain. Carbamazepine, phenytoin, and lamotrigine exert their membrane-stabilizing properties by blocking sodium channels; therefore, they are likely to reduce neuronal firing in sensitized C-nociceptors. Gabapentin interacts with GABA receptors or GABA metabolism, and binds to an $\alpha 2\delta$ subunit of voltage-gated calcium channels on neurons. Pregabalin is a higher-potency and higher-effective analogue of gabapentin.

We will briefly analyze the most common anticonvulsants currently used for neuropathic Pain, addressing the most common uses and side effects.

11.2.1 Carbamazepine

Carbamazepine has a well established efficacy both in trigeminal neuralgia – where it remains the treatment of choice (FDA approved in 1962) – with an NNT = 1.8 (1.3–2.2) (Sindrup and Jensen, 2002; Wiffen et al., 2005b) and painful diabetic neuropathy (Rull et al., 1969) as well as other kind of neuropathic Pain (Wiffen, 2005b). Recommended starting dosage is 100–200 mg × 2/day, to increase up to 800–1,200 mg in divided doses. A low initial daily dosage with a gradual increase is advised. Carbamazepine is ~70–80% protein-bound, is metabolized by the hepatic P450 enzymatic system, and eliminated by the liver. It induces its own metabolism within the first month of treatment and also may influence the metabolism of several hepatically metabolized drugs due to the induction of microsomal enzyme system. Like other tricyclic compounds, carbamazepine has a moderate anticholinergic action which is responsible for some of its adverse effects, the most common including dizziness, drowsiness, unsteadiness/ataxia, gum-hypertrophy, vomiting, and nystagmus. Skin rash, low white blood cell count, platelets, and sodium, and liver function abnormalities are also frequent. Therefore cell blood count, sodium, and liver functionality must be evaluated at baseline and should be monitored for at least 1 year of treatment. In clinical practice, intolerance to the side effects of carbamazepine limits its use, especially in the elderly. Oxcarbazepine, a keto-acid analogue of carbamazepine, is much better tolerated, and its efficacy on trigeminal neuralgia is similar to that of carbamazepine (Beydoun and Kutluay, 2002), and it is now preferred for its more favorable side effects profile. Moreover, different from carbamazepine, oxcarbazepine does not induce enzymatic activity and has low incidence of cutaneous manifestations. However, sodium levels must be carefully monitored, especially in the elderly. Oxcarbazepine has also been evaluated as monotherapy in painful diabetic neuropathy (Dogra et al., 2005).

11.2.2 Gabapentin

Gabapentin, first developed for the treatment of epilepsy, has a structure similar to the neurotransmitter GABA, but it does not seem to act on the GABA receptors. It does not interact in vitro with sodium channel, whereas binds to the α2δ subunit of voltage-dependent Ca2+ channel likely playing a role in modulation of central sensitization. Experimentally, gabapentin has been shown to inhibit the release of glutamate in the spinal cord dorsal horn in neuropathic rats suggesting that reduction of neuropathic Pain symptoms may be mediated by the inhibition of glutamate release in the spinal cord dorsal horn (Coderre, 2005). Gabapentin has ideal pharmacokinetics properties. Well absorbed orally, it circulates mostly unbound in the plasma, it does not induce hepatic enzymes and is excreted unchanged by the kidneys, so excretion is decreased in patients with renal impairment. Gabapentin has been shown to be effective in placebo-controlled studies in diabetic neuropathy (Backonja et al., 1998), postherpetic neuralgia (Rowbotham et al., 1998; Rice and Maton, 2001), Guillain-Barré syndrome (Pandey et al., 2002), phantom limb Pain (Bone et al., 2002), and Pain after spinal cord injury (Tai et al., 2002). A head to head study versus amitriptyline (Morello et al., 1999) showed no significant differences between the two drugs, but gabapentin had much lower adverse effects. The use of gabapentin for acute and chronic Pain has been recently extensively analyzed in a Cochrane Review (Wiffen et al., 2005c). The most common side effects of gabapentin are somnolence, dizziness, fatigue, peripheral edema, and dry mouth. To minimize possible side effects, gabapentin needs slow individual titration with initial dosages of 300 mg × 3/day (100–300 mg/day in the elderly), gradually increasing to a 3,600 mg/day or to significant Pain relief. Reduction in neuropathic Pain generally requires doses higher than 1,600 mg per day. As recently stated, this is an important consideration, since many patients are given doses that are too small (Mendell and Sahenk, 2003). Consistently, a study in diabetic neuropathy at lower dose (900 mg) showed no efficacy (Gorson et al., 1999).

11.2.3 Pregabalin

Pregabalin is a higher-effective analogue of gabapentin that binds to and modulates α2δ subunit of voltage-gated calcium channels in the central nervous system. The affinity for and the modulation of the α2δ subunit of voltage-gated calcium channels seem to be responsible for the analgesic actions of pregabalin (Field et al., 2006). Pregabalin is the second of only two medications that are FDA approved for the treatment of neuropathic Pain associated with diabetic peripheral neuropathy (Lesser et al., 2004; Rosenstock et al., 2004; Richter et al., 2005); it is also approved for the treatment of postherpetic neuralgia (Dworkin et al., 2003; Sabatowski et al., 2004; Freynhagen et al., 2005). The efficacy of pregabalin has been established in six double-blind, placebo-controlled trials, three involving patients with painful diabetic neuropathy, and three involving patients with postherpetic neuralgia (Sonnett et al., 2006). Similar to gabapentin, pregabalin is completely absorbed, it does not bind to plasma proteins, it is not metabolized, and is eliminated unchanged through the kidneys. Doses must be adjusted in patients with renal insufficiency. The most common adverse effects compared with placebo included dizziness, somnolence, dry mouth, peripheral edema, weight gain, difficulty with concentration/attention, and blurred vision. A slow tapering is recommended before discontinuation.

11.2.4 Lamotrigine

Lamotrigine has been shown to act by blocking voltage-dependent sodium channels and by inhibiting presynaptic release of excitatory neural transmitters, mainly glutamate. Lamotrigine has shown significant efficacy in diabetic painful peripheral neuropathy (Eisenberg et al., 2001) and, as add-on therapy, in refractory trigeminal neuralgia (Zakrzewska et al., 1997). In two randomized controlled trials in HIV, sensory neuropathy lamotrigine (300–600 mg/day) was effective only in a subgroup of patients receiving antiretroviral therapy (Simpson et al., 2000, 2003). A recent paper, however, reporting two large, replicate, randomized controlled trials on lamotrigine on painful diabetic neuropathy gave inconsistent results (Vinik et al., 2006). Lamotrigine has been shown to be effective in central poststroke Pain (Vestergaard et al., 2001) at a dose of at least 200 mg/day. It has also been used in patients with spinal cord injury (Finnerup et al., 2002) where it showed efficacy only in a subgroup of patients at relatively high dosage (up to 400 mg/die). Lamotrigine and amitriptyline are the only oral drugs proven to be effective in the treatment of central poststroke Pain in a placebo-controlled studies (Frese et al., 2006). Lamotrigine is well absorbed orally, with up to 98% bioavailability. Approximately 55% of the drug is protein bound; therefore, clinical interaction with other protein-bound drugs is unlikely. Ninety percent of the drug undergoes glucuronic acid conjugation in the liver, with the conjugate and the remaining 10% of unmetabolized drug excreted in the urine. Lamotrigine most common side effects are skin reactions, which may be severe, and can be minimized by using a very slow dose titration. Therapy should start with 25 mg daily and increased by 25 mg every other week. Other side effects include headache, nausea, and dizziness. Generally, doses of 200–400 mg/day are required for an analgesic effect.

11.3   Narcotic and Nonnarcotic Analgesics

11.3.1 Tramadol

Tramadol is an analgesic acting both through monoaminergic (norepinephrine and serotonin reuptake inhibitor) and opioid (mild μ opioid agonist) mechanisms which has been shown to relieve diabetic (Harati et al., 1998) and neuropathic Pain of various causes (Sindrup et al., 1999). A recent Cochrane review concluded that tramadol is effective in neuropathic Pain with a NNT compared to placebo of 3.8 (95% confidence interval 2.8 to 6.3) (Hollingshead et al., 2006). Starting dose is 50 mg once or twice daily, increasing by 50 mg every 1–3 days up to 400 mg daily. It is well tolerated and less likely than other opioid to cause dependence. The adverse effects include dizziness/vertigo, constipation, somnolence, and orthostatic hypotension. Caution should be used in the presence of history of seizures, and concurrent therapy with drugs which reduce seizure threshold (e.g., neuroleptics, antidepressants, opioids). Dosage adjustment may be required in old people and in patients with renal/hepatic failure. Its use is not advisable if there is tendency to drug abuse.

11.3.2 Opioids

Although neuropathic Pain has commonly been thought to be associated with a poor opioid response (McQuay, 2002), increasing evidence points to a possible role of opioids in controlling refractory neuropathic Pain.

A recent Cochrane review on opioids in neuropathic Pain concluded that only intermediate-term studies showed significant efficacy of opioids when compared to placebo (Eisenberg et al., 2006). Oxycodone (average doses 37–60 mg/day, range 10–99 mg/day), has been shown to be effective in painful diabetic neuropathy (Gimbel et al., 2003; Watson et al., 2003). Oxycodone, morphine, and methadone have also been shown to be effective in postherpetic neuralgia (Watson and Babul, 1998; Raja et al., 2002). The opioid agonist levorphanol, in double-blind, randomized controlled trials for refractory chronic peripheral and central neuropathic Pain, resulted to be more efficacious on peripheral neuropathic Pain. Patients with central poststroke Pain were the least likely to report benefit (Rowbotham et al., 2003). Gilron et al. (Gilron et al., 2005) compared the efficacy of gabapentin and morphine, alone or in association, in patients with painful diabetic neuropathy or postherpetic neuralgia showing that the combined use of gabapentin and morphine achieved a better analgesia at lower doses of each drug than either as a single agent. The most common side effects of opioids are constipation, sedation, nausea, dizziness, and vomiting. Dosages of opioids should be titrated individually balancing efficacy and side effects. Recently, in a randomized placebo-controlled trial smoked cannabis has been shown to relieve pain in patients with painful HIV-sensory neuropathy (Abrams et al., 2007). There are still several open questions regarding analgesic narcotics, among which are the long-term efficacy of the drugs, identification of factors influencing response to opioids, and the choice of which opioids are most appropriate in different kind of Pain. While ongoing and future studies will help address these issues, opioids (longer-acting agents preferred) should be reserved for refractory patients.

11.4 Topical Agents

Five percent lidocaine gel or patches have been shown to provide relief in blinded controlled studies in postherpetic neuralgia (Rowbotham et al., 1995, 1996; Galer et al., 1999, 2002; Wasner et al., 2005), for which it is FDA approved. Lidocaine is thought to act by reducing ectopic neural discharges in superficial nerves. Other topical agents are usually used as an adjunctive treatment for localized Pain (toes/feet). Capsaicin, which is extracted from hot chili peppers, exerts its analgesic effect by depleting substance P from sensory nerves in the skin. It has moderate efficacy in diabetic neuropathy whereas inconsistent results have been reported in postherpetic neuralgia. No Pain relief was reported in HIV painful neuropathy.

A recent systematic review (Mason et al., 2004) of topical capsaicin for the treatment of chronic Pain identified six double blind randomized controlled trials for a total of 656 patients with neuropathic Pain. The relative benefit from topical capsaicin 0.075% compared with placebo was 1.4 (95% confidence interval 1.2 to 1.7) and the NNT was 5.7 (4.0 to 10.0). The authors concluded that topical capsaicin may be useful as an adjunct or sole therapy for a small number of patients who are unresponsive to, or intolerant of, other treatments.

Capsaicin relatively favorable side-effects profile include burning, stinging, redness, sneezing. It is recommended to use it in well-ventilated rooms and to avoid rubbing in the eyes.

12 Conclusions

When dealing with neuropathic Pain, physicians should pursue the goal of finding effective agent with tolerable side effects. It is important to listen to the patients, to discuss together the possible therapeutic options and to provide the patients with reasonable expectations. It is crucial to explain that responses vary in different agents and different doses. Pharmacological therapy should be initiated with one drug at low doses and titrated upwards. Polypharmacotherapy should be considered only when higher doses of a partially effective drug cannot be tolerated, and drugs with different and complementary mechanism of action should be used, so to target different Pain mechanisms.

Treatment of neuropathic Pain remains challenging with considerable variability in individual response to the various agents and even to different drugs in the same class.

Using a NNT methodology based on data from randomized, placebo-controlled, double-blinded clinical trials, TCA have been recommended as first-line therapy for neuropathic Pain (Sindrup and Jensen, 2000).

Other authors (Mendell and Sahenk, 2003) suggest gabapentin as first choice therapy, on the basis of the proven efficacy and relatively favorable side-effects profile. The antiepileptics with best evidence for efficacy to date in diabetic painful peripheral neuropathy are gabapentin (1,200–3,600 mg/day) and prega-balin (150–600 mg/day) (Backonja et al., 1998; Simpson, 2001; Lesser et al., 2004; Rosenstock et al., 2004). The recent EFNS guidelines on pharmacological treatment of neuropathic Pain (Attal et al., 2006) suggest as first line therapy TCA, gabapentin, and pregabalin, and as second line therapy the SNRI (venlafaxine and duloxetine), lamotrigine and tramadol. Opioids should be considered in selective nonresponsive patients with chronic noncancer Pain (Kalso et al., 2003). An evidence-based algorithm for treatment of neuropathic pain has also been proposed (Finnerup et al., 2005).

References

Abrams DI, Jay CA, Shade SB, Vizoso H, Reda H, et al. 2007. Cannabis in painful HIV-associated sensory neuropathy: A randomized placebo-controlled trial. Neurology 68: 515-521.

Amadio PC. 1993. Outcomes measures. J Bone Joint Surg 5A: 1583-1584.

Attal N, Cruccu G, Haanpää M, Hansson P, Jensen TS, et al. 2006. EFNS guidelines on pharmacological treatment of neuropathic pain. Eur J Neurol 13: 1153-1169.

Backonja MM, Krause SJ. 2003. Neuropathic pain questionnaire—short form. Clin J Pain 19: 315-316.

Backonja M, Beydoun A, Edwards KR, Schwartz SL, Fonseca V, et al. 1998. Gabapentin for the symptomatic treatment of painful neuropathy in patients with diabetes mellitus: A randomized controlled trial. JAMA 280: 1831-1836.

Baron R. 2006. Mechanisms of disease: Neuropathic pain – a clinical perspective. Nat Clin Pract Neurol 2: 95-106.

Bennett DL. 2001. Neurotrophic factors: Important regulators of nociceptive function. Neuroscientist 7: 13-17.

Bennett GJ, Xie YK. 1988. A peripheral mononeuropathy in rat that produces disorders of pain sensation like those seen in man. Pain 33: 87-107.

Bennett M. 2001. The LANSS pain scale: The Leeds assessment of neuropathic symptoms and signs Pain 92: 147-157.

Beydoun A, Kutluay E. 2002. Oxcarbazepine Expert. Opin Pharmacother 3: 59-71.

Bone M, Critchley P, Buggy DJ. 2002. Gabapentin in postamputation phantom limb pain: A randomized, double-blind, placebo-controlled, cross-over study. Reg Anesth Pain Med 27: 481-486.

Bouhassira D, Attal N, Alchaar H, Boureau F, Brochet B, et al. 2005. Comparison of pain syndromes associated with nervous or somatic lesions and development of a new neuropathic Pain diagnostic questionnaire (DN4). Pain 114: 29-36.

Bouhassira D, Attal N, Fermanian J, Alchaar H, Gautron M, et al. 2004. Development and validation of the neuropathic pain symptom inventory. Pain 108: 248-257.

Bromm B, Treede RD. 1984. Nerve fibre discharges, cerebral potentials and sensations induced by CO2 laser stimulation. Hum Neurobiol 3: 33-40.

Campbell JN, Meyer RA. 2006. Mechanisms of neuropathic pain. Neuron 52: 77-92.

Chen Y, Sommer C. 2006. Nociceptin and its receptor in rat dorsal root ganglion neurons in neuropathic and inflammatory pain models: Implications on pain processing. J Peripher Nerv Syst 11: 232-240.

Chen Y, Sommer C. 2007. Activation of the nociceptin opioid system in rat sensory neurons produces antinociceptive effects in inflammatory pain: Involvement of inflammatory mediators. J Neurosci Res 85: 1478-1488.

Cleeland C. 1991. Pain assessment in cancer. Effect of Cancer on Quality of Life. Osoba D, editor. Boca Raton, FL: CRC Press; pp. 293-305.

Coderre TJ. 2005. Evidence that gabapentin reduces neuropathic pain by inhibiting the spinal release of glutamate. J Neurochem 94: 1131-1139.

Cohen S, Abdi S. 2002. Central pain. Curr Opin Anaesthesiol 15: 575-581.

Deschamps M, Band PR, Coldman AJ. 1988. Assessment of adult cancer pain: Shortcomings of current methods. Pain 32: 133-139.

Dogra S, Beydoun S, Mazzola J, Hopwood M, Wan Y. 2005. Oxcarbazepine in painful diabetic neuropathy: A randomized, placebo-controlled study. Eur J Pain 9: 543-554.

Dubinsky RM, Kabbani H, El-Chami Z, Boutwell C, Ali H. 2004. Quality standards subcommittee of the American academy of neurology. Practice parameter: Treatment of postherpetic neuralgia: An evidence-based report of the quality standards subcommittee of the American academy of neurology. Neurology 63: 959-965.

Dworkin RH, Corbin AE, Young JP Jr, Sharma U, La Moreaux L, et al. 2003. Pregabalin for the treatment of postherpetic neuralgia: A randomized, placebo-controlled trial. Neurology 60: 1274-1283.

Eisenberg E, McNicol E, Carr DB. 2006. Opioids for neuropathic pain. Cochrane Database Syst Rev 3: CD006146.

Eisenberg E, Lurie Y, Braker C, Daoud D, Ishay A. 2001. Lamotrigine reduces painful diabetic neuropathy: A randomized, controlled study. Neurology 57: 505-509.

Field MJ, Cox PJ, Stott E, Melrose H, Offord J, et al. 2006. Identification of the alpha2-delta-1 subunit of voltage-dependent calcium channels as a molecular target for pain mediating the analgesic actions of pregabalin. Proc Natl Acad Sci USA 103: 17537-17542.

Finnerup NB, Otto M, McQuay HJ, Jensen TS, Sindrup SH. 2005. Algorithm for neuropathic pain treatment: An evidence based proposal. Pain 118: 289-305.

Finnerup NB, Sindrup SH, Bach FW, Johannesen IL, Jensen TS. 2002. Lamotrigine in spinal cord injury pain: A randomized controlled trial. Pain 96: 375-383.

Frese A, Husstedt IW, Ringelstein EB, Evers S. 2006. Pharmacologic treatment of central post-stroke pain. Clin J Pain 22: 252-260.

Freynhagen R, Strojek K, Griesing T, Whalen E, Balkenohl M. 2005. Efficacy of pregabalin in neuropathic Pain evaluated in a 12-week, randomised, double-blind, multicentre, placebo-controlled trial of flexible- and fixed-dose regimens. Pain 115: 254-263.

Galer BS, Jensen MP. 1997. Development and preliminary validation of a pain measure specific to neuropathic pain: The neuropathic pain scale. Neurology 48: 332-338.

Galer BS, Rowbotham MC, Perander J, Friedman E. 1999. Topical lidocaine patch relieves postherpetic neuralgia more effectively than a vehicle topical patch: Results of an enriched enrollment study. Pain 80: 533-538.

Galer BS, Jensen MP, Ma T, Davies PS, Rowbotham MC. 2002. The lidocaine patch 5% effectively treats all neuropathic pain qualities: Results of a randomized, double-blind, vehicle-controlled, 3-week efficacy study with use of the neuropathic pain scale. Clin J Pain 5: 297-301.

Gilron I, Bailey JM, Tu D, Holden RR, Weaver DF, et al. 2005. Morphine, gabapentin, or their combination for neuropathic Pain. N Engl J Med 352: 1324-1334.

Gimbel JS, Richrds P, Portenoy RK. 2003. Controlled-release oxycodone for pain in diabetic neuropathy. A randomized controlled trial. Neurology 60: 927-934.

Gold MS. 2000. Spinal nerve ligation: What to blame for the pain and why. Pain 84(2–3): 117-120.

Goldstein DJ, Lu Y, Detke MJ, Lee TC, Iyengar S. 2005. Duloxetine vs. placebo in patients with painful diabetic neuropathy. Pain 116: 109-118.

Gorson KC, Schott C, Herman R, Ropper AH, Rand WM. 1999. Gabapentin in the treatment of painful diabetic neuropathy: A placebo controlled, double blind, crossover trial. J Neurol Neurosurg Psychiatry 66: 251-252.

Guyatt GH, Feeny DH, Patrick DL. 1993. Measuring health-related quality of life. Ann Intern Med 118: 622-629.

Harati Y, Gooch C, Swenson M, Edelman S, Greene D, et al. 1998. Double-blind randomized trial of tramadol for the treatment of diabetic neuropathy. Neurology 50: 1842-1846.

Hollingshead J, Duhmke RM, Cornblath DR. 2006. Tramadol for neuropathic Pain. Cochrane Database Syst Rev 3: CD003726.

Hu P, McLachlan EM. 2002. Macrophage and lymphocyte invasion of dorsal root ganglia after peripheral nerve lesions in the rat. Neuroscience 112: 23-38.

Kalso E, Allan L, Dellemijn PL, Faura CC, Ilias WK, et al. 2003. Recommendations for using opioids in chronic non-cancer pain. Eur J Pain 7: 381-386.

Katz J, Pennella-Vaughan J, Hetzel RD, Kanazi GE, Dworkin RH. 2005. A randomized, placebo-controlled trial of bupropion sustained release in chronic low back pain. J Pain 6: 656-661.

Kim SH, Chung JM. 1992. An experimental model for peripheral neuropathy produced by segmental spinal nerve ligation in the rat. Pain 50: 355-363.

Krause SJ, Backonja MM. 2003. Development of a neuropathic pain questionnaire. Clin J Pain 19: 306-314.

Lesser H, Sharma U, La Moreaux L, Poole RM. 2004. Pregabalin relieves symptoms of painful diabetic neuropathy. Neurology 63: 2104-2110.

Marchand F, Perretti M, McMahon SB. 2005. Role of the immune system in chronic pain. Nat Rev Neurosci 6: 521-532.

Marshall FJ, Kieburtz K, McDermott M, Kurlan R, Shoulson I. 1996. Clinical research in neurology. From observation to experimentation. Neuroepidemiol 14(2): 451-466.

Mason L, Moore RA, Derry S, Edwards JE, McQuay HJ. 2004. Systematic review of topical capsaicin for the treatment of chronic pain. BMJ 328: 991.

Max MB, Lynch SA, Muir J, Shoaf SE, Smoller B, et al. 1992. Effects of desipramine, amitriptyline, and fluoxetine on pain in diabetic neuropathy. N Engl J Med 326: 1250-1256.

McLachlan EM, Janig W, Devor M, Michaelis M. 1993. Peripheral nerve injury triggers noradrenergic sprouting within dorsal root ganglia. Nature 363: 543-546.

McQuay HJ. 2002. Neuropathic pain: Evidence matters. Eur J Pain 6(Suppl. A): 11-18.

McQuay HJ, Tramer M, Nye BA, Carroll D, Wiffen PJ, et al. 1996. A systematic review of antidepressants in neuropathic pain. Pain 68: 217-227.

Melzack R. 1975. The McGill pain questionnaire: Major properties and scoring methods. Pain 1: 277-299.

Melzack R. 1987. The short-form McGill pain questionnaire. Pain 30: 191-197.

Mendell JR, Sahenk Z. 2003. Clinical practice. Painful sensory neuropathy. N Engl J Med 348: 1243-1255.

Merksey H, Bogduk N. 1994. Classification of chronic pain. Descriptions of chronic pain syndromes and definitions of pain terms. 2nd ed. Seattle: IASP Press.

Moalem G, Xu K, Yu L. 2004. T lymphocytes play a role in neuropathic pain following peripheral nerve injury in rats. Neuroscience 129: 767-777.

Morello CM, Leckband SG, Stoner CP, Moorhouse DF, Sahagian GA. 1999. Randomized double-blind study comparing the efficacy of gabapentin with amitriptyline on diabetic peripheral neuropathy pain. Arch Intern Med 159: 1931-1937.

Moskowitz MA. 1993. Neurogenic inflammation in the path-ophysiology and treatment of migraine. Neurology 43 (6 Suppl. 3): S16-S20.

Pandey CK, Bose N, Garg G, Singh N, Baronia A, et al. 2002. Gabapentin for the treatment of pain in Guillain–Barre syndrome: A double blinded, placebo-controlled, crossover study. Anesth Analg 95: 1719-1723.

Perkins FM, Werner MU, Persson F, Holte K, Jensen TS, et al. 2004. Development and validation of a brief, descriptive Danish pain questionnaire (BDDPQ). Acta Anaesthesiol Scand 48: 486-490.

Pert CB, Snyder SH. 1973. Opiate receptor: Demonstration in nervous tissue. Science 179: 1011-1014.

Raghavendra V, Tanga F, De Leo JA. 2003. Inhibition of microglial activation attenuates the development but not existing hypersensitivity in a rat model of neuropathy. J Pharmacol Exp Ther 306: 624-630.

Raja SN, Haythornthwaite JA, Pappagallo M, Clark MR, Travison TG, et al. 2002. Opioids versus antidepressants in postherpetic neuralgia. Neurology 59: 1015-1021.

Raskin J, Pritchett YL, Wang F, D'Souza DN, Waninger AL, et al. 2005. A double-blind, randomized multicenter trial comparing duloxetine with placebo in the management of diabetic peripheral neuropathic pain. Pain Med 6: 346-356.

Reynolds DV. 1969. Surgery in the rat during electrical analgesia induced by focal brain stimulation. Science 164: 444-445.

Rice ASC, Maton S. 2001. Postherpetic neuralgia study group. Gabapentin in postherpetic neuralgia; a randomised, double-blind, controlled study. Pain 94: 215-224.

Richter RW, Portenoy R, Sharma U, Lamoreaux L, Bockbrader H, et al. 2005. Relief of painful diabetic peripheral neuropa-thy with pregabalin: A randomized, placebo-controlled trial. J Pain 6: 253-260.

Rosenstock J, Tuchman M, La Moreaux L, Sharma U. 2004. Pregabalin for the treatment of painful diabetic peripheral neuropathy: A double-blind, placebo-controlled trial. Pain 110: 628-638.

Rowbotham MC, Davies PS, Fields HL. 1995. Topical lidocaine gel relieves postherpetic neuralgia. Ann Neurol 37: 246-253.

Rowbotham MC, Davies PS, Verkempinck C, Galer BS. 1996. Lidocaine patch: Double-blind controlled study of a new treatment method for post-herpetic neuralgia. Pain 65: 39-44.

Rowbotham MC, Goli V, Kunz NR, Lei D. 2004. Venlafaxine extended release in the treatment of painful diabetic neu-ropathy: A double-blind, placebo-controlled study. Pain 110: 697-706.

Rowbotham M, Harden N, Stacey B, Bernstein P, Magnus-Miller L. 1998. Gabapentin for treatment of postherpetic neuralgia. JAMA 280: 1837-1843.

Rowbotham MC, Twilling L, Davies PS, Reisner L, Taylor K, et al. 2003. Oral opioid therapy for chronic peripheral and central neuropathic pain. N Engl J Med 348: 1223-1232.

Rull JA, Quibrera R, Gonzalez-Millan H, Lozano Castaneda O. 1969. Symptomatic treatment of peripheral diabetic neuro-pathy with carbamazepine (Tegretol): Double blind crossover trial. Diabetologia 5: 215-218.

Saarto T, Wiffen P. 2005. Antidepressants for neuropathic pain. Cochrane Database of Systemic Reviews 20: CD005454.

Sabatowski R, Galvez R, Cherry DA, Jacquot F, Vincent E, et al. 2004. Pregabalin reduces pain and improves sleep and mood disturbances in patients with post-herpetic neural-gia: Results of a randomised, placebo-controlled clinical trial. Pain 109: 26-35.

Seltzer Z, Dubner R, Shir Y. 1990. A novel behavioral model of neuropathic pain disorders produced in rats by partial sciatic nerve injury. Pain 43: 205-218.

Semenchuk MR, Sherman S, Davis B. 2001. Double-blind, randomized trial of bupropion SR for the treatment of neuropathic Pain. Neurology 57: 1583-1588.

Simpson DA. 2001. Gabapentin and venlafaxine for the treat-ment of painful diabetic neuropathy. J Clin Neuromuscular Dis 3: 53-62.

Simpson DM, McArthur JC, Olney R, Clifford D, So Y, et al. 2003. Lamotrigine for HIV-associated painful sensory neuropathies: A placebo-controlled trial. Lamotrigine HIV Neuropathy Study Team 60: 1508-1514.

Simpson DM, Olney R, McArthur JC, Khan A, Godbold J, et al. 2000. A placebo-controlled trial of lamotrigine for painful HIV-associated neuropathy. Neurology 54: 2115-2119.

Sindrup SH, Jensen TS. 2000. Pharmacologic treatment of pain in polyneuropathy. Neurology 55: 915-920.

Sindrup SH, Jensen TS. 2002. Pharmacotherapy of trigeminal neuralgia. Clin J Pain 18: 22-27.

Sindrup SH, Otto M, Finnerup NB, Jensen TS. 2005. Anti-depressants in the treatment of neuropathic pain. Basic Clin Pharmacol Ther 96: 399-409.

Sindrup SH, Bach FW, Madsen C, Gram LF, Jensen TS. 2003. Venlafaxine versus imipramine in Painful polyneuropathy: A randomized, controlled trial. Neurology 60: 1284-1289.

Sindrup SH, Gram LF, Brosen K, Eshoj O, Mogensen EF. 1990. The selective serotonin reuptake inhibitor paroxetine is effective in the treatment of diabetic neuropathy symptoms. Pain 42: 135-144.

Sindrup SH, Andersen G, Madsen C, Smith T, Brosen K, et al. 1999. Tramadol relieves pain and allodynia in polyneuropathy: A randomised, double-blind, controlled trial. Pain 83: 85-90.

Sindrup SH, Bjerre U, Dejgaard A, Brosen K, Aaes-Jorgensen T, et al. 1992. The selective serotonin reuptake inhibitor citalopram relieves the symptoms of diabetic neuropathy. Clin Pharmacol Ther 52: 547-552.

Sommer C. 2003. Painful neuropathies. Curr Opin Neurol 16: 623-628.

Sonnett TE, Setter SM, Campbell RK. 2006. Pregabalin for the treatment of painful neuropathy. Expert Rev Neurother 6: 1629-1635.

Tai O, Kirshblum S, Chen B, Millis S, Johnston M, et al. 2002. Gabapentin in the treatment of neuropathic pain after spinal cord injury: A prospective, randomized, double-blind, crossover trial. J Spinal Cord Med 25: 100-105.

Tasker RR. 1984. Deafferentation. Textbook of Pain. Wall PD, Melzack R, editors. Edinburgh, UK: Churchill Livingstone; pp. 119-132.

Torebjork E. 1993. Human microneurography and intra-neural micro-stimulation in the study of neuropathic pain. Muscle Nerve 16: 1063-1065.

Truini A, Cruccu G. 2006. Pathophysiological mechanisms of neuropathic pain. Neurol Sci 27 (Suppl. 2): S179-S182.

Uceyler N, Tscharke A, Sommer C. 2007. Early cytokine expression in mouse sciatic nerve after chronic constriction nerve injury depends on calpain. Brain Behav Immun. Jan 2. [Epub ahead of print]

Vestergaard K, Andersen G, Gottrup H, Kristensen BT, Jensen TS. 2001. Lamotrigine for central poststroke pain: A randomized controlled trial. Neurology 56: 184-190.

Vinik AI, Tuchman M, Safirstein B, Corder C, Kirby L, et al. 2006. Lamotrigine for treatment of pain associated with diabetic neuropathy: Results of two randomized, double-blind, placebo-controlled studies. Pain 128: 169-179.

Vogel C, Stallforth S, Sommer C. 2006. Altered pain behavior and regeneration after nerve injury in TNF receptor deficient mice. J Peripher Nerv Syst 11: 294-303.

Wagner KJ, Sprenger T, Kochs EF, Tolle TR, Vallet M, et al. 2007. Imaging human cerebral pain modulation by dose-dependent opioid analgesia: A positron emission tomography activation study using remifentanil. Anesthesiology 106: 548-556.

Wall PD. 1991. Neuropathic pain and injured nerve: Central mechanisms. Br Med Bull 47: 631-643.

Wall PD, Devor M. 1983. Sensory afferent impulses originate from dorsal root ganglia as well as from the periphery in normal and nerve injured rats. Pain 17: 321-339.

Wall PD, Waxman S, Basbaum AI. 1974. Ongoing activity in peripheral nerve: Injury discharge. Exp Neurol 45: 576-589.

Wang H, Ehnert C, Brenner GJ, Woolf CJ. 2006. Bradykinin and peripheral sensitization. Biol Chem 387: 11-14.

Wasner G, Kleinert A, Binder A, Schattschneider J, Baron R. 2005. Postherpetic neuralgia: Topical lidocaine is effective in nociceptor-deprived skin. J Neurol 252: 677-686.

Watson CP, Babul N. 1998. Efficacy of oxycodone in neuropathic pain: A randomized trial in postherpetic neuralgia. Neurology 50: 1837-1841.

Watson CP, Moulin D, Watt-Watson J, Gordon A, Eisenhoffer J. 2003. Controlled-release oxycodone relieves neuropathic pain: A randomized controlled trial in painful diabetic neuropathy. Pain 105: 71-78.

Wernicke JF, Pritchett YL, D'Souza DN, Waninger A, Tran P, et al. 2006. A randomized controlled trial of duloxetine in diabetic peripheral neuropathic pain. Neurology 67: 1411-1420.

Wiffen PG, McQuay HJ, Moore RA. 2005b. Carbamazepine for acute and chronic pain. Cochrane Database Syst Rev 3: CD005451.

Wiffen PG, McQuay HJ, Edwards JE, Moore RA. 2005c. Gabapentin for acute and chronic pain. Cochrane Database Syst Rev 3: CD005452.

Wiffen P, Collins S, McQuay H, Carroll D, Jadad A, 2005a. Anticonvulsant drugs for acute and chronic pain. Cochrane Database Syst Rev 3: CD001133.

Woolf CJ. 2004. Dissecting out mechanisms responsible for peripheral neuropathic pain: Implications for diagnosis and therapy. Life Sci 74: 2605-2610.

Woolf CJ, Salter MW. 2000. Neuronal plasticity: Increasing the gain in pain. Science 288: 1765-1769.

Yoshimura M, Furue H. 2006. Mechanisms for the anti-nociceptive actions of the descending noradrenergic and serotonergic systems in the spinal cord. J Pharmacol Sci 101: 107-117.

Zakrzewska JM, Chaudhry Z, Nurmikko TJ, Patton DW, Mullens EL. 1997. Lamotrigine (lamictal) in refractory trigeminal neuralgia: Results from a double-blind placebo controlled crossover trial. Pain 73: 223-230.

Zelman DC, Gore M, Dukes E, Tai KS, Brandenburg N. 2005. Validation of a modified version of the brief pain inventory for painful diabetic peripheral neuropathy. J Pain Symptom Manage 29: 401-410.

Index

Printing: Krips bv, Meppel, The Netherlands
Binding: Stürtz, Würzburg, Germany